THE ARRL RFI Book

Second Edition
Edited by: Mike Gruber, W1MG

Production Staff

Michelle Bloom, WB1ENT, Production Supervisor

Jodi Morin, KA1JPA, Compositor

David Pingree, N1NAS, Senior Technical Illustrator: Technical Illustrations

Sue Fagan, Graphic Design Supervisor: Cover Design

Cover Photo: Mel, W4MEL, at the Low Country Contest Club, NU4SC

Published by:
ARRL The national association for AMATEUR RADIO
225 Main Street
Newington, CT 06111-1494 • www.arrl.org

Printed in USA

ISBN: 0-87259-989-2

Second Edition

Editor's Note

ARRL wishes to acknowledge and thank the following people for their contributions to, and help with, updating the RFI Book and bringing this Second Edition to you:

Ron Hranac, NØIVN	Chapters 6, 7 and 8
Mike Martin, K3RFI	Chapter 11
Hartley Gardner, W1OQ	Chapters 13 and 14
Ghery Pettit, N6TPT	Chapter 15
Mark Steffka, WW8MS	Chapter 16
Terry Ryback, W8TR	Chapter 16
Mike Gruber, W1MG	Chapters 3, 5, 8, 11, 14, 17, 18 and 19
Ed Hare, W1RFI	Chapter 17

Contents

Foreword

If you're like most hams, radio-frequency interference has caused you headaches on more than one occasion during your ham career. Odds are, you're not anxious to repeat these RFI experiences! Despite our best efforts to lay RFI to rest over the past nine decades or so, it remains very much with us.

Part of the reason is that there are as many kinds of RFI as there are kinds of electronic devices. Add to that the variety of environments in which these devices can be installed, and you've got a thorny problem. Just when it looks like one type of interference problem has been licked, a new electronic device that is susceptible to RF energy enters the marketplace.

The best defense in this ongoing battle? Quite simply, drawing on the experiences of RFI experts who know the best way to tackle RFI problems. That's where *The ARRL RFI Book* comes in. We have asked experts in the various types of RFI to provide insights and solutions to the types of problems they know best. These experts, both amateur and professional, provide their expertise clearly and concisely.

Our thanks to these authors, as well as the hundreds of other hams who have unselfishly shared their RFI solutions with the ham community over the years. Our goal is to remove RFI as a problem for hams. Although we may never fully reach this goal, this book is a major step in that direction. You can help, too. Have an RFI tip that's worked for you and that doesn't appear in the book? Let us know about it. There's a handy Feedback Form at the back, or you can send an e-mail message to **pubsfdbk@arrl.org**.

David Sumner, K1ZZ
February 2007

Chapter 1

First Steps

Every RFI case involves both technical and personal issues. Knowing all of the technical solutions won't do you any good if your neighbor won't let you in the house to try them! Always solve the personal and emotional RFI problems first.

By Ed Hare, W1RFI
ARRL Laboratory Supervisor and
Bob Schetgen, KU7G (SK)

The first part of this chapter explains just what interference is and talks a bit about the scope of the RFI problem. Next, a discussion of "personal diplomacy" helps all those who are involved in an RFI problem successfully interact with each other. In some cases, a third party might be needed to mediate between parties that are no longer seeing eye to eye; ARRL Field Organization volunteers may be that needed third party. The chapter closes with a description of the Field Organization as it relates to RFI cases and tells you how you can often find local help.

RFI OR EMI?

Engineering professionals often talk about *electromagnetic interference* (EMI) —a very broad term that covers a lot of territory, ranging from electrostatic discharge (ESD), dealing with dc static fields, to the term more familiar to amateurs—*radio frequency interference* (RFI). It also encompasses extremely low frequency (ELF) electromagnetic fields (EMFs) such as those found near power lines.

Earlier editions of this book introduced the term EMI to amateurs and it was used extensively throughout the book. In spite of that, most hams still refer to all interference problems as RFI and rather than continue to fight the tide, this book uses *RFI* to describe most amateur interference. Either term is correct and either will be understood in most circles.

The term *electromagnetic compatibility* (EMC) may also be encountered. This is another general term that covers a large territory, implying that devices are (or should be) compatible with each other. Naturally, it is a term that is favored by those marketing electronic products.

Although the other terminology may be important to professionals who deal with a rather broad field, most hams are concerned mainly with interference from their transmitters or to their radio receivers. Both topics deal with RF signals, so the term RFI is well suited to most discussions about amateur interference.

Regardless of what terminology is used, when it happens to you, the ultimate translation is *trouble*!

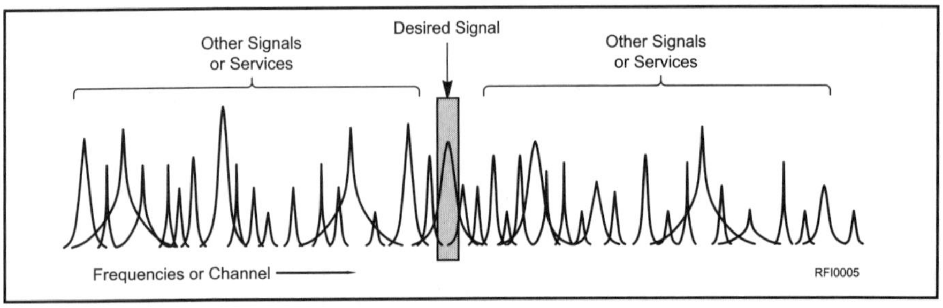

Figure 1.1—Every electronic appliance must select the signal it needs from all those present in the radio spectrum. The desired signal (in box) may not be as strong as the signals that must be rejected.

RFI Facts

All licensed transmitters must meet specifications set by the FCC. Those specifications are set in order to prevent interference to other communications services.

- Transmitter owners are responsible for the proper operation of their transmitters and conformance to the FCC regulations.
- Transmitter owners must reduce spurious (unwanted) emissions to whatever level is necessary to prevent interference to other *communications* services. (This provision does *not* apply to fringe-area TV reception.)
- Transmitter (including Amateur Radio) operators are obligated to conform to FCC regulations. Transmitter operators are not obligated to help consumers with RFI complaints that do not involve their transmissions (although they may *elect* to do so).
- FCC considers audio equipment (telephones, CD players, alarm systems and so on) that receives RFI to be *improperly functioning* as a radio receiver. Such improper function is a design inadequacy that should be corrected by the manufacturer. The FCC recommends that owners return such equipment (to the seller or manufacturer) as defective.
- The RFI susceptibility of consumer electronic equipment is limited only by the manufacturers' voluntary compliance with industry committee developed standards.
- The voluntary standards (and therefore manufacturers' designs of consumer equipment) do not provide for equipment to operate in close proximity to strong communications transmitters. Transmitter operators are not responsible for RFI in such situations.
- In general, all equipment owners are responsible for the proper operation of their equipment. *If the equipment is operating as intended and RFI persists, it is the responsibility of the equipment owner to modify the equipment as necessary in order to achieve the owner's operating goals.*

Example 1: RFI occurs as a result of the proper and legal operation of a licensed transmitter. The owner of the equipment *receiving* the RFI is responsible for corrective measures, which may include, but are not limited to added filters, shields and equipment modifications by qualified service personnel.

Example 2: Consumer equipment, such as a scanner, accepts RF energy from a nearby transmitter and produces RFI. The equipment owner is required to eliminate the RFI or remove the equipment from service.

Example 3: Amateur Radio transmissions prevent reception of "fringe area" TV stations. Since the TV receiver is in a fringe area (not in the intended TV-coverage range), the FCC does not protect the user. The user may either accept the interference or modify the TV system as needed to receive the desired channel (that may not be possible).

WHAT IS INTERFERENCE?

Let's start by defining the term *interference*. To some people, it implies action and intent. The statement, "You are interfering with my telephone," can sound like an accusation, even if it is intended as a statement of fact. From both a technical and social point of view, interference is defined as any unwanted interaction between electronic systems—period. No fault. No blame. It's just a condition.

There are three common types of interference:

Noise: Interference can be caused by an electromagnetic noise source. Defective neon signs, bug zappers, thermostats, electrical appliances, switches or computer systems are just a few of the possible sources of noise interference. Both you and your neighbors may be suffering from its effects. In some cases, the noise may be the result of a dangerous arc in electrical wiring or equipment and may provide warning of an unsafe condition that should be immediately located and corrected.

Overload: On the other hand, the world is filled with radio signals. Many cases of interference result from the inability of consumer equipment to reject strong (usually nearby) signals. This is known as *fundamental overload.* This fundamental RF energy is the signal from your transmitter that is intended and licensed to be there—the one

Figure 1.2—Apparently German amateurs can call on some *special* help in EMI cases.

you are trying to make as strong as possible at the receiving station on the other end of your transmissions. It is, by far, the strongest signal emitted from your station. You are not legally responsible for interference resulting from your fundamental signal. Any piece of consumer electronics equipment should be able to respond only to signals it is designed to receive (see Figure 1.1). Even if a nearby radio signal is being transmitted on its assigned frequency, the filtering and shielding in neighboring equipment may be inadequate to reject your strong fundamental signal. This condition is commonly called fundamental overload.

Spurious emissions: A radio transmitter could be inadvertently transmitting weak signals on a frequency not assigned to that transmitter. These signals are called *spurious emissions*. FCC regulations concerning spurious emissions are very clear. If interference is caused by spurious emissions, the operator of the transmitter must take whatever steps are necessary to reduce the spurious emissions as required by FCC regulations. Fortunately, modern transmitting equipment is manufactured to meet stringent regulations, and many radio operators are examined and licensed by the government. These federally licensed operators often have the technical skill to resolve interference problems that originate from their radio stations. FCC regulations are quite clear about spurious emissions: they must not cause interference to other services. As the operator of a transmitter, you must take whatever steps are necessary to eliminate interference from this cause. This almost always involves additional filtering, grounding or shielding of the transmitting equipment.

With all of these possibilities, it is difficult to guess which type of problem is causing your interference. Usually, only a technical investigation can pinpoint the cause and suggest a solution. This is where a spirit of cooperation and trust will pay off!

SCOPE OF THE PROBLEM

It's always important to remember our place as Amateur Radio operators in the overall scheme of things. Consider for a moment that the national governments extend to amateurs the privilege to operate in valuable radio spectrum. In a metropolitan area, amateur frequencies can have a commercial value of millions of dollars per MHz. The governments of the world have extended these privileges to amateurs because the world benefits from our existence. In addition to the emergency communications services often provided by amateurs, the world gains a reservoir of self-trained radio operators, skilled in operating practice and electronics technology. The amateur station may or may not be at fault in any particular case of interference (only a technical investigation can determine that for certain), but the amateur is *involved* in the problem. There is no better place to apply your technical skill, which has been gained from license study (and perhaps years of experimentation).

Present Engineering Resources

There are an estimated 300,000 electrical engineers in the United States. They are competent and dedicated people designing systems and products that must compete in a difficult marketplace. Unfortunately, most colleges don't offer courses in electromagnetic compatibility.

These engineers are asked to design systems that are compatible with other systems (including radio transmitters). The lack of EMC training and experience, coupled with the ever-increasing complexity and scope of electronic systems, can result in unexpected problems. Of 300,000 engineers, perhaps 6000 of them are actively engaged in EMC-related engineering. With so much engineering to be done, and such a relatively small percentage of engineers specifically familiar with EMC issues, the potential for incompatibility is fairly large. This is a worldwide problem, as demonstrated by a QSL card from Germany (see Figure 1.2). RFI obviously affects amateurs all over the world.

Increased Electronic Population

The electronics revolution has resulted in the proliferation of all sorts of home electronic devices. Thirty years ago, there were only TVs, telephones and various types of audio amplifiers. The opportunity for incompatibility was small, and the circuitry was unsophisticated (by today's standards).

Figure 1.3—The ultimate RFI problem! Sometimes the best solution is to temporarily cease operation! *(photo by John Frank, N9CH)*

The 1980s and '90s brought many new electronic technologies and associated new equipment. For example, one of the ARRL Lab Engineers saw an advertisement for a computer-controlled ac-power outlet strip. He jokingly noted that now a station might cause interference to a neighbor's extension cord! It was a joke, but it could also be reality. The use of microprocessors in such devices as toasters and set-back thermostats has multiplied the interference potential many times.

As our lives become filled with more technology, the likelihood of unwanted electronic interference increases. Every lamp dimmer, hair dryer, garage-door opener, radio transmitter, microprocessor-controlled appliance or remote-controlled new technical "toy" contributes to the electrical noise around us. Many of these devices also "listen" to that growing noise and may react unpredictably to their electronic neighbors.

Complex electronic circuitry is found in many devices used in the home. This creates a vast interference potential that didn't exist in earlier, simpler decades. Your own consumer electronics equipment can be a source of interference, or can be susceptible to interference from a nearby noise source. Interference can also result from the operation of nearby amateur, citizens band, police, broadcast or television transmitters.

Interference to consumer electronic equipment affects amateurs and their neighbors alike, although possibly in different ways. RFI can affect hams more directly. Hams and their families own consumer electronics equipment, too. Imagine the "personal diplomacy" problem that might occur if your station interferes with your teenager's phone call!

RFI can have far reaching implications. Figure 1.3 shows an interference problem no one wants to have! Fortunately, most RFI problems are not this explosive.

It is a Busy Street

Amateur Radio is only one of many possible interference sources. What about broadcast stations, taxicabs and police and fire services? What about cable TV leakage, unlicensed Part 15 devices (baby monitors, computers and so on)? Add power lines and electric motors to the list as well. They're all potential interference sources. Your neighbor's TV can even interfere with you!

Most hams have also experienced RFI *to* their stations, too. All of the above interference sources can also affect our station's operation. That "pop" in your station receiver when someone turns on a room light is RFI, too. Interference that happens to hams' own consumer devices is just as annoying to a ham as it would be to his or her neighbor.

It's a People Problem, Too!

Every RFI case involves people, and sometimes people compatibility is far more difficult to achieve than electromagnetic compatibility (EMC). You may have already been confronted by an unhappy neighbor with an RFI complaint. If so, take a few deep breaths, relax and read this chapter. It will take you step-by-step through the important aspects of working together to solve the common problem of RFI.

RESPONSIBILITIES

An Example

You, as an amateur, have purchased a transmitter that meets all of the FCC requirements for proper operation. You have installed it in a well-engineered station with proper grounding and filtering. You know your station is clean because you don't interfere with your own equipment. You have done nothing wrong.

The manufacturer of your neighbor's TV has designed and built the best possible product, constructed to meet hundreds of regulations set by dozens of federal regulatory agencies. (None of these regulations are directly related to EMC.) The product has probably met a few voluntary standards set by independent associations as well. Within the constraints of the law, the manufacturer has done nothing wrong.

Your neighbor has gone to the electronics store and has purchased a piece of equipment that has a fine reputation for quality and service. He has every right to expect his equipment to function as advertised. Clearly, your neighbor has done nothing wrong. Even so, when he turns on his set and you go on the air, you both have an interference problem.

So who is at fault? It should be obvious that no single individual is to blame. Everyone has done everything correctly, but the system has failed! Most cases of interference can only be resolved if all involved parties address their responsibilities fairly. No amount of wishful thinking (or demands for the "other guy" to solve the problem) will result in a cure for interference. Each individual has a unique perspective on the situation —and a different degree of understanding of the technical and personal issues involved. On the other hand, each person may have certain responsibilities toward the other and should be prepared to address those responsibilities fairly.

Two Erroneous Beliefs

Sometimes consumers believe that amateur stations spew out-of-band signals across the radio spectrum, and ruin reception for all around them. On the other hand, hams sometimes think consumer-equipment manufacturers deliberately leave out filters and shields needed to eliminate interference problems. Such blame-oriented viewpoints hinder the RFI-resolution process. In truth, manufacturers and hams both work diligently to prevent RFI problems.

The Radio Operator

It may be natural for some to assume that the transmitter operator is responsible for the interference [We have all heard, "It only happens when you use that transmitter!"—*Ed.*], but most interference is not caused by any rules violations by the transmitter operator (high-power 11-meter operation is one notable exception, but in some cases, the interference that results is exactly the same as would occur from legal 10-meter operation by a licensed ham).

FCC regulations require that amateur transmitters not emit spurious signals that interfere with other radio services. This is the sole *regulatory* responsibility of the radio operator. This doesn't apply in the case of interference to non-radio devices. In this regard, the FCC's own material is pretty clear—this type of interference is caused by inadequacies in the design of the non-radio equipment, not the transmitter.

Any individual who operates a radio transmitter, either commercial or private, is responsible for the proper operation of that radio station. All radio transmitters or sources are regulated by the FCC. The station should be properly designed and installed. It should have a good ground and use a low-pass filter, if needed. If consumer electronics equipment at the station is not suffering the effects of interference, you can be almost certain that the problem does not involve the radio station or its operation. However, if the interference is caused by a problem at the station, the operator must eliminate the problem there.

The manufacturers of amateur equipment are aware of the FCC requirements. Some commercially built amateur equipment must meet the FCC criteria ("certification") before it may be sold. All modern amateur equipment meets these federal standards. The ARRL regularly tests sample amateur transmitters as part of their Product Review process and advertising-acceptance policies. This helps ensure that modern amateur transmitters do not create spurious emissions that could cause interference.

You'll sometimes find that a manufacturer or utility company is willing to address an interference problem responsibly, but lacks the experience and training necessary to apply the correct solutions. You may need to apply your skills (and those of your ARRL Section Technical Coordinator or local club RFI committee) to help their personnel understand the technical issues.

It is not required by FCC rules, but the FCC hopes amateurs will cooperate with their neighbors, utility or repair people or the FCC, if necessary, to work toward a common solution. You may very well be an important part of finding a solution! Assume, for example, that your neighbor calls the cable company to service a cable leak that is letting your 2-meter signal into cable channel 18. If you do not make yourself available to put your station on the air for troubleshooting purposes, the cable company will not know if they have corrected the problem.

Keep Your Own House Clean

It is not exactly a responsibility, but you should first ensure that there are no interference problems in your own home! It may take a bit of work—installing high-pass filters on the TV, debugging telephone interference, getting the voices out of your stereo—but it will be time well spent.

There are several good reasons to start in your own home. You'll be able to demonstrate that RFI cures are not only effective, they cause no harm to the consumer equipment's proper operation. It is also a powerful diplomatic tool to tell your neighbor that you are sorry that they can't use their consumer equipment while you are on the air, but they can come over to your house and use *your* equipment while you are on the air! That usually helps them understand quickly that the problem might just be on their end!

The Role of the FCC

Lately it might appear that the FCC is "out" of the interference business. This is not completely the case, however. The FCC regulates radio transmitters. They have found, however, that with modern transmitting equipment, the vast majority of interference problems are *not* caused by any rules violations by the transmitter operator. The real problem is the susceptibility of consumer electronics equipment.

This often means that when RFI occurs, there is nothing for the FCC to enforce! The FCC rightfully believes that if consumer equipment is not working properly because it is receiving nearby radio transmitters, it should not become a federal case! This is a customer-service problem between the consumer and the manufacturer of the consumer device.

Nonetheless, the FCC realizes it will continue to receive inquiries from the public about RFI. They offer help to transmitter operators and consumers, primarily as an authoritative information source on self-help cures. Over the past several years, the FCC has made tremendous strides in creating unbiased, factual information about all types of interference problems, especially interference to non-radio devices such as telephones. The FCC clearly explains that the burden of resolution of most interference problems is *not* upon the transmitter operator, but with the manufacturer of the affected consumer equipment! Hams knew this all along, but in the past, the FCC offered scant material on the topic. Thanks to a major effort a few years ago by a national team of FCC staff, the RFI information the FCC offers is very useful and complete. (The ARRL was pleased to serve as a consultant to the FCC as this information was being developed.)

The former FCC Compliance and Information Bureau (CIB)[1] had created a publication, "Interference to Home Electronic Entertainment Equipment Handbook—May 1995 edition." This was released as Bulletin CIB-2. Although now out of print, a copy can be downloaded from the ARRL Web page (at **www.arrl.org/fcc/tvibook.html**). The former CIB had also written a bulletin specifically on telephone interference "What To Do If You Hear Radio Communications on Your Telephone," released as Bulletin CIB-10. A copy can be downloaded from the ARRL Web page (at **www.arrl.org/fcc/fcc_rfi_CIB-10.pdf**) Both bulletins have been reproduced in Appendix B and Appendix C. Other interference information is also available on the FCC Web site (at **www.fcc.gov/cgb/consumerfacts/interference.html**).

The FCC will do what it can to help consumers and radio operators resolve their interference problems. They expect everyone involved to cooperate fully. Experience has taught them that solutions imposed from the outside are not usually the best solutions to local problems. Instead, they provide regulatory supervision of radio operators and manufacturers. To help consumers, basic information concerning interference solutions is now available on the Internet through the FCC Compliance and Information Bureau Home Page. This basic information includes the *CIB Interference Handbook* and the *CIB Telephone Interference Bulletin*. The *CIB Interference Handbook* includes a list of equipment manufacturers who provide specific assistance with interference problems. The list also is available through the Commission's Fax on Demand at 202-418-2830. Callers should request document number 6904.

In many cases, the FCC serves as a mediator rather than a regulatory agency. Although the FCC Field Offices have solved a lot of RFI cases, mediation is not their primary purpose. The process is often slow, and the FCC solution may please neither the ham nor the neighbor.

If the FCC determines that a particular case of interference *is* caused by rules violations, they can and do avail themselves of the appropriate remedies. These usually consist of contacting the amateur and requiring that any violations be corrected. If necessary, the FCC could impose quiet hours on the amateur station, shutting down the station during a good part of the day. They could also fine an amateur who willfully violates any FCC regulations, although this is rare, especially in interference cases.

Manufacturer Responsibilities

The manufacturers also shoulder some responsibility for RFI problems. Public Law 97-259, enacted in 1982, gave the FCC the authority to regulate the susceptibility of consumer electronic equipment sold in the United States. The FCC, working with equipment manufacturers, decided to allow them to develop voluntary standards for RFI immunity and implement their own voluntary compliance programs. No system is perfect, especially a voluntary system, but the ARRL Laboratory staff has noted that RFI involving TVs, for example, seems to be decreasing.

Manufacturers of consumer electronics equipment are competing in a difficult marketplace. To stay competitive, most of them place a high priority on service and customer satisfaction. For example, many manufacturers have service information that can be sent to a qualified service dealer. Most manufacturers are willing to assist you in resolving interference problems that involve their products. The manufacturers are making some progress and we feel confident that they will continue to do so.

However, from reports received by the ARRL Lab, this is where the process often breaks down. The immunity standards are voluntary; there are no *regulations* that require them to help their customers. Some manufacturers and importers incorporate RFI immunity into their designs; others do not. Some manufacturers do not deal properly with consumers who report problems through the manufacturers' customer service representatives. In some cases, the *only* answers given by the customer service representatives who initially field the inquiry is to tell the consumer that the transmitter operator is running illegal power and that the consumer should contact the FCC!

This adds to the frustration of the people whose consumer electronics are being interfered with, adding fuel to an already hotly burning fire. This often puts the *social* burden back on the ham again. Neighbors of hams may not realize it at first, but the local amateur who has access to the information in this book and possible local technical advice probably represents the best source of help to find a solution to an interference problem! The ham is certainly the *closest* source of help.

Non-Radio Devices

When there is interference to audio devices (CD players, tape decks, most intercoms, burglar alarms, and so on) you are most certainly dealing with fundamental overload. *The FCC Interference Handbook* pamphlet indicates that when these devices are subject to interference, they are improperly functioning as radio receivers. The FCC clearly puts the responsibility for interference to audio devices on the manufacturer.

Contact the Manufacturer

The ARRL and FCC both encourage all people who have an interference problem to contact the manufacturer of the susceptible device. The Consumer Electronics Association, an arm of the Electronic Industries Association, can often help you locate a manufacturer. (See the Resources chapter for

[1]The functions of the former Compliance and Information Bureau (CIB) were enforcement and consumer information. The enforcement functions are now the responsibility of the Enforcement Bureau. Consumer information functions are now handled by Consumer & Governmental Affairs Bureau (CGB).

contact information.) Contacting the manufacturer helps ensure that the manufacturers (and their national association) are aware of the interference problems their customers are experiencing. Remember: interference that isn't reported officially doesn't exist!

Consumer Responsibilities

Consumers have responsibilities, too. They, too, must cooperate with the manufacturer, the radio operator and, if necessary, the FCC as they try to determine the cause of the problem. The FCC has gone to great pains to offer factual information designed to help consumers understand the issues surrounding RFI and to proceed correctly toward locating a cure. Consumers should educate themselves by reading the FCC's information and taking their advice. A consumer who refuses to discuss the problem or who demands that the transmitter operator correct the problem at the radio station is not doing his or her part toward finding a solution.

You are the Best Judge of a Neighborhood Situation

In general, hams should consider themselves as locators of solutions, not providers of solutions! The FCC requires that you operate your station legally. You are not required to filter your neighbors' equipment nor to buy them the necessary filters. In most cases, you do *not* want to fix their equipment at all! If you do, you may be walking into troubled legal waters. Some states require you to hold a repair license to perform even the simplest services—free or otherwise. Consider the future consequences of your actions as well. ARRL has reports of well-meaning amateurs who installed a high-pass filter inside a neighbor's TV, right at the tuner. When the picture tube on the old clunker suddenly went bad, the neighbor claimed that the filter caused the failure!

This doesn't mean you should never offer a helping hand, but it does mean you should look before you leap. You are the best judge of your neighborhood situation. Only you can decide what kind of assistance and diplomacy is appropriate.

It is sometimes counterproductive to be too accommodating. A helpful attitude sometimes gives people the idea that you will do anything they wish. They may wish that you stop operating permanently, or at least when they want to watch TV. You can best judge what constitutes the optimum balance between accommodation and cooperation. As long as your transmitter is clean, you have a federally granted privilege to operate. If you decide to curtail operation until a TVI problem is solved, make certain it is clearly understood that you are taking a voluntary, temporary measure until the necessary filters can be ordered or repairs can be made.

Remember that many people take advantage of others in any way they can. We live in a litigation-prone society. There is an unknown degree of liability (and risk) when dealing with strangers. Even installing a high-pass filter in a neighbor's external feed line can create problems. Each amateur must decide the best approach for each individual case. It is a good idea to have the neighbor install any external filters used to cure TVI. This minimizes liability, and helps show the neighbor that fundamental overload and its cures are the problems of the equipment manufacturer and owner. Amateurs can provide assistance and information about solutions.

ARRL-affiliated clubs with ARRL club-liability insurance can have the *official* activities of a club-authorized TVI committee covered if that activity has been properly

Local Help

What's a TC?

Technical Coordinators (TC) and Technical Specialists (TS) are ARRL field volunteers who provide technical support and guidance to amateurs in their Section. They should be contacted at the beginning of every RFI problem. Most TCs prefer to be called before tempers have flared and a situation has become difficult.

How to Locate Your ARRL Section TC

Grab a copy of *QST*; make it a recent one. Find the Section Manager's (SM) name for your ARRL Section. Ask the SM for the name and address of your Section TC.

Did you say you haven't got a *QST* handy? (For shame!) Well, all isn't lost. This information is available on the ARRL Web site. See **www.arrl.org/sections/** to find Section Web pages and Section appointments, including TCs. Don't have a computer? Give ARRL HQ a call or send an e-mail message; ask for Field and Educational Services, and inquire about the TC for your Section.

So What Do They Do?

Technical Coordinators are volunteers. As such, they each have certain capabilities and areas of understanding about Amateur Radio technology. Here are some general guidelines for TCs.

The Technical Coordinator:

1) Supervises and coordinates the work of the Section Technical Specialists (TSs).

2) Encourages amateurs in the Section to share their technical achievements with others through the pages of *QST*, at club meetings, hamfests and conventions.

3) Promotes technical advances and experimentation at VHF/UHF and with specialized modes, and works closely with enthusiasts in these fields within the Section.

4) Serves as an advisor to radio clubs that sponsor training programs for obtaining amateur licenses or upgraded licenses in cooperation with the ARRL Affiliated Club Coordinator.

5) In times of emergency or disaster, functions as the coordinator for establishing an array of equipment for communications use and is available to supply technical expertise to government and relief agencies to set up emergency communications networks, in cooperation with the ARRL Section Emergency Coordinator.

6) Refers amateurs in the Section who need technical advice to appropriate local TSs.

7) Encourages TSs to serve on RFI and TVI committees in the Section for the purpose of rendering technical assistance as needed, in cooperation with the ARRL OO Coordinator.

8) Is available to assist local technical program committees in arranging suitable programs for ARRL hamfests and conventions.

9) Conveys the views of Section amateurs and TSs about the technical contents of *QST* and ARRL books to ARRL HQ. Suggestions for improvements should also be called to the attention of the ARRL HQ technical staff.

10) Works with the appointed ARRL TAs (Technical Advisors) when called upon.

11) Is available to give technical talks at club meetings, hamfests and conventions in the Section.

In addition to the above duties, TCs and/or TSs in a given ARRL Section often serve as advisors for radio frequency interference (RFI) issues. Since RFI can drive a wedge into neighborhood relations, it is frequently a TC/TS with a cool head who resolves such problems.

TCs and TSs may also be asked to work with other ARRL officials, or represent the ARRL at technical symposiums in industry. They may serve on CATV advisory committees, or advise municipal governments on technical matters relating to the Amateur Radio Service.

described on the insurance application form. Contact ARRL HQ for details. A homeowner's or renter's insurance policy may also protect you from liability. Check with your insurance agent to find out if you have this coverage, or if a rider can be written.

Costs

Don't forget money issues. Repairing consumer electronic equipment usually isn't free. Buying RFI filters requires a bit of wallet-digging, too! These issues should be discussed up front, *before* filters are ordered or that call is made to the service department. You're *not* responsible for purchasing filters or repairing defects in your neighbors' consumer equipment. If you want to be neighborly and buy a filter for a neighbor, that's your choice. However, doing so may set a precedent. If you live in an apartment building where there are hundreds of telephones to contend with, you may have to re-evaluate your generosity! What if there are other neighbors in the area experiencing similar problems? You may be setting yourself up to spend a lot of money on filters! Other than problems that originate from your station, you should consider yourself as an advisor, not a service technician or parts supplier!

Reporting

You may be surprised to know that the number of reported cases of interference to consumer electronic equipment in recent years has been very small. As a radio service, this is our fault! Amateurs are notorious for not reporting RFI problems. Even if you come up with a solution, contact the manufacturer! Working with manufacturers makes them aware of the need to continue to develop better shielding and filtering methods. It also demonstrates to your neighbor that the manufacturer should receive a little of his anger and frustration too!

PERSONAL DIPLOMACY

You can't overestimate the importance of personal diplomacy when you're trying to solve a problem that involves two or more people. Everyone who is involved in an interference problem should remember that the best solutions are built on cooperation and trust. This is a view shared by electronic equipment manufacturers, the FCC and the ARRL.

The first contact between a ham and a neighbor with an RFI problem is often the most important; it is the start of all future relations between the parties. The way you react and behave when you first discuss the problem with other individuals, such as a neighbor or representative of a utility or manufacturer, can set the tone for everything that follows. If one party approaches the other with threats, demands and hostility, human nature urges a response in kind. From an amateur perspective, if the ham reacts to a neighbor by spouting off about the "right to operate" and doesn't offer either empathy or assistance, the neighbor may refuse to cooperate in any future investigation that may be required. This is really not what either party wants!

Many ARRL Field Organization volunteers and RFI professionals provided input for this book. Many agree that personal diplomacy is an important factor in the solution of RFI problems. This applies to *all* affected parties. In some of the less successful cases in the ARRL HQ files, poor (or hostile) communication often resulted in unsatisfactory resolutions. Both hams and their neighbors have been guilty of bad attitudes, hostile conversations and sometimes just plain bad manners.

That First QSO!

Imagine the following scenario: Joe Hamm has just passed his FCC examination and awaits the arrival of his ticket so he can have his first amateur contact. The big day finally arrives and Joe rushes into his newly assembled station to make use of the privileges he has worked so hard to earn. After about an hour, he finally gets over his initial nervousness and establishes contact with another ham—in Ireland! After about 15 minutes of excited conversation, Joe hears a loud knock on his door.

When Joe opens the door he is confronted by Sam Neighbor, who tells him that he is causing interference because Sam can hear Joe on his telephone. If Joe is lucky, Sam is pleasant and understanding. If Joe is not so lucky, Sam might be a bit more forceful, spouting demands and threats of FCC (or personal) intervention. Joe, like most hams first confronted with this situation, doesn't know what to do. Maybe it is Joe's fault? Should Joe get off the air? Who can he turn to for help? Joe and Sam are not alone. Thousands of hams and consumers have faced this situation since radio was first developed.

If this happens to you, begin by accepting the fact that your neighbor doesn't enjoy having his or her lifestyle hampered by RFI. Put yourself in his shoes. Admit that RFI is highly annoying. (Both of you can quickly agree on that point!) Calmly explain that you are responsible—by law—for the proper operation of your station. Assure him that you'll check your equipment right away and make any necessary corrections.

With any luck, the situation will begin to calm down. Now is the time to explain to your neighbor that the root cause of the problem could also be from a source other than your station. Perhaps his own equipment is to blame. Before he has a chance to misunderstand the last point, tell him that you're willing to help him locate a solution to the problem—even if it's not your fault.

Be Cool!

An argument never solves anything, so keep your cool when dealing with people involved in an RFI problem. This advice applies to the ham, the neighbor and any others helping solve the problem on behalf of these two key parties in the RFI problem.

By the time your neighbor becomes angry, he or she has probably decided that *you* are the cause of the problem. To your neighbor, your station is an imposition and you simply

Communications Service Transmitters
CB
Amateur Radio
Police
Cable TV (CATV)
Business or other two-way
Aircraft (near airports only)

Electrical Noise Sources
Doorbell transformers
Toaster ovens
Electric blankets
Fans
Heating pads
Light dimmers
Appliance switch contacts
Aquarium or waterbed heaters
Sun lamps
Furnace controls
Smoke detectors
Smoke precipitators
Computers (and video games)
Ultrasonic pest-control devices
Lights: fluorescent, mercury vapor and touch-controlled
Neon signs
Power Company Equipment
- Defective line insulators
- Loose or unbonded hardware
- Discharges from defective lightning arrestors
- Defective transformers

Electric fences
Alarm systems
Loose fuses
Sewing machines
Electrical toys (such as trains)
Calculators

Figure 1.4—A few possible RFI sources. Each transmitter or noise source can interfere with nearby equipment. The Amateur Radio Service is *not* the only possible source of RFI!

Interference Classifications and ARRL Assistance

Local ARRL volunteers are able to assist with the many types of RFI you might encounter. For them to best serve you, do some homework before calling. For example, is the interference best handled through assistance by a technical volunteer? Should you call the FCC and can and will they do anything? There are different kinds of interference and different ARRL volunteers to deal with them. Your ARRL RFI team can include Technical Coordinators and Technical Specialists, Amateur Auxiliary members (Official Observers and Official Observer Coordinators), Local Interference Committees and RFI/TVI Committees as well as Section Managers. All of these volunteers work together to help you solve your RFI problem.

It is important to understand that RFI problems can have technical or regulatory solutions (or both). Let's assume that you're already clear on amateur-to-nonamateur interference. If you're causing interference to your neighbor's brand new big screen television set, that is amateur-to-nonamateur interference and it is best handled with assistance from your ARRL Technical Coordinator (TC). What about the other kinds of interference? There are two general areas of "other" interference. They are amateur-to-amateur and nonamateur-to-amateur. The regulatory aspect of interference will be discussed in detail in Chapter 17.

AMATEUR-TO-AMATEUR INTERFERENCE

If you determine the source of the interference to be amateur, you must define it even more closely:

Inadvertent Interference

The first type of amateur-to-amateur interference is inadvertent interference. It is unintentional interference. The amateur may not know that another amateur is receiving interference from his or her station. This is not a major problem and education seems to be the best and most logical solution.

Careless Interference

The second kind of amateur-to-amateur interference is careless interference that is intentional but not premeditated or recurring. While this type of interference is deliberate, it is usually short-lived and caused by temporary eruptions of temper. Again, this is not a major problem in the Amateur Radio Service.

Harassment

The third kind of interference is classified as harassment. This is a more serious form of intentional interference. A harassing offender is likely to do anything to make life miserable for his victims. While inadvertent and careless interference is usually isolated and of short duration, harassment is recurring. It ranges from a "kerchunker" on a repeater to a jammer who uses foul language or any other means to disrupt the operation of others.

Malicious Interference

The fourth type of interference is malicious interference, which includes harassment and more. Malicious interference is recurring and usually meets the criteria in this example: when two or more amateurs are in a conversation and are interrupted by an interferer, they move to another frequency and the interfering station follows them to continue interfering. That's malicious interference. It is the most serious form of amateur-to-amateur interference and is clearly actionable by the FCC.

So, What Now?

In instances of intentional, harassing and especially malicious amateur-to-amateur interference, assistance is available from ARRL.

Members of the Amateur Auxiliary to the FCC's Enforcement Bureau are authorized to formally assist the FCC with data gathering for amateur-to-amateur interference problems. The Amateur Auxiliary is comprised of ARRL Official Observers who spend many hours of their time each week listening on the air for Part 97 rule violators. An ARRL Official Observer Coordinator oversees the Amateur Auxiliary in a particular section. The complaint must constitute a specific Part 97 rule violation. Contact your ARRL Section Manager for information about the Amateur Auxiliary and for the name of an OO or

don't care. If things have already gotten out of hand, all parties may have contributed some of the ill will that led to the impasse. If both parties want a satisfactory resolution (and they usually do), they must put aside resentment, pride and bad feelings and get back to the business of locating a solution. Apologies never hurt anyone!

At this point, it is very important to build a foundation of cooperation so that the RFI problem can be resolved. If your neighbor is already angry, it is doubly important that you keep calm. If you cannot do so, contact one of the volunteers listed later in this chapter. Sometimes a third party can build a bridge between warring neighbors.

If you are willing to build that bridge (translate: if you want to solve your RFI problem), begin by expressing empathy, understanding and concern. You have an equal desire for a satisfactory solution. The simple truth is that you both own equipment that was purchased, and is operated, with the best of intentions. Some of that equipment interacts with some other equipment in an undesired fashion. It is possible to reduce that interaction, but teamwork is needed. By working together, you can determine how the RFI is happening and select the most practical remedy.

Understanding

First, ensure that your neighbor (or family member) understands the issues. When you discuss it with them, be diplomatic. RFI can be a stressful situation and tempers *can* flare. First, help your neighbor understand the issues behind the interference. The FCC material explains things authoritatively.

Do ensure that things get started on the right foot. Your neighbor probably won't understand the complexities of RFI causes and cures. Be prepared to do a good job explaining things in non-technical terms. The ARRL and the Consumer Electronic Association jointly publish a pamphlet designed to explain interference to your neighbor. The text of this pamphlet has been reproduced in Appendix D, although an actual copy of the pamphlet is usually more effective with your neighbor. A copy can be downloaded from the ARRL Web site at **www.arrl.org/news/rfi/neighbors.html**.

Start With You!

When you first discuss RFI with your neighbor, start by talking about FCC regulations. If you start telling your neighbor that the FCC has regulations that do not permit your transmitter to emit signals that interfere with other radio services, your neighbor *will* be listening—you are talking about *your* responsibilities. From there, you can explain that you have put the necessary filters on your transmitter. If you have followed the sage advice to keep your own house interference free, you can then explain that your own consumer electronics equipment works well. Demonstrate this to your neighbor, and show off any filters you have installed. You can then offer to help your

OOC in your section. Some ARRL Affiliated Clubs have Local Interference Committees. LICs deal only with amateur-to-amateur operational difficulties. LICs can often provide direction-finding equipment and expertise to locate local caused-by-amateur problems. ARRL HQ can also refer you to the appropriate sources of assistance. Contact ARRL by mail at 225 Main St, Newington, CT 06111; by telephone at 1-860-594-0200; by fax at 1-860-594-0259 or by e-mail at **reginfo@arrl.org**. The ARRL Web page can be found at **www.arrl.org**.

The FCC should not be contacted since amateur-to-amateur interference problems are expected to be solved among the amateur community and through the Amateur Auxiliary. If the problem involves a non-amateur or "bootlegger," the Amateur Auxiliary can also help. This provision is meant to cover the unauthorized use of an amateur transceiver.

NONAMATEUR-TO-AMATEUR INTERFERENCE

Much of the information presented above deals with nontechnical matters. They are often best resolved by amateur-to-amateur peer pressure and groups such as the Amateur Auxiliary. A growing number of difficulties can be solved by individuals who have a clear understanding of the technical reasons behind certain kinds of interference. It's no secret that as more poorly designed consumer devices enter the marketplace, the potential for EMI problems to (and from!) Amateur Radio equipment increases dramatically.

Technical Problems

Amateurs sometimes receive interference caused by non-amateurs; these generally require technical solutions. Interference to amateurs from non-amateur devices should be reported to your ARRL Technical Coordinator as detailed earlier in this chapter. Assistance can also be sought from a club RFI or TVI Committee. Examples include a neighbor who uses a touch lamp, the local cable company with leaky cable and a local power company with a bad insulator on a line. All of these are capable of causing interference to the Amateur Radio Service. While a TVI or RFI committee will likely assist with EMI or RFI, the purpose of an LIC is more regulatory in nature. Your ARRL Section Manager can refer you to your TC and to other local sources of assistance. The ARRL Technical Information Service can also refer you to appropriate sources of assistance. It can be contacted by mail at: ARRL, 225 Main St., Newington, CT 06111; by telephone at 1-860-594-0200; by fax at 1-860-594-0259 or by e-mail at **tis@arrl.org**. The ARRL Web page at **www.arrl.org/tis/** contains a great deal of technical assistance.

Regulatory Problems

Non-amateur to amateur problems can also be regulatory problems. As mentioned earlier, it isn't always possible to know whether over-the-air interference is being caused by an amateur or a non amateur. If the interferer is found to be a non-amateur, the Amateur Auxiliary may continue to gather information on the matter since the Amateur Auxiliary has been expanded to cover these instances. This is sometimes the case when, for example, an unlicensed individual illegally uses the call of an amateur (a "bootlegger.") If the interference is caused by a non-amateur transmitter using a non-amateur device in the Amateur Radio Service, it is the responsibility of the Federal Communications Commission to solve these types of problems. An example is a leaky cable system. The FCC can be contacted at 1-888-CALL FCC. The FCC Web page can be found at **www.fcc.gov**.

In some cases, even the FCC doesn't have authority over certain types of interference that originates outside the US. Two examples are foreign broadcast interference and interference caused by a strong foreign military radar. The ARRL Monitoring System reports instances of non-amateur intrusion on amateur frequencies each month. This information is relayed to the FCC Treaty Branch in Washington, DC, for appropriate foreign governmental action.

IN CONCLUSION

ARRL programs can help assist with each kind of interference you're likely to experience. Each Amateur Radio operator is encouraged to do his or her fair share of the job by determining the kind of interference before making contact with the appropriate individual. Once you have determined whether the problem is technical or regulatory (or both) in nature, the assistance you require is only a phone call away.—*John C. Hennessee, N1KB (SK), former ARRL Regulatory Information Specialist*

neighbor find similar solutions.

Once you get your neighbor to understand that some interference can't be cured by making changes to your station, you can become the neighborhood "good guy" by locating cures for the interference, even though it is not directly your fault. Your neighbor may or may not believe your explanations, but if you are willing to help anyway, that is often enough for good neighborly relations to take over where technical misunderstandings might otherwise get in the way.

No Fault RFI!

Good RFI-problem diplomacy removes the need to determine who is at fault. Positive conflict resolution requires that both parties relinquish adversarial roles and realize that they share a common problem. Working together resolves the conflict, to the benefit of both parties. Your neighbor may not understand the technical issues or know about helpful ARRL programs. He or she usually is not sure how to approach you about the problem. A neighbor usually reacts positively if you express a willingness to help solve the problem regardless of fault.

Neighbors usually cooperate if you offer assurance that solutions are possible and outline how you can help. If you are willing to locate solutions to the problem, the neighbor should be willing to put those solutions into place.

COOPERATION, INVESTIGATION AND ACTION

Once all who are involved in an RFI problem have decided to work together, they are closer to a solution. That makes it possible to get started on the technical aspects of curing RFI problems. Effective RFI solutions require cooperation, investigation and action. This chapter tells how to establish the needed cooperation. The rest of this book contains the information you need to resolve the vast majority of all RFI cases satisfactorily.

Education

RFI problems can be complex. Stock answers do not always result in solutions. It is important for all concerned parties—the amateur, Technical Coordinator, Technical Specialist, TVI Committee members and the neighbor—to learn as much as possible about the subject of RFI. Books about RFI, Amateur Radio magazine articles, local clubs and other hams are all possible sources of information and education.

A good step toward solving an RFI problem is to learn about RFI. Some hams take this first step toward solving an RFI problem before they ever get on the air. The material and books hams study to get their licenses discuss the concepts of interference and some of the basic remedies. Some RFI cases call for more information than is presented in license-study material. Amateur Radio magazines have covered the subject since the 1920s. This book explains in detail how basic RFI problems occur and what can be done to cure them. Even so, RFI can present complex problems, and not all of these problems have standard solutions. Learn the fundamentals of RFI. Study troubleshooting techniques and the application of technical solutions.

If you are involved with the interference problem, organize yourself. Obtain local help, if possible, and, if necessary, the help of trusted, diplomatic friends. Obtain the information and materials necessary to work toward a solution.

Keep a Log

If you have gotten this far, hopefully you and your neighbor have agreed to stop blaming each other and work together to find a solution. But what if your neighbor thinks you were causing interference on a day you weren't even home?

Even though it's not a legal requirement, it's a good idea to keep a detailed station log. Now that you're involved in an interference issue, consider it a necessity! You and your neighbor should keep written records. Your station log will show when you are on the air, and on which bands and modes. If your neighbor writes down information every time interference occurs (dates, times, severity, etc.) you can quickly determine if you are involved in the interference problem, and what bands, modes, power and antenna combinations cause the problem.

Your neighbor should note which piece of equipment experienced the interference, what channels or frequencies were involved, the date and time the interference occurred and a description of the interference and its severity.

If you are lucky, a comparison between your log and his log will indicate that the interference isn't coming from your station. On the other hand, if your signal is the source of the problem, your neighbor is the lucky party—although he may not see it that way at first. As an Amateur Radio operator, you have access to the technical resources necessary to solve the problem (either from your own knowledge and experience, or with the help of other hams like your Technical Coordinator or local RFI expert). This is not necessarily true if the source of the problem is a business-band or citizens-band transmitter, for examples.

"It Ain't Me?"

After discussing things calmly with your neighbor, one of the first troubleshooting steps is to determine if the interference is really related to your transmissions. Conduct on-the-air tests to determine if the interference is coming from your station and what bands, modes, antenna and power levels are causing the problems.

The world is filled with radio transmitters and noise sources. When these operate in proximity to susceptible equipment, interference often results. Figure 1.4 shows just a few possible RF sources. Each transmitter or noise source can interfere with nearby equipment. The Amateur Radio Service is not the only possible source of RFI.

The station log is one valuable tool to prove that a station is not involved in a particular interference case.

Sometimes, noninterference problems can appear to be actual interference. Telephone RFI problems can be related to power lines or crosstalk within the telephone-company circuitry, most people have heard others talking faintly in the background on the phone. Electronic equipment can become noisy, due to internal component failures. If the RFI does *not* follow transmission patterns, and you can demonstrate that to the complainant, your role is over. (This is an example of my favorite interference problem—"It ain't me!"—*Ed.*)

When interference begins and ends with RF transmissions, RF is somehow involved in the problem. This does not mean that the transmitter is *causing* the problem. Since consumer electronic equipment is not intended to receive Amateur Radio signals, the affected equipment is not functioning properly. Continue troubleshooting to discover the exact device that causes the problem and determine possible solutions.

In a recent case reported to ARRL HQ, the husband said that they could hear voices but they were not intelligible. The wife said "Oh, we could hear a few words . . . and by the way what is this ten-four and good-buddy stuff they are always talking about?" The ham had been 98% convinced that he was not part of the problem, and this conversation strengthened that conclusion.

It may be natural to laugh at this point, but tact is required. Give a short explanation of the difference between municipal communications, business communications, Citizens Band operations and Amateur Radio. Many citizens lump all of these activities into one category—"ham radio." (Of course, you may also be able to show your neighbors enough about Amateur Radio to persuade them to get their licenses! If you can't lick 'em, join 'em, as the old saying goes!)

Testing

Offer to arrange a test. Ask your neighbor to invite a friend to visit your shack during the test. In addition, ask you neighbor if it would be possible for one of your friends to monitor the test at your neighbor's home. Having impartial participants will make you and your neighbor more comfortable with the outcome—whatever it may be. Be sure to choose your assistant carefully. Select someone who is known for diplomacy and tactfulness.

Your test must be thorough. Transmit on each band and mode you normally operate. If you have a beam antenna, aim it in different directions while you are transmitting. Try various power levels, too. Ask your friend to keep detailed notes of the results. A radio or telephone link between you and your friend is almost a necessity.

This testing will help demonstrate whether your station is involved, and, if so, just what bands, modes and frequencies are part of the problem. Even if your test proves that your station is not at fault, don't just drop the problem in your neighbor's lap and say "Good luck!" It is a neighborly gesture to offer to help find a solution, even if the interference is really caused by something other than your station.

HOW TO FIND HELP

Most consumers do not have the technical knowledge to resolve an interference problem. Even so, it's a comfort to know that help is available. Gather information about interference. The FCC and ARRL have self-help information packages or books. If the problem involves an electrical-power, telephone or cable-television system, contact the appropriate utility company. They usually have trained personnel who can help you and your neighbor pinpoint the cause of the problem.

The amateur community has programs to help hams and consumers with RFI problems. Start with the ARRL Section Technical Coordinator (TC). The TC knows the best help available in your area. He or she may refer you to an assistant Technical Specialist, a local TVI Committee, the Amateur Auxiliary or a local club that has an RFI expert or two. The technical needs and resources of each ARRL Section are different, and each Section Manager and TC have determined what works best for their own Section. If you don't know the name of your Section TC, contact your Section Manager (SM). Any recent *QST* gives you the name, address and usually the telephone number of your SM. ARRL HQ can also supply the name and address of your TC. This book refers to such help collectively as the Technical Coordinator or TC (to remind you that the TC is the first place to turn for help with RFI problems).

Most consumer electronics manufacturers are willing to help you. The owner's manual, or a label on the equipment, may give you information about interference immunity or tell you who to call about interference problems. If not, the Consumer Electronics Association will be able to give you the address of your equipment manufacturer's general customer service personnel. The manufacturers know their equipment better than anyone else and will usually be able to help you.

EVEN THOUGH IT'S NOT YOUR FAULT . . .

Many interference problems are not the fault or direct responsibility of the amateur. Although it is not required by FCC regulations, however, sometimes amateurs may need to make some operating changes at their stations to reduce RFI. (The ham who lives in an apartment building may be told to fix the problem or stop operating.) A station licensee may *elect* to make changes to improve the situation, but that decision does not establish cause or responsibility.

One obvious solution is to reduce the transmitter's output power. Many a QRP operator got started with QRP to eliminate problems caused by apartment living! Different operating modes also have different RFI potential, especially in cases involving audio rectification. SSB might cause objectionable interference, while CW results in only a few minor pops and thumps. Under some circumstances, RTTY might create almost no interference at all!

One of the most effective cures can also improve station effectiveness! Moving that antenna high and in the clear will maximize your signal strength at that coveted DX station while minimizing your signal at neighboring electronic devices. Not a bad trade-off! If moving an antenna a few feet can solve a neighborhood (or family) dispute, that may be the best solution. We amateurs create our own devil, so to speak, with antennas near the house, on the house and sometimes in the house. This results in strong RF fields in nearby homes. It is much better to send the signal skyward (to the other end of the QSO) than to neighboring equipment.

Even a slight increase in the distance between antenna and house could make a big difference. It is not practical to move a tower to solve an RFI problem, but moving one leg of an inverted-V might be an easy way out. This is not an ideal solution because it does not cure the fundamental problem, which is a often found in the consumer equipment or installation, but it is mentioned here so a ham can consider all alternatives.

Temporary Measures

Until the problem can be resolved, you can make some goodwill gestures. As a temporary measure, reduce your output power. (You may discover that you didn't need all those watts, anyway!) If you have a beam antenna, don't point it at your neighbor's house. Above all, try to gain some perspective on the situation. Amateur Radio may be your passion, but it doesn't mean a thing to your neighbor. Attempting to justify RFI by saying "There was a rare DX station on 10 meters and I just had to work him" may just sound like "ham lingo" to him and probably won't get you very far.

Hams Can Help

Your neighbor may not see it this way at first, but in cases of RFI that involve Amateur Radio, he or she is actually fortunate. At least hams have some technical ability (as demonstrated by an FCC Amateur Radio license). They can help determine the causes of RFI and suggest cures. Interference is caused by all radio services and many other unintentional sources of radio noise. If the problem were caused by a CB operator or the local fire department transmitter, there might not be an easy source of technical help. Not only is Amateur Radio noted for being helpful, but the nearby ham is also the *nearest* source of possible help.

Hams have access to the expertise of other hams, through the American Radio Relay League (ARRL) Field Organization. The ARRL Section Technical Coordinator (TC), Technical Specialists (TS) and local RFI Committees have all volunteered to help with RFI trouble. Other hams in the area may be willing to help as well.

MOST RFI CAN BE CURED!

The application of education, personal diplomacy, manufacturer contact and technical solutions (such as external filters and shields) can cure most RFI cases. It takes cooperation between the consumer, the manufacturer and the radio operator. With a little bit of work, you and your neighbor can both enjoy your favorite activities in peace.

Chapter 2

EMC Fundamentals

To learn any subject well, you have to start with the basics. This chapter covers the fundamentals of electromagnetic compatibility (EMC), laying the foundation for the practical RFI cures presented in the chapters that follow.

By Bryan Bergeron, NU1N
27 Sterns Rd #8, Brookline, MA 02141
and Ed Hare, W1RFI,
ARRL Laboratory Supervisor

Most consumer equipment uses electronic or electromagnetic signals, as do all radio transmitters. There is really no difference between the normal signals used by consumer electronics equipment and the ones used by radio transmitters, except that one signal is wanted and the other is unwanted. Which signal is which is a matter of perspective.

Knowledge about the fundamentals of EMC and RFI is one of the most valuable tools for solving RFI problems. RFI is any electrical interaction at RF that affects normal equipment operation. RFI is present when unwanted RF signals adversely affect the performance of a device. The term RFI covers interference from radio-frequency sources. A more generic term, *electromagnetic interference (EMI)* covers all electrical interactions, ranging from radio-frequency to audio-frequency sources, such as induced hum from a transformer, and electrostatic sources (ESD—electrostatic discharge). These are all electromagnetic phenomena, so EMI applies across the entire spectrum. Most amateurs are concerned with interference that affects or results from the operation of their radio stations. Although the terms "EMI" and "RFI" are usually interchangeable, RFI has become the more popular term among amateurs.

WHERE TO BEGIN—A BRIEF OVERVIEW

Resolving RFI is a three-step process, and all three steps are equally important:

- Identify the problem.
- Diagnose the problem.
- Determine what steps are necessary in order to eliminate the problem.

An accurate diagnosis, regardless of the problem, requires familiarity with relevant technology and troubleshooting procedures. This chapter lays the theoretical foundation toward those solutions. The principles of RFI diagnosis are explained in the Troubleshooting chapter.

MANIFESTATIONS OF RFI

Is it really RFI? Before solving a case of suspected RFI, verify that the symptoms actually result from external causes. A variety of equipment malfunctions can look like interference. It is important to be able to differentiate between RFI and equipment malfunction. Characterizing RFI is difficult because the symptoms depend on the nature of both the emitter and the susceptible device.

The basic causes of RFI can be grouped into several major categories:

- Spurious emissions from a transmitter.
- Fundamental overload effects.
- Intermodulation or externally generated spurious signals.

Spurious Emissions

All transmitters generate some (hopefully few) RF signals that are outside their allocated frequency bands. These out-of-band signals are called *spurious emissions*, or *spurs*. Spurious emissions can be discrete signals or wideband noise. They can be harmonics, which are signals at exact multiples of the operating (or *fundamental*) frequency or other discrete spurious signals caused by the superheterodyne mixing process used in most modern transmitters.

These emissions can cause interference to other services. Any emission from a transmitter other than the fundamental signal, including mixing byproducts and harmonics, is referred to as a spurious emission or signal. All transmitters produce some spurious signals. The FCC regulations specify the permitted levels of spurious emissions for different power levels and operating frequencies. (See the RFI Regulations chapter.) An amateur is responsible for harmful interference caused by spurious emissions from an amateur transmitter. A definition of harmful interference is found in the Regulations chapter.

Spurious emissions include harmonics, mixing products and parasitic oscillations. Parasitic oscillations usually bear no direct relationship to the fundamental frequency. The FCC requires that all commercially made amateur HF transmitters installed (FCC terminology for put into operation) ***after*** January 1, 2003 suppress spurious energy at least 43 dB below the RF output level. For example, all spurious energy must be 5 mW or less for a 100-W HF transmitter. At VHF, for transmitters with 25 watts or more power, the spurious energy must be at least 60 dB below the output power. Although all transmitters generate harmonics, they can be reduced to the required levels by proper transmitter design and filters.

Harmonics

The most common spurious emissions from transmitters are harmonics. Harmonics are caused by nonlinear operation of oscillator and amplifier stages—no amplifier design can avoid generating some harmonics. Harmonics are always exact multiples of the operating frequency. For example, the second harmonic (×2) of a 14.2 MHz signal is 28.4 MHz; the fourth harmonic is 56.8 MHz and so on. The 56.8-MHz harmonic from the last example falls within TV channel 2. In some rare cases, a circuit can generate subharmonics, which are exact divisions of the operating frequency, one-half, one-third, etc.

Harmonics are a common cause of interference, usually to televisions. The filters associated with circuitry usually ensure that the harmonics are attenuated or are kept confined to the circuitry generating them. If the filters are inadequate, the result can be the radiation of the harmonics and interference.

Mixing Products

Other discrete signals are inadvertently produced as part of the superheterodyne mixing process used in most modern amateur transmitters. Such mixing products can appear above and below the operating frequency. While harmonics are produced in the mixing process, and harmonics play a part in the production of other signals, the term "mixing products" usually is not applied to harmonics. Figure 2.1 shows the output of an amateur transmitter, with harmonics and mixing products present.

Although the products shown in this figure are typical of an amateur transmitter, the levels at which they are shown exceed the FCC limit and therefore are not typical.

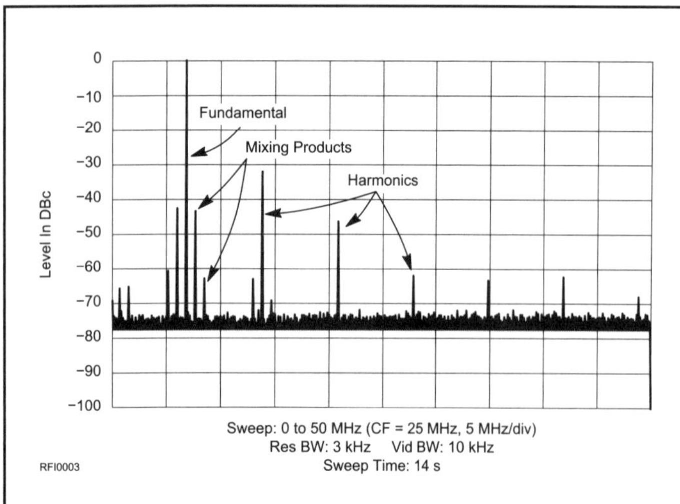

Figure 2.1—A spectral photo from a not so typical amateur transceiver. The fundamental (14.2 MHz) appears at left. Harmonics are visible at 28.4 and 42.6 MHz. Amateur stations are subject to stringent FCC spectral-purity regulations (this transceiver does not comply).

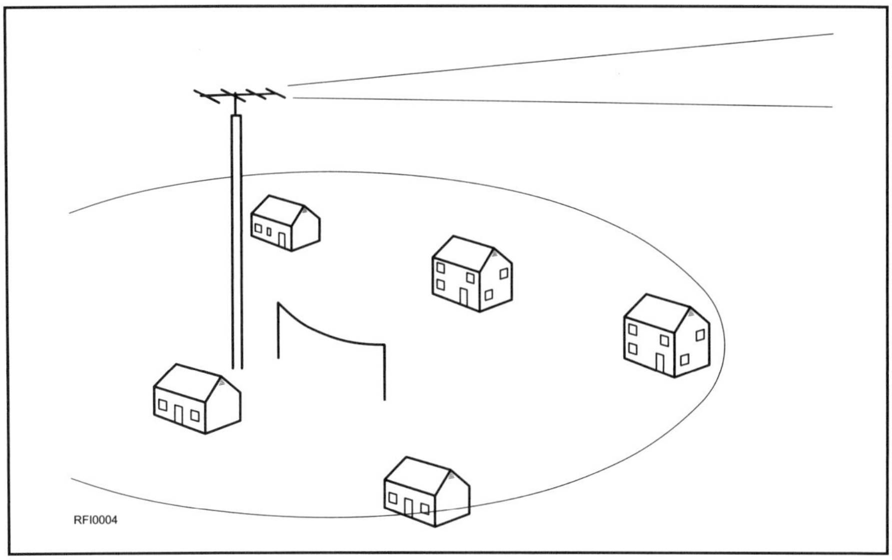

Figure 2.2—Some towns try to limit antenna height thinking they will prevent RFI problems. It's easy to see that a higher amateur antenna is better, from both amateur and RFI viewpoints. Here the higher antenna directs energy away from houses and equipment. The lower antenna places its field at ground level.

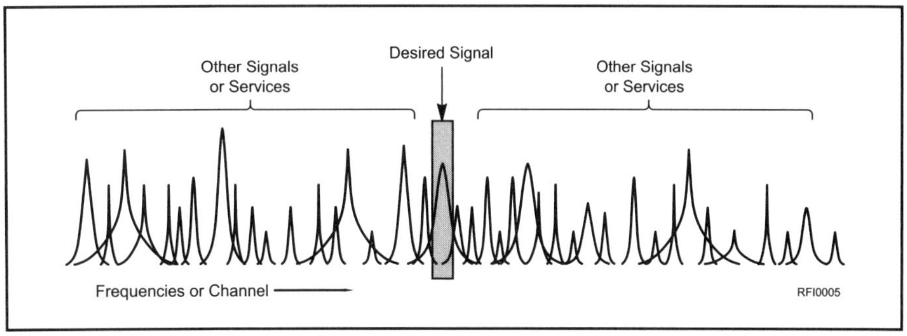

Figure 2.3—Every electronic appliance must select the signal it needs from all those present in the radio spectrum.

Noise

Transmitters may also produce noise and/or "parasitic" oscillations. (Parasitic oscillations are discussed in the Amplifiers chapter.) Noise may be broadband composite noise, keying transients or intermodulation products occurring in the linear amplifier stages. Transmitter noise can also cause interference problems, especially to frequencies that are close to the transmit frequency (such as 6 meters and TV channel 2).

FCC Regulations about Spurious Emissions

If any transmitter emitted unwanted signals cause interference to another radio service, FCC regulations require the owner to correct the problem.

FCC regulations set two different limits for all spurious emissions: an absolute worst-case limit and an interference limit. The worst-case limits set maximum spurious levels for transmitters. The allowed emission is expressed relative to the output power. For example, a spurious emission might be characterized as being –58 dBc; that is, 58 dB "below" (weaker than) the carrier or mean output power. The worst-case limits range from –43 dBc for all HF transmitters, to –60 dBc for 50-222 MHz transmitters with more than 25 watts of mean output power. There are no absolute limits for transmitters operating above 225 MHz, although the general requirements for good engineering practice would still apply. FCC regulations require transmitter operators to correct spurious emissions that exceed the absolute limits even if they don't cause interference.

FCC regulations also require attenuation of any spurious emissions that interfere with other radio services. For example, if a "legal," weak spurious emission from a VHF repeater interferes with an FAA communications channel, the emission must be reduced as necessary to eliminate the interference.

These regulations apply to amateur transmitters, so this topic is discussed in detail in the Transmitters chapter. It is also, of course, covered from a different perspective in the Regulations chapter.

Fundamental Overload

Most cases of interference are not caused by spurious emissions from transmitters; they are caused by overload of the affected consumer electronics equipment by the transmitter's fundamental frequency. The world is filled with RF signals. Properly designed electronic equipment should be able to select the desired signal, while rejecting all others. Unfortunately, because of design deficiencies such as inadequate shields or filters, some equipment is unable to reject strong undesired signals.

A strong fundamental signal can enter equipment in several different ways. Most commonly, it is conducted into the equipment by wires connected to it. Possible conductors include antennas and feed lines, interconnecting cables, power lines and ground wires. TV antennas and feed lines, telephone or speaker wiring and ac power leads are the most common points of entry.

Many active electronic devices (such as transistors, ICs and vacuum tubes) respond to *all* electrical signals at their inputs. When the total input signal is too large for a device to handle, the device is overloaded. When the signal overloading a device is the desired signal (fundamental) of a lawfully operating transmitter, the condition is called "fundamental overload." Many electronic devices exhibit a "low-pass" response, meaning that they are more susceptible to lower frequencies than higher ones.

The effect of an interfering signal is directly related to its strength. As an approximation, the strength of a radiated signal diminishes with the square of the distance from the source: When the distance from the source doubles, the strength of the electromagnetic energy decreases to one-fourth of its strength at the original distance from the source. This characteristic can often be used to help solve RFI cases. You can often make a significant improvement by moving the susceptible equipment and the transmitting antenna farther away from each other. Unfortunately, especially near antennas, a number of factors make it impossible to determine the distance to the transmitter from the field strength. See Figure 2.2.

When fundamental overload occurs, one or more signals at a device input are strong enough to change the operating characteristics of the circuit. The strong signal causes "gain compression" (the affected amplifier loses the ability to amplify), which may also be called *desensitization* or simply *desense*. *Saturation* is another term related to overloaded devices. A saturated device has more drive than it can faithfully reproduce. In other words, the device is overwhelmed by its input and does not perform properly. Thus, fundamental overload is a basic disruption of proper circuit operation because the affected device is incapable of rejecting strong unwanted signals.

Direct Radiation Pickup

The internal wiring, circuit-board traces and components of some poorly shielded equipment can pick up signals directly. This is known as *direct radiation pickup*. This is a form of fundamental overload, in most cases. In many cases of direct pickup, it is not practical to add shielding, so there is little an amateur can do to fix the problem. In these cases, contact the equipment manufacturer.

The term describes cases where a radiated signal is received by conductors *within* the affected equipment. RFI that arrives via radiation can be blocked by shielding the affected device (if that is possible; antennas, for example cannot be shielded). When shielding is not possible, direct radiation pickup can be treated as a conducted signal at the circuit-component level (modifications should be left to the manufacturer or a qualified technician).

Direct radiation pickup generally involves an intermediate element that supports re-radiation of the unwanted signal. For example, a fundamental emission could be picked up by a resonant length of ac wiring, which would concentrate and re-radiate it for pick-up by consumer equipment.

The amateur is not responsible for the effect of the amateur-band fundamental signal, but, of course, sometimes an amateur wants to help out anyway. (Practical considerations must sometimes be considered.) Figure 2.3 shows how properly designed consumer equipment should function.

If the construction of the equipment is faulty, it may be practical to improve the shielding. Direct pickup may result from something as simple as excess paint or anodizing—removal may fix the problem.

Fundamental Overload vs Spurious Emissions

Most receivers, including consumer-electronic receivers, have some form of band-pass filter in the front end. Strong signals that lie outside of the passband may, however, pass through to the first stage (fundamental overload). Once there, they may: (1) be passed directly to the output, (2) overload the receiver and make it inoperative (desensitization), or (3) combine with other signals in a nonlinear element to produce spurious signals within the receiver passband. Fundamental overload may occur when the affected device is inadequately shielded, has unfiltered leads, or is in close proximity to an RF source. Fundamental overload is the result of design deficiencies in the affected system, not improper transmitter operation or design.

In a high-density RF environment such as a repeater tower, one transmitter can overload a second repeater's receiver. A third transmitter can mix with the first signal, causing intermodulation. (Naturally, each user tends to blame the others for the problem.)

Intermodulation and External Rectification

Most amateurs have heard of "intermod" as applied to repeater operation. Intermodulation is an unwanted mixing of two or more signals in a nonlinear device (mixer). Because repeaters and repeater users are often located very close to other strong transmitters, intermod is a common problem. In some urban areas, intermod can make many handheld transceivers virtually unusable. Handheld transceivers are often susceptible to intermod in urban areas due to compromises made in receiver design. The Intermod chapter discusses this topic in detail.

While "intermod" usually refers to unwanted mixing in electronic equipment, mixing that occurs in nonelectronic components is called *external rectification*. The problem is so perplexing that there is a chapter dedicated to it later in this book. Although the terms are different, the action and results are essentially identical.

Most any semiconductor junction may act as a "nonlinear device" (mixer). Naturally occurring semiconductor junctions are very common in the form of poor electrical connections. House ac outlets with "push in" house-wiring connections and corroded connections on TV antennas are two examples of the types of installations that can have nonlinear junctions.

Complete intermod "systems" are common in ac-line wiring, telephone systems and TV antenna systems. Several systems have been particularly troublesome in recent years:

- VHF transmitters or receivers hooked up to antennas that are near transmitting antennas
- Rotator control boxes
- Lightning arrestors (telephone lines, TV receive antennas, transmitters)
- Corrosion in antennas, towers or guy wires
- Antenna-switching diodes and feed lines in VCRs, scanners, CBs and other radios.
- LEDs fed by long wires in SWR meters and other accessories.
- AC wiring that uses "push in" rather than screw-terminal connections for connection to the house wiring.

Nonlinear Junctions

Harmonic and intermodulation interference can be caused by RF current flowing through any nonlinear junction. Colloquially referred to as the "rusty bolt effect," the nonlinear characteristics of corroded joints (oxidized guy wires, coaxial connectors, antennas, feed lines and such) can cause harmonic, cross-modulation and intermodulation products that result in RFI. Even when the transmitter output is clean, RFI may result if there is a poor electrical joint somewhere in the antenna or feed line. A poor solder or mechanical joint can act as a rectifier, generating harmonic currents that may be radiated by the antenna. Poorly conducting joints in nearby conductors (rain gutters, for example) can generate and radiate harmonics when excited by RF energy.

RF current flowing through nonlinear electronic components, including transistors, diodes and ICs, can also result in RFI.

Electric and Magnetic Fields

The fields radiated by an antenna are fundamentally different far away (*far field*) from the antenna than they are close to it (*near field*). In the far field (approximately greater than about ½ λ from the source), the electric (E) and magnetic (H) components of the electromagnetic field always have the same relative amplitude. The ratio E/H is known as the *impedance of free space*, which is equal to 376.7 Ω.

In the near field, however, E/H ratio depends on the radiator. High-impedance sources generate mostly E fields. An example is a short rod or unterminated wire. Low-impedance sources generate mostly magnetic (H) fields, a small loop, for example.

The subject of electric and magnetic fields is treated in more detail in *The ARRL Antenna Book* and *RF Exposure and You*. These books are available from ARRL Headquarters Publication Sales Department.

Semiconductor junctions in such equipment as audio amplifiers, mast-mounted preamplifiers and SWR bridges may all rectify RF energy to produce spurious signals. Receivers can be RFI sources. When the first stage overloads, the resulting very weak intermodulation signals may be radiated from the receiving antenna. This will not cause widespread interference, but if another receive antenna is located very close by, IMD-generated interference can be the result. Sensitive preamplifiers, with limited dynamic range and poor selectivity, are especially vulnerable to overload problems.

Intermodulation products can be generated by transmitters when other strong signals are coupled to the final amplifier. There they mix with the fundamental and its harmonics. The resulting signals may be coupled to the antenna and radiated. Intermodulation problems are especially common in the VHF and UHF bands because amateur repeaters tend to cluster around commercial and government VHF and UHF radio services. Wide-band, solid-state amplifiers with low-Q circuits are more likely to generate spurious signals than are narrow-band, high-Q configurations.

External Noise

Electrical noise is usually a matter of interference *to* electronic equipment. Many hams have problems with electrical noise and reception at their station receiver. Most cases of interference reported to the FCC involve some sort of external noise source. The most common of these noise sources are electrical. External "noise" can also come from transmitters or from unlicensed RF sources such as computers, video games, video modulators hooked up to TV sets (especially if they are put in parallel with the TV antenna!), electronic mice repellers and the like.

Electrical noise is fairly easy to identify by looking at the picture of a susceptible TV or listening on an HF receiver. On a receiver, it usually sounds like a buzz, sometimes changing in intensity as the arc or spark sputters a bit. If you determine the problem to be caused by external noise, it must be cured at the source. Refer to the Electrical chapter of this book.

RFI CAUSES

There is rarely a single, well-defined cause for a given case of RFI. In many cases, there is a complex system of interactions and interdependencies. As described below, these include:

- Various mechanisms by which undesirable RF energy is generated, propagated and then received by susceptible devices
- Inadequate shields and filters that make some electronic devices especially vulnerable to RFI
- Improper filters, shields or operation of transmitters and other RF energy sources.

THE SOURCE-PATH-SUSCEPTOR CONCEPT

All cases of RFI involve a *source* of electromagnetic energy, a device that responds to this energy (susceptible equipment) and a transmission path (either conducted or radiated) that allows energy to flow from emitter to the susceptor. (This book often refers to the susceptor as the "susceptible equipment.")

Sources

Hams tend to think of Amateur Radio as the major source of interference problems. This is because we are so close to Amateur Radio. Hams are actually only a small part of the problem. Table 2.1 shows a number of possible sources of RF and noise signals. Emitters include radio transmitters, receiver local oscillators, computers and computer

Table 2.1
Possible Interference Sources

Communications Service Transmitters
Amateur Radio
Police
Cable TV (CATV)
Business or other two-way
Aircraft (near airports only)
Cellular phones
Land mobile transmitters and repeaters
Pagers

Electrical Noise Sources
Doorbell transformers
Toaster ovens
Electric blankets
Fans
Heating pads
Light dimmers
Appliance switch contacts
Aquarium or waterbed heaters
Sun lamps
Furnace controls
Smoke detectors
Smoke precipitators
Computers (and video games)
Ultrasonic pest-control devices
Lights: fluorescent, mercury vapor and touch-controlled
Neon signs
Electric fences
Alarm systems
Loose fuses
Sewing machines
Electrical toys (such as trains)
Calculators
Cash registers
Lightning arrestors
Cable television leakage

Power Company Equipment
Defective line insulators
Loose or unbonded hardware
Discharges from defective lightning arrestors
Defective transformers
(possible but surprisingly rare)

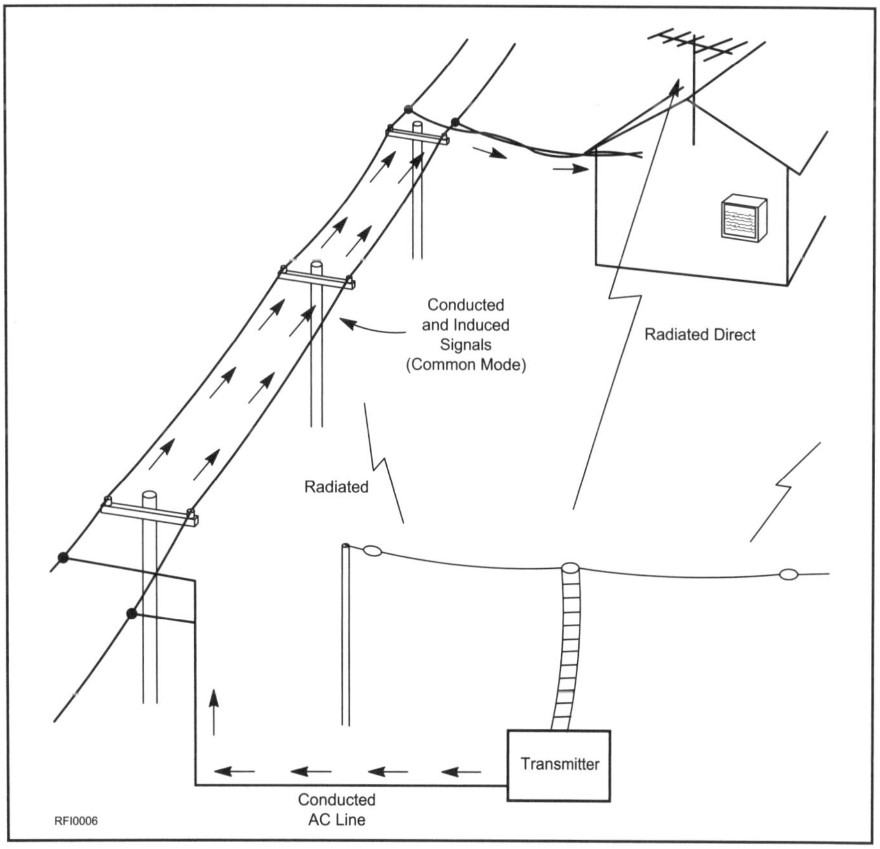

Figure 2.4—Conducted and radiated interference.

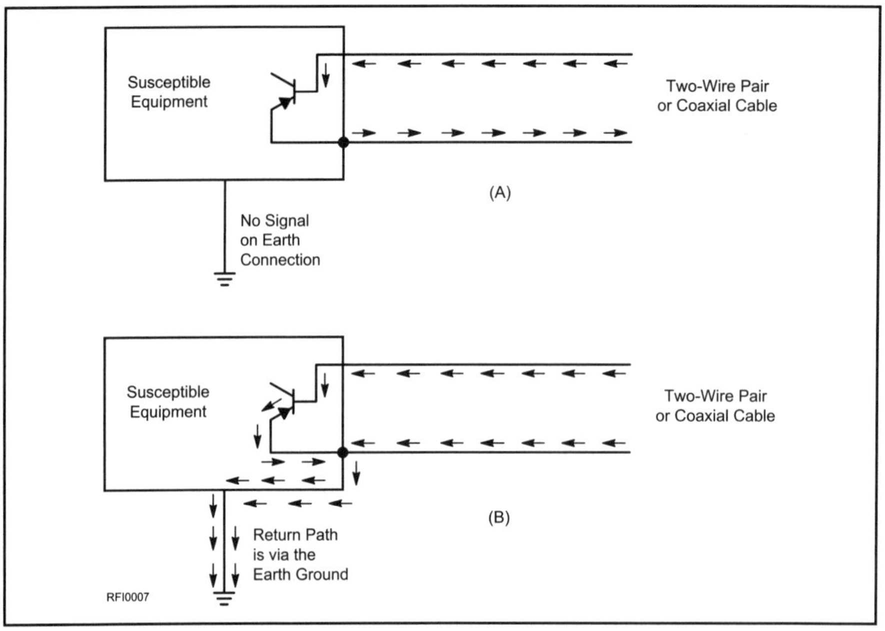

Figure 2.5—(A) Differential-mode signals are conducted between two wires of a pair. This signal is independent of earth ground. (B) A common-mode signal is in phase on all wires that form the conductor (this includes a coaxial cable). All wires act as if they are one wire. The ground forms the return path, as with a long-wire antenna.

> **Baluns**
>
> The balun is one of the most misunderstood components in an amateur station. It can help minimize feed-line radiation. Ferrite transformer baluns can saturate under high-power or high-SWR conditions. When ferrite saturates it causes distortion that can generate harmonic energy. An air-core balun is not subject to saturation.
>
> The common question about baluns is, "Do I need one?" The best answer is "Maybe." If the affected equipment is much closer to the feed line than the antenna, radiation received from the feed line might be stronger than the energy received from the antenna. This might be an important consideration in some installations, especially for amateurs who live in apartments.—*Ed Doubek, N9RF, Former Illinois Section Technical Coordinator (SK)*

peripherals, lightning, electrostatic discharge and other natural sources.

Paths

There are three ways RFI can travel from the source to the susceptible equipment: electromagnetic radiation, conduction and induction. These often exist in various combinations. The coupling path between an emitter and receptor can be extremely complicated. For example, there may be conduction from the emitter to a radiator, radiation to another conductor and conduction again to the susceptible equipment. There might be several transitions between radiation and conduction in the RFI path. Figure 2.4 shows an example of different paths that can exist between the source and susceptible equipment.

Radiated RFI propagates by electromagnetic radiation from the source, through space to some conductor in the susceptible equipment or system.

A conducted signal travels over wires connected to the source and the victim. Possible conductors include antennas and feed lines, interconnect cables, power lines and ground leads.

Induction occurs when two circuits are magnetically coupled. When magnetic induction occurs, the magnetic field of an inductor (such as a power transformer) produces an unwanted signal in a nearby conductor (such as an ac branch circuit or telephone cord). Capacitance can also cause interaction between two circuits as a result of electric field coupling.

Relevant characteristics of emitters and susceptible equipment are their emission and reception spectra and their susceptibilities. That is, what undesired signals do they emit or receive, and how do those unwanted signals leave or enter the equipment?

The Path and Troubleshooting

Identification of the RFI path is a key step in resolving RFI. The frequencies common to both the emitter and susceptible equipment are critical. The times of emitter operation and the relative location of emitter and susceptible equipment are also important. Only with this data in hand is it possible to select the most effective means of combating the RFI. In most instances, a combination of techniques must be used. The effectiveness of any one method can be enhanced when additional methods are also employed. For example, a low-pass filter does not operate effectively unless the transmitter is well shielded.

Conducted vs Radiated Emissions

Most RFI occurs via conduction (especially at HF) or some combination of radiation and conduction. For example, a signal is radiated by the source and picked up by a conductor attached to the victim (or directly by the victim's circuitry) and is then conducted into the victim.

Conducted emissions originate from a variety of sources, including relay and switch contacts, fan motors, oscillators, analog devices that are operated over nonlinear parts of their design curves, digital devices with short rise and fall times and high-speed switching devices. Radio transmitters can originate conducted emissions. Conducted interference arrives by wire (or other conductor). Conducted interference may be controlled by filtering the conductor. Seldom is conduction the sole path of an interfering signal. The incoming signal is usually radiated to a conductor near the affected device, which then passes the signal to the affected device. In those cases, interference may be reduced by shielding, or somehow detuning, the conductor. (Detuning a conductor might include changing the length, or inserting RF chokes to break its apparent length at RF.)

Conducted interference can be minimized by using filters to channel energy away from sensitive devices, and by using bypass capacitors to decouple devices from the power bus. Although the spectrum of conducted interference can exceed 1 GHz, higher-frequency conduction currents are heavily attenuated by resistive losses, wire inductance and shunt capacitance. In addition, the higher-frequency signals have a tendency to be radiated and coupled to nearby wiring.

The undesired signal is conducted by wires between the source and the susceptible equipment (see Figure 2.4). This includes ac wiring and shared ground leads. An induced signal (see the next paragraph) is also a conducted signal by the time it reaches the susceptible equipment.

Differential vs Common-Mode

There are two modes of conduction. Conducted interference may appear as either common-mode (all conductors except ground act in common; that is, as one) or differential-mode (the signal arrives on a pair of conductors, with a 180° phase difference between the pair) signals.

Common-mode signals are most prevalent, but have both common-mode and differential-mode filters on hand for testing. Some cases require both kinds of filter. A differential-mode signal arrives on one wire of a two-wire transmission line, and returns via the other wire. A common-mode signal arrives, in phase, on one or more wires of a system and returns via the power supply leads to ground.

It is important that you fully understand the difference between differential-mode and common-mode signals because the cures are different. Figure 2.5 shows differential-mode and common-mode signals.

As we shall see, each of these conduction modes requires different RFI cures. Differential-mode cures (the typical high-pass filter, for example) do not attenuate common-mode signals. On the other hand, a typical common-mode choke does not affect interference resulting from a differential-mode signal.

Many cables used in electronic equipment include two or more wires. Examples are coaxial cable, rotor cable, ac-line cords and microphone wires. Interfering signals can flow in these cables in two ways:

Differential-mode currents have equal strength and opposite phase in some of the wires. This is the mode used by the desired signal on a transmission line.

Common-mode currents are equal in strength and phase in all of the wires. The return path for common-mode signals usually includes ground. Common-mode currents on a transmission line cause feed-line radiation.

Differential Mode

Differential-mode currents usually have two easily identified conductors. Differential-mode currents occur between two conductors with no ground reference. In a two-wire transmission line, for example, the signal leaves the generator on one line and returns on the other. When the two conductors are in close proximity, they form a transmission line and there is a 180° phase difference between the signals in the two conductors. A differential-mode signal is created by a source, such as a transmitter, signal generator or antenna.

In a differential circuit, it is relatively simple to build a filter that passes desired signals and shunts unwanted signals to the return line. Most *desired* signals, such as the TV signal inside a coaxial cable are differential-mode signals.

Common Mode

In comparison, common-mode currents are in phase on each conductor of a multi-wire cable. In a common-mode circuit, many wires of a multiwire system act as if they were a single wire. The result can be a good antenna, either as a radiator or as a receptor of unwanted energy. Common-mode currents return to their source through some conductor common to both the source and affected circuit. The return path is usually earth ground. Since the source and return conductors are usually well separated, there is no reliable phase difference between the conductors and no convenient place to shunt unwanted signals. Toroid chokes are the answer to common-mode interference. (The following explanation applies to rod cores as well as toroids, but since rod cores may couple into nearby circuits, use them only as a last resort.)

Since common-mode currents flow in the same direction through all conductors in a cable, little field cancellation takes place. The result can be a good radiator. For example, a coaxial transmission line can act as a long-wire antenna "worked" against earth ground.

The magnitude of the common-mode current induced in each wire of a cable is a function of the cable design. In a balanced two-wire system, such as 300-Ω twin-lead, the conductors carry equal induced currents, just as though the two conductors were wired in parallel. In an unbalanced system, such as coaxial cable, the induced current magnitude is different in each conductor. The induced current is much greater in the coax shield than in the center conductor.

If not recognized, RFI from common-mode currents can be especially troublesome to eliminate. The usual fundamental-overload and harmonic-radiation RFI cures (a low-pass filter for the transmitter, a high-pass filter for the receiver, a ground connection for the transmitter coax shield and power-line filters for the transmitter and TV) do not reduce common-mode interference. The conductors carrying unwanted common-mode currents usually carry desired differential-mode currents as well. Therefore common-mode cures must block in-phase currents without affecting differential-mode signals. Fortunately, this is easily accomplished with common-mode chokes.

Differential vs Common-Mode Cures

Differential-mode cures (the typical high-pass filter, for example) do not work on common-mode signals. On the other hand, a typical common-mode choke does not affect interference resulting from a differential-mode signal.

The cure for differential-mode interference is a filter that passes the desired signal but blocks the unwanted signal. Examples are a low-pass filter on an HF transmitter and a high-pass filter on a TV set. These filters work only for differential-mode signals. If the cause is feed-line radiation from a transmitter, or large common-mode signals entering a TV tuner, use a different technique.

A cure for unwanted common-mode signals must somehow present a high impedance in the common mode, without obstructing differential-mode signals. This can be done by inserting a transformer (with isolated windings) in series with the cable.

A common-mode choke is one solution. A choke is formed by wrapping all conductors around a ferrite core (toroid or cylinder) or placing them through one or more ferrite beads. (All conductors must pass through the same beads, not a bead for each conductor.) Since differential-mode signals have equal-strength and opposite-phase currents in the wires, their fluxes cancel within the core, and no inductance results. Figure 2.6 illustrates a toroid used as a common-mode choke for an ac power zip-cord.

Toroids work differently, but equally well, with coaxial cable and paired conductors. A common-mode signal on a coaxial cable is usually a signal that is present on the *outside* of the cable *shield*. When we wrap the cable around a ferrite-toroid core the choke appears as a reactance in series with the outside of the shield, but it has no effect on signals inside the cable because their field is (ideally) confined inside the shield. With paired conductors such as zip-cord, signals with opposite phase set up magnetic fluxes of opposite phase in the core. These "differential" fluxes cancel each other, and

Figure 2.6—A common-mode choke. This technique can be used to suppress common-mode currents on any type of wiring.

there is no net reactance for the differential signal. To common-mode signals, however, the choke appears as a reactance in series with the line.

Toroid chokes work less well with single-conductor leads. Because there is no return current to set up a canceling flux, the choke appears as a reactance in series with *both* the desired and undesired signals.

Radiated

In this mode, the undesired signal is radiated directly from the amateur station to the affected devices and circuitry. Radiated interference arrives at the susceptible equipment in the same fashion as any other radio transmission. There is a radiating antenna, a transmission path and a receiving antenna. The antenna of the susceptible equipment device is often the interconnecting cables between various elements of the system.

The undesired signal is usually the fundamental signal of a transmitting station, which is radiated by the antenna (see Figure 2.4). This signal may, however, include spurious emissions and signals radiated by the feed line, ground lead or station equipment. It can also include any noise that is radiated by any source, such as a neon sign or an electrical storm.

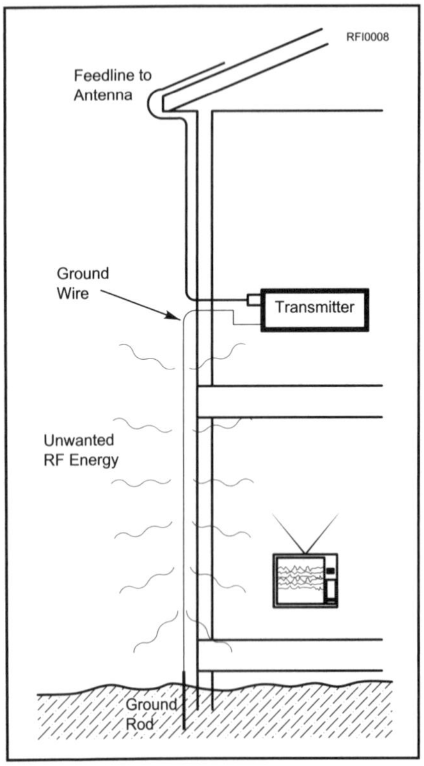

Figure 2.7—When a transmitter is located on an upper floor, the ground lead may act as an antenna for VHF/UHF energy. Such stations may be better off without a normal ground.

Radiated emissions can also result from leakage through coaxial cable shields, corroded surfaces, insulation breakdown or discontinuities in component housing. Whereas the same conditions that cause conducted interference may also result in radiated interference, the source of radiated interference is much more difficult to diagnose. The direction of maximum radiated energy may not correspond to the exact beam heading of the source antenna because reflections mask the direct path.

Induction

Induction is a combination of radiation and conduction. An induced signal is a radiated signal that is picked up by wiring and conducted to the susceptible equipment. Most induced interference occurs in the common mode.

The amount of coupling depends on the amount of capacitance, the impedance levels of the emitter and susceptible equipment circuits and the amplitude and frequency of the signal involved. High-impedance, high-frequency circuits favor capacitive coupling. Most amateurs who have ever tried to use an antenna and feed line system that placed a high-voltage point in the shack have experienced capacitive coupling. "RF in the shack" results when RF energy capacitively couples from the transmitter and transmission line into metal objects at the operating position.

The degree of inductive coupling between adjacent wires is a function of the frequency and magnitude of current flowing in one of the wires, the spacing of the wires and their common length. A good example of inductive coupling is the Yagi antenna. The current in the driven element induces currents in the parasitic elements that produce a directive radiation pattern.

Magnetic Induction

Magnetic-induction interference requires very close proximity of the interfering device (a coil or transformer) to its "susceptible equipment." Hams are not normally confronted with this situation, even in their own homes. Telephones can be prone to magnetic induction; see the Telephone chapter for more information. Large transformers or switching power supplies can also cause magnetic-induction in nearby equipment. This is sometimes seen in computer monitors. Magnetic induction can be reduced by physically separating or magnetically shielding (as by an enclosure of specially treated nickel-iron alloy, commonly called mu metal) the interacting devices.

Susceptible Equipment

Susceptible equipment includes radio and TV receivers, VCRs, telephones, amplifiers, computers and even devices such as pacemakers or alarm systems. Nearly any electronic device near a transmitter has the potential for being interfered with.

CURING RFI

Now that you have learned the theory, you can learn about the cures for RFI. This chapter gives an overview of various techniques; the following chapters in the book apply those cures to specific RFI problems.

When is Ground not a Ground?

In many stations, it is impossible to get a good RF connection to earth ground. A good RF ground is difficult to attain. Even an 8-foot ground rod can have a few dozen ohms of contact resistance in poor soil. Most ground systems use a shorter rod. (Have you ever driven an 8-foot ground rod into even the softest soil?)

Most practical installations require several feet of wire between the station ground connection and an outside ground rod. In amateur stations above ground level, the ground lead becomes longer and may, itself, radiate energy. A long ground lead can actually make interference worse. Many troublesome harmonics are in the VHF range. At VHF, a ground wire length can be several wavelengths long — a very effective long-wire antenna! Any VHF signals that are put on a long ground wire will be radiated. This is usually not the intended result of grounding.

Take a look at the station shown in Figure 2.7. In this case, the ground wire could very easily contribute to an interference problem in the downstairs TV set.

While a station ground may cure some transmitter RFI problems — either by putting the transmitter chassis at a low-impedance reference point or by rearranging the problem so the "hot spots" are farther away from susceptible equipment — it is not the cure-all that some literature has suggested. A ground is easy to install, and it may reduce stray fundamental or harmonic currents on your antenna lead; it is worth a try.—*Ed Hare, W1RFI, ARRL Laboratory Supervisor*

Grounds

This chapter deals with the EMC aspects of grounding. The most important reason to incorporate a good ground in the amateur station is for the safety of the operator and the building housing the station. Virtually all local building codes and the National Electrical Code require this. While grounding is not a cure-all for RFI problems, ground is an important safety component of any electronics installation. It is part of the lightning protection system in your station and a critical safety component of your house wiring. Any changes made to a grounding system must not compromise these important safety considerations. Although not always required for RFI control, the FCC generally expects you to have a good station ground.

Some amateur stations have several grounds: a safety ground that is part of the ac-wiring system, another at the antenna for lightning protection and perhaps another at the station for RFI control. These grounds can interact with each other in ways that are difficult to predict. In most cases, such multiple grounds are not in compliance with electrical building codes.

An electrical ground is not a huge sink that somehow swallows noise and unwanted signals. An improperly done ground can make an RFI problem even worse!

Grounding is the establishment of an electrically conductive path between two electrical systems or between an electrical system and a reference point (ground plane). Ground is a *circuit* concept, whether the circuit is small, like a radio receiver, or large, like the propagation path between a transmitter and cable-TV installation. Ground forms a universal reference point between circuits.

The ideal reference point (a zero-potential, zero-impedance body) can only be approximated. In practice, non-ideal ground (perhaps better called a universal reference) often consists of a metallic automobile or building structure, plumbing, earth ground, steel-reinforced concrete floors, station ground, signal-control cables or telephone lines.

Improper grounding can introduce unwanted audio hum and may violate the National Electrical Code or local regulations. An added ground for an audio system could create a big ground loop, which could act as an antenna and worsen the RFI.

A ground is supposedly a "zero-potential" surface or point, but most float at some small voltage in relation to other grounds in the system, or even to other parts of the same ground. In reality, our best hope is that there is little circulating current (particularly RF). Some households have multiple ground reference points: the ac power ground, the plumbing system and sometimes separate grounds for telephone and cable.

The RFI or station ground is used for: (1) cable-shield and equipment grounding, (2) RFI filter referencing, (3) noise and interference control (by providing a low-impedance sink for noise currents) and (4) circuit referencing (by allowing signals between equipment to be properly interpreted).

Grounding is Not a Cure All

At higher frequencies, however, the impedance of the ac ground lead may be much greater than that provided by stray capacitive paths between cables, circuit boards and equipment enclosures. Power ground is especially ineffective as an RF ground, especially when the ground wire approximates an odd multiple of ¼ λ for unwanted signals. Then, the ground lead transforms a low-impedance ground to a high impedance at the grounded equipment.

Any ground wire that has physical length can be thought of as a "ground wire antenna." This can be seen in Figure 2.7. Any RF from the transmitter that ended up on the ground wire would be radiated strongly toward the antenna-connected TV set right near the ground wire. In this case, the RFI would be less if there were no ground at all. Of course, you may be in trouble either way: if the RF ends up on the shield of the antenna feed line, it may be radiated over a wide area.

Multiple Grounds—A No-No

Many amateur installations make use of two ground systems: an electrical safety ground and an RFI ground (a few also use a system of conductors and ground rods for lightning protection). Safety ground is in the ac-line lead, which is connected to earth at the service entrance. At power frequencies, the safety ground provides a low-impedance path for fault currents. Multiple grounds, however, are not a good practice; the National Electrical Code requires that all grounds be connected to a single point.

The common practice of using house plumbing for RFI ground may not be a good idea. If you can tie into the plumbing system right where it enters the building *and* the rest of the building grounds are also tied to this point, the plumbing system can make an excellent ground. However, if these conditions are not met, using the plumbing ground may not be in compliance with local electrical building codes. Use of the plumbing system may create multiple grounds, making it likely that the entire plumbing system will act as a big antenna.

Grounding Techniques

Three fundamental grounding techniques are floating, single-point and multipoint ground. Each has its applications in circuit and station design.

Floating Ground

A floating ground is used to isolate circuits electrically from a common ground plane or from common wiring that might introduce circulating currents (ground loops). Floating ground is not used much in RFI prevention because it is ineffective at frequencies greater than approximately 1 MHz. At higher frequencies, capacitive coupling paths bypass isolation transformers and other isolation mechanisms. This allows RF currents to flow from one point to another on the ground plane. In addition, the electrical isolation may allow static charges to accumulate, resulting in random discharges of RFI-producing current.

Single-Point Ground

Single-point grounds rely on a single physical point in each circuit that is defined as the ground reference point. All ground connections are made directly to that point. In a single-point ground system with mul-

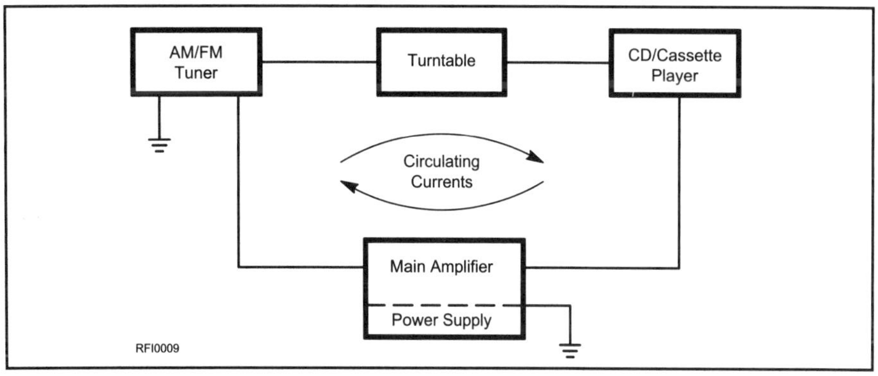

Figure 2.8—Ground loops in stereo systems. This diagram shows a system with multiple grounds, with inter-component ground connections made by cable shields. Circulating currents may flow through the loop formed by the ground connections between the tuner and phonograph, phonograph and CD player. There is also a loop between the tuner's separate earth ground and the main amplifier earth ground.

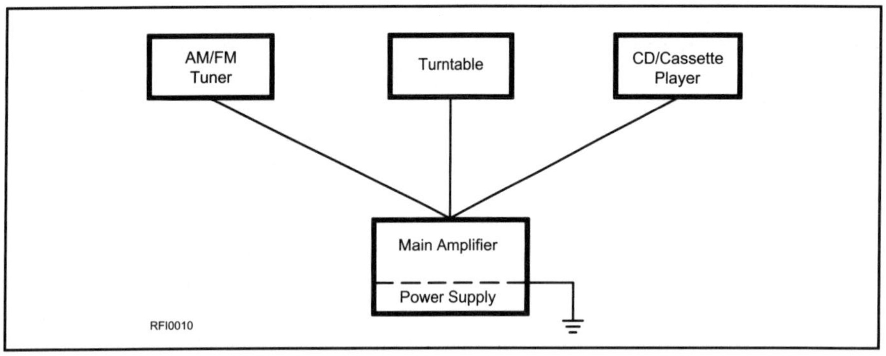

Figure 2.9—The proper way to ground a multiple-component system.

tiple cabinets, cabinet and electronic ground are kept separate, and a single ground is used inside of each cabinet. Cabinet grounds are in turn connected at a single reference point. Single-point grounds are effective at higher frequencies than floating grounds. As ground lead length approaches ¼ λ at the unwanted frequency, however, the impedance of the ground connection increases to an unacceptably high level. At the station level, single-point grounding is the only grounding that should be used.

Multipoint Grounding

In multipoint grounding, each ground connection is made directly to the ground plane at the closest available point. The advantages of this approach include easier circuit construction and higher operating frequencies. In addition, multipoint ground reduces electrostatic coupling between shielded cables. Unfortunately, multiple-point grounding permits ground loops, a potential source of interference.

The relative advantages and limitations of the three basic grounding techniques should be considered whenever a ground connection is required. For example, if ground lead length approaches ¼ λ for any potentially interfering signals carried on the lead, multipoint grounding should be used. Other good grounding practices include:

- Keep ground leads as short as possible.
- Ground all equipment to safety ground for shock protection.
- Avoid use of twisted-wire grounds on cables, especially on those carrying signals above 1 MHz.
- Insulate cable shields to prevent undesired grounding. Random contact between shield braid and chassis can result in noise.
- Cable shields should be grounded, typically as a minimum, at both ends. Some cases however, require the shield be grounded at only one end of the cable. You may wish to eperiment if in doubt.
- Shields should not be used for signal return, unless the shield is part of a coaxial cable carrying that signal.
- Use floating ground if interference caused by ground loops are a problem.
- Use separate circuit grounding systems for signal returns, signal shield returns, the power system and chassis grounds. Tie all of these grounds together at a single reference point.
- Isolate the grounds of low-level signals.
- Use multipoint grounding of the shield of coaxial cables used for high-frequency circuits.

Ground Loops

In nature, it is very unlikely that any two grounds are at exactly the same potential. Therefore, when a conductor is connected to two different ground points, a "circulating" current flows through the (supposedly grounded) wire. By connecting the wire to a second ground, a ground loop was created. The currents circulating in ground loops can cause or worsen RFI problems. Multipoint grounds are not only often in violation of the NEC or local electrical codes, they can create ground loops that add to RFI problems.

To avoid ground loops, there should be one (and only one) path from each point in the system to earth ground. (Visualize the ground system as a tree, with one trunk and many branches.) A "single point" grounding system is usually necessary to avoid circulating currents between components of a system. [Figures 2.8 and 2.9 show how to avoid ground loops between system components. In Figure 2.8, a turntable that has a shielded interconnect cable should not have a separate ground lead to the amplifier. Other components such as AM/FM tuners and CD or cassette players should be treated likewise.]

When each component has its own three-wire ac cord, it may be difficult to avoid ground loops. It may be satisfactory to connect them all to the same ac outlet. If not, try a ferrite choke on each ac cord. The ground loop will still exist at ac, but the ferrite will eliminate the RF ground, and the loop along with it. There are no hard and fast rules, because each station is unique. Try all combinations of potential cures; some are bound to work. Don't give up! When the final solution is achieved, it will hold lessons that can help in future cases.

All of these station grounds can form a large ground loop. This loop can act as a large loop antenna, with increased susceptibility to lightning or RFI problems. Figure 2.9 shows a proper single-point ground system

What does this mean for a station located on a second story, or in a far corner of the house? Establish a reference ground at the system itself. Then connect the reference ground to the single-point main building ground with as short a lead as possible. (Do not use the ground screw on an ac convenience outlet as an RF ground!) This may not be easily done, but it may be the only solution in severe RFI cases.

Common Ground

The coupling of interfering signals from emitter to susceptible equipment may involve common wiring, mutual capacitance or inductance, or direct radiation. Common wiring includes shared ground and power-supply leads as well as signal cables. For example, consider a 40-meter transmitter grounded to the telephone-entrance ground. If the ground point is 12 feet from the effective earth ground, it presents a high impedance to the third harmonic at 21 MHz. The third-harmonic energy then enters the telephone system, is rectified by a transistorized phone and appears as audio throughout the telephone system. An example of a common-ground connection is shown in Figure 2.10.

AC-Outlet Grounding Practices

Many homes have three-wire grounded outlets near water sources, while the rest of the home has only two-wire outlets. In such cases the home is usually wired with three-wire cable, but the ground wire is not connected at the two-wire outlets (see Figure 2.11).

Most building codes permit that the two-wire outlets can be replaced by three-wire outlets. If so, the ground wire must be connected to its proper terminal or a ground-fault, current-interrupter circuit or outlet must be used. For safety reasons, all work done on an electrical system must comply with local codes. The code may require that the work should be done by a professional electrician.

In many cases, electrical RFI "grounding" problems could well be caused by RF signals conducted from the transmitter to the susceptible equipment. In these cases, it may be helpful to filter the transmitter or susceptible equipment with an ac-line filter. You should also use a separate ac circuit,

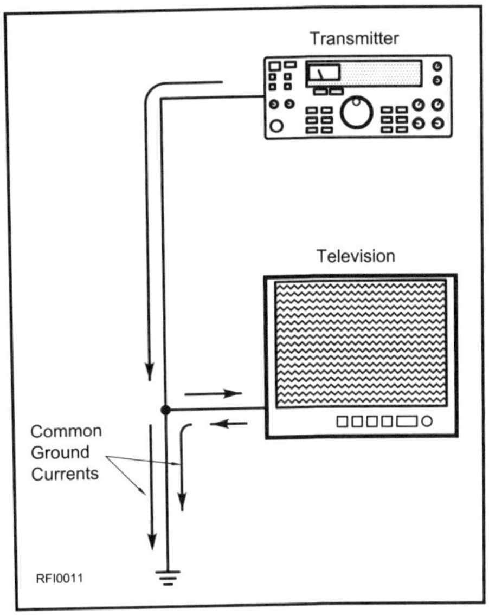

Figure 2.10—The ac-power supply electrical ground shares common ground currents. These currents can couple between the transceiver and television. This could result in interference to both devices. The cure in this case might be to use separate circuits for the transceiver and television.

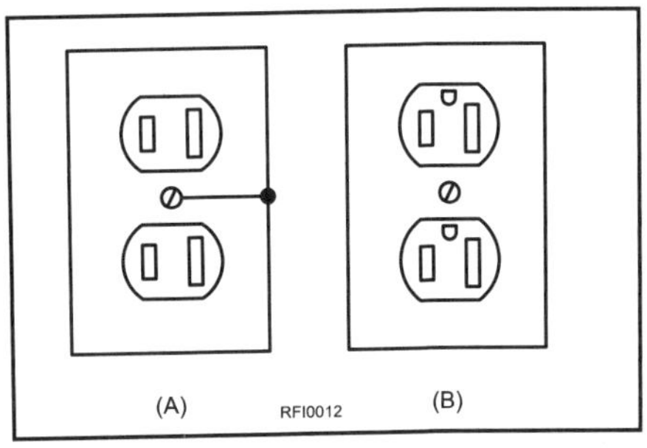

Figure 2.11—Old and new ac-outlet wiring styles. The old style, at A, has the colored wire (hot) connected to the small blade and the white wire (neutral) connected to the large blade. The green (ground) wire, if present, *should be* (but may not be) connected to the metal outlet box via the outlet center screw. The new style, at B, has the colored and white wires connected to the small and large blades as on the old style, but the green (ground) wire is connected to a D-shaped hole in each outlet. The outlet box of the new style may be plastic.

with its own circuit breaker, for higher-power amateur transmitters.

Shields

Shields are used to set boundaries for radiated energy. A low-pass filter can only reject harmonics that are exiting through the coaxial cable. Filters won't help if interference radiates from the chassis. Just as water passes through a sieve, RF passes through a leaky chassis. Good shielding can be used to plug the holes. Shields work by either attenuating the signal as it tries to pass through or by reflecting it. High conductivity metals such as copper or aluminum generally work by reflecting the unwanted energy, keeping it confined to the circuitry that is generating it.

Shielding Materials and Techniques

Magnetic fields below a few hundred kHz are best shielded by ferromagnetic materials, such as steel or mu metal (a specially heat-treated iron-nickel alloy). This is only true below a few hundred kilohertz. At those frequencies, the primary shielding mechanism is absorption. Absorptive loss is easily calculated: it is 8.7 dB times the thickness in skin depths (more about skin depth is found later in this chapter). Since ferromagnetic materials have shallower skin depths, attenuation is greater.

At HF, however, reflection is more important than absorption, so once again, copper and aluminum make better shields. To summarize: use steel shielding for low-frequency magnetic fields (such as stray radiation from a 60-Hz power transformer.) Aluminum or copper is better in other cases.

Solid Shielding

Most shielding is made from solid materials, usually aluminum or copper at RF; iron, steel or mu metal at LF. Solid shielding usually works by a combination of reflection and absorption.

Maximum shield effectiveness at RF generally demands solid sheet metal that completely encloses the emitter or susceptible equipment. Solid materials, such as sheet aluminum or copper, attenuate RF signals through both reflection and absorption. The shield surface reflects incident energy because of the impedance discontinuity at the air-shield boundary. Similarly, there is internal reflection of RF energy that reaches the opposite face of the shield. The RF energy is further attenuated by absorption as it passes through the shield material.

Cable Shielding

Cable shields are typically a compromise between effectiveness and practical considerations. Flexibility, low-weight and low-cost requirements lead owners to choose cables that are not completely shielded. (The main exception is rigid or flexible conduit.) Most cables are shielded with copper braid. The shield effectiveness of braid increases with weave density and decreases with increasing frequency. When braid shielded coaxial cables are bundled together, leakage from one cable can cause interference to adjacent cables. Leakage of electromagnetic fields through the braid causes a current to flow on the outside of the braid. This produces currents on the outside of other cables in the bundle. These are generally common-mode currents.

Thin Shielding

Because of the expense of solid shields and the popularity of plastic enclosures and composite materials, many modern electronic devices rely on thin film shielding to control RFI. Thin film shielding typically consists of a metallic (silver, copper or zinc) film deposited on the nonconductive support via vacuum metallization, flame spraying, plating or metal-filled paints. Pressure-sensitive foils and laminates can also add shielding capabilities to nonconductive materials. Unlike solid materials, thin films provide negligible absorption loss; attenuation is primarily from reflection.

Conductive Spray

It is possible to purchase conductive paint that can be used to create shielding on otherwise nonconductive materials. GC/Thorsen sells spray-on shielding that is readily available and easy to use. You can sometimes spray-paint this shielding material onto plastic enclosures and create a reasonably good shield.

This material should not be used indiscriminately, however! Remember, it is both paint and a good conductor. If some of the spray gets on to the electronics circuitry, the resultant short circuits can ruin the equipment. Like any paint, the materials being painted must be compatible with the paint, and clean. First, ensure that the paint will adhere well by thoroughly cleaning the material to be painted, using a suitable cleaner or solvent. Then, test-spray a piece of the material to make sure the paint adheres well. Make this assessment carefully; if any of the spray paint should ever come off, the conductive paint flakes could cause catastrophic equipment failure.

The spray-on shielding must follow all of the shielding "rules." All of the shielding material must be electrically connected to all other shields. It must also be tied in properly to the chassis and/or equipment ground. It may be necessary to disassemble various plastic parts and ensure that the shield spray will be connected across the width of all seams, etc. If necessary, ensure that the shield spray surface is electrically connected to the chassis. It may be necessary to run some sort of conductor to do this.

Shielding Spray Pitfalls

There are some important safety aspects to consider before using shielding spray, however. The first is a shock or electrical hazard. The equipment manufacturer may have used plastic to prevent shock to the equipment user or service technician. Some consumer-electronic equipment is "transformerless," meaning that the internal circuitry may be connected directly to the ac line. In this case, the plastic case is all that may stand between the consumer and a very dangerous electrical shock. If the plastic is made conductive, a potentially fatal shock could result.

The equipment manufacturer may have also used plastic parts near high voltage. Adding a conductive surface to these parts may result in internal arcs or short circuits. It is also possible that a service technician could be shocked by touching a plastic part "known" to be insulated. In some cases, the spray may also upset the thermal balance inside the circuitry, by reflecting heat back toward components.

Shielding sprays can be used, but they should be used with caution. A good rule is that if you are not *sure* a spray (or any RFI cure) is safe and appropriate, don't use it.

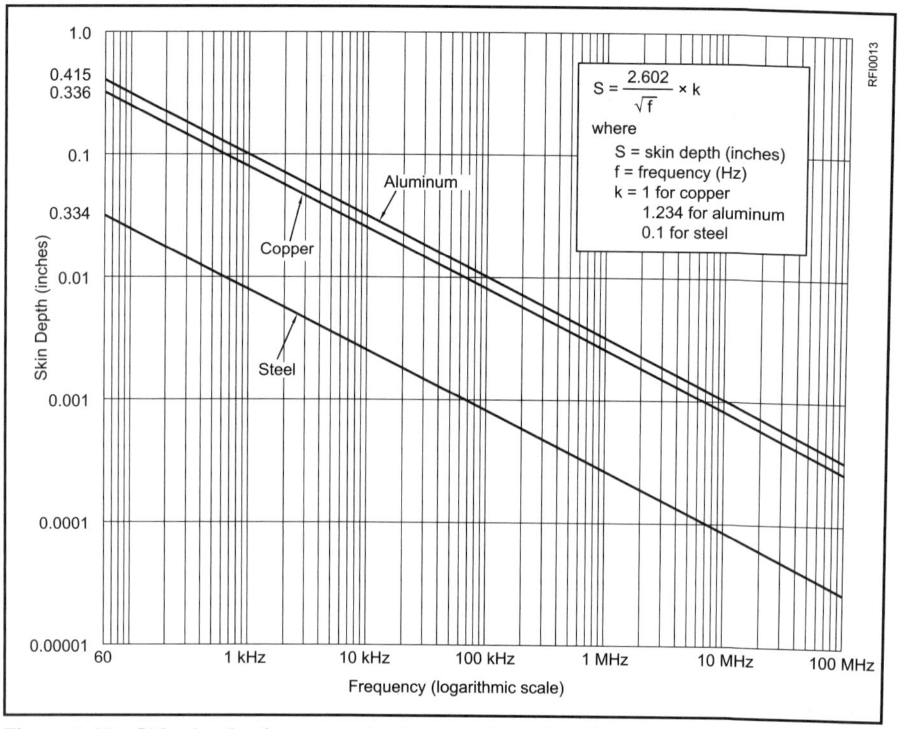

Figure 2.12—Skin depth of copper, aluminum and steel vs frequency. Ideally, shield thickness should be at least 3-10 times the skin depth. The skin depth of a particular steel varies greatly depending on the alloy used.

Shielding Effectiveness

Shield effectiveness is expressed as the number of decibels by which a shield reduces the field strength of radiated energy. Effectiveness is a function of shield composition and thickness, the frequency of the radiation and the quantity and shape of any shield discontinuities (seams or holes). For example, RG-59 with 51% braid coverage provides a relative isolation of 18 dB; a cable with 98% braid coverage provides 52 dB of relative isolation.

Discontinuities in shield materials decrease shield effectiveness. For this reason, shield seams should be bonded so that the RF impedance of the seam is the same as the material being joined. Otherwise, RF voltages may develop across the seam and allow RF energy to penetrate the shield. Pressure-sensitive foils, also called RFI tape (a form of thin film shielding), can be used across shield breaks to maintain shield effectiveness.

Panel openings (for displays, controls and ventilation) often compromise shield effectiveness. Panel openings for meters and other displays can be protected with conductive glass or a wire mesh across viewing surfaces. A honeycomb construction can be used for features like ventilation ducts, where it is not necessary for the shield to be transparent. Thin film or solid shields behind displays, combined with feedthrough capacitors, also minimize the flow of RF energy through panel openings.

Skin Depth

Shielding effectiveness is related directly to "skin depth," the depth to which RF penetrates shielding material. Shielding material should be chosen so that the actual material is at least 3 to 10 times the skin depth. Figure 2.12 shows the skin depth of common shielding materials vs frequency.

At RF, most of the current induced by the impinging signal flows near the surface of the shield. In fact, 63% of the current flows within one *skin depth* of the surface, falling off rapidly as depth increases. Ferromagnetic materials, such as steel, have a much shallower skin depth, and thus a much higher RF resistance than copper or aluminum.

Slots and Holes

Actually, the above discussion is not completely correct. While it is true that steel is a poorer RF shield than aluminum, it doesn't matter in most practical cases. Most RF leakage is not through the shielding material itself, but through openings in the shield.

The trouble caused by a particular hole is determined more by its maximum dimension than by its area. A round hole passes less radiation than a long, thin slot of the same area. Pay special attention to seams around shield covers. Surprisingly, a long gap that is too tight to pass a piece of paper can radiate almost as though the shield were not present. As a "rule of thumb" space mounting screws no more than $\frac{1}{20}\lambda$ (at the highest frequency of concern). In order to shield harmonics that fall within the North American VHF TV channels (up to 216 MHz), space shield screws no more than about 7 cm (2.75 inches) apart.

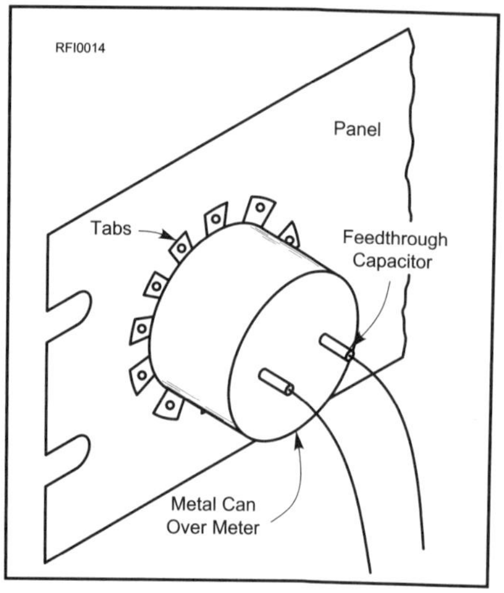

Figure 2.13—A shielding technique to prevent radiation from meter mounting holes.

Meters are a source of shield leakage, because they usually require a large hole in the front panel. Shield round meter openings with part of a discarded tin can (an old trick; see Figure 2.13). Find a can slightly larger than the meter. Cut off a portion (including an unopened lid) somewhat deeper than the meter (leave enough extra depth to form tabs that can be bent and drilled for mounting screws). Bypass the meter leads to the can at the exit point. Fabricate shield enclosures for square or edgewise meters from PC-board material.

Using Shields

Shield the Entire Unit

It is possible to shield an entire affected device, its enclosure, or cabinet. Although it is not often a practical solution, direct radiation pickup may be treated by placing the amplifier inside a home-built shielded enclosure. Perforated aluminum and metal screen are suitable materials. Be sure that the enclosure provides enough ventilation to prevent overheating of power amplifiers. Alternatively, the manufacturer or a qualified technician can add shielding or a conductive spray inside the amplifier cabinet or make circuit modifications to reduce RF susceptibility.

The owner may make (or have made) a shielded enclosure to hold the affected device. A suitable enclosure is reasonably easy to find or construct. For example, a stereo system might be placed inside a metal entertainment-center cabinet. The main drawbacks are appearance and convenient access to the equipment inside the enclosure.

A suitable enclosure may be made of metal, or a nonconductive enclosure may be lined with metal or a conductive spray. Copper and brass offer ease of connection; aluminum is cheaper (use lock-washer solder lugs for electrical connections to aluminum). A small metal case should be perforated (such as decorative screening) to allow adequate ventilation. Use a piano hinge for good shielding at openings. Opening covers should fit closely and overlap the opening edge. If the enclosure cures the RFI without a ground connection, leave it that way. A good earth ground helps in some situations.

Internal Shielding

Internal shielding is a design issue. The owner or ham should generally not attempt to install shields *inside* an affected device. There are safety, regulatory and other issues involved. Conductive material may create short circuits in the equipment or present a shock hazard (especially with some transformerless designs). Adding shielding may also change the thermal balance inside the equipment, making it more prone to long or short-term failure. Internal shields should be installed only by qualified technicians with the manufacturer's approval. The manufacturer or a qualified technician may be able to install rigid (often foil bonded to cardboard) or spray shielding inside the case.

Internal Construction

Although it is not usually a good idea to add internal shielding, it *is* a good idea to ensure that the shielding that is present is working properly. All shields and materials should be electrically and mechanically bonded to each other and the equipment chassis ground. Gaps, seams or large holes can leak a lot of RF energy. If the equipment has seams, ensure that they are well bonded. If there are gaps, fix them, using some of the techniques that follow. If fasteners are used, they should be spaced less than 0.05 λ at the frequency being shielded.

Paint is an insulator. Remove all paint from seam joints for the best connection. Where screw mounting is not practical (such as around an access door), use finger stock or one of the other commercial RF-gasket materials (Figure 2.14).

The orientation of interference-carrying wires inside the chassis is important. Obviously, it is best if they are far from any holes in the shield, but it's even more important to orient them parallel to the seams. Wires at right angles to slots cause maximum leakage (because the magnetic field is parallel to the slot, at 90° to the wire).

Tape

Treat shield seams with "RFI tape." 3M makes copper adhesive tape in several widths. Copper tape is often sold at hobby/craft stores. Florists also use it in preparing floral arrangements.

True "duct" tape, sold in the heating-supply departments of major building-supply stores, is made from aluminum. It can also be used in shielding.

Simply apply the tape across shield seams to seal them effectively. Most tapes will not have good electrical conductivity through the glue. Generally, the capacitance between the tape and the rest of the shielding is enough to have a good RF connection at VHF (where most of the troublesome transmitter harmonics are).

Holes in shield enclosures may be covered with conductive screen to prevent radiation. Copper screen can be easily attached to steel enclosures by soldering. Attach conductive screen to aluminum enclosures with stainless-steel hardware.

Grounding and Shields

How important is grounding to shield effectiveness? If "ground" means "earth ground" the answer is not much at all! A well shielded transmitter causes no RF cur-

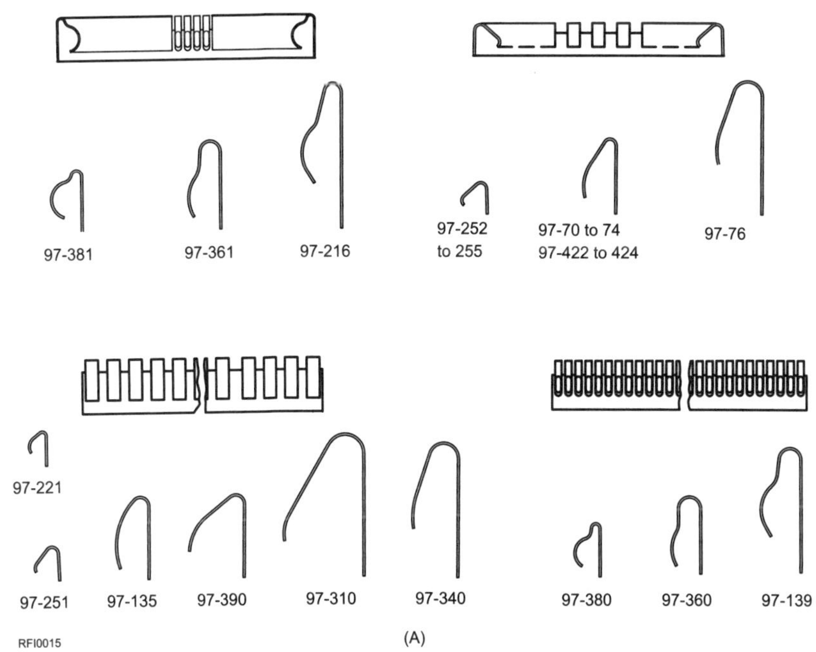

(A)

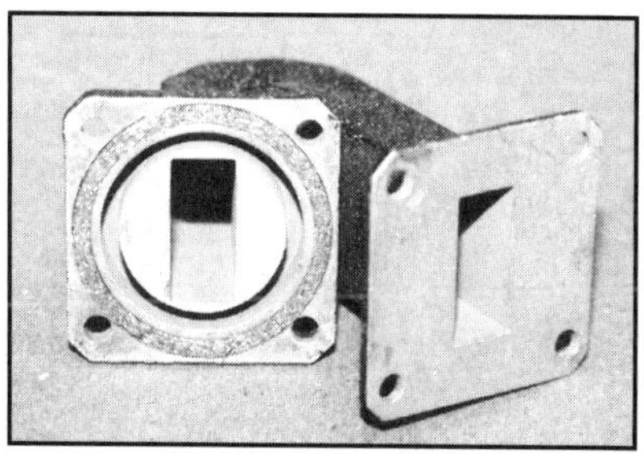

(B)

Figure 2.14—Several types of "finger stock" (A) are useful for shielding seams that must be opened and closed frequently. See Instrument Specialties Co and Richardson Electronics in the Suppliers List. Wire-mesh RF gaskets (B) can give an adequate RF-tight seal right up into the UHF and microwave range.

rents to flow on the outside of the shield, so the presence or absence of an earth ground can have no effect. If poor shielding does allow RF currents to flow on the outside of the chassis, a ground wire will not cure the condition. As discussed in the grounding section of this chapter, it is difficult to get a good RF ground with physically long ground wires.

What *is* important is internal chassis grounding. Internally, chassis grounding is used to help keep RF contained within the appropriate circuits. Missing or resistive internal ground connections can sometimes permit the RF energy to appear where it isn't supposed to be, perhaps in an unshielded circuit that normally wouldn't radiate RF. Part of the internal considerations should also be to ensure that the chassis grounding is functioning as the manufacturer designed it.

Two Heads Are Better Than One

One final shielding tip: Two poor shields are usually better than one good one. A good shield might have 100 dB of isolation. Two 60 dB shields add up to 120 dB, a good 20 dB (100 times) better than the single "good" shield.

FILTERS

Filters separate signals based on their frequency differences. Generally, filters make this possible because they can offer little opposition to certain frequencies while blocking (by reflection and/or absorption) others. Filters vary in attenuation characteristics (high-pass, low-pass, band-pass and notch), power-handling capabilities and in their passband and stopband frequencies. The names given to various filters are based on their uses.

A properly designed and installed filter can reduce the levels of conducted interference, as long as the spectral content of the interference is different from that of the desired signal. Filters are often used to resolve problems resulting from design compromises. For example, harmonic filters would not be needed if circuit linearity were perfect.

The application of standard filter configurations is the same, regardless of operating frequency. Low-pass filters are used at the output of any device that generates unwanted harmonics. Depending on the desired frequencies, however, receiver front ends may require low-pass, high-pass, band-pass or notch filters at the antenna input.

From the perspective of RFI control, the most important attribute of a filter is its frequency characteristic—the relationship between insertion loss and frequency. In choosing a filter, both the desired and unwanted frequencies must be considered. If the frequencies are relatively close, then a filter with a large attenuation-vs-frequency slope may be needed. Such filters are expensive to construct, however, because of the precision component values and mechanical accuracy required.

Filter Types

The following describes a number of filter types helpful in RFI control. Other than the ferrite filters, these are all differential-mode filters: the filters work on the differential-mode signal inside the conductors being filtered.

Low-Pass Filters

Low-pass filters pass frequencies below some cutoff frequency, while attenuating frequencies above that cutoff frequency. They are typically used as transmit filters for HF transmitters. A typical low-pass filter curve is shown in Figure 2.15. A schematic is shown in Figure 2.16. These filters can be difficult to construct properly and suitable parts can be hard to find, so most amateurs buy them. Many retail Amateur Radio stores that advertise in *QST* stock low-pass filters. Low-pass filters are discussed in more detail in the Transmitters chapter.

High-Pass Filters

High-pass filters pass frequencies above some cutoff frequency while attenuating frequencies below that cutoff frequency. They are typically used as receive filters for VHF/UHF broadcast television. A typi-cal high-pass filter curve is shown in Figure 2.17. Figure 2.18 shows a schematic of a typical high-pass filter. Again, it is best to buy one of the commercially available fil-

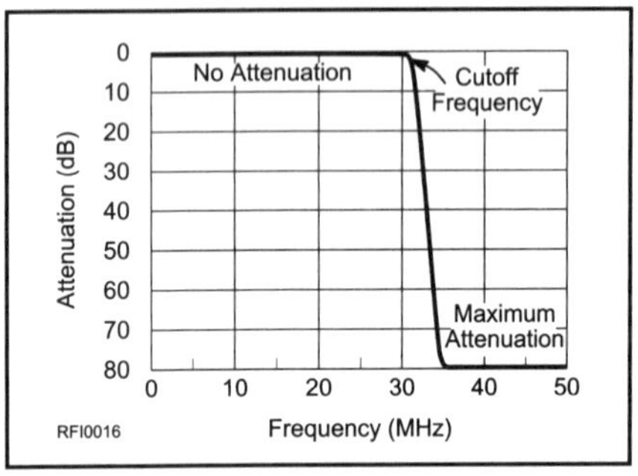

Figure 2.15—An example of a low-pass filter frequency response curve.

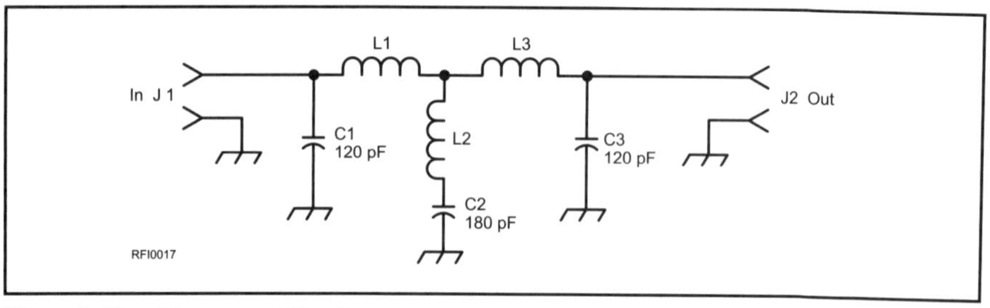

Figure 2.16—A low-pass filter for amateur transmitting use.

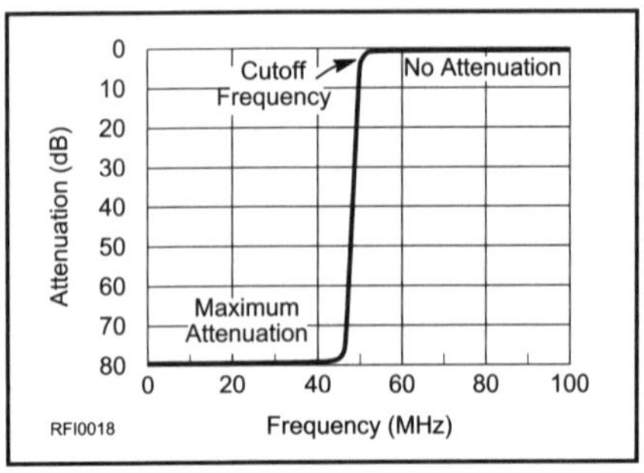

Figure 2.17—An example of a high-pass filter frequency response curve.

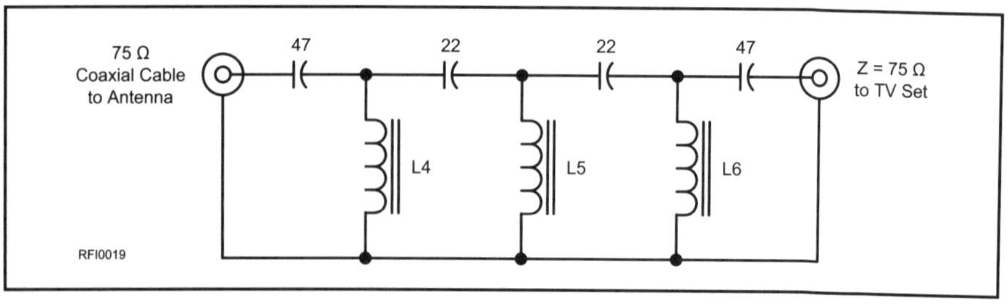

Figure 2.18—A differential-mode high pass filter for 75-ohm coax. It rejects HF signals picked up by a TV antenna or that leak into a cable-TV system. It is ineffective against common-mode signals because the ground connection is continuous between input and output. All capacitors are high-stability, low-loss NP0 ceramic discs. Values are in pF. The inductors are made with #24 enameled wire on T-44-0 toroid cores. L4 and L5 are each 12 turns (0.157 µH). L5 is 11 turns (0.135 µH).

ters. High-pass filters are discussed in more detail in the Televisions chapter.

Band-Pass Filters

High-pass and low-pass filter designs can be combined to form a band-pass filter. A band-pass filter passes a band of frequencies, typically an entire amateur band, and rejects signals significantly above and below that frequency band. An example of a band-pass filter response is shown in Figure 2.19.

Notch Filter

Notch filters can also be helpful. A notch is the opposite of a band-bass filter. A notch filter eliminates a single frequency, or narrow-band of frequencies. If a notch filter eliminates a wider band of frequencies, it may be called a band-stop filter. The frequency characteristics of a notch filter are shown in Figure 2.20.

Reflection vs Absorption

Most people think that filters remove unwanted energy. In reality, most filters are reflective—they reflect unwanted RF energy back toward its source, where presumably it can't be radiated efficiently. Reflective filters function as reflectors in their stopband. These filters usually consist of a capacitor-inductor combination configured to present a mismatch in the stopband (a high series impedance with a low shunt impedance, for example) and an impedance match in the passband.

Absorptive Filters

Since reflective filters simply divert undesired signals, the potential for interference remains. When undesired signals must actually be eliminated, a filter that absorbs the unwanted RF energy must be used. These filters provide attenuation via absorption. These are called "absorptive filters." Absorptive filters are especially effective when combined with reflective filters; the combination is capable of providing steep cutoff slopes and high stopband attenuation. Most absorptive-filter designs include a resistor to absorb the unwanted energy being filtered. An example of an absorptive low-pass filter is featured in the Transmitters chapter.

AC-Line Filters

Power-line filters are low-pass filters that provide little attenuation to 60-Hz energy but substantial attenuation to RF energy. Power-line filters are useful for suppressing conducted emissions that may enter equipment from the power line and vice versa.

AC-line filters, sometimes called "brute-force" filters, are used to filter RF energy from power lines. A schematic is shown in Figure 2.21. Use ac-rated components as specified. We *strongly* recommend UL-listed, commercially made ac-line filters; the ac-power lines are no place for homebrew experimentation!

Ferrites and Common-Mode Chokes

Common-mode chokes may be the best-kept secret in Amateur Radio. The differential-mode filters described earlier are *not* effective against common-mode signals. To eliminate common-mode signals properly, you need common-mode chokes. They may help nearly any interference problem, from cable TV to telephones to audio interference caused by RF picked up on speaker leads.

Common-mode chokes usually have ferrite core materials. These materials are well suited to attenuate common-mode currents. Several kinds of common-mode chokes are shown in Figure 2.22.

The optimum size and ferrite material are determined by the application and frequency. For example, an ac cord with a plug attached cannot be easily wrapped on a small ferrite core. The characteristics of ferrite materials vary with frequency, as shown by the graph in Figure 2.23.

Ferrite is a ceramic containing granulated iron compounds. It comes in toroids, beads and bars, that can be used to absorb RF energy or used as a low-pass filter. When current flows in a wire passing through a ferrite bead, magnetic flux circulates inside the bead. In ferrites formulated for RFI control, most of the magnetic flux is dissipated in the material as heat. The bead forms an absorptive filter with energy absorption proportional to frequency and bead length. The optimum bead size and composition is determined by the application.

Ferrite cores are uniquely suited for attenuating the flow of common-mode currents. Wrapping a cable through a lossy toroid (that is, one with a high-permeability mix) forms an RF choke for common-mode currents, without attenuating differential-mode signals. With unbalanced cables (coax), common-mode current in the shield sets up a field that dissipates the energy as described for beads. With balanced cables (speaker cables, twin lead and such), the differential-mode currents produce magnetic fields of opposite polarity, which cancel so there is no net field and negligible signal loss. Common-mode currents are dissipated as described for beads.

Ferrite cores are effective filters at higher frequencies than wire-wound inductors because of their low associated capacitance. Ferrite cores are also easy to use. By simply slipping one or more cores onto an antenna lead, speaker cable, pick-up lead, power cable or multiwire cable, RF current on the wire is attenuated. A single core slipped over

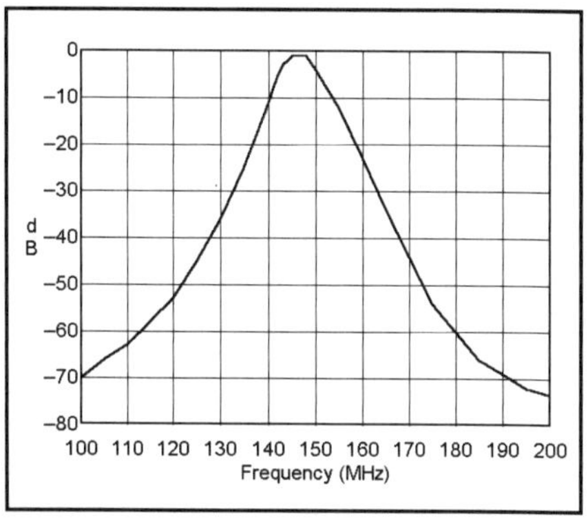

Figure 2.19—This bandpass filter passes a band of frequencies, rejecting frequencies above and below the passband.

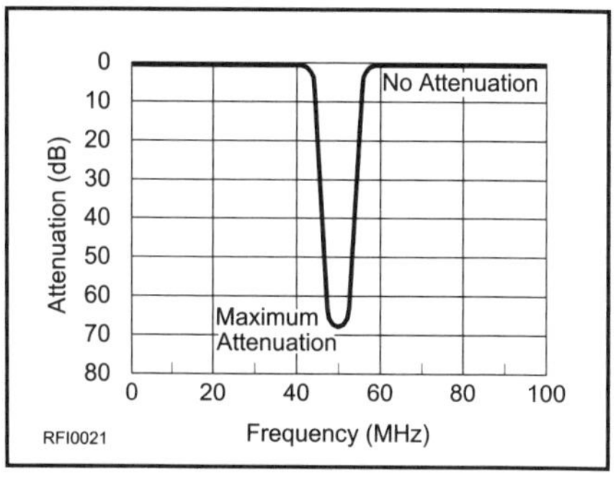

Figure 2.20—An example of a notch filter frequency response curve. If the notch is deep, it is often narrow; if the notch is wide, it is not often very deep.

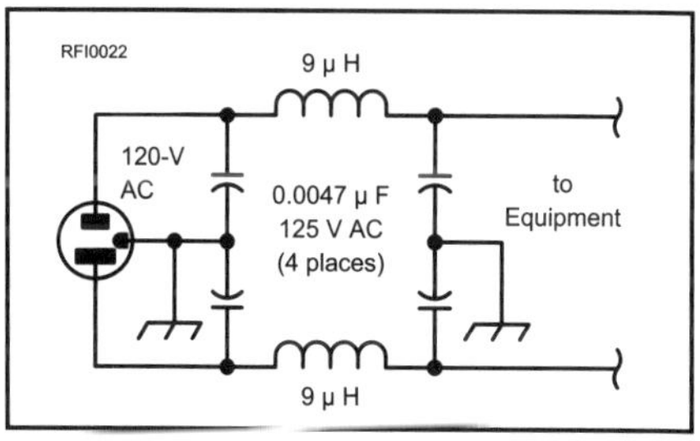

Figure 2.21—A "brute-force" type ac-line filter. It is often not advisable to construct one of these filters. In a safe design, the inductors will be able to handle the full-rated ac current that might go through the filter. The capacitors must be rated to be placed across 120 volt ac lines.

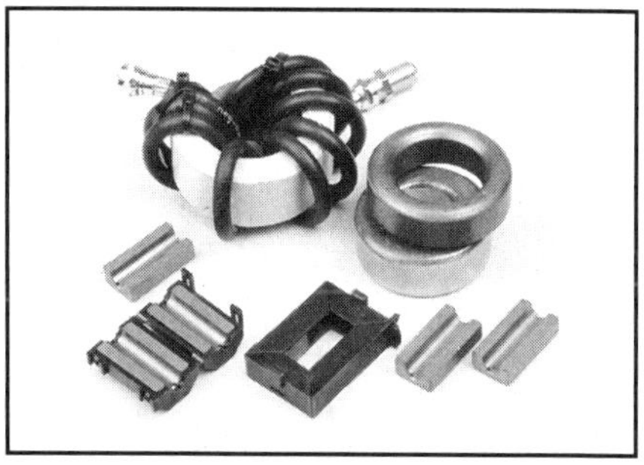

Figure 2.22—Several styles of common-mode chokes.

a wire is equivalent to a single-turn RF choke. Increasing the series inductance is simply a matter of adding more turns, more or longer beads, in direct proportion to the additional inductance required.

Speaker leads may be wrapped through a ferrite toroid core to reduce RF energy entering an audio amplifier via the speaker leads. Similarly, common-mode RF current flowing through an ac-line cord can be attenuated by winding a single layer of the cord on a ferrite rod. When both the antenna and power leads carry common-mode current, the attenuation afforded by a toroid can be multiplied by winding both leads (in opposite directions) on the same core.

Ferrite cores are made by several suppliers. There are two groups of ferrites. The first group acts inductive over its specified frequency range. Ferrites in the second group act like a resistor in parallel with an inductor over the specified frequency range. For these reasons, it is difficult to buy ferrite material at hamfests; the material and supplier are frequently unknown. Unfortunately, there is no standard or color code for marking ferrites; each manufacturer uses their own. When misapplied, ferrite seldom works; this breeds mistrust of a valuable RFI suppressor.

In general, a ferrite filter formed with a toroid is more effective than one formed with beads. Ferrite filters, especially those formed with beads, are usually only effective in low-impedance circuits. Of course, the impedance of any particular circuit (at the interference frequency) is often difficult to predict.

Ferrite Beads

Ferrite material can also be shaped into beads. These dowel-like ferrite cores have one or more holes through them. When placed on wires, they act as RF chokes. The beads are made from several different ferromagnetic materials, in order to vary their permeability. Beads work where their larger brothers won't fit. In most cases, ferrite beads do not provide much inductance, so they do not work as well as ferrite chokes as do toroids with multiple turns of wire. If possible, use a few turns of wire in conjunction with a ferrite bead to help improve its effectiveness.

Beads come in various sizes and per-meabilities, so manufacturers' literature should be consulted to determine their size, permeability and optimum frequency range. Manufacturers usually publish this data, along with an "impedance factor" (related to the permeability), which relates impedance to bead size and the number of turns.

The impedance coupled into a circuit by a bead is a function of the bead material, the signal frequency and the bead size. As

Ferrite Materials

The following types of ferrite materials are the most useful:

Material 43: μ = 850. Wide-band transformers to 50 MHz. Optimum frequency attenuation from 14 MHz to 450 MHz. This material is a good choice common-mode filter against VHF as well as both VHF and HF signals. It is also used for ferrite beads, with a peak impedance of about 30 Ω per bead at 200 MHz.

Material 61: μ = 125. Wide-band transformers to 200 MHz. Optimum attenuation above 200 MHz. This material makes a good common-mode filter against VHF signals.

Material 64: Ferrite bead material. Its peak attenuation is approximately 40 Ω per bead at 400 MHz. This is the best bead for both VHF and UHF amateur signals.

Material 73: Ferrite bead material. Its peak attenuation is approximately 30 Ω per bead at 25 MHz. This is the best bead for the upper end of the HF range and a good choice for general-purpose HF coverage.

Material 75: Ferrite bead material or wide-band transformers. Its peak attenuation is approximately 30 Ω per bead at 6 MHz. This material is a good choice for AM broadcast interference and MF/lower-frequency HF amateur signals.

J material: This is similar in characteristics to 75 material.

Ferrite material is available from Amidon, Palomar and many electronic-parts distributors. See the ARRL Parts Suppliers list in the References chapter of *The ARRL Handbook* for an up-to-date list.

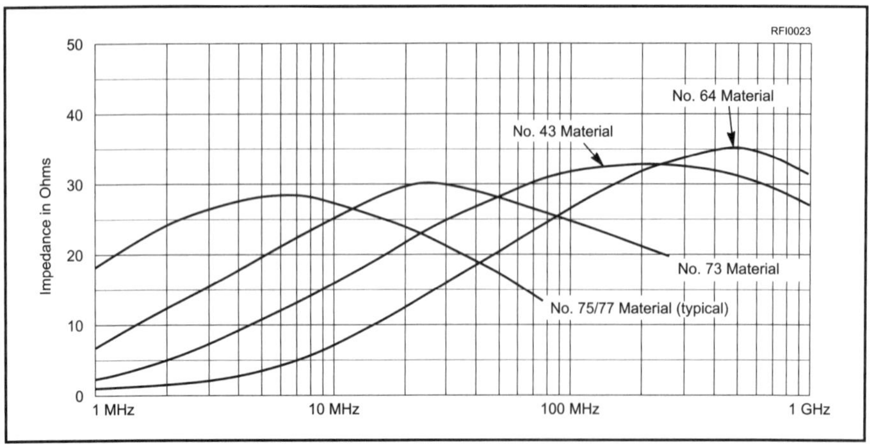

Figure 2.23—A plot of attenuation vs frequency for "101" size ferrite beads. Larger beads have higher impedance. Adding additional beads on a wire increases the impedance proportional to the number of beads.

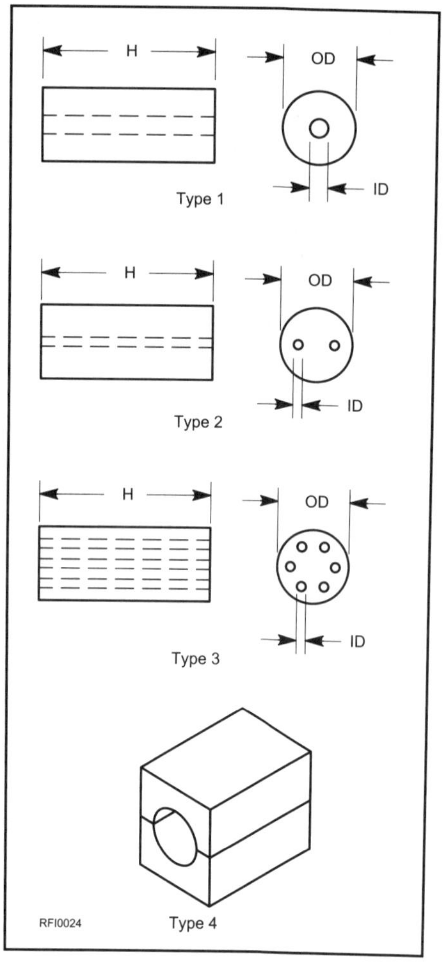

Figure 2.24—Typical ferrite bead configurations. Types 1 through 3 are one-piece beads. These are often used in circuit designs to suppress VHF signals. Type 4 is a split bead for assembly around cables or wire bundles. Some hams try to use these as common-mode chokes for HF signals. They usually do not work well unless a number of turns of wire can be used.

Table 2.2
Ferrite Bead Data

Part #†	Bead Type	Dimensions (in.) OD	ID	Ht	A_L v Mix No. (nH_t††) 43	64	73	75	77	Impedance Factor
FB-(xx)-801	1	0.296	0.094	0.297	1300	390	3900	—	—	2.5
FB-(64)-901	2	0.250	0.050	0.417	—	1130	—	—	—	†††
FB-(xx)-1801	1	0.200	0.062	0.437	2000	590	5900	—	—	3.9
FB-(xx)-2401	1	0.380	0.197	0.190	520	—	1530	—	—	1.1
FB-(xx)-5111	3	0.236	0.032	0.394	3540	1010	—	—	—	††††
FB-(xx)-5621	1	0.562	0.250	1.125	3800	—	—	—	9600	7.4
FB-(xx)-6301	1	0.375	0.194	0.410	1100	—	—	—	2600	2.1
FB-(43)-1020	1	1.000	0.500	1.112	3200	—	—	—	—	6.2
FB-(77)-1024	1	1.000	0.500	0.825	—	—	—	—	5600	3.7
2X-(43)-151	4	1.020	0.500	1.125	split bead, no. 43 only					
2X-(43)-251	4	0.590	0.250	1.125	split bead, no. 43 only					

†Complete part no. by substituting material no. for "xx."
††Based on low-frequency measurements.
†††Based on a single "U-turn" winding.
††††Based on a 2½-turn side-to-side winding.
Information courtesy of Fair-Rite Corp and Amidon Associates.

frequency increases, permeability decreases, resulting in a band-reject response. Figure 2.23 shows the variation of impedance with frequency for several materials, based on a "101" bead size. Mix 43, 73 and 75 ferrite beads are most commonly used for RFI suppression.

Impedance is directly proportional to the length of the bead, and it increases as more beads are used. The magnetic field is totally contained within the beads, and it does not matter whether they touch one another or not. Ferrite beads need not be grounded; in fact, take care with these high-permeability materials because they are semi-conductive. They should not touch uninsulated wires or ground.

Figure 2.24 shows four different bead configurations. As the figure shows, one need only thread one or more beads on a lead, or wind one or more turns through them (Type 2 and 3 beads) to impede unwanted RF signals. Table 2.2 shows design data for many popular ferrite beads. Bead type 4 is a split bead that can be used over cable assemblies (¼ to ½ inch diameter) without disassembly of the cable. The split bead is mounted around the cable to enclose it. Flat, split beads are available for use on multiconductor flat ribbon cable.

To install a ferrite bead in a transistor amplifier built on a PC board, unsolder the base lead, lift the lead clear of the board, place a ferrite bead over it, reinsert the lead and solder it. This places an inductively coupled impedance in the signal path without adding resistance. Remember you are trying to eliminate RF, and the desired signal is AF.

Ferrite beads typically work best in low-impedance circuits. They provide little (if any) attenuation when used in high-impedance circuits.

BYPASS CAPACITORS

In some cases, bypassing techniques can be used to eliminate unwanted RF from circuits, without affecting desired signals. Bypassing is a form of low-pass filtering. It is also generally a differential-mode filtering method. Capacitors are installed between a signal line and ground (or across two signal lines) to short circuit unwanted RF signals. Bypassing can be used where there is a large difference in frequency between the desired signal and unwanted RF. For examples, bypassing works well on audio lines or on dc meter leads.

Bypassing generally provides a low-impedance path to ground for RF signals. Normally, a bypass capacitor will have an impedance at the RF signal frequency that is 20% or less of the circuit impedance. This forms an effective low-pass filter. By shunting higher frequencies to ground, bypass capacitors offer an inexpensive means of reducing RF energy in a circuit. Bypass capacitors (typically 0.001 to 0.1-µF ceramic discs) can be used at panel meters to reduce emissions, from each side of the ac line to the chassis to reduce ac-line pickup, and on speaker connections (on vacuum-tube amplifiers only, more on this later) to shunt the RF energy to ground before it enters circuitry.

Conventional bypassing techniques (as discussed in the RFI Fundamentals chapter) can work against many RFI causes. Over the years, *QST* "Hints and Kinks" has featured audio-RFI cures that involve placing capacitors across audio inputs or outputs. In many cases, these can also work together with inductors to form a very effective filter. Several examples are shown in the Stereos chapter.

Bypassing does have its side effects, however. A 150 pF capacitor has an impedance of about 100 ohms at 10 MHz, making it an effective RFI filter at that frequency if used in a high-impedance audio circuit. It has an impedance of about 50,000 ohms at 20 kHz, however, and if used in a typical audio high-impedance circuit, it will roll off the 20 kHz response by about 3 dB. This is tolerable in some applications, such as a house intercom, but will probably be unacceptable to most high-end audio users.

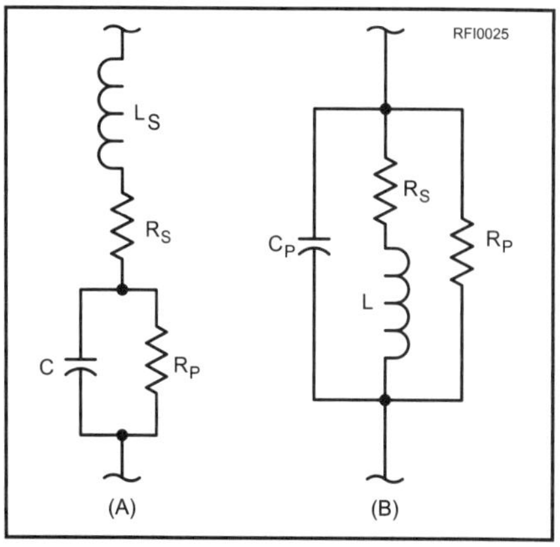

Figure 2.25—A, every capacitor includes stray series inductance and resistance. B, every inductor has stray shunt capacitance and series resistance. Losses in high-permeability cores appear as a parallel resistance, Rp.

When Not to Bypass

Do not apply bypass capacitors to the output of transistor amplifiers! The reasons not to do this are discussed in detail in the Stereos chapter.

Components Aren't Perfect

The reliability, stability, size, weight, efficiency and effectiveness of a filter is ultimately defined by the nature of the components used in its construction. For example, filter performance is limited because perfect components do not exist. Capacitors provide not only capacitance but also resistance and inductance. The resistance presented by a capacitor can be attributed to dielectric losses, foil resistance and the lead-to-foil contact. The inductance is present in the capacitor plates and leads. Because of self-inductance, capacitors exhibit self-resonance at a frequency where the inductive and capacitive reactances are equal. At frequencies above resonance, a capacitor behaves more like an inductor than a capacitor.

The magnitudes of inductance and resistance presented by a capacitor, and therefore the capacitor's suitability for filtering, is a function of capacitor type and construction. Metallized-paper capacitors are poorly suited for use in RF filters because of high contact resistance and a tendency to create RF noise. Large tantalum capacitors are resonant at 2 to 5 MHz, depending on construction and capacitance. In comparison, aluminum-foil capacitors can be used up to 20 MHz, depending on capacitance and lead length. Even low-inductance mica and ceramic-disc capacitors, which are effective up to 200 MHz, are limited by lead inductance. Feedthrough capacitors offer the highest operating frequencies. With their reduced lead inductance, feedthrough capacitors self-resonate above 1 GHz.

Just as there are no perfect capacitors, there are no perfect inductors. In addition to inductance, inductors exhibit resistance, capacitance and self-resonance. At resonance, the reactance of the interwinding capacitance is equal to that of the inductance. To increase the self-resonant frequency (and therefore the useful range of the inductor), inductors can be wound on separate cores and connected in series. This decreases the total interwinding capacitance.

Even the best components and filter designs are fruitless without good shielding and grounding practices. Improper shielding and grounding of a transmitter cabinet allows RFI to radiate around the filter. In addition, improper shielding and grounding make it possible for unfiltered RF currents to flow unimpeded on the outside of the coaxial cable braid and around low-pass filters and traps.

Finally, filters must be properly terminated. The input impedance of a filter, at any frequency, depends on its load impedance. Since the ability of a filter to pass or reject energy results from its input impedance, a suitable load impedance is critical to filter performance.

Capacitors

Every capacitor you buy comes with an inductor and resistor included free of charge. (Figure 2.25A). The "equivalent series resistance" (ESR) results primarily from losses in the dielectric. The "equivalent series inductance" (ESL) results from self-inductance in the lead wires. As frequency increases, the ESL dominates; the capacitor acts like an inductor, which means the impedance increases with frequency. Hence it is important to keep bypass capacitor leads short.

The inductance present in a capacitor can be used to some advantage. A capacitor and inductor in series form a series-tuned circuit, which has very low impedance at its resonant frequency. Bypass capacitors work best near their series resonance. This fact can be used to advantage in a trick called *series-resonant bypassing*. To bypass a single band of frequencies, choose a capacitor value that self-resonates with its own ESL. A typical ceramic capacitor mounted with short (1/16 inch) leads has around 3-4 nH (0.003-0.004 μH) of ESL. Table 2.3 lists the optimum series-resonant-bypass capacitor values for various amateur bands.

Capacitor Types

Electrolytic capacitors, both the aluminum and tantalum versions, pack a lot of capacitance into a small space, but relatively high ESR and ESL limit their usefulness at high frequencies. Broadband bypassing requires a large-value electrolytic (for low frequencies) in parallel with a ceramic capacitor (to handle RF).

Ceramic capacitors have low ESL, and they are widely used in RF-bypass and filter applications. Those with larger values use high dielectric-constant materials, with high losses (high ESR). This can cause overheating in high-power RF circuits. The capacitance tends to be unstable as well. Thus, ceramic capacitors are primarily used as power-supply bypass capacitors, where stability is not as important. The dielectric type is indicated by the temperature-coefficient code printed on the side of the capacitor body. Z5U, Y5V and Z5P are lossy. Low-loss styles list the temperature coefficient (TC) in parts per million (PPM): like N750 (–750 PPM/°C), P150 (+150 PPM/°C) or NP0 (zero nominal TC).

Mica-dielectric capacitors have little loss, and they are stable, but they are not available in values greater than about 0.01 μF. Poly-film capacitors (including polyester or mylar, polypropylene, polystyrene and polycarbonate) are good low-ESR, medium-ESL styles with excellent stability. They are available up to a few μF, but they are physically large.

Figure 2.26 illustrates a special capacitor called a *feedthrough*. Feedthrough capacitors are useful for filtering leads that exit a shield enclosure. The exiting lead forms one side of the capacitor, with the other side connected to the shield. When the feedthrough capacitor is installed in the enclosure wall, there is almost no inductance from the "hot" side to ground. The result is excellent high-frequency performance.

For broadband bypassing, however, use the largest ceramic bypass capacitor that fits. It does a good job at low frequencies, and all ceramic capacitors have about the same reactance (ESL) at high frequencies anyway.

Table 2.3
Series-Resonant Bypass Capacitors for Various Amateur Bands

Frequency (MHz)	*Capacitance*
7	0.15 μF
14	0.039 μF
28	0.01 μF
50	0.0033 μF
144	390 pF
220	150 pF
440	39 pF

Values are based on measurements made using ceramic-disc capacitors mounted with minimum lead length on a 0.060-inch-thick PC board with zero effective trace length. Above 300 MHz or so, the resonance bandwidth becomes so narrow that the technique is not very useful. At such frequencies, it is better to use feedthrough or chip capacitors to reduce lead inductance.

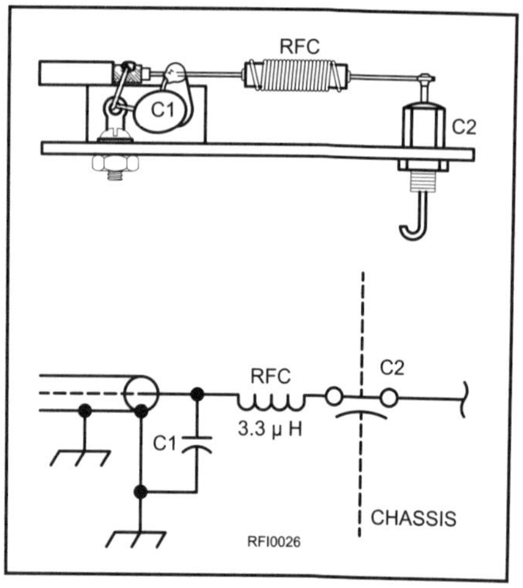

Figure 2.26—A feedthrough capacitor filtering a lead where it exits a shield enclosure. This filter rejects frequencies in the high-VHF TV band (174-216 MHz).

C1—0.001 μF disc ceramic.
C2—500 or 1000 pF feedthrough capacitor.
RFC—14 inches, #26 enameled wire close-wound on a 3/16-inch form (3.3 μH).

Unknown "Ferrite" Material

Hams often dig into their junk- boxes to find toroidal cores from which to construct common-mode "ferrite" chokes. In many cases, the results can be somewhat disappointing. This unknown material could be ferrite, powdered iron or even plastic! Even if it is ferrite, it may not be a material suitable for the frequency range involved. Other hams have used the ferrite from an old TV flyback or yoke. This material was designed for 15 to 20 kHz; it may or may not work effectively at RF.

It is okay to try unknown material; if it works, it is okay to use. If it doesn't work, however, don't falsely conclude that a common-mode choke will not solve your RFI problem. It is very possible that making a choke out of the correct material for the job at hand would have worked handily. Skilled amateurs may be able to characterize ferrites by measuring their attenutation vs. frequency with an antenna analyzer.—*Ed Hare, W1RFI, ARRL Laboratory Supervisor*

Ferrite Beads

The ARRL Lab staff often advise hams to try ferrite common-mode chokes to solve RFI problems. The advice is usually to use and FT-240-43 or FT-140-43 ferrite toroidal core and wrap about 10 turns of wire onto it. Many hams have used this sage advice to solve RFI problems. Others have listened to the advice and concluded that the cable is too thick or its connector is too large so some of those "split-bead" type ferrites should work just as well. While this can be true if you are trying to choke a VHF signal, at HF this arrangement just doesn't have enough inductance to be a good common-mode choke. It would take 20 to 40 of these split beads to be effective on 3.5 MHz, for example. At about $5 each, this would be an expensive proposition. They will work for marginal cases of RFI, especially at the upper end of the HF range. They will work better if you can get a few turns of wire on them. However, for the low end of HF, there is nothing that beats the right material (#43, #75, J, etc) with enough turns to be effective. — *Ed Hare, W1RFI, ARRL Laboratory Supervisor*

Inductors

Just as every capacitor includes stray inductance and resistance, so every inductor has stray capacitance and resistance (Figure 2.25B). The shunt capacitance results from distributed capacitance between turns, and the ESR is from the RF resistance of the wire. In addition, inductors with high-permeability cores exhibit core losses, which are modeled as a parallel resistance.

Inductor Types

Air-wound inductors are used in high-power RF circuits or where good stability is required (as in VFOs). Their disadvantage is large physical size.

Inductors wound on ferromagnetic forms pack higher inductance in less space. One disadvantage is core saturation: at some current level, the magnetic flux reaches the maximum density that the core can support. Inductance drops rapidly above this level, and the nonlinear response may cause distortion and harmonics.

Two magnetic materials are often used in RF inductors: ferrite and powdered iron. Ferrite usually yields greater inductance, but it saturates more easily, and the inductance value is less stable. Both ferrite and powdered iron come in various mixes that are optimized for different frequency ranges and power levels.

Molded chokes look like resistors: they have a cylindrical body with pigtail leads on each end, and they are often color coded. Small-value chokes are usually wound on a simple insulating form, while the larger values use a ferromagnetic core to increase the inductance. Be sure to observe the maximum dc current rating, which can be limited either by core saturation or by wire heating.

Toroid inductors nearly always use a magnetic core. The "doughnut" shape confines the lines of flux within the inductor. Toroids are often used to reduce unwanted radiation (or susceptibility to incoming radiation).

A *ferrite bead* is a small cylinder of ferrite with one or more holes through the center for wire. Figure 2.23 charts impedance vs frequency for several ferrite mixes. For more impedance, use a bigger bead, string several beads in series on the same wire, or use a multihole bead.

Split beads are available for wires or cables with connectors that would not fit through the hole. For best results, be sure that the mating bead surfaces make good solid contact.

OPERATING PRACTICES

Although it is no substitute for proper shielding, grounding and filtering, modified operating practices can be an effective means of avoiding RFI. Abstain from transmitting, or operate at relatively low power levels, during prime-time TV viewing hours. Also avoid operating on frequencies with a high potential for RFI. For example, the second harmonic of 10 meters falls within TV channel 2 (54-60 MHz). Since the lowest harmonic frequency is 2 MHz above the low edge of the TV channel, fine cross-hatching of the TV picture can be avoided by restricting operation to 29 MHz and above. The only restriction is that the second harmonic must be kept 200-500 kHz from the TV sound carrier, which is up 5.75 MHz from the low end of the TV channel. Thus, avoid 29.875 MHz (±250 Hz) as well.

Other operating practices that can reduce the potential of RFI include properly adjusting the drive and tuning of final-amplifier circuits. Mistuning the output of a tube-type amplifier or overdriving the final amplifier can result in very high harmonic output. Similarly, an improperly neutralized transmitter final amplifier may produce parasitic oscillations. A Transmatch can increase the selectivity of a transmitter output circuit, thereby reducing harmonics and parasitics at the antenna.

Antenna Management

Most amateurs select an antenna based on power gain figures, radiation pattern, beamwidth, space requirements, weight, wind survivability and cost. Many of us fail to recognize that antenna polarization, physical configuration, height and orientation affect not only communications, but also RFI. For example, single-band antennas, tuned and resonant at a single band, are much less efficient radiators of harmonics than are multiband antennas.

Often, what is best for communications is also optimum from an RFI-prevention standpoint. For example, a narrow beam- width is not only effective against QRM, but it also restricts the area potentially affected by RFI. A pair of stacked antennas generally compress the vertical radiation pattern, relative to a single larger one. A four-bay antenna array, with four antennas stacked two high and two wide, compresses the radiation in both the vertical and horizontal planes.

Increased antenna height not only improves communications, but also reduces the field intensity of transmitted signals at nearby homes. Place the transmit antenna as high above ground as possible to increase the separation between the antenna and neighbors' equipment. Keep the transmit antenna away from CATV and power lines. Energy coupled into these lines may be conducted into neighboring houses. On the way, rectification may occur at corroded connections and junctions of dissimilar metals, resulting in RFI. A beam directed at a CATV system may result in RFI, simply because the connectors or housings of amplifiers and taps are inadequately shielded.

The optimum antenna polarization for RFI control depends on the situation. A switch from vertical to horizontal polarization has been effective in some TVI cases. Vertically polarized transmit antennas, however, induce stronger common-mode signals in nearby cables than do their horizontally polarized counterparts. Experiment to determine what is best for a particular case. Also consider the physical configuration of the susceptible equipment. Horizontally oriented CATV wiring, for example, picks up more RF energy from horizontally polarized antennas.

Most of us strive for a low SWR in our antenna systems. This is reasonable for increased communications efficiency, but a high SWR is often erroneously presented as an intrinsic cause of RFI. While several factors in the antenna system have specific RFI significance, high feed-line SWR does not create RFI problems. For example, it is important that any transmit filters have an appropriate load. That end, however, is easily achieved by placing the filter between the transmitter and Transmatch. An antenna system that is unmatched at the fundamental frequency, however, does not significantly increase feed-line radiation and has little or no bearing on RFI.

DESIGN PRACTICES

Amateurs have an obligation to reduce harmonics and spurious emissions in accordance with good engineering practice. This responsibility not only encompasses the operation of commercial equipment, it should also govern the design and construction of home-built equipment. Because of the typical amateur's limited resources, it is much better to provide for RFI control in the design stages than to correct RFI problems once they appear. For example, although a spectrum analyzer is the best way to ascertain that transmitter output is clean, few amateurs have access to one. In contrast, the following design guidelines for minimizing RFI can be followed by virtually all amateurs:

- Avoid long cable runs in proximity to one another (to minimize crosstalk).
- Minimize wire length whenever possible to reduce the potential for RF radiation and pickup.
- Isolate return lines of noisy components from the return lines of sensitive components.
- Use thick wire insulation to reduce capacitive coupling between wires.
- Choose minimum signal levels that are consistent with the needed signal-to-noise ratio.
- If signal and power leads cross, make the crossing perpendicular to minimize coupling.
- Use decoupling capacitors or a voltage regulator to decouple circuits from the power supply.
- Use wired returns rather than structure returns to ensure a clean, single-point ground that is free of ground loops. Structure returns (for example, chassis ground) may be more convenient, but voltage drops may occur in the structure used for the return, which may in turn induce voltages into sensitive circuits. (This does not apply to PC-board ground planes.)
- When working with digital circuits, use slow, low-power ICs where possible. The RFI potential of a device is directly proportional to its operating frequency and output current, and inversely proportional to its rise and fall times. When the output transition times are much shorter than required by the load input, the transition can be slowed by adding a small series resistor (25 Ω) at the output pin, and a small capacitor (50 pF) between the output pin and ground.
- Minimize radiated RFI from PC-board traces by positioning each RF load as close to its signal source as possible. Radiation is a direct function of the RF current path length.
- If a noisy line must cross a PC board, locate the input and output connectors as close to each other as possible, and run separate ground traces for each line.
- Connect unused IC inputs and outputs to ground or V_{cc}, according to the manufacturer's recommendations.
- The ideal PC-board aspect ratio is 1:1. The less a printed circuit board approximates a square, the more difficult it will be to design a good RFI-proof layout.
- Lay out op-amp circuits so that the input and output traces are as far from each other as possible (to prevent amplifier oscillations).
- Don't cover unused circuit board areas with ground plane. The area enclosed by return currents constitutes a radiating loop, and additional ground plane area may increase RFI by increasing the effective loop areas.
- Route all conductors subject to common-mode pickup tightly in cabinet corners or otherwise close to a ground plane (to minimize radiation and coupling).
- Shielded twisted-pair wiring can help protect susceptible circuits.

SIDE EFFECTS OF RFI CURES

Some might argue that a good working knowledge of the manifestations and root causes of RFI is unnecessary, that a random trial-and-error approach can be equally effective. As an illustration, examine the following list of actions known to be effective against RFI:

- Install a filter in the ac line to the affected device.
- Install a common-mode choke in the speaker leads.
- Connect the affected device to a good RF ground.
- Use grounded, shielded cable for speaker leads.
- Install a 0.01-µF capacitor between the affected device chassis and ground.
- Ground the device coax shield.
- Install an RF choke in the antenna coax.
- Install a shielded, grounded high-pass filter on the device antenna terminals.

All of these suggestions would likely control most cases of fundamental overload. However, each of these actions may also have unplanned and undesirable effects on the equipment involved. The more troublesome side effects, including parasitic oscillations and high-frequency attenuation in audio amplifiers, hum in phonograph preamplifiers and the "hot- chassis" syndrome, may cripple or destroy the affected device.

Bypass capacitors installed on speaker leads of transistor amplifiers may cause inaudible feedback. The result is oscillations that may destroy the audio amplifier. The Stereos chapter fully discusses this problem and appropriate cures.

Reduced high-frequency response can result when a phonograph preamplifier for a magnetic cartridge is bypassed. This happens because the bypass increases capacitance in the signal path beyond the maximum load capacitance of the cartridge. Similarly, RF chokes in phonograph preamplifiers (which are high-gain, high-impedance circuits) may cause hum when nearby magnetic fields induce voltage in the choke. Shield the choke or replace it with ferrite bead (over the input lead of the first amplifying transistor in the preamplifier circuit) to eliminate the hum.

An RF choke in series with the speaker leads not only breaks the RF path, but also attenuates higher audio frequencies. Even if the fidelity loss is not noticeable, the choke may cause the amplifier load to vary dynamically. Since some receivers and amplifiers can't tolerate loads of varying impedance, a ferrite bead on each speaker lead, or a toroid common-mode choke may be a better solution.

Bypass capacitors at the primary of a power transformer prevent RF from entering via the ac line, but they may create a shock hazard. The capacitors form a voltage divider that places the chassis above ground potential. Underwriter's Laboratories specifies that the maximum leakage current through any such capacitor should be less than 0.5 mA. Hence, use 0.01-µF capacitors or less. Also, any capacitors installed in ac-line circuits should be rated for that use (1.4 kV is the typical ac rating). An external power filter is a better alternative in most cases. External filters usually work, and anyone can install them.

Given these and other possible side effects of RFI control, it is best to use only minimum control methods after a detailed examination of all emitters, transmission paths and susceptible equipment involved. Examine each in terms of possible RFI control techniques and potential side effects. While certain exploratory methods can and should be used in determining these techniques, the most appropriate method of RFI control varies from one situation to the next. That is where knowledge of RFI fundamentals becomes indispensable.

GLOSSARY

BCI—broadcast interference; interference to broadcast receivers.

Bonding—the establishment of a low-impedance path between two metal surfaces; the physical implementation of grounding.

Broadband emission—an emission that has a spectral energy distribution sufficiently broad, uniform and continuous so that the response of the measuring receiver does not vary significantly when tuned over a specified number of receiver bandwidths.

Bypass capacitor—a capacitor used to provide a comparatively low-impedance ac path around a circuit element.

Common-mode signals—signals that are in phase on both (or several) conductors of a signal lead (often the antenna lead).

Conducted emission—a signal that propagates through an electrical conductor or any conductive structure. The level of conducted emissions is usually expressed in terms of voltage or current (for example, V or dBV).

Counterpoise—the reference-plane portion of an unbalanced antenna.

Crosstalk—an electromagnetic disturbance introduced by unwanted coupling between conductors.

Cross modulation—modulation of a desired signal by an undesired signal.

Decibel—a logarithmic unit of relative power measurement that is used to express the ratio of two power levels. It is equal to 10 times the base 10 logarithm of this ratio.

Desensitization—a reduction in receiver sensitivity caused by RF overload from a nearby transmitter or noise source.

Electromagnetic compatibility (EMC) — the capability of electronic equipment or systems to be operated with a defined margin of safety in the intended operational environment, at designed levels of efficiency, without degradation from interference.

Electromagnetic interference (EMI)—any electrical imposition that may interfere with the normal operation of equipment. RFI is said to be present when undesirable voltages or currents adversely influence the performance of a device. This term encompasses interference from radio-frequency sources (RFI), audio-frequency sources (that is, induced hum from a transformer) and electrostatic sources (ESD—electrostatic discharge).

Electrostatic discharge (ESD)—a flow of current that results from static electrical charges. This term covers events from doorknob shocks to lightning strikes.

Emission—electromagnetic energy propagated from a source by radiation or conduction.

Emitter—a source of electromagnetic energy.

Filter—a network of resistors, inductors, capacitors or transmission lines that offers comparatively little opposition to certain frequencies while blocking or attenuating other frequencies.

Functional interference—occurs when the normal functions of one system part directly interfere with functions of another part. Functional interference is generally easy to resolve because the frequencies and power levels are established by design.

Fundamental overload—unwanted desensitization or generation of spurious responses in a receiver that is caused by large amounts of RF energy from a nearby transmitter fundamental output signal.

Grounding—the establishment of an electrically conductive path that connects electrical and electronic elements of a system to one another, or to some reference point that may be designated as "ground."

Ground plane—a metal sheet or plate used as a common circuit return or reference point for electrical or signal measurements.

High-pass filter—a filter designed to pass all frequencies above a certain cutoff frequency, while rejecting those below the cutoff frequency.

Intersystem interference—when the source-coupling-receptor routes of interference include two or more separate and discrete systems.

Intrasystem interference—when the source-coupling-receptor routes of interference are located within a system.

Intermodulation distortion (IMD)—the undesired mixing of two or more frequencies in a nonlinear device, which produces additional sum and difference frequencies.

Low-pass filter—a filter designed to pass all frequencies below a certain cutoff frequency, while rejecting those above the cutoff frequency.

Narrowband emission—an emission that has its principal spectral energy within the passband of the measuring receiver.

Noise—anything that interferes with the exchange of intelligence in electronic communications.

Nonlinear—having an output that does not rise or fall in linear proportion to the input.

Notch filter—a filter that rejects or suppresses a narrow band of frequencies within a wider band of desired frequencies.

Passband—the band or range of frequencies that a filter conducts or is intended to pass. In most cases, the attenuation is minimal. Some designs, however, incorporate attenuation inside the passband to improve stopband rejection.

Peak envelope power—the average power supplied to the antenna transmission line by a transmitter during one RF cycle at the crest of the modulation envelope, taken under normal operating conditions.

Radiated emission—RF energy that is coupled between circuits, equipment or systems via electromagnetic fields. Radiated energy leaves the source and spreads out in space according to the laws of wave propagation. The strengths of radiated emissions are usually expressed in terms of power density or field strength (for example, V/m and dBV/m).

Receptor—the generic class of devices, equipment and systems which, when exposed to conducted or radiated electromagnetic energy, either suffer from performance, degradation or malfunction.

RFI—radio frequency interference. Interference to radios and televisions as well as to various appliances, such as computers, audio systems and telephones, from a source of RF energy. This is a specific case of EMI.

Selectivity—the ability of a receiver to reject unwanted signals. Selectivity, a measure of equipment quality, is a critical factor in most interference cases.

Service entrance—the point where some utility, usually ac power or telephone, enters a building.

Spurious emission—any electromagnetic emission, from the intended output terminal of an electronic device, that is outside of the designed emission bandwidth.

Spurious response—any response of an electronic device to energy outside its designed reception bandwidth through its intended input terminal.

Stopband—that part of the frequency spectrum that is attenuated by a filter.

Susceptibility—the characteristic of electronic equipment that permits undesirable responses when subjected to electromagnetic energy.

Susceptor—a device that responds to unwanted electromagnetic energy.

TVI—interference to television receivers.

Chapter 3

RFI Troubleshooting Techniques

Consider for a moment the various ways an RF signal can get from your transmitter to a neighboring piece of equipment. By the time you factor in all of the variables and the possible cures, the possible combinations can number in the millions. A systematic approach to troubleshooting is a must to identify the problem and to determine the correct cure.

By Ed Doubek, N9RF (SK)
Former Illinois Section Technical Coordinator

Many RFI cases are complex. They involve a source, a path and susceptible equipment. Each of these main components has a number of variables: Is the problem caused by harmonics, conducted emissions, radiated emissions or a combination of all of these factors? Are harmonics being generated outside the amateur station? Should it be fixed with a low-pass filter, high-pass filter, common-mode chokes or ac-line filter? How about shielding, isolation transformers, a different ground or antenna configuration? Is it a case of fundamental overload? Is the RF energy conducted along the power lines or picked up directly by the affected equipment? These are the kinds of questions that you must answer as you troubleshoot a case of RFI.

These are only a *few* of the questions you might need to ask. Any information you gain about the systems involved will help find the RFI cause and cure. With all of the variables involved, the possibilities could number in the millions. There are just too many possible variations, so you probably will not see your exact problem and cure listed in this book or any other. You must diagnose the problem! An accurate diagnosis, regardless of the problem, requires skill with the relevant technology and troubleshooting procedures.

This book has chapters on a number of different RFI problems. Each contains specific troubleshooting information for their particular RFI specialty. This chapter covers RFI troubleshooting fundamentals that apply to all types of RFI problems. It also covers the general area of checking and troubleshooting one area involved in most RFI problems—the amateur station.

The first two chapters in this book talked about EMC fundamentals and theory. Now that the basics are covered, it is time for lesson no. 3—troubleshooting techniques that will help you locate the source of a specific problem.

The Major Steps to Troubleshooting

Troubleshooting and curing an RFI problem is a three-step process, and all three steps are equally important. In many cases, the line between troubleshooting and curing RFI is not clear. The major components to an RFI problem are:

- Identify the problem
- Diagnose the problem
- Cure the problem

Identify the Problem

Is It Really RFI?—Before trying to solve a suspected case of RFI, verify that the symptoms actually result from external causes. A variety of equipment malfunctions or external noise can look like interference. "Your" RFI problem might be caused by another ham or a radio transmitter of another radio service, such as a local CB or police transmitter.

Is It Your Station?—If it appears that your station is involved, operate your station on each band, mode and power level that you use. Note all conditions that produce interference. If no transmissions produce the problem, your station *may* not be the cause. (Although some contributing factor may have been missing in the test.) Have your neighbor keep notes of when and how the interference appears: what time of day, what station, what other appliances were in use, what was the weather? You should do the same whenever you operate. If you can readily reproduce the problem with your station, you can start to troubleshoot the problem.

Diagnose the Problem

Look Around—Aside from the brain, eyes are a troubleshooter's best tool. Look around. Installation defects contribute to many RFI problems. Look for loose connections, shield breaks in a cable TV installation or corroded contacts in a telephone installation. Fix these first.

If your station is held together with baling wire and chewing gum (not that *yours* would be, mind you, but if it were) this could directly contribute to RFI problems. The same is true for the installation of the consumer equipment. Curing installation defects will often diagnose the RFI by curing the problem.

Look for the obvious; look for the unexpected. You know what good wiring should look like; you know what good engineering practices should apply to your station or consumer installations. Look to see that these practices are being followed, and that nothing unexpected can be found. If you find, for example, that the cable TV is installed in parallel with a TV antenna, you have probably found the TVI problem.

Where to Begin?

Many things other than amateur transmission can cause RFI. Fluorescent lamps, neon signs, electric motors, and transmitters licensed and unlicensed in other services can all contribute to the problem. A TV set can even interfere with itself. One of the first troubleshooting steps is actually "diplomatic": determine if you and your station are involved in the problem at all.

Keep an RFI Log

As discussed in the First Steps chapter, if you and your neighbor keep RFI logs, you should be able to quickly determine if you are involved. The station log is one valuable tool to prove that a station is not involved in a particular interference case. The FCC has greatly relaxed logging requirements over the past few years, but it is still important to maintain a logbook. The log can quickly relate any reported interference to station operation.

If you are involved, the logs will provide valuable troubleshooting clues about bands, modes, power, antenna pointing, etc. If you are not involved, the logs will demonstrate that, too. You may still choose to help, especially because you may be experiencing the same interference source as your neighbor.

If It Is You!

It is important to understand amateur responsibilities. There are two ways a station can affect electronic equipment. All transmitters generate small signals outside of the intended bands. If these spurious emissions exceed the FCC limitations for out-of-band signals or cause interference to another service, it is the amateur's responsibility to reduce these signals as necessary to eliminate the interference.

Many cases of interference result from the inability of consumer equipment to reject strong out-of-band signals. This is known as *fundamental overload* (see the first two chapters for more information). This fundamental RF energy is the signal from your transmitter that is supposed to be there—the one you are trying to make as strong as possible. It is, by far, the strongest signal emitted from your station. You are not legally responsible for interference resulting from your fundamental signal (explaining this to a neighbor can be a real challenge!). How this occurs is demonstrated in Figure 3.1.

Non Radio Devices

You should understand what kind of problems can, and cannot, be caused by the improper operation of your station. When there is interference to audio devices (CD players, tape decks, most intercoms, burglar alarms, and so on) you are most certainly dealing with fundamental overload. The FCC *Interference Handbook* pamphlet indicates that when these devices are subject to interference, they are improperly functioning as radio receivers. Thus you need not troubleshoot the station, look for harmonics or other causes. In the case of fundamental overload, the only things you can do at your station are those that reduce your field strength at the affected equipment. This includes reducing power and relocating antennas as far as possible from the susceptible equipment. The cures to fundamental overload problems are found at the other end—with the faulty equipment. The necessary filtering, shielding or grounding must be applied there.

RFI Troubleshooting

The process of RFI elimination is composed of two phases. Phase one identifies the root cause of the RFI problem. Once the source and susceptible equipment victim are identified, and the means of interference is understood, you can select and implement a solution. This chapter deals mostly with identifying the problem. Several easily performed tests are described that make the troubleshooting process easier. More detailed explanations of specific cures appear in other chapters of this book. The main steps to dealing with any RFI problem are outlined in Table 3.1.

Types of Interference

There are a number of different types of interference you should look for. Each has different causes and cures. Some of them are not really interference at all!

- Spurious emissions
- Fundamental overload
- External noise
- External rectification
- Self-generated interference
- Defective consumer equipment
- Interference caused by someone else
- Multiple causes

Spurious Emissions

Since modern transmitters must meet FCC regulations (see the chapter on RFI Regulations), the probability of spurious radiation RFI in a broadcast service area is low if the transmit equipment is properly operated. RFI caused by spurious emissions is the direct responsibility of the transmitter operator. If it occurs, the FCC rules are clear—the transmitter operator must fix it. Interference from spurious emissions generally applies *only* to interference to other radio services, such as over-the-air (not cable) television or broadcast radio. Fortunately, with modern transmitters, spurious emissions are rare. If they do cause RFI, a low-pass or band-pass filter at the transmitter is usually all it takes to cure it.

If the installation of a transmit filter doesn't reduce the RFI, it is unlikely that spurious emissions are the cause. One possible exception might be spurious emissions directly radiated by the transmitter chassis or wiring. To diagnose this condition, operate the transmitter into a dummy load. If the interference is still present, it is likely that it is coming directly from the transmitter chassis. (In that case, see the EMC Fundamentals and Transmitters chapters for more information.) Install a brute-force ac-line filter and ferrite common-mode choke on the transmitters ac supply line. If the interference goes away, it was being coupled into the ac line.

The likelihood of interference caused by spurious radiation increases with power level, so it may be necessary to add filtering, usually a commercially available low-pass filter, to high-power amplifiers. Some exciters exhibit a relatively constant spurious output regardless of the drive-control setting. They can cause a problem if drive is reduced to accommodate an amplifier.

Fundamental Overload

This is the most common interference problem. It can be difficult to diagnose, however. In most cases, apply the appropriate RFI cure to the affected device. If the RFI is eliminated, the problem was probably fundamental overload. Fundamental overload can also happen when electronic equipment's internal circuitry and internal wiring directly pick up interference. This is known as "direct pickup." In this case, there is little that can be done externally; the appropriate RFI cure must be applied internally by qualified service personnel.

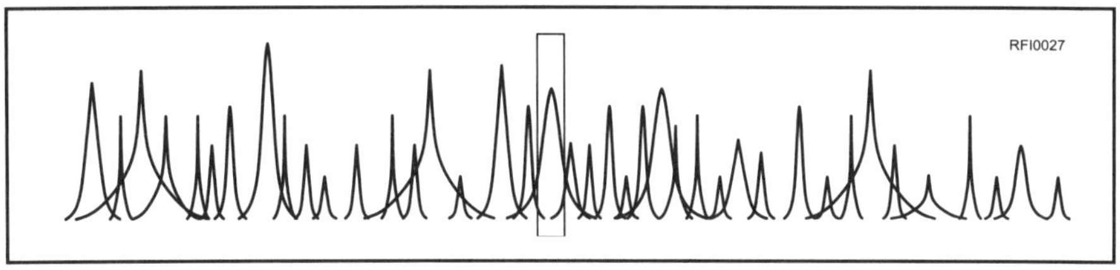

Figure 3.1—Fundamental overload results when equipment fails to reject signals it was never intended to receive.

The Telephone Troubleshooter

In some installations, unexpected problems can crop up. When I was a lad, I was, of course, avidly interested in radio and electronics. I started my "career" by building the obligatory crystal-radio receiver. I strung a wire in my room and picked up an AM broadcast station a whole mile away! (The signal strength improved when I actually added a ground.) Like every youngster, I, of course, wanted more, but the realities of my city backyard just didn't permit an effective antenna system.

One day, however, as I gazed at the phone line, the light came on! I had miles and miles of antenna that came right into my house! The phone wire should make a great antenna. When I ran the antenna connection on my crystal radio to one of the terminals on the phone line, the difference was spectacular! Not only could I hear the local station louder, but I could hear four others faintly in the background at the same time. I then tried running the ground connection to the other wire. It worked even better!

About an hour later, there was a knock at the door. I opened the door and my eyes must have lit up in awe—right before my eyes stood a telephone repair man, complete with tools, test equipment and all of the trappings. He explained that our party line had called in a problem—our shared line was dead. He immediately descended into the basement, and I followed reverently, watching every move and every wire. He explained that he had measured a strange short circuit across the phone line and was tracing it through the house. Suddenly, I turned white as a sheet as I realized just where and what the short circuit was. I politely excused myself, ran up the stairs and ripped the wires out of the phone socket. (He had stepped over them as he walked into the basement, but never noticed.) As I sat in my room in terror, he came up the stairs, looking quite puzzled, indeed. He told my father that he was following the short, when it suddenly disappeared. I said. "Golly, that's sure strange," or words to that effect, and did my best not to look guilty. I must not have done a great job of playing innocent, because the phone repair man kept giving me odd looks, but somehow, I got away with it.

I can just imagine how unexpected the diode across the phone line would be, and how much RFI it would have caused. Look for the obvious; look for the unexpected. Doing so will find a lot of RFI problems quickly.—*Ed Hare, W1RFI, ARRL Laboratory Supervisor*

Consumer Equipment Provides Valuable Clue

If an amateur is having interference involving another radio service (such as TV, broadcast radio or even interference with a local scanner enthusiast) and can demonstrate that a device tuned to the same frequency and located close to the transmitter has no interference at the same time that the affected device shows interference, the amateur station emission is clean. (Any spurious emissions affect all receivers within range.) Nothing further can be done at the transmitter to help.

The example equipment need not be the ham's own, but it is a good idea for demonstration purposes. Many devices are quite susceptible to fundamental overload, so it is a good idea to carry a portable receiver or TV for tests.

Filters should be installed on all of the amateur's own consumer electronics equipment, whether needed or not. This demonstrates that filters are effective and do no harm.

Corollary

If the amateur station is interfering with a known clean receiver or TV set, the station must be cleaned up before anything is done with the affected equipment. The susceptible equipment may also need work of some sort, but the transmitter must be clean before any sensible diagnosis is possible.

These principles may seem obvious, but often hams whose stations are already "clean" put wasted effort into adding more transmit filters, shields, grounds and so on. Of course the results are often "no improvement." Sometimes even skilled troubleshooters neglect the obvious.—*Dave Heller, K3TX, Former ARRL Assistant Director Atlantic Division, Yardley, Pennsylvania*

Table 3.1
An RFI Troubleshooting Procedure

The Amateur Station

Learn about RFI.
Keep a log.
Obtain outside assistance.
Check the amateur station.
Install a transmit filter at the amateur station.
Try an ac-line filter and common-mode choke at the amateur station.
Install a good RF ground at the amateur station.
Check the integrity of station connections. Corrosion and arcs can cause interference.
Cure interference to home electronics equipment in the amateur household.
Maintain an antenna-connected TV monitor at the amateur station.
Simplify the problem—disconnect (completely) all accessories and all but one transmitter and determine if the problem still exists. Reconnect one unit at a time.
Change one, and only one, thing at a time.
Keep test conditions (frequency, power level) constant.
Consider ways to reduce the signal strength at the effected equipment: Reducing transmit power and altering antenna placement are two possibilities.
Eliminate any problems from the amateur station first before working on the consumer equipment.

The Consumer Equipment

Maintain a spirit of cooperation between the amateur and neighbor. Nothing gets solved with anger and bad feelings.
Simplify the affected system as much as possible.
Install appropriate filters on the consumer equipment. This can include differential-mode high-pass filters, common-mode chokes for signal and power-supply leads and ac-line filters.
Try one, and only one, thing at a time.
Disconnect the equipment from long leads to check for direct radiation pickup.
Inspect the consumer-electronics ground connection. (Obey all necessary safety precautions.) This may include grounding the equipment or feed lines or breaking ground loops with common-mode chokes.
Ensure that the consumer-equipment is installed in accordance with good RFI engineering practice.
If possible, increase the strength of the desired signal. A better TV antenna or good distribution amplifier may help in some cases.
CATV does wonders for TVI!

External Noise

Some cases of interference can be caused by externally generated noise. This commonly comes from power-line problems or devices connected to the ac line, such as computers, electric motors, arc welders, etc. In many cases, amateurs are blamed for these types of interference problems. Your troubleshooting steps should identify external noise.

Self-Generated Interference and Defective Equipment

In some cases, electronic equipment can interfere with itself. TVs can have internal arcing or oscillation problems that can appear to be interference. (It technically *is* interference, but it has an internal source.) In other cases, people may suspect interference where consumer electronics equipment has become defective. A noisy transistor in an audio amplifier is one example; another might be a TV that has had a failure in the horizontal hold circuitry.

Interference Caused by Someone Else

One of the simplest troubleshooting steps is to determine that you and your station are not involved in the interference problem. As described earlier, you and your neighbor should keep logs on the dates, times, frequencies and circumstances of the interference. Using the log, you may discover that the interference is not at all related to the way you use your station. It may help you determine that the interference is really caused by power-line noise, a neighboring computer, the CBer up the street, the ham next door or some device in your neighbor's own home.

Ensure that your own House is Clean!

You can make a good start on troubleshooting if you ensure that your own house is clean! To show that your station is free of spurious emissions that cause harmful interference, fix your own consumer-electronics equipment. If the consumer equipment in your own house is not subject to interference, you have cut the size of your troubleshooting problem in half: You know that your station is not the cause of the problem. The FCC will see it that way, too!

Simplify the Problem!

The *number one* troubleshooting step is to simplify the problem. Every RFI problem is a puzzle, and puzzles with fewer pieces are easier to solve. When you begin troubleshooting keep an open mind and don't be afraid to experiment. Make your approach to the problem a logical one, rather than a hit or miss proposition. See Figure 3.2.

For example, when troubleshooting a case of telephone interference, first disconnect all but one line and one telephone, preferably near the telephone service entrance. Each line, instrument or accessory in a system may cause RFI, and several causes may interact. It is much easier to fix a simple system. Once a minimum system is fixed, reconnect components one at a time. That way causes and effects are more easily recognized and problems can be handled individually.

At the transmitter, remove all nonessential accessories. Disconnect all but one transmitter feed line and antenna (remember to include VHF transceivers, scanners and so on). See if the problem still exists. If it does, troubleshoot and eliminate the problem and reconnect equipment one unit at a time, eliminating RFI as you go.

In the case of television interference (TVI), temporarily disconnect any extra TVs, VCRs, games or splitters. (In cable systems, use only one cable-ready TV or the converter supplied by the cable company.) Once the interference is out of the single TV, reconnect equipment one unit at a time, debugging RFI as it recurs. With luck, you will cure all of the problems. If not, at least you will find the specific equipment that causes the problem.

With an audio amplifier, disconnect all separate components from the main amplifier. Disconnect the speakers as well, and listen with headphones on a short cord. If a radio tuner is part of the main unit, disconnect the radio antenna. If the interference persists, it is entering the main amplifier through either the power cord or by direct radiation pickup.

One At a Time

Try one (and only one) thing at a time. If you change too many variables at once, you won't know which change cured the problem. It is useful to know specific causes if you want to make future changes to your station. (You will know the most prevalent mechanism of RFI at your location.)

When you are testing to determine the cause of interference, or to see if an attempted cure has had any effect, keep the test conditions constant. It is difficult to notice improvements if you are constantly changing power levels, frequency or other test conditions. If you need to test at multiple operating frequencies and power levels, eliminate the problem at one set of conditions before you move on to the next set.

Multiple Causes

"Most RFI problems have multiple causes. The problem won't be fixed until you find them all!"— *John Norback, W6KFV.*

TVI, for example, may be caused by a combination of transmitter harmonics, an overloaded TV and an unused mast-mount TV preamp generating harmonics. When you put the low-pass filters on the transmitter, the other causes are still in place, so the interference remains. If you also filter the TV, the preamp is still generating harmonics, so the TVI still remains. The TVI won't go away until all of the causes have been properly dealt with. Although you may want to try cures one at a time, to determine exactly what worked, it is a good idea to leave a cure in place once it is tried, to ensure that all of the causes are taken care of to cure the problem.

Try the Easy Things First

"The problem is fixed when your neighbor stops complaining."—*Mike Tracy, KC1SX.*

Sometimes, the easiest solutions are the best. Many cases of interference can be resolved without the need for technical investigations or knowledge. Often the best troubleshooting technique is to determine what remedy eliminates or reduces the interference. Thus, the distinction between

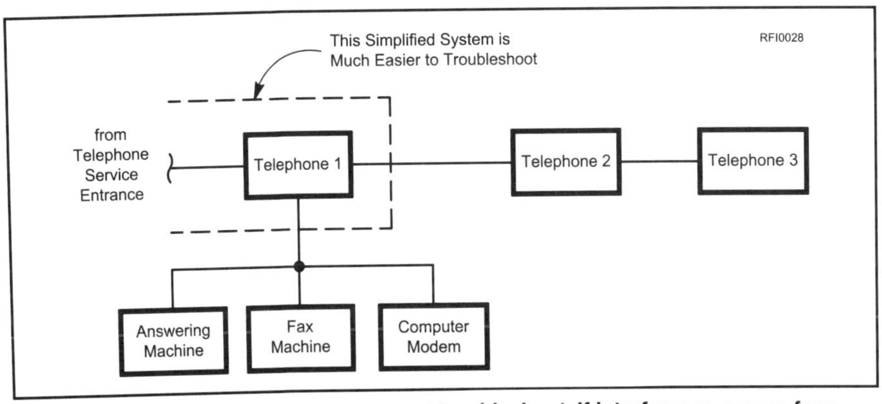

Figure 3.2—Simplify the systems you must troubleshoot. If interference comes from a system accessory, it might take months to find the culprit with all accessories connected.

Another Harmonic Test

Here is another test to confirm the presence of a harmonic interference problem. Temporarily connect a TV set to the amateur beam antenna normally used with the HF transmitter. Connect a dc voltmeter to the TV AGC line, or disable the TV AGC and use an attenuator between the antenna and TV. Plot a curve of the amateur antenna gain on the channel exhibiting TVI. With an HF beam, you should see a pattern such as that shown in Figure A. The pattern shows almost constant gain (loss) in any direction. (If you have an old 6-meter converter you can convert it to channel 2 and run a plot of your antenna gain.)

A TV field-strength meter or a VHF/UHF receiver with an S meter can be used as well. Once you have this plot, connect your amateur antenna to the transmitter and transmit while rotating the amateur antenna and watching a monitor TV. You will see a TVI level corresponding to either the plot of the harmonic response of your amateur antenna or an interference level corresponding to the HF response of your amateur antenna. The presence of an interference level that corresponds to the normal HF directivity (gain) of your ham antenna indicates that TVI does not come from transmitter harmonics.—*Ed Doubek, N9RF, Former Illinois Section Technical Coordinator*

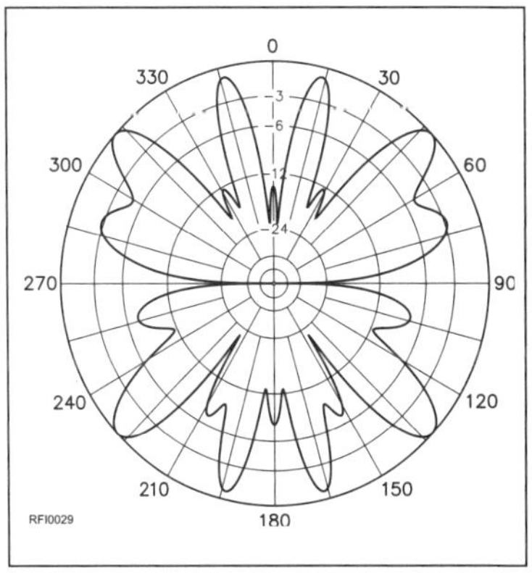

Figure A—An amateur HF antenna is many wavelengths long at VHF. Here is a typical VHF radiation pattern for an HF Yagi antenna. If interference results from VHF spurious signals created in the transmitting system, the interference magnitude should display similar peaks and nulls as the HF antenna is rotated while transmitting.

troubleshooting and an actual cure is often blurred. This chapter concentrates on determining the interference causes, but must also (by necessity) discuss techniques that are covered in more detail in other chapters.

As first steps, you might check the equipment's wiring for damage, for open outer wire shields, or for loose terminal connections. Try removing any added devices, such as video games, or even relocating the equipment or reorienting the device's antenna and power cord.

Stock Cures

Another diagnostic trick is to install appropriate filters on the equipment that is receiving interference. If you try the stock cures—a high-pass filter (TVs), common-mode choke, telephone filter or three-wire ac-line filter with a separate common-mode ferrite-core choke, bypassing, etc—and it fixes the problem, your job is done. You may still want to do some troubleshooting to *know* just what caused the problem, but the "stock" cures can usually be applied quickly and often result in an instant cure!

Testing 1, 2, 3

If you suspect that a ham neighbor may be causing interference, you can help find the source of the problem with a transmitting test. Your neighbor may want to ask a ham friend to participate in the test at your home, and you may want to ask a ham friend of yours to attend the test at your neighbor's home. Having witnesses can make you and your neighbor more comfortable with the outcome–whatever it turns out to be. They should keep detailed written notes while the tests are being conducted.

Transmit on each band and mode you normally operate. If you have a beam antenna, aim it in different directions while you are transmitting. Try various power levels, too. Your neighbor should then repeat the same test at his or her station. Keep in touch with your friend on VHF or via the telephone. Once you and your neighbor have determined which frequencies and power levels cause the problem, you'll be one step closer to finding a solution.

THE AMATEUR STATION

Although it is not exactly a "troubleshooting" step, testing and configuring an amateur station so it is not likely to cause interference, and curing any problems associated with the station, are important to good RFI control. Eliminate *all* station problems before you even have an RFI problem. If you *know* that your station is not the direct cause of interference, you have cut the size of your troubleshooting problem in half. The Transmitters chapter also contains information about the transmitters used in amateur stations.

Transmit Filters

If you have an RFI problem, you should use a transmit filter whether you think you need one or not. It will help ensure that there are no spurious emissions coming from your station to cause RFI problems. It is not a cure-all for interference, but if an interference case is ever brought before the FCC they will want to know that a transmit filter has been used. When confronted with interference, an amateur can point to the filter with pride. Showing this filter to a neighbor may help him or her understand more about the root causes of an RFI problem (usually the consumer electronics equipment).

The transmit filter for an HF transmitter is usually a low-pass filter, rejecting frequencies above its cut-off frequency (40 MHz, or so). An antenna tuner can also function as a transmit filter to a limited degree, but under some circumstances it may offer only a few dB of attenuation to spurious VHF signals.

Commercial transmit filters for VHF (and up) are available, too. Digital Communica-

Trouble in Paradise

As a rule, RF circuitry requires extensive, multi-point grounding to keep circulating ground currents from affecting circuits. On printed circuit boards, a good ground plane is mandatory for this purpose. If the system consists of several large racks of equipment, good low-inductance ground strapping is necessary not only between racks, but also to the best possible "earth" ground. More grounds, however, are not always better.

Such was the case a number of years ago when a fellow engineer and I were field testing a spread-spectrum system in Hawaii. The spread-spectrum system was a prototype that occupied three cabinet racks. Since it was a prototype, the inter-chassis cabling had not been optimized (although the cables were shielded). This might not seem a problem, but the transmitter boosted our low-power, frequency-hopping signal to 8 kW (peak) output.

No special shielding had been added to the prototype so we expected to have problems. In addition, we were operating inside an FAA transmitter site on Oahu, with row upon row of multi-kilowatt transmitters. Yes, we definitely expected problems. So when we initially "fired up" the equipment, their appearance was no surprise.

The Hawaiian islands are a combination of volcanic lava and coral. Neither is a good conductor, but the extensive station ground system appeared to be quite good. Our problem was to make a low-inductance connection with minimum ground loops. We decided to apply a little "overkill" by running half-inch-wide copper-braid straps between all racks, the transmitter and the station ground system.

The transmitter was a 10-channel Collins Autotune HF rig modified for pulse operation. Sure enough, all 10 HF channels showed RFI problems that distorted and smeared the pulse train. While my associate switched channels and watched the oscilloscope for RFI changes, I went about removing one end of the various ground straps. Where no effect was noticed by my cohort, I removed the other end of the braid and discarded it.

Finally, I removed one end of a 15-foot strap, and suddenly all channels were clear of RFI. After a recheck of all channels, it appeared we were set to begin testing. When I removed the other end of the braid, my associate immediately asked me what I had done. I told him, and he said "Put it back." I did, and he said "Leave it there." The RFI had reappeared when I removed the second end of the strap.

This is similar to using $\frac{1}{4}$-λ wires as a "counterpoise," except for two things: We had to stay near the strap, and it couldn't lie loose on the floor. First, I coiled it up on the floor next to the rack, and that worked fine! Since the braid was not insulated, each turn of the coil shorted out against the next one; that should change the counterpoise or ground radial effect, but it did not. Next I loosely arranged it on the floor and put a blank rack panel over it with the same result: no RFI on any channel. In the end, we simply stuffed the braid under the rack it "grounded" and ran our "round-the-clock" tests for a full week with no RFI problems at all.

What's the moral to all of this? Don't be afraid to experiment. Ground-loop currents are difficult to predict, and it may take an unusual approach to eliminate them. Good layout and circuit design can only go so far; it cannot anticipate every RFI problem. Have the courage to use a little ingenuity; go with what works; it can make all the difference in the world.—*James G. Lee, W6VAT, San Jose, California* tions, Inc sells a broad line of VHF and UHF band-pass filters. VHF/UHF band-pass filters are not often needed, but they can be helpful in some cases. VHF/UHF transmit filters are generally required only when there are symptoms of harmonic interference, transmitted composite phase noise or IMD in the transmitter output stage.

As a troubleshooting step, insert a transmit filter between the transmitter output and the feed line or antenna coupler. If the interference level is reduced via a low-pass filter at the HF transmitter, spurious radiation is indicated. If no apparent difference is observed, spurious signals may be present, but generated after the filter. Even more likely, the interference is caused by something other than transmitter spurious energy.

A transmit filter should be installed with short leads, between the transceiver and the antenna or Transmatch. (Filters require a matched load to function properly, so use a Transmatch if needed.) A right-angle connector with a single double-male connector at one end does a good job. If not installed this way, some spurious energy may be radiated from the coaxial cable between the transmitter and filter. If an amplifier is used, the filter should be installed after the amplifier, between the amplifier and antenna tuner. (In some cases a second filter between the exciter and amplifier helps.)

There are many myths about the effect of SWR on interference. By itself, a high SWR on a feed line will not cause interference. However, if you install a coaxial transmit filter, it is imperative that a low SWR (1.5 to 1 or so) exists on your feed line at the filter.

A typical transmit filter may reduce the spurious output reaching the antenna by 60 to 70 dB when used in a matched transmission line. With a high SWR, the filter effectiveness may be reduced. In addition to increased spurious output, high SWR can cause filter failure from increased feed-line voltage. You may want to read "SWR and TVI," Jan 1954 *QST*, pp 44, 45, 128, 130. Information about line conditions with high SWR appears in *The ARRL Antenna Book*.

If there is any suspicion that the transmitter might be coupling harmonic or fundamental energy into the ac-power lines, install an ac-line filter at the transmitter. This may cure the RFI problem, and it eliminates the transmitter as a direct source of power-line RF energy. This is discussed again later in this chapter and in the Power Lines and Electrical Devices chapter.

Transmitter Chassis Radiation

To determine whether the interference is the result of spurious radiation from the

"RF Sniffer"

RFI0034

The "Sniffer"

It is fortunate that the FCC doesn't require us to measure chassis radiation levels accurately, as this requires sophisticated equipment. The home-built "sniffer" illustrated in Figure B can give a qualitative measure of chassis leakage.

The sniffer does not respond strongly to electric fields, but favors magnetic fields, which are prevalent in radiation from poor RF shields. Connect the sniffer to a device capable of receiving the interfering signal (such as a TV set) and move the loop around the suspect chassis. Maximum pickup should occur with the loop wire at right angles to any slots or seams and parallel to any wires exiting the shield enclosure.

The "sniffer" is used to measure unwanted chassis leakage of harmonics and other spurious emissions. Construct as follows:

1) Remove 12 inches of outer insulation from the end of the coaxial cable.

2) Remove 11 inches of shield from the end of the cable, leaving 1 inch of exposed shield.

3) Remove ½ inch of inner insulation from the end of the cable.

4) Wrap the end of cable into a coil with three or so loops as shown in Figure B.

5) Wrap and solder the center conductor to the exposed shield.

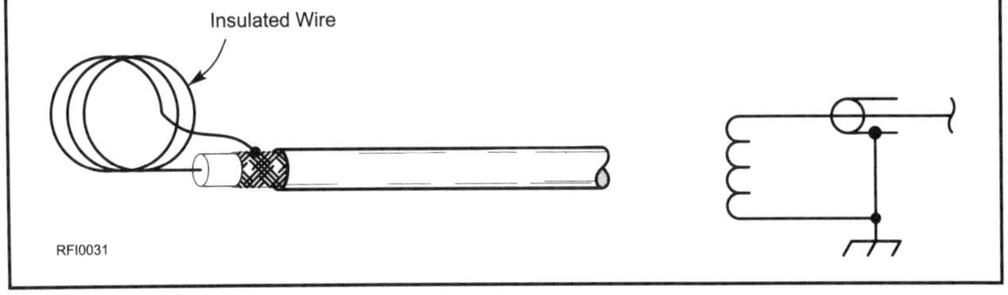

Figure B—A pickup loop for detecting cabinet radiation and IMD from external rectification. Use the loop with an appropriate radio or television receiver.

RFI0030

"No ground problems here!"

transmitter chassis, connect a dummy load to the transmitter output (in place of the antenna) using coaxial cable. If any interference is present when transmitting into a shielded dummy load, it indicates that the transmitter design (notably shielding and lead filtering) is not adequate. Improved shields and filters or replacement of the transmitter will be required to eliminate the interference. The techniques for properly shielding and bypassing transmitter key and microphone lines are covered in the chapter about transmitters. That chapter also discusses techniques used to filter ac-power leads, as does the Power Lines and Electrical Devices chapter.

Spurious emissions can leak out of a transmitter cabinet. Even if there are no shielding defects that result in leaks, the equipment cabinet itself can radiate undesired signals. Meter holes, seams and large gaps are possible sources of cabinet leakage.

While you are transmitting into a dummy load, cabinet leakage can be located by means of a sensing loop and receiver. See the sidebar on the RF sniffer. As the loop passes near any opening (such as a meter, control knob or wire that exits the transmitter enclosure), the signal level measured on the receiver S meter, or seen on the TV screen may increase. This indicates hot spots (shield deficiencies).

Accessories

Many accessories such as SWR indicators, external electronic TR switches and monitor scopes all rectify a small amount of RF energy in order to operate. This rectification can generate harmonics. The test is straightforward—temporarily remove each accessory from the system. When the RFI goes away, you have found the culprit. You can also insert a dummy load at each point, sequentially as RF flows through the system. When the interference returns, the last device added to the system was probably creating harmonics and radiating them directly. In some cases a transmit filter may be needed between the offending device and the antenna.

A good rule to remember: Use as few accessories in the transmitting feed line as possible. Each accessory can pump both spurious energy and fundamental into the ac line (although this is not very likely).

Antennas and Towers

Here are some guidelines to use when installing a tower:

Make sure all mechanical joints are both mechanically and electrically secure! If the tower is new, the bright metal makes a good connection. If the tower is old, shine up the galvanizing with very fine sandpaper at the joints (don't forget the insides). Just brighten it up, don't sand through the coating! Guy lines should be tight, so that they don't move or rub where they are attached. Rohn specifies that guy wires should be tightened to about 10% of their ultimate strength. Tree branches, wire antennas and other material should not touch the tower as they may arc and cause noise.

Joints between dissimilar metals should be eliminated. For example, never allow bare copper to contact the zinc (galvanized) coating on a tower or guy wire. Not only will the joint quickly form a diode (a rectification interference source), but the zinc will be removed, allowing the steel to rust. Use either bronze or stainless-steel clamps to attach metal to a galvanized tower (or weld the connection).

For lightning and noise protection (and also to comply with the electrical code) ground the tower to a good ground system (Figure 3.3). The ground system should be specifically designed for soil conditions at the site. If the soil conductivity is good and the lightning risk small, a single 8-ft by ⅝-in. ground rod may be adequate. Drive it close to the base of the tower, and connect it with #6 (or larger) solid copper wire. In areas with poor soil conductivity or high lightning risk, more rods or a radial system may be necessary. Connect the ground system to your station with #6 copper wire; make all runs as short as possible, with smooth bends and never in metallic conduit. For additional lightning protection, bond the shields of all cables to the tower where they leave it for the shack. (Be sure to weatherproof the connections.) Avoid braid as an outdoor station ground. If you must use it, "flood" it with noncorrosive caulk (such as Permatex blue gasket compound) to prevent water from wicking into the braid.

There is some controversy about insulating guy wires. If there is concern about resonant guys disturbing the pattern of an HF antenna, use nonconductive guys. If resonances are not a problem (say, for a VHF/UHF-only array), leave the guys continuous

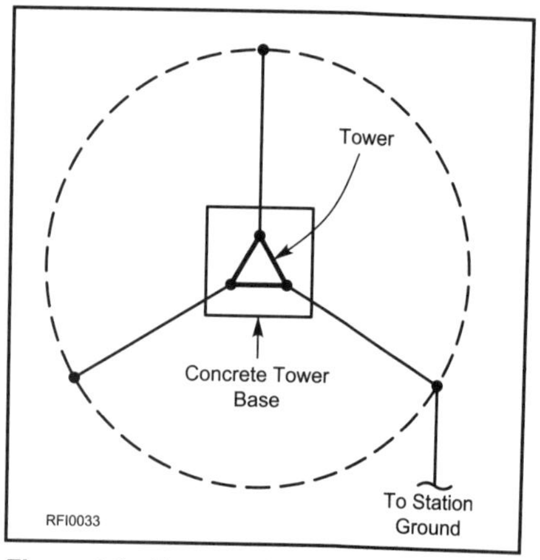

Figure 3.3—Towers should be properly grounded. Tower sections should be electrically bonded.

and ground them at the anchors for good lightning protection. When using metal guys broken by insulators, ensure that the joints are in good condition without touching or rubbing metal. Keep tree limbs, wire antennas and such well away from guy lines.

Loose hardware such as clamps, bolts and screws can also contribute to interference problems because they can contribute to arcs. Dissimilar-metal junctions can also cause rectification (harmonics). Aluminum oxidizes and can cause arcing and high SWR in a system. The near-field intensities of an amateur antenna that is radiating high power can induce voltages in other wires such as guys and metal clotheslines. These can, in turn, arc to tree branches, thus causing interference. This is an uncommon, but easily overlooked, RFI problem.

All types of antennas may require occasional maintenance. Tighten hardware and trim trees, as needed (at least annually). Damage from thunderstorms must also be repaired. The ground for an antenna support structure also needs an annual (at minimum) inspection.

RFI0032

"Y'know OM, you really ought to check that line sometime"

These are general guidelines. In any case, be sure that all aspects of station and antenna installation comply with local electrical codes.

External Station Problems

Harmonics can be generated outside an amateur station, after the RF energy passes through the transmit filter. This subject is covered in the chapter on External Rectification. These external harmonics can happen at any system hardware or connections that are loose or corroded. For example, if a commercial or home-built spark-gap lightning protector is inserted in the line, bugs, corroded copper or brass pieces can generate harmonics via rectification. Other troublesome items include rotator controllers, corroded tower joints and guy

Planning a New House

People who are building a new house often ask if RFI should be considered during the planning and construction. The answer is yes, and here is a list of suggestions.

Checklist for the New Home

Plan to place the ham shack at (or below) ground level (makes grounding easier).

Place remote speaker, telephone, burglar alarm, TV cables and intercom wires inside metal conduit (separate conduit runs are better).

Provide electrical insulation where pipes and conduits cross. As an alternative, electrically bond the systems together.

Run all wiring at the lowest possible height. Conduit and other wiring in an attic can pick up relatively large RF voltages and conduct them into appliances.

Electronic doorbells and computerized thermostats may create more trouble than they are worth.

Plan the position of ham and TV antennas (see Figure C).

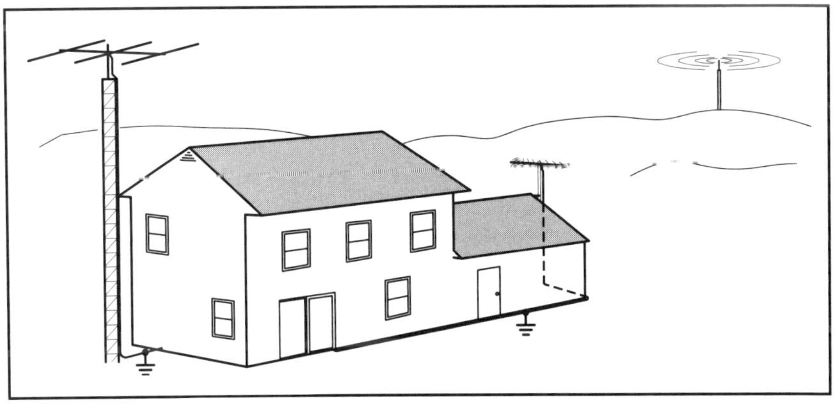

Figure C—RFI is minimized when antenna and feed-line locations are based on sound RFI thinking. Note the RFI advantages.

A) TV antenna located far from ham antenna
B) TV feed line is run vertically to ground level
C) TV distribution amplifier mounted at ground level
D) TV feed-line shield connected to ground rod
E) TV feed-line horizontal run at ground level
F) Ham antenna does not block TV-reception path
G) Ham feed line runs vertically to ground level
H) Ham feed-line shield connected to ground rod
I) Ham shack located close to station ground

Additional RFI Hints

Here are a few stock cures that have solved a lot of RFI cases:

When using a transmit filter, use a split ferrite bead at each end of the filter. Common-mode signals are then effectively attenuated.

Even if there is no transmit filter in the line, a couple of split beads should be installed. If coax is used to feed a balanced antenna, it is a good idea to place a couple of ferrite toroids at the antenna end too.

AC power cords should also have toroid chokes, with the wire wrapped through the core several times (in the same direction).

Keep all cables (microphone, key and so on) short. If a particular cable must be extended, install a toroid choke in it (close to the rig).

Disconnect all unnecessary items from the feed line. Unnecessary items include antenna relays and switches, SWR bridges and monitors. If an SWR bridge must be in the line it should be on the transmitter side of the transmit filter.

Ground leads should be short, direct and thick (large surface area means low inductance).

Commercially available ground rods are usually copper-plated steel, and the simple act of driving them into the earth removes the copper. It is far better to use standard (approximately 1-in.) pipe, available from plumbing supply stores. A ground wire is then easily soldered to the pipe for a corrosion-free connection.

Filters must only be used in lines that are terminated in their characteristic impedance. If not, a Transmatch must be installed between the filter and load.—*Howard Liebman, W2QUV, Former NYC-Long Island Section Technical Coordinator, New Hyde Park, New York*

wires, and unpowered solid-state equipment (such as a VHF transmitter connected to an antenna).

Metallic structures in the near field of your antenna are also subject to these effects. In an RF system, corrosion means nonlinearity, which means harmonics or other undesired mixing effects (intermodulation—IMD). Include a check of these things in your troubleshooting procedures, especially in difficult cases. (Corroded connections in your station installation should be fixed anyway, as part of good engineering practice.) For example, when copper water pipes touch any other metal object (including other pipes or the electrical system ground), corrosion can cause rectification at the points of contact, thus generating harmonic energy after the illuminating RF energy leaves the amateur antenna. If significant harmonic radiation is identified, either bond the pieces of metal together or separate them, as convenient.

Wire antennas are often connected to trees instead of towers. If the antenna arcs to tree branches or the trunk, the arcs can disrupt home-electronic equipment. The effect of these arcs, however, normally shows up as a series of black dots on all TV channels, or as broadband noise on audio devices. You are responsible for interference caused by arcs that occur in your station equipment, so ensure that the wires of your transmitting antenna cannot arc to nearby objects.

The tuners in many TVs and VCRs can generate harmonics and other mixing effects when overloaded by strong signals. These harmonics look the same on the TV screen as harmonics generated by the transmitter. Try a differential-mode high-pass filter, with a separate ferrite-core common-mode choke at the TV set. If the harmonics are self-generated, the filter will give a marked improvement.

And in Conclusion

If the troubleshooting methods and cures described in this book are applied, most cases of interference will be reduced or eliminated. There is one other final suggestion, however: when the Super Bowl or some similar big-audience event is broadcast, join the ranks of viewers, or try QRP, just in case one or two of the TV sets in your neighborhood remains unfixed.

Chapter 4

Radio Direction Finding

The other chapters in this book tell you what RFI is. This chapter tells you how to find it.

By Joe Moell, KØOV, and
Bob Schetgen, KU7G (SK)

Far more than simply finding the direction of an incoming radio signal, radio direction finding (RDF) encompasses a variety of techniques for determining the exact location of a signal source. The process involves both art and science. RDF adds fun to ham radio, but has serious purposes, too.

Among them is tracking down sources of interference. The FCC's former Compliance and Information Bureau created an Amateur Auxiliary, administered by the ARRL Section Managers, to deal with interference matters. In many areas of the country, there are standing agreements between Local Interference Committees and district FCC offices, permitting volunteers to provide evidence leading to prosecution in serious cases of malicious amateur-to-amateur interference. RDF is an important part of the evidence-gathering process.

In support of the growing interest in RDF in the US, in 1998 the ARRL Board of Directors established the new position of ARDF Coordinator to help promote direction finding in the US and elsewhere. In addition, 1998 saw the US field its first team of US fox hunters at the ARDF World Championships in Hungary.

The most basic RDF system consists of a directional antenna and a method of detecting and measuring the level of the radio signal, such as a receiver with signal strength indicator. RDF antennas range from a simple tuned loop of wire to an acre of antenna elements with an electronic beam-forming network. Other sophisticated techniques for RDF use the Doppler effect, which measures the time-of-arrival difference of the signal at multiple antennas.

All of these methods have been used from 2 to 500 MHz and above. RDF practices vary greatly between the HF and VHF/UHF portions of the spectrum, however. For practical reasons, high gain beams, Dopplers and switched dual antennas find favor on VHF/UHF, while loops and phased arrays are the most popular choices on 6 m and below. Signal propagation differences between HF and VHF also affect RDF practices. But many basic transmitter hunting techniques, discussed later in this chapter, apply to all bands and all types of portable RDF equipment.

Some sources of RFI can affect relatively large areas. For example, touch-controlled lamps have placed RFI on power lines, where it affected broadcast receivers several miles away. External-rectification sources are often many yards from the RF source and RFI receptor. Similarly, intermodulation distortion (IMD) contributors may be some distance from the affected devices. In treating these and similar situations, direction finding (DF) skills are useful.

An RFI investigator is likely to confront two kinds of signals for DFing; broad-band and narrow-band noise. Electric motors, arcs, sparks, external rectifiers and digital equipment all produce broadband noise. Sources of fundamental overload, harmonics and IMD produce narrow-band signals. Let's consider broadband sources first.

BROADBAND DF

Broadband noise is usually an AM phenomenon. Current flow in sparks and arcs varies widely. Therefore, AM receivers are best for noise DFing.

Broadband sources release energy all across the spectrum. The strength of that energy, however, varies with frequency. Although there may be peaks at certain frequencies (resulting from resonance of the conductor that radiates the energy), the energy is generally stronger at lower frequencies. Hence, signal strength decreases as frequency increases. This characteristic is useful in DF work.

Distant RFI sources are best detected on a battery powered AM-broadcast receiver. Tune the receiver to a clear frequency somewhere near the low end of the band and listen for the interference. Most such receivers use a ferrite-loop (loopstick) antenna oriented along the longest dimension of the radio case.

Happily, loopstick antennas are directional. Figure 4.1 shows the field pattern of a typical loopstick antenna. When the noise is faint, use the loopstick lobes to indicate the source along a line perpendicular to the antenna core. As the search progresses, the signal should strengthen. Then use the antenna nulls to indicate the direction of the RFI source.

If the RFI becomes too strong for DF bearings in the loopstick nulls, switch to a higher frequency and continue the search. A pocket scanner that receives the aircraft band is suitable. (Aircraft use AM for communication.) A monopole or dipole antenna has maximum response perpendicular to the antenna axis, and minimum response off the end. Use the maximum response mode when the signal is weak, and switch to the minimum response technique as the signal grows stronger. This technique should be sufficient to locate typical building powering electrical noise sources such as thermostats, fluorescent lights and touch-controlled lamps.

Within a Building

Continue listening to the RFI and use the main fuse or circuit breaker to switch the power off and then on, so that the entire building is momentarily without electricity. Likewise, disconnect batteries in backup systems such as alarm systems. If the interference disappears, the source is probably in the building.

If the source is in the building, a process of elimination can locate the offending device. Remain at the electrical service box, and cut power to one branch circuit at a time. The RFI should start and stop as power to one of the branches is switched. When the source branch has been determined, unplug each electronic or electrical device on the branch while monitoring the receiver. Once the RFI source is located, refer to the appropriate chapter of this book for suitable treatments.

On the Power Lines

If DFing leads to a power line, the situation gets a little more complicated. RFI radiated from power lines is limited to the general area of the source, while RFI induced into power lines can travel significant distances.[1] It may be more appropriate to continue the search by car.

If the car has an AM radio, tune it to a clear frequency at the high end of the band and listen for the noise. Listen to the noise while driving around the general area. Be concerned with general signal-strength trends, not normal variations at down leads, pole grounds and hardware. When the line with the strongest noise is located, drive along that line.

The strength of RFI radiating from a single source connected to a power line usually varies in a pattern of peaks and nulls. The distance between peaks (or between nulls) is minimum at the source.[2] By noting the peak/null pattern, the source can be found. (The signal strength is not significant at this point.)

Do the final location at VHF. Listen for a noise peak with an AM VHF receiver.

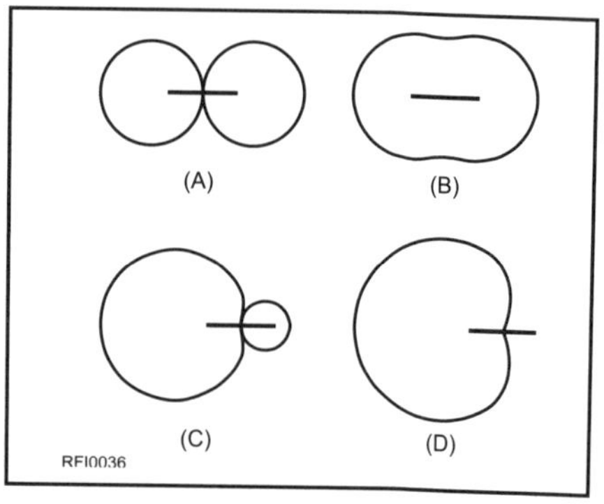

Figure 4.1—Small loop field patterns with varying amounts of antenna effect—the undesired response of a loop acting merely as a mass of metal connected to the receiver antenna terminals. The horizontal lines show the plane of the loop turns.

Once the maximum noise is found, drive past it until the noise becomes very weak. Then, turn around and drive in the opposite direction until it is again weak. Perform this check by driving on both sides of the suspect line if possible.

If the source is at a building, speak with the owner and perform the service-panel tests described earlier in this chapter ("Within a Building"). If there is no building nearby, note the pole number (usually stamped on a metal plate attached to the pole) and call the power company. When reporting the problem, give the pole number, describe the noise and the time of day that it was heard. *Do not, under any circumstances, tamper with the power lines or their attached hardware or equipment!* It is dangerous to strike poles with hammers or automobiles. Doing so could down live wires and cause broken hardware or insulators to fall.

Cable TV DF With a Handheld Transceiver

If you believe that you are experiencing cable TV interference, you can perform your own cable TVI hunt with a handheld 2-meter transceiver, a detachable antenna and some basic transmitter-hunting skills. First, tune your receiver to 145.250 MHz. If you hear a loud buzzing (the TV sync pulse) or your receiver quiets, start sweeping the area looking for the strongest signal. (It helps if your receiver has an S meter, but it is possible to gauge signal strength by ear alone.) When you locate the strongest signal, loosen the antenna connector (to reduce the received signal strength) and continue to hunt for the strongest signal. (It sometimes helps to use your body as a shield for increased directivity.) When the signal strength again reaches the point of receiver overload, remove the antenna entirely, and probe with the open receiver antenna connector for the strongest signal. (You'll probably locate the leak this way.)

With an FM Broadcast Receiver

It is possible to locate distant, strong cable leaks with a mobile FM broadcast receiver.[3] Some CATV systems have a "cuckoo" signal near the high end of the FM broadcast band. The signal is often a double-tone signal, but it may be a single tone. At its strongest, the cuckoo is about 6 dB above the other FM-band signals. The strength of the cuckoo is usually stepped periodically. These steps should be visible on the signal-strength meter of receivers so equipped.

If the cable TV system does not use a cuckoo, use a weak FM station that is on the cable, but not received over the air in your area. Since broadcast stations do not use stepped power levels, DFing the station will not work as well as DFing the cuckoo.

Once in the vehicle tune the mobile receiver to the cuckoo. If you are far from the source, only the strongest steps will be audible, and those may not fully quiet the receiver. Drive the area while listening to the strength steps. This method can indicate the general area of leaks. The technique is not completely reliable, however. The cable TV shield, telephone cables, guy wires and other nearby conductors can echo the leak. Once you suspect an area, note the pole number and call the cable company. Let them find the actual leak and fix it.

With a Portable TV

It is also possible to DF cable TV leakage with a portable, battery powered TV and a directional antenna. The directional patterns of commercial TV antennas vary, so experiment with a known over-the-air station to determine the pattern. If the pattern is too bad, construct a directional antenna for an appropriate frequency. Suitable directional antennas appear in this chapter, *The ARRL Antenna Book* and *The ARRL Handbook.*

Once the equipment is assembled, tune the TV to the weakest channel that is on the cable TV system, but not received over the air. DFing with a TV is easy because multipath signals are visible as "ghosts." TV ghosts are caused by the time delays inherent in multipath propagation. Therefore, the left-most image on the TV screen is always the image produced by the most-direct signal path. Simply follow the path indicated by the maximum strength of the left-most image.

External Rectification Source DF

First, determine the power threshold at which rectification begins. Reduce transmit power gradually until the problem disappears. TVI with less than 5 W indicates a problem with the transmit system. Give the transmit antenna, connectors and ground system a detailed inspection. 10 or 20 W is a more common threshold (higher on 80 and 160 meters, where the average bit of metal around the house isn't of significant length) for nearby rectifiers.

It is difficult to exactly locate a rectifier. If the affected device is a TV with a rotatable antenna, make a "first pass" by rotating the TV antenna and looking for a peak in the interference strength. Be careful; it is easy to mistake a TV-signal null for an interference peak. Rotate the transmit antenna (where possible), and look for a TVI peak in the direction of the rectifier. (Do this test with the minimum power required to cause interference when the transmitting antenna is pointed in the most sensitive direction.)

A portable TV or scanner is required to "home in" on the actual interference source. Check suspicious metallic objects by proximity, one at a time. Don't confuse a drop in legitimate signals with a "hot spot" of interference. The usual clanging, banging, twisting, "torquing" and pushing may produce recognizable interference changes that lead to the source. Above all else, check the transmitting and receiving antenna systems thoroughly before beginning a rectifier hunt.

If the transmitter is a commercial broadcast station, the hunt follows a slightly different course. Since the transmitter is always on, and the antenna can't be rotated, the rectifier must be "DFed" from the receiver end. This isn't too difficult—the signal is coherent, narrow band and easily identified. After DFing the general direction, set out on foot with a portable receiver and search by proximity. If the interference seems to be farther afield, use a mobile rig or portable receiver in a car to locate the general vicinity of the source.

When rectification occurs in rain gutters, rusty water pipes and so on, second and third harmonics are usually evident, as well as third- and high-order IMD with broadcast stations. More often than not, the audio is remarkably clear, but scrambled by the simultaneous presence of two sources.

Hum on an interference signal is a possible sign of power-line rectification. A mobile rig and a quick drive around the neighborhood should indicate one or more poles. Report them to the utility company for repair. (Sometimes you can actually hear the broadcast audio while standing near the offending pole.)

Arcs are notorious for producing a wide variety of RF products. Arcs can generate second and third harmonics of the fundamental, as well as second- and third-order IMD of many broadcast signals near and far.

Rectification may occur in passive conductors (grounds, guys and so on), yet a strong hum component may be evident. This is caused by 60-Hz energy in close proximity; it modulates the current flowing in the rectifying junction.

If IMD products are intermittent but seemingly periodic (seconds), suspect a long expanse of wire. Power lines in residential areas swing like pendulums, with a period of 1 to 2 seconds. As a line sways in the breeze, it tightens and loosens or gently rocks the insulators back and forth. This action may be just sufficient to repetitively break or short the offending rectifier, thus switching the broadcast interference on and off.

Any lengthy conductor can be checked with a portable receiver. Couple the internal loop antenna to the conductor under test. The

BCI always gets louder (the metal object is an antenna); but in BCI sources it gets disproportionately louder. Test several objects in the area, using an attenuator (or proximity) to control receiver sensitivity. It doesn't take long to get an intuitive feel for levels that are "normal" and those that aren't.

If the interference is generated in residential wiring or plumbing, check for ground return currents in the water pipe between the street (meter) and dwelling. In temperate climates (where pipes are buried relatively close to the surface) return currents radiate quite well. By simply "sweeping" the sidewalk near the water entrance with a portable receiver, the relative interference level can be detected. Signals are strongest at the house with rectification problems.

A search by foot and by car is usually required to locate IMD sources. If the IMD is weak, it may be difficult to detect on a portable receiver. Check other related frequencies for a stronger IMD product. The second and third harmonics should be strongest, followed by third-order IMD with other local stations. The search for TVI and BCI is more art than science. A few simple experiments, however, and some practical experience with a portable receiver quickly makes an expert out of a beginner.

NARROW-BAND DF

Narrow-bandwidth signals appear on a small set of frequencies. Possible sources are two-way radios as used for Citizens Band, police, fire, aircraft, radar, land-mobile, cellular telephone, cordless telephone and Amateur Radio.

Hams often DF narrow-band signals (repeater interference, aircraft emergency-locator transmitters—ELTs—and transmitter hunts) for public service and for fun. Hence, narrow-band DF techniques have been discussed in many magazine articles and a few books.[4]

Is a Search Necessary?

Because the energy of narrow-band signals is concentrated in one area of the spectrum, a narrow-band noise source may be located several miles from the affected device. It is unlikely that distant narrow-band sources will interfere with consumer electronic equipment such as telephones and cassette players. Nonetheless, distant sources may affect sensitive receivers.

Before beginning a search, listen to the target signal on the affected device. Can you discern intelligence? If a voice or Morse code is evident, try to copy the station identification. (Nearly all licensed transmitters are required to identify periodically.) Some digital modes (such as amateur AMTOR, RTTY and ASCII) are permitted to identify in the digital code. If the target signal sounds like a digital code, try to get equipment to read the intelligence. If you can discern a call sign, DF techniques may not be needed. Check the FCC's Universal Licensing System (**http://wireless.fcc.gov/uls/index.htm**). Then contact the owner and discuss the situation. Remember: Cooperation is the key to most RFI problems.

RDF ANTENNAS FOR HF BANDS

Below 50 MHz, gain antennas such as Yagis and quads are of limited value for RDF. The typical installation of a tribander on a 70-foot tower yields only a general direction of the incoming signal, due to ground effects and the antenna's broad forward lobe. Long monoband beams at greater heights work better, but still cannot achieve the bearing accuracy and repeatability of simpler antennas designed specifically for RDF.

RDF Loops

An effective directional HF antenna can be as uncomplicated as a small loop of wire or tubing, tuned to resonance with a capacitor. When immersed in an electromagnetic field, the loop acts much the same as the secondary winding of a transformer. The voltage at the output is proportional to the amount of flux passing through it and the number of turns. If the loop is oriented such that the greatest amount of area is presented to the magnetic field, the induced voltage will be the highest. If it is rotated so that little or no area is cut by the field lines, the voltage induced in the loop is zero and a null occurs.

To achieve this transformer effect, the loop must be small compared with the signal wavelength. In a single-turn loop, the conductor should be less than 0.08 λ long. For example, a 28-MHz loop should be less than 34 inches in circumference, giving a diameter of approximately 10 inches. The loop may be smaller, but that will reduce its voltage output. Maximum output from a small loop antenna is in directions corresponding to the plane of the loop; these lobes are very broad. Sharp nulls, obtained at right angles to that plane, are more useful for RDF.

For a perfect bidirectional pattern, the loop must be balanced electrostatically with respect to ground. Otherwise, it will exhibit two modes of operation, the mode of a perfect loop and that of a nondirectional vertical antenna of small dimensions. This dual-mode condition results in mild to severe inaccuracy, depending on the degree of imbalance, because the outputs of the two modes are not in phase.

The theoretical true loop pattern is illustrated in Figure 4.1. When properly balanced, there are two nulls exactly 180° apart. When the unwanted antenna effect is appreciable and the loop is tuned to resonance, the loop may exhibit little directivity, as shown in Figure 4.1B. By detuning the loop to shift the phasing, you may obtain a useful pattern similar to Figure 4.1C. While not symmetrical, and not necessarily at right angles to the plane of the loop, this pattern does exhibit a pair of nulls.

By careful detuning and amplitude balancing, you can approach the unidirectional pattern of Figure 4.1D. Even though there may not be a complete null in the pattern, it resolves the 180° ambiguity of Figure 4.1A. Korean War era military loop antennas, sometimes available on today's surplus market, use this controlled-antenna-effect principle.

An easy way to achieve good electrostatic balance is to shield the loop, as shown in Figure 4.2. The shield, represented by the dashed lines in the drawing, eliminates the antenna effect. The response of a well constructed shielded loop is quite close to the ideal pattern of Figure 4.1A.

For 160 through 30 meters, single-turn loops that are small enough for portability are usually unsatisfactory for RDF work. Multi-turn loops are generally used instead. They are easier to resonate with practical capacitor values and give higher output voltages. This type of loop may also be shielded. If the total conductor length remains below 0.08 λ, the directional pattern is that of Figure 4.1A.

Ferrite Rod Antennas

Another way to get higher loop output is to increase the permeability of the medium in the vicinity of the loop. By winding a coil of wire around a form made of high-permeability material, such as ferrite rod, much greater flux is obtained in the coil without increasing the cross-sectional area.

Modern magnetic core materials make compact directional receiving antennas practical. Most portable AM broadcast receivers use this type of antenna, commonly called a *loopstick*. The loopstick is the most popular RDF antenna for portable/mobile work on 160 and 80 meters.

As does the shielded loop discussed earlier, the loopstick responds to the magnetic field of the incoming radio wave, and not to the electrical field. For a given size of loop, the output voltage increases with increasing flux density, which is obtained by choosing a ferrite core of high permeability and low loss at the frequency of interest. For increased output, the turns may be wound over two rods taped together. A practical loopstick antenna is described later in this chapter.

A loop on a ferrite core has maximum signal response in the plane of the turns, just as an air core loop. This means that maximum response of a loopstick is broadside to the axis of the rod, as shown in Figure 4.3. The loopstick may be shielded to eliminate the antenna effect; a U-shaped or C-shaped channel of aluminum or other form of "trough" is best. The shield must not be closed, and its length should equal or slightly exceed the length of the rod.

Sense Antennas

Because there are two nulls 180° apart in the directional pattern of a small loop or loopstick, there is ambiguity as to which null indicates the true direction of the target station. For example, if the line of bearing runs east and west from your position, you have no way of knowing from this single bearing whether the transmitter is east of you or west of you.

If bearings can be taken from two or more positions at suitable direction and distance from the transmitter, the ambiguity can be resolved and distance can be estimated by triangulation, as discussed later in this chapter. However, it is almost always desirable to be able to resolve the ambiguity immediately by having a unidirectional antenna pattern available.

You can modify a loop or loopstick antenna pattern to have a single null by adding a second antenna element. This element is called a sense antenna, because it senses the phase of the signal wavefront for comparison with the phase of the loop output signal. The sense element must be omnidirectional, such as a short vertical. When signals from the loop and the sense antenna are combined with 90° phase shift between the two, a heart-shaped (cardioid) pattern results, as shown in Figure 4.4A.

Figure 4.4B shows a circuit for adding a sense antenna to a loop or loopstick. For the best null in the composite pattern, signals from the loop and sense antennas must be of equal amplitude. R1 adjusts the level of the signal from the sense antenna.

In a practical system, the cardioid pattern null is not as sharp as the bidirectional null of the loop alone. The usual procedure when transmitter hunting is to use the loop alone to obtain a precise line of bearing, then switch in the sense antenna and take another reading to resolve the ambiguity.

Phased Arrays and Adcocks

Two-element phased arrays are popular for amateur HF RDF base station installations. Many directional patterns are possible, depending on the spacing and phasing of the elements. A useful example is two ½-λ

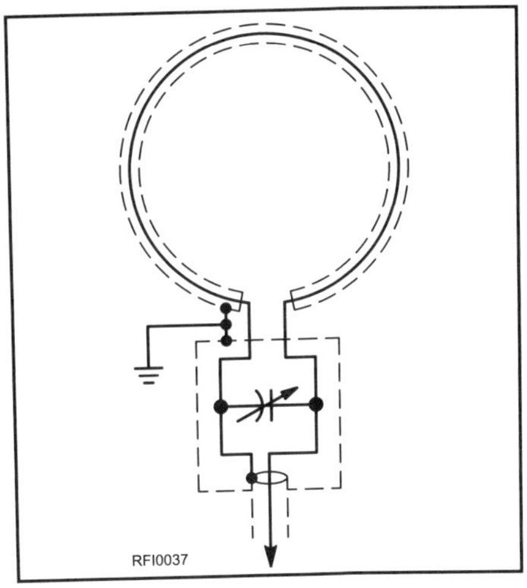

Figure 4.2—Electrostatically shielded loop for RDF. To prevent shielding of the loop from magnetic fields, leave the shield unconnected at one end.

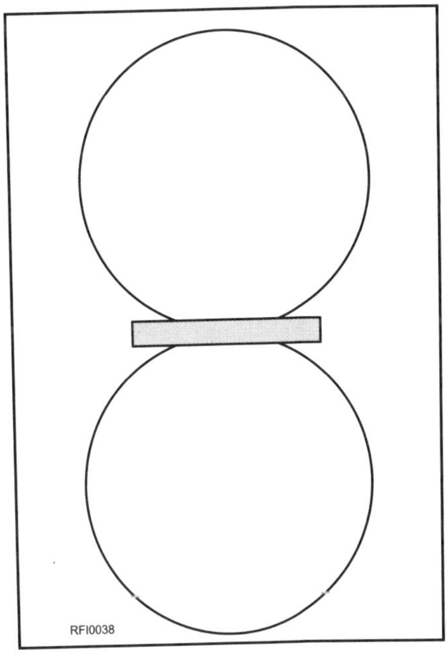

Figure 4.3—Field pattern for a ferrite rod antenna. The dark bar represents the rod on which the loop turns are wound.

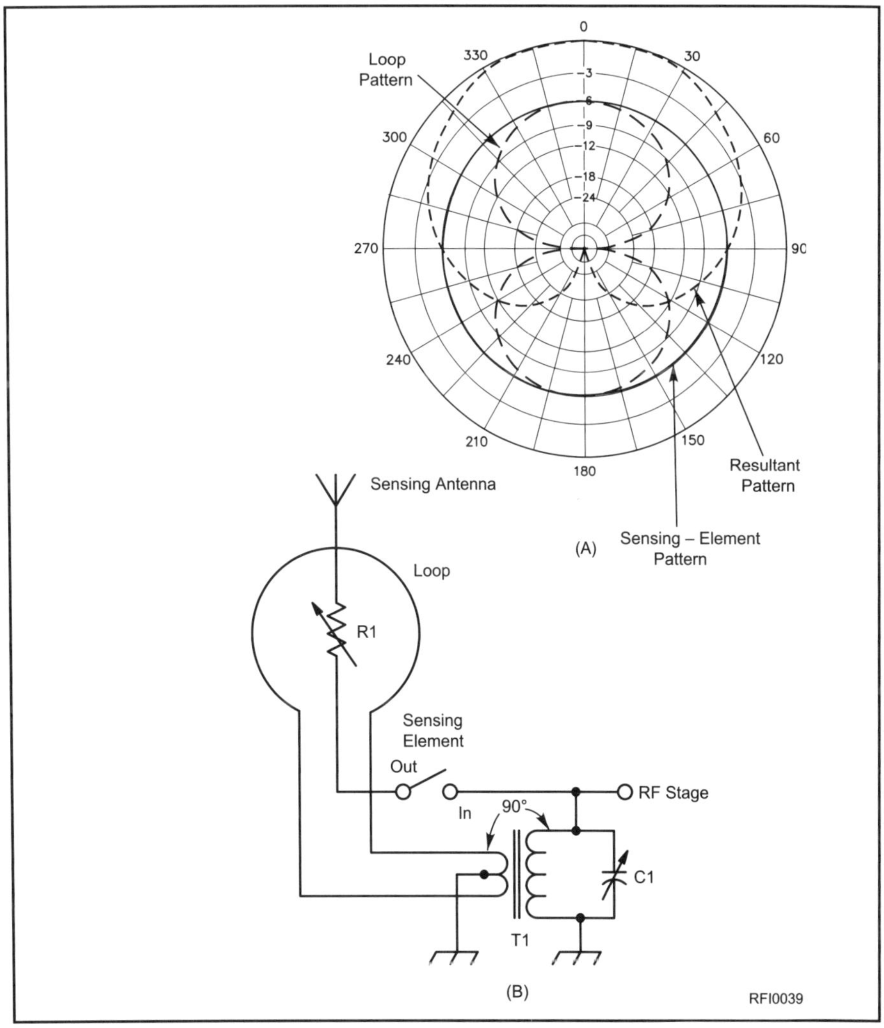

Figure 4.4—At A, the directivity pattern of a loop antenna with sensing element. At B is a circuit for combining the signals from the two elements. Adjust C1 for resonance with T1 at the operating frequency.

elements spaced ¼ λ apart and fed 90° out of phase. The resultant pattern is a cardioid, with a null off one end of the axis of the two antennas and a broad peak in the opposite direction. The directional frequency range of this antenna is limited to one band, because of the critical length of the phasing lines.

The best-known phased array for RDF is the Adcock, named after the man who invented it in 1919. It consists of two vertical elements fed 180° apart, mounted so the array may be rotated. Element spacing is not critical, and may be in the range from 1/10 to ¾ λ. The two elements must be of identical lengths, but need not be self-resonant; shorter elements are commonly used. Because neither the element spacing nor length is critical in terms of wavelengths, an Adcock array may operate over more than one amateur band.

Figure 4.5 is a schematic of a typical Adcock configuration, called the H-Adcock because of its shape. Response to a vertically polarized wave is very similar to a conventional loop. The passing wave induces currents I_1 and I_2 into the vertical members. The output current in the transmission line is equal to their difference. Consequently, the directional pattern has two broad peaks and two sharp nulls, like the loop. The magnitude of the difference current is proportional to the spacing (d) and length (l) of the elements. You will get somewhat higher gain with larger dimensions. The Adcock of Figure 4.6, designed for 40 meters, has element lengths of 12 feet and spacing of 21 feet (approximately 0.15 λ).

Figure 4.7 shows the radiation pattern of the Adcock. The nulls are broadside to the axis of the array, becoming sharper with increased element spacings. When element spacing exceeds ¾ λ, however, the antenna begins to take on additional unwanted nulls off the ends of the array axis.

The Adcock is a vertically polarized antenna. The vertical elements do not respond to horizontally polarized waves, and the currents induced in the horizontal members by a horizontally polarized wave (dotted arrows in Figure 4.5) tend to balance out regardless of the orientation of the antenna.

Since the Adcock uses a balanced feed system, a coupler is required to match the unbalanced input of the receiver. T1 is an air-wound coil with a two-turn link wrapped around the middle. The combination is resonated with C1 to the operating frequency. C2 and C3 are null-clearing capacitors. Adjust them by placing a low-power signal source some distance from the antenna and exactly broadside to it. Adjust C2 and C3 until the deepest null is obtained.

While you can use a metal support for the mast and boom, wood is preferable because of its nonconducting properties. Similarly, a mast of thick-wall PVC pipe gives less distortion of the antenna pattern than a metallic mast. Place the coupler on the ground below the wiring harness junction on the boom and connect it with a short length of 300-Ω twin-lead feed line.

Loops vs Phased Arrays

Loops are much smaller than phased arrays for the same frequency, and are thus the obvious choice for portable/mobile HF RDF. For base stations in a triangulation network, where the 180° ambiguity is not a problem, Adcocks are preferred. In general, they give sharper nulls than loops, but this is in part a function of the care used in constructing and feeding the individual antennas, as well as of the spacing of the elements. The primary construction considerations are the shielding and balancing of the feed line against unwanted signal pickup and the balancing of the antenna for a symmetrical pattern. Users report that Adcocks are somewhat less sensitive to proximity effects, probably because their larger aperture offers some space diversity.

Skywave Considerations

Until now we have considered the directional characteristics of the RDF loop only in the two-dimensional azimuthal plane. In three-dimensional space, the response of a vertically oriented small loop is doughnut-shaped. The bidirectional null (analogous to a line through the doughnut hole) is in the line of bearing in the azimuthal plane and toward the horizon in the vertical plane. Therefore, maximum null depth is achieved only on signals arriving at 0° elevation angle.

Skywave signals usually arrive at non-zero wave angles. As the elevation angle increases, the null in a vertically oriented loop pattern becomes more shallow. It is possible to tilt the loop to seek the null in elevation as well as azimuth. Some amateur RDF enthusiasts report success at estimating distance to the target by measurement of the elevation angle with a tilted loop and computations based on estimated height of the propagating ionospheric layer. This method seldom provides high accuracy with simple loops, however.

Most users prefer Adcocks to loops for skywave work, because the Adcock null is

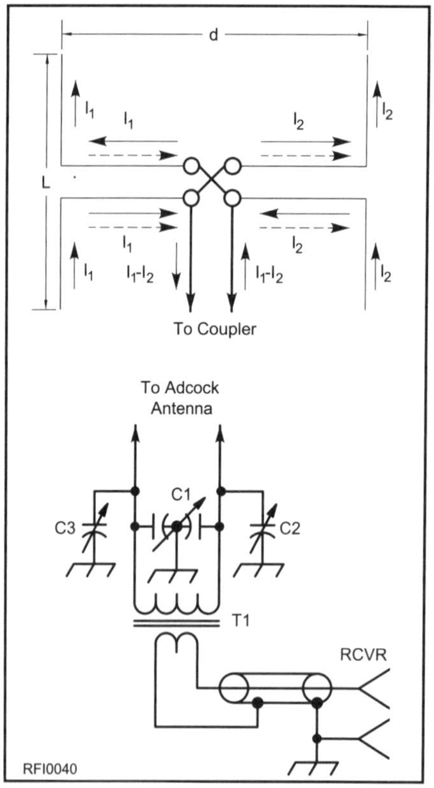

Figure 4.5—A simple Adcock antenna and its coupler.

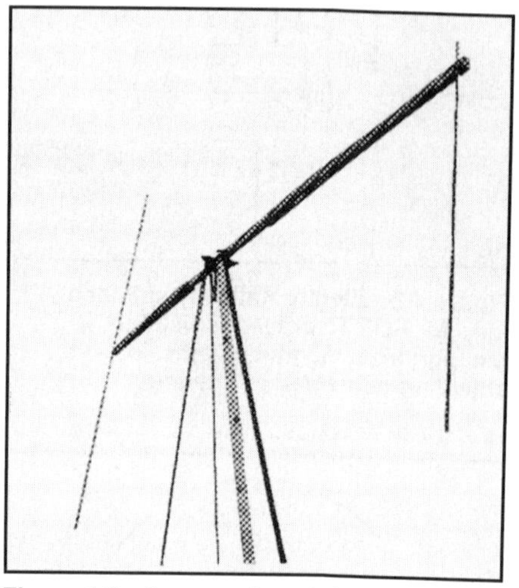
Figure 4.6—An experimental Adcock antenna on a wooden frame.

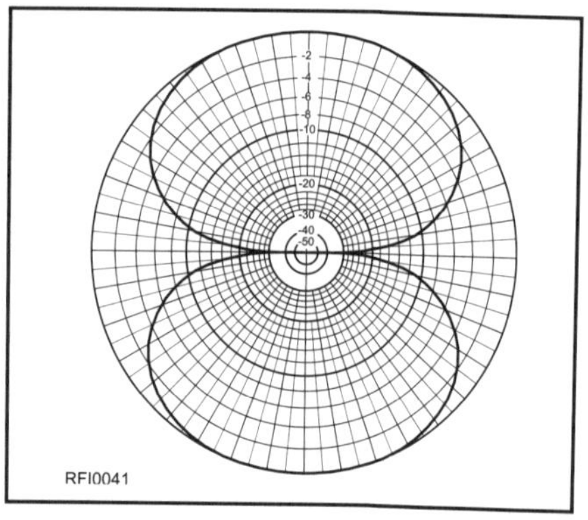

Figure 4.7—The pattern of an Adcock array with element spacing of ½ wave-length. The elements are aligned with the vertical axis.

present at all elevation angles. Note, however, that an Adcock has a null in all directions from signals arriving from overhead. Thus for very high angles, such as under-250-mile skip on 80 and 40 meters, neither loops nor Adcocks will perform well.

Electronic Antenna Rotation

State-of-the-art fixed RDF stations for government and military work use antenna arrays of stationary elements, rather than mechanically rotatable arrays. The best known type is the Wullenweber antenna. It has a large number of elements arranged in a circle, usually outside of a circular reflecting screen. Depending on the installation, the circle may be anywhere from a few hundred feet to more than a quarter of a mile in diameter. Although the Wullenweber is not practical for most amateurs, some of the techniques it uses may be applied to amateur RDF.

The device which permits rotating the antenna beam without moving the elements has the classic name *radiogoniometer*, or simply *goniometer*. Early goniometers were RF transformers with fixed coils connected to the array elements and a moving pickup coil connected to the receiver input. Both amplitude and phase of the signal coupled into the pickup winding are altered with coil rotation in a way that corresponded to actually rotating the array itself. With sufficient elements and a goniometer, accurate RDF measurements can be taken in all compass directions.

Beam Forming Networks

By properly sampling and combining signals from individual elements in a large array, an antenna beam is electronically rotated or steered. With an appropriate number and arrangement of elements in the system, it is possible to form almost any desired antenna pattern by summing the sampled signals in appropriate amplitude and phase relationships. Delay networks and/or attenuation are added in line with selected elements before summation to create these relationships.

To understand electronic beam forming, first consider just two elements, shown as A and B in Figure 4.8. Also shown is the wavefront of a radio signal arriving from a distant transmitter. The wavefront strikes element A first, then travels somewhat farther before it strikes element B. Thus, there is an interval between the times that the wavefront reaches elements A and B.

We can measure the differences in arrival times by delaying the signal received at element A before summing it with that from element B. If two signals are combined directly, the amplitude of the sum will be maximum when the delay for element A exactly equals the propagation delay, giving an in-phase condition at the summation point. On the other hand, if one of the signals is inverted and the two are added, the signals will combine in a 180° out-of-phase relationship when the element A delay equals the propagation delay, creating a null. Either way, once the time delay is determined by the amount of delay required for a peak or null, we can convert it to distance. Then trigonometry calculations provide the direction from which the wave is arriving.

Altering the delay in small increments steers the peak (or null) of the antenna. The system is not frequency sensitive, other than the frequency range limitations of the array elements. Lumped-constant networks are suitable for delay elements if the system is used only for receiving. Delay lines at installations used for transmitting and receiving employ rolls of coaxial cable of various lengths, chosen for the time delay they provide at all frequencies, rather than as simple phasing lines designed for a single frequency.

Combining signals from additional elements narrows the broad beamwidth of the pattern from the two elements and suppress unwanted sidelobes. Electronically switching the delays and attenuations to the various elements causes the formed beam to rotate around the compass. The package of electronics that does this, including delay lines and electronically switched attenuators, is the beam forming network.

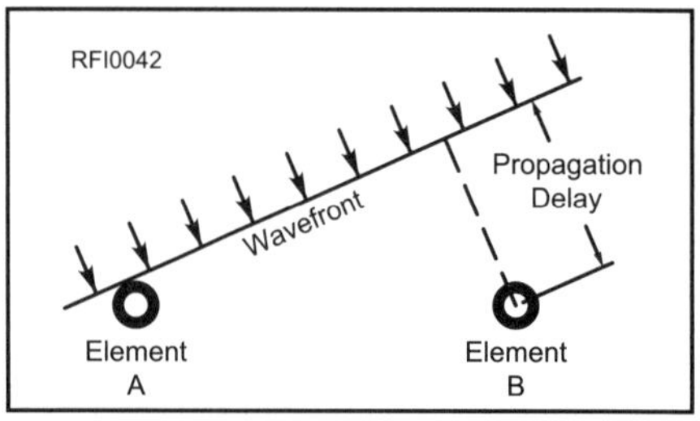

Figure 4.8—One technique used in electronic beam forming. By delaying the signal from element A by an amount equal to the propagation delay, two signals are summed precisely in phase, even though the signal is not in the broadside direction.

METHODS FOR VHF/UHF RDF

Three distinct methods of mobile RDF are commonly in use by amateurs on VHF/UHF bands: directional antennas, switched dual antennas and Dopplers. Each has advantages over the others in certain situations. Many RDF enthusiasts employ more than one method when transmitter hunting.

Directional Antennas

Ordinary mobile transceivers and handhelds work well for foxhunting on the popular VHF bands. If you have a lightweight beam and your receiver has an easy-to-read S meter, you are nearly ready to start. All you need is an RF attenuator and some way to mount the setup in your vehicle.

Amateurs seldom use fractional wavelength loops for RDF above 60 MHz because they have bidirectional characteristics and low sensitivity, compared to other practical VHF antennas. Sense circuits for loops are difficult to implement at VHF, and signal reflections tend to fill in the nulls. Typically, VHF loops are used only for close-in sniffing where their compactness and sharp nulls are assets, and low gain is of no consequence.

Phased Arrays

The small size and simplicity of 2-element driven arrays make them a common choice of newcomers at VHF RDF. Antennas such as phased ground planes and ZL Specials have modest gain in one direction and a null in the opposite direction. The gain is helpful when the signal is weak, but the broad response peak makes it difficult to take a precise bearing.

As the signal gets stronger, it becomes possible to use the null for a sharper S-meter indication. However, combinations of direct and reflected signals (called *multipath*) will distort the null or perhaps obscure it completely. For best results with this type of antenna, always find clear locations from which to take bearings.

Parasitic Arrays

Parasitic arrays are the most common RDF antennas used by transmitter hunters in high competition areas such as Southern California. Antennas with significant gain are a necessity due to the weak signals often encountered on weekend-long T-hunts, where the transmitter may be over 200 miles distant. Typical 144-MHz installations feature Yagis or quads of three to six elements, sometimes more. Quads are typically home-built, using data from *The ARRL Antenna Book* and *Transmitter Hunting* (see Bibliography).

Two types of mechanical construction are popular for mobile VHF quads. The model of Figure 4.9 uses thin gauge wire (solid or stranded), suspended on wood dowel or fiberglass rod spreaders. It is lightweight and easy to turn rapidly by hand while the vehicle moves. Many hunters prefer to use larger gauge solid wire (such as AWG 10) on a PVC plastic pipe frame (Figure 4.10). This quad is more rugged and has somewhat wider frequency range, at the expense of increased weight and wind resistance. It can get mashed going under a willow, but it is easily reshaped and returned to service.

Yagis are a close second to quads in popularity. Commercial models work fine for VHF RDF, provided that the mast is attached at a good balance point. Lightweight and small-diameter elements are desirable for ease of turning at high speeds.

A well-designed mobile Yagi or quad installation includes a method of selecting wave polarization. Although vertical polarization is the norm for VHF-FM communications, horizontal polarization is allowed on many T-hunts. Results will be poor if a VHF RDF antenna is cross-polarized to the transmitting antenna, because multipath and scattered signals (which have indeterminate polarization) are enhanced, relative to the cross-polarized direct signal. The installation of Figure 4.9 features a slip joint at the boom-to-mast junction, with an actuating cord to rotate the boom, changing the polarization. Mechanical stops limit the boom rotation to 90°.

Figure 4.9—The mobile RDF installation of WB6ADC features a thin wire quad for 144 MHz and a mechanical linkage that permits either the driver or front passenger to rotate the mast by hand.

Parasitic Array Performance for RDF

The directional gain of a mobile beam (typically 8 dB or more) makes it unexcelled for both weak signal competitive hunts and for locating interference such as TV cable leakage. With an appropriate receiver, you can get bearings on any signal mode, including FM, SSB, CW, TV, pulses and noise. Because only the response peak is used, the null-fill problems and proximity effects of loops and phased arrays do not exist.

You can observe multiple directions of arrival while rotating the antenna, allowing you to make educated guesses as to which signal peaks are direct and which are from nondirect paths or scattering. Skilled operators can estimate distance to the transmitter from the rate of signal strength increase with distance traveled. The RDF beam is useful for transmitting, if necessary, but use care not to damage an attenuator in the coax line by transmitting through it.

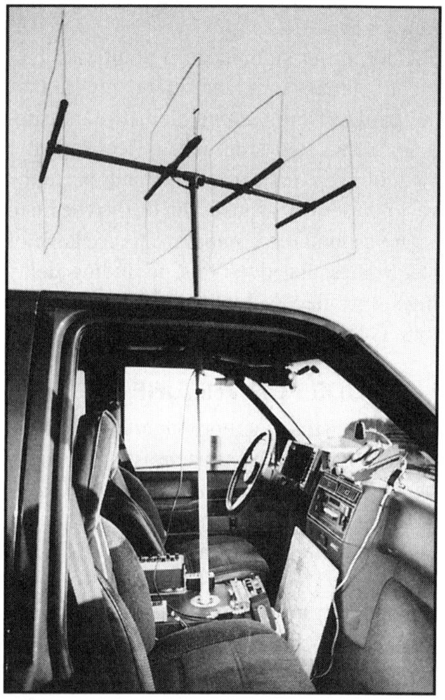

Figure 4.10—KØOV uses this mobile setup for RDF on several bands, with separate antennas for each band that mate with a common lower mast section, pointer and 360° indicator. Antenna shown is a heavy gauge wire quad for 2 meters.

The 3-dB beamwidth of typical mobile-mount VHF beams is on the order of 80°. This is a great improvement over 2-element driven arrays, but it is still not possible to get pinpoint bearing accuracy. You can achieve errors of less than 10° by carefully reading the S meter. In practice, this is not a major hindrance to successful mobile RDF. Mobile users are not as concerned with precise bearings as fixed station operators, because mobile readings are used primarily to give the general direction of travel to "home in" on the signal. Mobile bearings are continuously updated from new, closer locations.

Amplitude-based RDF may be very difficult when signal level varies rapidly. The transmitter hider may be changing power, or the target antenna may be moving or near a well-traveled road or airport. The resultant rapid S-meter movement makes it hard to take accurate bearings with a quad. The process is slow because the antenna must be carefully rotated by hand to "eyeball average" the meter readings.

Switched Antenna RDF Units

Three popular types of RDF systems are relatively insensitive to variations in signal level. Two of them use a pair of vertical dipole antennas, spaced ½ λ or less apart, and alternately switched at a rapid rate to the input of the receiver. In use, the indications of the two systems are similar, but the principles are different.

Switched Pattern Systems

The switched pattern RDF set (Figure 4.11) alternately creates two cardioid antenna patterns with lobes to the left and the right. The patterns are generated in much the same way as in the phased arrays described above. PIN RF diodes select the alternating patterns. The combined antenna outputs go to a receiver with AM detection. Processing after the detector output determines the phase or amplitude difference between the patterns' responses to the signal.

Switched pattern RDF sets typically have a zero center meter as an indicator. The meter swings negative when the signal is coming from the user's left, and positive when the signal source is on the right. When the plane of the antenna is exactly perpendicular to the direction of the signal source, the meter reads zero.

The sharpness of the zero crossing indication makes possible more precise bearings than those obtainable with a quad or Yagi. Under ideal conditions with a well-built unit, null direction accuracy is within 1°. Meter deflection tells the user which way to turn to zero the meter. For example, a negative (left) reading requires turning the antenna left. This solves the 180° ambiguity caused by the two zero crossings in each complete rotation of the antenna system.

Because it requires AM detection of the switched pattern signal, this RDF system finds its greatest use in the 120-MHz aircraft band, where AM is the standard mode. Commercial manufacturers make portable RDF sets with switched pattern antennas and built-in receivers for field portable use. These sets can usually be adapted to the amateur 144-MHz band. Other designs are adaptable to any VHF receiver that covers the frequency of interest and has an AM detector built in or added.

Switched pattern units work well for RDF from small aircraft, for which the two vertical antennas are mounted in fixed positions on the outside of the fuselage or simply taped inside the windshield. The left-right indication tells the pilot which way to turn the aircraft to home-in. Since street vehicles generally travel only on roads, fixed mounting of the antennas on them is undesirable. Mounting vehicular switched-pattern arrays on a rotatable mast is best.

Time of Arrival Systems

Another kind of switched antenna RDF set uses the difference in arrival times of the signal wavefront at the two antennas. This narrow-aperture Time-Difference-of-Arrival (TDOA) technology is used for many sophisticated military RDF systems. The rudimentary TDOA implementation of Figure 4.12 is quite effective for amateur use. The signal from transmitter 1 reaches antenna A before antenna B. Conversely, the signal from transmitter 3 reaches antenna B before antenna A. When the plane of the antenna is perpendicular to the signal source (as transmitter 2 is in the figure), the signal arrives at both antennas simultaneously.

If the outputs of the antennas are alternately switched at an audio rate to the receiver input, the differences in the arrival times of a continuous signal produce phase changes that are detected by an FM discriminator. The resulting short pulses sound like a tone in the receiver output. The tone disappears when the antennas are equidistant from the signal source, giving an audible null.

The polarity of the pulses at the discriminator output is a function of which antenna is closer to the source. Therefore, the pulses can be processed and used to drive a left-right zero center meter in a manner similar to the switched pattern units described above. Left-right LED indicators may replace the meter for economy and visibility at night.

RDF operations with a TDOA dual antenna RDF are done in the same manner as with a switched antenna RDF set. The main difference is the requirement for an FM receiver in the TDOA system and an AM receiver in the switched pattern case. No RF attenuator is needed for close-in work in the TDOA case.

Popular designs for practical do-it-yourself TDOA RDF sets include the Simple Seeker (described elsewhere in this chapter) and the W9DUU design (see article by Bohrer in the Bibliography). Articles with plans for the Handy Tracker, a simple TDOA set with a delay line to resolve the dual-null ambiguity instead of LEDs or a meter, are listed in the Bibliography.

Performance Comparison

Both types of dual antenna RDFs make good on-foot "sniffing" devices and are excellent performers when there are rapid amplitude variations in the incoming signal. They are the units of choice for airborne work. Compared to Yagis and quads, they give good directional performance over a much wider frequency range. Their indications are more precise than those of beams with broad forward lobes.

Dual-antenna RDF sets frequently give inaccurate bearings in multipath situations, because they cannot resolve signals of nearly equal levels from more than one direction. Because multipath signals are a combined pattern of peaks and nulls, they appear to change in amplitude and bearing as you move the RDF antenna along the bearing path or perpendicular to it, whereas a non-multipath signal will have constant strength and bearing.

The best way to overcome this problem is to take large numbers of bearings while moving toward the transmitter. Taking bearings while in motion averages out the effects of multipath, making the direct signal more readily discernible. Some TDOA RDF sets have a slow-response mode that aids the averaging process.

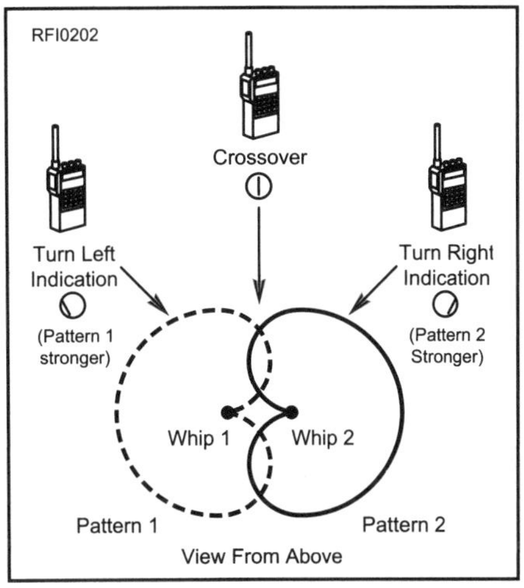

Figure 4.11—In a switched pattern RDF set, the responses of two cardioid antenna patterns are summed to drive a zero center indicator.

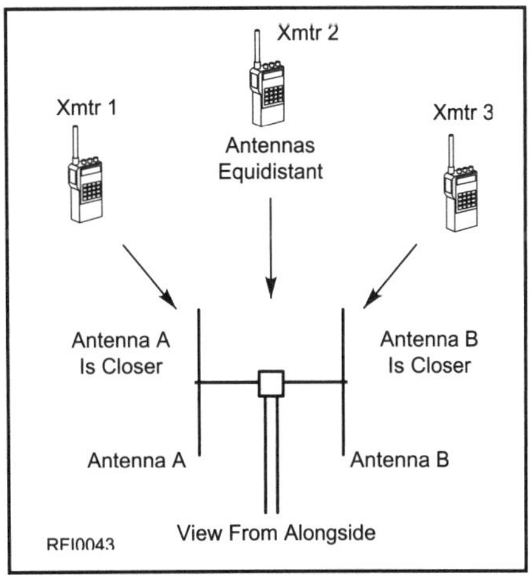

Figure 4.12—A dual-antenna TDOA RDF system has a similar indicator to a switched pattern unit, but it obtains bearings by determining which of its antennas is closer to the transmitter.

Switched antenna systems generally do not perform well when the incoming signal is horizontally polarized. In such cases, the bearings may be inaccurate or unreadable. TDOA units require a carrier type signal such as FM or CW; they usually cannot yield bearings on noise or pulse signals.

Unless an additional method is employed to measure signal strength, it is easy to "overshoot" the hidden transmitter location with a TDOA set. It is not uncommon to see a TDOA foxhunter walk over the top of a concealed transmitter and walk away, following the opposite 180° null, because there is no display of signal amplitude.

Doppler RDF Sets

RDF sets using the Doppler principle are popular in many areas because of their ease of use. They have an indicator that instantaneously displays the direction of the signal source relative to the vehicle heading, either on a circular ring of LEDs or a digital readout in degrees. A ring of four, eight or more antennas picks up the signal. Quarter-wavelength monopoles on a ground plane are popular for vehicle use, but half-wavelength vertical dipoles, where practical, perform better.

Radio signals received on a rapidly moving antenna experience a frequency shift due to the Doppler effect, a phenomenon well known to anyone who has observed a moving car with its horn sounding. The horn's pitch appears higher than normal as the car approaches, and lower as the car recedes. Similarly, the received radio frequency increases as the antenna moves toward the transmitter and vice versa. An FM receiver will detect this frequency change.

Figure 4.13 shows a ¼-λ vertical antenna being moved on a circular track around point P, with constant angular velocity. As the antenna approaches the transmitter on its track, the received frequency is shifted higher. The highest instantaneous frequency occurs when the antenna is at point A, because tangential velocity toward the transmitter is maximum at that point. Conversely, the lowest frequency occurs when the antenna reaches point C, where velocity is maximum away from the transmitter.

Figure 4.14 shows a plot of the component of the tangential velocity that is in the direction of the transmitter as the antenna moves around the circle. Comparing Figures 4.13 and 4.14, notice that at B in Figure 4.14, the tangential velocity is crossing zero from the positive to the negative and the antenna is closest to the transmitter. The Doppler shift and resulting audio output from the receiver discriminator follow the same plot, so that a negative-slope zero-crossing detector, synchronized with the antenna rotation, senses the incoming direction of the signal.

The amount of frequency shift due to the Doppler effect is proportional to the RF frequency and the tangential antenna velocity. The velocity is a function of the radius of rotation and the angular velocity (rotation rate). The radius of rotation must be less than ¼ λ to avoid errors. To get a usable amount of FM deviation (comparable to typical voice modulation) with this radius, the antenna must rotate at approximately 30,000 r/min (500 Hz). This puts the Doppler tone in the audio range for easy processing.

Mechanically rotating a whip antenna at this rate is impractical, but a ring of whips, switched to the receiver in succession with RF PIN diodes, can simulate a rapidly rotating antenna. Doppler RDF sets must be used with receivers having FM detectors. The DoppleScAnt and Roanoke Doppler (see Bibliography) are mobile Doppler RDF sets designed for inexpensive home construction.

Doppler Advantages and Disadvantages

Ring-antenna Doppler sets are the ultimate in simplicity of operation for mobile RDF. There are no moving parts and no manual antenna pointing. Rapid direction indications are displayed on very short signal bursts.

Many units lock in the displayed direction after the signal leaves the air. Power variations in the source signal cause no difficulties, as long as the signal remains above the RDF detection threshold. A Doppler antenna goes on top of any car quickly, with no holes to drill. Many Local Interference Committee members choose Dopplers for tracking malicious interference, because they are inconspicuous (compared to beams) and effective at tracking the strong vertically polarized signals that repeater jammers usually emit.

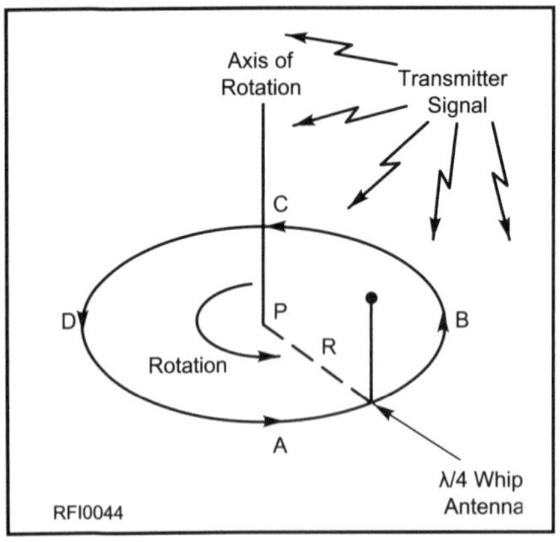

Figure 4.13—A theoretical Doppler antenna circles around point P, continuously moving toward and away from the source at an audio rate.

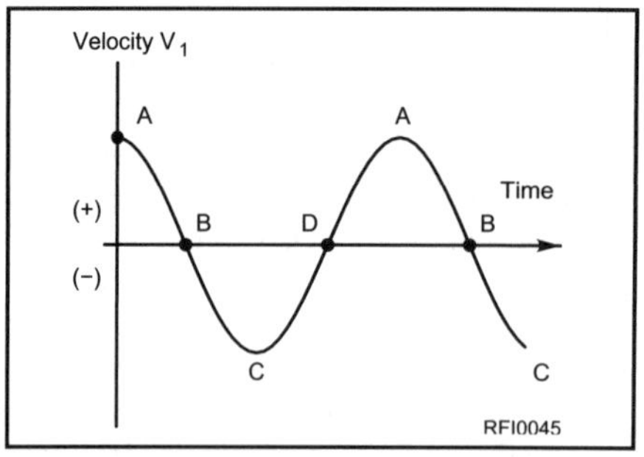

Figure 4.14—Frequency shift versus time produced by the rotating antenna movement toward and away from the signal source.

A Doppler does not provide superior performance in all VHF RDF situations. If the signal is too weak for detection by the Doppler unit, the hunt advantage goes to teams with beams. Doppler installations are not suitable for on-foot sniffing. The limitations of other switched antenna RDFs also apply: (1) poor results with horizontally polarized signals, (2) no indication of distance, (3) carrier type signals only and (4) inadvisability of transmitting through the antenna.

Readout to the nearest degree is provided on some commercial Doppler units. This does not guarantee that level of accuracy, however. A well-designed four-monopole set is typically capable of ±5° accuracy on 2 meters, if the target signal is vertically polarized and there are no multipath effects.

The rapid antenna switching can introduce cross-modulation products when the user is near strong off-channel RF sources. This self-generated interference can temporarily render the system unusable. While not a common problem with mobile Dopplers, it makes the Doppler a poor choice for use in remote RDF installations at fixed sites with high power VHF transmitters nearby.

Mobile RDF System Installation

Of these mobile VHF RDF systems, the Doppler type is clearly the simplest from a mechanical installation standpoint. A four-whip Doppler RDF array is easy to implement with magnetic mount antennas. Alternately, you can mount all the whips on a frame that attaches to the vehicle roof with suction cups. In either case, setup is rapid and requires no holes in the vehicle.

You can turn small VHF beams and dual-antenna arrays readily by extending the mast through a window. Installation on each model vehicle is different, but usually the mast can be held in place with some sort of cup in the arm rest and a plastic tie at the top of the window, as in Figure 4.15. This technique works best on cars with frames around the windows, which allow the door to be opened with the antenna in place. Check local vehicle codes, which limit how far your antenna may protrude beyond the line of the fenders. Larger antennas may have to be put on the passenger side of the vehicle, where greater overhang is generally permissible.

The window box (Figure 4.16) is an improvement over through-the-window mounts. It provides a solid, easy-turning mount for the mast. The plastic panel keeps out bad weather. You will need to custom-design the box for your vehicle model. Vehicle codes may limit the use of a window box to the passenger side.

For the ultimate in convenience and versatility, cast your fears aside, drill a hole through the center of the roof and install a waterproof bushing. A roof-hole mount permits the use of large antennas without overhang violations. The driver, front passenger and even a rear passenger can turn the mast when required. The installation in Figure 4.10 uses a roof-hole bushing made from mating threaded PVC pipe adapters and reducers. When it is not in use for RDF, a PVC pipe cap provides a watertight cover. There is a pointer and 360° indicator at the bottom of the mast for precise bearings.

PRACTICAL DIRECTION-FINDING TECHNIQUES AND PROJECTS

The ability to locate a transmitter quickly with RDF techniques is a skill you will acquire only with practice. It is very important to become familiar with your equipment and its limitations. You must also understand how radio signals behave in different types of terrain at the frequency of the hunt. Experience is the best teacher, but reading and hearing the stories of others who are active in RDF will help you get started.

Verify proper performance of your portable RDF system before you attempt to track signals in unknown locations. Of primary concern is the accuracy and symmetry of the antenna pattern. For instance, a lopsided figure-8 pattern with a loop, Adcock or TDOA set leads to large bearing errors. Nulls should be exactly 180° apart and exactly at right angles to the loop plane or the array boom. Similarly, if feed-line pickup causes an off-axis main lobe in your VHF RDF beam, your route to the target will be a spiral instead of a straight line.

Perform initial checkout with a low-powered test transmitter at a distance of a few hundred feet. Compare the RDF bearing indication with the visual path to the transmitter. Try to "find" the transmitter with the RDF equipment as if its position were not known. Be sure to check all nulls on antennas that have more than one.

If imbalance or off-axis response is found in the antennas, there are two options available. One is to correct it, insofar as possible. A second option is to accept it and use some kind of indicator or correction procedure to show the true directions of signals. Sometimes the end result of the calibration procedure is a compromise between these two options, as a perfect pattern may be difficult or impossible to attain.

The same calibration suggestions apply for fixed RDF installations, such as a base station HF Adcock or VHF beam. Of course it does no good to move it to an open field. Instead, calibrate the array in its intended operating position, using a portable or mobile transmitter. Because of nearby obstructions or reflecting objects, your antenna may not indicate the precise direction of the transmitter. Check for imbalance and systemic error by taking readings with the test emitter at locations in several different directions.

The test signal should be at a distance of 2 or 3 miles for these measurements, and should be in as clear an area as possible during transmissions. Avoid locations where power lines and other overhead wiring can conduct a signal from the transmitter to the RDF site. Once antenna adjustments are optimized, make a table of bearing errors noted in all compass directions. Apply these error values as corrections when actual measurements are made.

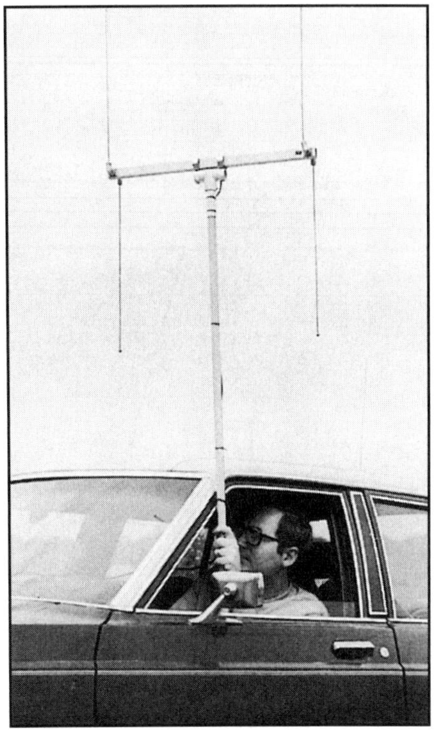

Figure 4.15—A set of TDOA RDF antennas is lightweight and mounts readily through a sedan window without excessive overhang.

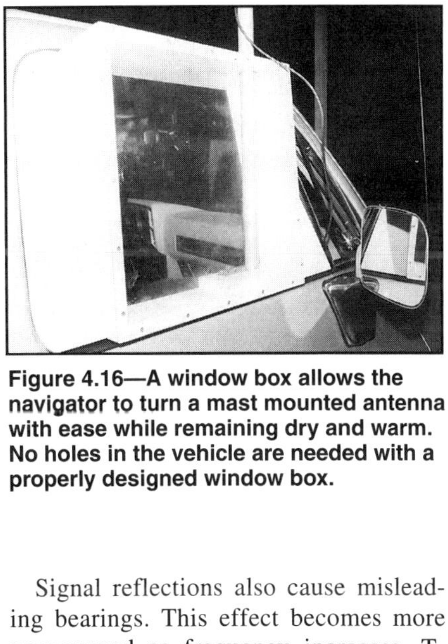

Figure 4.16—A window box allows the navigator to turn a mast mounted antenna with ease while remaining dry and warm. No holes in the vehicle are needed with a properly designed window box.

Preparing to Hunt

Successfully tracking down a hidden transmitter involves detective work—examining all the clues, weighing the evidence and using good judgment. Before setting out to locate the source of a signal, note its general characteristics. Is the frequency constant, or does it drift? Is the signal continuous, and if not, how long are transmissions? Do transmissions occur at regular intervals, or are they sporadic? Irregular, intermittent signals are the most difficult to locate, requiring patience and quick action to get bearings when the transmitter comes on.

Refraction, Reflections and the Night Effect

You will get best accuracy in tracking ground wave signals when the propagation path is over homogeneous terrain. If there is a land/water boundary in the path, the different conductivities of the two media can cause bending (refraction) of the wave front, as in Figure 4.17A. Even the most sophisticated RDF equipment will not indicate the correct bearing in this situation, as the equipment can only show the direction from which the signal is arriving. RDFers have observed this phenomenon on both HF and VHF bands.

Signal reflections also cause misleading bearings. This effect becomes more pronounced as frequency increases. T-hunt hiders regularly achieve strong signal bounces from distant mountain ranges on the 144-MHz band.

Tall buildings also reflect VHF/UHF signals, making midcity RDF difficult. Hunting on the 440 MHz and higher amateur bands is even more arduous because of the plethora of reflecting objects.

In areas of signal reflection and multi-path, some RDF gear may indicate that the signal is coming from an intermediate point, as in Figure 4.17B. High gain VHF/UHF RDF beams will show direct and reflected signals as separate S-meter peaks, leaving it to the operator to determine which is which. Null-based RDF antennas, such as phased arrays and loops, have the most difficulty with multipath, because the multiple signals tend to make the nulls very shallow or fill them in entirely, resulting in no bearing indication at all.

If the direct path to the transmitter is masked by intervening terrain, a signal reflection from a higher mountain, building, water tower, or the like may be much stronger than the direct signal. In extreme cases, triangulation from several locations will appear to "confirm" that the transmitter is at the location of the reflecting object. The direct signal may not be detectable until you arrive at the reflecting point or another high location.

Objects near the observer such as concrete/steel buildings, power lines and chain-link fences will distort the incoming wavefront and give bearing errors. Even a dense grove of trees can sometimes have an adverse effect. It is always best to take readings in locations that are as open and clear as possible, and to take bearings from numerous positions for confirmation. Testing of RDF gear should also be done in clear locations.

Locating local signal sources on frequencies below 10 MHz is much easier during daylight hours, particularly with loop antennas. In the daytime, D-layer absorption minimizes skywave propagation on these frequencies. When the D layer disappears after sundown, you may hear the signal by a combination of ground wave and high-angle skywave, making it difficult or impossible to obtain a bearing. RDFers call this phenomenon the *night effect*.

While some mobile T-hunters prefer to go it alone, most have more success by teaming up and assigning tasks. The driver concentrates on handling the vehicle, while the assistant (called the "navigator" by some teams) turns the beam, reads the meters and calls out bearings. The assistant is also responsible for maps and plotting, unless there is a third team member for that task.

Maps and Bearing-Measurements

Possessing accurate maps and knowing how to use them is very important for successful RDF. Even in difficult situations where precise bearings cannot be obtained, a town or city map will help in plotting points where signal levels are high and low. For example, power line noise tends to propagate along the power line and radiates as it does so. Instead of a single source, the noise appears to come from a multitude of sources. This renders many ordinary RDF techniques ineffective. Mapping locations where signal

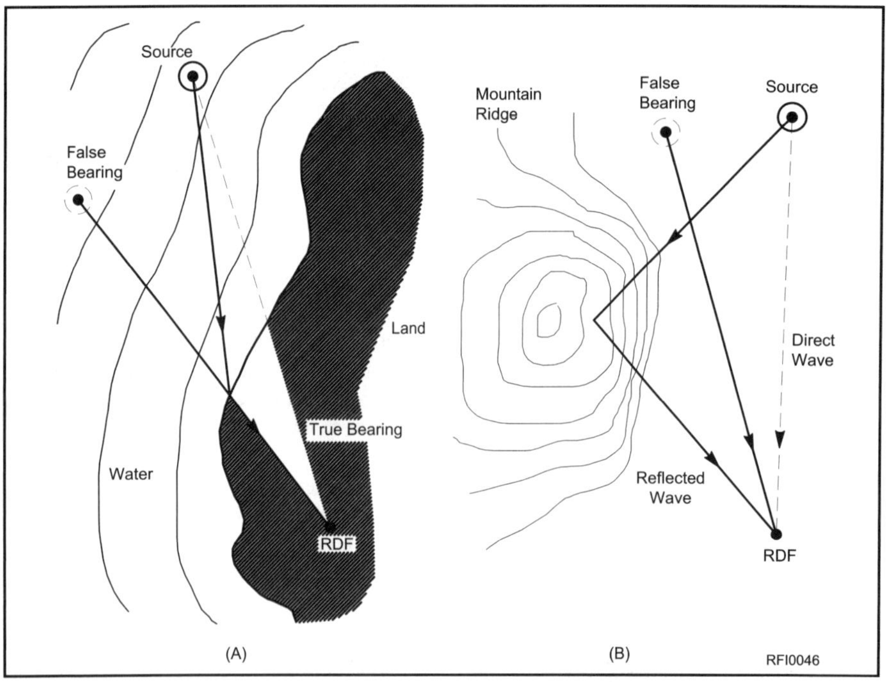

Figure 4.17—RDF errors caused by refraction (A) and reflection (B). The reading at A is false because the signal actually arrives from a direction that is different from that to the source. At B, a direct signal from the source combines with a reflected signal from the mountain ridge. The RDF set may average the signals as shown or indicate two lines of bearing.

amplitudes are highest will help pinpoint the source.

Several types of area-wide maps are suitable for navigation and triangulation. Street and highway maps work well for mobile work. Large detailed maps are preferable to thick map books. Contour maps are ideal for open country. Aeronautical charts are also suitable. Good sources of maps include auto clubs, stores catering to camping/hunting enthusiasts and city/county engineering departments.

A *heading* is a reading in degrees relative to some external reference, such as your house or vehicle; a *bearing* is the target signal's direction relative to your position. Plotting a bearing on a hidden transmitter from your vehicle requires that you know the vehicle location, transmitter heading with respect to the vehicle and vehicle heading with respect to true north.

First, determine your location using landmarks or a navigation device such as a GPS (or loran) receiver. Next, using your RDF equipment, determine the bearing to the hidden transmitter (0 to 359.9°) with respect to the vehicle. Zero degrees heading corresponds to signals coming from directly in front of the vehicle, signals from the right indicate 90°, and so on.

Finally, determine your vehicle's true heading, that is, its heading relative to true north. Compass needles point to magnetic north and yield magnetic headings. Translating a magnetic heading into a true heading requires adding a correction factor, called *magnetic declination*, which is a positive or negative factor that depends on your location.

Declination for your area is given on US Geological Survey (USGS) maps, though it undergoes long-term changes. Add the declination to your magnetic heading to get a true heading.

As an example, assume that the transmitted signal arrives at 30° with respect to the vehicle heading, that the compass indicates that the vehicle's heading is 15°, and the magnetic declination is +15°. Add these values to get a true transmitter bearing (that is, a bearing with respect to true north) of 60°.

Because of the large mass of surrounding metal, it is very difficult to calibrate an in-car compass for high accuracy at all vehicle headings. It is better to use a remotely mounted flux-gate compass sensor, properly corrected, to get vehicle headings, or to stop and use a hand compass to measure the vehicle heading from the outside. If you T-hunt with a mobile VHF beam or quad, you can use your manual compass to sight along the antenna boom for a magnetic bearing, then add the declination for true bearing to the fox.

Triangulation Techniques

If you can obtain accurate bearings from two locations separated by a suitable distance, the technique of *triangulation* will give the expected location of the transmitter. The intersection of the lines of bearing from each location provides a *fix*. Triangulation accuracy is greatest when stations are located such that their bearings intersect at right angles. Accuracy is poor when the angle between bearings approaches 0° or 180°.

There is always uncertainty in the fixes obtained by triangulation due to equipment

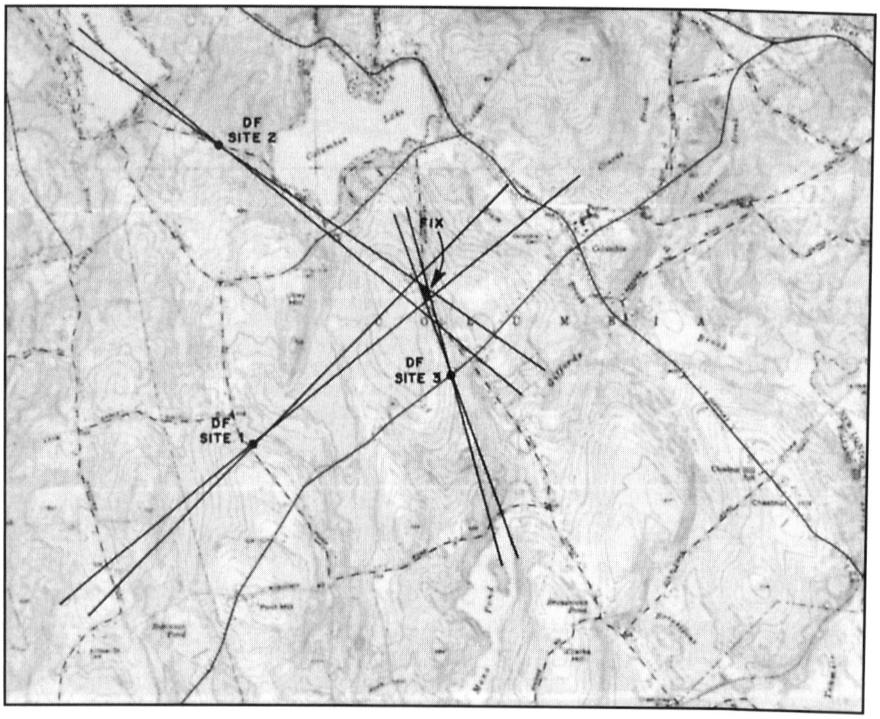

Figure 4.18—Bearing sectors from three RDF positions drawn on a map for triangulation. In this case, bearings are from loop antennas, which have 180° ambiguity.

limitations, propagation effects and measurement errors. Obtaining bearings from three or more locations reduces the uncertainty. A good way to show the probable area of the transmitter on the triangulation map is to draw bearings as a narrow sector instead of as a single line. Sector width represents the amount of bearing uncertainty. Figure 4.18 shows a portion of a map marked in this manner. Note how the bearing from Site 3 has narrowed down the probable area of the transmitter position.

Computerized Transmitter Hunting

A portable computer is an excellent tool for streamlining the RDF process. Some T-hunters use one to optimize VHF beam bearings, generating a two-dimensional plot of signal strength versus azimuth. Others have automated the bearing-taking process by using a computer to capture signal headings from a Doppler RDF set, vehicle heading from a flux-gate compass, and vehicle location from a GPS receiver (Figure 4.19). The computer program can compute averaged headings from a Doppler set to reduce multipath effects.

Provided with perfect position and bearing information, computer triangulation could determine the transmitter location within the limits of its computational accuracy. Two bearings would exactly locate a fox. Of course, there are always uncertainties and inaccuracies in bearing and position data. If these uncertainties can be determined, the program can compute the uncertainty of the triangulated bearings. A "smart" computer program can evaluate bearings, triangulate the bearings of multiple hunters, discard those that appear erroneous, determine which locations have particularly great or small multipath problems and even "grade" the performance of RDF stations.

By adding packet radio connections to a group of computerized base and mobile RDF stations, the processed bearing data from each can be shared. Each station in the network can display the triangulated bearings of all. This requires a common map coordinate set among all stations. The USGS Universal Transverse Mercator (UTM) grid, consisting of 1×1-km grid squares, is a good choice.

The computer is an excellent RDF tool, but it is no substitute for a skilled "navigator." You will probably discover that using a computer on a high-speed T-hunt requires a full-time operator in the vehicle to make full use of its capabilities.

Skywave Bearings and Triangulation

Many factors make it difficult to obtain accuracy in skywave RDF work. Because of Faraday rotation during propagation, skywave signals are received with random polarization. Sometimes the vertical component is stronger, and at other times the horizontal. During periods when the vertical component is weak, the signal may appear to fade on an Adcock RDF system. At these times, determining an accurate signal null direction becomes very hard.

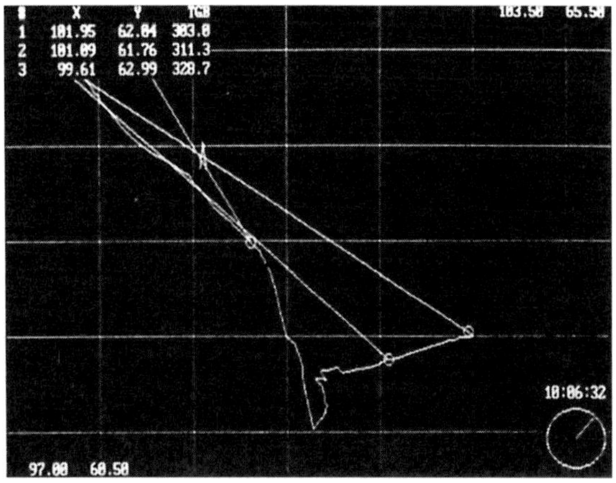

Figure 4.19—Screen plot from a computerized RDF system showing three T-hunt bearings (straight lines radiating from small circles) and the vehicle path (jagged trace). The grid squares correspond to areas of standard topographic maps.

For a variety of reasons, HF bearing accuracy to within 1 or 2° is the exception rather than the rule. Errors of 3 to 5° are common. An error of 3° at a thousand miles represents a distance of 52 miles. Even with every precaution taken in measurement, do not expect cross-country HF triangulation to pinpoint a signal beyond a county, a corner of a state or a large metropolitan area. The best you can expect is to be able to determine where a mobile RDF group should begin making a local search.

Triangulation mapping with skywave signals is more complex than with ground or direct waves because the expected paths are great-circle routes. Commonly available world maps are not suitable, because the triangulation lines on them must be curved, rather than straight. In general, for flat maps, the larger the area encompassed, the greater the error that straight line triangulation procedures will give.

A highway map is suitable for regional triangulation work if it uses some form of conical projection, such as the Lambert conformal conic system. This maintains the accuracy of angular representation, but the distance scale is not constant over the entire map.

One alternative for worldwide areas is the azimuthal-equidistant projection, better known as a great-circle map. True bearings for great-circle paths are shown as straight lines from the center to all points on the Earth. Maps centered on three or more different RDF sites may be compared to gain an idea of the general geographic area for an unknown source.

For worldwide triangulation, the best projection is the *gnomonic*, on which all great-circle paths are represented by straight lines and angular measurements with respect to meridians are true. Gnomonic charts are custom maps prepared especially for government and military agencies.

Skywave signals do not always follow the great-circle path in traveling from a transmitter to a receiver. For example, if the signal is refracted in a tilted layer of the ionosphere, it could arrive from a direction that is several degrees away from the true great-circle bearing.

Another cause of signals arriving off the great-circle path is termed *sidescatter*. It is possible that, at a given time, the ionosphere does not support great-circle propagation of the signal from the transmitter to the receiver because the frequency is above the MUF for that path. However, at the same time, propagation may be supported from both ends of the path to some mutually accessible point off the great-circle path. The signal from the source may propagate to that point on the Earth's surface and hop in a sideways direction to continue to the receiver.

For example, signals from Central Europe have propagated to New England by hopping from an area in the Atlantic Ocean off the northwest coast of Africa, whereas the great-circle path puts the reflection point off the southern coast of Greenland. Readings in error by as much as 50° or more may result from side-scatter. The effect of propagation disturbances may be that the bearing seems to wander somewhat over a few minutes of time, or it may be weak and fluttery. At other times, however, there may be no telltale signs to indicate that the readings are erroneous.

Closing In

On a mobile foxhunt, the objective is usually to proceed to the hidden T with minimum time and mileage. Therefore, do not go far out of your way to get off-course bearings just to triangulate. It is usually better to take the shortest route along your initial line of bearing and "home-in" on the signal. With a little experience, you will be able to gauge your distance from the fox by noting the amount of attenuation needed to keep the S meter on scale.

As you approach the transmitter, the signal will become very strong. To keep the S meter on scale, you will need to add an RF attenuator in the transmission line from the antenna to the receiver. Simple resistive attenuators are discussed in another chapter.

In the final phases of the hunt, you will probably have to leave your mobile and continue the hunt on foot. Even with an attenuator in the line, in the presence of a strong RF field, some energy will be coupled directly into the receiver circuitry. When this happens, the S-meter reading changes only slightly or perhaps not at all as the RDF antenna rotates, no matter how much attenuation you add. The cure is to shield the receiving equipment. Something as simple as wrapping the receiver in foil or placing it in a bread pan or cake pan, covered with a piece of copper or aluminum screening securely fastened at several points, may reduce direct pickup enough for you to get bearings.

Alternatively, you can replace the receiver with a field-strength meter as you close in, or use a heterodyne-type active attenuator. Plans for these devices are at the end of this chapter.

The Body Fade

A crude way to find the direction of a VHF signal with just a handheld transceiver is the body fade technique, so named because the blockage of your body causes the signal to fade. Hold your handheld close to your chest and turn all the way around slowly. Your body is providing a shield that gives the handheld a cardioid sensitivity pattern, with a sharp decrease in sensitivity to the rear. This null indicates that the source is behind you (Figure 4.20).

If the signal is so strong that you can't find the null, try tuning 5 or 10 kHz off frequency to put the signal into the skirts of the IF passband. If your handheld is dual-band (144/440 MHz) and you are hunting on 144 MHz, try tuning to the much weaker third harmonic of the signal in the 440-MHz band.

The body fade null, which is rather shallow to begin with, can be obscured by reflections, multipath, nearby objects, etc. Step well away from your vehicle before trying to get a bearing. Avoid large buildings, chain-link fences, metal signs and the like. If you do not get a good null, move to a clearer location and try again.

Air Attenuators

In microwave parlance, a signal that is too low in frequency to be propagated in a waveguide (that is, below the *cutoff frequency*) is attenuated at a predictable logarithmic rate. In other words, the farther inside the waveguide, the weaker the signal gets. Devices that use this principle to reduce signal strength are commonly known as *air attenuators*. Plans for a practical model for insertion in a coax line are in *Transmitter Hunting* (see Bibliography).

With this principle, you can reduce the level of strong signals into your handheld transceiver, making it possible to use the body fade technique at very close range. Glen Rickerd, KC6TNF, documented this technique for *QST* (Jan 1994). Start with a pasteboard mailing tube that has sufficient inside diameter to accommodate your handheld. Cover the outside of the tube completely with aluminum foil. You can seal the bottom end with foil, too, but it probably will not matter if the tube is long enough. For durability and to prevent accidental shorts, wrap the foil in packing tape. You will also need a short, stout cord attached to the handheld. The wrist strap may work for this, if long enough.

To use this air attenuation scheme for body fade bearings, hold the tube vertically against your chest and lower the handheld into it until the signal begins to weaken (Figure 4.21). Holding the receiver in place, turn around slowly and listen for a sudden decrease in signal strength. If the null is poor, vary the depth of the receiver in the tube and try again. You do not need to watch the S meter, which will likely be out of sight in the tube. Instead, use noise level to estimate signal strength.

For extremely strong signals, remove the "rubber duck" antenna or extend the wrist strap with a shoelace to get greater depth of suspension in the tube. The depth that works for one person may not work for another. Experiment with known signals to determine what works best for you.

THE SIMPLE SEEKER

The Simple Seeker for 144 MHz is the latest in a series of dual-antenna TDOA projects by Dave Geiser, WA2ANU. Figure 4.12 and accompanying text shows its principle of operation. It is simple to perform rapid antenna switching with diodes, driven by a free-running multi-vibrator. For the best RDF performance, the switching pulses should be square waves, so antennas are alternately connected for equal times. The Simple Seeker uses a CMOS version of the popular 555 timer, which demands very little supply current. A 9-V alkaline battery will give long life. See Figure 4.22 for the schematic diagram.

Pin diodes are best for this application because they have low capacitance and handle a moderate amount of transmit power. Phillips ECG553, NTE-555, Motorola MPN3401 and similar types are suitable. Ordinary 1N4148 switching diodes are acceptable for receive only use.

Off the null, the polarity of the switching pulses in the receiver output changes (with respect to the switching waveform), depending on which antenna is nearer the source. Thus, comparing the receiver output phase to that of the switching waveform determines which end of the null line points toward the transmitter. The common name for a circuit to make this comparison is a *phase detector*, achieved in this unit with a simple bridge circuit. A phase detector balance control is included, although it may not be needed. Serious imbalance indicates incorrect receiver tuning, an off-frequency target signal or misalignment in the receiver IF stages.

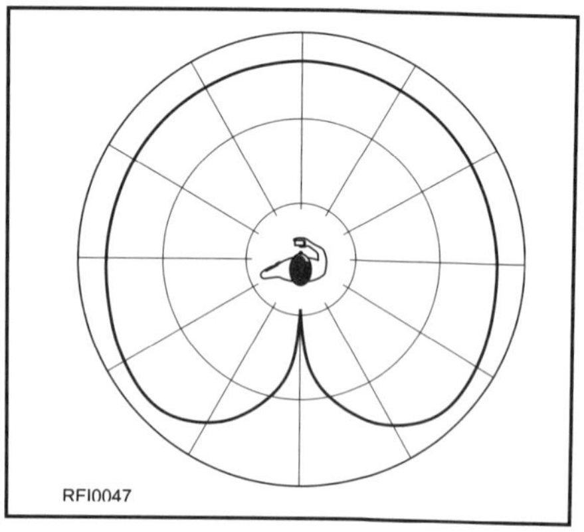

Figure 4.20—In the body-fade maneuver, a handheld transceiver exhibits this directional pattern.

Figure 4.21—The air attenuator for a VHF handheld in use. Suspend the radio by the wrist strap or a string inside the tube.

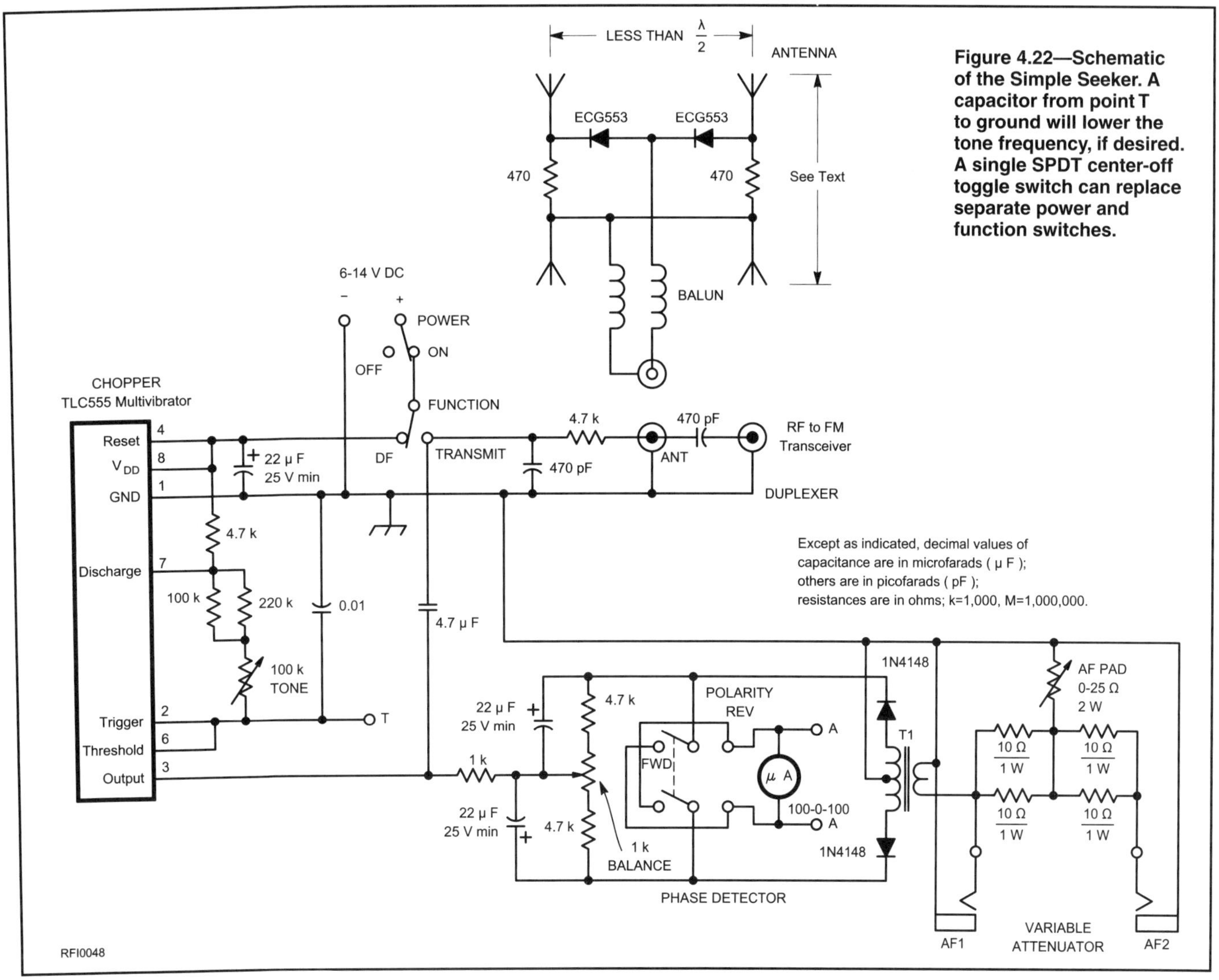

Figure 4.22—Schematic of the Simple Seeker. A capacitor from point T to ground will lower the tone frequency, if desired. A single SPDT center-off toggle switch can replace separate power and function switches.

Almost any audio transformer with approximately 10:1 voltage step-up to a center-tapped secondary meets the requirements of this phase detector. The output is a positive or negative indication, applied to meter M1 to indicate left or right.

Antenna Choices

Dipole antennas are best for long-distance RDF. They ensure maximum signal pickup and provide the best load for transmitting. Figure 4.23 shows plans for a pair of dipoles mounted on an H frame of ½-inch PVC tubing. Connect the 39-inch elements to the switcher with coaxial cables of *exactly* equal length. Spacing between dipoles is about 20 inches for 2 meters, but is not critical. To prevent external currents flowing on the coax shield from disrupting RDF operation, wrap three turns (about 2-inch diameter) of the incoming coax to form a choke balun.

For receive-only work, dipoles are effective over much more than their useful transmit bandwidth. A pair of appropriately spaced 144-MHz dipoles works from 130 to 165 MHz. You will get greater tone amplitude with greater dipole spacing, making it easier to detect the null in the presence of modulation on the signal. But do not make the spacing greater than one-half free-space wavelength on any frequency to be used.

Best bearing accuracy demands that signals reach the receiver only from the switched antenna system. They should not arrive on the receiver wiring directly (though an unshielded case) or enter on wiring other than the antenna coax. The phase detecting system is less amplitude sensitive than systems such as quads and Yagis, but if you use small-aperture antennas such as "rubber duckies," a small signal leak my have a big effect. A wrap of aluminum foil around the receiver case helps block unwanted signal pickup, but tighter shielding may be needed.

Figure 4.24 shows a "sniffer" version of the unit with helix antennas. The added RDF circuits fit in a shielded box, with the switching pulses fed through a low-pass filter (the series 4.7-kΩ resistor and shunt 470-pF capacitor) to the receiver. The electronic switch is on a 20-pin DIP pad, with the phase detector on another pad (see Figure 4.25).

Because the phase detector may behave differently on weak and strong signals, the Simple Seeker incorporates an audio attenuator to allow either a full-strength audio or a lesser, adjustable received signal to feed the phase detector. You can plug headphones into jack AF2 and connect receiver audio to jack AF1 for no attenuation into the phase detector, or reverse the external connections, using the pad to control level to both the phones and the phase detector.

Convention is that the meter or other indicator deflects left when the signal is to the left. Others prefer that a left meter indication indicates that the antenna is rotated too far to the left. Whichever your choice, you can select it with the DPDT polarity switch. Polarity of audio output varies between receivers, so test the unit and receiver on a known signal source and mark the proper switch position on the unit before going into the field.

PIN diodes, when forward biased, exhibit

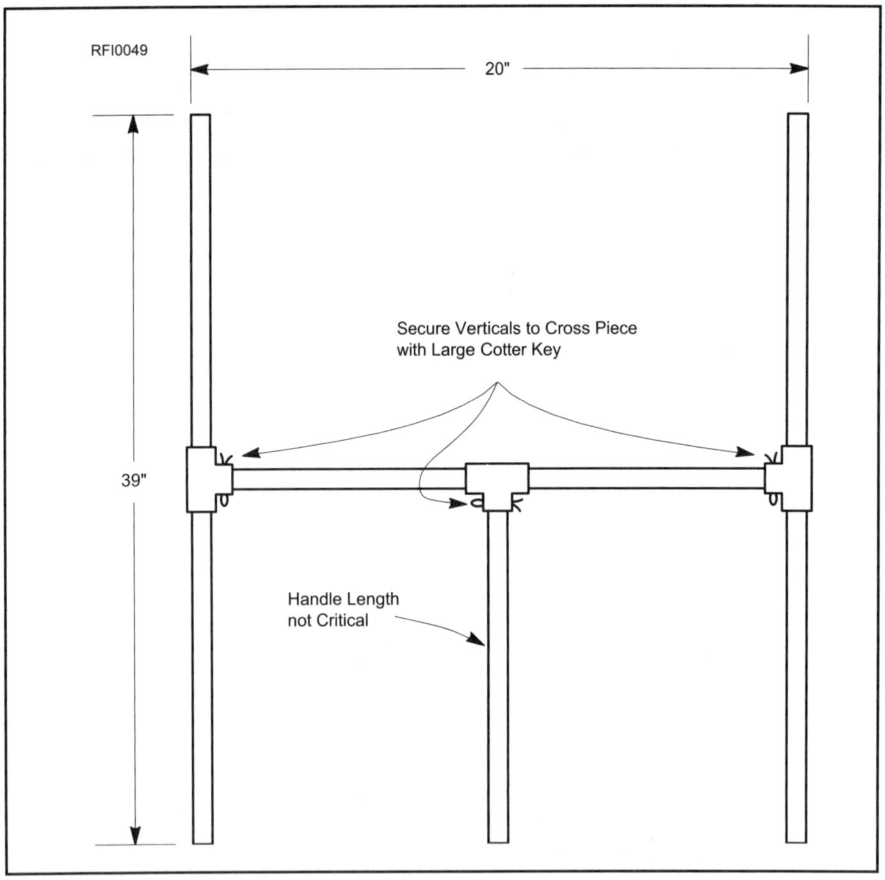

Figure 4.23—"H" frame for the dual dipole Simple Seeker antenna set, made from ½-inch PVC tubing and Ts. Glue the vertical dipole supports to the Ts. Connect vertical Ts and handle to the cross piece by drilling both parts and inserting large cotter pins. Tape the dipole elements to the tubes.

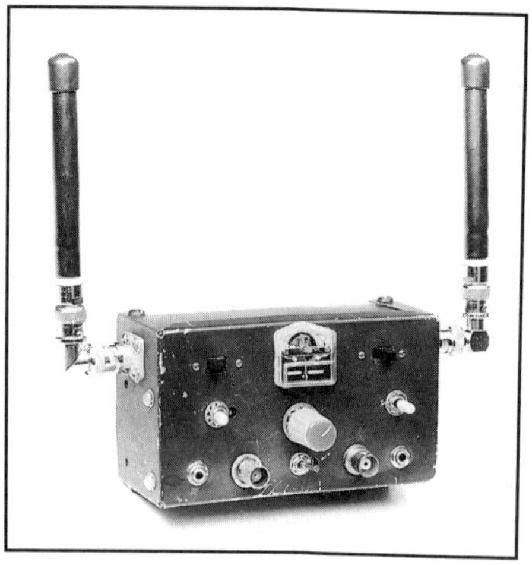

Figure 4.24—Field version of the Simple Seeker with helix antennas.

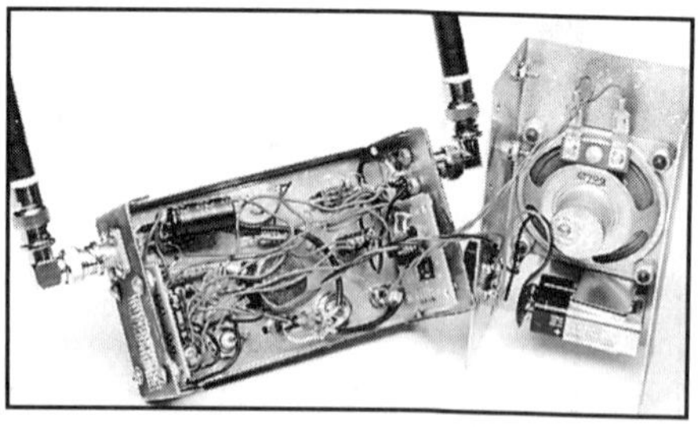

Figure 4.25—Interior view of the Simple Seeker. The multivibrator and phase detector circuits are mounted at the box ends. This version has a convenient built-in speaker.

low RF resistance and can pass up to approximately 1 W of VHF power without damage. The transmit position on the function switch applies steady dc bias to one of the PIN diodes, allowing communications from a handheld RDF transceiver.

AN ACTIVE ATTENUATOR FOR VHF-FM

During a VHF transmitter hunt, the strength of the received signal can vary from roughly a microvolt at the starting point to nearly a volt when you are within an inch of the transmitter, a 120-dB range. If you use a beam or other directional array, your receiver must provide accurate signal-strength readings throughout the hunt. Zero to full scale range of S meters on most handheld transceivers is only 20 to 30 dB, which is fine for normal operating, but totally inadequate for transmitter hunting. Inserting a passive attenuator between the antenna and the receiver reduces the receiver input signal. However, the usefulness of an external attenuator is limited by how well the receiver can be shielded.

Anjo Eenhoorn, PAØZR, has designed a simple add-on unit that achieves continuously variable attenuation by mixing the received signal with a signal from a 500-kHz oscillator. See "An Active Attenuator for Transmitter Hunting," Nov 1992, *QST*, p 28. This process creates mixing products above and below the input frequency. The spacing of the closest products from the input frequency is equal to the local oscillator (LO) frequency. For example, if the input signal is at 146.52 MHz, the closest mixing products will appear at 147.02 and 146.02 MHz.

The strength of the mixing products varies with increasing or decreasing LO signal level. By DFing on the mixing product frequencies, you can obtain accurate headings even in the presence of a very strong received signal. As a result, any handheld transceiver, regardless of how poor its shielding may be, is usable for transmitter hunting, up to the point where complete blocking of the receiver front end occurs. At the mixing product frequencies, the attenuator's range is greater than 100 dB.

Varying the level of the oscillator signal provides the extra advantage of controlling the strength of the input signal as it passes through the mixer. So as you close in on the target, you have the choice of monitoring and controlling the level of the input signal or the product signals, whichever provides the best results.

The LO circuit (Figure 4.26) uses the easy-to-find 2N2222A transistor. Trimmer capacitor C1 adjusts the oscillator's frequency. Frequency stability is only a minor concern; a few kilohertz of drift is tolerable. Q1's output feeds an emitter-follower buffer using a 2N3904 transistor, Q2. A linear-taper potentiometer (R6) controls the oscillator signal level present at the cathode of the mixing diode, D1. The diode and coupling capacitor C7 are in series with the signal path from antenna input to attenuator output.

This frequency converter design is unorthodox; it does not use the conventional configuration of a doubly balanced mixer, matching pads, filters and so on. Such sophistication is unnecessary here. This approach gives an easy to build circuit that consumes very little power. PAØZR uses a tiny 1.4-V hearing-aid battery with a home-

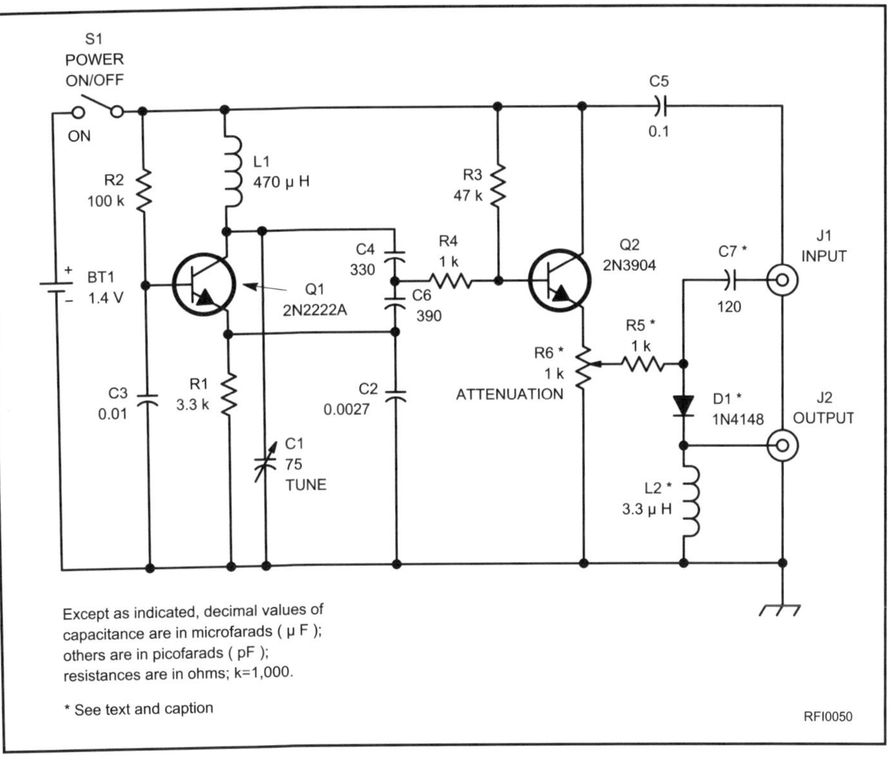

Figure 4.26—Schematic of the active attenuator. Resistors are ¼-W, 5%-tolerance carbon composition or film.

BT1—Alkaline hearing-aid battery, Duracell SP675 or equivalent.
C1—75-pF miniature foil trimmer.
J1, J2—BNC female connectors.
L1—470-μH RF choke.
L2—3.3-μH RF choke.
R6—1-kΩ, 1-W linear taper (slide or rotary).
S1—SPST toggle.

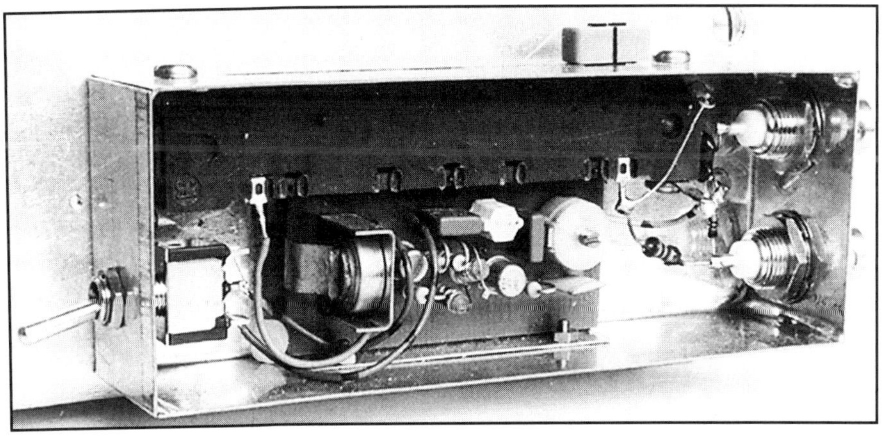

Figure 4.27—Interior view of the active attenuator. Note that C7, D1 and L2 are mounted between the BNC connectors. R5 (not visible in this photograph) is connected to the wiper of slide pot R6.

made battery clip. If your enclosure permits, you can substitute a standard AAA-size battery and holder.

Construction and Tuning

To order a template for this project, including the PC board layout and parts overlay, contact the Technical Department at ARRL HQ. A circuit board is available from FAR Circuits. The prototype (Figure 4.27) uses a plated enclosure with female BNC connectors for RF input and output. C7, D1, L2 and R5 are installed with point-to-point wiring between the BNC connectors and the potentiometer. S1 mounts on the rear wall of the enclosure.

Most hams will find the 500-kHz frequency offset convenient, but the oscillator can be tuned to other frequencies. If VHF/UHF activity is high in your area, choose an oscillator frequency that creates mixing products in clear portions of the band. The attenuator was designed for 144-MHz RDF, but will work elsewhere in the VHF/UHF range.

You can tune the oscillator with a frequency counter or with a strong signal of known frequency. It helps to enlist the aid of a friend with a handheld transceiver a short distance away for initial tests. Connect a short piece of wire to J1, and cable your handheld transceiver to J2. Select a simplex receive frequency and have your assistant key the test transmitter at its lowest power setting. (Better yet, attach the transmitter to a dummy antenna.)

With attenuator power on, adjust R6 for mid-scale S-meter reading. Now retune the handheld to receive one of the mixing products. Carefully tune C1 and R6 until you hear the mixing product. Watch the S meter and tune C1 for maximum reading.

If your receiver features memory channels, enter the hidden transmitter frequency along with both mixing product frequencies before the hunt starts. This allows you to jump from one to the other at the press of a button.

When the hunt begins, listen to the fox's frequency with the attenuator switched on. Adjust R6 until you get a peak reading. If the signal is too weak, connect your quad or other RDF antenna directly to your transceiver and hunt without the attenuator until the signal becomes stronger.

As you get closer to the fox, the attenuator will not be able to reduce the on-frequency signal enough to get good bearings. At this point, switch to one of the mixing product frequencies, set R6 for on-scale reading and continue. As you make your final approach, stop frequently to adjust R6 and take new bearings. At very close range, remove the RDF antenna altogether and replace it with a short piece of wire. It's a good idea to make up a short length of wire attached to a BNC fitting in advance, so you do not damage J1 by sticking random pieces of wire into the center contact.

While it is most convenient to use this system with receivers having S-meters, the meter is not indispensable. The active attenuator will reduce signal level to a point where receiver noise becomes audible. You can then obtain accurate fixes with null-seeking antennas or the "body fade" technique by simply listening for maximum noise at the null.

Notes

[1]W. Nelson, *Interference Handbook* (Wilton, CT: Radio Publications, 1981), p 105.
[2]Ibid, p 106.
[3]J. Moell and T. Curlee, *Transmitter Hunting* (Blue Ridge Summit, PA: TAB Books, 1987).
[4]Happy Flyers, *Radio Direction Finding.* Copies are available from Hart Postlewaite, WB6CQW, 1811 Hillman Ave, Belmont, CA 94002; E-mail **HartPost@aol.com**.

Chapter 5

Transmitters

According to the FCC, most interference from amateur transmitters is not the fault of the amateur equipment. Transmitters can and do sometimes malfunction, however. If you suspect a problem with the rig, you'll want to read this chapter.

By Alan Bloom, N1AL,
San Francisco Section Technical Coordinator,
1528 Los Alamos Rd,
Santa Rosa, CA 95409-3308

Modern amateur transmitters are almost always operating in compliance with FCC regulations. Most cases of interference are caused by fundamental overload. This has been confirmed by the experiences of many amateurs over the years. Nonetheless, it is important that amateurs know that their transmitter is working properly and what to do about it when it is not.

SPURIOUS EMISSIONS

All transmitters emit *some* harmonics or other spurious frequencies. The transmitter may be operating well within specifications, but because it's in a TV fringe area, or because the ham antenna is close to the TV antenna, the specified performance is just not good enough.

The spurious emission levels listed in the FCC amateur regulations [§97.307 (d) and (e)] are not stringent. Below 30 MHz, a properly designed vacuum-tube amplifier can meet these requirements with no more low-pass filter than the π network used for output impedance matching. For broadband solid-state amplifiers, a simple two-section low-pass filter is sufficient.

Below 30 MHz, current FCC rules specify a minimum spurious output attenuation of an amateur transmitter of 43 dB below the mean power of the fundamental emission – regardless of output power. The maximum spruious radiation at the full legal limit of 1,500 watts is, therefore, 75 milliwatts. Now, this may seem like real QRP to the avarage ham, but that much power close to TV-channel carrier frequencies can wipe out fringe-area TV sets for blocks. See Figure 5.1 and caption for FCC spurious output requirements above 30 MHz.

Even worse, harmonic attenuation is specified (and tested) into a broadband 50-Ω dummy load. Actual amateur antennas are *not* 50 Ω and resistive at TV frequencies. This can cause filter networks to provide less harmonic attenuation than specified.

The point is, a perfectly legal transmitter is quite capable of causing television interference (TVI). You can calculate the interference level if you like. See the sidebar "Taking the Mystery Out of TVI" for the details.

The calculations show that if you run high power with nearby TV antennas in an area of deep-fringe TV reception, the interference may be difficult or impossible to eliminate. A partial solution is to operate on frequencies where harmonics do not fall within any local TV channels. Several examples of this can be seen in the Television chapter. As a harmonic approaches the video carrier, picture interference worsens.

ISOLATING THE PROBLEM

It Ain't Me

In cases of interference to non-radio devices, such as telephones, computers, alarm systems, audio amplifiers, etc. troubleshooting the problem is easy: This type of interference is *not* caused by transmitter spurious emissions or other transmitter rules violations. Other than reducing power or relocating the transmitting antenna, there is nothing that can be done at the transmitter to eliminate this type of interference. The interference is caused when the affected equipment is overloaded by your strong, *fundamental* signal. Refer to the appropriate chapters for cures for the affected consumer equipment.

Transmitter Chassis Radiation

Let's assume that you have read the Troubleshooting chapter and have proved that the interference is caused by transmitter harmonics or other spurious frequencies. A simple test can determine whether the unwanted signals are radiated directly from the transmitter chassis and/or cables.

First, disconnect the coax from the back of the transmitter or amplifier, and connect the rig to a shielded dummy load. Use a short piece of good quality (95% braid coverage) coaxial cable with properly installed connectors. (This means the coax shield is soldered to each connector around the entire circumference of the connector shell.)

Transmit with full power into the dummy load. Use all bands, and a selection of frequencies on each band. "Play" with the

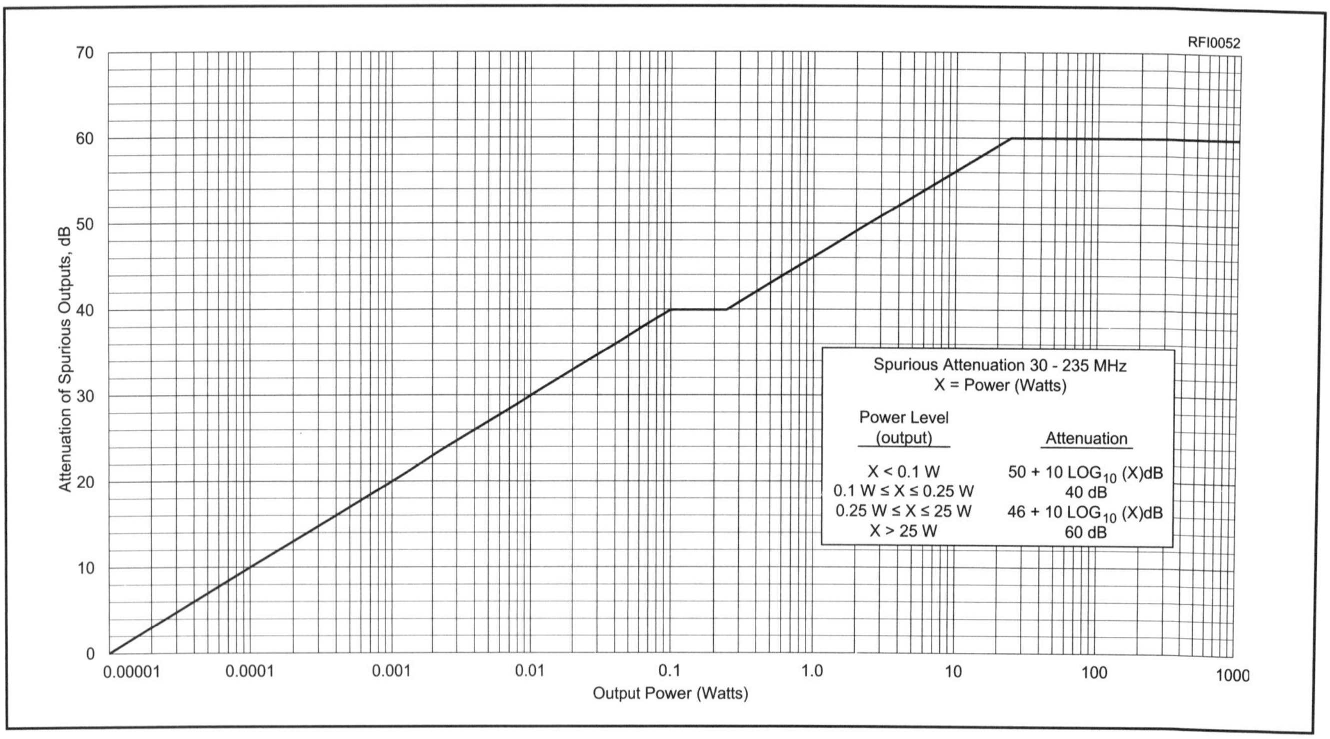

Figure 5.1—FCC specifications for minimum spurious output attenuation of amateur transmitters operating between 30 and 235 MHz. There are no absolute limits for transmitters operating above 235 MHz, although general requirements for good engineering practice still apply. Current Amateur transmitters operating below 30 MHz must have a spurious output attenuation of at least 43 dB below mean power of the fundamental emission—regardless of output power. This includes HF QRP transmitters.

Taking the Mystery Out of TVI

Why do some hams have no trouble with television interference when other amateur stations, seemingly just as "clean," turn the neighbors' TV screens into multi-color light shows? The mystery can be solved with a single equation:

$$H_R = (T_P + T_A + G_T + G_R + S_A) - S_S + S_R$$

H_R is the additional harmonic attenuation (in dB) required for the transmitter. T_P is peak transmitter power in dBm, T_A is harmonic attenuation (dB) of the "bare" transmitter, G_T is transmit antenna gain (dB) at the TV frequency, G_R is TV antenna gain (dB), S_A is path loss between the transmitter and TV antennas (dB), S_S is TV signal level (dBm), and S_R is the signal-to-interference ratio required for an acceptable picture (dB). Here's how to estimate these numbers:

Transmitter power, T_P, is measured in dBm, or decibels referenced to 1 mW. One watt is +30 dBm, 100 W is +50 dBm, and 1500 W is +62 dBm.

Transmitter harmonic attenuation, T_A, is typically somewhat better than required by FCC regulations, especially for the higher-order harmonics. The specifications in the transmitter manual should give a conservative estimate. Older transmitters, built before FCC harmonic regulations, may have only –30 to –35 dB suppression of the second and third harmonics.

The transmitting antenna may have up to –10 dB or so of gain, G_T, at TV frequencies, depending on the type of antenna and its orientation. A conservative estimate is 0 dB.

A good (fringe-area) TV antenna may give up to +10 dB of gain, G_R, in its main lobe. Figure about –10 dB at right angles to the main lobe and 0 dB off the rear.

Path loss can be estimated from the procedure described below, or calculated using the following equation:

$$S_A = 10\log_{10}\left(\frac{\lambda^2}{(4\pi D)^2}\right)$$

D is the distance between antennas in the same units as the wavelength, λ. To estimate S_A, note that if D is 10 meters (33 feet), then S_A varies from –27 dB on TV channel 2 to –31 dB on channel 6. Add –6 dB each time the distance doubles.

The received TV signal level, S_S, depends on the TV-station effective radiated power, distance to the TV antenna, terrain, and other factors. Proper equipment can measure the signal level right at the TV end of the coax. Otherwise, figure about –60 dBm for fringe-area reception (more than 50 miles) and –35 dBm for local stations.

The signal-to-interference ratio, S_R, required for a decent TV picture varies. It's about 40 dB for interference close to the video carrier. For signals more than 2 MHz from the video carrier and not too close (a few hundred kHz) to the sound and color subcarriers, it's about 20 dB. 35 dB is a reasonable value.

Sample Calculation

Let's perform the calculation for an amateur station operating on 15 meters with 1.5 kW of output power. The third harmonic (in TV channel 3) is down 45 dB, just within FCC specifications. The TV antenna is 40 meters (130 ft) away, pointing at the ham antenna. The TV station is more than 50 miles away. The numbers are:

T_P = +62 dB,
T_A = –45 dB,
G_T = 0 dB,
G_R = +10 dB,
S_A = –27 –12 = –40 dB,
S_S = –60 dBm,
S_R = +35 dB.

$$H_R = (+62) + (-45) + (0) + (+10) + (-40) - (-60) + (+35) = 82 \text{ dB}$$

It would probably take two low-pass filters in series to get this kind of attenuation. Other factors, such as power- and control-wire radiation and shielding, not to mention fundamental overload of the TV set, would also be tough "nuts" to crack. Fortunately, few of us are faced with such a worst-case scenario. This information was derived from "Harmonic TVI: A New Look at an Old Problem," by D. Rasmussen, W6MCG, and D. Gerue, K6YX (Sep 1975 *QST*, pp 11-14).—*Alan Bloom, N1AL, San Francisco Section Technical Coordinator, Santa Rosa, California*

transmitter tuning controls and attempt to generate TVI. If there is *any* interference on the TV set, there is radiation either from the transmitter chassis (poor shielding) or from the power cord or other leads. If the transmitter is "clean" into the dummy load, but not into the antenna, and if the problem is not in the TV set or caused by external rectification, then the transmitter must be emitting signals outside the amateur bands via the coax feed line or the equipment is experiencing fundamental overload.

If there is no interference with the dummy load connected directly to the transmitter output, try moving the load to the antenna end of the coax. If this point is not accessible, move it to the point where the coax leaves the shack, after any low-pass filters, wattmeters, antenna tuners, and so on. If the interference returns, examine the connecting coax cables for good solid connections between the cable shields and connector bodies. This is especially important for the cable between the transmitter and low-pass filter. It should be as short as possible and have the shield solidly soldered around the full circumference of the connector shell. Even better: use a double-male coaxial adapter and mount the filter directly to the transmitter output connector. Coax-shield problems may only show up with the antenna feed line connected, because the feed line can act as an antenna for the interference.

Another good test is described in the Troubleshooting chapter. If you operate a transmitter into a rotary beam antenna, if interference is caused by transmitter harmonics, the interference will show several lobes as the antenna is rotated. Refer to the Troubleshooting chapter for details.

Poor Shielding

If there is interference with a dummy load connected, look for poor shielding in the transmitter. Are any shields missing? Are there any missing or loose screws? Corrosion or paint at the seams can let a surprising amount of VHF energy escape. Make sure all joining surfaces are clean and bright. A salty or acidic atmosphere can corrode plated metal surfaces. So can cigarette smoke. If the radio has been used much by a heavy smoker, a major cleaning of all contacting surfaces may be necessary to restore good connections.

The ac-line cord is another common source of incidental radiation. Wrap several turns of the cord around a ferrite rod or through a ferrite toroid core. If there is *any* change in interference, it is a sure sign that there is RF on the line cord. Other connecting leads, such as the microphone or key cords, merit the same attention. A good device for the purpose is a toroid with a permeability greater than 800 such as an FT-82-43.

Spurious Signals Out the Coax

If the interference only occurs with the feed line actually connected to the antenna, the transmitter output signal may contain interference that is reaching the antenna. (Interference to TVs or other radio services can be caused by spurious emissions, but it is more often caused by fundamental overload, too.)

First, try a low-pass filter. Locate the filter directly at the transmitter output, before any SWR meter, Transmatch or other accessories. If there is already one filter, add a second in series. One may not be enough in areas of weak TV reception.

Check used transmitters for modifications or improper alignment by a previous owner. The proper alignment procedures appear in the manufacturer's service manual (often an extra-cost item, separate from the owner's manual).

Vacuum-tube transmitters are especially susceptible to misadjustment because the operator adjusts the tuning controls from the front panel. Be sure that the final amplifier is not overdriven, especially with older rigs that do not have automatic level control (ALC). Reduce the drive to the lowest level that yields full power output. Also be sure that the final amplifier bias is set correctly (check for the proper *resting current* with the transmitter on, but with no RF output), as specified in the owner's manual.

Neutralization

Some tube amplifiers include a special circuit that prevents oscillation by *neutralizing* the tube's internal (feedback) capacitance. If the neutralization circuit is not adjusted correctly, there may be evidence of oscillation: erratic jumps in plate or grid current while tuning. In a properly operating amplifier, maximum power output, minimum plate current, and maximum control and screen-grid currents should all occur at about the same control settings. If not, the transmitter needs neutralization. This procedure should be described in the transmitter service manual, or use the procedure described in the *ARRL Handbook* (look in the index under "Neutralization").

Don't overlook the transmitter manufacturer as a source of help for spurious-emission problems. If there is a stubborn problem, the service department may be willing to offer suggestions.

TRANSMITTER MALFUNCTIONS

This section covers transmitters that are operating outside their design limits: There is something wrong with the rig. The simple fixes mentioned above didn't work, and now it's time to dive into the circuitry to find out what is wrong. The two most common problems are excessive harmonics and spurious oscillations.

Final amplifier overdrive is a common cause of harmonics. Most modern transmitters have an ALC circuit to limit the peak power level. If a slight reduction in drive level causes a dramatic reduction in interference, the ALC circuit may be faulty. There is usually an internal adjustment to set the ALC level, and readjustment of this control may be all that is needed. A previous owner may have "optimized" this setting to increase power output.

Both solid-state and tube transmitters can have defective tank-circuit or low-pass filter components that cause excessive harmonic generation. This problem typically causes low power output as well. Check for signs of overheating, mechanical damage, or poor solder connections.

VHF PARASITIC OSCILLATIONS IN HF VACUUM-TUBE AMPLIFIERS

Although it doesn't show on the schematic diagram, most capacitors have some series inductance from self inductance of the connecting leads. Similarly, every inductor has some parallel capacitance from distributed capacitance between the turns of wire in the coil. These stray (or parasitic) reactances become important at high frequencies. Figure 5.2 illustrates a typical kilowatt amplifier tuned for the 40-meter band. Parasitic inductances L1 and L2 form a high-Q resonance with C1 and C2 around 200 MHz, which causes the tube to have high voltage gain at this frequency. If enough feedback exists between output and input, the amplifier may oscillate.

How can you spot parasitic oscillations? They often reveal themselves through erratic plate-current fluctuations when tuning the amplifier output or input tuned circuits. Also try reducing the drive level to zero. If there is *any* change in grid or plate current when tuning with no drive, it is a sure sign of spurious oscillations.

It is helpful to determine the approximate frequency of the oscillation. The best way is to feed the transmitter output (through a high-power attenuator) into a spectrum analyzer. The traditional alternative (for those of us who don't have several thousand dollars to invest in test equipment) is a tuned wavemeter or a grid-dip oscillator in its wavemeter mode. Use the wavemeter coil to "sniff" near the amplifier tank coil while transmitting, and tune the wavemeter dial for maximum meter indication. Be *very* careful if you try this—there is high voltage nearby! A safer alternative is to listen for the oscillation on a nearby receiver. A general-coverage shortwave receiver can detect oscillations in the HF region, and a TV set is a good detector of VHF parasitics.

Once the approximate frequency of the parasitic oscillation is determined, try to pin down which part of the tank circuit exhibits the spurious resonance. With the amplifier tuned for maximum interference, turn off the power, remove the covers and short the high-voltage capacitors to ground with an insulated screwdriver. Place the dip meter in its oscillator mode, and slowly tune it around the parasitic frequency while holding the dip-

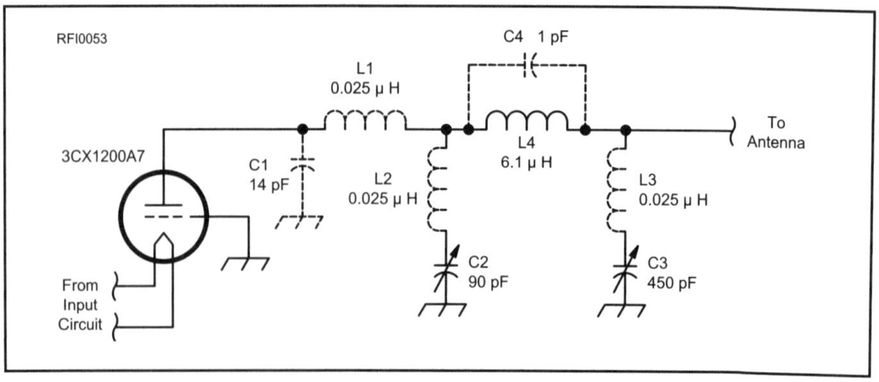

Figure 5.2—Schematic of a typical pi-output matching network in an HF vacuum-tube amplifier. C2 and C3 are the tuning and loading capacitors; L4 is the tank coil. C1, L1, L2, L3 and C4 are not actual components, but represent small stray reactances that are important at VHF frequencies. (The plate choke and coupling capacitor have been omitted for clarity.)

"Don't overload the transmitter"

meter coil close to various circuit components. A sharp dip in meter current indicates the circuit with unwanted resonance.

VHF parasitic oscillations can often be reduced or eliminated by moving the spurious resonances higher in frequency. This is done by reducing the stray inductance of all final-amplifier RF connections: Make leads as short and fat as possible. Pay special attention to screen bypass capacitors in tetrode and pentode tube circuits. Sockets with built-in low-inductance screen bypass capacitors are available for some high-power tubes. In grounded-grid amplifiers, connect the grid pins directly to the chassis with a short, wide copper strap. If the grid is not grounded directly, use the shortest possible leads on the grid bypass capacitor(s). Also, several capacitors in parallel have less inductance than one.

The plate-tuning capacitor should be bolted directly to the grounded chassis. The connection from tube plate through the coupling capacitor to the tuning capacitor should be as short and fat as possible. It is especially important to use low-inductance connections between tubes connected in parallel—otherwise they can act as a push-pull VHF oscillator.

Don't overlook sources of stray feedback from amplifier output to input. Make sure all power-supply leads are well bypassed and filtered, including the filament supply. The filament choke and input matching network should be well shielded from the output circuitry.

PARASITIC SUPPRESSORS

Eliminate a VHF parasitic resonance by installing a resistor in series with either the inductance or capacitance of the undesired VHF tuned circuit.[1] Unfortunately, this tends to "kill" the efficiency of the desired HF tuned circuit as well. The answer is to connect a small coil in parallel with the resistor, forming a circuit known as a *parasitic suppressor*. Since inductive reactance is proportional to frequency, the coil shorts out the resistor at HF frequencies, while presenting a high reactance at VHF.

If the VHF parasitic is too low in frequency, it may be difficult to design a parasitic suppressor capable of suppressing the oscillation without also degrading efficiency at the highest desired operating frequency. This is one reason to first move all parasitic resonances as high in frequency as possible.

Traditionally, parasitic suppressors are used in series with the plate, but they may also be useful in the control- and screen-grid circuits of grounded-cathode amplifiers. The resistance and inductance values are found by experimentation. If the inductance is too high, the resistor gets hot and efficiency at the high end of the HF range suffers. If the inductance is too small, the oscillation may not be suppressed (which may also cause resistor overheating!). A typical suppressor for the plate circuit of a 150-W transmitter consists of a half-dozen turns of wire in parallel with a 50 to 100-Ω, 1-W carbon or metal- film resistor. For a kilowatt amplifier, a good starting point would be three or four 220-Ω, 2-W resistors in parallel with 1 or 2 turns of no. 12 or 14 wire.

Parasitic suppressor resistors often overheat, not only from the RF current through them, but also because of the nearby hot final-amplifier tube. Replace any resistors that have become discolored.

PARASITIC OSCILLATIONS IN SOLID-STATE AMPLIFIERS

VHF parasitic oscillations are not normally a problem in HF solid-state amplifiers. Since the high-frequency current gain of a bipolar transistor drops 6 dB every time the frequency doubles, there is usually not enough gain at VHF to sustain an oscillation.

Low-frequency oscillations are another matter. Since transistors have so much gain at low frequencies, they oscillate readily if unwanted feedback is present. Detect low-frequency spurs by listening to the transmitted signal with a nearby receiver. Slowly tune back and forth several hundred kilohertz either side of the carrier frequency, while the transmitter is transmitting a series of CW dits. Try several different power levels. A modulation monitor or oscilloscope capable of displaying the RF signal will show any "fuzz" on the CW carrier.

There is a common low-frequency feedback path from the transistor collector, through the dc power supply and into the base bias network. Power supply decoupling must be effective not only at the RF frequencies the amplifier is designed for, but at low frequencies as well. A 0.05-µF ceramic capacitor in parallel with a 220-µF electrolytic should serve to bypass the +12 V power supply to the final amplifier of a 100-200 W HF transmitter. The base bias network should be either supplied from an IC voltage regulator or decoupled with an RC low-pass filter.

Remember that every amplifier stage in the transmitter RF chain needs power supply decoupling. Failure of any decoupling components can cause low-frequency oscillations. Solder connections can degrade over time, especially in high-temperature environments like around high-power RF amplifiers. Resolder any suspicious solder joints.

Many amplifiers include negative-feedback networks to stabilize gain. The network often consists of a capacitor and a resistor (and sometimes an inductor) series-connected from the collector to the base. Often a resistor is connected from the base to ground in order to lower the low-frequency input impedance. If low-frequency oscillations develop in a solid-state amplifier, check to see that all parts still function and that the solder joints are good.

SHIELDS—PLUGGING THE LEAKY SIEVE

A low-pass filter can only reject harmonics that are exiting through the coaxial cable. Filters won't help if interference radiates from the chassis. Just as water passes through a sieve, RF passes through a leaky chassis. The object is to plug the holes.

How important is grounding to shield

effectiveness? If "ground" means "earth ground" the answer is not at all. A well shielded transmitter causes no RF currents to flow on the outside of the shield, so the presence or absence of an earth ground can have no effect. If poor shielding does allow RF currents to flow on the outside of the chassis, a ground wire will not often cure the condition because at VHF, most ground wires are also ground-wire "antennas." In fact, they may radiate the unwanted energy more efficiently than the transmitter chassis, so grounding may be counterproductive.

The FCC amateur rules do not directly regulate incidental radiation. Paragraph 97.307(c) states in part, however: "If any spurious emission, *including chassis or power-line radiation*, causes harmful interference to the reception of another radio station, the licensee of the interfering amateur station is required to take steps to eliminate the interference, in accordance with good engineering practice." (Emphasis added.)

Slots and Holes

Actually, the above discussion is not completely correct. While it is true that steel is a poorer RF shield than aluminum, it doesn't matter in most practical cases. Most RF leakage is not through the shielding material itself, but through openings in the shield.

The trouble caused by a particular hole is determined more by its maximum dimension than by its area. A round hole passes less radiation than a long, thin slot of the same area. Pay special attention to seams around shield covers. It is amazing, but true: A long gap that is too tight to pass a piece of paper can radiate almost as though the shield were not present. As a "rule of thumb" space mounting screws no more than 1/20 λ (at the highest frequency of concern). In order to shield harmonics that fall within the North American VHF TV channels (up to 216 MHz), space shield screws no more than about 7 cm (2.75 inches) apart.

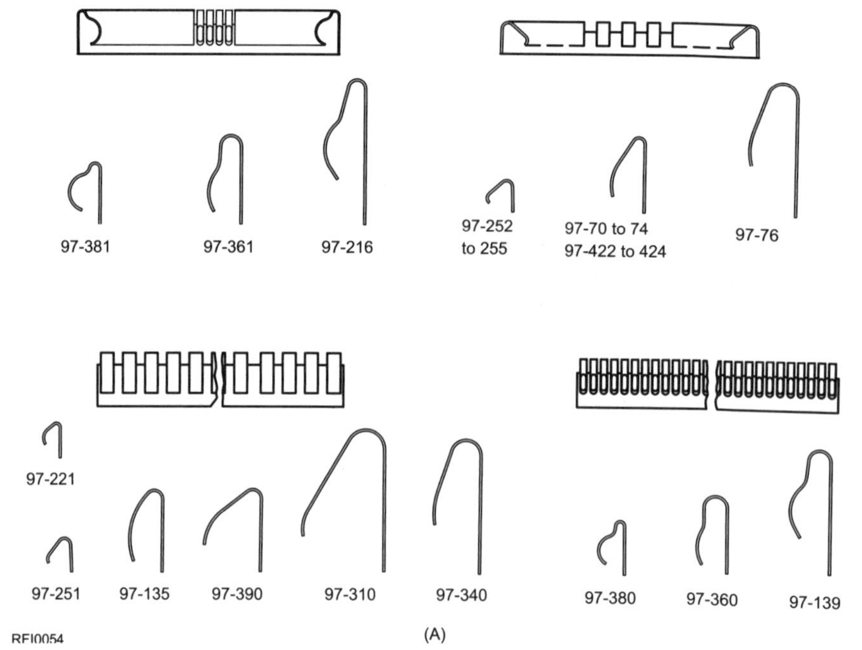

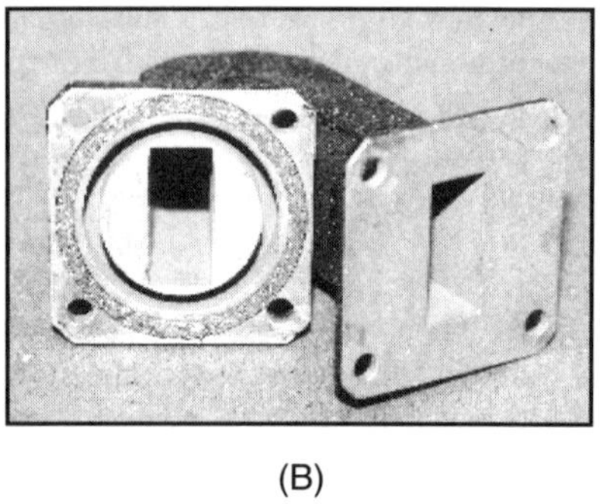

(B)

Figure 5.3—Several types of "finger stock" (A) are useful for shielding seams that must be opened and closed frequently. (See Instrument Specialties Co and Richardson Electronics in the Suppliers List.) Wire-mesh RF gaskets (B) can give an adequate RF-tight seal up into the UHF and microwave range.

Improving Shields

Treat shield seams with "EMI tape." 3M makes copper adhesive tape in several widths. Simply apply the tape across shield seams to effectively seal them. Holes in shield enclosures may be covered with conductive screen to prevent radiation. Copper screen can be easily attached to steel enclosures by soldering. Attach conductive screen to aluminum enclosures with stainless-steel hardware.

Unshielded equipment may need an added shield. Plastic enclosures can be shielded by the addition of metal foil or conductive paint (see GC/Thorsen in the Suppliers List) either inside or outside the cabinet. While internal shields usually look better, they have several disadvantages. Conductive material may short circuits in the equipment and present a shock hazard. Internal shields should be installed only by qualified technicians with the manufacturer's approval.

Despite aesthetic disadvantages, most anyone can fabricate a suitable metal enclosure for equipment. Use copper or aluminum sheet. Make sure that the enclosure provides adequate ventilation via small holes, or large ones covered with screen. "Piano hinge" provides better shielding than small hinges where openings are needed. If the affected equipment is installed in a metal cabinet, such as an entertainment center, no fabrication is needed.

The orientation of interference-carrying wires inside the chassis is important. Obviously, it is best if they are far from any holes in the shield, but it's even more important to orient them parallel to the seams. Wires at right angles to slots cause maximum leakage

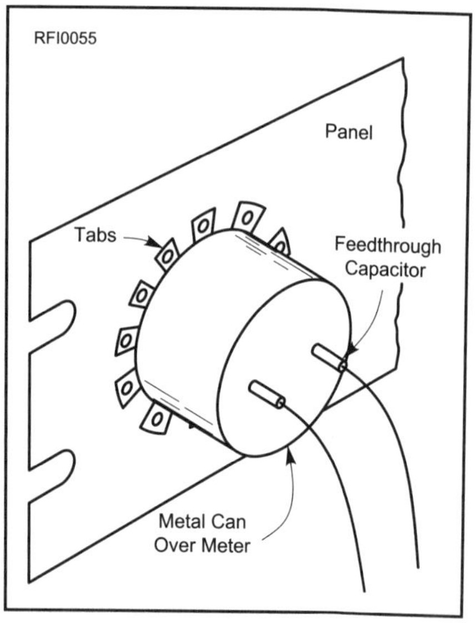

Figure 5.4—A shielding technique to prevent radiation from meter mounting holes.

(because the magnetic field is parallel to the slot, at 90° to the wire).

Paint is an insulator. Remove all paint from seam joints for the best connection. Where screw mounting is not practical (such as around an access door), use finger stock or one of the other commercial RF-gasket materials (Figure 5.3).

Meters are a source of shield leakage, because they usually require a large hole in the front panel. Shield round meter openings with part of a discarded tin can (an old trick, see Figure 5.4). Find a can slightly larger than the meter. Cut off a portion (including an unopened lid) somewhat deeper than the meter (leave enough extra depth to form tabs that can be bent and drilled for mounting screws). Bypass the meter leads to the can at the exit point. Fabricate shield enclosures for square or edgewise meters from PC-board material.

One final shielding tip: Two poor shields are usually better than one good one. A good shield might have 100 dB of isolation. Two 60 dB shields add up to 120 dB, a good 20 dB (100 times) better than the single "good" shield.

DECOUPLING AND BYPASSING

The Maginot line was France's defense against a German invasion before World War II. It was the strongest, most impregnable defensive structure ever built. The French high command was convinced that the Germans would never breach it. They didn't. They simply bypassed it by attacking around the end of the line, through Belgium and The Netherlands.

Similarly, the best shielding in the world won't work if there is a path through it. Any wire that pierces the shield wall acts as a receiving antenna inside the shield and a transmitting antenna outside. There are two ways to prevent this: by shielding the wire, or by filtering it at the point where it pierces the shield.

Shielded Wire

Shield the wire on the inside of the chassis, on the outside—or both. The most common method uses coaxial cable. Ground the coax shield right where it passes through the chassis. For best results, don't twist the braid into a pigtail and solder it to a ground lug. It is much better to *connect the braid around the entire circumference* of the hole or connector, so that the center conductor is completely surrounded.

If interference is generated by a current flowing in a wire, it is magnetic (low impedance) in nature, and the H field is stronger. In that case, ground the coax shield at both ends. If the interference results from capacitive (voltage) coupling, then the E field is stronger. It may be better to ground only one end of the coax shield. This avoids a *ground loop*, where magnetic fields can cause current to flow in the loop formed by the coax shield and chassis. If the cable length is greater than $\frac{1}{20}\ \lambda$, however, ground both ends.

If shielding a wire is impractical, use the "poor man's coax"—a "twisted pair." Twisting the hot wire together with a ground-return wire reduces both electric and magnetic pickup. Ground the return wire according to the rules for coax shields, above. Any interference on the pair will be a common-mode signal. Install common-mode chokes at one end of the wire (or both ends if needed).

Try to route the wire tight against the chassis for its entire length—in the angle between bottom and sides, if possible. The shielding effect is better when the wire is close to the chassis. This technique is even useful for coax, to prevent ground loops.

For the same reason, any PC board with potential radiation or susceptibility problems should have a ground plane on at least one side. To reduce radiated emissions, computer PC boards often use four or more layers: one or two ground plane(s) and the others for circuitry.

FILTERS

If interference does manage to find its way onto a wire, the wire must be filtered before it exits the shield enclosure. When the desired signal is lower in frequency than the undesired signal, a low-pass filter is called for. (A low-pass filter at a transmitter RF output connector is a special case that is covered later.)

A Low-Pass Filter

The simplest low-pass filter is a *bypass capacitor* connected between the wire and ground. Install it with shortest possible leads directly where the wire exits the shield. Use additional bypass capacitors inside the shield, as close to the interference source(s) as possible. In a computer or other digital device, include a bypass capacitor from the power-supply lead to ground near each IC. In a transmitter, bypass the power-supply lead near each amplifier stage. Place the capacitor close to the RF source, to minimize the wire length carrying (and radiating) unwanted RF energy.

If a simple bypass capacitor doesn't work, add one or more resistors or inductors in series with the signal lead. Resistors are cheaper than inductors, and they work over a broader frequency range. They are not suitable in all applications, however, because they consume power from the filtered circuit. Shunt capacitors should have low impedance, and the series resistors or inductors should have high impedance at the lowest frequency to be filtered. (An R:X_C ratio of 5:1 or 10:1 should be sufficient.) Use the standard reactance formulas:

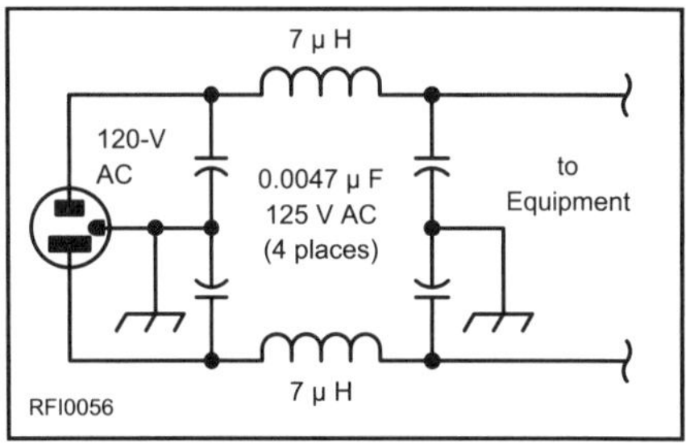

Figure 5.5—A "brute-force" ac-line filter. The inductors are 47 turns of no. 18 enameled wire on a 0.5-inch diameter form (7 μH). The 0.01-μF, UL-recognized capacitors are available from Digi-Key Corporation, part no. P4601.

$$X_L = 2\pi\ fL \qquad \text{(Eq 5.1)}$$

$$X_C = \frac{1}{2\pi fC} \qquad \text{(Eq 5.2)}$$

where

f = frequency (MHz)
L = inductance (μH)
C = capacitance (μF)
X_L= inductive reactance (Ω)
X_C= capacitive reactance (Ω).

Typical values for power-supply decoupling of RF signals in the 3.5 to 200-MHz range would be 1 μH and 0.05 μF.

Be careful with LC filters as they suffer from a resonance effect. At frequencies near

$$f = \frac{1}{2\pi\sqrt{LC}} \qquad \text{(Eq 5.3)}$$

they can *increase* the circuit response over that with no filter at all! One solution is to make L and C large enough that the resonant frequency falls well below the lowest frequency of interest. Another solution is to load the resonant circuit, by adding a resistor in series with the coil or capacitor. The resistance should equal twice the reactance of either L or C at resonance.

Try "Brute Force"

As an example of interference filtering, consider the "brute-force" ac-line filter in Figure 5.5. Safety is an important consideration with ac-line filters. Three major organizations that promote safety standards are UL (Underwriters Laboratories) in the USA, CSA (Canadian Standards Association) in Canada, and IEC (International Electrotechnical Commission) in Europe. All capacitors connected to the ac line should be certified by one of these organizations for the line voltage in use. Certified parts are labeled with an agency symbol or *mark*. UL uses a backwards "R" attached to an "L." The CSA mark is "SA" inscribed in a large "C." Each European country has its own standards agency. For example, the mark of Germany's Verband Deutscher Elektrotechniker is "VDE" inscribed in a triangle.

Some references recommend using 1000-V dc ceramic-disc capacitors. Unfortunately, these parts are intended for dc-bypass applications, and their performance with 60-Hz ac is unspecified. UL-recognized parts are inexpensive and readily available in mail-order parts catalogs, so there is no reason not to use them.

In an ac-line filter, capacitor size is limited by allowable 60-Hz leakage current. A 0.1-µF bypass capacitor connected from the 120-V ac line to a chassis would let up to 4.5 mA of 60-Hz current flow in anyone unlucky enough to touch both the chassis and ground at the same time. Any value less than 0.01 µF keeps leakage current within the UL limit of 0.5 mA. (Be cautious about installing ac-line filters in any home entertainment equipment, because the same 0.5-mA limit applies.)

The inductor value is limited by physical size (bigger is better). The coils described in Figure 5.5 are about 7 µH each, which results in a self-resonant frequency near 600 kHz. Install the filter inside the transmitter chassis if there is room, or in a small metal box bolted to the rear panel.

Figure 5.6—A common-mode choke to suppress shield current in coaxial cable. Reversing the windings as shown allows more turns with less shunt capacitance.

AC-line filters are also available commercially, both as external "add ons" and for inclusion inside equipment. The internal units often include a line cord receptacle and a fuse holder. Manufacturers include Corcom, Cornell-Dubilier (CDE), Schaffner and Sprague. (See the Suppliers List in the appendices of this book.)

Capacitors Are Not Capacitors

No electronic component is perfect. Capacitors also exhibit inductance, inductors exhibit capacitance, and both have resistance. These factors limit the frequency range over which a particular component can be used. The EMC Fundamentals chapter discusses this topic in detail.

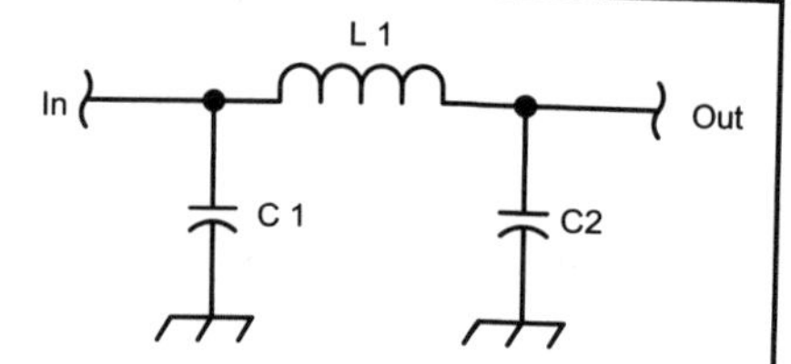

Design formulas are:

$$C1 = C2 = \frac{1}{(2\pi f (50))}$$

$$L1 = \frac{50}{(2\pi f)}$$

where,

C1 and C2 are in µF,

L1 is in µH,

f is the pass frequency in MHz.

RFI0057

Figure 5.7—The "quarter-wave" low-pass filter uses elements of 50-Ω reactance. Cascade as many sections as desired.

LOW-PASS TVI FILTERS

Before we get to the "nitty-gritty" of low-pass filter design, let's get one thing straight: A low-pass filter only acts upon signals coming through the feed line. If any VHF harmonics escape before reaching the filter, it won't help.

The connection between the transmitter

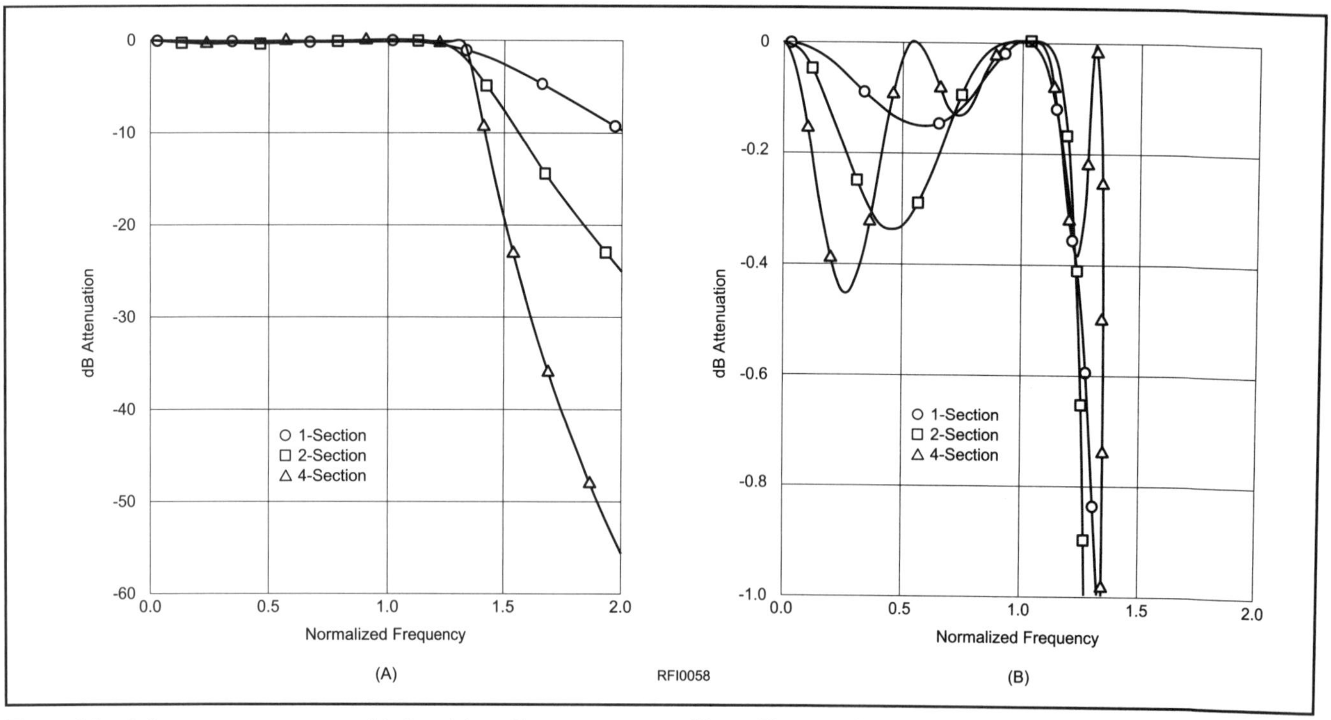

Figure 5.8—A, frequency response of 1, 2 and 4-section quarter-wave filters. The pass frequency has been normalized. B shows the same curves with the attenuation scale expanded to illustrate passband ripple.

and filter is critical! Use only good-quality coax with at least 95% braid coverage. Be warned that "RG-8" is no longer a valid military specification for coaxial cable. Coax bearing that label may not be of high quality. RG-213 is the same size and impedance as RG-8, but it has guaranteed specifications. Pay special attention to the connection between the shield and the connectors at each end of the cable. The braid should be soldered around the entire connector-shell circumference. PL-259 "UHF" connectors have several solder-access holes: make sure that the braid is well-soldered through each hole. See the sidebar "How to Install a PL-259 Connector."

The connection between a transmitter and low-pass filter should be as short as possible. This not only reduces the chances of stray radiation, but also ensures that the transmitter "sees" the proper load impedance at VHF frequencies (above the filter cut-off frequency). Even better: use a double-male connector between the filter and transmitter.

Antenna SWR should be low at the operating frequency. High SWR can cause overheating and arcing in a low-pass filter. If necessary, install a Transmatch between the filter and antenna.

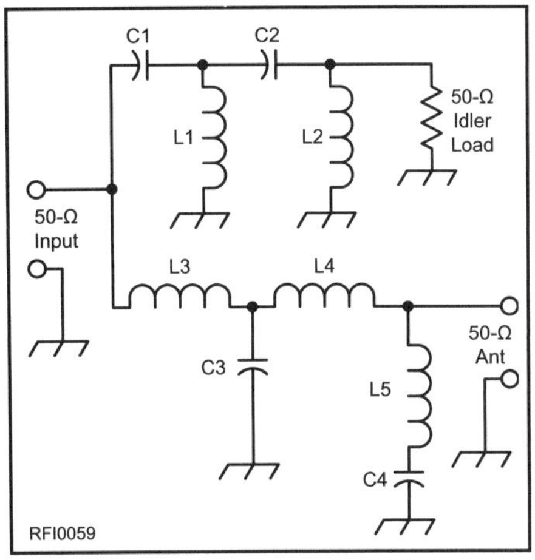

Figure 5.9—An absorptive low-pass TVI filter. The following values are for a cutoff frequency of 40 MHz and a rejection peak in TV channel 2:

C1—52 pF.
C2—73 pF.
C3—126 pF.
C4—15 pF.
L1—0.125 µH.
L2—0.52 µH.
L3—0.3 µH.
L4—0.212 µH.
L5—0.55 µH.

A Transmatch can act as a filter, especially for low-order harmonics. Attenuation of high-order harmonics in the VHF TV range is limited, however: perhaps 5 to 20 dB, compared to 50-80 dB for a good low-pass filter. HF attenuation is best with tuners that use a low-pass circuit (with coils in series and capacitors shunting the load).

Designs

Low-Pass filter designs appear in *The ARRL Handbook*. *Chebyshev*, *Butterworth* or *elliptic* filters are covered. Simply select a suitable design, look up the component values in a table, and scale the values for the desired impedance and frequency. The details are well covered in any recent *Handbook* as well as many other references.[3] Computer programs are also available to generate the component values.

Figure 5.7 illustrates a *quarter-wave* filter, so called because it electrically resembles a ¼-λ transmission line below the cutoff frequency. This filter is easy to design because each of the three elements (C1, C2 and L1)

How To Install a PL-259 Connector

Install PL-259s to RG-8 or RG-213 coax like this, and you will have good RF-tight connections that should not work loose. First, remove the coupling ring from the connector and slide it over the cable. (If you forget this step, you'll be sorry!) Now *carefully* score the outer insulation (jacket) around its circumference 1⅛ inches from the end (A). If you nick the braid, cut the cable and start over. Remove the jacket without disturbing the braid. Tin the braid with rosin-core solder using a high-wattage soldering iron or gun (B). (Try not to melt the center insulation, and don't use too much solder.) Use a tubing cutter or sharp knife to cut through the tinned braid (and about half-way into the center insulator) ¾ inches from the end (C). Twist off the cut braid and insulation. Clean and tin the exposed center conductor. Smooth any bumps or rough edges on the tinned braid with a file. Screw the connector body onto coax jacket (D). (You may need to hold the connector body with pliers.) Solder the braid to the connector body through each access hole, and solder the center conductor to the pin (E). Trim the center conductor even with the end of the pin and file off any solder build-up. Finally, scrape away any solder flux on outer surface of the pin.—*Alan Bloom, N1AL, San Francisco Section Technical Coordinator, Santa Rosa, California*

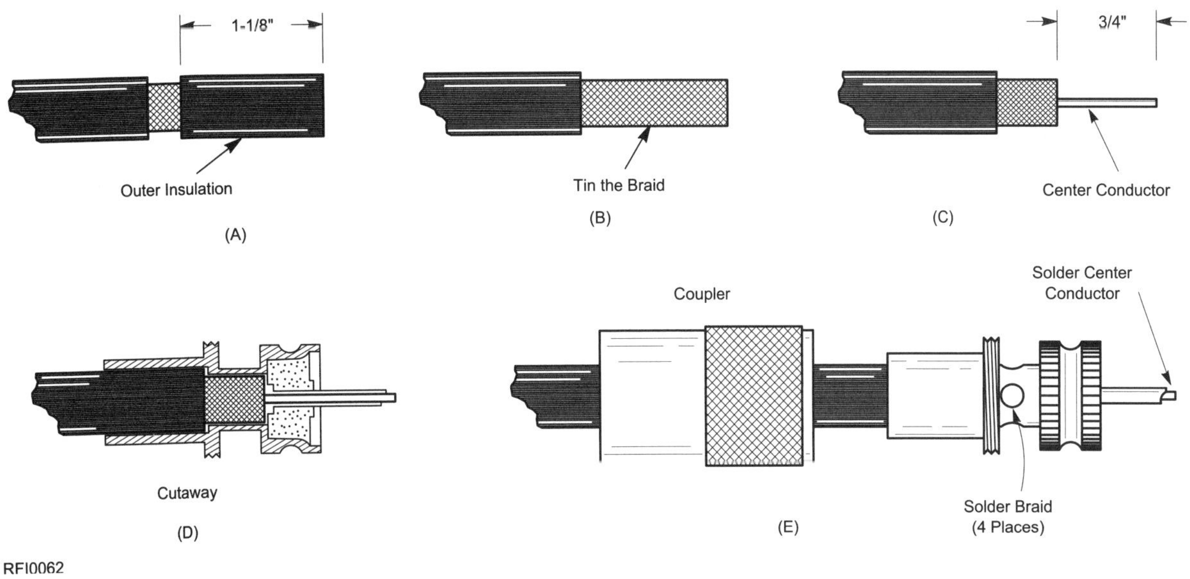

have 50 Ω of reactance at the design frequency. Like image-parameter designs, cascade as many sections as needed to get additional high-frequency rejection. Unfortunately, the low-frequency ripple is not well controlled, especially for multisection designs (Figure 5.8). This kind of filter is only good for single-band applications.

Absorptive Filters

Most filters are designed for a 50-Ω impedance at both terminations. In a typical amateur installation, this is only true at the transmit frequency. Most antennas are not 50 Ω outside the band. The same is true for the transmitter output circuit.

Unfortunately, filter response into a highly reactive load is unspecified. Harmonic attenuation may be degraded at some frequencies (and improved at others). Take the 50-Ω frequency-response plots as rough approximations of performance in actual installations.

An *absorptive filter* (Figure 5.9) is one partial solution. In addition to a low-pass filter, this circuit includes a matching high-pass filter connected to a 50-Ω load. The filter provides a matched, 50-Ω resistive load at VHF frequencies. Connect this filter after a good conventional low-pass filter to guarantee excellent harmonic suppression.

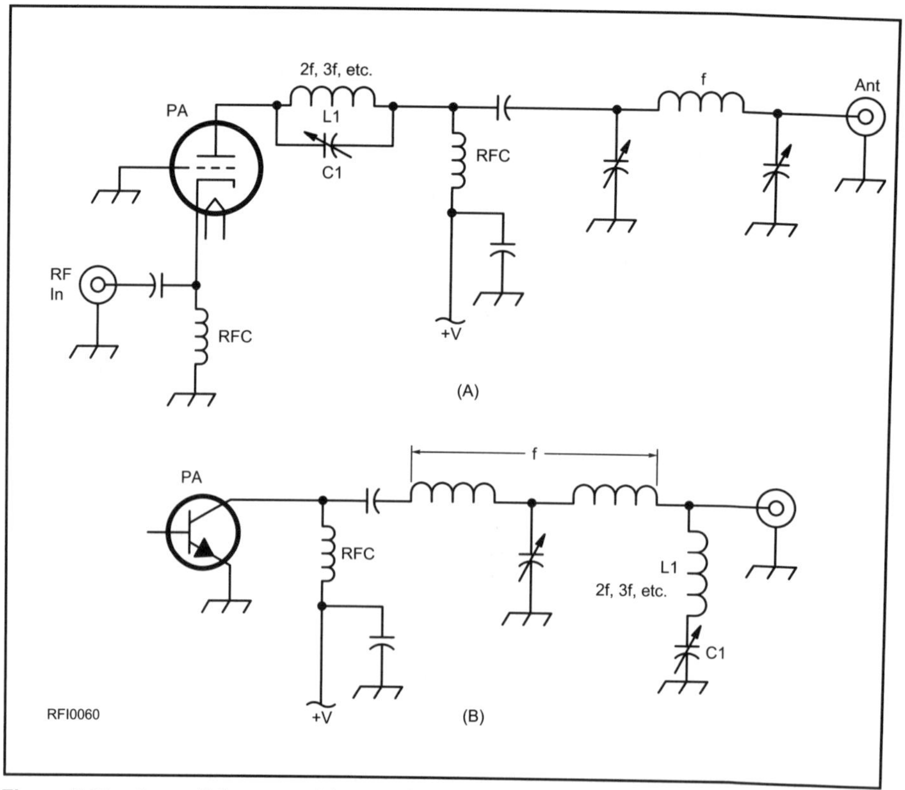

Figure 5.10—A parallel-resonant harmonic trap is shown in the plate lead of the circuit at A. The example at B uses a shunt connected series-resonant trap.

BANDSTOP (TRAP) FILTERS

For interference problems resulting from a single spurious frequency, a series- or parallel-tuned *wave trap* within the transmitter, or near the antenna terminal, can help. (See Figure 5.10.) Each LC circuit is resonant at the interfering frequency.

With vacuum-tube transmitters, install a parallel-tuned trap in the plate lead of the final amplifier. This might cure mild TVI cases. The final amplifier output impedance is too low for this technique in solid-state amplifiers, but a series-tuned trap in parallel with the antenna connector may help. For either type of trap, the inductor and capacitor should each have a reactance in the neighborhood of 200 Ω. Tune the trap to the frequency of the affected TV channel. *Beware the high voltage present on the tube plate!* Use a plastic alignment tool to adjust the capacitor, or use a dip meter to align the trap with the transmitter switched off. To trap out TV channel 2, for example, suitable values would be about 0.56 µH and 15 pF (use a 20-pF variable capacitor.)

A tuned *transmission-line stub* is similar to the LC trap. A ¼-λ transmission line (open at the far end) acts as a series-tuned circuit. Place such a ¼-λ stub (cut for the

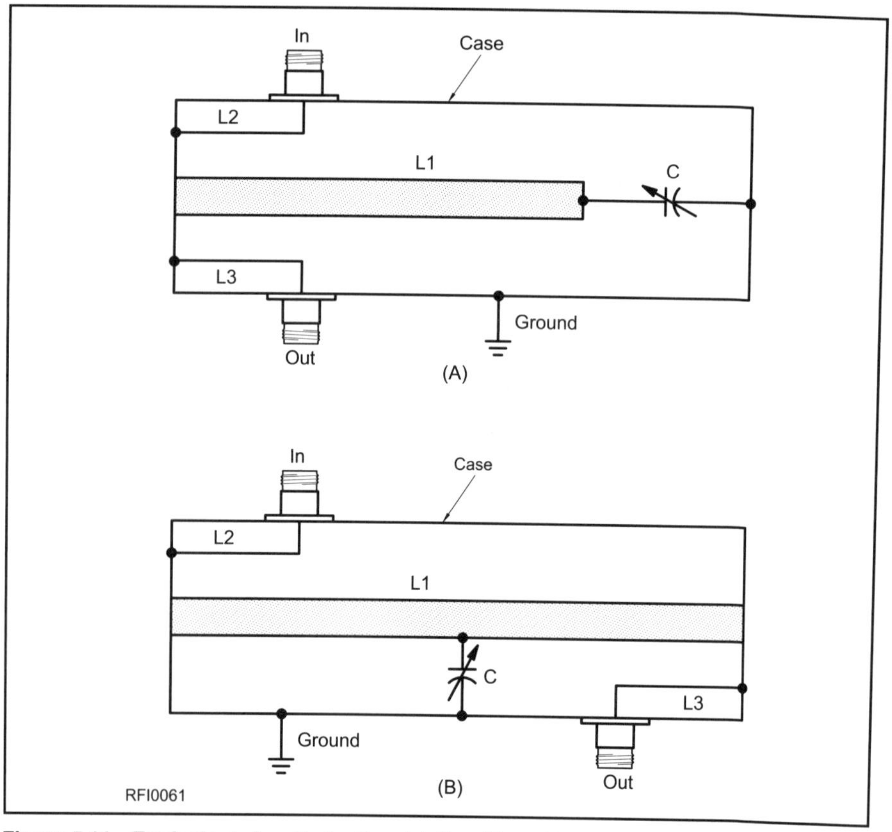

Figure 5.11—Equivalent circuits for the stripline filters. The circuit at A applies to the 6 and 2-meter designs. L2 and L3 are the input and output links. At B is the circuit for the 220 and 432-MHz filters. All of the filters are bilateral (the input and output terminals may be interchanged).

50-kHz CW Sidebands

I once received an ARRL Official Observer report for transmitting a T-6 (hum-modulated) note just outside the bottom of the 20-meter band. My logbook revealed that I was operating about 50 kHz higher at the time. The transmitter I was using included a transistor amplifier stage with 0.01-µF coupling capacitors and 1-mH shunt-feed inductors at the input and output. A quick calculation revealed that these values resonate at about 50 kHz. The engineers who designed this rig had accidentally designed a tuned-collector, tuned-base oscillator! The 50-kHz oscillation amplitude modulated the transmitted signal, which caused sidebands to appear 50 kHz above and below the carrier frequency.—*Alan Bloom, N1AL, San Francisco Section Technical Coordinator, Santa Rosa, California*

There's No Such Thing As a Free Lunch!

Some of the most severe interference is caused by unwanted oscillations in transmitter final amplifiers. I once owned a popular make of solid-state 2-meter FM transceiver that I retuned for maximum power (into a wattmeter). After vociferous complaints from my neighbors, I repeated the tuning procedure, but this time with my TV set on. Not all of the power registering on the wattmeter was within the 2-meter band! At certain tuning conditions, the amplifier would break into oscillation, causing severe TV interference (and a higher reading on the wattmeter).—*Alan Bloom, N1AL, San Francisco*

Figure 5.12—Interior of the 50-MHz stripline filter. The aluminum inner conductor is bent into a "U" shape to fit inside a standard 17-inch chassis.

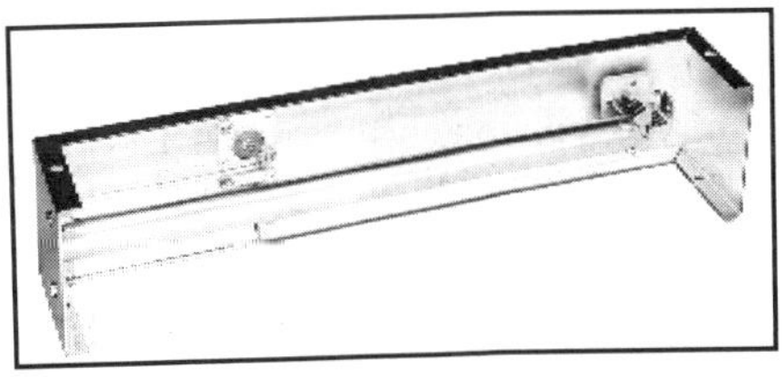

Figure 5.13—The 144-MHz filter has an inner conductor of ½-inch (⅝ inch OD) copper tubing, 10 inches long. It is grounded to the left end of the case and supported at the right end by the tuning capacitor.

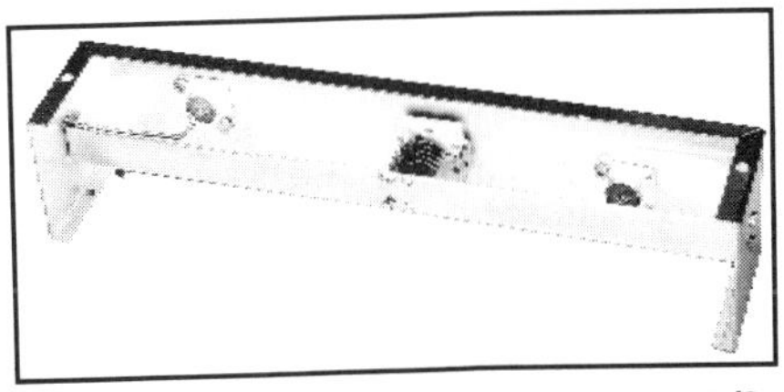

Figure 5.14—The 220-MHz filter uses a $\lambda/2$ stripline. It is grounded at both ends and tuned at the center.

interfering frequency) in parallel with the transmitter output; it acts like the circuit in Figure 5.10B.

Parallel-conductor transmission lines present a special problem: balanced low-pass filters are not readily available. Use a shunt trap, either LC or a tuned stub. Generally, the best solution is a standard coaxial low-pass filter at the transmitter output followed by either a balun or a balanced-output Transmatch.

INTERFERENCE FROM VHF TRANSMITTERS

Some amateur VHF and UHF transmitters use an LF oscillator that is multiplied to the transmit frequency. In the process, unwanted oscillator harmonics can leak through, causing interference to other services. Of course, VHF and HF transmitters are equally likely to have output-frequency harmonics and unwanted mixer products. For any of these problems, add a band-pass filter between the transmitter output and the antenna.

VHF Band-pass Filter Designs

A *cavity* or *stripline* filter is a tuned transmission line. Instead of using coaxial cable, a high-Q line is built out of wide strap or tubing for the center conductor and some kind of chassis or enclosure for the outer conductor.

Figures 5.11 through 5.14 illustrate stripline filters for the amateur VHF and UHF bands from 50 through 450 MHz. Construction is easy, and the cost is low. Standard boxes are used for ease of duplication.

The filter of Figure 5.12 is selective enough to pass 50-MHz energy and attenuate the seventh harmonic of an 8-MHz oscillator that falls in TV channel 2. With an insertion loss (at 50 MHz) of about 1 dB, it can provide up to 40 dB of attenuation to energy at 58 MHz.

The filter uses a folded transmission line to keep it within the confines of a standard 6×17×3-inch chassis. The aluminum partition down the middle of the assembly is 14 inches long. The inner conductor is 32 inches long and 13/16 inch wide, made of 1/16-inch brass, copper or aluminum. In the prototype, two pieces of aluminum were spliced together to provide the 32-inch length. The sides of the "U" are 2⅞ inches apart, with the partition at the center. The line is supported on ceramic standoffs shimmed with sections of hardwood or Bakelite rod to give the required 1½ inch height.

The tuning capacitor is a double-spaced variable (Hammarlund HF-30-X, 4.9-30 pF, 1600 V) mounted 1½ inches from the right end of the chassis. The input- and output-coupling loops are made from no. 10 or 12 wire, 10 inches long. They are spaced about ¼ inch from the line.

The 144-MHz model shown in Figure 5.13 is housed in a 2¼×2½×12-inch box. One end of the tubing is slotted (¼-inch deep) with a hacksaw. This slot takes a brass angle bracket 1½ inches wide and ¼-inch high, with a ½-inch mounting lip. The ¼-inch lip is soldered into the tubing slot, and the bracket is bolted to the end of the box to position the tubing at the center of the end plate. The tuning capacitor (Hammarlund HF-15-X, 3.6-15 pF, 1600 V) is mounted 1¼ inches from the other end of the box with the two stator bars soldered to the inner conductor.

The two SO-239 coaxial connectors are 11/16 inch in from each side of the box, 3½ inches from the left end. The coupling loops are no. 12 wire, bent so that each is parallel to the center line of the inner conductor and about ⅛ inch from its surface. Their "cold" ends are soldered to the brass mounting bracket.

The 220-MHz filter (Figure 5.14) uses

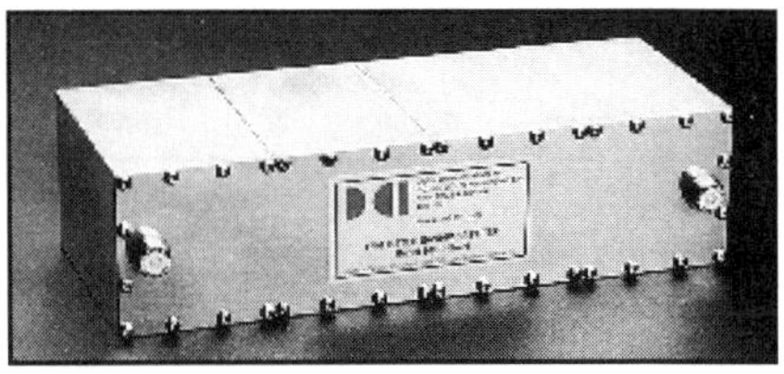

Figure 5.15—This commercially available filter can be used as an effective VHF transmit filter. It offers up to 100 dB of attenuation on transmit and receive. It can also help in cases of receiver intermod.

the same size box as the 144-MHz model. The inner conductor is 1/16-inch brass or copper, 5/8-inch wide, just long enough to fold over at each end and bolt to the box. It is positioned so there is a 1/8-inch clearance between it and the rotor plates of the tuning capacitor. The latter is a Hammarlund HF-15-X (3.9-15 pF, 1600 V), mounted slightly off-center in the box, so that its stator plates connect to the exact mid-point of the line. The 5/16-inch mounting hole in the case is 5½ inches from one end. The SO-239 coaxial fittings are 1 inch in from the opposite sides of the box, 2 inches from the ends. The coupling links are no. 14 wire, spaced 1/8 inch from the inner conductor.

The 420-MHz filter is similar in design, using a 1⅝×2×10-inch box. A half-wave line is used, with a tuning capacitor (fabricated from brass discs) at the center. The two discs are each 1/16 thick and 1¼ inch in diameter. The fixed disc is centered on the inner conductor, the other is mounted on a no. 6 brass lead-screw. This passes through a threaded bushing (which can be taken from the end of a discarded slug-tuned coil form). Such bushings usually include a tension device for the screw. If there is none, use a lock nut.

The 420-MHz model uses N connectors. They are located ⅝ inch from each side of the box, and 1⅜ inches from the ends. The coupling links are no. 14 wire, spaced 1/16 inch from the inner conductor.

To adjust the filters, simply connect them between the station SWR meter and the antenna; tune for minimum SWR. The SWR should be close to 1:1 into a VHF dummy load or a well-matched antenna. All coaxial cable between the transmitter and filter should be good quality (95% braid coverage), with the shield properly soldered to the connector shells.

Commercially Available Band-pass Filters

Digital Communications, Inc (DCI) makes a wide range of commercially avail-

A Low Pass TVI Filter

Here's a low-pass filter that you can build. It's almost as good as commercial products. It should be adequate for all but the most severe TVI caused by transmitter harmonics.

Construction

The filter is constructed in an aluminum box measuring 3½×2⅛×1⅝ inches. Input and output connectors are mounted at the center of each end. Use 5%-tolerance (2% would be better) capacitors. The 500-V capacitors specified should be more than adequate for a 200-W transmitter.

The coils are space wound from no. 18 enameled copper wire. Wind L1 and L3 using a ¼-inch drill bit as a form, use an ⅛-inch bit for L2. To space the windings, wind two pieces of wire in parallel. To remove the extra winding, just grasp one wire and pull (with the windings still on the drill bit). Solder all parts to a two-terminal-with-ground solder strip as illustrated.

Tuning

Adjust the coils by spreading or squeezing the turns. Set L2 first for maximum rejection of TV channel 2 (55.25 MHz). Then tune L1 and L3 for lowest SWR, or minimum insertion loss, at 28.5 MHz.

Here are two ways to adjust L2: Method 1 requires a grid-dip oscillator or solid-state dip meter. Temporarily short L2 (the end that connects to L1 and L3) to the grounded end of C2 with a *very short, fat* conductor. (A piece of tinned coax braid should do nicely.) Set the dip meter frequency by placing it near a TV set tuned to channel 2. Tune the dip meter to produce broad horizontal interference bars on the TV. Loosely couple the dip meter to L2, and adjust the coil for a dip at 55.25 MHz.

Method 2 requires a strong channel-2 signal and a TV fed with coax. Cable TV service should suffice. Connect the filter between the TV antenna connector and the feed line. Adjust L2 for maximum "snow" in the channel-2 picture. It may be necessary to temporarily short L1 and L3 to yield a strong enough signal.

Performance

The filter attenuates all VHF TV frequencies a minimum of about 48 dB. It has about 70 dB of rejection at the channel-2 carrier frequency (often one of the worst trouble spots). The worst-case insertion loss is about 0.3 dB, and the SWR is less than 1.3:1 below 29.0 MHz. The loss rises to over 0.4 dB at 29.7 MHz. If you operate at the high end of 10 meters, you may want to peak L1 and L3 at 29.0 MHz instead of 28.5 MHz.

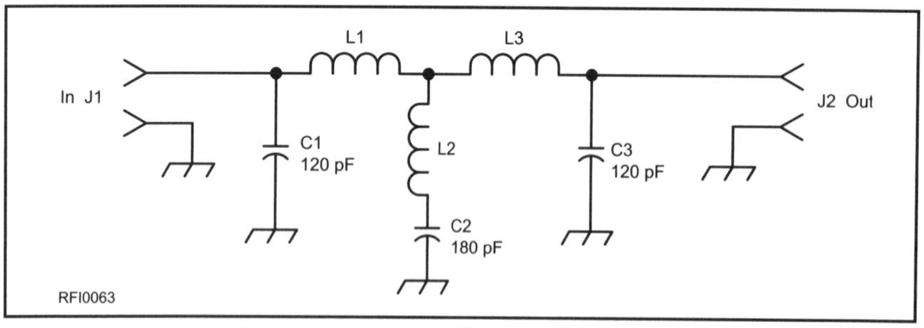

(A)

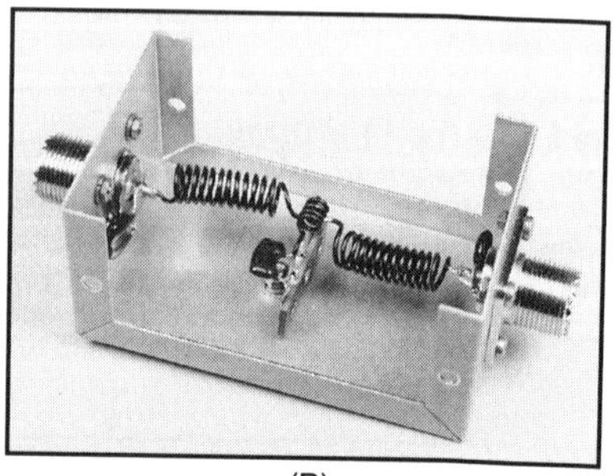

(B)

A low-pass filter for amateur transmitting use. The circuit (A) is constructed on a two-terminal-with-ground terminal strip (15TS003), which is fastened to the chassis with no. 4 solder lugs and hardware. (Part numbers in parentheses are catalog numbers from Mouser Electronics; see Suppliers List for address.) B is a photo of the completed filter.

C1, C3—120 pF, 5% tolerance, 500 V silver mica (232- 1500-120).
C2—180 pF, 5% tolerance, 500 V silver mica (232-1500- 180).
L1, L3—13 turns no. 18 wire, space wound ¼-inch ID.
L2—4 turns no. 18 wire, space wound ⅛-inch ID.
J1, J2—SO-239 "UHF" connectors (16SO239) or RF connectors of the builder's choice.
Aluminum chassis—3½×2⅛×1⅝ inches (537-TF-773).

able VHF and up band-pass filters. These can offer up to 100 dB of stop-band attenuation, although their cavity design doesn't permit that much attenuation of odd-ordered harmonics. They have models for any frequency between 50 and 3000 MHz, using 4 to 10 poles. They can also supply filters in nearly any bandwidth. An example of one of these filters is shown in Figure 5.15. See the Suppliers List in the appendices for information on DCI.

Parasitics in VHF/UHF Transmitters

Parasitic oscillations far above the operating frequency should not be a problem in VHF or UHF vacuum-tube transmitters. If short, low-inductance connections are used throughout, the frequency of any parasitic resonances in the final amplifier tank circuit are so high that the tube won't have enough gain to sustain oscillation.

Oscillation near the operating frequency can be a problem, however. Grounded-grid amplifiers should be stable if the input and output circuits are well isolated from each other. Grounded-cathode triode amplifiers usually need to be neutralized. Some tetrode or pentode circuits require neutralization as well. Series-resonant screen bypassing is a possibility here: Above the tube self-resonant frequency, the correct bypass capacitor can tune out the self-inductance of the screen.

Solid-state VHF amplifiers can oscillate at either low or RF frequencies. For low-frequency oscillations, apply the same solutions previously mentioned for HF solid-state amplifiers.

To determine whether high-frequency oscillations are present, retune the final amplifier while watching its collector current (or total power-supply current). Any erratic current fluctuations during tuning indicate unwanted oscillation, probably in the VHF/UHF range. The problem may be caused by simple misadjustment. If, after proper alignment, all tuning controls can be slightly detuned (near their optimum setting) without causing oscillations, the transmitter is probably operating correctly.

If there is no stable tuning condition, check all circuitry around the affected stage for poor solder joints and damaged or overheated components. If there are no defects, circuit modifications may be in order.

One quick-and-dirty solution: Add a low-value resistor (5-50 Ω) in series with a 0.001-µF ceramic disc capacitor from the base of the affected stage to ground. With luck, a resistor value that is low enough to kill the oscillation will not reduce output power excessively. If there is a coil or RF choke from the base to ground, try adding a series ferrite bead. The same trick sometimes works with a choke in the collector power-supply lead.

BE AN OPTIMIST

When troubleshooting interference problems in a transmitter, keep in mind that there is almost always a technical solution. After all, there are literally millions of transmitters around the world that work, day in and day out, without interference-causing spurious emissions. With proper shields, filters and bypassing, any transmitter can be clean.

Notes

[1]R. Measures, "Parasitics Revisited," Sep 1990 *QST*, pp 15-18 and Oct 1990 *QST*, pp 32-35.

[2]Ferrite beads, rods and toroid cores are available from several sources: Amidon Associates, Palomar Engineers, and others. See the Suppliers List at the end of this book.

[3]Filter-design references, arranged in order of ease of use:

C. Hutchinson and L. Wolfgang, *The ARRL Handbook for Radio Amateurs* (Newington, CT: ARRL, 1991 and earlier editions), Chapter 2. These tables are the easiest to use, and some are set up for standard capacitor values. The selection of filter types is limited, however, and only low-pass and high-pass filters are covered.

E. Jordan, *Reference Data for Engineers* (Indianapolis, IN: Howard W. Sams & Co, Seventh Edition, 1985), Chapter 9. This resource is more difficult to use, but more inclusive than the *Handbook*. It includes band-pass and bandstop filter transformations.

A. Zverev, *Handbook of Filter Synthesis* (New York, NY: John Wiley & Sons, 1967). This is "everything you ever wanted to know about filter design." It contains 203 pages of design tables, plus charts and graphs and extensive mathematical theory. For the engineer.

Chapter 6

Antenna-Connected Televisions

Although the number of reported cases has been declining, Television Interference (TVI) is still with us today. When TVI strikes you or your neighbor's antenna-connected TV, the techniques discussed in this chapter will help find the cause and a cure.

By Ed Hare, W1RFI,
ARRL Laboratory Supervisor and
Bob Schetgen, KU7G (SK)

Several chapters in this book are devoted to various aspects of television interference (TVI). There are substantial differences between the causes and cures for TVs connected to antennas, TVs connected to cable TV systems and VCRs. This first "TVI" chapter covers TVs connected to antennas.

Other TVI chapters are:

Chapter 7—Cable Television Interference: The approaches used to diagnose and cure cable TVI are different than those used to cure TVI to TVs connected to antennas. This chapter covers the basics, and includes a section designed specifically to help you teach your cable company everything they need to know about cable TVI.

Chapter 8—DVDs and VCRs: DVDs and VCRs are much like TVs, but they, too, have some unique characteristics that make them a separate RFI topic.

WHAT IS TVI?

In simple terms, "television interference," or TVI, describes a condition where a television receiver fails to operate properly in the presence of unwanted RF energy. An amateur transmitter in perfect working condition (with no harmonics or other spurious radiation) can still interact with TVs. This occurs because the TV manufacturer did not build the product to reject high-level signals that are a part of its normal environment.

There are several "types" of TVI, each with its own separate causes and cures. When caused by a radio transmitter, TVI may appear as crosshatching or bars in the picture, a change in contrast or color, complete blanking of the screen, audio interference or any combination of these effects. The manifestation in a specific case depends on a number of factors, such as the frequency, bandwidth and intensity of the interfering signal and the design of the affected TV equipment.

PRELIMINARY STEPS

Amateurs involved with or assisting with a TVI problem should take a few preliminary steps before starting to diagnose and fix the TVI problem. Read the First Steps, EMC Fundamentals and Troubleshooting chapters. This chapter has only a very brief summary of the detailed information in those chapters.

RFI problems can be complex. Stock answers do not always result in solutions. It is important for all concerned parties—the amateur, Technical Coordinator, Technical Specialist, TVI Committee members and the neighbor—to learn as much as possible about the subject of RFI.

TVI CAN BE CURED

TVI is not the oldest RFI problem facing hams, but with the widespread introduction of broadcast TV in the 1950s, it became one of ham radio's worst problems. Although its incidence seems to be decreasing, TVI remains one of the more common RFI problems. It is probably a dead tie between TVI and the other "T" problem—telephone interference.

When TVs were first introduced, amateur regulations didn't require any specific amount of attenuation of harmonics from amateur stations (although the rules did require that "good amateur practice" be followed). In TVI terms, most amateur transmitters of that day were, to put it bluntly, terrible! In addition, TVs were not well filtered. Further, neither hams, their neighbors nor, to some extent, the FCC, knew much about this new TVI monster. The result was that TVs and amateur stations were not compatible—a reputation the Amateur Radio Service is still trying to leave behind.

Fortunately, things have improved a lot! FCC regulations require that harmonics and other spurious emissions from radio transmitters be attenuated enough not to cause TVI. Hams are a lot smarter about interference than they were when TVI first became a problem. In addition, modern industry standards for immunity are resulting in more and more manufacturers considering immunity when they design consumer electronics equipment. As can be seen in Figure 6.1, most TVI can be cured!

DIPLOMACY

It May Be Your Fault!

The topic of responsibility comes up in every chapter, but it is important, so it is worth repeating: The technical parts of a TVI problem are really the second half of the problem. The human part of the problem usually must be dealt with first. Review the First Steps chapter for some important guidelines on how you and your neighbor can work together to find a solution.

There is something about the relationship between TVI and personal diplomacy that doesn't apply to most interference problems. Broadcast TV is a radio service, so it is possible that TVI *could* be caused by unwanted spurious emissions from your amateur transmitter. Translation: TVI could be your fault and it may be your responsibility to fix it! Keep this in mind when you talk to your neighbor.

It May Not Be Your Fault

Take heart, though; it may not be your fault. Like any other type of interference,

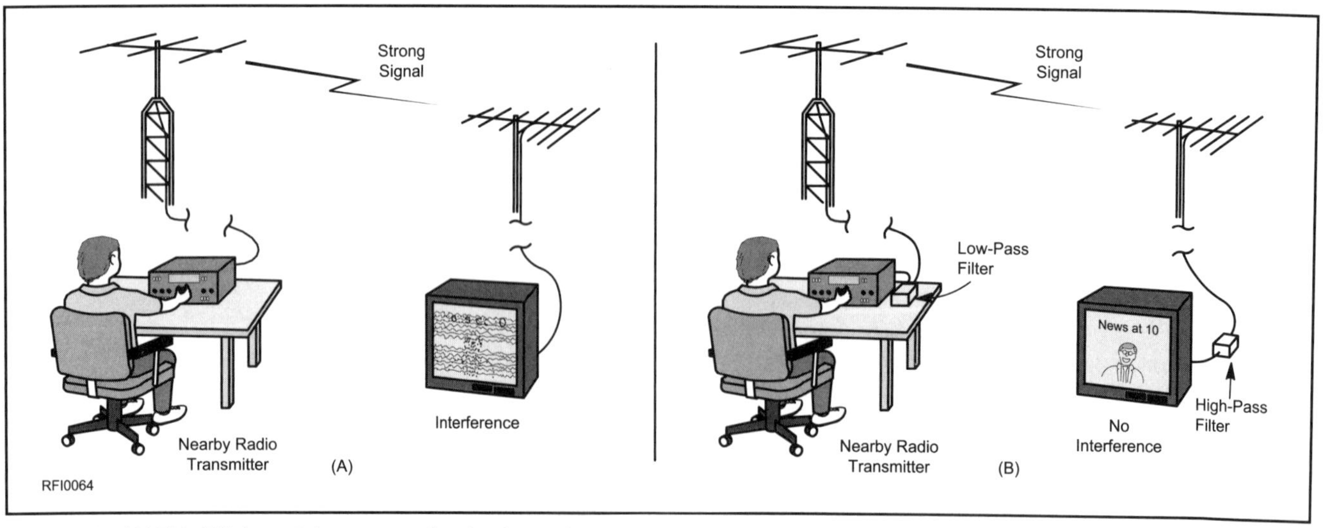

Figure 6.1—(A) This TV doesn't have enough rejection to keep out an amateur fundamental signal. The transmitter may also be emitting some VHF spurious signals that fall inside the TV channel. (B) The appropriate cures have been installed and everyone is happy again.

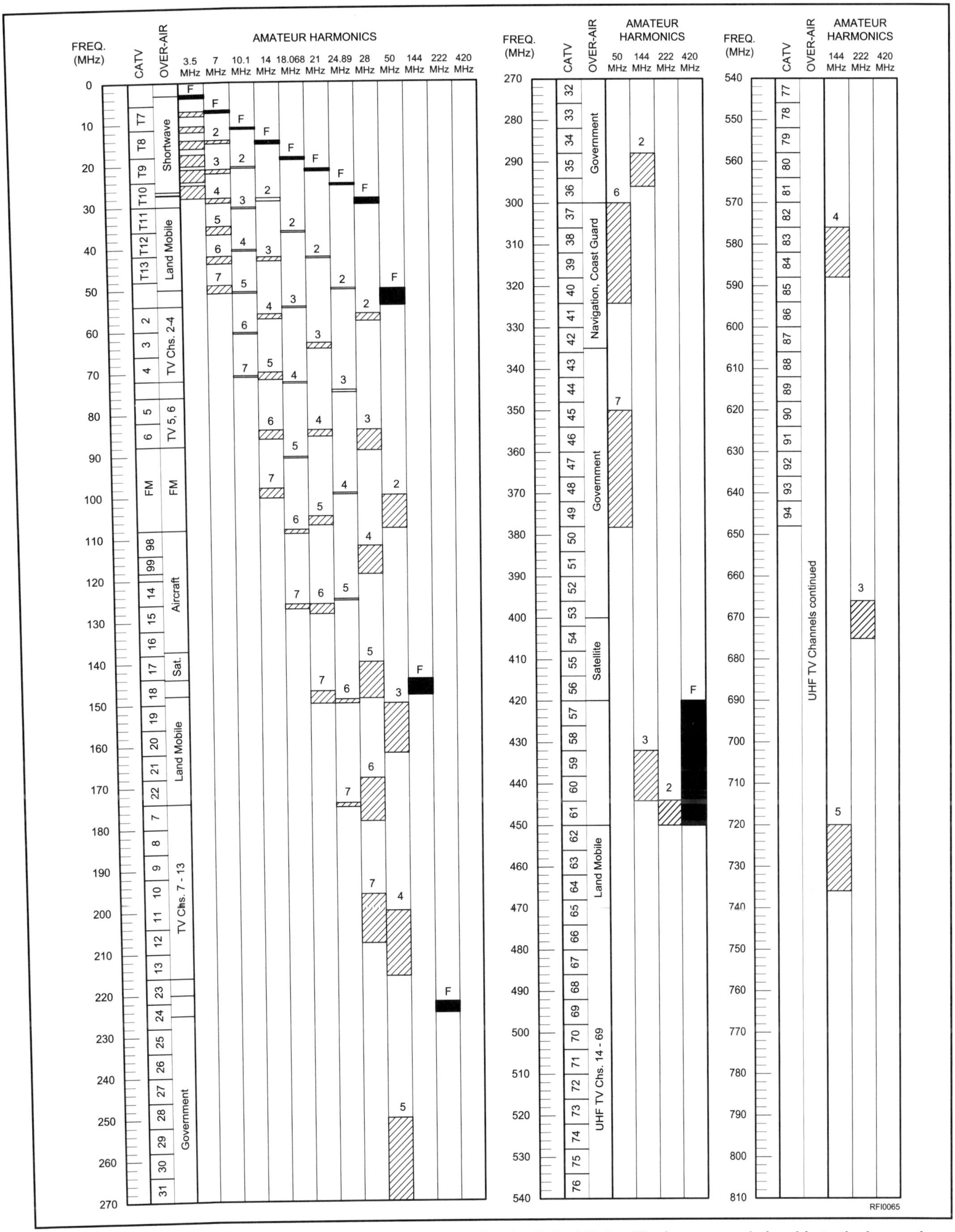

Figure 6.2—This chart shows CATV and broadcast channels used in the United States. The frequency relationship to the harmonics of HF, VHF and UHF amateur bands can be readily seen. This will help identify those channels that are harmonically related to the transmit frequency. Use this chart to determine if there is a harmonic relationship between operating frequency and any TV channels affected by station operation. If the interference is only on channels that are multiples of your operating frequency, you clearly have interference from harmonics. (It is not certain that these harmonics are coming from your station, however.)

fundamental overload of a susceptible TV may be the root cause of the problem. The TVI could be coming from another source altogether, such as power-line noise or another transmitter in the area. It could also be coming from a VCR recording another channel, or from a video game that is generating interference to the whole system.

ABOUT TV CHANNELS AND TV SETS

Frequencies Allocated to Broadcast TV

Broadcast television is becoming more varied. In addition to the standard VHF and UHF analog TV signals, broadcasters now transmit standard and high-definition digital TV signals and various satellite analog and digital signals. Each of these can have their own TVI potential, causes and cures.

Figure 6.2 shows the frequencies allocated to over-the-air and cable TV in the VHF and UHF ranges.

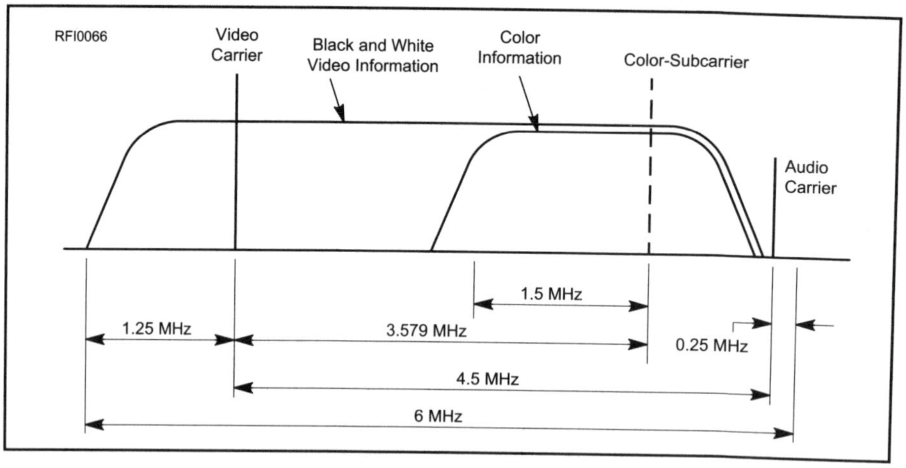

Figure 6.3—The frequency layout of a television channel.

Figure 6.4—This "herringbone" pattern is typical of a TV receiving interference from a radio transmitter.

The TV Channel

Figure 6.3 shows the frequency layout of a television channel. Each TV channel is 6 MHz wide. The left side of Figure 6.3 is the lower channel edge. The lower sideband of the video signal is found from 0 to 1.25 MHz from the lower edge of the channel. This is a "vestigial" sideband, meaning that only part of the sideband is transmitted; the rest is filtered out at the transmitter. The visual (video) carrier is 1.25 MHz up from the low-frequency edge. The upper video sideband extends to about 5.75 MHz from the lower channel edge. The color subcarrier and aural (sound) carrier and sidebands are also found as part of the upper sideband. The sound carrier is 4.5 MHz higher than the visual carrier. The color subcarrier frequency is 3.579 MHz from the visual carrier. The color information is found near the color subcarrier frequency.

As the interference frequency approaches that of the visual carrier, the interference bars become thicker and more objectionable. As the interference moves away from the visual-carrier frequency, the bars become finer until they become a "herringbone" pattern on the screen. This pattern sometimes becomes so fine that it appears as a slight reduction of picture contrast. When these horizontal or diagonal lines are modulated by a voice, they are usually called "sound bars." Figure 6.4 shows a typical case of TVI.

The TV

Figure 6.5 shows a typical color TV block diagram. The major systems are designed to work together. Study the way that the various circuits interact. As you read through the rest of this chapter, this fundamental knowledge about how TVs work will help you under-

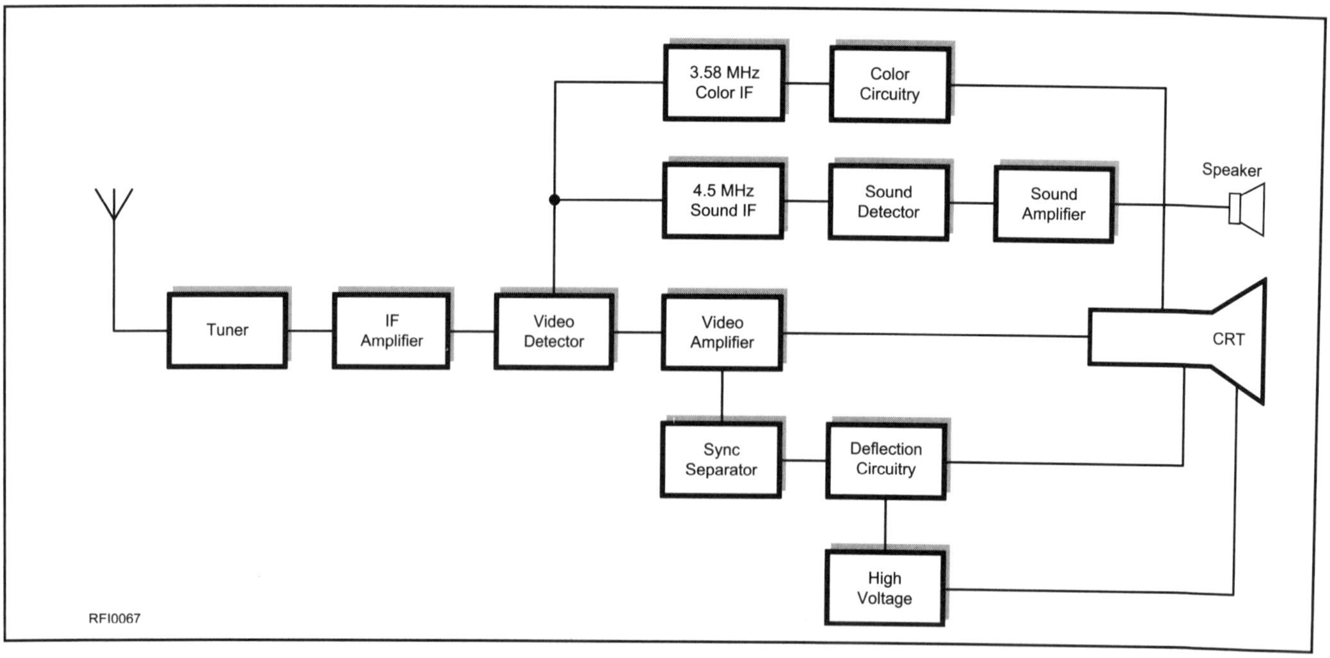

Figure 6.5—Block diagram of a typical color-television receiver.

stand how interfering signals can enter each circuit and affect the picture or sound. TVs are by necessity broadband devices and are therefore relatively susceptible to interference from external RF signals. The susceptibility problem is compounded because interference makes a much greater impression on the eyes than it does on the ears.

TV sets receive their signals from over-the-air broadcast, cable or a satellite dish. Over-the-air broadcast TV (because of the TV antenna system) is more susceptible to TVI than are cable and satellite TV. Broadcast TV is frequently the only TV service available in remote areas, and TV signals there are weak. When the TV signal is weak, the chance for TVI is greatly enhanced. A good TV signal and typical antenna usually result in a TV signal of about –45 dBm at the receiver terminals. Interfering signals that are 40 to 60 dB weaker can cause perceptible interference.

Like any other piece of home electronics equipment, a TV set can be overloaded by a strong, usually nearby, signal. This is known as *fundamental overload*. Fundamental-overload TVI results from the inability of a TV to reject an amateur's transmitted signal. Anechoic-chamber tests show that some older TVs have only 30 dB of 20-meter signal rejection, although some manufacturers have improved newer TVs quite a bit beyond this level. There may be even less rejection of 10 meter and VHF/UHF signals. When compared to a received TV signal, this sensitivity to amateur frequencies is potentially one of the most severe RFI problems (see Figure 6.6).

Although the internal circuitry of consumer-electronics equipment is fairly immune to strong local signals, the equipment is connected to 117-V ac lines, and antenna or cable systems, which can couple a tremendous signal into the set. This makes it more likely that a strong enough signal will be coupled into the TV to cause fundamental overload. Fundamental overload usually generates a herringbone pattern similar to that caused by transmitter harmonics. In serious cases, the picture can be blanked out, often on all channels. Fundamental overload is the most common form of TVI.

Amateur Station TVI Checklist

Have proof that your station is clean. Be sure to separate wishful thinking from reality. Thinking or saying that station output is clean may sound good, but it doesn't help the neighbors. Children have become the innocent victims of neighborhood TVI disputes. The following check list should make it easier to objectively examine a station and determine the source of a problem.

- ❒ Is the station grounded with a short low-inductance ground lead?
- ❒ Is the SWR low at the transmit filter?
- ❒ Does TVI increase when the transmitting antenna points toward the TV (fundamental overload)?
- ❒ Do you always use the same frequency for TVI tests.
- ❒ Is the TVI level constant as the antenna rotates (harmonics)?
- ❒ Does TVI cause the TV raster to change with no TV antenna connected (direct radiation pickup)?
- ❒ Are the drive level and any narrow-band stages tuned for minimum harmonic output?
- ❒ Does TVI decrease when any VHF radio equipment is disconnected from its antenna?
- ❒ Do you observe interference on a monitor TV while transmitting?
- ❒ Is all station equipment bonded to a common ground?
- ❒ Are all station cables and connectors in good shape?

After completing the amateur station TVI checklist, proceed to the TV receiver checklist. Then make a complete analysis of the transmitter, affected equipment and interference path to solve the EMI problem.

TV Receiver Checklist

Complete the following checklist for each TV with interference.

- ❒ Is the received TV signal STRONG?
- ❒ Do you use common-mode, high-pass filters?
- ❒ Is there an audio-rectification problem?
- ❒ Does the TV antenna have good low-channel gain?
- ❒ Could the TV system benefit from a properly installed distribution amplifier?
- ❒ Is the TV antenna and coaxial feed line properly grounded?
- ❒ Have you checked the area for rectifying devices?
- ❒ Is there a line filter at the power plug?
- ❒ Are the TV antenna and feed line in good condition?
- ❒ Is the TV-antenna feed-line horizontal-run at ground level?

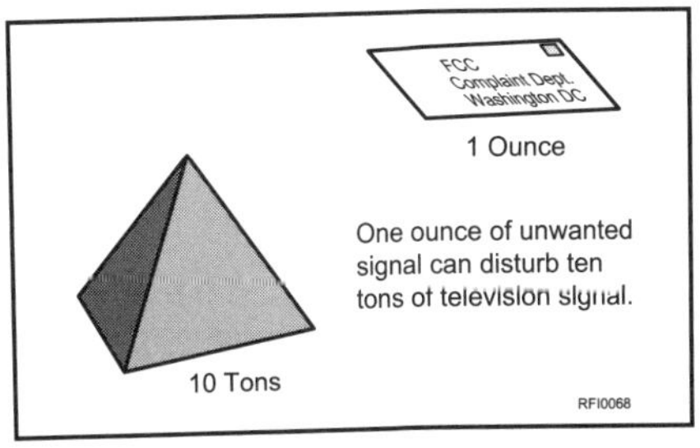

Figure 6.6—For good quality, a television signal must be much stronger than any competing signal on the same frequency.

The TV itself may be a direct cause of an RFI problem. It may have a defective component (although this is not likely in a well maintained set). If a defect does exist in the TV set, one can spend a lot of time looking for problems outside the set when the actual problem is inside the set. Broken shields, defective AGC circuits or improper IF alignment can all contribute to TVI. In some instances, a "rabbit ears" antenna on top of a TV set can receive interference from the set's own switching power supply!

TV-set circuitry is also subject to blocking—a reduction of sensitivity. The effect can range from a just-perceptible change in color or contrast, to a totally blanked-out picture, color or sound.

The TV audio circuitry can envelope-detect an RF signal and amplify the resultant audio signal. This effect is known as *audio rectification*. The resultant audio signal may or may not be affected by the TV volume control. If audio interference is not affected by the volume control, audio rectification is clearly indicated. CW audio interference appears as clicks and thumps, or a reduction in TV audio as the transmitter is keyed. SSB audio interference sounds distorted.

If only one neighbor has TVI problems, the TV set and its installation are suspect. If the external solutions described in the rest of this chapter do not cure the problem, contact the TV manufacturer (through the Consumer Electronics Manufacturers Association) to obtain additional assistance. Some set manufacturers produce service bulletins that describe TVI cures for their models.

If it is not practical to contact the manufacturer or dealer, the amateur and set owner can work with a TV-service technician who can install the appropriate cures given in this chapter. These can be effective for fundamental-overload interference. Additional shielding such as EMI tape may eliminate most of the chassis pickup. Leave the installation of internal shielding, especially in a neighbor's TV, to qualified service personnel. When shielding is added, devices in a set may overheat if air flow around the devices is changed. Added shielding may also cause short circuits; it is best to turn to the equipment manufacturer for assistance.

Leave TV repair to professionals. Safety, licensing and liability issues make it inappropriate for amateurs to repair a neighbor's equipment.

The TV Preamp

In theory, a TV preamplifier at the TV antenna should boost the signal in relation to the interference picked up by the feed line. In practice, this does not work because the selectivity of these preamps is poor, and they eagerly amplify amateur signals (especially at VHF) as well. In general, avoid the use of such preamps.

TV preamplifiers are notoriously susceptible to strong RF fields. They usually lack front-end selectivity and are easily overloaded. They can also oscillate; when they do, they may radiate an interfering signal over many blocks. Many times, TVI lessens when the preamplifier is removed. Check the TV installation carefully for a preamp; sometimes they have been installed in the attic or a closet and forgotten. Preamplifiers such as the one shown in Figure 6.7 should be used only when necessary in weak-signal areas.

The TV Antenna and Feed Line

The television antenna and TV installation may be involved in the RFI problem. Check the TV and antenna installation. The TV antenna is usually strapped to a chimney, where it has been subjected to years of rain, wind and contaminants. The twin-lead feed line has alternately rattled in the icy wind of winter and deteriorated in the summer sun. Defects in the antenna system cause problems in several different ways—a poor antenna system picks up very little desired TV signal. Feed line problems can also result in leakage or excessive common-mode pickup on the feed line itself. Corrosion in the antenna system can also generate harmonics (see the External Rectification chapter.) Defects in the TV antenna system should be identified and replaced.

Perform a quick check on the overall condition of the television antenna system by temporarily substituting an indoor antenna and selecting a weak station. If the signal is not stronger on the outside antenna, there is clearly something wrong with the outdoor TV antenna system.

In some cases, correcting installation or antenna defects may result in an instant TVI cure. If the antenna is old, broken or corroded, it needs to be replaced. Likewise, if the feed line is old, broken or corroded, it needs to be replaced. In many cases, these simple steps will result in the installation receiving more TV signal and less ham signal, curing the TVI problem.

If the TV antenna feed line is made from 300-Ω twin-lead, it may be helpful to replace the feed line with 75-Ω coaxial cable or shielded 300-Ω twin-lead. Most modern TVs have a 75-Ω input, but some older TVs or small portables may have only a 300-Ω bal-

Satellite TVI

Although satellite TV receiving equipment is sometimes regarded as immune to interference from Amateur Radio transmissions, it can be interfered with under some circumstances.

Satellite TV antennas and receivers are generally less likely to be interference victims than most other consumer grade television equipment because of the frequencies and the highly directional antennas being used. However, satellite TV reception can be interfered with if there are deficiencies in the system. These deficiencies can include poor grounding, loose connections, low quality coaxial cable and interference entering the satellite receiver via the ac line or as common mode interference.

Is the dish antenna properly grounded? An inadequate ground may lead to interference problems as well as safety hazards. Make sure the dish antenna is properly grounded in accordance with the manufacturer's instructions and local electrical codes.

Many satellite dish antennas use F connectors on the coaxial cables. These screw-on connectors sometimes come loose after exposure to temperature extremes and movements of the antenna as it is moved from one satellite to another. Slight movements of the antenna caused by wind can also contribute to F connectors becoming loose. Loose, improperly installed, or poorly weatherproofed connectors are an invitation to interference.

A carefully planned and maintained installation keeps Amateur Radio signals out of the satellite TV system. *—Ron Hranac Jr, NØIVN*

Ferrite beads and/or toroids can be placed on the coaxial cable leading from the antenna to the satellite receiver to suppress common mode interference resulting from the coax picking up the Amateur Radio signals.

Interference could also enter the satellite receiver via the ac line. If this is the case, an ac line filter located as close to the satellite receiver as possible may be beneficial. Alternatives to the ac line filter would include wrapping the line cord around a ferrite rod or using a snap-together ferrite core (RadioShack catalog number 273-104 or equivalent) and winding several turns of the line cord through it.

Most direct broadcast satellite receivers rely upon a phone line connection for pay-per-view and other services. Interference could enter the satellite receiver via the phone line connection. A common mode choke should be tried first, installed in the phone line as close to the satellite receiver as possible. If the interference is differential mode, use an appropriate differential-mode phone line filter. See the Telephone RFI chapter for more information.

Poorly shielded patch cords between the satellite receiver and TV receiver or VCR can often increase susceptibility to interference. Some consumer grade satellite receivers use phono jacks (also called RCA jacks) as the connectors for the video and audio outputs. Although these are not the best connectors available, they are adequate if they are kept clean and used with good quality connectors. When buying patch cords, look for gold plated connectors. If you are making your own patch cords, use gold plated connectors and coax with 100% shield coverage. In either case, keep patch cords as short as possible.

anced input. Balun transformers can be found at most retail stores selling TV accessories. Figure 6.8 shows the basic hookup for using 75-Ω coaxial cable with a 300-Ω antenna and TV tuner input. (Of course, if cable ser-vice is available in your area, you can always hope that your neighbor will consider subscribing. A properly maintained cable system is well shielded and will deliver a strong TV signal to the set. Both factors significantly lessen the potential for interference.)

The feed line between the TV receiver and the antenna usually picks up a great deal more energy from nearby HF transmitters than does the TV antenna. These induced feed-line currents are common-mode signals. (The line acts like two wires connected together to operate as one.) If the TV input circuit is perfectly balanced, it rejects common-mode signals and responds only to true transmission-line (differential-mode) currents. That is, only signals from the TV antenna (and not the feed line) would cause a receiver response.

The feed line is much larger than the antenna in most TV receiving installations. Thus, the feed line picks up more interfering signals than does the antenna. To reduce this problem: (1) shield the feed lines, (2) use a common-mode filter at the TV antenna terminals and (3) relocate and/or reposition the transmitting and/or receiving antenna to take advantage of directive effects.

When installing a TV antenna system or connecting a TV to a cable system, it is best to run the TV feed line as directly to ground level as possible and then run the line horizontally in the basement, crawl space or near the ground surface. This should increase the vertical separation between the amateur antenna and feed line. When a feed line is shorted at one end it acts as a long-wire antenna. The TV antenna and feed line are not a good termination for HF and therefore act as a relatively good antenna at amateur frequencies. The TV feed line can pick up a tremendous number of amateur signals, usually in the common mode.

No receiver is perfect in this respect, however, and many TV receivers respond strongly to such common-mode currents. The result is that signals from nearby transmitters are much more intense at the TV front end than they would be if the TV were well designed.

A fringe-area (over 150 miles) TV antenna increases received TV signal strength, especially at channels 2 to 4. Even though a smaller TV antenna may provide an acceptable picture, there is no signal-to-noise margin to cope with near-field amateur transmissions. A larger TV antenna may also offer some directivity to reject a nearby amateur signal, especially if the ham signal is on VHF or UHF. It's best to buy a long-range TV antenna once to avoid replacing a smaller antenna later. In most cases, a higher-gain TV antenna will give more improvement in reception than a TV preamplifier, offering some interference rejection at the same time.

Splitters and Proper Terminations

Consumers often use TV splitters to use a single antenna (or cable) feed with multiple TVs. These work well if there is ample TV signal available, but each splitter introduces some loss. Each doubling of the splitter's capability adds a little more than 3 dB of loss—a 2-way splitter has about 3.5 dB of loss; a 4-way splitter about 7 dB, etc. The use of multiple splitters or a splitter that splits more than is needed can result in excessive signal loss. Many of these splitters are poorly shielded (plated plastic). Splitters may be adequate for a two-TV installation, but larger splitters introduce too much loss.

All connections on a coaxial system should be properly terminated. If multi-tap splitters or distribution amplifiers are used, all inputs or outputs should be connected through quality feed line to matched loads. Unused inputs or outputs should have a suitable resistive termination. RadioShack sells 75-Ω terminating resistors for this purpose. See Figure 6.9.

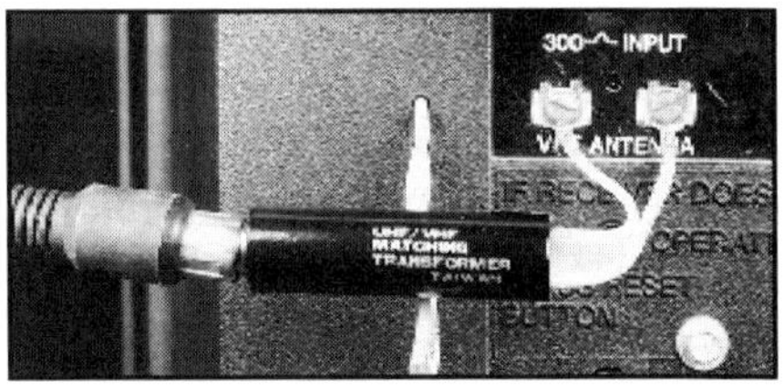

Figure 6.8—This is how to use 75-Ω coax with a 300-Ω TV antenna and television.

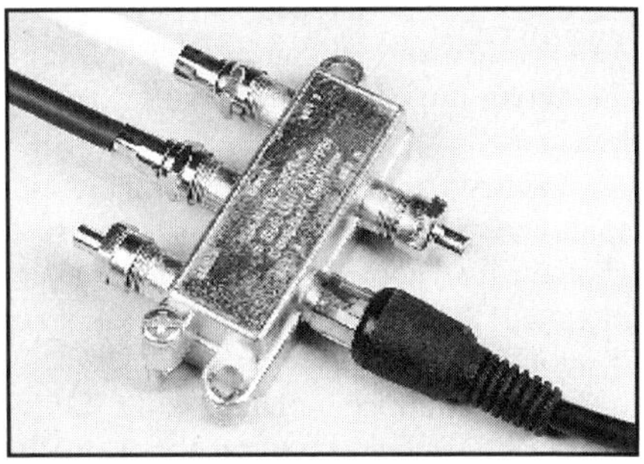

Figure 6.9—All multi-tap splitters should be properly terminated. 75-Ω resistors terminate unused taps of this multi-tap splitter.

An Oscillating TV Preamp

Uncommon causes of TVI can really make a ham dig to find a cause and solution. I live in a fringe area where there were only three snowy TV channels (until cable service finally arrived).

To keep peace at home, I made many attempts to improve the TV signal. My inverted-V antennas were removed from the 55-foot telephone pole, and a new high-gain TV antenna with a transistorized booster were installed in their place. The TV picture went from snow to medium fog, and everything looked rosy on the home front.

Several nights later, after the dipole was installed at about 20 feet, I fired up on 40-meter CW. Everything was great until the next morning, when my neighbors let me know that I am not as interesting as their favorite programs—and it would be okay with them if I didn't play with my interference generator on Friday night.

A check on my TV set confirmed the problem. All three channels were wiped off the screen when I keyed the transmitter. After the booster was removed, everything settled down to normal. The booster became a wide-band oscillator when it met up with a little RF from my rig. For my family that meant—back out of the fog into the snow.

There are other devices that can generate TVI: garage-door openers, receivers, wireless intercoms and audio preamps. There are bound to be others that are more difficult to identify.

If you find an unusual solution to a TVI problem, let ARRL HQ know what you found and the conditions surrounding the case. Fill out the ARRL EMI Report Form described elsewhere in the book. Maybe your report will help someone who has come up against a blank wall.—*Art Block, W3YK*

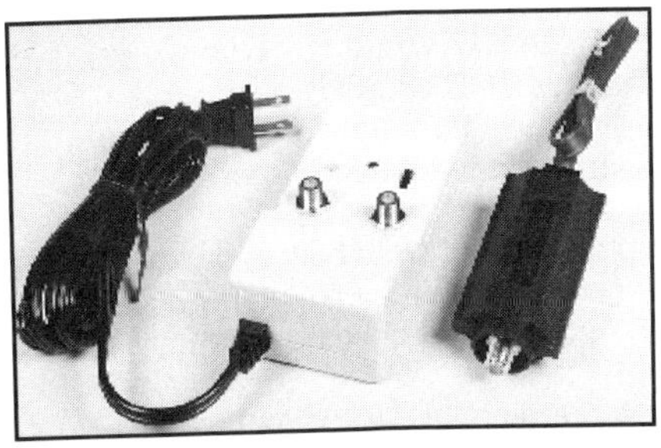

Figure 6.7—A TV preamp. These devices are not necessary in strong signal areas, and in many cases are the cause of TVI.

CAUSES OF TVI

Most cases of TVI are caused by one or more of the following conditions:

- Spurious signals within the TV channel coming from your transmitter or station. These spurious emissions can be discrete signals (harmonics or mixing products) or broadband noise.
- There could be factors in your station operation or configuration that increase the likelihood of interference.
- A television set can be overloaded by the strong fundamental signal of a nearby transmitter.
- Signals within the TV channel from some source other than your station, such as electrical noise, an overloaded mast-mounted TV preamplifier or a transmitter in another service.
- There could be problems in the TV installation or antenna system that increase the likelihood of interference.
- The TV set might be defective or misadjusted, making it look like there is an interference problem.

TVI can originate from many sources other than radio transmitter operation. External noise sources such as power lines, electric motors, household appliances and computer systems (including some video games!) can all interfere with televisions. A complete list of potential interference sources is found in Table 2.1 in the EMC Fundamentals chapter. The list in this table may help put the problem into perspective. Obviously, Amateur Radio is only one of many possible RF or noise energy sources. Although Amateur Radio related RFI is *our* biggest concern, remember that amateurs do not comprise the bulk of the RFI problem.

Interference can result from several different causes—electrical noise (such as a vacuum-cleaner motor), spurious signals from your transmitter, or the inability of the TV set to "tune out" a strong local signal. As an analogy to the last point, if you were to tune a TV set to channel 3, and saw channel 13 instead, you would probably conclude that the TV was broken. If you tune your TV to channel 3, and see a local HF transmitter instead, you could reach the same conclusion. That is, the TV is "broken" in the sense that it needs additional filtering or shielding to work near a strong RF source.

The Amateur Station

Amateur transmitters must meet stringent FCC requirements for spectral purity, and most TV manufacturers design TVs to meet the ANSI standards for rejection of unwanted signals. These two factors make it less likely that the TVI is caused directly by the amateur transmitter.

Nonetheless, in cases of interference to other radio services, such as broadcast TV, the amateur station itself *must* be considered as a possible cause of the interference. If, for example, a transmitter were emitting spurious signals that were inside the frequency allocated to a local TV station, if those signals were strong enough, TVI would result. This can happen even if the spurious signals are low enough to be in compliance with the absolute limits on spurious emissions (see the Regulations chapter for more information). When this happens, the interference is the direct result of the transmitter operation. Under FCC regulations, the transmitter operator is responsible for reducing the spurious emissions to whatever extent is necessary to eliminate the interference. If a spurious

Digital TV

Television broadcasters are in the process of converting to digital transmissions to support standard- and high-definition TV. On the consumer side of things, a variety of display technologies support the new digital transmissions: conventional cathode ray tube (CRT), digital light processing (DLP), liquid crystal on silicon (LCOS), liquid crystal display (LCD) and plasma. All of these can potentially cause interference to amateur communications, and of course, may be susceptible to interference from amateur transmissions.

When interference does occur, try the usual fixes first: common mode chokes, high-pass filters, ac-line filters—brute force and common mode—and so forth. If these don't work, try contacting the set's manufacturer for assistance.

Plasma displays in particular have been found to be sources of interference to HF communications, highlighted in this question from "The Doctor is IN" (June 2006 *QST*).

Q. Tom Bickley, K4NA, asks: I just bought a plasma TV. When I turned on my HF transceiver I got a rude awakening. I heard lots of noise across almost every band. When I turned off the TV, it all went away. I probably will return the set for a refund. The question I have now is: will an LCD model do the same thing? Any recommendations would be appreciated.

A. Almost any television can cause harmful interference to equipment operating in the amateur service. In fact, the problem isn't just confined to televisions. Many appliances, such as computers, air purifiers, electric blankets and others can cause interference. In many cases the problem is with ancillary equipment, such as switching power supplies, rather than the appliance itself. Thus the interference is often more a function of how well the equipment is designed than what type of equipment it is.

One way to improve your chances would be to bring a battery operated portable shortwave receiver with you to listen for potential problems before you buy another unit. You may have to ask the sales person to turn off nearby gear to avoid confusion and be sure that you're listening to the set you're considering. Use the AM mode and tune across all frequencies of interest.

Most TVI per Dollar!—War Stories from W6KFV

Accessory equipment is often overlooked in TVI cases. This includes things like unshielded A/B switches, remote TV tuners (with and without built-in cable converters), video games, preamps and VCRs. These devices are known to contribute directly to RFI. As can be found in an old Electronic Industries Association publication, "Consumers Should Know Something About Interference":

Anything connected to your TV is a suspected source of interference.

An older RFI case involved a Magnavox VCR that was rated by *Consumer Reports* to have the most features per dollar. This VCR caused severe audio interference to the TV set even when the TV set was turned off! No amount of external filtering on the cable leads and ac-line reduced the interference. The manufacturer said that it could not be fixed.

Another case involved an automatic remote TV tuner (with built-in cable converter). No degree of external filtering would solve the audio-interference problem. The manufacturer said, "What can I do that you have not already done? It cannot be fixed."

In another case involving an unshielded A/B switch the switch had to be removed from the circuit to solve the associated interference problem.—*John Norback, W6KFV*

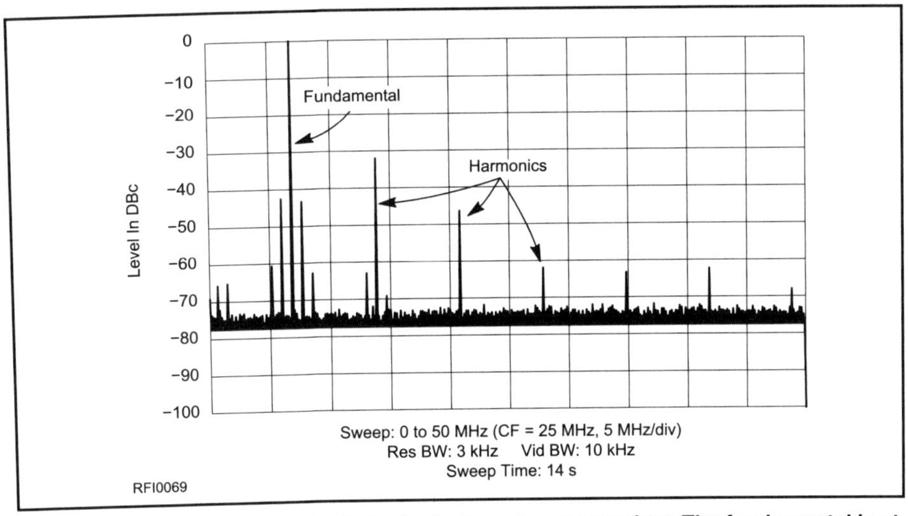

Figure 6.10—The spectral output of a typical amateur transmitter. The fundamental is at 7 MHz. There are visible harmonics at 14, 21 and 28 MHz. The unlabeled lines close to the fundamental are non-harmonic spurious emissions. This transmitter complies with the FCC spectral-purity regulations for HF transmitters with less than 5 W output.

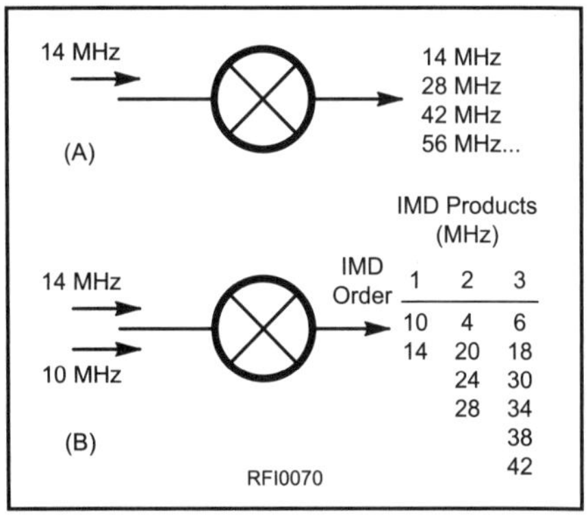

Figure 6.11—Mixers produce many spurious signals. When there is a single input signal, harmonics result (A). Multiple input signals produce both harmonics and many mixing products known as IMD (B).

emission falls within the passband of a TV channel and is not at least 40 dB below the received TV signal, it may cause TVI. This is usually seen as a herringbone pattern that is superimposed on the intended TV picture. An example of a herringbone pattern can be seen in Figure 6.4 and in Appendix B, the FCC's *Interference Handbook.*

In many cases, these spurious emissions are *harmonics* (exact multiples) of the operating frequency (see Figure 6.10). Transmitters that use a mixing process to obtain the desired output frequency (as do most modern amateur transmitters) can also produce unwanted outputs as a result of the mixing process (see Figures 6.10 and 6.11). These emissions can cause interference to other services. Any emission from a transmitter other than the fundamental signal, including mixing byproducts and harmonics, is referred to as a *spurious emission* or signal. All transmitters produce some spurious signals. The FCC regulations specify the permitted levels of spurious emissions for different power levels and operating frequencies. (See the Regulations chapter.) An amateur is responsible for interference caused by spurious emissions from the amateur transmitter.

These spurious signals are usually harmonics of the operating frequency, although they can also be various mixing products resulting from the superheterodyne and phase-lock-loop design of most modern amateur equipment. Figure 6.2 shows the way various amateur bands relate to over-the-air and cable TV channels.

Make Sure Your Own House is Clean

You are responsible for spurious signals produced by your station. If your station is generating any interfering spurious signals, the TVI problem must be cured at your station before you attempt to cure any TVI at the TV receiver. So, if the problem occurs only when you transmit, go back and check your station. Refer to the Troubleshooting chapter, the Transmitters chapter and the sections later in this chapter on troubleshooting and specific cures. It is especially important that you check your station if it appears that you have TVI only on those channels harmonically related to your transmit frequency. (Harmonics can be generated in an overloaded TV or in some other system external to the amateur station.) In some cases, you can cure an RFI problem by avoiding the transmit frequencies that have harmonics that fall inside of local TV channels.

In all cases of TVI, use a TV set to monitor TVI from the shack operating position. If any antenna-connected TV set in the immediate vicinity of the station is interference-free on all channels, then the station is not generating TVI from harmonics or other spurious emissions, or from external effects related to the station (see the chapter on externally caused EMI). If an interference case is ever brought before the FCC, they are able to quickly determine that the interference is not caused by the station if a station monitor is clean. If the station has been determined to be clean, then continue troubleshooting at the equipment that is experiencing interference.

Other Station Problems

It is easy for even the best operator to develop habits that can inadvertently contribute to TVI. The operator can become careless when tuning the rig for the millionth time. An overdriven final stage usually increases spurious output without any significant increase in output power. ALC and speech-compressor settings are important considerations that are frequently overlooked.

The location of your antenna, feed line and station ground can also be a factor. If it is possible for you to locate your antenna farther away from the TV and its antenna (this usually means higher), you may be able to cure a TVI problem.

The feed line and station ground are not *usually* directly involved with an interference problem, but if the feed line and/or ground wire are *much* closer to the TV than is the antenna, it is possible that feed line or ground wire radiation are to blame. In this case, the susceptible equipment may receive more energy from the ground lead than from the antenna. (See the EMC Fundamentals chapter for more information.)

Equipment grounding has long been considered a first step in eliminating interference. This is not necessarily true, however. While the method is very effective in the MF range and below, it is less useful in suppressing VHF energy; even short lengths of wire have considerable reactance at VHF. The delay effects along the wire are similar to those on the surface of an antenna. As the electrical length approaches ¼ λ, the wire appears as an open circuit rather than a low-impedance ground. It may also radiate like an antenna, as shown in Figure 6.12.

You may have a defect in your station, such as a defective trap in your antenna system that is arcing when you transmit, generating very strong RF noise. If your station is held together with baling wire and chewing gum (not that yours would be, mind you, but if it were), these types of installation defects could be a very strong contributor to your RFI problem.

If your transmitter and station check "clean," then you must look elsewhere. The

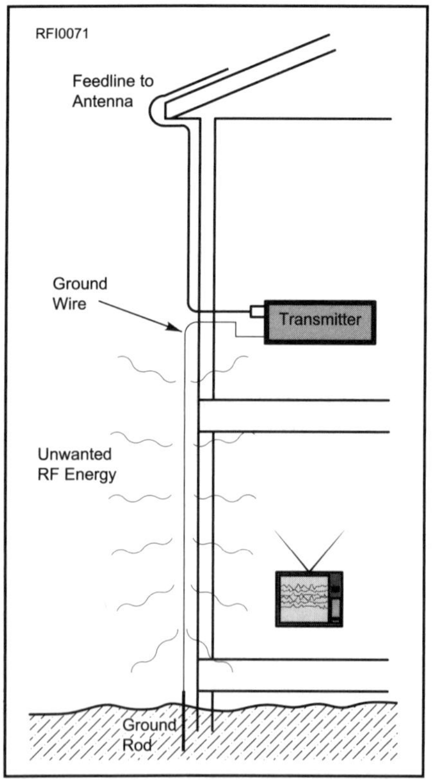

Figure 6.12—A second-floor apartment presents a nearly impossible grounding situation. The ground lead acts as a VHF long-wire antenna that can radiate unwanted energy. This station might be better with no ground connection. A lossy ground plane, described in "Eliminate TVI with Common-Mode Current Controls," May 1984, *QST*, pp 22-25, offers a better alternative in cases where a good RF ground is required.

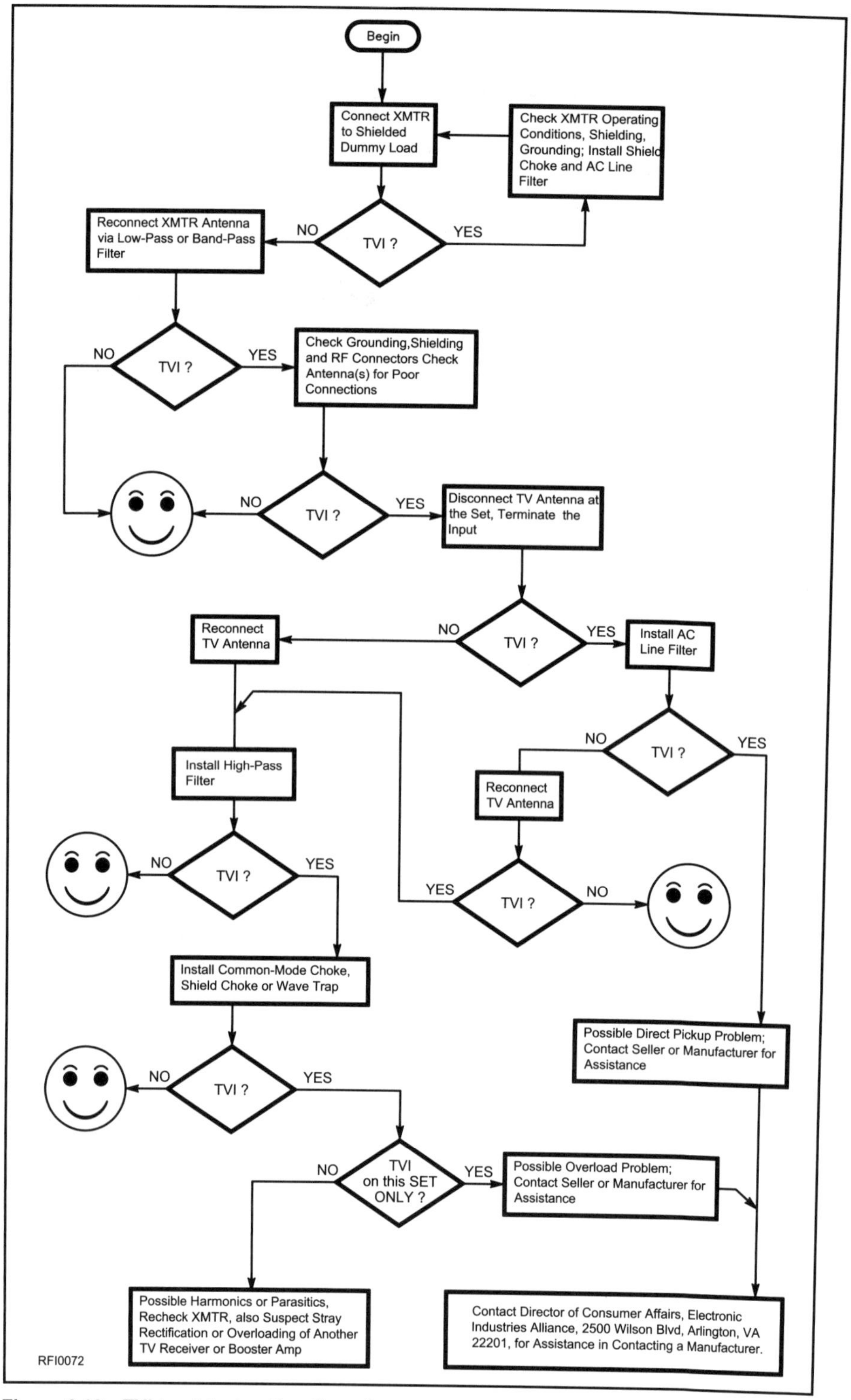

Figure 6.13—TVI troubleshooting flow chart.

most likely cause is TV susceptibility—fundamental overload. This is usually manifest by interference to all channels, or at least all VHF channels.

Fundamental Overload

Although the transmitter *could* be the cause of TVI, most cases are caused by fundamental overload. This term means that the TV circuitry is not able to function in the presence of strong RF energy and is being overloaded by your fundamental signal. A television set can be overloaded by a strong, local RF signal; this is *not* the fault of the transmitter! Fundamental overload happens because the manufacturer did not install the necessary filters and shields to protect the TV set from other signals present on the air. Fundamental overload is usually cured at the TV or its installation, usually by applying external filters that correct the design deficiencies that contribute to the problem.

Direct Pickup

In some rare cases, the TV set itself may directly pick up the RF signal. This is known as *direct pickup*. The term refers to interference that results when fundamental RF energy is picked up directly by the circuitry of susceptible equipment. As fundamental energy is picked up by the circuitry of a TV (or other piece of equipment), it can result in severe interference. This generally involves the internal wiring or printed-circuit traces. This type of interference is rare in modern TVs, but it may still be a problem with older equipment. External shielding may help, but it is not always practical. In cases of direct radiation pickup, contact the set manufacturer.

Measurements in an anechoic chamber indicate that several kilowatts are required to generate this type of interference at the normal distances between an amateur antenna and a TV set. It is highly unlikely that direct radiation pickup by TV circuitry will be encountered unless operating in an apartment building with an indoor or attic amateur antenna. However, this mode of EMI may affect TV-control remote circuitry. Such interference to a remote control is usually eliminated by shielding the leads to the infrared detector on the front of the set.

To test for this type of interference to the picture, replace the TV antenna with an appropriate 300 or 75-Ω termination and note the level of interference to the set raster. If the raster shows no flicker or drastic change and the IR controller doesn't false, there is not a direct-radiation-pickup problem. There must be some other mode of interference such as fundamental overload. With the TV signal removed from the TV set, the TV gain drastically increases (because of the TV circuitry). Hence, this test is a sensitive indicator.

A second test for the presence of direct radiation pickup can be made at amateur stations that have rotatable antennas. Temporarily remove the antenna from the TV set and rotate the amateur antenna while transmitting. If the interference level increases while the amateur antenna is pointing at the house (especially the TV and its feed line), it is likely that the set has a direct-radiation-pickup problem.

If direct-pickup interference is indicated, contact the TV manufacturer through the Consumer Electronics Manufacturers Association (CEMA).

IF Interference

Some TV receivers do not have sufficient selectivity to reject strong signals at the TV intermediate frequency (IF). The third harmonic of 14 MHz and second harmonic of 21 MHz fall at the television IF (41.0-47.0 MHz), as do some local-oscillator frequencies used in heterodyne transmitters or transceivers. If these frequencies break through the TV tuner, a high-pass (cutoff above 40 MHz) filter and common-mode choke can significantly improve the situation.

There is a form of IF interference that is peculiar to 50-MHz operation near the low edge of the band. Some TVs with the standard 41-47 MHz IF (sound carrier = 41.25 MHz; picture carrier = 45.75 MHz) pass a 50-MHz signal into the IF stages. The 50-MHz signal beats with the IF picture carrier to give a spurious response at (or near) the IF sound carrier, even though the interfering signal is not actually in the normal passband of the IF amplifier.

IF interference is easily identified because it affects all channels (although the intensity sometimes varies from channel to channel) and the crosshatch pattern it causes rotates as the TV fine-tuning control is moved. With harmonic, fundamental-overload and IMD interference, the orientation of the interference pattern does not change (its intensity may change) as the fine-tuning control is varied.

THE SUSCEPTIBLE TV

Fundamental overload (which can also affect FM receivers) is cured by preventing undesired signals from entering the affected receiver. The remedies are simple and well established. Use a combination of high-pass filters, ac-line filters and common-mode chokes on the feed line and ac power leads.

When a television receiver is located close to a non-TV transmitter, the intense RF signal from the transmitter's fundamental may overload one or more of the receiver circuits to produce spurious responses or blocking that cause interference.

When overload is moderate, the interference is much like harmonic interference (the harmonics are generated in the early stages of the receiver). Since it occurs only on channels harmonically related to the transmit frequency, it is difficult to distinguish light overload from transmitter harmonics. In such cases, additional harmonic suppression (transmit filters) at the transmitter does no good, but measures to reduce amateur signal strength at the TV improve performance. When fundamental overload is severe, interference affects all channels.

All of these potential problems are made more severe because the TV set is hooked up to two antenna systems: (1) the incoming antenna and its feed line and (2) the ac power lines. These two "long-wire" antennas can couple a lot of fundamental or harmonic energy into the TV set. The TVI Troubleshooting Flowchart, Figure 6.13, is a good starting point.

Start by determining if the interference is affecting the video, the sound or both. If it is present only on the sound, it is probably a case of audio rectification. (See the Stereos section of this chapter.) If it is present on the video, or both, it could be getting into the video circuitry or affecting either the tuner or IF circuitry.

The first line of defense for an antenna-

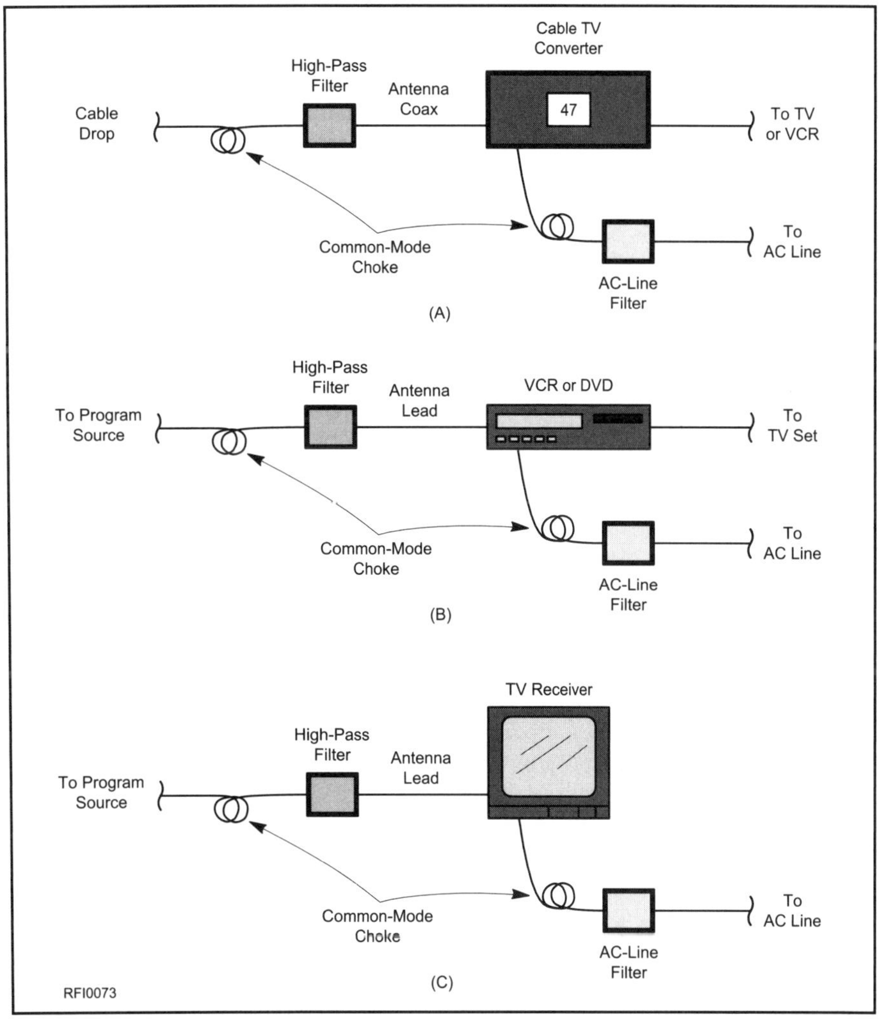

Figure 6.14—This sort of installation should cure any kind of conducted TVI. It will not cure direct-pickup or spurious-emission problems.

connected TV is a high-pass filter. Install a high-pass filter directly on the back of the TV set. You may also have a problem with common-mode interference. The second line of defense is a common-mode choke on the antenna feed line, also placed as close to the set as possible—try this first in a cable-television installation. These two filters can probably cure most cases of TVI!

Figure 6.14 shows a "bulletproof" installation. If this doesn't cure the problem, the TV circuitry is picking up your signal directly. In that case, don't try to fix it yourself—it is a problem for the TV manufacturer.

Intermodulation (IMD)

Under some circumstances, fundamental overload results in IMD, or mixing of the overloading signal with others (such as those from local FM or television stations). The most common place for this to happen is in the front end of a TV set or VCR.

For example, a 14-MHz signal can mix with a 92-MHz FM station to produce a beat at 78 MHz, the difference between the two frequencies (92 – 14 = 78). Since 78 MHz is within TV channel 5, TVI would occur on that channel. The affected TV channel has no harmonic relationship to 14 MHz. Both signals must be on the air for interference to occur. Eliminating either at the receiver will eliminate the interference.

There are many possible IMD combinations, depending on the band in use and the frequency assignments of local stations. The interfering frequency is equal to the sum or difference of the stations involved. Their harmonics may enter into the mixing process as well. Whenever interference occurs at a frequency that is not harmonically related to the transmitter frequency, investigate IMD possibilities.

Electrical Noise

Electrical noise is fairly easy to identify by looking at the picture or listening on an HF receiver. Electrical noise on a TV screen is shown in Figure 6.15. On a receiver, it usually sounds like a buzz, sometimes changing in intensity as the arc or spark sputters a bit. If you have a problem with electrical noise, go to the electrical-interference chapter.

The FCC rules do not protect fringe-area reception. If you are outside that station's primary coverage area, you will just not get enough TV signal to be usable. If you or your neighbors are having problems receiving that station, it is not surprising.

VHF TVI CAUSES AND CURES

Television interference has traditionally been mostly a problem of HF operators. This results from a number of factors. HF operation is more apt to use high power. (Even a 100-W "barefoot" rig uses more power than the typical VHF FM base station.) The FCC requirements for spectral purity at VHF are more stringent than those at HF. (No spurious emission from a 25-W VHF station can exceed 25 μW.) This results in a significantly reduced TVI potential.

This doesn't mean, however, that TVI doesn't happen to VHF operators, too! A VHF signal is within the passband of a TV high-pass filter. A VHF signal picked up by a TV antenna passes through the high-pass filter to the TV set. A band-reject filter can help reduce VHF fundamental energy.

Many of the techniques discussed in this chapter are most appropriate to an HF transmitter. This is not meant to exclude VHF (and UHF) transmitters. Most VHF operation uses 25 W or less power, which reduces the likelihood of problems. If a VHF transmitter reached the market since 1978, the FCC VHF spectral-purity regulations (spurious emissions less than –60 dBc at power levels of 25 W or greater), provide greater protection than is afforded HF transmitters. In addition, only a few TV channels are harmonically related to the VHF fundamental signal.

All these factors tend to make television interference a smaller problem at VHF than at HF. Of course, a 6-meter operator who lives in an area with channel-2 over-the-air TV service, or a 2-meter operator who lives in an area with cable channel 18 might see things differently. Most TVs don't have enough front-end selectivity to reject a 6-meter fundamental signal, and cable channel 18 includes frequencies allocated to the amateur 2-meter band.

Table 6.1 lists common causes of VHF TVI. There are other possibilities. Nearly all can be corrected completely; the rest can be substantially reduced.

Items 1, 4 and 5 are receiver faults; no amount of filtering at the transmitter can eliminate them. (A reduction of transmit power can help, but that is a voluntary concession, not a cure.) The cure for these problems must be applied to the TV receiver. In mild cases, increasing the antenna separation may reduce or eliminate the problem. Item 6 is also a receiver fault; it may be minimized by using FM or CW instead of SSB.

Treat harmonic troubles (items 2 and 3) with standard methods given in the Transmitters chapter. Builders of VHF equipment should become familiar with TVI-prevention techniques and incorporate them in projects.

UHF TELEVISION

Harmonic TVI in the UHF-TV band is far less troublesome than in the VHF band. Harmonics from transmitters operating below 30 MHz are normally weak. In addition, the components, circuit conditions and construction of HF transmitters tend to prevent strong VHF harmonics. This is not true of amateur VHF transmitters, particularly those working in the 144 MHz and higher bands. There the problem is similar to that of the low-VHF TV channels with respect to HF transmitters.

Multiple Causes

Most interference problems have multiple causes. The problem won't be fixed until you identify *all* of the causes and apply the appropriate cures. If, for example, TVI were being caused by harmonics from the transmitter, fundamental overload of the TV, harmonics being created in an overloaded TV preamplifier *and* there was electrical noise present, the TVI wouldn't go away until everything had been taken care of. Consider the impact of multiple causes as you troubleshoot TVI.

FILTERS

Most TVI is cured by the proper application of the appropriate TVI filter(s). The characteristics of several types of filters are

Figure 6.15—Two examples of TVs experiencing electrical noise.

Table 6.1

The principal causes of TVI from VHF transmitters are:

1. Fundamental overload from the transmitted fundamental signal
2. Fourth harmonic of 50 MHz in channels 11, 12 or 13, depending on the operating frequency
3. Radiation of transmitter local-oscillator or multiplier stage harmonics
4. IF-image interference in channel 2 from 144 MHz transmitters (in TVs with a 45-MHz IF—see Figure A)
5. Sound interference (picture is clear in some cases) resulting from RF pickup by the TV audio circuits

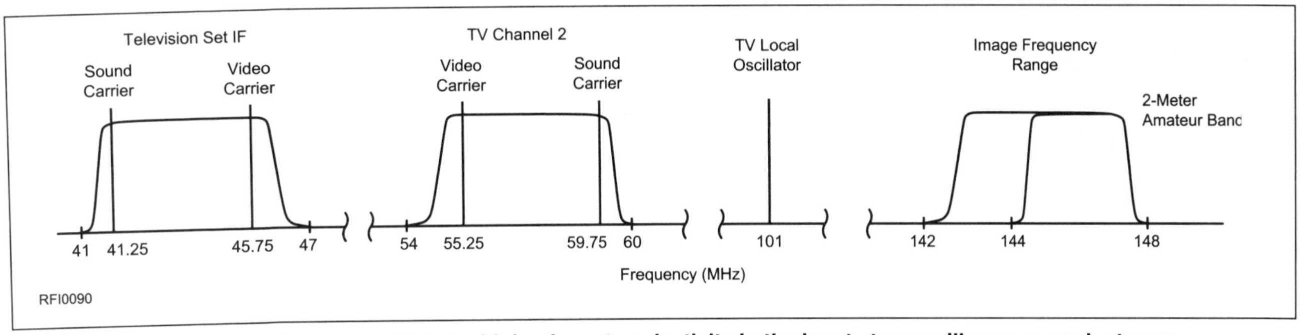

Figure A—A receiver tuned to channel 2 but with inadequate selectivity in the input stages will pass enough strong 144-MHz band signal to cause image interference with reception. Adding a 144-MHz trap in the antenna feed line improves the selectivity of the TV receiver and prevents image interference from 144-MHz band signals.

Intermodulation Problems and TVI—A Few Case Histories

In most cases intermodulation (IMD) problems that cause TVI are seen on one of the higher TV channel numbers. These problems frequently involve commercial paging transmitters running high power and located in metropolitan areas.

In one Chicago case, both a 220-MHz repeater and a paging transmitter were located in a hotel. When the 35-MHz paging transmitter and the 220-MHz repeater came on at the same time, there was severe channel-9 interference in the whole area. Signals from the two transmitters were mixing in the front ends of many TV sets. The solution was to move the paging transmitter. It was possible to accomplish this because the hotel management recognized that their severe interference problem started when the paging transmitter was installed.

In another case, an amateur was having a problem when his packet station transmitted while he was operating on the HF bands. In this case, the packet transmitter had IMD in the final RF transistor. A TV high-pass filter at the VHF transmitter output protected it from the HF energy. In this case the filter was able to handle the power level of the packet transmitter. At high power levels a custom-built filter may be required. (Even though TV high-pass filters are designed for 75-Ω feed lines, they work reasonably well in a 50-Ω transmitting system up to the 50-W level.) High-power filters are discussed at length in the *ARRL Handbook*.

Another amateur had a problem during HF transmissions: Interference to TV channel 2 came and went. The interference seemed random until a careful log of household activities showed that the problem occurred only when another member of the family was home. The son always parked his truck, which contained a CB transceiver, next to the tower. The CB receiver was protected by a diode across the input. When there was enough RF energy, the diode generated harmonics that were radiated from the CB antenna. Simply parking the truck away from the tower solved the problem.

A similar scenario occurred when a garage-door opener receiver converted amateur HF energy to harmonic energy. This caused a channel-2 TVI problem. The problem was cured by shortening the opener antenna.

One amateur had been working on a TVI problem for a year with no success. The culprit turned out to be a weather monitoring receiver with a wall-outlet power supply. An unbypassed diode in the power module was the cause. The solution was quite humorous: The amateur picked up the monitor receiver and threw it as hard as possible across the basement—instantly generating a Heathkit.

Finally, a ham had a severe case of TVI that was generated by his DX-spotting 2-meter transceiver. This unit had an antenna mounted on the same mast that supported the HF beam. Once again, energy picked up by the VHF antenna was delivered to the VHF transceiver, where it was rectified by protective diodes and radiated. This problem was also solved with the help of a TV high-pass filter mounted on the 2-meter transceiver. It is encouraging to see that newer VHF rigs place the protection diodes after the input filter, so that HF energy can no longer reach them.—*Ed Doubek, N9RF, Illinois Section Technical Coordinator, Naperville, Illinois*

discussed in the EMC Fundamentals chapter. This chapter offers an overview of the types of filters that are generally helpful in curing TVI.

Resonances in the TV feed line may affect filters. Sometimes filters work better at a low-impedance point in the line. It is often difficult to locate a low-impedance point, so experiment with filter position when necessary.

Types of Filters

The types of filters that help cure TVI are:

- Transmit low-pass or bandpass filters
- VHF high-pass filters
- VHF band-reject or notch filters
- Common-mode chokes
- AC-line filters

Every TVI troubleshooter should keep a variety of filters on hand for test purposes. Several filters appear in Figures 6.16 through 6.19. ARRL HQ receives many RFI letters that indicate there are no filters installed on

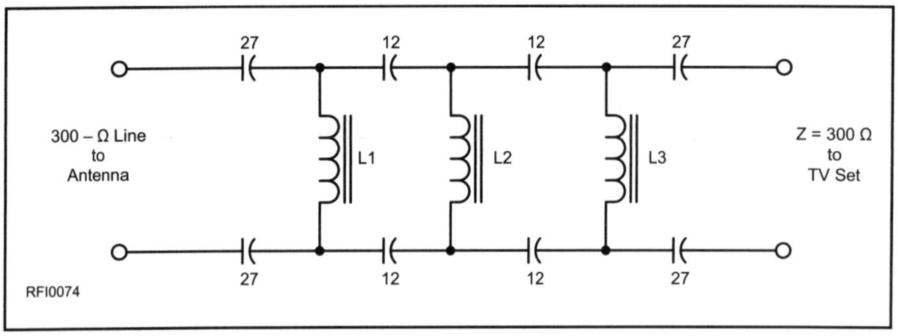

Figure 6.16—A differential-mode high-pass filter for 300-Ω twin-lead. When installed in the feed line at a TV or accessory input, it rejects differential-mode HF signals picked up to the TV feed line or antenna. Capacitor values are in pF.

L1, L3—13 turns no. 24 enameled wire on T-44-10- toroid core (0.67 μH)

L2—12 turns no.24 enameled wire on T-44-10 toroid core (0.57μH)

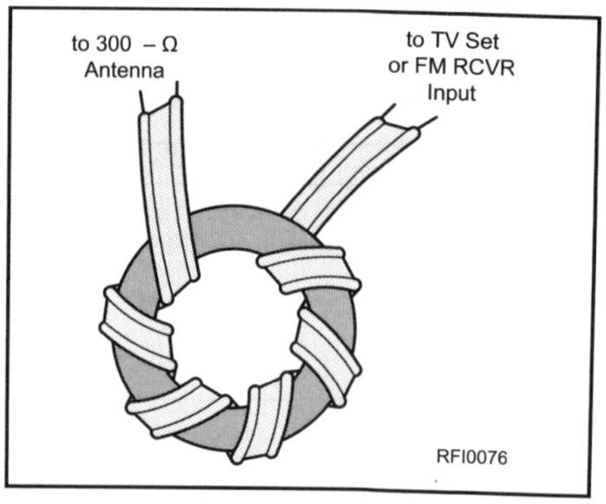

Figure 6.18—Several turns of TV twin-lead cable on a toroid ferrite core can eliminate HF and VHF common-mode signals from a twin-lead TV system.

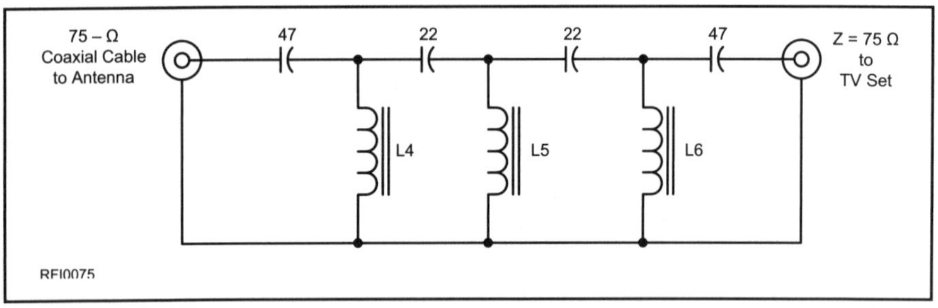

Figure 6.17—A differential-mode high-pass filter for use with 75-Ω coax. It rejects HF signals picked up by a TV antenna or signals that leak *into* a CATV system. It will *not* reject common-mode signals, HF or VHF, on the outside of the coax. Capacitor values are in pF.

L4, L6—12 turns no. 24 enameled wire on T-44-0 toroid core (0.157 μH)

L5—11 turns no. 24 enameled wire on T-44-0 toroid core (0.135 μH)

Figure 6.19—Several turns of coax on a ferrite core eliminate HF and VHF signals from the outside of a coaxial feed-line shield.

the consumer equipment. Always try the easy things first—if they are real easy, try them twice!

The Transmit Filter

Every amateur who expects to encounter interference problems, and this includes nearly every amateur, should have a properly installed transmit filter on the transmitter or transceiver antenna lead (see Figures 6.20 and 6.21). It is not a cure-all for interference, but it can be considered "good engineering practice," which is required by the FCC. When confronted with interference, an amateur can point to the filter with pride.

For an HF transmitter this filter is usually a commercially available low-pass filter. For VHF transmitters, a band-pass filter is the most useful. Figure 6.22 shows a commercially available VHF band-pass filter made by DCI. A cavity-type filter is represented in Figure 6.23.

Notch filters can also be used as transmit filters. A few examples are shown in Figure 6.24.

High-Pass Filters

Most cases of TVI to antenna-connected

Fringe Area Reception

For a TV signal to look good, it must have about a 40 to 45 dB or better signal-to-noise ratio. This requires a good signal at the TV antenna-input connector. This brings up an important point: to have a good signal, you must be in a good signal area. FCC regulations define the service area of a TV station. The minimum acceptable signal area is defined as the Grade-B signal contour. In general, this is typically about 50 miles or so from the TV station, although the exact distance can be affected by factors such as the height of the TV antenna, the surrounding terrain, etc. The TV station operator should be able to tell you where the Grade B contour is. Reception outside of this defined service area *is* possible, but the TV signals are not strong enough to guarantee acceptable reception. This weak-signal reception is known as "fringe-area" reception.

The FCC does not provide protection to fringe-area TV signals. *The FCC Interference Handbook* says, "Although fringe area signals may be satisfactory, they are weak signals that are highly susceptible to interference and are not protected by the Commission." You may want to help your neighbors if you live in a weak-signal area, but if you are outside the television service area, this is a voluntary action on your part.

Fringe-area reception can be a two-edged sword. TVI is more of a problem in TV fringe areas, since even low-level RF signals can disrupt a weak TV signal. The good news is that the FCC does not offer protection to fringe-area reception. TV coverage areas are divided into "contours" according to signal strength. Fringe areas are those outside of the "Grade-B contour" area. This means that fringe-area viewers are not in the intended audience of the station, and the FCC does not protect them from interference. Of course, hams usually want to be good neighbors and are willing to offer some help in cases of fringe-area TVI.

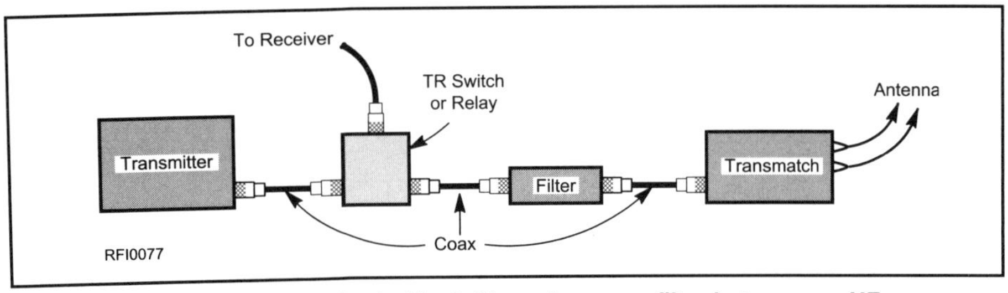

Figure 6.20—The proper method of installing a low-pass filter between an HF transmitter and Transmatch. The transmitter and filter must be well-shielded and grounded. If a TR switch is used, install it between the transmitter and low-pass filter. (TR switches can generate harmonics.) If the antenna is well matched (SWR = 1.5:1 or less), the Transmatch can be eliminated. A mismatch degrades filter performance.

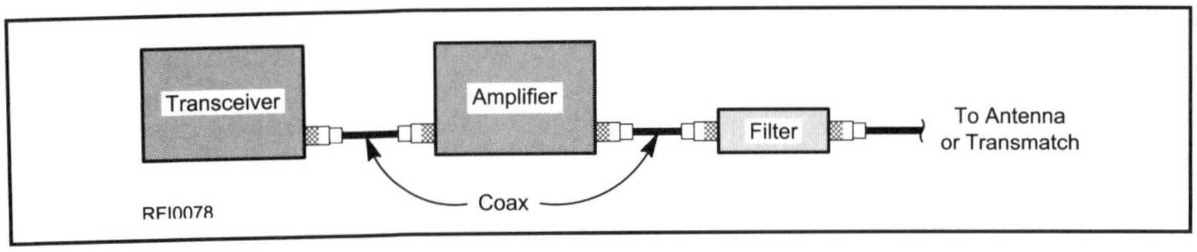

Figure 6.21—This is the proper location for a low-pass filter in stations that use a linear amplifier. Some hams insert a low-pass filter between the transmitter and amplifier, but it isn't necessary in most cases.

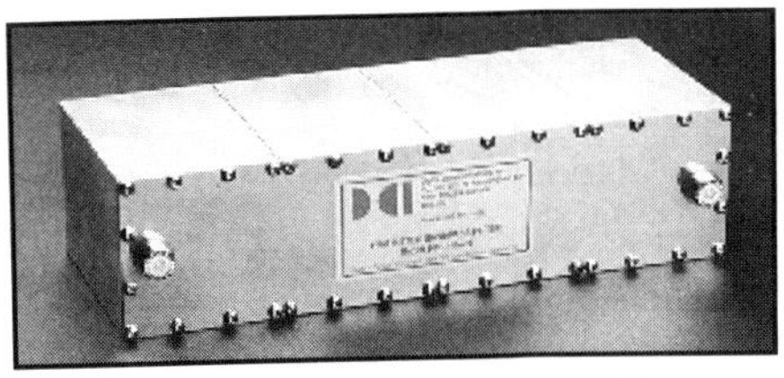

Figure 6.22—This VHF bandpass filter can eliminate spurious emissions on VHF transmitters.

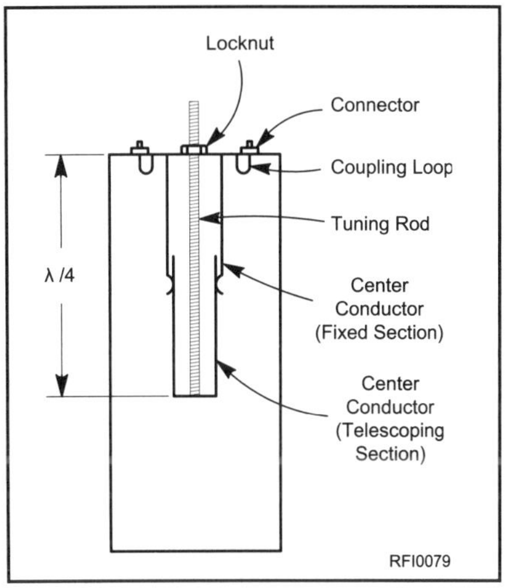

Figure 6.23—A cutaway view of a typical cavity. Further discussion of cavity filters and construction details for a 144-MHz duplexer (an assembly of cavities) appear in *The ARRL Antenna Book*.

6-Meter TVI

Very little information has been written lately about TVI caused by operation in the 6-meter band, where frequencies are in proximity to TV channels 2 and 3. Six-meter signals are some of the most difficult to isolate from these two TV channels. Normal high-pass filters do not effectively filter out strong VHF amateur signals. Other steps must be taken.

To reduce, or preferably eliminate, TVI caused by a 50-MHz signal, a resonant circuit of some sort is needed at the TV receiver. One such circuit is a resonant stub installed at the TV antenna terminals and tuned to the 6-meter operating frequency. There are two drawbacks to the use of a stub: It only notches out one operating frequency, and it also notches out all odd harmonics of the operating frequency (it's probably unsuitable for cable use). If the amateur station operates on a frequency much removed from that of the stub, interference returns. The amateur can only operate over the limited frequency range for which the stub is tuned.—*Roger Laroche, N6FOU, Santa Barbara Section Emergency Coordinator, Arroyo Grande, California*

TVs from HF transmitters can be cured by installing a high-pass filter at the TV set. Figure 6.25 shows a commercially available high-pass TVI filter.

Simple high-pass filters cannot always be applied successfully in the case of 50-MHz transmissions, because most filters do not have sufficiently sharp cutoff characteristics to give both good attenuation at 50-54 MHz and no attenuation above 54 MHz. A more elaborate design capable of giving the required sharp cutoff has been described (F. Ladd, "50-MHz TVI—Its Causes and Cures," *QST*, June 1954, pp 21-23, 114, 116 and July 1954, pp 32-33, 124 and 126.) This article also contains other information useful in coping with the TVI problems peculiar to 50-MHz operation.

Install a commercially made differential-mode high-pass filter in the TV feed line, as close to the TV as practical. Such filters are sold by RadioShack and many department stores that sell TV accessories. High-pass filters are available for 300-Ω twin lead or 75-Ω coax.

The filter may not work well with some newer TVs. Sets with a 75-Ω coaxial input and an "ac/dc" chassis (no isolation transformer) use a capacitor in the shield lead to isolate the coaxial shield from the ac line. This provides a low-impedance path for VHF energy, while presenting a high impedance at 60 Hz, and some moderate impedance at HF. Because of this poor HF ground, high-pass filters at the antenna input terminals may not be as effective as those at the TV tuner.

High-pass filters are available commercially at moderate prices. All should understand however, that while an amateur is responsible for harmonic radiation from the station transmitter, the amateur is not responsible for filters, wave traps or other devices needed to protect the receiver from interference caused by the fundamental frequency. The question of cost should be settled between the set owner and the manufacturer or seller. Don't overlook the possibility that the TV manufacturer may supply a high-pass filter free of charge.

Common-Mode Filters

If a stock high-pass filter doesn't cure the problem, suspect common-mode interference. A common-mode choke is another possible cure for common-mode interference. The filter arrangement shown in Figure 6.26 can be used. A photograph of a common-mode choke and a differential-mode high-pass filter installed on a 75-Ω cable is seen in Figure 6.27. Additional cures for common-mode problems appear in Figures 6.28 and 6.29. Any of these methods should effectively eliminate the common-mode signal picked up by twin-lead or coaxial cable.

If the TV is fed by a 75-Ω cable, ensure that

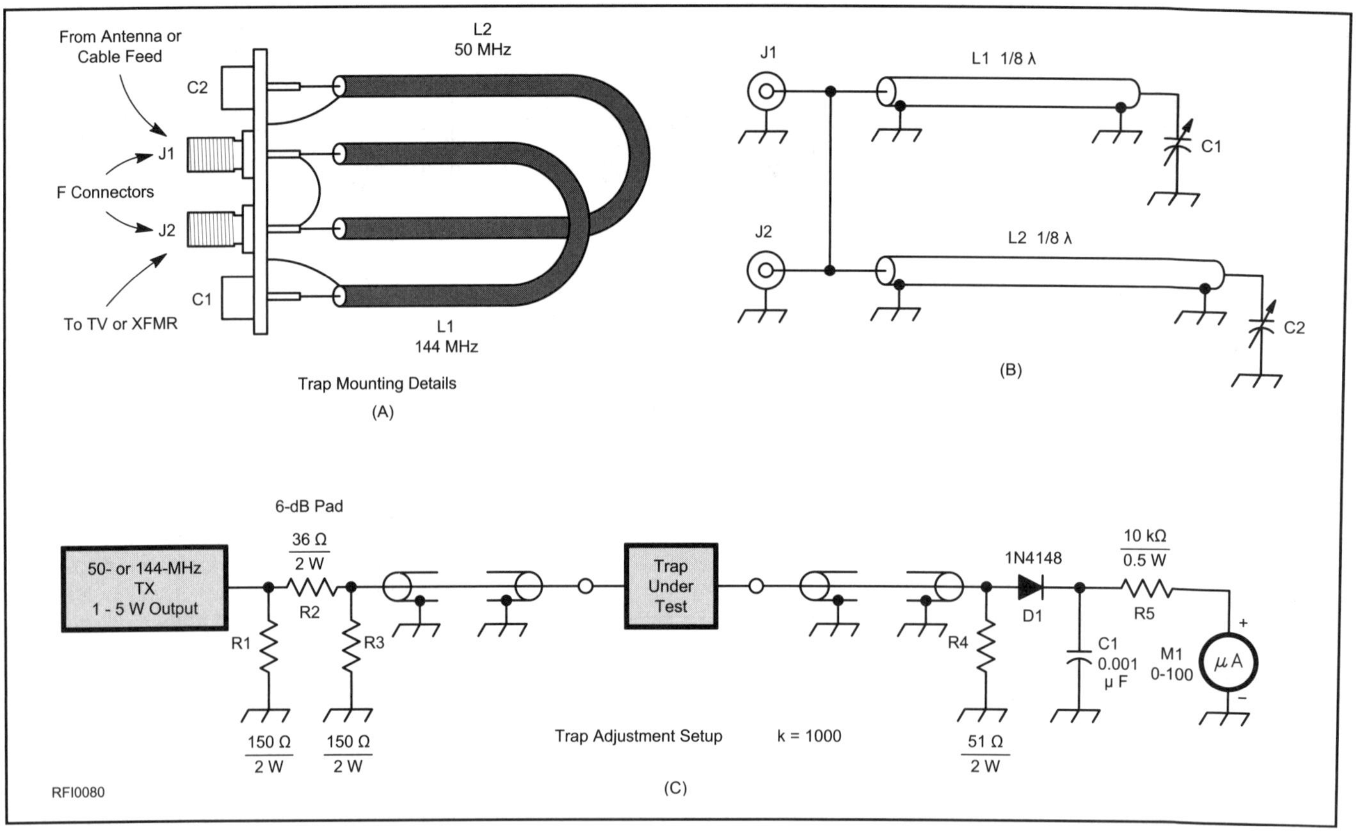

Figure 6.24—A pictorial diagram of the mounting details for the coaxial traps is shown at A. The F connectors, variable capacitors and cable loops are mounted on a scrap of double-sided PC-board material. At B, the schematic diagram of the trap circuit is shown. The filter is bilateral; either connector can be used for input or output. C shows a test setup to pretune the coaxial traps. Adjust the capacitors for minimum meter deflection at the desired frequency. Some retuning is often necessary after the traps are connected to the TV.

C1—3-12 pF ceramic trimmer.
C2—7-45 pF ceramic trimmer.
L1 (144 MHz)—6.7 inches for cables with 0.66 velocity factor (VF) (RG-8, RG-58); 8.2 inches for cables with 0.80 VF (semi-rigid and foam-dielectric cables).
L2 (50 MHz)—18.7 inches for cables with 0.66 VF; 22.7 inches for cables with 0.80 VF.

the shield of the cable is properly grounded where it enters the house. This may shunt most of the common-mode signal harmlessly away from the TV. Grounds on cable-TV installations should be installed only by qualified cable service personnel.)

Notch Filters

VHF notch band-reject filters are available to amateurs. Notch filters are most often used to notch out a strong VHF signal on antenna connected TVs. Notch filters should not be used in a cable TV installation; most cable systems use the VHF and UHF ham bands for active channels (see Figure 6.2); a notch filter would remove one or more of these channels. Figure 6.30 shows the schematic of a home-brew notch filter.

A notch filter should normally be installed right at the TV antenna input. If a TV preamplifier is used, however, the notch filter usually needs to be installed at the preamplifier input. In the case of mast-mounted preamps, installing the notch filter is *not* an easy job.

Even though a common-mode filter cannot reduce the pickup of your amateur signal by an antenna and feed line (only reducing the total length of the feed line can), it can reduce the coupling of the amateur signal to the TV. Remember that the TV antenna is not effective at HF, but the feed line to the TV is an effective HF antenna. The TV antenna can pick up amateur VHF/UHF signals (usually in a differential mode), and the TV feed line can pick them up in the common mode.

Figure 6.25—This commercially available TVI high-pass filter is the first line of defense for RFI to antenna-connected TVs.

Figure 6.26—This is how to wind an antenna-lead common-mode choke. Wrap ten to fifteen turns of feed line through the core. Use no. 73 core material for interference from HF signals. Number 43 core material works well for common-mode interference from VHF signals. In some installations there is not enough slack feed line to allow the required number of turns, so it is a good idea to have a common-mode choke prepared, using either 75-Ω cable with the appropriate connectors (two crimp-on male connectors, along with a double female "barrel" connector work well) or 300-Ω twin lead, as appropriate.

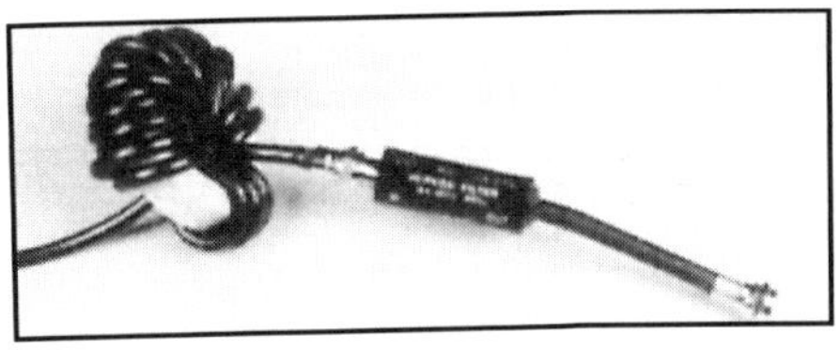

Figure 6.27—A 75-Ω high-pass filter has been installed with a common-mode choke. This combination will eliminate nearly all TVI from signals that are entering the set by way of the feed line.

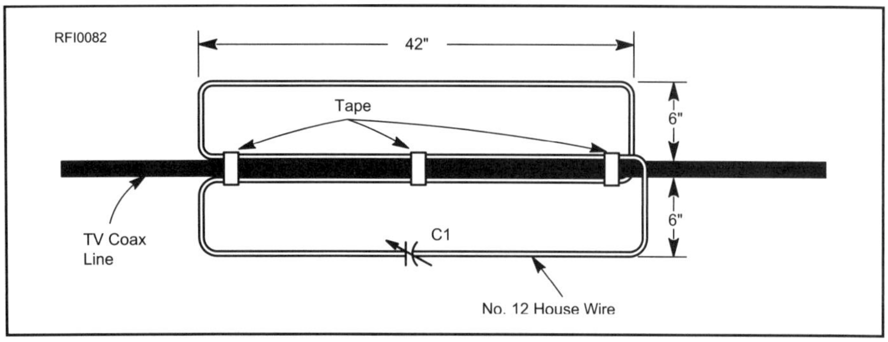

Figure 6.29—A breaker loop for 80 and 40 meters. The loops should be taped securely to the coaxial cable at several locations. Gaps between the loops and the outer surface of the coaxial cable greatly lessen the coupling effectiveness of the assembly. C1 should be adjustable and have a maximum capacitance of 150 pF or more. Either air-variable or compression-trimmer capacitors are satisfactory.

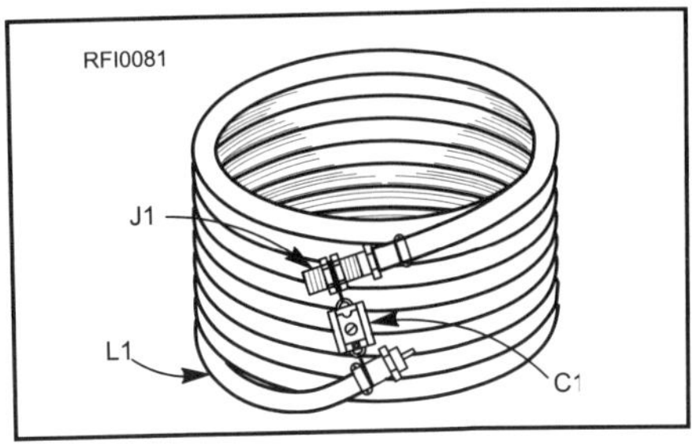

Figure 6.28—A compact shield-breaker assembly. This is another way to eliminate common-mode interference. It forms a high-impedance, parallel-resonant circuit at the operating frequency. It is effective, but only for interference resulting from one HF band.

C1—150 pF ceramic trimmer
J1—F-81 bushing (Radio Shack 278-213)
L1—8 feet (2.4 meters) 75-Ω coaxial cable with an F-59 connector on each end (RadioShack 15-1530).

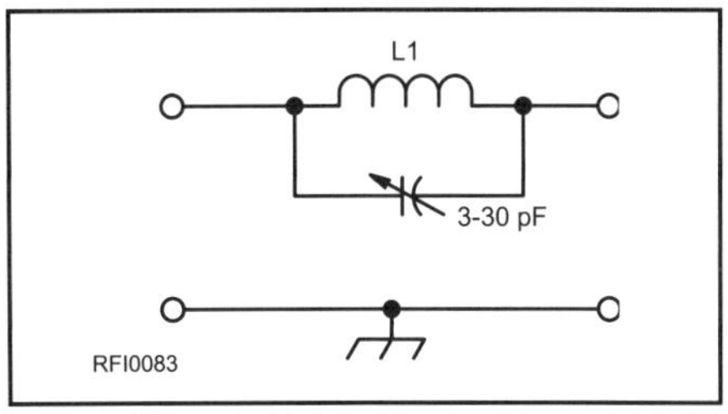

Figure 6.30—A tunable VHF notch filter. For 50-MHz use, L1 is 9 turns of no. 16 enameled wire close wound on a ½-inch-diam form (air core). For 144 MHz, L1 is 6 turns of no. 16 enameled wire close wound on a ¼-inch-diameter form (air core).

Figure 6.31—An ac-line common-mode choke. This eliminates most forms of ac-line related RFI that results when power lines pick up a radio signal. Wrap about 10 turns of the ac-line cord through a ferrite toroid or around a ferrite rod. Use no 43 (VHF and upper HF) or no. 75 (lower HF) material.

AC-Line Filters

An ac-line filter with an additional common-mode choke should be installed on the TV set power cord. This reduces the possibility of RF energy entering the ac power system by that route. Examples of common-mode and ac-line filters that can be used on transmitters or home-electronic equipment are shown in Figures 6.31 and 6.32. Figure 6.33 shows a "brute-force" ac-line filter.

An ac-line filter and common-mode ac-line choke eliminate the possibility that the interfering signal enters through the ac line. Most ac-line filters are effective only against differential-mode signals present between the hot and neutral leads; they do not filter common-mode signals that may also be present. In most cases, an ac-line common-mode choke is needed as well.

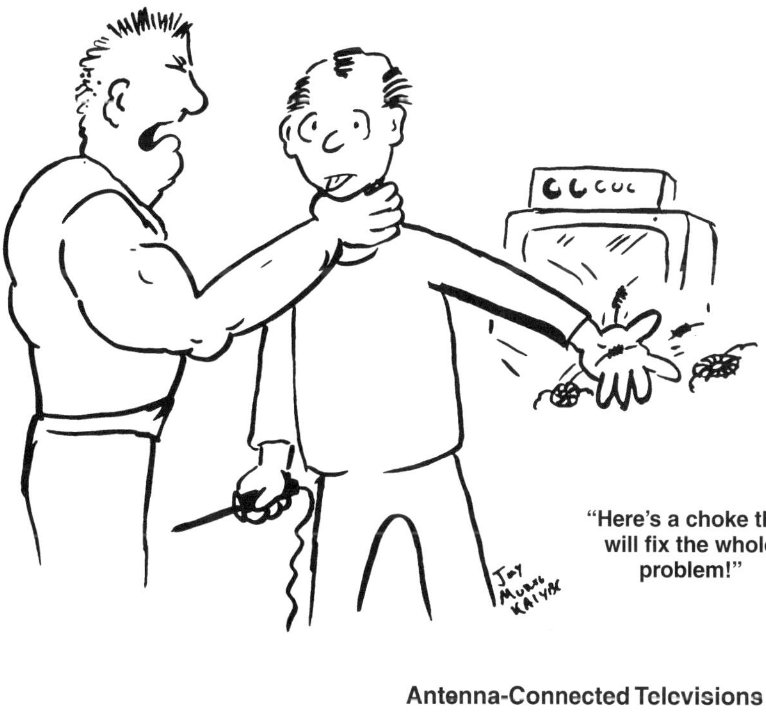

"Here's a choke that will fix the whole problem!"

FILTER CONSTRUCTION

TV High-Pass Filters

In most TVI cases, interference can be eliminated if the fundamental signal strength is reduced to a level that the receiver can reject. To accomplish this with signals below 30 MHz, the most satisfactory device is a

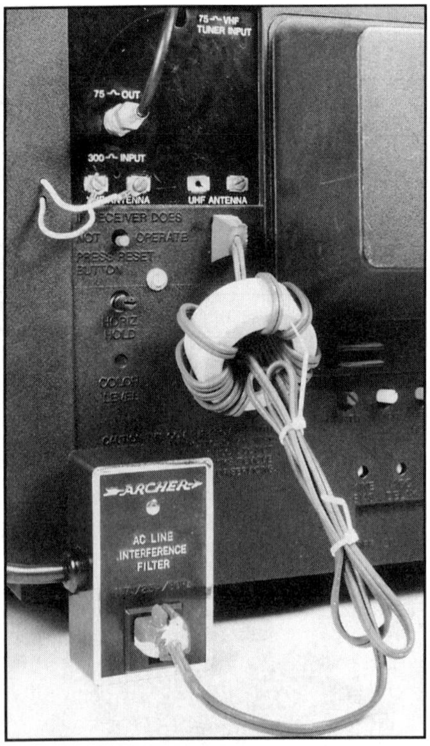

Figure 6.32—A commercial ac-line filter installed with a common-mode choke. Note that the excess line cord has been coiled around the common-mode choke, minimizing the "capture area" of the ac-line cord "antenna."

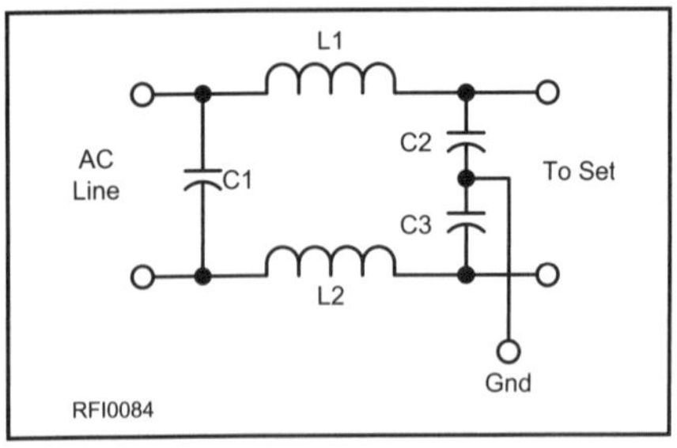

Figure 6.33—A "brute force" ac-line filter. C1, C2 and C3 can be any value from 0.001 to 0.01 µF, rated for ac-line service (1.4 kV dc). L1 and L2 are each 2-inch-long windings of #18 enameled wire on a ½-inch diameter form. If installed outside the equipment cabinet, enclose the filter to eliminate shock hazard. This filter is similar to commercial ac-line filters.

high-pass filter (with a cutoff frequency just below 54 MHz) installed at the tuner-input terminals of the TV.

Refer to Figures 6.34 and 6.35. Slice off the extruded insulation around the solder pins on two type-F coaxial connectors (RadioShack 278-212). Butt the connectors directly against the PC board. Solder the connector shells to the bottom ground plane and the center pins to the microstrip line. Cut the microstrip line in four equally spaced places. The capacitors should be mounted across the spaces; inductors can be connected between the capacitor junctions and the ground plane on the top of the board. Use NP0-ceramic or silver-mica capacitors. Inductors are wound on toroidal powdered-iron cores; winding details are given with the schematic.

Figure 6.36 shows the schematic and pictorial diagrams of a 300-Ω balanced elliptical high-pass filter that uses PC-board capacitors. Use double-sided ¹⁄₃₂-inch FR-4 glass-epoxy PC board. Thicker board will require more area for the desired capacitances. C2, C4 and C6 should be NP0-ceramic or silver-mica capacitors. It is easier to strip away rather than etch copper to form the series of capacitive elements. Mark the edges by cutting with a sharp knife; soldering-iron heat helps lift the strips more easily. Top and bottom views of the filter are shown in Figure 6.37.

Neither of the high-pass filters described requires a shielded enclosure. For mounting outside a receiver, some kind of protective housing is desirable, however. These filters were first presented in "Practical 75- and 300-Ohm High-Pass Filters," *QST*, February 1982, pp 30-34, by Ed Wetherhold, W3NQN.

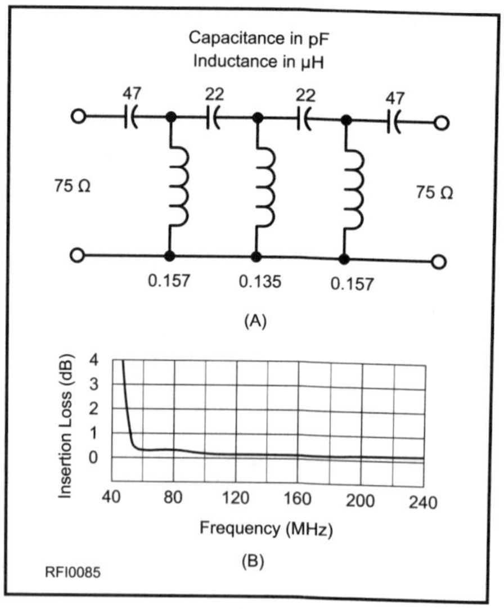

Figure 6.34—The schematic diagram of a 75-Ω Chebyshev filter assembled on PC board is shown at A. At B, the passband response of the 75-Ω filter. Design inductances: 0.157 mH: 12 turns #24 wire on T-44-0 core. 0.135 mH: 11 turns #24 wire on T-44-0 core. Turns should be evenly spaced, with approximately ¼ inch between the ends of the winding. If T-37-0 cores are used, wind 14 and 12 turns, respectively.

A High-Q Wave Trap

As an alternative to such a filter, a high-Q wave trap (tuned to the transmitting frequency) may be used. It suffers only the disadvantage that it is quite selective and therefore protects a receiver from overloading over only a small range of transmitting frequencies. A trap of this kind is shown in Figure 6.38. These "suck-out" traps absorb energy at the trap frequency, but do not otherwise affect receiver operation. The assembly should be mounted near the input terminals of the TV tuner and its case should be RF grounded to the TV set chassis by means of a small capacitor. The traps should be tuned for minimum TVI at the transmitter operating frequency. An insulated tuning tool should be used for adjustment of the trimmer capacitors, since they are at a hot point and will show considerable body-capacitance effect.

Figure 6.35—Photo showing construction of the 75-Ω unbalanced filter. Double-sided ¹⁄₁₆-inch FR-4 epoxy-glass PC board is used as a base for the filter components. A section of copper on the top is stripped away on both sides of center to approximate a 75-Ω microstrip line about ³⁄₃₂-inch wide. Both sides of the top copper foil (at the edges) are connected to the ground plane foil underneath.

Shielding

Additional shielding may help in a few cases of direct radiation pickup, when the manufacturer can't supply a cure. This is not always a practical solution: Envision telling an irate, nontechnical neighbor that a brand new, expensive, rather attractive TV must be placed inside a sheet-metal box. Just the thought of the resultant conversation would give most hams nightmares for weeks!

There are sometimes less-objectionable ways to improve shielding. Sometimes it is possible to add shielding to a home-entertainment-center cabinet. It may also be possible to relocate the TV so there are more walls between the TV and the amateur antenna. A qualified service technician can sometimes add shielding (either metal or a spray-on material) inside the TV. Con-

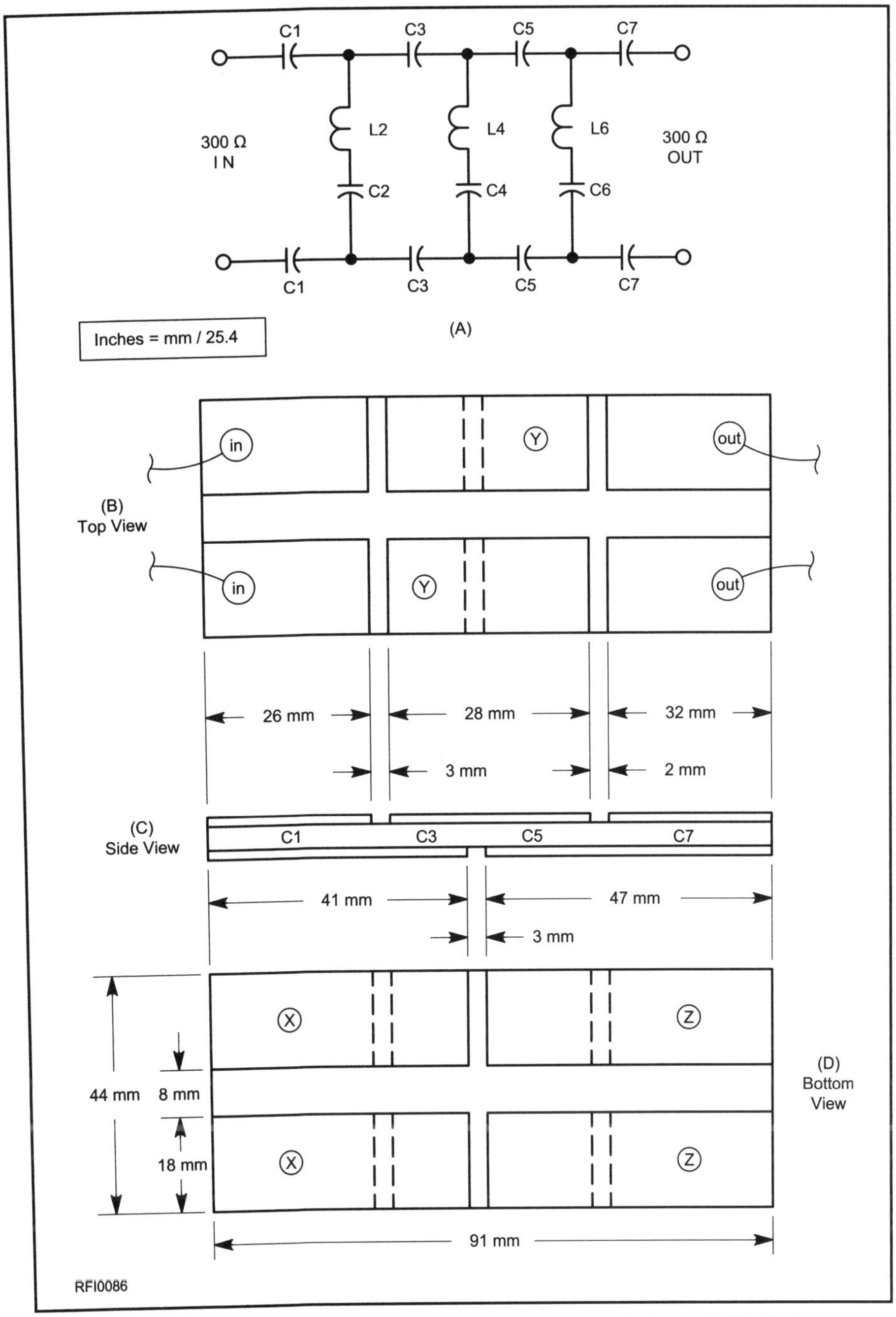

Figure 6.36—Schematic and pictorial diagrams of the 300-Ω balanced elliptical high-pass filter with PC-board capacitors. Shaded areas indicate where copper has been removed. Dimensions are given in millimeters for ease of measurement. L2-C2 connects between the points marked "X," L4-C4 connects between the points marked "Y" and L6-C6 connects between the points marked "Z" on the pictorial. C1=28.0 pF, C3 = 14.0 pF, C5 = 14.8 pF, C7 = 34.2 pF, C2 = 162 pF, C4 = 36.0 pF, and C6 = 46.5 pF. Design inductances: L2 = 0.721 µH: 14 turns #26 wire evenly wound on a T-44-10 core. L4 = 0.766 µH: 14 turns #26 wire bunched as required on a T-44-10 core. L6 = 0.855 µH: 15 turns evenly wound on a T-44-10 core. These coils should be adjusted for resonance at 14.7, 30.3 and 25.2 MHz.

ductive paint is available from GC/Thorsen (see the Resources chapter). Take note, however, that added shielding can cause short circuits if not installed properly, and it may upset the air flow around heat-generating components. Internal shielding is a task best left to the manufacturer or qualified service personnel.

Troubleshooting and Curing TVI—Step By Step

The Troubleshooting chapter covers the general aspects of troubleshooting RFI problems. Read that chapter before you tackle the troubleshooting of any type of RFI problem. In most cases, you "troubleshoot" a TVI problem by trying one or more of the cures. A flow chart of TVI troubleshooting procedures is shown in Figure 6.13.

After you have covered the "interpersonal" bases (First Steps chapter), you can start to do a technical investigation to pinpoint the cause of the TVI problem. Many cases of RFI are fairly easy to troubleshoot. In cases of interference to non-radio devices, for example, the troubleshooter knows that the root cause of the problem is something at the consumer-equipment end; harmonics and other spurious emissions from the transmitter are *not* involved in some types of interference. The size of the troubleshooting problem has been cut in half.

This is *not* necessarily true for interference involving other radio services. Over-

Troubleshooting Rules

If an amateur can demonstrate that a TV set reasonably close to the transmitter has no TVI at the same time that the affected device shows interference, the amateur station emission is clean. (Any spurious emissions affect all TVs within range.) Nothing further can be done at the transmitter to help.

The example TV need not be the ham's own set, but it is a good idea for demonstration purposes. Many TVs are quite susceptible to fundamental overload, so it is a good idea to carry a portable TV for tests.

Filters should be installed on all of the amateur's TV sets, whether needed or not. This demonstrates that filters are effective and do no harm.

Corollary: If the amateur station is interfering with a known clean TV set, the station must be cleaned up before anything is done with the TVs. The susceptible TVs may also need work of some sort, but the transmitter must be clean before any sensible diagnosis is possible.

These principles may seem obvious, but often hams whose stations are already "clean" put wasted effort into adding more transmit filters, shields, grounds and so on. Of course the results are often "no improvement." Sometimes even skilled trouble-shooters neglect the obvious.—*Dave Heller, K3TX, ARRL Assistant Director Atlantic Division, Yardley, Pennsylvania*

(A) (B)

Figure 6.37—A top view of the 300-Ω elliptic filter using PC-board capacitors is at A. Twin-lead should be soldered to the left and right ends of the board. At B, bottom view of the filter.

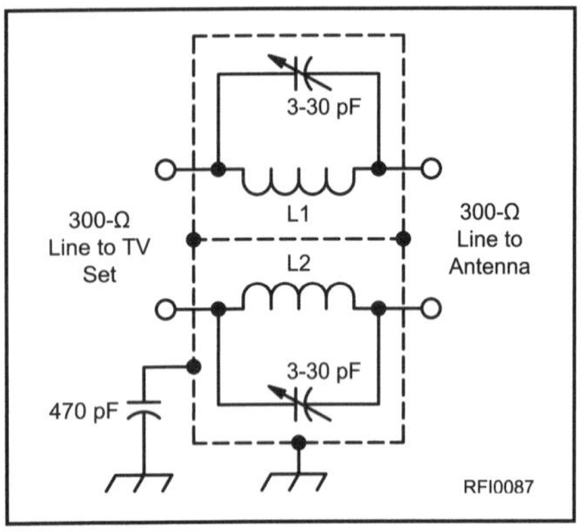

Figure 6.38—Parallel-tuned traps for installation in the 300-Ω line to a TV set. Mount the traps in a shielded enclosure with a shield partition as indicated. For 50-MHz use, L1 and L2 are each 9 turns of #16 enameled wire close wound on a ½-inch-diameter form (air core). For 144-MHz use, L1 and L2 are each 6 turns of no. 16 enameled wire close wound on a ¼-inch-diameter form (air core). This trap technique can be used to over-come HF fundamental overload as well.

the-air television, including satellite and other "direct" TV, *is* another radio service and it is quite possible that TVI involves spurious emissions from your station and is thus your direct responsibility. (Refer to the RFI Fundamentals chapter for information on spurious emissions. The Transmitters chapter discusses ways to cure problems in your transmitter and station.)

It Ain't Me

Before you even start to troubleshoot, you should first determine if *you and your station* are involved in the TVI problem. If there is no correlation between the TVI and your operating time, the TVI could be caused by something or someone else. Perhaps it is a misadjusted TV, a defective TV, VHF tropospheric propagation, electrical noise or that other ham or CBer up the street. If you know that your station is not involved at all, you have certainly cut the problem down to size. You may, of course, choose to help out anyway, but if the TVI does not involve your station, that help is entirely voluntary.

The TVI Log

The advice given in the First Steps chapter is for you to keep a log of your station operation and for your neighbor to keep a log of the RFI. A quick comparison of the logs will often yield valuable clues. This is especially true if you have verified that everything at your station is working properly. The TVI log is especially important when you are not sure of the cause of the interference. If, for example, the interference is caused by harmonics of a nearby 11-meter (CB) transmitter, much time could be wasted on various fundamental-overload cures. With any luck, the TVI log will give some clues to the nature of the problem. You can also compare the TVI log with the operating times of other amateur or CB operators in the area.

The TVI log should contain the date and time, the channel selected and a description of the interference. If the TV system includes accessories (other TVs, VCRs or FM receivers), ask the neighbor to note what other equipment was in use at the time.

Your Station

When troubleshooting TVI, start by considering your own station. This advice is given, not because your station is the most likely source of interference—because it is not, but because this is the easiest part of the problem for you to control.

The number one rule of finding and fixing RFI is: *ensure that the amateur's own house is clean*! Fix any and all station problems before proceeding on to troubleshoot the TV installation and system. If you have a case of TVI that involves your station, one important step is for you to be *sure* that your station is not *creating* the problem. One of the best ways to show that a station is not generating spurious emissions that cause TVI is to be sure that your own home television equipment functions properly. If an antenna-connected television next to the station is not subject to interference on any channel during amateur transmissions, the amateur station is certainly not the source of the problem. The FCC sees it that way, too!

The next step is to demonstrate to yourself that there are no spurious emissions coming from your station to cause TVI. If there is no interference to the antenna-connected TVs in your own household, it is almost certain that your transmitter is *not* producing spurious emissions. This is an ideal situation. If you *do* have TVI in your own house, you may be experiencing fundamental overload *or* you may have a transmitter problem. You should switch your TV through all channels, verifying that each channel is interference-free while the amateur operator transmits. The test frequencies, bands and modes should be those the operator uses most frequently.

If you do have TVI, as a troubleshooting step, operate your transmitter into a dummy load. If you still have some interference, your transmitter may be radiating spurious emissions from the case and/or power lines. You can install an ac power-line filter on your transmitter to eliminate the power lines as a source. If you still have interference after you have installed the ac-line filter, the transmitter may have shielding or circuit defects that are causing it to radiate spurious emissions directly. (Refer to the Transmitters chapter of this book and/or contact the

Interference to DVD Devices

Depending on whose nomenclature you use, DVD can mean digital video disc or digital versatile disc. The latter is often the preferred meaning because the disc is capable of much more than just storing video. The DVD can be used to store data, computer programs, audio and video. DVD equipment is less prone to interference than some other types of home electronics apparatus, but it is not totally immune to interference from nearby transmitters.

The DVD uses technology similar to that used in CD players. The DVD uses a shorter wavelength laser than that used in CD equipment and has smaller "pits" on the surface read by the laser. These smaller pits allow more information to be held on each DVD.

Many consumer quality DVD devices are built in metal cases, although low cost DVD players are often housed in plastic cases. The metal provides some degree of shielding and reduces the likelihood of interference due to direct penetration of the case by unwanted signals. However, the ac line cord and the patch cords that connect the DVD device to other electronic equipment can still provide a path for interference to enter the system.

If you suspect the interference is entering the DVD device via the ac line cord, a filter can be installed in the ac line as close as possible to the DVD device. Wrapping the line cord around a ferrite rod or toroid can also be helpful in eliminating this type of interference.

Inexpensive patch cords can sometimes allow interference to enter a system. Inadequate shielding on inexpensive patch cords allows interference to enter whatever piece of equipment the DVD device is connected to. The remedy to this type of interference is well shielded patch cords with good quality connectors.

Common mode interference involves the offending signal traveling on the shield of the patch cords. Running the patch cord through one or more ferrite beads can sometimes cure this type of interference. If this is impractical, you may be able to wind several turns of the cord through a snap-together ferrite core.—*John Frank, N9CH*

transmitter manufacturer.)

If you have TVI only when the antenna is connected, install a transmitter filter permanently—whether you need one or not! (Generally, you should use a low-pass filter for HF transmitters; use a band-pass filter for VHF or UHF transmitters.) You can always point to it with pride, explaining that you have installed a filter on your end. It is also a good idea to install a common-mode choke at the transmitter, to eliminate the possibility that spurious emissions are unintentionally getting onto the shield of your feed line.

If you still have TVI, a high-pass filter, ac-line filter and a common-mode choke should be installed on the amateur's TV sets, whether they need them or not. This demonstrates that filters are effective and do no harm.

QRP

Another method that may improve the situation is to use an exciter alone or QRP power levels. The few dB lost at the fundamental is not noticed in most contacts. A station power-output increase from 100 to 1000 W typically changes the received signal typically by only 1 to 2 S units, depending on the calibration of the receiver. That same change, however, may increase the interference potential ten-fold. Some amateurs use an amplifier when making the first call and then switch it to bypass to continue. This mode of operating fulfills the FCC regulations as well, and reduces QRM levels on the bands. QRP is not the best RFI-reduction solution, but hams should understand the relationship between transmitter power and RFI problems such as fundamental overload.

In general, the combination of these cures will work to cure your own TVI. Once you get the bugs out of your own house, you can tell your neighbor that you are sorry that they can't watch TV at their house while you are on the air, then offer to let them watch TV at your house while you are on the air! This latter offer usually gets the point across pretty quickly that the problem might be on their end. You can then point to the cures you have used on your TV installation and offer to help them find similar cures for their equipment. Not surprisingly, the cures that were just outlined are the very steps you will need to follow if you choose to help your neighbor solve his or her TVI problem.

TVI Troubleshooting Primer

One thing is worth noting up front: For years amateur literature has not emphasized the difference between common-mode and differential-mode signals, nor has it fully explained the different cures needed for each mode. There are critical differences! It can't be said too many times—*a differential-mode high-pass filter is not enough!* Common-mode cures are usually required as well. Remember this as you read the rest of this book. The RFI Fundamentals chapter contains an explanation of differential- and common-mode signals.

Complete elimination of TVI is not often a simple process. A single measure, such as a high-pass filter at the TV, does not always cure the problem. Sometimes, a number of methods must be applied. An important factor in any TVI case is the ratio of TV signal strength to interference level (also called signal-to-noise ratio, S/N). This includes interference of all kinds, such as ignition noise, random or thermal noise (which sets the minimum signal required for snow-free reception) and unwanted signals that fall within the TV channel. An S/N ratio greater than approximately 40 dB is required for good picture quality.

Anything connected to the TV is a possible source of interference. It is impossible to efficiently troubleshoot a system with multiple TVs, video games, VCRs, splitters, amplifiers, remote TV tuners, cable TV converters, A/B switches and long cable runs. Many cases of TVI have one easily located cause and cure. When you are confronted with a case caused by multiple factors, however, much time can be wasted trying different things that apparently don't work. Again (and again if necessary), the first important step in troubleshooting a case of TVI is to simplify the problem. Simplify things to one TV and one short feed line (close to where the feed line enters the house). Once the bugs are out of the simple system, reconnect accessories one at a time. Eliminate problems as you go along.

This principle applies to the amateur station as well as to the affected equipment, and especially to VCRs. They (especially the older ones) are notoriously susceptible to EMI problems.

Try one, and only one thing at a time. If too many variables change at once, you will never know which cure actually solved the problem.

Do You Want to Service Your Neighbor's Equipment?

Don't forget that you generally should *not* service your neighbor's equipment. Once you have determined that the problem is not your radio station, you want to be a good neighbor. In general, you should be a *locator* of solutions, not a *provider* of solutions. If you take the back off the TV, you may be blamed years later when the 25-year old

RF-Sensitive Safety Circuits in TVs

A local ham was having a problem when he operated on 20-meter CW. As soon as he keyed his 100-W transceiver, the TV set in the next room would switch off! A safety circuit in the TV set was obviously doing more than its job, but what was triggering it?

The TV was connected to the cable service, so we disconnected the cable. This had no effect on the problem. There was a 300-Ω twin-lead cable (from an abandoned TV antenna in the attic) hanging out of the wall with its end near the TV terminals. Moving this wire away from the TV set cured the problem. Its length, about 16.5 feet (a good ¼-λ 20-meter antenna), coupled enough RF into the TV to cause the problem.—*Howard Liebman, W2QUV, NYC-LI Section Technical Coordinator*

All modern TVs contain overvoltage sense circuitry that trips if the TV power-supply voltages exceed a factory-set limit. These are critical safety circuits that have been installed to ensure that a power-supply malfunction will not cause the picture tube to emit Xrays.—*Ed Hare, W1RFI, ARRL Laboratory*

NTSC Channel Impairment Graph

Approximate signal-to-discrete interference limits for NTSC analog TV channels. As long as the level of discrete interference relative to the level of the visual carrier remains in the shaded area on the graph, the picture will usually not be affected. Note that as the frequency of the interference approaches either the visual carrier or color subcarrier, its amplitude must be greater than about 56 dB down to not be seen in the picture. (Source: Motorola)

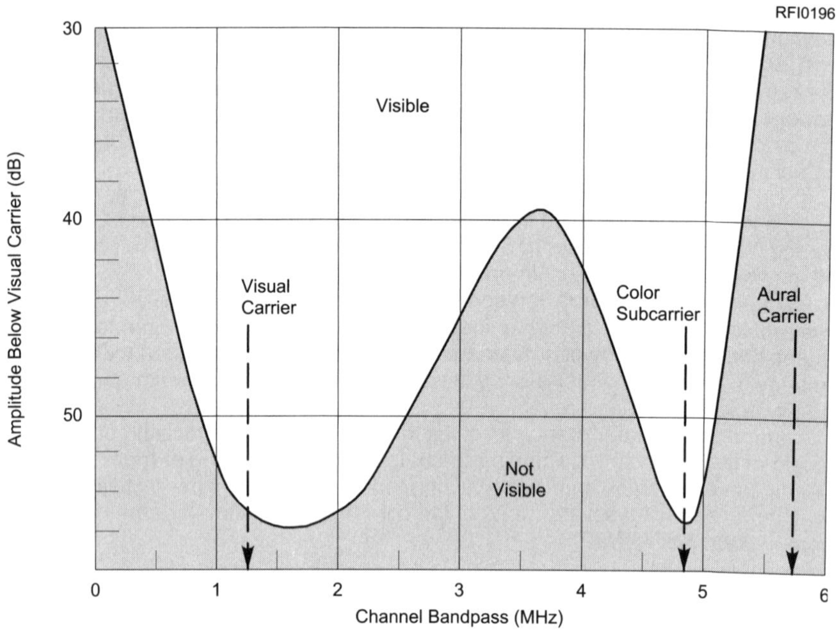

clunker gives up the ghost. Some hams have learned this the hard way—they have been held responsible for defects that occurred years after they modified a neighbor's set. Allegedly, the failure was caused by the modifications. In some states, a state-issued electronic-service license is required to work on consumer-electronics equipment (your amateur license will not do!).

If you *do* have occasion to dig into a TV, remember that high voltages remain after the set is turned off. Also remember that in some sets the circuitry and some hardware is dc connected to one side of the ac line (these sets contain no isolation or power transformer). The 117-V ac line can kill! Please be careful—the Silent Key is not an award. Leave the repair of consumer equipment to qualified service personnel!

Figure 6.13 shows a TVI troubleshooting flow chart.

TVI Cures at the TV and Its Installation

If you and your neighbor decide that the TVI problem is something on her end, your help is voluntary. FCC regulations require *only* that your station not emit spurious emissions that interfere with other radio services. You may, however, choose to be a good neighbor and help your neighbor find a solution.

Start by conducting tests. Operate your transmitter on each band, mode, power and antenna-pointing combination you usually use. If you own more than one transmitter, try the tests with each transmitter. Since the TV set may exhibit varying sensitivity to TVI from different amateur frequencies, keep the transmit frequency constant when testing cures. This frequency-dependent TVI sensitivity can be caused by several factors: The presence of a comb filter in a TV set causes variable gain across the channel, which can reduce interference at some frequencies. When a harmonic is near either the picture or sound carrier, or near the color subcarrier, the beat frequency is lower and appears as a more coarse (and more visible) interference pattern. As the frequency of the HF transmitter increases relative to the TV visual carrier, the video interference pattern is normally less objectionable. Color or audio interference may increase, however. Because of this behavior, video interference caused by operating on phone frequencies is not normally as visible as when operating in the CW portion of a band (but color or audio EMI may increase). Remember, when testing a station for TVI, always listen before transmitting to prevent QRM, and always identify properly.

Then, check the TV and its installation. Much like at your station, chewing gum and baling wire, or more likely, broken or corroded antenna or feed line, too many (or unterminated) splitters or other obvious installation defects could be part of the problem.

The following steps can sometimes help improve a TV installation

- Install a TV antenna with greater gain, hopefully with improved directivity.
- Increase the TV antenna height, thus increasing the desired signal.
- Increase the height of the amateur antenna, thus moving it farther away from the susceptible equipment.

Indoor TV antennas are potential problems. Move the TV antenna above the roof from under the roof. (Many roofing materials contain aluminum and carbon and attenu-

Figure 6.39—RFI is minimized when antenna and feed-line locations are based on sound RFI-prevention thinking. Note the advantages:
A) TV antenna located far from ham antenna
B) TV feed line is run vertically to ground level
C) TV distribution amplifier mounted at ground level
D) TV feed-line shield connected to ground rod
E) TV feed-line horizontal run at ground level
F) Ham antenna does not block TV-reception path
G) Ham feed line runs vertically to ground level
H) Ham feed-line shield connected to ground rod
I) Ham shack located close to station ground

ate TV signals by up to 20 or 30 dB when wet.)

Try Installing or Improving the TV Grounding System

You may also want to consider the consumer-electronic equipment ground. A good RF ground may not always be practical, especially in a neighbor's house, but an understanding of the basic principles can sometimes help. If coaxial feed line is used (either in an antenna installation or cable TV system) it can be connected to a good earth ground, preferably outside the house. This can shunt unwanted RF energy on the outside of the coax shield (common-mode) to ground. The electric-power mains and telephone lines coming into the house also have grounds that should be inspected. All utility grounds should be serviced only by qualified personnel.

The consumer-electronics equipment is usually also "grounded" through the ac-power system. In most cases, this is not a good RF ground. An interference problem can sometimes be cured by improving this ground with an actual earth connection, either to a ground rod or a cold-water pipe that is continuous to earth. Note that the ground must comply with National Electrical Code or other relevant local requirements. In other cases the situation can be improved by RF isolating the equipment ground connection, using 10-20 turns of the ac-line cord on a ferrite toroid or rod, thus breaking the ground loop. Remember that some equipment has dangerous voltages on parts of the chassis. If you are not *sure* it is safe to connect an earth ground to any piece of equipment, contact the manufacturer.

The amateur signal at the TV can be reduced greatly by proper location of the amateur and TV installations. When planning your antenna placement, consider the horizontal and vertical distances between the amateur and TV antennas and feed lines (see Figure 6.39). Place the amateur antenna as far as possible from the TV system. If you can locate the ham antenna so that there are no TV antennas or feed lines between it and favorite operating directions, the field strength at the TV will be reduced.

With proper antenna placement and orientation, including antenna polarization, harmonic and fundamental overload conditions can be reduced. You may even be able to give nearby neighbors a break with this trick. Since harmonic energy will be picked up to a lesser degree off the back side of a TV antenna, about 20 to 40 dB less signal will be received if the ends of each antenna can be pointed away from each other.

Vertical separation between antennas or between the amateur antenna and horizontal runs of the TV feed line is more effective than horizontal separation alone. Raising the height of the amateur antenna not only reduces take-off angle, but also the field strength at the TV (thereby reducing fundamental overload).

You can try a TV distribution amplifier to increase the strength of the desired signal. *Some* distribution amplifiers have better RFI immunity than a typical TV or FM receiver. They are often available from electronic distributors who sell components to television-service shops. Be wary however; some poor-quality distribution amplifiers can worsen the problem. Like preamplifiers, distribution amplifiers boost unwanted, as well as wanted, VHF signals.

These signal-increasing fixes all seek to increase the desired-signal to undesired-signal ratio received by the TV set or radio receiver.

If the installation checks out okay, it is time to start trying filters.

TVI FILTERS

High-Pass

The first line of defense against TVI to antenna-connected TVs is the installation of a high-pass filter for interference from HF transmitters. If an interfering signal enters the TV through the feed line, a filter will eliminate or minimize the problem. Each TV may require its own filter. This should be installed at the TV, as close as possible to the TV's antenna terminals. These are available from RadioShack (catalog #15-579 or 15-582). Several other sources are listed in the "EMI/RFI Package."

For VHF and UHF transmitters, a high-pass filter will do no good. VHF and UHF signals are located between various TV channels, so the high-pass filter does what it is designed to do and lets VHF and UHF amateur signals pass unimpeded. To prevent fundamental overload of TVs from VHF and UHF signals, a band-reject or notch filter must be used. These are commercially available. Those who operate in the VHF weak-signal parts of the band (50.2 MHz, 144.2 MHz, etc) should consider a notch filter. Look for filters that offer performance optimized for the weak-signal operator. One of these filters is shown in Figure 6.40.

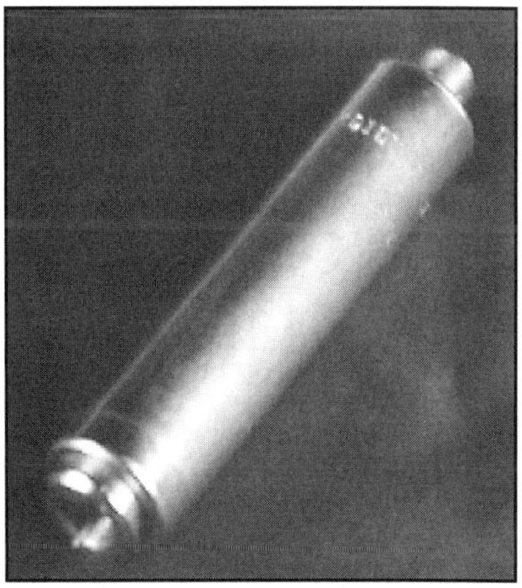

Figure 6.40—This notch filter is optimized for the weak-signal part of 2 meters (144.2 MHz). For other parts of the band, a tunable notch filter should be used.

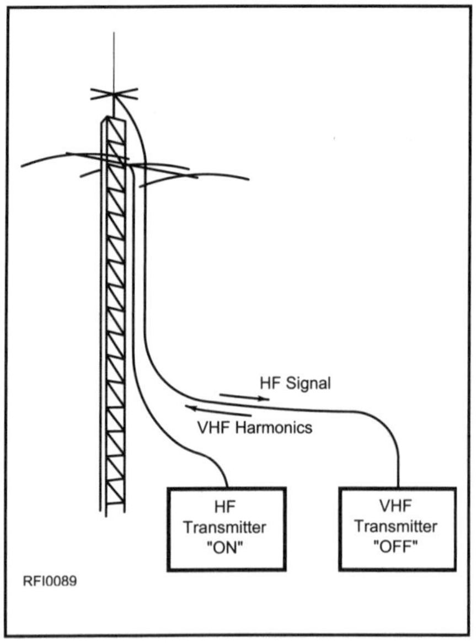

Figure 6.41—The 2-meter transceiver in this example is powered off, but it can still cause RFI problems. The ground-plane antenna near the HF Yagi can pick up a lot of HF energy. This is then conducted down the VHF transceiver's feed line. The unpowered VHF transceiver can act in a non-linear fashion, turning some of the HF signal into VHF harmonics. These are then conducted back up the VHF feed line, where they will be radiated efficiently by the VHF antenna.

Cures Summarized

TVI Cures

TVI Problem	*Action*
HF or VHF spurious emission	Cure problem at radio transmitter or station
HF or VHF direct pickup	Contact set manufacturer through the CEMA
HF or VHF electrical differential mode	Install ac-line filter
HF or VHF common mode	Common-mode filter on feed line and/or ac line
HF differential mode	Install high-pass filter
HF cure-all	Common-mode filter and high-pass filter and electrical cure-all
VHF differential mode	Notch filter or tuned stub.
VHF cure-all	Combine notch filter or tuned stub with the VHF common-mode choke, plus the cures from electrical cure-all
Electrical differential mode	AC-line filter
Electrical common mode	Common-mode choke
Electrical cure-all	AC-line filter with common-mode choke

Electrical

RF may be getting into the TV through the power line. An ac-line filter should be installed on the TV-line cord, as close to the TV as possible. If the interfering signal is conducted via the ac power system, this filter will eliminate or reduce the severity. If the problem is still evident, the television circuitry may be picking up signals directly. See the section on direct radiation pickup. (This could also be a case of locally generated harmonics or some form of intermodulation.)

Install the filter as close to the antenna terminals as is possible. If you own the TV, you may want to install it inside the cabinet, at the tuner. If the interference is to an audio device, such as a stereo or alarm system, install the filters in the speaker leads or audio input cables, as close to the amplifier as possible. Telephone interference filters should be installed as close to the telephone as possible. Refer to the appropriate chapters in this book for more details about filter selection. Also, don't forget that dangerous voltages are present inside most consumer-electronics equipment. Figure 6.32 shows a power-line filter that is available from RadioShack.

Common-Mode

A high-pass filter will attenuate the HF signal that's picked up by the antenna and sent to the TV inside the TV's feed line. But you can also pick up a lot of signal on the outside of the feed line. The signal that is not inside the feed line is called a common-mode signal.

If the installation of a high-pass or notch filter does not cure the TVI, it is possible that the TV is responding to a common-mode signal, either on the feed line or on the ac lines. By now, you have probably learned enough about RFI to know that a common-mode choke, made by wrapping about 10 turns of wire onto to a suitable ferrite toroidal core or ferrite rod, can be used to eliminate unwanted common-mode signals. See the EMC Fundamentals chapter for information on selecting the correct ferrite material.

Harmonics

When TVI occurs on channels harmonically related to the transmit frequency, *harmonics* are involved. These *can* come from the transmitter, and must be corrected by the transmitter operator. They can, however, be generated in the front end of an overloaded TV. (The TV cannot distinguish between transmitter harmonics and those generated inside the TV, so the interference can look exactly the same.) This problem is usually solved when the fundamental overload is cured with the proper filters. In some rare cases, however, harmonics can be generated externally from the TV or the transmitter. This is known as "external rectification" or, more colloquially, the "rusty bolt effect." An entire chapter is devoted to this topic.

Many hams have heard about TVI caused by a rusty downspout or bad aluminum siding. These specific cases rarely happen. When external rectification does occur, suspect the following causes (pretty much in the likely order of occurrence):

- Arcing in the antenna system
- Corrosion in the transmit antenna tower or guy wires
- An unpowered solid-state device near the antenna (such as an unpowered VHF transmitter—see Figure 6.41)
- An overloaded, sometimes unused, TV preamplifier
- Last, and probably least, a rusty downspout

Harmonics are likely in cases where TVI is experienced only when the transmitter is operated on specific bands. Harmonics can be generated by the transmitter or external devices, such as an unpowered VHF transmitter connected to a VHF antenna. They can also be generated inside the TV tuner, a form of fundamental overload.

IN CONCLUSION

The "Cures Summarized" sidebar lists the cures used to troubleshoot and fix most types of TVI problems.

HF Differential Mode

A differential-mode high-pass filter will usually help with this type of interference. Keep in mind that a differential-mode high-pass filter will not work against a common-mode signal that may also be present.

HF Common Mode

A common-mode choke is easy to build using a ferrite core or rod.

HF "Cure-All"

Many times the interfering signal will have both common and differential-mode components. This means the filter must be able to eliminate both types of interference, or two filters must be used. Use a commercially available high-pass filter along with a common-mode choke. The cures described under electrical cure-all should also be applied.

VHF/UHF Differential Mode

A commercially available high-pass filter will not work to eliminate interference from a VHF transmitter because the VHF signal falls within the filter passband. It would be necessary to use a filter with a sharp notch tuned to the fundamental, or a tuned stub.

VHF/UHF Common Mode

A conventional high-pass filter is effective only against HF signals. It does not offer significant attenuation to VHF common-mode signals. A common-mode choke rejects VHF common-mode interference.

VHF "Cure-All"

A VHF notch filter used in conjunction with a common-mode choke will sometimes help against interference from a VHF signal. For a number of reasons interference from VHF transmitters is not as common as interference from HF transmitters. When it does occur, however, it is more difficult to eliminate because TV tuners don't offer any attenuation to VHF signals. Cable systems use several amateur bands to carry TV signals. The cures described under electrical cure-all should also be applied.

Electrical Differential Mode

This is the kind of interference generated by electric motors. A commercial ac-line filter will usually reduce this kind of interference.

Electrical Common Mode

Many noise sources place a common-mode signal on the ac power line. When ac wiring picks up an amateur fundamental signal, that signal is nearly always carried in the common mode. On an ac line, the common-mode signal is carried on the hot, neutral and ground leads. Most ac-line filters do not filter the ground lead, so they are ineffective against the most common forms of ac-line interference. A common-mode choke does a good job with ac-line interference. It is not effective against ac-line differential-mode interference.

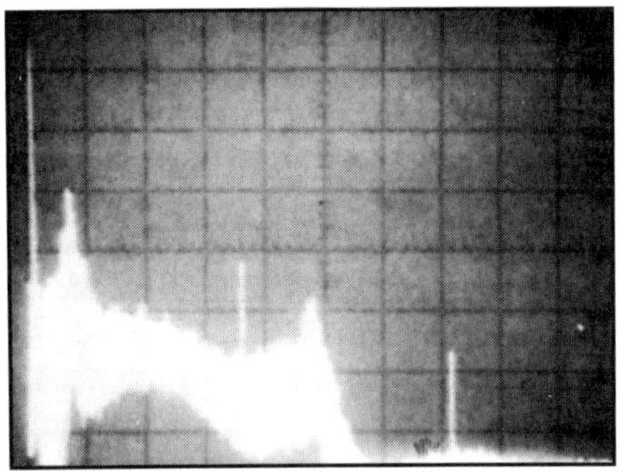

Figure 6.42—A spectral display of a TV-receiver output as measured at the antenna terminals. Horizontal divisions are 1 MHz; vertical divisions are 10 dB. The top line represents –30 dBm (7000 μV).

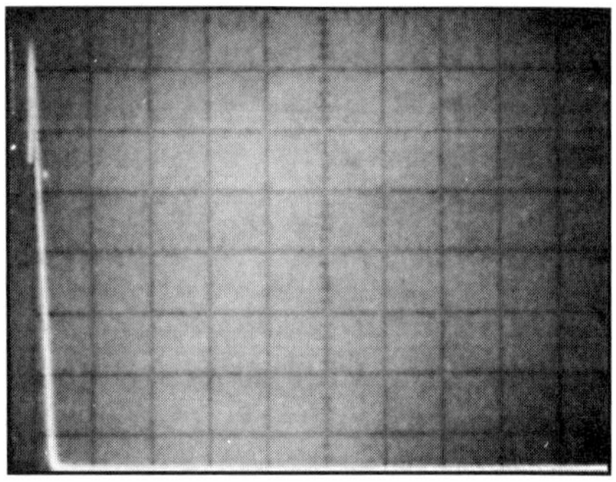

Figure 6.43—A spectral display of the filtered TV output. The interference is gone!

Electrical "Cure All"

Most noise sources, and some RF pickup, on the ac line contains both differential- and common-mode signals. Use a commercial ac-line filter in conjunction with the common-mode choke shown in Figure 6.31 to ensure that all interfering ac-line signals are properly filtered. This book has a chapter on electrical interference.

Interference From Television Sets

As mentioned earlier in this chapter, a TV can cause EMI. The horizontal-sweep and color burst oscillators are among the causes. Switching power supplies in TVs can also be a source of interference. Figure 6.42 shows the spectral output as measured at the antenna terminals of a currently manufactured set. Each horizontal division is equal to 1 MHz with the "pip" at the far left representing zero frequency. Each vertical division is 10 dB with the top line equal to –30 dBm (7000 μV). This energy is radiated from the antenna system. Since the level is great, the chances for interference are good.

The definitive indicator of horizontal-oscillator interference from TVs is the presence of raspy groups of interference about every 15 kHz, especially on lower HF. In addition, most modern TVs use switched-mode power supplies, which can radiate broadband noise. As if this weren't bad enough, most modern TVs use microprocessors that can radiate unwanted energy as well.

To cure this interference, add a high-pass filter at the antenna terminals of the set. The filter attenuates frequencies below 40 MHz by at least 60 dB. The spectrum-analyzer photograph at Figure 6.43 shows the difference after the filter was installed. Note the absence of interference.

Interference caused by the oscillators can also be conducted along the ac line. Most sets have built-in line filters. Their effectiveness is questionable, however, because an additional line filter often eliminates the interference. A ferrite common-mode choke may help as well. (Use 10-15 turns of the ac cord on an FT-240-43.)

SUMMARY

TVI can be cured! It takes a combination of technical knowledge, troubleshooting skills, tried-and-true cures and a good dose of personal diplomacy. Most difficult cases of TVI have involved problems with common-mode signals on a coaxial or twin-lead feed line. Those cases were only difficult because cures for common-mode signals were not well known. Armed with cures for most types of TVI, amateurs should be able to operate in peace.

Chapter 7

Cable Television Interference

This chapter discusses interference to televisions connected to cable-television systems. Although there is much in common between TVs connected to cable-distribution systems and antenna-connected TVs, the differences sometimes require different approaches to RFI trouble-shooting and cures.

By Robert V. C. Dickinson, W3HJ
President, Dovetail Systems Corp
125 Goodman Dr
Bethlehem PA 18015

This chapter picks up where the last one left off—by discussing RFI issues particular to cable-television installations. It is divided into two sections. The first offers an overview of the cable industry and specific information about cable TV interference (CATVI). The second part is adapted from a three-part article published in the July, August and September 1993 issues of Communications Technology, the journal of the Society of Cable Telecommunications Engineers. This has been included in this book for your cable operator. If you have a problem with cable-related interference, show this information to your cable company repair personnel. Although it duplicates some information from other parts of this book, it offers complete and authoritative information that the cable company can use to cure the RFI problem. This work represents a good example of voluntary cooperation between the cable industry and Amateur Radio

WHAT IS CABLE TV?

The details of a cable system are covered later in this chapter. In a nutshell, however, a cable operator delivers a number of local and network television channels, digital video, high-speed data and in some cases cable telephony to individual homes by direct connection between the home and a coaxial cable or hybrid fiber coax (HFC) distribution system. A properly maintained cable television system can be both a blessing and a curse for Amateur Radio RFI problems. A cable system is ideally self-contained. On the plus side, the cable delivers a strong, consistent signal to the TV receiver or other customer premises equipment. It is also in theory a shielded system, so external signals shouldn't leak in and cause trouble, and signals inside the cable shouldn't leak out and interfere with over-the-air services. On the minus side, the cable forms a large, long-wire antenna that can pick up lots of external signals on its shield in the common mode. Many TVs and VCRs and even some cable set-top boxes are easily overloaded by such common-mode signals. These factors combine to make cable systems relatively free from interference, but when CATVI strikes, some of the standard cures, such as high-pass filters, don't usually work.

By design, a cable system is configured so that all users can receive approximately the same TV signal levels over a broad range of frequencies. Since a cable network is considered to be a closed system, any frequency within the transmission capabilities of the network can be used inside the cable. Cable networks can, therefore, use frequencies allocated to other services. The cable industry uses frequencies ranging from 5 MHz to over 860 MHz. This is the fundamental problem in CATVI: cable systems use frequencies allocated to other services. In a case of CATVI, a legal over-the-air fundamental signal may fall within a cable channel. Once inside the cable network, such signals cannot be removed by filtering. Obviously, if a cable network is not a closed system, signals can enter the network and degrade TV reception, while cable signals can leave the network and interfere with over-the-air services.

CABLE TV HISTORY

In order to gain a proper perspective on the subject, let's briefly review the history of the cable industry. In the late 1940s there were the early signs of what is now a pervasive technology. With the advent of television broadcasting, many viewers were far from the television transmitters or shielded by terrain. A number of innovators began businesses providing television signals to the remote viewers. The normal procedure was to locate a high vantage point with reasonable reception and then carry the signals by wire to viewers in a community where direct reception was poor or impossible. These early operations were called CATV (community antenna television); today, CATV is generally accepted to mean cable television.

Cable TV is a nearly 60-year-old industry that really took off in the 1970s. By now, it is as pervasive as any other technology, available in almost all areas of the country. While not nearly as ancient as broadcast radio or even television, it has been estimated that cable operators serve more than 65 million homes (and still growing) throughout the United States and multimillions more overseas. While cable TV is an outgrowth of modern technologies, the leakage problems we encounter stem from well understood phenomena that can be addressed by modern forms of traditional remedies.

TVI began early in the history of broadcast television. Amateurs were particularly concerned with TVI prevention in the 1950s and early 1960s, but the problem has decreased as a result of improved transmitter and TV performance. With the arrival of cable TV, instances of TVI have reduced considerably because television signals are delivered to viewers via coaxial cable at a much greater signal strength than provided by broadcast TV. This method of transmission is designed to be a "closed system," and it does provide sufficient shielding to reduce the occurrence and severity of TVI caused by radio transmissions. In practice, however, cable systems are often only partially closed. To some extent, they are susceptible to radio transmissions and can cause interference to other communications services, including Amateur Radio.

Early cable systems varied in their approaches. It is said that some used barbed-wire fences to carry the signal, although more traditional methods were required for reasonable signal quality. These methods include balanced open-wire line, twin lead, G-line (an approach where the signal was launched down a single wire) and, of course, coaxial cable. Coaxial cable won out and is currently the method of choice. Modern cable networks generally use a combination of fiber optics and coaxial cable to deliver signals to subscribers, although a few fiber-to-the-home installations transport signals all the way to the customer using glass fiber.

By the early 1970s cable TV was becoming more sophisticated. Rather than the original 12 over-the-air channels, cable systems were beginning to carry 20, 30, 36 channels and even more. In those days, little formal attention was paid to cable signal leakage, except when: (1) leaky systems began interfering with over-the-air signals (which gave some people free cable TV service by way of local leaks); or (2) a radio transmitter interfered with cable TV service.

NCTA Sees the CATVI Problem

In the early 1970s the National Cable Television Association (NCTA, now known as the National Cable and Telecommunications Association, is the principal trade association of the U.S. cable television industry) became aware of numerous complaints of signals leaking from cable networks. NCTA felt the growing concern that this leakage might interfere with other services, particularly aircraft-navigation and radio communication (we'll call them NAVCOM) circuits.

Department of Commerce (DoC) Tests CATV vs NAVCOM

At the same time, the DoC ran susceptibility tests on typical navigation receivers and pointed out the possible consequences of excess cable TV signal leakage. Other work explored the modes of leakage from coaxial cable, including probable levels, radiation patterns and so on.

The results of the navigation receiver testing showed that it is virtually impossible for a random signal(s) to yield improper navigational information that would lead a pilot in the wrong direction. It was demonstrated, however, that a significant interfering signal could make a navigational service unusable. Such a situation would, fortunately, notify a pilot that the navigational system was inoperative or unreliable. This took away the fear that leakage would cause someone to fly the wrong course. Nonetheless, it left the possibility that cable TV leakage could obstruct instrument approaches in bad weather.

A NAVCOM Crisis—FAA and FCC Step In

A situation in Harrisburg, Pennsylvania was probably the most significant instance of real-life cable TV leakage. After the FAA made a new air-traffic-control communications-frequency assignment, pilots in the area heard various tones on the FAA channel. Although the tones did not cover the controller's voice, they were annoying and sometimes held the receiver squelch open. The FAA was unable to locate the cause of the problem over a period of weeks and called the FCC. The FCC Field Operations Bureau decided that the problem had something to do with the local cable TV system.

The cable operator was cooperative and the final explanation was rather surprising: The cable system had four separate "headends." The AGC for the cable amplifiers was derived from a pilot signal that coincidentally was on the same frequency as the FAA. The four headends had separate pilot generators, on slightly differing frequencies. Although a single interfering frequency would simply cause a squelch break and a little background noise, leakage from two (or more) cable network sectors was heard in some areas, and AF beats between the pilot carriers appeared as audio tones. Once the cause was found, the cable system was shut down, and the AGC system was reconfigured to operate on TV-carrier frequencies. (This was an old and large system that was slated for rebuild.)

The Harrisburg situation was the proverbial "straw that broke the camel's back." A huge uproar in aviation circles followed, and great efforts were made to prevent cable systems from using any frequencies not used for standard over-the-air TV broadcasting. Had those efforts been successful, the industry and the public interest would have been severely affected. As it worked out, the FCC threatened operators of leaky cable systems with severe consequences. This edict got the attention of some, who began to seriously address the problem.

The Advisory Committee on Cable Signal Leakage was formed by the FCC and the FAA. It included industry participants who studied the situation and made recommendations to the FCC for further rule making. In 1979, the committee recommended measures that ultimately became part of FCC Rules and Regulations Part 76. One of the most significant FCC actions (even before the report of the committee) requires continuous monitoring of cable systems for leaks and timely repair of any leaks. A few cable operators took the situation very seriously and began to clean up their act. Others virtually ignored the problem.

Cable TV Leakage Controls

Cable-industry signal-leakage control programs have a long history, with a great deal of success and some notable failures. The FCC rules enacted in 1985 not only require an ongoing monitoring program to find and fix leaks, but also annual qualification of nearly all cable systems to demonstrate compliance by actual measurements. (This part of the rules did not take effect until July 1, 1990.) At present, all cable systems using aeronautical frequencies (108-137 MHz and 225-400 MHz) must qualify under §76.611 annually. A brief description of the qualification requirements is given in the "Cable Leakage" sidebar.

RESPONSIBILITIES

What does all of this mean to the Amateur Radio community? Primarily, it means that cable TV is here to stay, and we amateurs must be cognizant of the facts and diligent to avoid or reduce the inherent problems of coexistence.

Cable Operator Responsibilities

Absolute Maximum

The cable industry has some comprehensive legal responsibilities under the FCC regulations. The FCC regulations, in Part 76, specify the maximum permitted level of leakage out of a cable system (also known as a cable "plant"). In cable-company terminology, leakage out of the system is known as "egress." If the cable network isn't leaking, there should be no external signals getting inside the cable.

The permitted signal leakage out of a cable system is limited according to frequency range as shown in Table 7.1. Note that 20 µV/m at 3 meters is a small, but not insignificant, signal. A 20-µV/m field indicates over S9 on an average amateur 2-meter receiver, if the receive antenna is located 3 meters from the source of the leak. This means (under these FCC rules) occasional squelch breaks while riding through town are possible, and interference with weak signals may be acceptable. In addition to these absolute levels, the FCC requires that cable operators conform to overall leakage limits for the entire plant, often verified by having an airplane fly over the area that the cable company serves, making a measurement of the "cumulative leakage" from the entire system.

Non-interference

There is a second responsibility of the cable operator (as well as anyone using FCC frequency assignments): systems may not cause "harmful interference." Harmful interference is defined under FCC Rules and Regulations Part 76, paragraph 76.613(a) as "any emission, radiation or induction which endangers the functioning of a radio navigation service or other safety service or seriously degrades, obstructs or repeatedly interrupts a radio communications service operating in accordance with this chapter." "This chapter" covers essentially all radio frequency use and therefore includes Amateur Radio operation.

The regulations also require, however, that leakage from the cable system not cause "harmful interference" to over-the-air radio services. It is possible that a leak could be below the absolute maximum signal levels, but still be very disruptive to the Amateur Radio Service (or other services). If this does occur, the cable operator is still responsible for the leakage.

The scope of §76.613 is broad, and the rule can be difficult to interpret. Statements from high-ranking FCC personnel indicate that resolution of interference under §76.613 is a cooperative venture between the interfered and interfering parties. Cases referred to the FCC may well be resolved in ways not favorable to either party. The bottom line is: Work with the cable company to find an acceptable solution, rather than calling in the FCC (for leaks that are not of major consequence). Although the FCC decides exactly what constitutes interference, amateurs should apply common sense to any potential interference situation. It is unreasonable to interpret merely breaking squelch as interference.

It is sometimes hard to know where to draw the line. If a leak is occurring on the input frequency of an amateur repeater, and the repeater cannot be used while the leak is present, this is clearly a case of harmful interference. Even this, however, may be a bit muddied by other factors. How would the FCC view the matter if the repeater frequency could be easily moved to another unused channel in the area? In addition, for there to be "harmful interference," both harm and actual interference must be present. The FCC has generally decided that merely hearing squelch breaks on unused channels is not harmful interference and they don't generally require cable companies to correct such problems unless the leak is greater than the maximum permitted level.

Table 7.1
CATV Leakage Limits

Frequency (MHz)	*Field Strength (µV/m)*	*Distance (ft)*
0-54	15	100
54-216	20	10
216+	15	100

Who is Responsible for Cable Interference Problems?

The Cable Operator

Many cable-interference problems are the direct result of a leak or other defect in the cable system. The cable operator must fix the leak to prevent signals from leaking into or out of the system. Interference caused by leaks is generally the responsibility of the cable operator.

The Amateur Operator

It would be easy (for amateurs) if we could blame all cable interference problems on the cable system. This is not always a correct assessment of responsibility, however. Unfortunately leaks are not the only cause of interference to a TV connected to a cable system. There are even ways the amateur may be responsible. If, for example, a spurious emission generated by the amateur station causes the interference, the amateur should eliminate the problem. This is true even when a cable leak contributes to the problem, because the spurious emission may interfere with antenna connected TVs or disrupt communication in other services.

The TV Owner or Manufacturer

Some TVs do not have adequate shielding and filtering. A strong radio signal, especially at VHF, may be picked up directly by the TV circuitry, resulting in interference. There is little a cable company can do when interference results from direct radiation pickup in TV or VCR circuitry. A cable company set top box (converter) installed at the input to a cable-ready TV can help alleviate many direct pickup problems.

The shield of the cable system forms a large long-wire antenna. This antenna can pick up large amounts of RF energy from nearby radio transmitters. (This does not indicate a leak or defect in the cable system.) Most of the time, the RF present on the cable shield does not cause a problem, however, some TVs and VCRs are quite susceptible to this common-mode signal. The cable company is not responsible for such common-mode interference.

Cable repair personnel should understand the various ways that interference can occur in their cable system. In order to determine that interference is not caused by a leak, it is usually necessary to diagnose how interference does occur. A common-mode choke should be installed in all cases to determine whether interference is caused by common-mode signals on the cable shield. The proper construction and use of a common-mode shield are discussed in other sections of this chapter.—*Ed Hare, W1RFI, ARRL Laboratory*

Proof of Performance

The regulations also require that the cable signal be of good quality. §76.605 stipulates minimum specifications for things like frequency stability, carrier and signal levels, and various carrier-to-noise and carrier-to-coherent interference parameters. These latter are of special interest to amateurs because when our signals leak *into* cable systems, we can represent noise or coherent interference on the desired signals. Our leaking into the cable plant, called "ingress" in cable parlance, is of concern to cable operators, both as a customer-service and as a regulatory issue.

Now that most cable operators carry high-speed data service and digital video (and sometimes also cable telephony), leakage and ingress have taken on a much higher priority. Ingress of over-the-air signals and even local noise from power lines, appliances and other electrical devices can disrupt data transmissions carried on cable networks.

Responsibility Misunderstood

Most cable companies understand their responsibilities about leakage. They understand that signal leakage must not exceed the maximum levels permitted by the FCC. Most of them also understand the requirements not to cause harmful interference. In some cases, however, cable companies have not been very concerned about leakage within the ham bands. If there is leakage in the ham bands, it is almost certain that the leakage also exists in the aeronautical bands. Use a scanning receiver to locate the video carrier for one of the aeronautical channels and identify this to the cable operator. This usually gets their attention pretty fast because the FCC will shut down a cable channel quite quickly if it is causing interference to critical safety services.

What cable companies don't always understand is that they may be responsible for keeping interfering signals out of their own systems. Leakage into the system is called "ingress" in the cable industry. §76.605 (a) (7) and (a) (8) require that the cable signal be delivered to the subscriber with a specific carrier-to-noise and carrier-to-coherent-disturbance ratio. The FCC has probably never determined whether a signal leaking into a cable is noise or a coherent disturbance, but if the cable operator tells you they are not responsible for leakage into the system, you should refer them to these rules.

Amateur Responsibilities

In all interference cases, amateurs are responsible only for interference caused by spurious emissions from their stations. Amateurs are not legally responsible for interference caused by fundamental overload, and interference to cable TV systems is almost always a case of fundamental overload. Nonetheless, you may want to help. Try basic TVI cures first: high-pass filters, common-mode chokes and ac-line filters. These devices often cure CATVI, and they are more convenient than securing an appointment and waiting for a service technician.

Some TVs and VCRs respond to signals picked up on the coaxial-cable shield or are subject to direct radiation pickup. The cable company is not responsible for these two kinds of problems. In cases of direct pickup, contact the TV or VCR manufacturer through the Electronic Industries Alliance (EIA). To eliminate problems from signals on the coaxial-cable shield, install a common-mode choke. Common-mode chokes are discussed in the TVI chapter. In many instances the installation of a cable set top box (converter) will eliminate direct pickup interference in affected cable-ready TVs and VCRs.

CABLE TV FUNDAMENTALS—A TECHNICAL DESCRIPTION

The Cable Plant

Let's look at some technical fundamentals of a cable network. (See Figure 7.1) A cable system is a very unusual network, and is quite interesting from a technical point of view. Its signals are transmitted from a central point normally called a headend. There, over-the-air signals are collected and combined with others (public access, satellite feeds and so on) to form a broad spectrum of television programming. This often includes digital video and cable modem signals, occasionally FM radio stations, and cable telephony signals.

From the headend, signals are routed throughout the system over a network of optical fibers and coaxial cables. Amplifiers in the system compensate for cable losses. As you may know, cable losses increase with frequency. Equalizers installed in the amplifiers compensate for the cable frequency response.

Transmission quality is governed by two general factors: Noise contributed by the amplifiers tends to degrade signal quality. Distortion products generated by the amplifiers cause interference that gradually increases toward the system end. Impairments caused by impedance mismatches, diplex filter group delay and so forth also can degrade signal quality.

Since noise is generated in each amplifier, the system noise is kept low by keeping a good carrier-to-noise ratio at each amplifier input (so that the amplifier noise contribution is relatively insignificant). The signal level must not be too high, however, or distortion may reduce picture quality. This combination of effects defines a window of acceptable operation. The window gets smaller as more amplifiers are cascaded. There is a finite limit to the number of cascaded amplifiers, which is known as the "cascade depth" of the cable system.

In a modern HFC network, the headend's cable signals are converted to light in optical transmitters. The entire 50-860 MHz downstream spectrum amplitude modulates each optical transmitter's laser output. Glass fiber transports the light—typically at 1310 nanometers wavelength—to fiber nodes, which convert the light back to RF. Each node's RF output is connected to hardline coaxial cables (feeder cables) that serve an area of a few hundred homes. The fiber node and amplifiers in the coaxial feeder portions of the HFC network are powered by 60 to 90 volts quasi-square-wave ac carried in the hardline coax along with the RF signals. A ferroresonant transformer-based power supply converts the electrical utility's 120 volts ac feed to 60 to 90 volts, which is coupled into the coax with a passive device called a line power inserter.

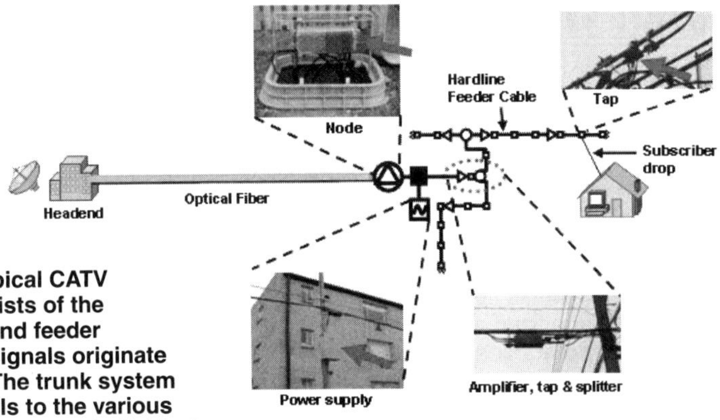

Figure 7.1—A typical CATV installation consists of the headend, trunk and feeder systems. CATV signals originate at the headend. The trunk system carries the signals to the various parts of the service area. Individual subscribers receive their signals from the feeder system.

As the RF signals pass through the feeder cables, the signals must be periodically amplified to compensate for cable and passive device attenuation. The nodes and amplifiers have their own internal power supplies that convert the hardline cable's ac voltage to dc voltage to power the various circuits. Cable networks are designed to minimize amplifier cascades, and RF operating levels are carefully controlled in order to manage noise and distortions.

Subscriber drop cables, which are usually 6-series coax (RG-6 type cable), connect to the hardline feeder cable via passive devices called taps or multitaps. Depending on housing density in the neighborhood being served, taps are installed in the feeder cable about every other lot line, in order to minimize the length of the smaller drop cable to each home.

At the side of the home, a bonding block (also called a ground block) is used to bond the shield of the drop cable to the utility company neutral conductor in accordance with National Electrical Code or similar requirements. The National Electrical Safety Code requires similar bonding of the hardline cable's shield (and support strand in the case

What to Do If The Cable Company Won't Fix the Problem

There is another problem with cable companies that crop up from time to time—although most cable companies have technical staff who understand the technical and regulatory issues, and policies that they will do all that FCC regulations and good customer-service principles require, these well-intentioned policies do not always get followed by the field repair people. Many a ham has encountered the problem of a cable-TV repair person arriving on the scene and, on seeing the amateur's tower, pointing in the general direction and saying, "There's *your* problem!"

There are a few steps you can take to help minimize the risk of this happening to you. First, discuss the matter directly with the cable company's plant manager, system engineer or chief technician. Often, the customer-service people who answer the telephones and schedule repair staff are *not* fully trained about RFI problems. The same can even sometimes be said for the field repair staff. They generally know a lot about cable-system maintenance, but have not usually had much experience with RFI problems. If the repair folks are coming out to fix a neighbor's problem, it is a good idea to make yourself available. This will serve two purposes: You can put your station on the air to help the cable company identify the problem. Secondly, if the cable repair staff doesn't understand the issues, you can help explain the responsibilities of the cable company, the TV manufacturer, and you. If you do have to correct a misunderstanding, be as diplomatic as you can.

You can also share the second part of this chapter with your cable company. This is a revision of an article printed in *Communications Technology,* a magazine well known to the cable industry. It is a complete package that starts at the beginning and goes step-by-step through the repair process.

If all else fails, the ARRL has some "big-gun" help. We maintain a liaison with the SCTE who is willing to help cable companies understand their regulatory and other responsibilities. If you need this help, contact ARRL HQ by mail (or e-mail), clearly explaining the problem, what you and other involved parties have done to resolve it and the final outcome with the cable company. Include any supporting documentation you think might be helpful. Contact:

Attn: RFI Desk
American Radio Relay League
225 Main Street
Newington, CT 06111-1494
Or e-mail tis@arrl.org
—Ed Hare, W1RFI, ARRL Laboratory Supervisor

CABLE TV Leakage

Are you experiencing what could be defined as harmful interference from CATV signal leakage? Not getting suitable support or cooperation from local CATV system personnel? Problems with CATV Channel 18 interfering with your local repeater at 145.250 MHz? If so, let Ron Hranac, NØIVN share some thoughts on the problem of CATV leakage.

The FCC's Signal Leakage Requirements

Part 76 includes some specific regulations that pertain to signal leakage from cable TV networks. In general, the rules comprise two major areas: Quarterly monitoring and an annual flyover or ground-based measurement. Regarding the former, cable operators are required to monitor essentially 100% of the network for leakage once per quarter, log any leaks found that are greater than 20 microvolts per meter (at a 10 ft measurement distance), and repair those leaks in a timely manner. Once per year, the cable operator must conduct either a flyover or ground-based measurement to come up with a "snapshot" of leakage performance. A flyover involves an aircraft-based measurement of the network's leakage from an average altitude of 450 meters above the community, and the network must be in the 90th percentile with regard to a 10-microvolt per meter field strength at the measurement altitude.

Alternatively, the cable operator may conduct a ground-based measurement of at least 75% of the plant, record leaks 50 microvolts per meter or greater, and calculate a "cumulative leakage index." The CLI is a figure of merit that shows leakage performance at a given point in time. The cable operator also is required to comply with the harmful interference clause of Part 76.

In theory, if the quarterly monitoring and repair have been conducted as required, the annual flyover or ground-based CLI should show no leaks. The reality is that leakage does occur—even in the best-maintained cable networks—often from consumer-installed devices such as poorly shielded cable-ready TVs, VCRs, etc. As well, the occasional craftsmanship issue crops up, as does cable damage from rodents, weather, and so forth. So, leakage monitoring and repair is an on-going battle for cable operators.

In the vast majority of cases, when CATV interference to amateur radio occurs, it's able to be resolved locally. Every now and then for whatever reason, the affected amateur is unable to get the interference resolved locally. Contact the ARRL for help in these cases.

Conversion of Analog TV Channels to all Digital

Now, let's talk about the cable industry's move from analog TV channels to digitally modulated signals.

This move is being done in part to "reclaim" RF spectrum. If you think about it, allocating a 6 MHz chunk of spectrum to a single TV channel is pretty inefficient, especially considering that digital compression and available digital modulation formats such as 64- and 256-QAM (quadrature amplitude modulation) allow the carriage of, say, 10 or 12 channels in the same 6 MHz bandwidth that one analog TV channel now occupies!

So, does the move to digital represent an interference nightmare ready to pounce on us hams? Not likely. Indeed, a 64- or 256-QAM digitally modulated signal has LESS probability of interfering with over-the-air communications such as ours. Here's why:

When an analog TV channel is the source of interference to, say, 2-meter voice communications, the usual culprit is the TV channel's visual carrier. The vast majority of the channel's RF power is in that carrier, so it represents the signal most likely to cause problems. Naturally, the aural carrier can be an issue, too, but its lower amplitude (typically about 15 dB lower than the visual carrier) often means the visual carrier is the one we have to worry about.

A downstream digitally modulated signal occupies the same 6 MHz bandwidth as an analog TV channel, but its power is spread across the entire channel bandwidth rather than concentrated in a discrete visual or aural carrier! The digitally modulated signal's average power also is intentionally set 6 dB to 10 dB below what an analog TV channel's visual carrier amplitude would be on the same frequency. The noise-like characteristic of the digitally

of overhead plant) to the utility company neutral on a periodic basis. The telephone company support strand also is required to be bonded to the utility company neutral. This is known as common bonding, and is done for safety of personnel and equipment, and to minimize ground potential differences among power, cable and telephone.

Once inside the home, the drop cable is connected to various subscriber terminal devices, including TVs, VCRs, set top boxes, and cable modems.

The TV-signal level at any point in the cable network seldom exceeds 1 mW (0 dBm), and the nominal level at a TV set may be on the order of –45 dBm. The cable system is designed to maintain a constant impedance (nominally 75 Ω). Hence, 75-Ω cable is used throughout the system. Most modern subscriber terminal devices have a 75-Ω coaxial input that allows direct connection to the drop cable.

In two-way cable systems, signals from the end user are transmitted back to the cable company's headend in the 5-42 MHz band. Because upstream signals are on different frequencies than downstream signals, most of the same cables and network components are used for simultaneous two-way signal transmission. See "Just What is a Two-way Cable System Anyway?"

Channelization Schemes

Several years ago, the EIA and NCTA collaborated to create a standard (EIA 542) for cable channelization. It defines cable channel plans from 54 to 1002 MHz.

There are three general schemes for channelization, which are known as Standard, HRC, and IRC. Standard channelization begins with the normal frequency assignments of the VHF-television bands and the extension of these channels (6n + 1.25 MHz) into the cable-only frequencies (the spectrum above FM and below channel 7, 108-174 MHz, and all channels above 13, 216 MHz and higher). HRC and IRC channelization schemes are not common today because of available amplifier technology.

The TV Channel Chart in chapter 6 shows which cable channels coincide with ham bands. If, for example, you have interference to cable channel 18 from amateur 2-meter operation, suspect cable ingress.

Harmonically Related Carriers (HRC)

During the development of cable-system amplifiers, amplifier-semiconductor distortion limited the number of channels and signal levels that could be carried. Efforts were made to "hide" distortion products by placing them on the frequencies of other-channel carriers, where they would be the less visible. The Harmonically Related Carriers (HRC) approach resulted. In this system, all channel frequencies are multiples of 6 MHz and locked to a precise 6.0003-MHz comb. (A graph of an array of equally spaced channels resembles a comb. See Figure 7.2.)

modulated signal is going to be less likely to cause interference if leakage exists, too!

If we look at a digitally modulated signal's noise power bandwidth, it is equal to the symbol rate. For 256-QAM, the symbol rate is 5.360537 million symbols per second (Msym/sec), which means the equivalent noise power bandwidth (–3 dB points) is 5.360537 MHz.

A 2 meter transceiver with an equivalent noise power bandwidth of, say, 15 kHz (the IF bandwidth), won't be affected by the entire 6 MHz of "noise" (OK, 5.36 MHz), but rather the noise power in the IF bandwidth: 15 kHz. This means the received signal strength will be 10log(15000/5360537) lower, or 25.53 dB below the noise power in a 5.36 MHz bandwidth. So, if the digitally modulated signal is carried –6 dBc relative to what an analog visual carrier would have been on the same channel, the received "noise" power from the digitally modulated signal will be 25.53 + 6 = 31.53 dB LOWER than the interference from a discrete visual carrier on the same channel.

If the cable operator is just complying with the FCC's 20-uV/m leakage limit (10 ft from the network), that limit will yield a dipole level of –43.67 dBmV (75 ohms), or –92.42 dBm (6.55 microvolts) at the dipole's terminals for an analog TV channel visual carrier at 145.25 MHz. If that analog signal were a digitally modulated signal at –6 dBc relative to the visual carrier, when you factor in bandwidth correction (25.53 dB), the dipole level will now be –75.2 dBmV (–123.95 dBm).

As well, rather than a carrier, the interference will be noise-like. At this low level, it's unlikely the interference could be differentiated from normal background noise, or at worst, a slight increase in background noise.

Interference to 145.250 MHz

This problem is typically associated with the visual carrier on CATV Channel 18. This carrier is at 145.250 MHz, which also happens to be a common repeater output frequency. In fact, we have this very scenario here in the Denver area. A local 2-meter repeater is on 145.25 MHz, and I can receive some localized—but fairly low level—interference from signal leakage. It appears to be coming from a nearby house, suggesting that the likely culprit is a poorly shielded cable-ready TV set connected directly to the cable. Fortunately the interference is low enough level that it's not an issue when the repeater transmits. Unfortunately, this is not always the case...

If severe enough to warrant attention, there is a potential work-around to this problem. The cable company's headend equipment that creates TV channels can in some cases be offset from the standard xxx.250 MHz visual carrier frequency. Many headend modulators include an offset switch that facilitates an offset of 12.5 kHz or 25 kHz from the nominal xxx.250 MHz frequency. Offsets are used in the aeronautical navigation and communication bands to comply with FCC offset requirements in Part 76 of the FCC Rules. Fortunately, most offset-capable modulators can be offset outside of the aeronautical bands, too, which allows continued operation on a given channel. TV set AFC circuits have no problem with a channel being offset 12.5 or 25 kHz, and it should be enough to eliminate the occasional interference problem that crops up on the repeater's 145.25 MHz output frequency.

Here is my suggestion if having a problem with interference to 145.250 MHz:

Contact the local cable system, and ask for the Plant Manager (may also be called "Chief Engineer," "Chief Tech," "VP of Engineering," or similar). Let him or her know what's been going on with leakage on Ch. 18. Add that you'd be appreciative of any cooperation to resolve specific leakage problems. Also let him or her know that offsetting Ch. 18 might go a long way toward reducing or eliminating interference on 145.25 MHz, and ask if it would be possible for them to do that.

As a side note, some cable systems have done this where a 145.25 MHz repeater was in use. The nice thing is it's a "free" fix that works quite well. Granted, it won't eliminate leakage (that still has to be kept in check), but it will substantially reduce the possibility of on-frequency interference.—*Ron Hranac, NØIVN, Technical Leader,Cisco Systems, Inc.*

Therefore, IMD products between channels also fall on a 6-MHz comb, and the distortion products fall at or near zero beat and are hidden by the carriers of other channels. This is a successful approach that is still used in some cable systems, although its use has diminished in recent years.

Incrementally Related Carriers (IRC)

The Incrementally Related Carriers (IRC) system was developed to achieve some of the HRC benefits without using nonstandard TV-channel frequencies. It uses the basic VHF comb based on 6n + 1.25 MHz for all channels; Channels 5 and 6 are shifted to slightly lower frequencies to accommodate this channel plan. In this system, all carriers are locked to a coherently generated comb of frequencies to stabilize the cable channels. IRC has some of the IMD-hiding effect of HRC while not displacing TV-carrier frequencies (except Channels 5 and 6) from those of over-the-air channels.

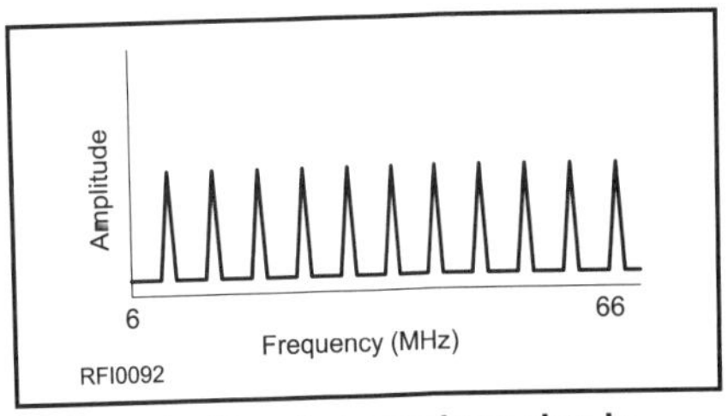

Figure 7.2—Equal channel spacing is called a "comb" because of a graph of amplitude v frequency resembles a comb.

Two-Way Cable Networks

Cable systems are capable of two-way operation, and the majority of today's cable systems support two-way signal transmission. Upstream signals originate at the user and travel back to the headend (upstream: opposite to the entertainment-signal downstream flow from the headend to the user). In a normal cable system, the frequency range from 5-42 MHz is available for upstream operations, although in some countries outside of North America the return path may operate on frequencies as high as 65 MHz.

Some parts of the cable network are used for non-entertainment purposes. These are generally included in what is known as an institutional network (I-Net). In such systems the number of upstream channels is expanded, while the number of downstream channels is reduced to provide approximately equal capacity in both directions.

The upstream channels (5 to 42 MHz) may be somewhat more susceptible to interference than those in the VHF range (susceptibility depends on a number of factors). It may be more important that amateur power levels in the HF range are typically higher than at VHF: More serious interference can be expected. When the cable system is known to have reverse transmissions, the energy from these signals can be monitored for leakage with HF equipment. This, however, involves the use of unwieldy or inefficient antennas and has no particular advantage over VHF monitoring. Things to remember: In general, a leak leaks at all frequencies, although not uniformly; where energy exits, energy can enter.

FCC Requires Offsets To Protect NAVCOM Frequencies

In the process of the FCC rulemaking about cable signal leakage, some extra caution was applied. Based on the possibility of a catastrophic leak (such as a severed cable) the FCC established a system of frequency offsets in §76.612. The offsets protect aircraft NAVCOM channels from strong carriers in cable systems. The offsets are generally 12.5 kHz offset (or odd multiples of 12.5 kHz) in aeronautical radio communication bands and 25 kHz offset (or odd multiples) in aeronautical radio navigation bands. These bands include the frequencies of 118-137, 225-328.6, 335.4-400, 108-118 and 328.6-335.4 MHz bands.

This is important to amateurs because interference from cable leakage may not appear exactly where indicated by channel plans. Note that offsets are not required in the amateur bands, but cable operators may choose to use them for consistency with the aeronautical channels.

For instance, 145.25 MHz is the nominal visual carrier frequency of Channel 18 in the standard and IRC channelization plans (144.0072 MHz in an HRC system). If this channel were offset like those in the 118-137 MHz band, Channel 18's visual carrier would be 145.2625 MHz (+12.5 kHz) rather than 145.25 MHz, or perhaps 145.2375 MHz (-12.5 kHz). (Actually, the offset can be any odd multiple of 12.5 kHz, but most cable operators choose ±12.5 kHz.)

CATVI Isn't Always the Fault of the Cable Company!

Many hams have the mistaken belief that all cable problems are the fault of the cable company. In reality, this is not the case! The *only* CATVI problems that are legally the fault and responsibility of the cable company are those caused by cable leakage, either ingress or egress. In many other cases, the fault that is the root cause of the interference is in the customer-owned equipment. You may find that cable companies are willing to address CATVI due to causes that are not their fault, but they do so as a customer-service issue.

CATVI to the picture and sound of a TV set from HF transmitters is rarely caused by a fault in the cable system. In most cases, it is caused by the common-mode signal that is picked up *on* the shield of the cable-system coaxial cable. This common-mode signal can get into a TV any number of ways. Some TVs are immune; others respond strongly. In either event, common-mode pickup problems do not stem from cable-plant defects. (In some cases, the common-mode signal can get into cable-company equipment, such as set-top converters. In most cases, the cable company probably should have responsibility for curing this.) Fortunately, the cure for HF common-mode CATVI is usually pretty easy—install a common-mode choke in the incoming cable, just where it connects to the first piece of electronic equipment.

More cable companies are making use of HF two-way channels than ever before, primarily for digital signals, but with some analog uses as well. HF signals can leak into the cable (for the same reasons they do on VHF) and cause interference. See the two sidebars on two-way cable TV channels.

On VHF, the situation is a bit more clear cut. If there is interference from an over-the-air service to a cable channel that uses frequencies allocated to that over-the-air service, such as 2 meters interfering with cable channel 18, there is clearly a leak somewhere. This leak can occur in the cable system or equipment, or it can occur in customer-owned, cable-ready TV equipment. While the cable company is responsible for its own equipment, it is *not* responsible for the customers' equipment. In fact, in most cases, company policy or state law prohibits them from working on the customers' TVs and VCRs. It is up to the manufacturer of the TV, or a local service shop, to resolve any problems with customer-owned equipment.

This can lead to a bit of finger pointing! The cable company can claim that the leak is in customer's equipment; the TV manufacturer can claim the cable equipment is at fault. Fortunately, there is an easy solution: Ask the cable company to demonstrate adequate performance using a set-top channel converter connected to the affected device. If they can, the leak is clearly in the customer's equipment, not in the cable system. This also offers a solution to the interference problem. If there is still interference, either the cable system or the converter box is leaky—either way, it's the cable company's problem.

Of course, you should suggest to the cable company that they buy a copy of this book and contact the ARRL "RFI Desk" (that's me!) here at ARRL HQ.—*Ed Hare, W1RFI, ARRL Laboratory Supervisor*

WHY CABLE SYSTEMS LEAK

Bottom Line: A Leaky System

When all is said and done, our attention focuses on one fact: the shielding integrity of a cable system is not perfect. Where cable-system RF energy can leak out, external RF energy can enter. Energy that leaks out can interfere with amateur communication, while energy that leaks in can interfere with viewer reception (which often results in unpleasantness for the radio amateur).

Cable systems leak for several reasons. The seamless aluminum jacket (the cable shield) has a relatively high coefficient of expansion: As air temperature rises, the shield lengthens. When the air temperature lowers (after sundown, for example), the shield shortens. This constant cycling may develop small cracks in the shield in systems not built properly (usually behind the connector compression fitting—see Figure 7.3A). The cracks present effective escape routes for signal leakage. Left untreated, cracks can develop into complete shield breaks, called "ring cracks," which expose the dielectric and center conductor (see Figure 7.3B). If this happens, ground continuity along the shield can be lost!

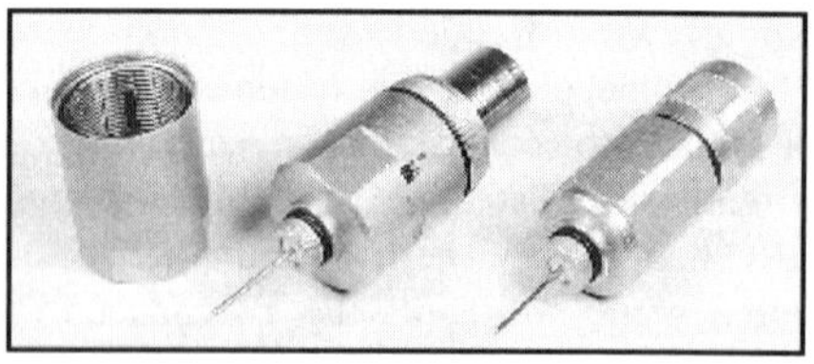

(A)

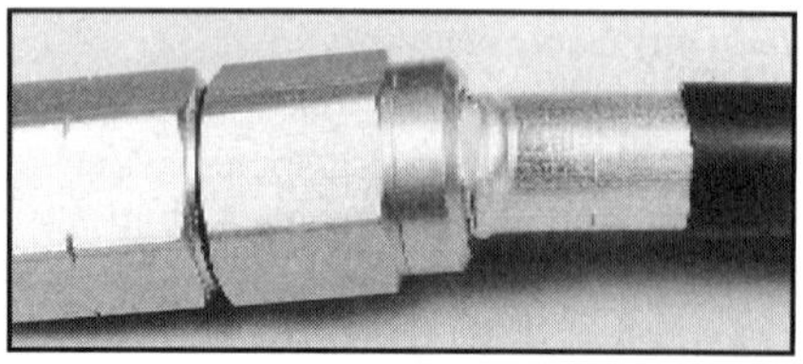

(B)

Figure 7.3—(A) A Hardline connector as used on CATV trunk lines. (B) A ring crack breaks the shield connection completely.

Rodent damage is another culprit that can cause signal leakage. It seems that squirrels in particular like to include coaxial cables in their regular diet.

This same condition can occur anywhere hardline is used. For amateurs, a worsening SWR is usually the first clue of a mechanical fault such as a crack. This is an excellent reason why all amateurs should keep detailed antenna-system records and check system operation several times each year. Measurement comparisons "flag" developing defects in the system. (They may also be of assistance in convincing the FCC that your neighbor's nasty RFI problem isn't caused by technical deficiencies at your station!)

There are a number of other leak sources. These can generally be easily located by cable-company repair personnel.

In the network or outside-plant part of the system, the cable company may find:

- Loose connectors
- Loose or warped amplifier housing covers
- Loose passive device (splitter or tap) face-plates
- Damaged cables

You may be able to fix some of the problems that might be found on the customer premises. In fact, in the customer area, you are apt to find nearly anything!

In the customer "drop" or at the customer premises, you may identify:

- Loose connectors
- Damaged cables
- Poorly shielded splitters
- Illegal connections
- Inferior, consumer-grade leaky interconnect cables (TV to VCR, for example)
- Improperly installed additional outlets
- Twin lead or telephone wire
- Poorly shielded TVs or VCRs.

Connector Continuity

Cable systems often leak because technicians haven't properly tightened the compression fittings used on hardline and subscriber-drop cables. As with ring cracks, connectors with poor ground continuity present excellent paths for signal leakage.

Illegal Connections

Not all leaks result from maintenance problems. Illegal connections to the cable system (made by technically inept people without the proper tools and parts) are a major source of leaks. As can be seen in Figure 7.4, it is not uncommon for an illegal hookup to use 300-Ω twin-lead cable or even zip cord for runs to additional outlets, or

Just What is a Two-way Cable System Anyway?

Most cable systems have, until recently, been one-way systems. That is, signals are carried in one direction only: From the headend "downstream" to the subscribers. In a typical system, downstream, or *forward signals* occupy frequencies from 50 MHz to as high as 860 MHz. Since the 1970s, cable companies have been designing their networks to be two-way capable, although the majority have not actually used the upstream, or *reverse signal* path. With the advent of cable modems and other interactive technology, operational two-way CATV systems are now fairly commonplace.

Upstream signals can simultaneously travel through the same cables and equipment as the downstream signals do because the upstream signals are in a different range of frequencies than the downstream signals. Most upstream operation occurs in the 5 to 42-MHz spectrum. When a cable operator decides to implement two-way services, it is necessary to install reverse amplifier modules and diplex filters in all of the distribution network amplifier housings. Figure A shows a block diagram of a typical two-way amplifier.

Fifty to 860-MHz signals enter the amplifier housing at the downstream input (the left side of the figure), and pass through the first diplex filter's high pass circuit to the forward amplifier stages. After amplification, the downstream signals pass through the second diplex filter's high-pass circuit to the housing's downstream output. Five to 42-MHz signals enter the housing on the downstream output side (right side of the figure), and pass through the second diplex filter's low-pass circuit to the reverse amplifier stage. After amplification, the upstream signals pass through the first diplex filter to the housing's downstream input.

When a cable network is configured for two-way operation as just described, it is known as a *sub-split network*. There also are *mid-split* and *high-split* networks, and while uncommon, they are used mostly for institutional applications rather than providing traditional cable services. The upstream spectrum in a mid-split network is 5 to 112 MHz, and the downstream spectrum starts at 150 MHz. In a high-split network, the upstream spectrum is 5 to 174 MHz, and the downstream spectrum starts in the 222 to 234-MHz range.—*Ron Hranac, NØIVN, Technical Leader, Cisco Systems, Inc*

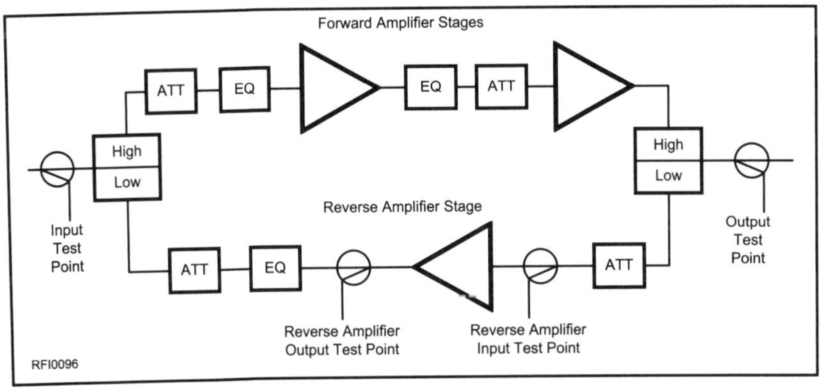

Figure A—Block diagram of a two-way amplifier.

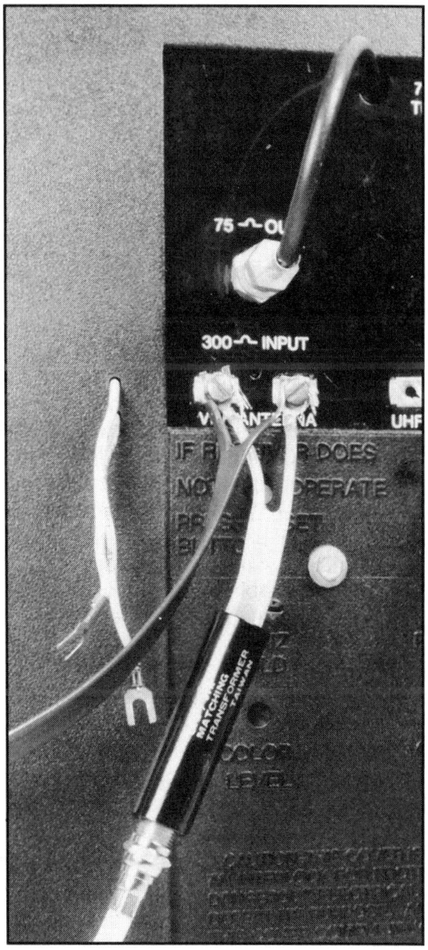

Figure 7.4—The 300-Ω twin lead used in this illegal CATV hookup certainly causes CATV leaks and results in interference to and from amateur stations. All CATV installation work should be done by CATV personnel.

even to a neighboring home. Needless to say, such hookups are sources of strong leaks.

Inferior Cable from 1960s and '70s

The drop cables used to wire homes during the 1960s and '70s are another major source of cable leaks. Before cable operators started programming on frequencies used primarily by other services, it was common practice to use drop cables with 67% braided shields (see Figure 7.5). While such cable does provide ground continuity, it does not always reduce leakage to levels acceptable to amateurs with 0.1-μV-sensitivity receivers! Modern drop coax has a composite shield consisting of one or more layers of foil tape and wire braid. Dual-shield drop cables, which are the most common, have the first layer of foil bonded to the dielectric, and that foil layer is covered by braid. Some installations near transmitters may benefit from tri- or quad-shield coax, which has a foil-braid-foil or foil-braid-foil-braid shield configuration. All of these modern cable designs efficiently keep signals within the cable system. (If you experience interference on 144.0072 MHz or 145.25 MHz, ask your cable operator to determine if your home/station is wired with the old style drop cable.)

Understanding and Identifying CATVI

Cable interference to Amateur Radio typically occurs on 2 meters and 222 MHz. The usual frequencies are at or near 144.00 MHz or 145.25 MHz (cable channel 18[E]) and 223.25 MHz (cable channel 24[K]). The interference is a constant "BUZZ" (the picture-sync buzz), or a steady carrier that gets stronger as you get closer to the cable leak.

Causes

Such interference happens when cable carriers leak out of the system. The reasons for leakage usually fall into several general categories:

Poor Cable Company Coax Shielding

Until the early 1980s, coaxial drop cables (which connect the TV to the cable company's feeder line) employed a single layer of braided wire as the shield. Some drop cables had less than 70% of the center conductor covered. This allows much more leakage than contemporary bonded foil-braid cables.

Damaged Hardline Cable Shields

The main distribution network coaxial cables used in cable systems are constructed with a copper-clad aluminum center conductor surrounded by hardened foam. An aluminum tube forms the electrical shield. A plastic jacket may provide an environmental barrier. Unfortunately, aluminum expands and contracts with heat and cold. Circumferential ("ring") cracks can develop at the point of maximum stress, usually at the back nut of a hardline connector. Leakage then occurs. Leakage can also occur if "smog" erodes unjacketed cables. Other causes of hardline cable leakage include rodent damage, bullet or pellet gun holes (yes, this does happen!), chafing to the shield where the cable passes through trees, and unintentional damage caused by landscaping or utility work.

Loose Connectors and Hardware

Another potential source of signal leakage is loose connectors on hardline cables, as well as loose or improperly installed drop cable connectors (commonly known as "F" connectors). Newer F connectors have an integral compression to improve shielding, but millions of older "two-piece" connectors are still in use, and are still sold at neighborhood electronics stores. Leaks also occur when amplifier housings are not properly closed or signal-tap face plates are not tightened properly.

Law of Reciprocity

If you can hear them, they can hear you. Not only do cracked or poorly shielded cables emit potential interference to your operations, leaks also permit amateur signals to enter the cable system. Ever wonder why your neighbors complain about TVI on their channels 18 and/or 24? Now you know!

Depending on the ingress point and the strength of your signal, you could cause herringbone (beat) patterns for dozens (or hundreds) of your neighbors! Whose responsibility is it? As long as you are ". . . a radio-communication service operating in accordance with [the FCC rules]" . . . it's probably the cable operator's responsibility. But for the sake of your neighborly relations and general sanity, work with your cable operator to help identify the leak! (See the FCC CATV rules §76.613.)

Use Transmitter-Hunt Skills to Find CATVI

Brush up on your transmitter-hunting skills and dial up 145.25 MHz (or 144.0072 MHz for "HRC" systems) on your hand-held radio. Set your squelch to just close. As you get closer to the leak, the squelch will break, the signal will get louder, and the background noise will quiet down. (You may hear the buzz of the television sync pulse.) Keep tightening the squelch as you get closer. When you can no longer squelch the radio, remove the antenna and then use the radio as a "near field" detector. When you've located the leak, or the approximate location, call your cable operator.

Don't try to repair cable leaks yourself. The lines are the property of the cable company. They are responsible for system condition, and they must ensure that leakage is within FCC specifications. Also, some hardline distribution cable lines carry dangerous voltages.

While you can try to describe the signal leakage problem to the customer service representative who answers the phone, it's better to speak directly with the "Technical Manager" or "Chief Technician." Always log the date, time, and names when you refer CATVI complaints or inquiries to your cable operator! You'll never know when that log will be needed to substantiate your claim of interference.

Finally

Cable operators do want to resolve service problems, especially leakage. The FCC requires every cable system to file an annual report of leakage control. When you call about a CATVI problem, it's usually just a matter of getting through to the "right person," who can understand you. Local radio clubs should initiate and maintain a liaison with local cable operators before problems occur.

In the event that you are unable to resolve a problem at the local level, report (in writing, see the Report Form chapter) your complaint to ARRL HQ. Mark it "Attention: Box RFI." For many years, the League and the SCTE (a cable-industry engineering association based in Exton, PA) have formally cooperated to resolve problems that couldn't be resolved on the local level.—*Jonathan Kramer, W6JLK, Principal Technologist, Kramer.Firm, Inc.*

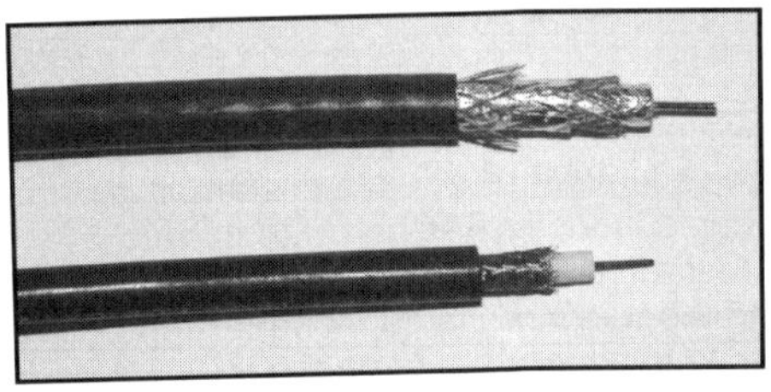

Figure 7.5—Old-style CATV drop cable has relatively poor shield characteristics. Newer quad-shield cable such as shown in the upper part of the photo has much better shielding. This cable is 6-series quad-shield CATV drop cable. It has an inner foil shield bonded to the dielectric, followed by additional layers of braid, foil and braid. The lower cable is copper braid 59-series coax, which is unsuitable for CATV use. Some older variations of copper braid cable have even less braid coverage than what is shown here.

LEAKAGE AT THE TV AND ACCESSORIES

Even if the hardline is free of cracks or other damage, the connectors are properly installed and the drop coax is modern foil-braid shield, there may still be leakage interference from a TV set. Many television receivers with 75-Ω "F" connectors on the back use a balun to convert the 75-Ω incoming signal to 300 Ω for the tuner. The short section of 300-Ω twin lead can act as an antenna for cable system leakage. As well, some cable-ready TVs and VCRs have less-than-optimum shielding, which may cause leakage. Your cable operator can't do anything about this leakage, except to disconnect the service until the leaky TV set is corrected. In some cases a cable set top box can be used to minimize cable-ready TV or VCR leakage, because the only signal from the set top box to the TV or VCR is Ch. 3 or 4 rather than the entire 50-860 MHz cable spectrum.

Some other leaks originate in the subscriber's home. These may result from fittings or extemporaneous connections and electronic devices. Electronic devices include poorly shielded TV sets, VCRs and FM tuners, switches for video games and the like, high output levels from distribution amplifiers used to feed extra drops within the dwelling and other similar sources. It is obvious that these are difficult for the cable operator to address because they are inside the subscriber's home. Note that the FCC allows cable operators to disconnect subscribers who have excessive leakage within the dwelling. The normal scenario is: Notify the subscriber and ask for entrance to find and repair the problem. If such is not granted, the drop may be disconnected.

HOW CABLE OPERATORS FIND LEAKS

FCC Requires Vigilance

The FCC requires all but the smallest cable operators to have and follow a signal-leakage patrol program. That requirement is part of the FCC rules (see "CATV Leakage"). The FCC requires that a cable operator continuously patrol the system looking for leaks, logging them and repairing them.

Most cable operators use truck-mounted commercial antennas and special commercial receivers to patrol the system for leaks. The receivers are usually meter-calibrated so the technician can read or compute leakage field strength. When a leak is heard, the technician narrows down and logs its general location and strength. In large systems, those logs are given to other technicians who pinpoint and repair the leak. In small systems, the same technician often patrols, finds and fixes leaks.

FIXING CATVI

In most cases, the cable company will have to fix CATVI problems. If the problem is caused by a leak in the cable, they are the *only* ones that can fix it because no one but qualified personnel should be working on the cable system. In some cases, however, the ham can help by offering advice to the cable repair personnel, or by directing the cable customer toward external solutions, such as common-mode chokes.

Later in this chapter, the subject of curing CATVI is treated in detail. This section offers a quick overview of the causes and cures for CATVI.

Finding Leaks

Leak location finding takes many of the same skills that amateur transmitter hunters use. The technician localizes the area of the leak by monitoring signal strength while driving. As signal strength increases, the technician may use an attenuator to avoid overloading the receiver. When this technique narrows down the leak location to within about 80 feet, the technician usually uses a hand-held receiver with a small antenna to "walk the area." When the leak is determined within several feet, the technician may use a highly directional slotted antenna with the hand-held receiver to find the exact point of leakage. (Because most cable systems use special heat-shrink boots over hardline connectors, cracks may not be visible.)

Sometimes a leak is difficult to locate because energy couples into the wire rope (messenger or support strand) holding up the cable. On the strand and outer surface of the cable's shield, the leak propagates as a series of peaks and nulls (see Figure 7.6). The largest peak occurs very near the leak. Unfortunately, it may be very difficult to determine which is the largest peak. The actual leak is found and corrected by repeated measurements (or sometimes by luck).

LOCATING CABLE INTERFERENCE NEAR YOUR SHACK

Safety

First follow basic safety rules: Don't ever touch your antenna or antenna connector to any cable system surface. You never know where you'll find ac power! Leave the actual repairs to trained cable technicians who are equipped with the proper tools and parts.

Interference From Cable

If you believe you are experiencing CATVI, you can perform your own CATVI-hunt with a hand-held 2-meter receiver, a detachable antenna and some basic transmitter-hunting skills. First, tune your receiver to 145.25 MHz or near 144.0072 MHz. If you hear a loud buzzing (the TV sync pulse) or your receiver quiets, start sweeping the area looking for the strongest signal. (It helps if your receiver has an S meter, but it is possible to gauge signal strength by ear alone.) When you locate the strongest signal, loosen the antenna connector (to reduce the received signal strength) and continue to hunt for the strongest signal. (It sometimes helps to use your body as a shield for increased directivity.) When the signal strength again reaches the point of receiver overload, remove the antenna entirely, and probe with the open receiver antenna connector for the strongest signal. (You'll probably locate the leak this way.)

Don't be surprised if you find yourself staring at a wall! It's not unusual for coax inside walls to deteriorate and begin leaking. Also, a careless installer may have accidentally pierced the cable with a staple or other fastener during installation. As the fastener rusts, the cable deteriorates and begins to leak.

Interference to Cable

Sometimes cable leakage is encountered by the fixed-station operator who uses a scanner to monitor the entire spectrum. The scanner may routinely stop at the cable-leakage frequency. Conversely, the fixed-station operator may find that his or her transmissions interfere with cable connected television sets. In such cases, follow the diagnostic and repair procedures set forth in the Troubleshooting chapter and in the first part of this chapter.

When all receivers in the same area are

Two-way CATV Systems and Amateur Radio

Most amateurs are familiar with the issues surrounding CATVI from one-way systems. The majority of interference problems are confined to VHF and UHF bands. When CATVI occurs, it normally involves continuously operating TV channels. The interference is easy to identify on a receiver, because of the steady visual carrier at frequencies such as 145.25 MHz or 223.25 MHz, telltale sync buzz from the video's horizontal sidebands spaced every 15.734 kHz, or perhaps even audio from the aural carrier.

With two-way cable systems, interference problems are more challenging to troubleshoot. First, signal leakage from a two-way system's reverse spectrum is likely to occur in the HF bands between 5 and 30 MHz. Second, many of the signals carried in a cable system's reverse path are digitally modulated signals that are very bursty in nature, and while they are discrete carriers, they resemble wideband noise occupying anywhere from a few hundred kilohertz to several megahertz bandwidth. In other words, the interference is likely to be very brief and intermittent, and hard to identify because of its noise-like characteristics.

There are no standardized frequencies for CATV reverse path operation, in part because of potential interference problems from over-the-air signals that may enter the network via a loose connector or even a poorly shielded TV set. Cable modems, for instance, are frequency agile to allow cable operators to choose relatively clean portions of the reverse spectrum for two-way operation. Some equipment will automatically switch to a clean frequency in the event of interference. In almost all cases, cable modems transmit upstream carriers somewhere in the 20-40 MHz range.

Perhaps of greater concern to amateurs is the potential interference that may occur to the cable reverse spectrum from an HF transmitter. While the digital modulation schemes used by reverse path data equipment are fairly robust, they are not completely immune to outside interference. If a strong interfering carrier appears within the digitally modulated signal's bandwidth, it will result in retransmissions and a decrease in data throughput. Severe interference, even at HF frequencies other than where the digitally modulated signal is operating, may cause reverse path amplifiers or fiber optics equipment to go into compression and clip (this phenomena is known as cross-compression). When this happens, data transmission will come to a halt, at least for the duration of the interference.

Why should this be of concern to amateurs? Because of the finger pointing that is likely to occur. Angry subscribers may blame "the ham next door" for cable modem Internet access problems, and cable technicians unfamiliar with troubleshooting reverse path interference may do the same thing. For these reasons, education and mutual cooperation are very important. Amateurs cannot automatically assume that random noise-like interference is coming from the cable network, and cable company personnel should not automatically assume that your transmitter is the source of reverse path interference.

Successful two-way cable system operation requires that the network be kept very tight—essentially leak-free. If signals can leak out of a cable system, then over-the-air signals also can leak in! Even if a cable system complies with the FCC's signal leakage regulations, that may not be enough to prevent problems with two-way operation. Low level leaks below the FCC's 20 microvolt per meter (µV/m) leakage limit (as measured 3 meters from the network components using a resonant halfwave dipole) still represent potential ingress points. Most cable operators have found lowering the leak threshold to 5 µV/m or less provides relatively problem-free two-way operation.

When outside signals leak "into" a cable system, it is called ingress. The cable industry has found that up to 95% of the ingress that affects the reverse path of a two-way system gets into the network via subscriber drops. Roughly 25% gets in somewhere between the telephone pole (or pedestal in underground installations) and the side of the house, and the remaining 70% or so gets in somewhere between the side of the house and the subscriber terminal devices. In fact, many cable-ready TVs and VCRs are notorious sources of interference, often because of poor shielding.

The most common causes of drop-related ingress problems include loose or improperly installed F connectors, damaged cable, poorly shielded splitters, illegal connections, extra outlets installed by the subscriber using substandard "consumer-grade" parts, and the previously mentioned consumer devices connected directly to the cable. If you suspect interference from your cable system, or suspect that you may be causing interference to the system, don't try to fix these problems yourself because of the potential for making the interference worse.

Instead, work with the cable company to resolve the problem. You might start by looking for downstream signal leakage with your 2 meter rig tuned to a midband VHF channel's visual carrier frequency. Remember, if there is a leak, that represents a potential ingress point. The following table highlights common midband visual carrier frequencies used in most cable systems.

Channel	*Standard visual carrier frequency (MHz)*
98	109.275*
99	115.275*
14	121.2625*
15	127.2625*
16	133.2625*
17	139.25
18	145.25
19	151.25
20	157.25
21	163.25
22	169.25

*Note—TV channels carried in the aeronautical navigation and communication bands are required to be offset in increments of 25 kHz and 12.5 kHz respectively from the standard xxx.25 MHz visual carrier frequency. The majority of cable operators use the frequencies shown, but in practice any offset may be in use.

When a potential point of subscriber drop-related interference has been identified, the cable company may have to replace the drop cable with a tri-shield or quad-shield variety; install new connectors and other components; and install high-pass filters in one-way portions of the drop. One effective tool for reducing or eliminating ingress getting in through TVs, VCRs and FM tuners is the common-mode choke. Simply coiling up 10 feet of cable into a five to six-inch diameter loop at the input of the offending device often will take care of most 5 to 30-MHz interference. For some TVs or VCRs where filters and chokes do not provide adequate interference suppression, the cable company may have to install a set-top box (converter) to provide the required amount of isolation between the network and problem device, or the subscriber may have to have the set repaired.—*Ron Hranac, NØIVN, Technical Leader, Cisco Systems, Inc.*

Two-way Cable Systems and Digital Services

The majority of North American cable systems have been upgraded to two-way operation, and now provide a variety of interactive services such as high-speed Internet access. RF transmissions to and from cable modems use digitally modulated signals: 64-QAM or 256-QAM in the downstream (headend to modem), and usually either QPSK or 16-QAM in the upstream (cable modem to headend). Digital video service also uses 64- or 256-QAM downstream signals.

Downstream 64- and 256-QAM signals occupy the same 6 MHz bandwidth as an analog TV channel. The QAM signal's RF energy is spread across nearly all of that 6 MHz—there is no carrier, per se, since quadrature amplitude modulation results in a double-sideband suppressed carrier signal—which on a spectrum analyzer resembles a flat-topped noise-like "haystack." The QAM signal's average power (digital channel power) is normally set 5 dB to 10 dB lower than what the visual carrier amplitude would be for an analog TV channel on the same frequency.

What does a downstream QAM signal leaking out of a cable network sound like on a typical amateur FM transceiver? Think very low-level noise, and difficult to distinguish from normal background noise. Let's first consider a 20 µV/m leak from a regular analog TV channel (Ch. 18, or 145.25 MHz visual carrier) that causes a 2-meter transceiver's S-meter to read S9. For this example, we'll assume the 2-meter antenna is a half-wave dipole placed 10 feet from the leak, and each S-unit on the transceiver equals 6 dB.

If the analog TV channel were replaced with a QAM signal whose digital channel power was set at the same level as the analog TV channel's visual carrier (as previously mentioned, the digital channel power of a QAM signal is set several dB less, but we'll assume a worst-case situation here), you might be inclined to think the interference would still indicate S9 on the transceiver. In fact, it would be substantially less, because of the noise-like characteristics of the QAM signal. Remember, a downstream QAM signal's digital channel power is spread across nearly all of the 6 MHz channel bandwidth, yet the 2-meter transceiver has an IF bandwidth of perhaps 15 kHz or so. Since we're not dealing with a discrete carrier, but rather a noise-like signal, the S meter would indicate a received signal strength that is about 26 dB (a little more than four S units) less than what a carrier would produce, or just under S5.

The receiver bandwidth correction we have to apply to this example is 10log(6,000,000/15,000), or 26.02 dB. That is, the receiver's 15 kHz IF bandwidth won't pass the entire 6 MHz-worth of "noise" from the QAM signal, only 15 kHz worth.

What About Upstream Signals?

Transmissions from the cable modem back to the headend occur in the 5-42 MHz spectrum, and the potential for leakage-related interference in the HF bands has generated concern among amateurs.

There have been some interference cases reported to ARRL in which noise-like interference in the lower part of the HF spectrum—for instance, 160 meters and 75-meters—was believed to be coming from the outside cable plant. In more than one instance, the affected amateur has found that the noise appears to be radiating from the cable company's equipment and cables. It's pretty easy to assume that since the interference appears to be coming from the cable plant, it's noise-like, and the cable operator has high-speed data signals in the upstream, it must be a leakage problem!

While today's cable networks use the 5-42 MHz spectrum for transporting upstream signals, operators generally don't carry any signals below about 10 MHz. The noise and other ingress in that frequency range make the spectrum essentially unusable. Some cable companies transmit set top box data on narrow-band FSK carriers in the 8 to 12 MHz range, but this would be fairly easy to identify. As well, cable modem upstream signals are typically either 1.6 MHz or 3.2 MHz bandwidth QPSK or 16-QAM digitally modulated signals. Like downstream QAM signals, upstream QPSK and 16-QAM signals are noise-like signals, but they occupy a defined channel bandwidth, not the entire 5-42 MHz upstream spectrum.

In most cases, noise that appears to be radiating from the outside cable plant in the lower part of the HF spectrum is not a leakage problem, but rather is noise from another source: Part 15 devices in consumers' homes, power line noise, and even telephone company DSL signals! Cable operators, as well as the phone company, are required to bond the support strand (the steel rope that holds up the hardline cables) and, in the case of the cable company, the cable shield, to the utility company neutral conductor. It's not unusual for noise and other non-cable interference to be coupled to the coax shield, support strand and equipment housings via the neutral bonds. The cable network then acts like a giant long-wire antenna, re-radiating the interference. When this happens, it is not a cable leakage problem.

The cable company should definitely be contacted so their service technicians can check for leakage, and confirm whether or not the interference is a cable network problem. If the cable plant is clean, it's time to look for other potential sources. Here's one example of such an interference case:

"The cable company turns out not to be at fault. They said I should go to the power company. I made a complaint to them on Friday last and they were out by Tuesday of the following week. The technician called me and asked me some very good questions about the interference. I told him about the cable vault and the transformer vault I localized the interference to. He went out and found no problems with the transformer but went to the neighbor whose house was right next to the vault. He had him turn off the power to his house and the interference went away! They used the breakers in the house to isolate it to a bad GFI in the bathroom. The neighbor repaired it and voila, I am back on the air. It astounded me that a bad GFI could generate that much noise and couple to the cable coax shield, the power lines and maybe the phone lines as well."

There have been some instances of interference from cable company set top boxes and cable modems. The usual fixes apply. If you are able to confirm a cable company in-home device is causing interference, contact the cable company. The device or its power supply (usually a switching power supply) may be defective. Remember, these are Part 15 devices, and even correctly functioning ones may cause some interference, especially to sensitive amateur gear. If a replacement device does the same thing, common mode chokes placed on all leads connected to the device may help. Some older cable modems may benefit from the use of a shielded CAT5 cable between the modem and router or PC.—*Ron Hranac, NØIVN, Technical Leader, Cisco Systems, Inc.*

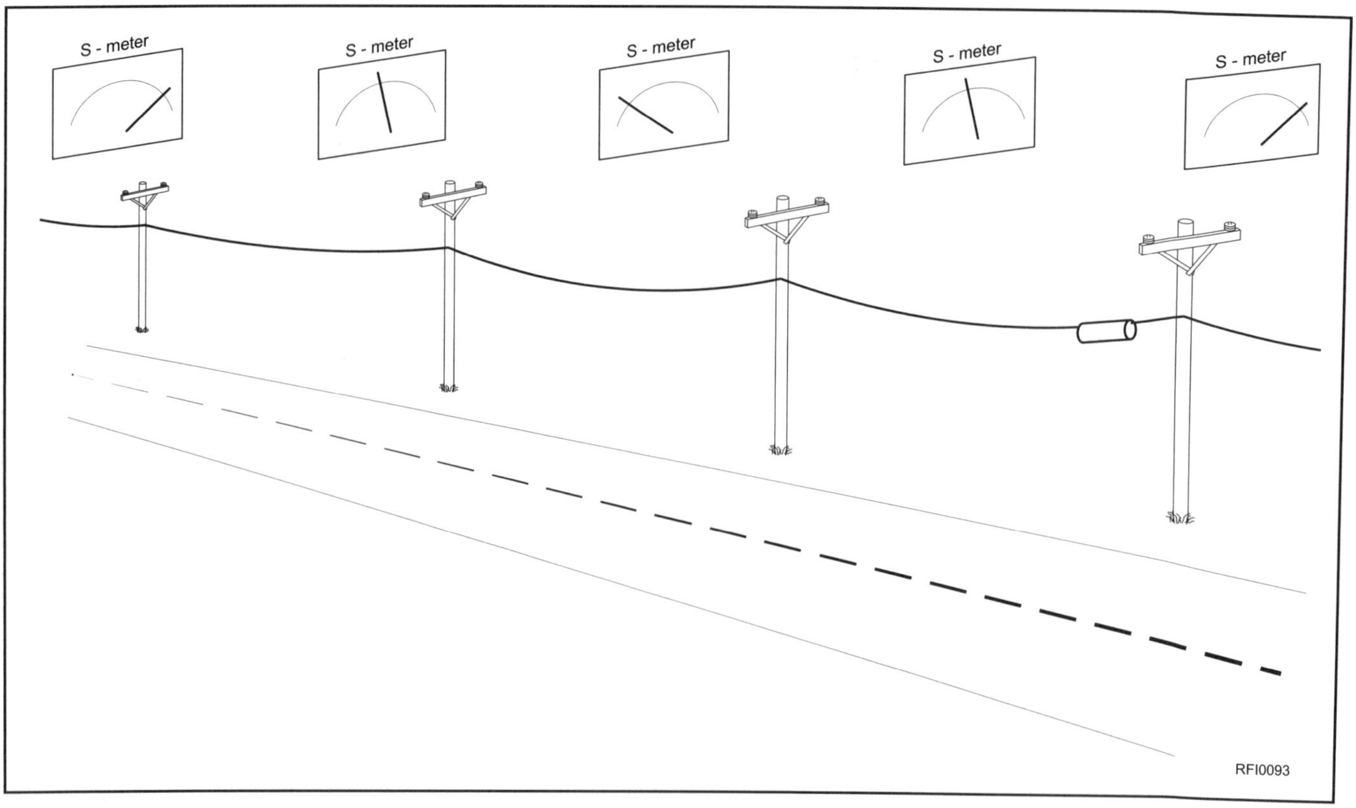

Figure 7.6—A CATV leak may set up a pattern of RF peaks and nulls on the metal-rope "strand" that supports the cable.

interfered with, it is possible that energy is entering the cable company's coax and propagating to other receivers on the system. As you can see, there are few situations to be dealt with, and the solutions to the problems have few options. When the problem is interference on the cable, it may disappear when the cable is disconnected and the TV set input connector properly terminated.

WHEN CABLE LEAKAGE IS FOUND

It is clear that the amateur has rights in regard to cable leakage. These rights come from the privileges and protections supported by FCC regulations. On the other hand, amateurs should be careful and thorough in working out solutions to specific problems and only involve the FCC as a last recourse.

The "political" procedure for gaining relief from cable leakage problems is generally as follows:

Contact the cable operator. It is best to contact a supervisor in the technical (preferably maintenance) side of the company. Normal approaches may not work efficiently here; if possible, find the name of the chief technician or plant manager and give him a call. (Your ARRL Section Technical Coordinator, TC, may know the best cable company contact. If you don't know your TC, ask your Section Manager.)

CATVI and High-Pass Filters

When most hams think of TVI, they almost immediately think of high-pass filters. Over the years, these have often been the first line of defense against TVI from HF transmitters. When an HF transmitter interferes with a cable system, it is not surprising that some hams mistakenly believe that a high-pass filter should help with CATVI, too.

A high-pass filter is a differential-mode filter. This means that it will filter the signal *inside* the coax. If the cable system is not leaking, however, there is no HF signal inside the cable to filter! In most cases, a high-pass filter is the wrong filter to use to cure CATVI. This is especially true in modern cable systems, which use HF for the two-way cable channels; installing a high-pass filter can make the system fail to work at all.

It is important for hams to understand this because if a neighbor has interference to a number of cable-connected TVs and the ham convinces neighbors to buy several high-pass filters, he or she may have to go back to the neighbors and explain that he or she specified the wrong filter! In most cases, the correct filter to try is the reliable common-mode choke. For a cable-connected TV, the first line of defense is a common-mode choke. Figure B shows a common-mode choke.—*Ed Hare, W1RFI, ARRL Laboratory Supervisor*

Figure B—A common-mode choke is the first line of defense against CATVI! This choke can be easily made with a few turns of wire onto an F(T)-240-43 or -75 ferrite core. See the EMC Fundamentals chapter for more information.

Discuss the problem with the chief technician in a matter-of-fact way. Give him the details of your observations. (The focus on cable leakage over the past several years has made most cable operators cooperative. In fact, good amateur relations can help with their leakage program.)

2 Meters and Channel 18

Cable channel 18 uses the same frequencies as the 2-meter amateur band! If the cable system is well shielded, this doesn't present a problem for the cable company or for nearby hams. But when things can go wrong, they will go wrong—most VHF operators have learned the hard way that cable TV and 2 meters are not always compatible.

Nearly all cable television systems make use of frequencies allocated to over-the-air services. Cable channel 18, or channel "E" in cable parlance, is from 144-150 MHz. The amateur 2-meter band falls within that channel. Your 2-meter signal can leak into the system. When it does, the hard part is finding out where.

Start by determining if the signal is leaking either in (ingress) or out (egress). In most systems, the video carrier will be on or near 145.25, plus or minus 25 kHz. If you hear a strong carrier (an FM receiver will break squelch, with full quieting) on that frequency, the cable system is probably leaking, although it might be customer-owned cable-ready TVs, VCRs or FM tuners connected directly to the cable system. The cable company will appreciate your calling this to their attention. This type of leakage can be expensive if they fail their cumulative leakage tests or if they're cited for violations of FCC rules. If their signal is leaking out, you can be sure your signal is leaking in. In the case of cable-plant leaks, it is the cable company's responsibility to cure the problem.

Even if the leak from the cable system is below the FCC leakage requirements, the interference could still be caused by a leak in the cable system. An interfering signal that's more than 40 dB below the visual carrier level can still cause interference! If the leak is not severe, the cable company may try to tell you that they don't have to fix it. Don't be fooled. The cable company must adhere to several regulations about picture quality. FCC regulations 76.605(a) state, among other things, that the carrier-to-noise ratio must be at least 43 dB and the carrier-to-coherent disturbances ratio must be at least 51 dB.

Unfortunately, it is not all black and white! The leak could be in the cable system, the cable company's set-top converter, or a customer-owned cable-ready TV or VCR. The cable company is responsible for the first two, but cannot be held responsible for the design of the TV. If the TV leaks, the only resolution may be with the TV-set manufacturer. In many instances a cable company set top box (converter) connected between the cable and affected device may alleviate direct pickup problems with cable-ready devices.

Sometimes, a common-mode choke does help. Try one on the incoming cable, just before it contacts the first piece of electronic equipment. Try a common-mode choke on the ac line connected to each piece of equipment, too. Other than that, there is no filtering you can do. If you filter the 2-meter signal, you will be filtering cable channel 18, too!—*Ed Hare, W1RFI, ARRL Laboratory Supervisor*

ARRL + SCTE = A Strong Hand

If there is no positive result from your initial and follow-up contact with a cable operator, the next step should involve the ARRL. The ARRL, in conjunction with the Society of Cable Telecommunications Engineers (SCTE), has developed a program of complaint processing designed to assist peaceful and constructive resolution of cable problems (and avoid involving the FCC). Begin by providing complete information on the incident as described in the chapter on EMI reporting.

If necessary, the HQ staff may become involved in some CATVI cases. In cases where the amateur and cable operator are unable to agree on the technical or regulatory issues, the League can call on the SCTE liaison. The liaison exercises his/her contacts with the particular cable operator in an effort to induce cooperation. Give the liaison a chance to operate, and help wherever possible by providing detailed documentation of your efforts to resolve the issue with the cable company.

Complaints should go to the FCC only as a last resort. When all else fails, do contact the FCC, but do so in cooperation with the ARRL. This procedure ensures that all possibilities for a peaceful solution have been exhausted and that the complaint is properly addressed at the Commission level.

This procedure has been in effect for a number of years, and it has been a substantial help in resolving leakage problems. The cable operator has real incentives to cooperate. Resolution of amateur leakage complaints (1) improves the cable system integrity, (2) reduces complaints from other services and (3) improves customer signal quality. The amateur incentive is that a cooperative cable operator better controls leakage. There will be fewer future problems.

This procedure has worked in many cases. When complaints have been made directly to the Commission, the FCC has rejected them and instead sent them through the SCTE and ARRL. Most of these were eventually resolved. A sensible, cooperative approach is best for all. It leaves a better image in the mind of the Commission, who licenses both amateurs and cable operators. And it leads to better performance in both the cable and Amateur Radio communities.

Cable Installation

Cable installation is a little simpler. In the ideal cable installation, the cable runs directly to the house, vertically to the ground, then into the house through the basement or crawl space. It should be connected to a bonding or ground block where it enters the house. This may shunt some common-mode signals harmlessly to earth ground.

In a typical installation, the coax comes to the outside of the house, drops vertically to a ground block, and enters the house through a wall. The ground block should be wired directly to the utility neutral conductor.

Some cable installations do not minimize RFI. In a large apartment building, there are too many mechanical and constructional constraints for RFI to be a main consideration during installation, although the installation is supposed to comply with FCC leakage regulations. High-pass filters and common-mode chokes remain the most effective tools against CATVI.

The cable installer usually uses crimp-on coax connectors throughout the installation. If the shield connections are not tightly crimped, the shield can float and act as a receiving antenna. Wear and tear can result in shield breaks at connector crimps. People who try to repair the damage themselves often do not crimp connectors properly and a leak results.

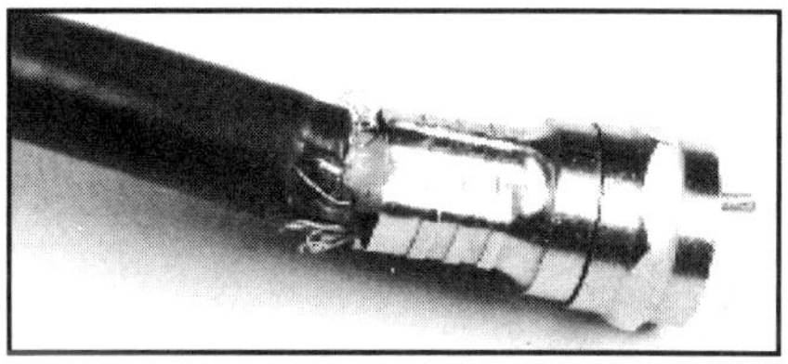

Figure 7.7—All connections in a CATV system should be tight and secure. This shield crimp connection has loosened, resulting in poor shield integrity, possible leaks and resultant TVI problems. Call CATV service personnel to repair such problems.

The result of this can be seen in Figure 7.7.

Illegal hookups are another major source of CATVI problems. People do the strangest things. Spliced-on 300-Ω twin-lead runs to another set (or another house!) are not inconceivable. Some people hook an outdoor TV antenna in parallel with the incoming cable line. That makes a real mess—radiated cable signals throughout the neighborhood! If a neighbor is reluctant to consult with the cable company about TVI problems, he or she may have something to hide.

Since cable systems are not readily accessible, it may be difficult to locate the point of failure. EMI may enter the system via the flexible drop cable at a subscriber's home, a poor connection, an unterminated coupler or splitter that feeds a TV, external devices or ac-power lines, to name just a few.

Many cable interference problems can be traced to poorly installed illegal hookups. These are often performed by untrained personnel. There is little regard for proper installation techniques, such as good shielding and secure connections. Some of the worst illegal installations even use twin-lead or zip cord instead of coaxial cable.

After making sure the antenna and feed line (or cable company's coax) are in good order, look at the TV set. The only parts you should check are the external connections on the back of the set. Do not open the case of equipment you do not own! Internal problems (such as broken connections and dirty tuners) should be left to qualified service personnel. Most states permit only state-licensed persons to repair electronic equipment. The license is required even if you do not charge for the service. A neighbor is more likely to hold you liable for future TV problems if you have opened the set. The ARRL Laboratory staff tell some real horror tales about hams who have not heeded this advice. Don't become part of the next story!

Let cable company repair personnel perform any and all repairs to the cable installation. Under FCC regulations, they are responsible for any leaks. If they are responsible for leaks, it's only fair that the entire system remain under their direct control.

Present and Future CATVI Hot Spots

Fortunately there are usually only a few cable signals in a given area that might cause trouble. One is a visual-carrier frequency that falls in the amateur 2-meter band (probably 145.25 MHz, and possibly the first sidebands located 15.734 kHz either side of the carrier). In HRC cable systems, Channel 18's visual carrier is at 144.0072 MHz, where it can interfere with amateur EME work. There are equivalent signals in the 222 MHz and the 70-cm bands. As the cable industry moves up toward 1 GHz, other frequencies may be involved.

NEXT . .

The next section of this chapter is a revision of a series of articles written by Ed Hare, W1RFI (then KA1CV) that appeared in the July, August and September 1993 issues of *Communications Technology.* It has been revised by the author in consultation with the editors at *Communications Technology.*

Reprinted with permission from July, August and September 1993 issues of *Communications Technology*, CT Publications, 50 S Steele St, Suite 500, Denver, CO 80209 (303) 355-2101.

Electromagnetic interference and the cable operator–Part 1

This is the first installment of a three-part series. The American Radio Relay League, a not-for-profit, membership-services organization representing the interests of Amateur Radio operators (hams), has long-standing relationships with groups and organizations such as the Federal Communications Commission, the National Cable Television Association and the Electronic Industries Association. All of these organizations have active interference-reduction programs.

A note about Amateur Radio: One no longer needs to learn Morse Code to obtain a Technician Class Amateur Radio License. A free information package that describes how to get started in Amateur Radio is available from ARRL Headquarters. Write them a letter at 225 Main St., Newington, CT 06111, asking for the "Prospective Ham Package," contact them at (800) 326-3942 or connect to **www.arrl.org/hamradio.html** *to download free information about Amateur Radio.*

By Ed Hare, W1RFI
Technical Department Laboratory Supervisor
American Radio Relay League, Inc.
225 Main Street
Newington, CT 06111
860-594-0200 voice
860-594-0259 fax
w1rfi@arrl.org
www.arrl.org/

"Interference (is) any unwanted interaction between electronic systems—period! No fault. No blame. It's just a condition."

Among the more difficult problems faced by the cable TV industry are those caused by leakage or ingress. The effects of these problems are felt from the front-office staff that has to deal with the paperwork associated with all of the applicable regulations, to the maintenance personnel in the field dealing with subscriber picture quality. Leakage is a two-way phenomenon; when signals can get out, over-the-air signals can also get in. The result of the latter is a condition known as electromagnetic interference (EMI), radio-frequency interference (RFI), TV interference (TVI) or ingress. No matter what you call it, the ultimate translation is trouble.

Ingress by over the-air signals can come from many sources. There are millions of regulated radio transmitters in the United States, belonging to utilities, municipalities, business users, Amateur Radio operators and military users, to name just a few.

In addition to these licensed sources of radio-frequency energy, there are even more sources of radio energy or noise. The FCC Part 15 regulations govern license-free transmitters used in walkie-talkies, garage-door openers, video games and modulators, the infamous VCR "Rabbit," and a rather long list of other unlicensed, low-power radio transmitters. Unintentional sources of noise include computer equipment, microprocessor circuits used in consumer electronics equipment or security-alarm systems, motors, neon signs, thermostats, touch-to-switch lamps, the electrical-power distribution system and even street lighting. (Newer street lamps can be radio-frequency powered.)

The interference problems that affect cable TV operation can be manifested as interference to the subscriber's TV reception (analog and digital video), high-speed Internet access, and voice service in those systems providing cable telephony. Cable systems make use of spectrum allocated to over-the-air services, relying on the inherent shielding of coaxial cable to allow cable operators to use spectrum assigned to broadcasting and other users. It must be stressed that FCC regulations require that a cable operator must not cause interference to the licensed users of this spectrum. When the shield integrity is compromised, in addition to the problems associated with egress, the result can be interference and an unhappy subscriber. The resultant service costs (or a lost subscriber) represent a financial loss to the cable operator.

The examples and FCC regulations cited in this article are specifically applicable to the Amateur Radio Service, and cable TV interference (CATVI) resulting from amateur operation. However, the principles discussed and the steps described are applicable to nearly any interference problem resulting from cable TV ingress. The FCC regulations that govern Amateur Radio have provisions that help control interference; all radio services have similar regulations.

DEFINITIONS

- Amateur Radio Service, FCC Part 97 regulations govern the Amateur Radio Service. They essentially define Amateur Radio operation as a non-commercial service comprised of people interested in radio operation and technology. The governments of the world have encouraged the growth of Amateur Radio because of the various forms of public service performed by hams. They provide message handling in times of disaster, and continuing contributions to electronics and radio technology.

An amateur is often "amateur" only by title. Many hams are professionals in the electronics industry. (Yes, the cable industry has more than its share of hams. SCTE maintains a list of licensed amateur radio operators in its annual membership directory.) Even amateurs in relatively "low-tech" occupations had to pass a highly technical examination to get their licenses. For example, the Amateur Extra Class examination is roughly equivalent to the test for the FCC commercial radiotelephone license.

- *Interference.* The term "interference" should be defined without emotion. To some people, it implies action and intent. The statement, "You are interfering with my television" sounds like an outright accusation. It is better to define interference as any unwanted interaction between electronic systems—period! No fault. No blame. It's just a condition.

RESPONSIBILITIES

Each person involved in an interference problem has individual needs, a unique perspective and a varying degree of understanding of the technical and personal issues involved. On the other hand, each of them may have certain responsibilities toward the other, and should be prepared to address those responsibilities fairly.

The transmitter operator is responsible for

the proper operation of the radio station. If the transmitting equipment is emitting out-of-band signals, the operator must take the necessary steps to ensure that the station is in compliance with FCC regulations. The operator also should cooperate with the subscriber and cable company repair staff to find a resolution to the problem. It may be necessary for the operator to participate in the tests to help them find and correct the fault. The cable company's repair crew will work in close cooperation with the radio transmitter operator.

The subscriber should extend the same cooperation. An interference problem can only be corrected at its source. If, for example, interference is due to leakage in the wiring located in the subscriber's home or in the subscriber's TV receiver, asking the transmitter operator to correct the problem by making changes to the radio station will not result in a cure. The cable company's repair crew may need access to a subscriber's home or want to try some interference filters on the input cable to the subscriber's TV receiver or VCR. The subscriber should help in every reasonable way possible.

The cable company must ensure that the cable system is in proper working order. Beyond that, most cable operators are willing to help a customer resolve a problem concerning the quality of the received service. It is costly to leave an interference problem unresolved. Dissatisfied viewers can cancel their subscriptions and, if the interference problem is caused by a cable system leak, the FCC can levy fines or shut down channels.

Consumer electronic equipment manufacturers also share in the responsibilities. In 1982, Public Law 97-259 became law. Among other things, this legislation gave the FCC the *authority* to regulate consumer electronic equipment susceptibility. The FCC chose to implement a program of voluntary immunity standards and allow the manufacturers to devise a program of voluntary compliance. The American National Standards Institute (ANSI) has accredited standards for TV sets and VCRs that do provide a fair degree of immunity.

SOME ELECTROMAGNETIC COMPATIBILITY (EMC) FUNDAMENTALS

• *The non-technical side—diplomacy.* There are occasional misunderstandings between the cable subscriber, the cable company service personnel, and the transmitter operator who is involved in the problem. The subscriber often feels that the transmitter is to blame. After all, if the radio transmitter wasn't on the air, there wouldn't be a problem. The station operator often feels that the problem is *always* the responsibility of the cable company, and that it can always be fixed if the cable system shielding is improved. The cable company repair personnel are often stuck in the middle.

This scenario is sometimes compounded by hostility, hurt feelings and a hostile posture adopted by everyone who is involved. This creates an atmosphere in which the problem cannot be solved. At this point, the subscriber, the ham and, if necessary, the cable company staff must decide to put aside bad feelings and approach the situation with an open mind.

• *Technical.* There are three components to any interference problem: (1) a source of radio-frequency energy or electrical noise, (2) a piece of susceptible equipment, and (3) a path over which the unwanted energy is propagated. Any solution to an interference problem is going to involve a change made to one or more of these three factors.

In most cases of cable systems ingress, the source is difficult to control. If the source is a transmitter of a licensed radio service (this includes Amateur Radio, among many others) operating in compliance with the appropriate regulations, there may be little justification to ask the operator to make any changes to the station. Under some circumstances, however, it may be possible to improve the situation by making changes to the source. Ensure that the transmitting station is installed and operated using good engineering practice. The equipment should be properly grounded and installed so that the transmitted signal is coming from the antenna, not the transmitter chassis or ground. Transmitter output power and the proximity of the antenna to the susceptible equipment or the cable distribution system can all affect the severity of an interference problem.

There is little that can be done by the cable repair personnel to improve susceptible equipment. Most states have laws and licensing requirements that regulate the electronic equipment repair industry. In addition, most cable companies have policies that prohibit the cable company service staff from repairing or modifying any subscriber-owned equipment. The best source of help with susceptible equipment is the equipment's manufacturer.

Interference can propagate via several different possible paths. Take a look at Figure 7.8. In the direct path, the interfering signal travels directly from the transmitting station's antenna to the susceptible equipment. Conducted interference travels from the source to the susceptible equipment by wires. Most of the time, however, the conducted signal is one that has been induced into the external or internal wiring of the susceptible equipment.

All electronic circuits require two wires

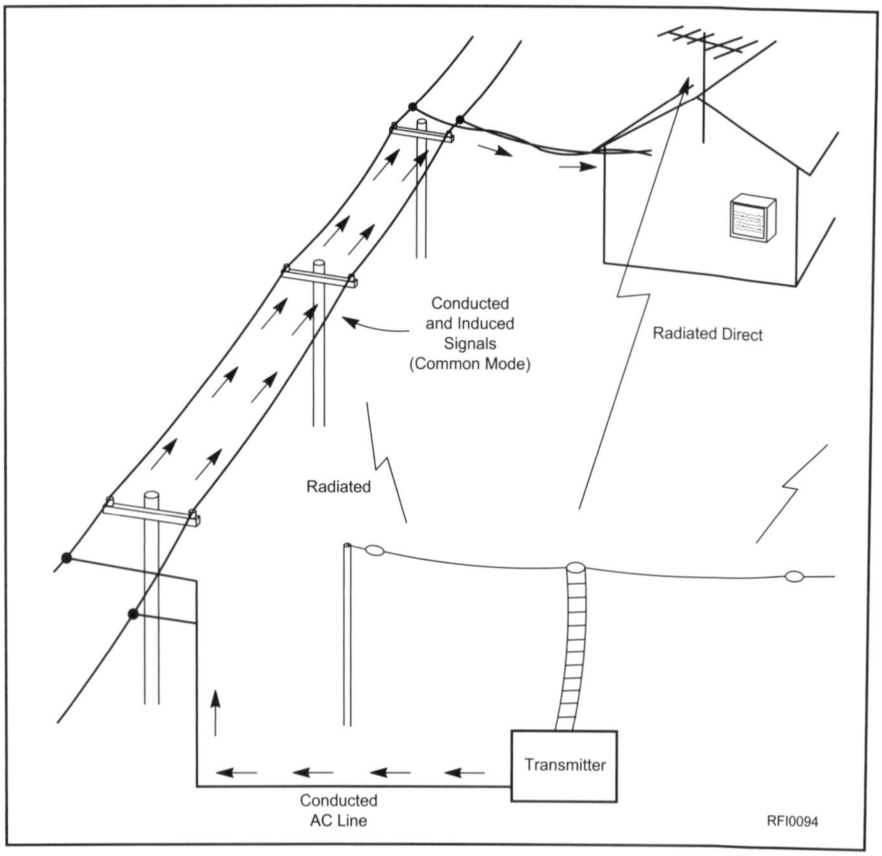

Figure 7.8—Conducted and direct interference.

(or their equivalents) - one wire for the signal to travel on and another for the signal return. In the *differential mode* a signal travels down the center conductor of a coaxial cable and uses the inside of the shield as its return path. In the *common mode*, the shield and center conductor act as if they were one wire, with the return path being earth ground (usually through the ac wiring). Figure 7.9 shows the difference between common-mode and differential-mode signals.

Most signal-leakage cures will be applied to the path. The most important of these cures involve filtering and shielding. If a shield is placed between the source and the susceptible equipment, the effect of the interfering signal can be significantly reduced. Unfortunately, shielding is not a practical solution in most cases. It is difficult to imagine convincing a subscriber that the entire video system (including the viewers!) must be put into a grounded, shielded box.

Another means of affecting the path is the use of filters to attenuate unwanted energy. Part 2 of this article will discuss the practical steps that should be taken to cure interference problems, concentrating on the use of the necessary filters.

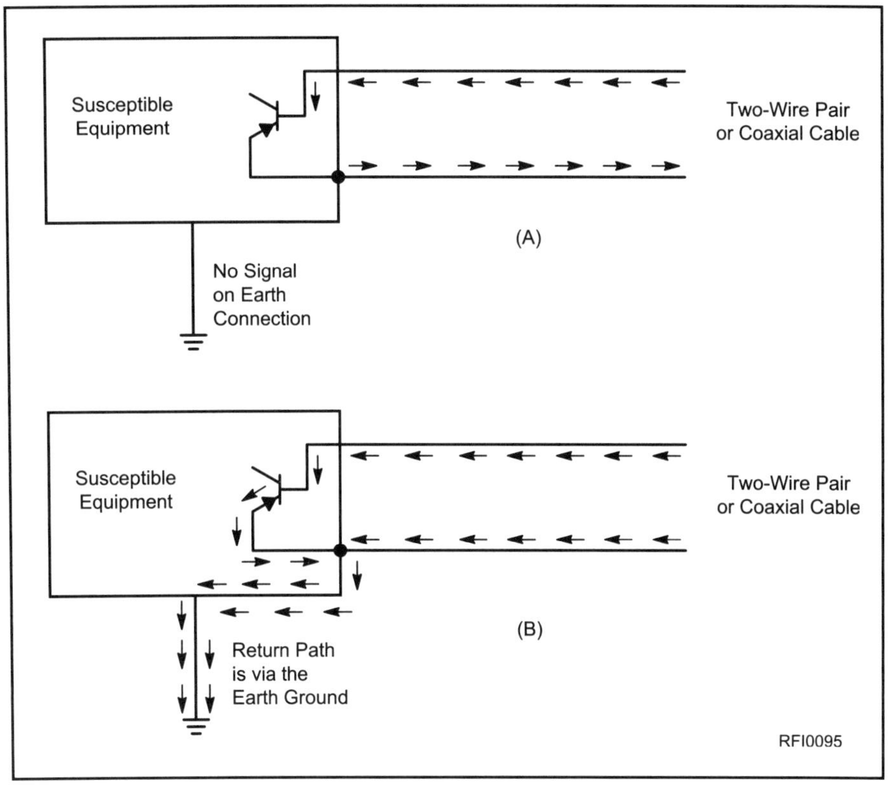

Figure 7.9—Differential (A) and common-mode (B) signals.

Electromagnetic interference and the cable operator—Part 2

Part 1 of this article appeared last month. It discussed the nature of the interference problem, ways to deal effectively with the subscriber and the radio operator, and electromagnetic compatibility technical fundamentals. It is time to outline some standard approaches that usually result in a solution to the problem.

By Ed Hare, W1RFI
Technical Department Laboratory Supervisor
American Radio Relay League

The first technical step toward finding a solution to an interference problem is for the cable company service personnel to determine if there is any leakage in the cable system wiring. If the interfering radio signal is leaking in, it will, in most cases, be impossible to filter.

Fortunately (for troubleshooting purposes), if a signal is leaking in it, others are leaking out. Cable operators have many good methods for measuring leakage, although a passing cumulative leakage index (CLI) figure on the Federal Communications Commission's Form 320 does not ensure that leakage and ingress are completely under control. Modern communications receivers have a lot more sensitivity than test conditions used by cable operators to find 20 µV/meter leaks.

The radio operator can do a quick check for leakage in the VHF ham bands by using a receiver designed for the Amateur Radio Service. (A local ham can often be of help here.) If a strong carrier is heard in the subscriber's home on the frequency of the visual carrier in the 144 or 222 MHz amateur bands, then leakage is indicated. (*Editor's Note: Be aware that some poorly shielded TV sets, VCRs or FM tuners may be a source of leakage.*) On the other hand, if no carrier is heard, then the cable system assuredly does not leak at that point. In some cases, it may be possible to hear a signal when the receiver's antenna is placed within inches of the cable. It may, however, be leaking elsewhere in the neighborhood and being delivered to the subscriber's drop. The accompanying Table 7-2 shows the frequen-

Table 7.2
Problem Channels for Amateur Radio Operators

*Channel**	*CATV*	*Video Carrier (in MHz)*	*Sound Carrier (in MHz)*	*Amateur Band*
Cable 2	2	55.25	59.75	50-54 MHz (6 meter)
Cable 17	D	139.25	143.75	144-148 MHz (2 meter)
Cable 18	E	145.25	149.75	144-148 MHz (2 meter)
Cable 23	J	217.25	221.75	222-225 MHz (1.25 meter)
Cable 24	K	223.25	227.75	222-225 MHz (1.25 meter)
Cable 57	UU	421.25	425.75	420-450 MHz (70 centimeter)
Cable 58	VV	427.25	431.75	420-450 MHz (70 centimeter)
Cable 59	WW	433.25	437.75	420-450 MHz (70 centimeter)
Cable 60	XX	439.25	443.75	420-450 MHz (70 centimeter)
Cable 61	YY	445.25	449.75	420-450 MHz (70 centimeter)
Cable 62	ZZ	451.25	455.75	420-450 MHz (70 centimeter)

*Note: Jerrold/General Instrument/Motorola converters use a different numbering scheme than the EIA convention for certain channels. For example, cable channel 57 (UU) is 65, 58 (VV) is 66 and so cies of analog TV channel visual and aural carriers that can be found in, or adjacent to, the amateur bands.

Another effective way of evaluating leakage is to use a spectrum analyzer. Spectrum analyzers are usually well-shielded and filtered, and can be relied upon to measure only signals inside the cable. Under some conditions, an in-band, spurious signal that is 60 dB weaker than the video carrier can cause just-perceptible interference. If the in-channel radio transmitter's fundamental signal is weaker than this, then cable system leakage is not contributing to the problem, so other causes should be investigated.

CHECK THE CABLE SYSTEM THOROUGHLY

Many cases of ingress can be traced to something the subscriber has done. It is not rare for a subscriber to add additional cable and splitters to the installation to save monthly fees for additional outlets. It is not unheard of for subscribers to run cables to a neighbor's house, to give away free cable, or to take advantage of free cable (with or without the other party's consent). As if this weren't bad enough, this is often done using substandard cable, splitters or installation techniques (check the crimps). Even worse, they may use wires such as twin-lead or zip cord, or even hook up an external TV antenna in parallel with the cable. Even if the subscriber has done none of these things, substandard cables may have been substituted for the quality jumper cables used at the time of installation, or the subscriber may have repaired a broken connector using parts and techniques not up to industry standards. Look for these things—you may fix your problem without having to go any further.

All of these defects can result in severe leakage. Not only can they be the direct cause of the interference problem, but they can have quite the impact on your system's leakage performance. Some of them also represent a significant revenue loss.

THE SUSCEPTIBLE TV SET

In many cases the ingress is occurring in the subscriber's TV set or VCR. This should be tested early in the troubleshooting process. Disconnect the TV set from the cable system and terminate its input with a 75 ohm terminator or 300 ohm carbon resistor (as appropriate for its input impedance). Select the channel the TV set is normally tuned to—usually Ch. 3 or 4 if a set-top converter or VCR is used to select channels. If the TV set is cable-ready, select the channel that is most prone to interference. If the snowy raster is *strongly* affected when the interfering signal is present, this indicates that the TV set is susceptible to direct pickup. If the TV set tests okay, repeat the process for the VCR.

The cable operator is not responsible if the TV set or VCR is susceptible to direct pickup of the interfering signal. However, many operators want to offer some help to any subscriber who is having reception problems. If there is definitely no ingress, and these steps do not effect a cure, then the problem is in the subscriber-owned equipment. This equipment should only be fixed by the manufacturer or an authorized service facility. Have the subscriber contact the manufacturer through the address given for the Electronic Industries Association. The EIA maintains a data base of manufacturers and key electromagnetic compatibility (EMC) personnel.

THE RADIO TRANSMITTING STATION AND OPERATOR

Because a cable system is (in theory) a closed system, it really shouldn't matter if the radio station is transmitting harmonics or other out-of-band spurious emissions; the cable system is not supposed to respond to any over-the-air signals. However, it is helpful if the radio station can be demonstrated to be "clean," especially if the interference is only on channels harmonically related to the transmitted fundamental frequencies. If the station is not causing interference to an antenna-connected TV set located on the premises (often at home), this usually indicates that the transmitter is not transmitting any interfering out-of-band signals.

COOPERATION

As mentioned in Part 1, secure the cooperation of the transmitter operator. If the transmitter is an Amateur Radio station, you and the subscriber are lucky—it is in the spirit of the Amateur Radio Service to perform public service. As a public service, most amateurs will be willing to help with an interference problem, even if it doesn't involve their station. In any event, the station operator can help ensure that there is a signal on the air while the repair crew is at the subscriber's house; this will prevent the "no problem found" syndrome. The station operator can vary the station power, frequency and beam antenna headings (all within the limits of the station authorization) to help diagnose the cause of the problem.

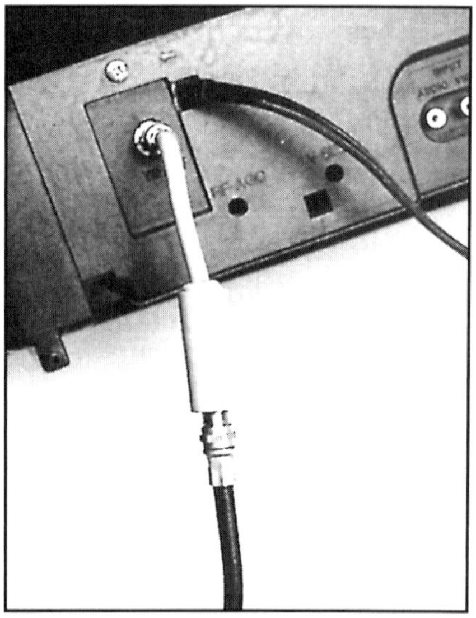

Figure 7.10

SOME MORE TECHNICAL "FIRST STEPS"

• *Look around and investigate other causes.* Although it is natural to think first of the local amateur or CB station with the 120-foot tower as the source of the interfering signal, there are many other potential sources of interference. Much time can be wasted blaming the local ham if the problem is being caused by a defective electrical system insulator on a near-by utility pole, or by a noisy appliance operating in the subscriber's own house.

• *Simplify the problem.* Many cases of interference have several causes. If the subscriber has multiple outlets, many of which have VCRs, video games, A/B switches and TV receivers connected to stereo systems with long speaker leads and who-knows-what attached, it may be the interference is entering the system. Anything connected to the cable system—by any means—is suspect.

To begin troubleshooting such a complex system, completely disconnect everything except one outlet and one TV set and its set-top converter. (More about the set-top converter later.) When the following steps have been employed to reduce the interference to this simplified installation, start adding things back, one at a time, eliminating interference one step at a time. If all goes well, all of the interference will be eliminated. If not, at least the defective piece of susceptible equipment will have been identified. Most subscribers will realize that if one piece of equipment is interference-free and another is not, the equipment that is having the problem must be deficient in some way.

Testing

Determine which transmitted frequencies cause the problem. It may be possible to obtain this information from the radio station log book. If not, have the station operator transmit on each frequency band for which the station has equipment, observing any interference present. If the operator has been able to secure the help of another qualified operator—for an Amateur Radio operator this is usually the ARRL Section Technical

Figure 7.11

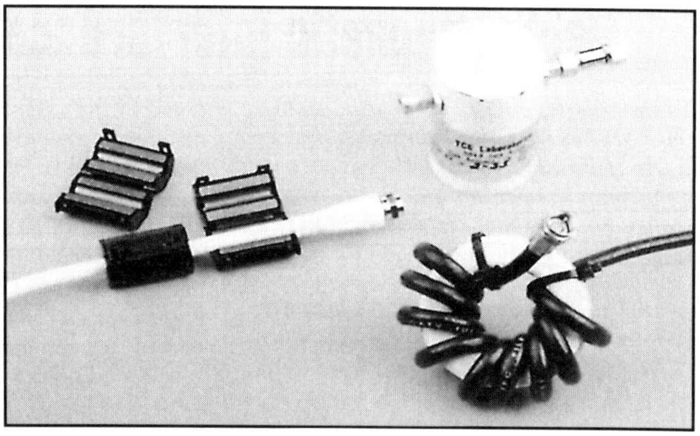
Figure 7.12

Coordinator (more on this very important person later in the article) – they may be able to coordinate the testing using hand-held, two-way radios. If not, a telephone link between the subscriber's home and the station operator will help things to run smoothly.

If the interference results from medium-frequency (MF) or high-frequency (HF) operation (signals below 30 MHz), the solution to the problem may be as simple as the installation of a differential-mode, high-pass filter in the cable that feeds the set-top converter. Figure 7.10 shows one way to do this. Even if this filter doesn't cure the interference, leave it in place, at least for the time being. Caution: this filter is not compatible with most two-way systems. It will filter out all signals from 5 to 42 MHz! Note that many modern digital set top boxes use the cable system's upstream for interactive communications with the headend. A standard high-pass filter will prevent this two-way communication, likely rendering the set top inoperable. For these applications, a windowed high-pass filter is recommended. This type of high-pass filter allows upstream signals from the set top box to pass through the filter, while rejecting other frequencies.

TIME TO GET DOWN TO BRASS TACKS

There are a number of cable channels that can be affected by cable system ingress. These channels make use of over-the-air frequencies that are allocated to other services, often located immediately adjacent to amateur, business or pager frequencies. The table earlier in this article (Table 7-2) lists the frequencies for a standard system with no offsets. Many systems use offsets, or one of the other channel schemes.

If the interference results to Ch. E, Ch. J or Chs. UU through YY (Ch. 18, Ch. 23-24 or Ch. 57-61) from amateur VHF or UHF operation, it will not be possible to filter the undesired signal from inside the cable without affecting the desired TV signal as well. All is not lost however. Sometimes the TV receiver, VCR or set-top converter will exhibit a strong response to the common-mode signal present on the outside (shield) of the cable, or to a lesser extent the common-mode or differential-mode RF signals that may be present on the ac electrical system wiring. After all leaks have been repaired, it may be possible to do something about these signals.

THE COMMON-MODE FILTER

The common-mode filter or choke may be one of the best-kept secrets in the Western world. Ingress problems that are not caused by cable leakage are usually the result of the common-mode response of the TV receiver, the VCR or the set-top converter. This common-mode response usually can be cured with the use of a common-mode choke.

The simple cable TV installation shown in Figure 7.11 is virtually bulletproof. The placements of the cable and ac line common-mode filters have been chosen to minimize the amount of undesired common-mode energy reaching the set-top converter. Two different cable common-mode filters are shown in Figure 7.12. Do not substitute unsuitable ferrite materials for those specified in the description. As an example, it might be possible to scrounge some ferrite from an old TV set's deflection yoke, but the ferrite material used was designed to perform at 15.734 kHz. It may, or may not, work well at the frequency of the radio station.

Common-mode filters for coaxial cable should be installed as close as possible to the cable input connector of the set-top converter, TV receiver or VCR. In extreme cases, it may be necessary to install a common-mode choke at each end of all interconnecting cables (VCR to TV set, for example) in the system.

AC LINE FILTERING

• *AC line common-mode choke.* The undesired signal also can be coupled into the subscriber's equipment through the ac power lines. These lines can function as antennas. The signal that is induced into ac power lines can be either common- or differential-mode. Install a common-mode choke on the ac line cord of the TV set, as close as possible to the point where the cord enters the set. This choke should be made by wrapping 10 to 20 turns of the ac line cord through the ferrite toroid (or onto a ferrite rod). This will reduce any common-mode signal that has been picked up by the ac lines.

• *The ac line differential-mode filter.* If the ac line common-mode choke doesn't help, add a differential-mode ac line filter. That is, the type of ac line filter commonly sold by electrical supply companies. (They also are sometimes called "brute force" ac line filters.) This filter should be installed as close as possible to the point where the ac line cord enters the TV set.

These "standard" cures will probably take care of 90% of the interference problems that plague your subscribers. But that other 10% can be the most difficult to pin down. Part 3 of this series discusses some of the more advanced troubleshooting techniques and cures, and lists several sources of help.

Electromagnetic interference and the cable operator—Part 3

Part 1 of this article appeared in July. It discussed the nature of the interference problem, ways to deal effectively with the subscriber and the radio operator, and electromagnetic compatibility technical fundamentals. Part 2 was in last month's issue. It discussed the common cures for cable TV interference. Now, we will pick up where we left off, delving into some troubleshooting techniques and cures.

By Ed Hare, W1RFI
Technical Department Laboratory Supervisor
American Radio Relay League

All of the cures covered in last month's installment should result in an interference-free installation. However, if interference is still present, there may be a problem with direct pickup by the TV receiver, VCR or set-top converter. If any of these devices are tuned to the frequency of a local Amateur Radio signal (Ch. E, Ch. 18 in most cable systems, for amateur 144 to 148 MHz signals, for example), it should be the prime suspect. If ingress (and most likely leakage as well) is taking place inside the subscriber's TV set or VCR, there is almost nothing the cable company service personnel can do to cure the problem. The only source of help is the product manufacturer or electronic service personnel.

VCR—FRIEND OR FOE?

If it appears that the subscriber's cable-ready TV set is the entry point for the ingress, the battle is not necessarily lost. Many subscribers have a VCR and, in some configurations, that VCR can be used as a channel selector. VCRs are often newer than the TV receiver, and they may have been designed

Multiple Problems—A Case History

The ARRL Technical Department staff often gets letters and telephone calls about ham radio-related EMI. I once received a telephone call from Chris Saunders, technical customer service manager for Cox Cable Greater Hartford, about a Ch. 18 ingress problem that was affecting one of the company's customers. They had thoroughly rewired the installation to correct a few identified leaks, but this did not completely fix the problem. Chris had heard about the help that can be supplied by the ARRL, so he gave me a call. He described the tests that had been performed by the cable repair crew. I agreed with his conclusion that the ingress was external to the cable system.

In addition to my duties here at ARRL Headquarters, I also serve as an unpaid field volunteer Technical Specialist for the Hartford area. I explained to Chris that I would be willing to go over to the subscriber's house sometime and see what I could do.

I made an appointment for a Sunday afternoon, and brought my survival kit with me—a collection of ferrite toroids, a few 75 Ω high-pass filters, 75 Ω terminators and some brute-force ac line filters.

The subscriber met me at the door. He appeared to be quite friendly (what a relief!) and we sat down to talk a bit before I got down to "business." I explained that I was an Amateur Radio volunteer who was offering to help with a problem that involved a local amateur. He then pointed out the 100-foot tower on an adjacent piece of property.

The Problem

We then switched on his TV set to Ch. 18 (Ch. E). Within a few minutes, the screen was filled with a broad cross-hatch interference pattern that lasted about 20 seconds. This repeated randomly about every minute or so. I had brought a 144 to 148 MHz handheld transceiver with me, so I was able to transmit a signal that easily duplicated the interference. I also was able to quickly locate the other station on the air. It turned out that the station was an amateur repeater, set up to automatically rebroadcast the signals of hundreds of area amateurs to increase their service area.

This explained the nearly non-stop nature of the interference. The subscriber had wondered how the ham was able to stay up and operate day and night!

The Cox repair crew had previously replaced the drop and rewired the entire house. The subscriber told me that this made quite an improvement, but that the interference was still objectionable. (I agreed.) Because there had been an improvement, I knew that there had been a leak. I turned my handheld receiver to the Ch. 18 visual carrier frequency. I could not hear any trace of the visual carrier. This meant that the cable installation was clean. Good! I could now see if we could locate the point where the ingress was happening.

The subscriber was using a cable-ready TV set to select channels, in conjunction with a VCR. I followed my own advice and disconnected the VCR. We still had interference. I then tried common-mode chokes on the incoming cable, to no avail. A common-mode choke on the TV set's ac line cord seemed to make a small difference, but not enough to really fix the problem. It was beginning to look like the problem might be inside the TV set!

I disconnected the TV set from the cable and put a 75 terminator on the incoming cable and on the TV set input. I keyed the handheld transmitter. Uh, oh ... the screen went 100% black! This was not good news. The TV set was picking up the interfering signal directly. I explained how Connecticut state law (and common sense) prohibited me from trying to repair his TV set for him.

The Hunch

The subscriber had a hunch, though, that saved the day! He suggested that we try the VCR, just to see if the VCR might work properly now that the common-mode chokes were in place. It worked! It seems that the whole problem resulted from a multitude of causes, all of which had to be fixed before the problem was solved. Originally, there was a leakage problem, resulting in interference that couldn't be filtered when the 145-MHz amateur signal leaked into the cable. The TV receiver was subject to direct pickup of over-the-air signals, a problem for the manufacturer. The VCR was sensitive to the common-mode signal on the shield of the coaxial cable, but when the leak was fixed and the common- mode chokes were installed, the VCR was able to work properly. When the VCR was used as a channel selector, the susceptible TV set was tuned to Ch. 4 instead of Ch. 18, thus eliminating the problem. Most set-top converters are well shielded, and, if common-mode chokes are installed, they will usually perform properly in the presence of strong, over-the-air signals. Another happy customer (and a happy ham). A neighborhood dispute had been resolved.

Information about the ARRL Technical Coordinator program is available from ARRL Headquarters. See "Useful addresses" in the "Sources of Information" sidebar. Write to our EMI specialist (currently that's me) here in Newington, CT. We are looking for the cable operator who wants to do something about EMI!—*Ed Hare* to meet the ANSI immunity standards. They are often better shielded than the TV receiver. With the proper installation of cable common-mode chokes and the appropriate ac line filters, it may be possible to successfully use the VCR as a channel selector. In that case, the susceptible TV set will be tuned to a low VHF channel (usually Ch. 3 or 4). This makes it much less susceptible to interference than it was when it was tuned to the radio station's transmitted frequency. Unfortunately, in some cases, the VCR can be more susceptible than the TV set, or can exhibit intermodulation problems that only worsen in the presence of strong over-the-air signals.

Most set-top converters are well shielded, although they can be susceptible to interference from common-mode signals. If used in conjunction with cable and, if needed, ac line electrical chokes, they often can be used as a channel selector for a susceptible TV receiver. A fully filtered set-top converter can be used as a diagnostic tool. Keep one available for use when troubleshooting all cases of interference. It will help demonstrate to the subscriber that the problem originates from a susceptible piece of subscriber-owned equipment.

THE UNEXPECTED

Be prepared for the unexpected. The science of electromagnetic compatibility (EMC), although exact and predictable (in retrospect), is sometimes thought of as "black magic." Unexpected factors can sometimes come into play. It may take a specific combination of subscriber equipment hookups, or a combination of two or more transmitters being on the air simultaneously to cause the problem. In some cases, even the weather can be a factor. The ingress may only occur when the ingress point is wet (or dry, or hot, or…), which is an almost sure sign that the leak is outdoors. In complex cases like these it may take the services of an EMC expert, or a cable repair person who has been forced into that role!

SOURCES OF HELP

Many cable companies have a regional or corporate engineer who can help with interference and leakage problems. He will certainly get involved if the problem escalates to a formal Federal Communications Commission interference complaint, so he would rather be informed early in the interference resolution process!

The station operator and service personnel are handy sources of help. If the station operator is a licensed radio amateur, the operator may have a good understanding of the applicable technical issues. Even if the radio amateur is not technically well-versed, ham help may be available. The ARRL has set up a system of technical volunteers to help with interference problems that involve the Amateur Radio Service. This system consists of the ARRL Section Technical Coordinators (TC), their assistant Technical Specialists (TS) and possibly local amateur-club TVI (TV interference) or RFI committees. Many cable operators already have standing relationships and active interference resolution programs with local radio amateur clubs. These are usually coordinated through the ARRL section TC.

The amateur may know how to get in touch

Sources of Information

The ARRL Headquarters staff has prepared several important web pages to help amateurs, consumers and utility companies win the battle against EMI. The ARRL Regulatory Information Branch offers a wide range of online assistance that describes rules and regulations that apply to RFI involving amateurs. These may be found at **www.arrl.org/FandES/field/regulations/rfi-legal/index.html**. You will also find links to a variety of resources related to amateurs and RFI. The ARRL Technical Department also has a web page[1] of information that lists parts suppliers for filters, ferrites and other EMI-control devices. It also talks about the technical and diplomatic aspects of interference problems. There is also a downloadable pamphlet[2] for the consumer that explains interference in non-technical terms. This is useful for your neighbor or any cable TV subscriber. The FCC Interference Handbook 1990 Edition[3] is also available on this site. There is no charge for downloading or printing any of these ARRL informational pamphlets.

In addition to publishing *The ARRL RFI Book*[4] (this book), the ARRL funds several programs besides those already described to help amateurs, consumers and utilities with interference problems.

For interference related to electric power lines, the book *AC Power Interference Handbook—2nd Ed*[5] by Marv Loftness, KB7KK is recommended. Loftness is an EMC engineer and true RFI pioneer with many years experience in the power industry.

The Electronic Industries Alliance (address below) represents manufacturers and major distributors in the United States. It has several brochures that it will send out upon request. In addition, it maintains a database listing specific people with each manufacturer who can be contacted to help resolve interference problems. Write to the EIA for additional information.—*Ed Hare, W1RFI*

References

The following may be ordered using the addresses that are listed below.

[1]ARRL EMI/RFI Web Pages," American Radio Relay League. **www.arrl.org/tis/info/rfigen.html**.

[2]A self-help guide for the consumer published jointly by the American Radio Relay League (ARRL) and the Consumer Electronics Association (CEA); this information can be downloaded for both amateurs and neighbors of amateurs at **www.arrl.org/news/rfi/neighbors.html**.

[3]FCC Interference Handbook 1990 Edition, Federal Communications Commission, Washington, DC. This publication is now out of print. It has been replaced by The FCC Interference to Home Electronics Entertainment Equipment Handbook, Bulletin CIB-2. This publication is also out of print but available on the ARRL Web Site at **www.arrl.org/fcc/tvibook.html**. Additional information from the FCC can be obtained from the Consumer & Governmental Affairs Bureau (CGB) at **www.fcc.gov/cgb/consumerfacts/interference.html**.

[4]The ARRL RFI Book, edited by Mike Gruber, W1MG. American Radio Relay League. Available from ARRL Publication Sales.

[5]AC Power Interference Handbook—2nd Ed, by Marv Loftness, Published by Percival Technology, PO Box 4172, Tumwater, WA 98501. Available from ARRL Publication Sales. Another useful publication no longer available from the ARRL is the Interference Handbook, by William Nelson, Radio Publications, Inc, 925 Sherwood Dr, Box 247, Lake Bluff, IL 60044.

Useful Addresses

• American Radio Relay League (ARRL), 225 Main St, Newington, CT 06111; (860) 594-0200 voice; (860) 594-0259 fax; e-mail: **hq@arrl.org**; Web: **www.arrl.org/**.

• Electronic Industries Alliance (EIA), Consumer Electronic Association (CEA), 2500 Wilson Blvd, Arlington, VA 22205; (703) 907-7600 (voice); Web: **www.ce.org/**.

• National Cable & Telecommunications, Science and Technology Department, 1724 Massachusetts Ave, NW, Washington, DC 20036; (202) 775-3669; Web: **www.ncta.com/**.

with the TC or local club interference committee. If not, the ARRL Section Manager for each Section will be able to supply the address of the Section Technical Coordinator. A listing of ARRL Section Managers is found near the front of any recent *QST* (the official journal of the ARRL). The ham probably has a copy. If not, ARRL headquarters will be able to supply the information.

These local volunteers are dedicated in every sense of the word. There will be no charge for their "services." The other side of that coin, however, is that there is a limit to what can be asked of them. It would not be appropriate for the amateur, the subscriber, or the cable company to make unrealistic demands upon any of these people. It only takes one 3 a.m. telephone call to ruin an otherwise fine relationship!

The June 1993 issue of *Communications Technology* lists hundreds of amateurs that are employed in the cable industry. This list is updated yearly by Steve Johnson, NØAYE, of Time Warner Cable, and is a "natural" resource for the cable operator. It may be possible to encourage any amateurs that work for your cable company to take an active role in local-club interference activities to act as a first line of defense against leakage and ingress. (Editor's Note: *Communications Technology* no longer publishes the ham operator list. This information is now available in SCTE's annual membership directory.)

SUMMARY

Cable TV interference problems can be solved. It takes a combination of diplomacy, expertise, perseverance and a thorough approach to troubleshooting. Contact the ARRL for assistance with any difficult interference problems that involve a radio amateur.

Chapter 8

DVDs and VCRs

In many ways, DVD devices and VCRs are much like TVs. The RFI cures are similar. This chapter explains some important differences.

By John Frank, N9CH
PO Box 5113
Madison, WI 53705

DVD devices, VCRs and TVs have a lot in common. They all deal with video, color and sound, but there are differences. They often use frequencies used by HF Amateur Radio transmissions. Fortunately, it is often possible to cure interference using the filtering and shielding techniques that apply to most RFI problems.

DVD DEVICES

Depending on whose nomenclature you use, DVD can mean digital video disc or digital versatile disc. The latter is often the preferred meaning because the disc is capable of much more than just storing video. The DVD can be used to store data, computer programs, audio and video. DVD equipment is less prone to interference than some other types of home electronics apparatus, but it is not totally immune to interference from nearby transmitters.

The DVD uses technology similar to that used in CD players. It uses a shorter wavelength laser than that used in CD equipment and has smaller "holes and slots" on the surface read by the laser. These smaller holes and slots allow more information to be held on each disk.

Many consumer quality DVD devices are built in metal cases. The metal provides some degree of shielding and reduces the likelihood of interference due to direct penetration of the case by unwanted signals. However, the ac line cord and the patch cords that connect the DVD device to other electronic equipment can still provide a path for interference to enter the system. Unfortunately, many newer DVD devices are housed in plastic cases, which may affect shielding quality.

If you suspect the interference is entering the DVD device via the ac line cord, a filter can be installed in the AC line as close as possible to the DVD device. Wrapping the line cord around a ferrite rod or toroid can also be helpful in eliminating this type of interference.

Inexpensive patch cords can sometimes allow interference to enter a system. Inadequate shielding on inexpensive patch cords allows interference to enter whatever piece of equipment the DVD device is connected to. The remedy to this type of interference is well shielded patch cords with good quality connectors.

Common mode interference involves the offending signal traveling on the shield of the patch cords. Installing a common mode choke on the patch cord is usually the cure of choice in this case. Use toroids if the interference source is HF. In the case of VHF or UHF, running the patch cord through one or more ferrite beads can sometimes cure this type of interference. If this is impractical, you may be able to wind several turns of the cord through a snap-together ferrite core. Don't forget the antenna coax cable connection either. This is typically the longest coaxial cable, and therefore the longest shield, connected to the DVD. In most cases, it is the only coaxial cable to be an appreciable portion of a wavelength at HF, and therefore more likely to be problematic at these lower Amateur frequencies. We'll talk more about common mode chokes in a little later in this chapter.

VCRS

Try to visualize the following scenario: your brother-in-law has been awarded a Nobel Prize, and the presentation will be on TV the same night that you hope to work a DXpedition to Outer Elbonia. If you watch the awards presentation on TV, you miss working Outer Elbonia on 80 meters. If you work Outer Elbonia, you incur the wrath of your family for decades to come.

Wait—a VCR can solve this problem for you! You can automatically record the historic event (the award, not the DX contact) while you are in the ham shack making a once-in-a-lifetime DX contact.

After working Outer Elbonia, however, you rewind the videotape and play it back. Your brother-in-law is barely recognizable: The interference is so bad you can't tell the Nobel-prize presentation from an old rerun. Now you incur your family's wrath because (1) you missed your brother-in-law's moment of glory and (2) you interfered with the VCR.

Fortunately, most VCR owners don't experience any interference from radio transmitters. The potential for interference is clearly present, however, because of the design and construction of VCRs. VCRs are highly susceptible to interference because their circuitry uses the frequencies of (and near) several heavily populated radio services. Additionally, older VCRs lack shielding needed to keep unwanted signals out of their video modulators and demodulators and other circuits.

Is the situation hopeless? Must VCR users endure interference? The answer to both questions is "No, not necessarily." Newer VCRs are better shielded than older models. Interference to VCRs can be reduced or eliminated with some relatively simple procedures.

Although some of the techniques used to curb interference to VCRs are similar to those used to eliminate interference to other home entertainment devices, others are quite different because of the frequencies involved in the video recording and playback process. Also, as the ham who worked Outer Elbonia discovered: Once interference has been recorded, it's on the tape as long as the program is there.

Why Are VCRs Prone To Interference?

In order to understand why VCRs are susceptible to interference, it is important to understand how video is recorded and how the process differs from audio recording. The audio spectrum is generally regarded as frequencies from 20 Hz to 20 kHz. It is relatively easy and inexpensive to obtain good frequency response by recording the audio signal directly onto the tape. Video signals, however, contain frequencies from 30 Hz to about 4.5 MHz.[1]

Although it is possible to obtain good frequency response over such a wide bandwidth with direct recording, the cost would be prohibitive for home entertainment equipment. The simplest way to record this bandwidth on magnetic tape (at a reasonable price) is to frequency modulate a sine wave with the baseband video signal. At first this may seem like technological overkill, but it works and offers good immunity to noise.

Recording the Video Signal

The problem for amateurs is that most VCRs use frequencies in or near amateur bands to record the FM video signal. For example, VHS video recorders use the frequencies from 3.4 to 4.4 MHz to record the luminance (brightness) portion of the video signal.[2] This frequency choice makes VCRs especially prone to interference from amateur stations operating in the 80-meter band.[3] It also makes harmonic suppression an absolute must for amateur stations using the 160-meter band. Remember that VCRs are high-gain, broadband devices that can suffer interference from other HF amateur bands.

The Super VHS system uses different frequencies and a wider bandwidth to deliver better video resolution. The FM video signal is recorded from 5.4 to 7.0 MHz.[4] This higher frequency and greater bandwidth provides almost twice the resolution of conventional VHS recording. The video signal, however, is adjacent to the 40-meter amateur band.

Recording the Audio Signal

The audio portion of a television program can be recorded onto tape as an audio track. Several other techniques are used as well.

In the case of stereo audio on videotapes, the audio is used to frequency modulate a carrier, which is then recorded on the tape. In the case of VHS stereo, the audio-channel carriers are at 1.3 and 1.7 MHz.[5] Unfortunately, the lower-channel carrier is in the AM-broadcast band, and the upper-channel carrier is perilously close to the 160-meter amateur band. Even though FM is less prone to interference than some other modes of recording might be, it is not immune to all interference. VCRs that record audio directly onto tape (as baseband audio rather than as FM signals) are susceptible to interference in much the same way as are audio tape recorders.

The relationships between frequencies used by VCRs and the amateur bands are shown in Figure 8.1.

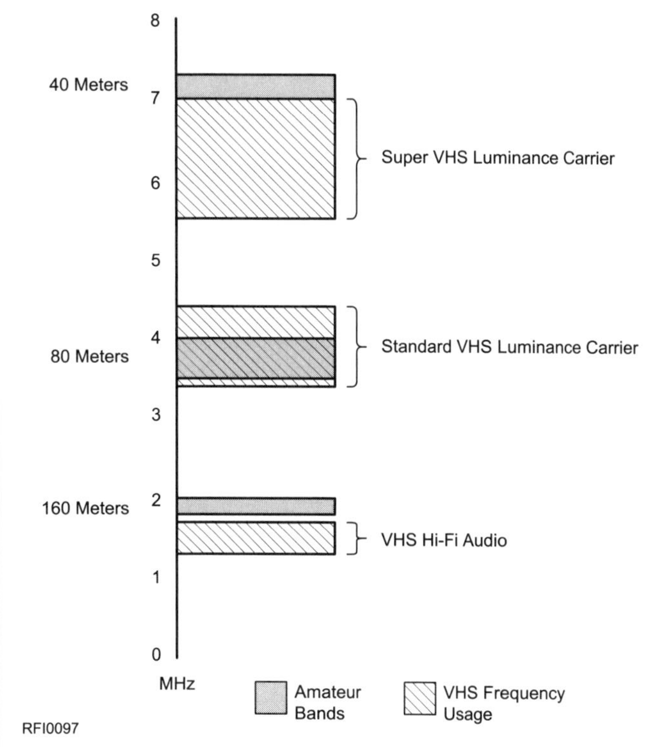

Figure 8.1—The VHS recording format uses frequencies in and near the US 160, 80 and 40-meter amateur bands. Beta recorders use similar, but not identical, frequencies.

CURBING INTERFERENCE TO DVD DEVICES AND VCRS

Although the technology of video recording and playback may make interference suppression seem a hopeless task, the key to curbing DVD and VCR interference is isolation. Isolation is achieved through the proper use of shields, filters and grounds. The VCR installation shown in Figure 8.2 is probably susceptible to interference from the transmitters of several different radio services. The proper use of filters (shown in Figure 8.3) usually eliminates DVD and VCR interference.

Some Stock Cures Are Unsuitable For DVDs and VCRs

Although interference to many home entertainment devices can be cured with the proper value of bypass capacitor or RF choke in the right place, interference to DVD devices and VCRs is a bit more complex. If the frequency of the interfering signal is within the bandwidth of the desired signal, conventional bypass and filter techniques can have disastrous effects on DVD and VCR performance.

Suppose that a CW signal on the 80-meter amateur band is interfering with a VCR. Although installing capacitors in the VCR might bypass the intruding signal to ground, the sync pulses of the video signal would be bypassed as well. Bypass capacitors in the rotating drum that contains the video heads might cause even greater problems if the balance of the drum were upset. Because the video signal occupies the same frequencies as some amateur HF bands, unwanted signals must be removed in ways that do not affect the desired signal: It is more appropriate to remove the 80-meter signal with a common-mode choke on the feed line or a filter on the power cord, or place the VCR in a shielded enclosure (as determined by the interference path).

Do not install metal foil or screen shields inside a VCR cabinet. While this technique is common with audio amplifiers and stereo receivers, they don't contain many moving parts. Since VCRs contain a multitude of moving parts, aftermarket shielding inside a VCR cabinet is extremely hazardous. If a wayward piece of foil or screen got tangled in the videotape, it could cause considerable damage. The EMI-RFI coatings mentioned elsewhere in this chapter are preferable shielding methods. The spray must be used with care, however, to reduce the likelihood of damage to the environment, individual or equipment.

Avoid the use of capacitors across the ac line as shown in Figure 8.4. Although this RFI cure might actually work on a DVD device or VCR, it has several disadvantages. C1 and C2 form an ac voltage divider that can hold the device chassis above ac ground. Since home DVD players and VCRs usually do not have three-wire line cords, there is a shock hazard should C1 short out. Furthermore, if the unit is connected to a properly grounded cable TV system, ac could show up on the coax shield. An ac-line common-mode choke (see Figure 8.5) is a much safer alternative.

It is common practice to connect a DVD device or VCR to the antenna with twin lead when cable TV service is not used. There is nothing wrong with this practice as long as interference cures are not improperly applied. Use a common-mode choke as shown in Figure 8.6.[6] *Do not place a ferrite bead over each wire of the twin lead as shown in Figure 8.7. The TV signal will be suppressed along with the interference!*

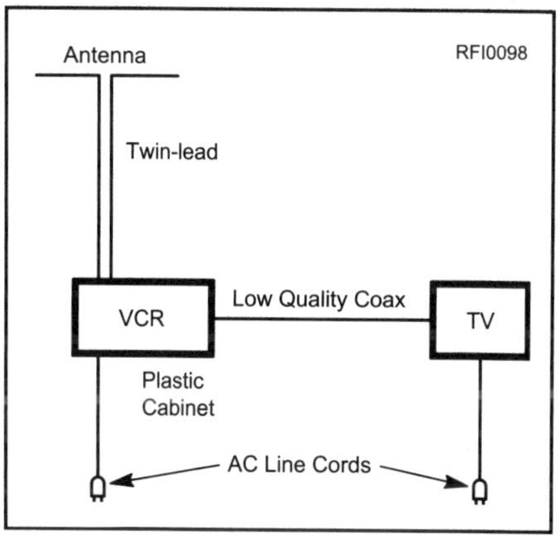

Figure 8.2—This is a typical home VCR installation. It has no protcotion against EMI.

Filter Shopping

When shopping for a high-pass filter, try to find a unit with appropriate connectors for the suffering DVD device or VCR. This eliminates the need for additional adapters or balun transformers. Additional components may degrade performance because they do not present a proper termination to the high-pass filter. Also, use filters with metal cases. The metal is an extra shield against unwanted signals. Finally, install the filter as close to the unit input as possible (as shown in Figure 8.8). High-pass filters are compact and easy to install. They are relatively inexpensive insurance against interference.

To Find a Cure, Understand the Interference Mechanism

Even though the design and construction of DVD players and VCRs makes them susceptible to interference from radio transmitters, it is generally possible to reduce or eliminate the effects of the interfering signal. The first thing to do is to determine how the offending signal is entering the unit.

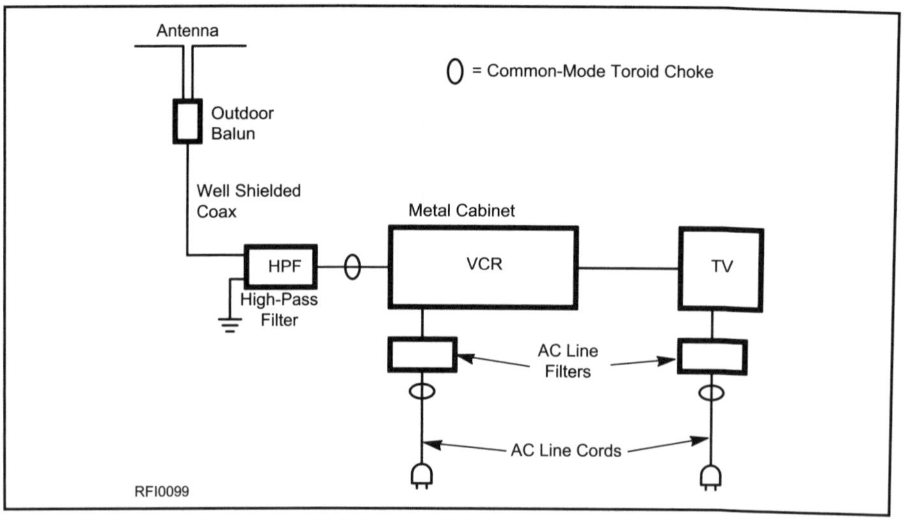

Figure 8.3—Proper filters and shields make VCRs more immune to interference. See Figure 8.5 for details of the common-mode chokes.

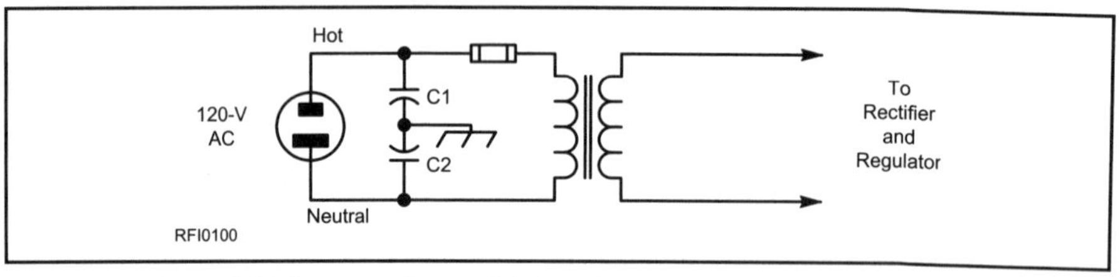

Figure 8.4—C1 and C2 form a voltage divider, which is a shock hazard in equipment that does not have a three-wire line cord.

Antenna/Feed Line

The TV antenna system is one of the most common DVD/VCR interference paths. DVD players and VCRs connected to cable TV systems can also accept interference from the cable; either as a differential-mode signal or a common-mode signal (refer to the cable television interference chapter). You can determine whether interference is entering through the antenna connection by disconnecting the feed line and viewing a DVD or prerecorded tape while the amateur transmits. The accepted cure for this interference path is a quality high-pass filter installed at the DVD device or VCR input. A common-mode choke is often necessary as well.

In many interference cases the unwanted signal enters the unit via coax shield.[8] This is called common-mode interference, and its cure is not the same as the cures for fundamental or harmonic interference. You can construct a ferrite toroid common-mode choke shown in Figure 8.9; more turns of cable through the core give more suppression. In some stubborn interference cases, it may be necessary to use more than one choke. In some cases, it may be necessary to use a normal (differential-mode) high-pass filter as well.

Common-mode interference can also occur in systems using twin lead feed line. When using twin lead, remember that both conductors must go through the core.

Figure 8.7 shows the wrong way to suppress common-mode signals on twin lead. Figure 8.6 shows the right way to suppress common-mode signals on the ac power cord. The same technique is applicable to twin lead.

AC Line

In some cases, interference reaches a DVD device or VCR through the ac line. Although most units have some filtering in their power supply circuitry, there is considerable room for improvement. Figure 8.10 is a simplified representation of a DVD or VCR power-transformer filter circuit. If you suspect the ac line as an interference path, wrap the ac-line cord around a ferrite rod as shown in Figure 8.11 or on a toroid as shown in Figure 8.6 (a common-mode choke).

To make the most of this simple filtering method, wrap the line cord around the ferrite rod in one smooth layer with no gaps between turns of wire. Use nylon cable ties to hold the line cord in place, and the filter is complete. When using this technique, place the ferrite rod close to the DVD unit or VCR as shown. This helps minimize line-cord RF pickup between the choke and the unit. Stubborn cases of ac-line interference may require a "brute force" ac-line filter (Figure 8.3). Only qualified electronics technicians should install components inside a DVD player or VCR. Most states have licensing and certification requirements for home-electronic service personnel.

Direct Radiation Pickup

Direct penetration of the cabinet by unwanted signals is not nearly the problem it once was. Many early top-loading VCRs were built in plastic cabinets. Newer video recorders are certified to comply with FCC Rules and Regulations, Part 15 (as a result of excessive radiation from RF modulators). When VCR manufacturers installed the shielding necessary to suppress modulator radiation, the shielding reduced direct radiation pickup.

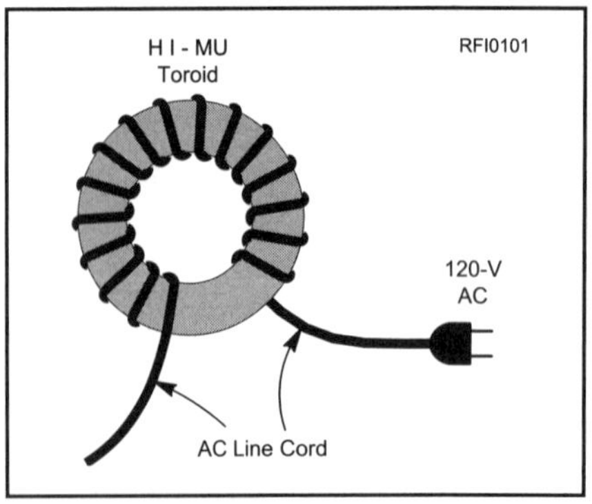

Figure 8.5—A common-mode choke for line cords. Wind 10-15 turns of the appropriate cord around an FT-240-43 ferrite toroid core.

Figure 8.6—A common-mode choke made with an FT-240-43 ferrite core. Install the filter as close to the VCR terminals as possible.

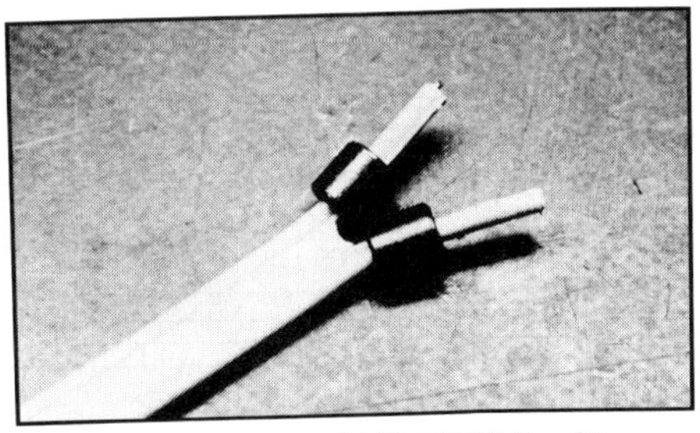

Figure 8.7—THE WRONG WAY! Ferrite beads installed this way will suppress the desired (differential-mode) signals. Even if used properly as a common-mode choke, these small ferrite beads just don't have enough inductance to work with HF signals.

Figure 8.12 shows the inside of a typical consumer-grade VCR. The metal box near the head drum contains much of the RF circuitry, and it is well shielded to prevent the exit and/or entrance of unwanted signals. The Phillips-head screw just left of the large PC board holds down one end of a springy metal piece; the other end of that metal presses up against the bottom of the cover and grounds it to the chassis.

VCRs and DVD players often have metal cabinets, but many plastic-cased units are in use. If direct radiation pickup is the prime suspect in an interference complaint, there are two ways to RF-proof a plastic cabinet. (Again, keep in mind that these internal cures should be performed only by qualified service personnel.)

The first involves careful use of an aerosol spray on the inside surfaces of the unit cabinet. There are several anti-RFI sprays on the market. One such spray advertises shielding ability of 35-50 dB depending on the frequency involved.[9] When using RFI shield sprays: (1) Exercise caution when removing the unit from its cabinet, (2) mask areas you don't want coated, and (3) allow adequate drying time before reassembly. Remember that the resulting coating is conductive. Make sure there are no component leads touching the coated areas (where they might cause a short circuit).

Figure 8.8—Commercial high-pass filters are compact and easy to install. Place them as close to the VCR terminals as possible.

A second, much less attractive, way of shielding plastic cabinets involves constructing a metal shield around the outside of the unit. Use perforated aluminum rather than solid sheets because DVD players and VCRs need ventilation. (The small-diameter holes do not significantly affect the shield quality.) If a DVD device or VCR shield enclosure is necessary, make it as RF tight as possible. Overlap the seams to help keep RF out, and use a "piano" hinge. Such exterior cabinet shields are a last resort, but serious interference sometimes calls for serious shields.

Accessory Connections

DVD players and some, but not all, VCRs have input and output jacks for audio and video signals. These jacks are often on the back of the unit, and videophiles sometimes leave patch cords attached when they are not being used. This practice is convenient, but it is also an invitation to interference. Cables can act as antennas and feed interference into the VCR.

Audio and video cables should be well shielded and equipped with proper connectors (adapters may degrade the connections). Patch cords should also be unplugged from the unit when they are not in use. A common-mode choke formed with about 10-15 turns of cable on a ferrite core will sometimes help. Audio cables can be bypassed as shown in the stereo interference chapter.

INTERFERENCE TO VCR AUDIO

Interference to the audio portion of videotape recordings is often more complex than interference to conventional audio tape recordings. VCR audio can be interfered with as part of the TV signal, after demodulation, during recording and during playback.

As discussed under "Recording the Audio Signal," some VCR formats record separate audio tracks. Others use the audio to frequency modulate carriers that are then recorded via audio heads mounted in the same drum that carries the video heads. To cure audio interference in VCRs, determine how the offending signal enters the VCR and which part of the circuit is affected.

Audio Examples and Cures

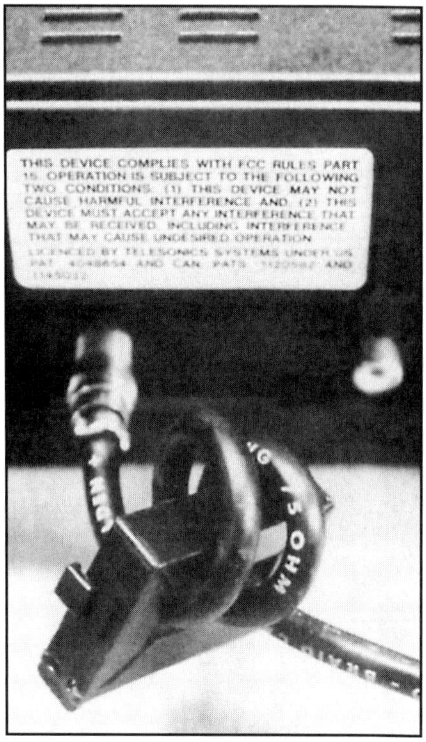

Figure 8.9—A common-mode choke using a snap-on core installed on coaxial cable. Install the filter as close to the VCR terminals as possible.

If audio interference occurs when a VCR is used with a microphone and camera but not during playback of prerecorded tapes, the interference is probably affecting the audio circuitry or microphone as it might an audio tape recorder. Such interference to the audio circuitry of VCRs can usually be cured with the same techniques used for audio tape recorders: ferrite beads, RF chokes and bypass capacitors (as described in the Stereos chapter).

On the other hand, VCRs that record audio as FM signals may also receive interference via record audio modulator or the playback audio demodulator: Suppose a VHS-stereo VCR picks up interference in one audio channel. Further suppose that there is a nearby AM-broadcast station at 1310 kHz. (VHS stereo uses FM audio with carriers at 1300 and 1700 kHz.) Even though FM is often considered immune to AM interference, the AM broadcast station is interfering with the VCR audio circuitry. To cure this kind of interference, use shields and filters to keep the unwanted signal from reaching the VCR by using techniques described elsewhere in this chapter.

When confronting audio interference in VCRs, determine which circuitry is affected and how the unwanted signal reaches that circuit. Remember that VCRs use frequencies in and near the AM broadcast band, international short-wave bands and amateur

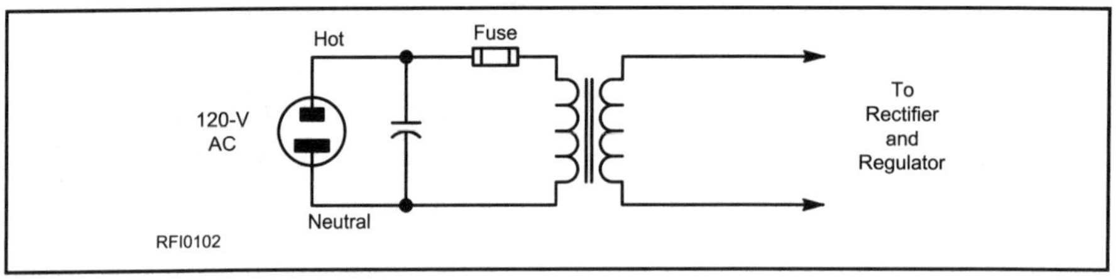

Figure 8.10—Most consumer-grade VCRs use a power supply with a two-wire line cord and a polarized plug. There is no on-off switch in the primary circuit of the power transformer.

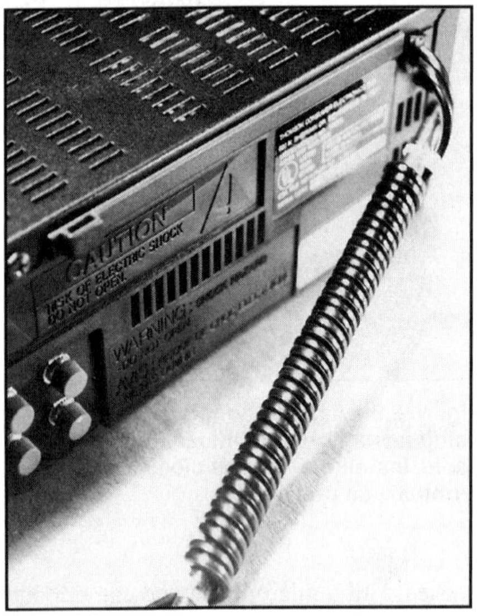

Figure 8.11—A common-mode choke made by winding the ac-line cord around a ferrite rod. Toroids are better (higher permeability and self shielding) than rods, but rods are easier to wind. They are adequate for some EMI cases.

bands. Place filter and bypass components carefully so that they do not disrupt the normal operation of the VCR.

ACCESSORIES AND VCR INTERFERENCE

High-quality baluns, splitters, switches, combiners and patch cords can make the difference between interference immunity and susceptibility. Some baluns are built in plastic boxes while others have metal enclosures. The same holds true for splitters, combiners and switches. Most good-quality video accessories are built in metal enclosures, which provide much better shielding than plastic enclosures. Don't save money by using low-quality accessories; they are an open invitation to interference.

Patch Cords

Poorly shielded patch cords contribute to DVD and VCR interference by allowing unwanted signals to enter the system. All 75-Ω coaxial cable is not the same: Some cables (with a single braided shield) have as little as 70% shield coverage. Better cables (such as RG-6) use a foil shield together with a braided shield for 100% coverage.

Cable TV Converters

The converters and decoders used on some cable systems may not be well shielded. Ideally, such devices should be built in RF-tight metal boxes. Unfortunately, many converters and decoders are built in plastic boxes, which offer no shielding. Although it might be possible to shield the inside of the plastic box with an anti-RFI spray, this is not advisable if the device is sealed to prevent tampering. Do not attempt to shield or modify cable-company owned converters or decoders to make them resistant to interference. That is the job of the cable system operator.

If cable equipment is the point of interference ingress, notify the cable system operator of the problem. The responsibilities of system operators are defined in FCC regulations and explained in the cable television interference chapter of this book.

Wireless VCR Remote Controls

The wireless remote controls that come with most DVD players and VCRs (Figure 8.13) are not RF devices, but rather infrared devices. Hence, they are generally not affected by strong RF fields. In a few rare cases, however, the DVD or VCR-remote logic circuitry is adversely affected by strong RF signals. For example, if a VCR switches from "playback" to "pause" or from "record" to "rewind" whenever a nearby transmitter is keyed, it is reasonable to assume that RF is affecting the logic circuits that control the VCR mode.

Again, it is easier to keep RF out of the DVD or VCR than to cure the problem at the affected circuitry. Proper shields and a good ac-line filter should eliminate most cases of interference to wireless remote controls.

Wired VCR Remote Controls

Wired remote controls require a different approach than their wireless counterparts. The cable between the control and the DVD player or VCR can act as an antenna and feed unwanted signals to the control circuitry. The simplest cure is to feed the cable through a ferrite bead, toroid, or ferrite core in much the same way that twin-lead feed line is treated to eliminate common-mode interference.

If the common-mode choke does not cure the problem, contact the unit's manufacturer for help. Other components should only be added as directed by the manufacturer's service bulletins. Only qualified personnel should perform service on home-entertainment electronic equipment.

Figure 8.12—The inside of a typical consumer-grade VCR.

THE CAMCORDER DILEMMA

Regrettably, camcorders do not respond well to the interference cures described in this chapter. Since there is usually no antenna or ac power connected to a camcorder, most interference results from direct radiation pickup. Camcorder construction and use generally preclude home-installed shields (see Figure 8.14).

Can anything be done to eliminate interference to camcorders? Yes. Contact the manufacturer. When a camcorder functions as a shortwave receiver (as it does when receiving interference), it is defective or unsuitable for the user's environment.

In desperate situations, an amateur can voluntarily alter amateur station operation. Careful selection of operating frequency and transmitter power can reduce or eliminate interference to camcorders.

AT THE AMATEUR STATION

Finally, the way in which an amateur station is assembled and operated can influence interference to DVD players and VCRs. Proper RF grounding is important, as is an ac-line filter to reduce RF on the power line. Use the minimum power necessary for communication to reduce the possibility of interfering with DVDs and VCRs.

Transmitter Treatments

Although the nature of the recording system used in VCRs makes them prone to interference, it is every amateur's duty to make sure that station transmitters don't radiate harmonics or spurious signals that could cause interference. Part 97 of the FCC Rules and Regulations states, "All spurious emissions from a station transmitter must be reduced to the greatest extent practicable. If any spurious emissions, including chassis or power line radiation, causes harmful interference to the reception of another radio service, the licensee of the interfering amateur station is required to take steps to eliminate the interference, in accordance with good engineering practice."[10]

A Transmit Filter May Not Help

Many amateurs think that a low-pass filter between the transmitter and antenna will solve all interference problems. It may solve cases of broadcast television interference, but it probably won't reduce DVD or VCR interference. Since amateur 160 and 80-meter fundamental signals are within the VCR passband for example, no transmit filter can help.

Consider an Amateur Radio station transmitting in the 160-meter band. The second harmonic is likely to fall in the luminance portion of a VCR record or playback signal. Since the harmonic is within the passband of the low-pass filter, it is not attenuated. How can harmonics below the transmit filter cut-off frequency be eliminated? A properly designed and carefully tuned matching network (Transmatch) can sometimes suppress harmonics by 20 dB or more. Information about the construction and use of matching networks can be found in *The ARRL Antenna Book*.

Transmitter AC-Line Filters

Install an ac-line filter as close to the transmitter or its power supply as possible. If the filter is located any appreciable distance from the transmitter, the wires between the transmitter and the filter might still radiate RF. Remember, the object is to keep the RF off the power line and out of the air (except at the antenna). RF signals must be directed to the antenna and nowhere else.

Figure 8.13—Typical VCR wireless remote controls operate with infrared, rather than RF. That's a blessing.

If a transmitter is used with a separate power supply, install the ac-line filter as close as possible to the power supply. If the dc leads between the supply and transmitter are long, a common-mode choke may be required on the dc leads.

SOME FINAL THOUGHTS

Proper selection of DVD and VCR accessories can be a factor in interference susceptibility. Top quality baluns, coaxial cable feedlines, interconnecting cables, splitters, combiners and switches help protect DVD players and VCRs from unwanted signals.

Some radio amateurs are under the impression that nothing can be done about interference to VCRs because of the frequencies they use to record the video. While it is true that VCRs use frequencies in and near amateur bands, interference to these devices can be controlled by the proper use of shielding and filtering to keep the unwanted signals out of the VCR.

The nature of FM video recording, and in some cases the FM audio recording, is such that interference control techniques used in other home entertainment devices like tape decks and tape players are not appropriate for

Figure 8.14—Will Eduardo, KA9YIT, experience interference to the camcorder if his neighbor (AC9J) calls "CQ" on 80-meter CW?

"My Cousin Hooked It Up That Way"

Several years ago, a friend of mine received a VCR as a gift. Before long the neighborhood was abuzz with stories of a pirate TV broadcaster showing racy movies late at night. The neighbors were more amused and entertained than shocked, but my friend was embarrassed and confused as to how the movies she and her husband were watching could be seen on their neighbors' TV sets.

This happened because the VCR output and the outdoor TV antenna were parallel connected at the antenna terminals of the TV set. Whenever the VCR was used to watch a videotape, the signal was not only fed to the TV set, but also to the antenna—and radiated around the neighborhood.

The cure was very simple: Connect the antenna to the VCR input and the TV to the VCR output. This shows the importance of proper VCR connections. Follow the instructions in the owner's manual! Who connected the VCR to the TV the first time? "My cousin hooked it up that way." The late night broadcasts stopped and none of the neighbors suspected a thing.—*John Frank, N9CH, Wisconsin Section Technical Specialist, Madison, Wisconsin*

VCRs. In some cases, improper interference control measures could actually degrade VCR performance or damage the unit.

Notes

[1]R. Goodman, *Maintaining & Repairing Videocassette Recorders*, Tab Books, Inc, 1983, p 1.

[2]Consumer Guide, *Video Buying Guide*, Publications International Limited, copyright 1985, p 23.

[3]W. Nelson, *Interference Handbook*, Radio Publications Inc, copyright 1981, second edition, fourth printing, 1990, p 247.

[4]J. Meigs, "Super Home Video," Jan 1988 *Popular Mechanics*, p 59.

[5]Consumer Guide, *Video Buying Guide*, Publications International Limited, 1985, p 23.

[6]Palomar Engineers, *RFI Tip Sheet*. See the References chapter for contact information.

[7]See the Suppliers List for contact information.

[8]*1992 ARRL Handbook for the Radio Amateur* (Newington: ARRL, 1991) p 40-14.

[9]A spray EMI-RFI coating is available from GC /Thorsen. See the Suppliers List for contact information.

[10]FCC Rules and Regulations Part 97.307 (c).

Chapter 9

Telephone RFI

In the FCC's own words, telephone interference is not caused by a rules violation by the transmitter operator: "Telephone interference generally happens because telephones are not designed to operate near radio transmitters and the telephone improperly operates as a radio receiver." With this in mind, most cases of telephone interference can be easily solved with the proper application of commercially available filters.

By Pete Krieger, WA8KZH
PO Box 82
Randolph, OH 44265

OVERVIEW

If you have a telephone interference problem, it is likely that you have skipped immediately to this chapter. The cures for telephone interference are usually fairly easy to understand and implement, but there are often other issues that need to be addressed first. Technical issues are really the *second* half of an interference problem. First and foremost you must deal with personal and political matters. After all, knowing all of the technical solutions will do no good if your neighbor won't let you into the house to try them!

The "First Steps" chapter helps you deal with your neighbor, the "EMC Fundamentals" chapter explains technical principles and terms and the troubleshooting chapter explains how to track down interference. Within the telephone-RFI chapter, specific fundamentals about telephones and telephone systems are covered, followed by the specific steps to take to diagnose and cure telephone interference.

Improvements in regulations and technology have actually reduced the likelihood of some types of interference. Modern transmitters no longer emit spurious signals that can easily interfere with other radio services. Cable TV has also helped—its high signal levels and shielding have actually reduced the likelihood of TVI. Unfortunately, the same regulatory and technical factors that have helped TVI have contributed to the likelihood of telephone interference! Telephone-industry regulations have been deregulated and modern telephone equipment contains sophisticated electronic circuitry that has more active devices and functions that can experience interference.

What is Telephone Interference?

Telephone interference has been around since the first time an Amateur Radio transmitter and a telephone had the misfortune of being located too close to each other. The result was interference—a problem for hams and their neighbors. Before we heat up the soldering iron or start trying to fix telephone equipment, let's get some perspective on the conditions surrounding telephone interference and the applicable FCC regulations.

To put it all in perspective, take a look at the telephone shown in Figure 9.1. This is what your telephone looks like to RF energy. In this phone, there is no shielding; there is no filtering. Inside the telephone there *are* a lot of active devices that can make this telephone very susceptible to RFI. However, telephone interference is like any other RFI problem—it usually can be cured by properly applying good troubleshooting techniques and the proper filters to the telephone or telephone device.

Telephone RFI occurs when telephones inappropriately listen and react to nearby radio signals. Well designed telephones do *not* pick up radio stations. Although it is not intentional, telephones can act as simple radio receivers when components rectify (AM detect) radio signals and produce unwanted audio.

All radio services can experience problems with telephone RFI. Amateur Radio is only one facet of the problem. (According to FCC statistics, Amateur Radio is a *small* facet!) People living near AM broadcast stations are familiar with interference from that source. Telephone RFI can result from any radio transmission of sufficient power on a frequency that telephones can detect.

There are exceptions, but the greatest likelihood of telephone RFI is from signals within the 0.5 to 30-MHz range. This results from several factors:

- Telephone components that detect RF usually have a frequency cutoff somewhere in the upper HF or lower VHF range.
- The resonances of the telephone wiring are usually more pronounced in the HF range. (Even small homes can contain hundreds of feet of telephone wiring.)
- The field strength of VHF and UHF stations is generally weaker than that of HF stations. VHF/UHF stations usually run less power, and the attenuation offered by buildings and trees decreases field strength even further.

Several components contribute to telephone RFI: Telephones and accessories that function improperly as radio receivers (some worse than others), telephone wiring that functions as a radio antenna, and radio transmitting RF sources. In some cases, improper RF reception (audio rectification) can occur outside telephone instruments, as in faulty wiring, for example.

History

Years ago, the telephone industry was heavily regulated. Before "deregulation," all telephone equipment in the US was owned, installed and maintained by the telephone company. Since the telephone system was the responsibility of the telephone company, there was no question who should diagnose and correct interference problems. The telephone company was clearly responsible for all phases of the telephone system and its proper operation, from the main office to the customer telephone. Procedures to fix RFI problems were developed in telephone-company laboratories and documented for use by field-service personnel. The *Bell Systems Practices Plant Series Manual* (now out of print) for example, provided detailed information on RFI treatment of telephones in use at the time. Some of those procedures are still valid, especially for older phones.

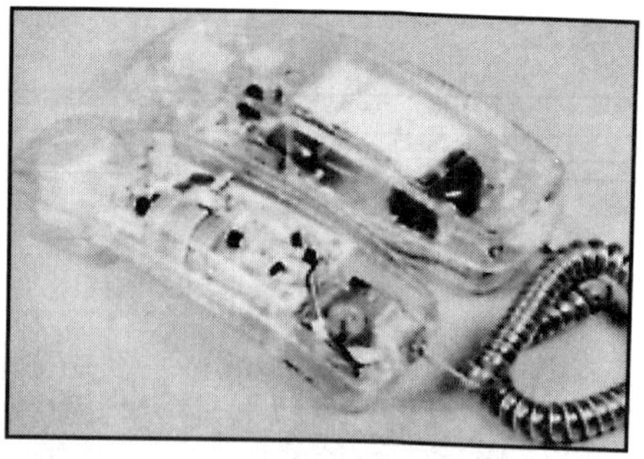

Figure 9.1—This telephone is transparent to light so we can see right through it. If we could see what RF sees, most telephones would look like this!

In those simpler times, it was easier to solve telephone RFI problems. The telephone equipment consisted primarily of rotary dial, electromechanical telephones that could be desensitized to RF with simple bypassing methods. RFI problems in such older telephones can still be cured by installing 0.001-µF ceramic-disc capacitors across the carbon microphone elements.

The telephone company produced filters that could be installed at the customer's home, although they were more effective at broadcast band frequencies than the ham bands. In stubborn cases, the telephone company provided replacement phones with RF-resistant transmission networks or used shielded twisted pairs for the line-drop and house wiring.

The Present Status

Unfortunately, those "golden days," like most golden days, are over. Hams and their neighbors usually have to find their own solutions to telephone RFI problems. Now, in most cases, the phone company no longer owns the telephones or the wiring in your home—you do. Because you own it, you are responsible for maintenance. (Some telephone companies have provisions for "inside wire maintenance" where, for a monthly fee, they agree to maintain the wiring in your home.) You are now responsible for telephone equipment that, not all that long ago, you weren't even allowed to touch!

RESPONSIBILITIES

The first step to solving a telephone interference problem is for all involved to understand the issue of responsibility. Most cases of interference can be resolved only if all involved parties address their responsibilities fairly.

The Radio Operator

It may be natural for some to assume that the transmitter operator is responsible for the interference (We have all heard, "It only happens when you use that transmitter!"—*Ed.*), but telephone interference is not caused by any rules violations by the transmitter operator (high power 11 meter operation is one notable exception, but the interference that results is exactly the same as would occur from legal 10 meter operation by a licensed ham).

FCC regulations require that amateur transmitters not emit spurious signals that interfere with other radio services. This is the sole *regulatory* responsibility of the radio operator. This doesn't apply in the case of interference to nonradio devices; the FCC's own material is pretty clear—telephone interference is caused by inadequacies in the telephone equipment, not the transmitter. From a regulatory point of view, this is all fine and good, but the neighbors of hams are not apt to understand this subtle point! It can certainly be a neighborly gesture for you to help your neighbor understand telephone interference and help your neighbor *locate* a solution. In most cases, this will involve helping your neighbor select the proper telephone filters and understand the simple instructions that come with them. Although the FCC doesn't *require* you to help, they certainly hope that the public service of Amateur Radio will be extended to a local neighborhood situation.

It is possible, however, to carry this too far! It is not usually a good idea for you to do repairs on your neighbors' equipment or telephone lines. In most cases, it is not legal for you to work on your neighbors' telephones or home wiring. You could also be held liable for any problems your neighbors ever have with their phones or phone system. In all RFI cases, you should be a locator of solutions, not a provider of solutions.

You'll sometimes find that a manufacturer or utility company is willing to address an interference problem responsibly, but lacks the experience and training necessary to apply the correct solutions. You may need to apply your skills (and those of your ARRL Section Technical Coordinator or local club RFI committee) to help their personnel understand the technical issues.

Personal Diplomacy Overview

The subject of the interpersonal aspects of RFI is covered in the "First Steps" chapter. This is must reading for the ham with an RFI problem! Do ensure that things get started on the right foot. Your neighbor won't understand the complexities of RFI causes and cures. Be prepared to do a good job explaining things in nontechnical terms. The ARRL and the Consumer Electronic Manufacturers Association jointly publish a pamphlet designed to explain interference to your neighbor. The text of this pamphlet has been reproduced in Appendix D, although an actual copy of the pamphlet is usually more effective with your neighbor. A copy can be obtained from the ARRL Web site at **www.arrl.org/news/rfi/neighbors.html**.

Once you get your neighbor to understand that telephone interference can't be cured by making changes to your station, you can become the neighborhood "good guy" by locating cures for the interference, even though it is not directly your fault. Your neighbor may or may not believe your explanations, but if you are willing to help anyway, that is often enough for good neighborly relations to take over where technical misunderstandings might otherwise get in the way.

Remember, too, the adage to keep your own house in order. You should first ensure that there are no interference problems in your own home. There are several good reasons to start in your own home. If you own your telephone wiring, you have the flexibility to try different troubleshooting steps easily.

There's another good reason to start with your own house: You'll be able to demonstrate that RFI cures are not only effective, they cause no harm to the telephone's proper operation. It is a powerful diplomatic tool to tell your neighbor that you are sorry they can't use their telephones while you are on the air, but they can come over to your house and use *your* telephones while you are on the air! That usually helps them understand quickly that the problem must be on their end.

The Role of the FCC

The FCC regulates radio transmitters. The Federal Communications Commission (FCC) (**www.fcc.gov/**) doesn't have any specific rules that govern interference to telephones. This is because telephones are not radio devices and, as such, are not supposed to pick up radio signals. Although they also regulate telephones (through Part 68 of the FCC rules), the FCC does not require that telephones have any interference protection. Radio transmitters do not transmit the audio signals that telephones use, so there are no applicable rules that govern radio transmitters interfering with telephones.

This means that when telephone RFI occurs, there is nothing for the FCC to enforce! The FCC rightfully believes that telephone interference is a matter to be resolved by the telephone equipment manufacturers as a customer service issue. Nonetheless, the FCC realizes it will continue to receive inquiries from the public about telephone RFI. They offer help to transmitter operators and consumers, primarily as an authoritative information source on self-help cures. The former FCC Compliance and Information Bureau (CIB) created a publication, "Interference to Home Electronic Entertainment Equipment Handbook—May 1995 edition." This was released as Bulletin CIB-2. A copy can be downloaded from the ARRL Web page at **www.arrl.org/fcc/tvibook.html**. The former CIB had also written a bulletin specifically on telephone interference, "What To Do If You Hear Radio Communications on Your Telephone," released as Bulletin CIB-10. A copy can be downloaded from the ARRL Web page (at **www.arrl.org/fcc/fcc_rfi_CIB-10.pdf**). These bulletins have been reproduced in Appendixes B and C. See Figure 9.2, which reproduces the first page of CIB-10.

Over the past several years, the FCC has made tremendous strides in creating unbiased, factual information about all types of interference problems, especially interference to nonradio devices such as telephones. The FCC clearly explains that the burden of resolution of telephone interference prob-

Phone Modifications

What about home-brew RFI treatments inside a phone? There are several good reasons for owners to avoid such modifications:

Many new telephones are mass produced as "throwaways" that are not easily disassembled. Others permit partial access, but the "guts" are encapsulated or extremely difficult to reach.

By law, only registered telephone refurbishers may repair or modify phones in ways that affect telephone operating characteristics or requirements for FCC registration.

Anti-RFI techniques used on older phones may not work. Internal measures such as bypass capacitors may adversely affect modern telephone operation.

External RF filtering works nearly as well as internal modifications. This is an alternative to replacing or modifying most telephones. Ferrite-core chokes can eliminate RF current on the wiring before it reaches the phone, or at least decrease the current to an insignificant level.

It is best to simply avoid internal modifications to telephones. Internal modifications should only be performed by registered telephone service personnel.

lems is *not* upon the transmitter operator, but with the manufacturer of the affected telephone instrument! Hams knew this all along, but in the past, the FCC offered scant material on the topic. Thanks to a major effort a few years ago by a national team of FCC staff, FCC material on interference is useful and complete. The ARRL was pleased to serve as a consultant to the FCC as this information was being developed.

The FCC *encourages* telephone manufacturers to consider RFI susceptibility in the design of their equipment. The Commission also encourages manufacturers to help customers resolve interference problems that occur after the telephones have been purchased. This encouragement is continued in their bulletins, which contain a sample letter that consumers can write to telephone manufacturers, and a list of addresses and telephone numbers. They also encourage hams and their neighbors to work together toward finding solutions to telephone interference problems.

Telephone Manufacturer Responsibilities

This is where the process often breaks down. Although the FCC points to the telephone company manufacturers as the root source of telephone interference problems and solutions, there are no *regulations* that require them to help their customers. Some manufacturers and importers design telephones with susceptibility in mind. RadioShack is known to consider susceptibility when they qualify manufacturers for their product line—their test procedures for telephones include several that measure susceptibility. (It is not surprising that the ARRL Lab has received numerous reports over the years praising the immunity of RadioShack telephone products.)

Companies like these, however, are still the exception. From reports received by the ARRL Lab, many manufacturers do not deal properly with consumers who report problems with their telephones. In some cases, customer service personnel tell the consumer that the transmitter operator is running illegal power and that the consumer should contact the FCC!

This adds to the frustration of the people whose telephones are not working, adding fuel to an already hotly burning fire. This often puts the *social* burden back on the ham. The neighbors of hams may not realize it at first, but the local amateur who has access to the information in this book and possible local technical advice, probably represents the best source of help to find a solution to an interference problem! The ham is certainly the *closest* source of help.

G. TELEPHONE EQUIPMENT INTERFERENCE:

Telephone interference generally happens because telephones are not designed to operate near radio transmitters and the telephone improperly operates as a radio receiver.

You may contact the nearest FCC Office and request FOB Bulletin No. 10, "Telephone Interference". You may also try the following.

1. Contact the telephone company if you are using a leased phone The telephone company may have responsibility for correcting interference to their leased phones.

2. Disconnect all of your telephones and accessories such as answering machines and take them to one telephone jack. Connect each instrument, one at a time, and listen for the interference. If you hear the interference through only one telephone, the interference is being generated in that unit.

3. Install a filter on the telephone line cord at the end nearest the telephone and/or at the telephone handset cord.

Filters are very selective. (See Section IV.) They must be designed for the type of interference you are experiencing or they will not work. For example, if your phone is reacting to an Amateur or CB radio transmitter, install a filter designed for that purpose. FM Broadcast interference requires a filter designed to reject FM broadcast stations. AM broadcast interference requires a filter designed to reject AM broadcast stations, etc.

4. Filter the incoming telephone line with ferrite beads and snap-together ferrite cores. You may need to experiment to find the best style of bead or core and the best location on the cord.

5. If you cannot eliminate the interference using the above techniques, consider purchasing an interference free telephone which has been specifically designed to be immune to interference.

Cordless telephones use radio frequencies and have no FCC protection from interference. If you are receiving nearby transmissions on a cordless phone your only recourse is to contact the manufacturer for assistance. The remedies above will not be of any use.

Figure 9.2—This page, reproduced from the FCC's Bulletin CIB-10 on telephone interference makes it clear that telephone interference is not caused by a violation of FCC rules.

The manufacturer is responsible for the proper operation and repair of the telephone. The ARRL and FCC both encourage all people who have an interference problem to contact the manufacturer of the susceptible device. The Consumer Electronics Association, an arm of the Electronic Industries Alliance, can often help you locate a manufacturer. (See the Resources chapter for contact information.) Contacting the manufacturer helps ensure that the manufacturers and their national association are aware of the interference problems their customers are experiencing. Remember: Interference that isn't reported officially doesn't exist!

Some manufacturers do give excellent customer service and a few supply filters free of charge. We've even heard a few reports of major telephone manufacturers giving refunds for defective telephones.

Telephone Utility Responsibilities

The redefined role of the telephone company (now broken up into many local, regional or national service providers) makes telephone RFI a problem some companies choose not to confront. In most cases, the telephone company is responsible for the proper operation of the telephone system only up to the point where the telephone wire is connected to the subscriber's home. The wire, whether overhead or underground, and the connection point at the house, are usually referred to as the "drop."

Although telephone-line amplifiers or other electronic equipment may also be the root cause of the problem, it's best to leave the telephone company equipment to their experts. There are important safety and reliability issues related to the maintenance of the lines and equipment that are the sole responsibility of the telephone company.

Even with the help of telephone company professionals, there's no guarantee of a cure for interference. The days of extensive laboratory testing and documentation of RFI remedies for company-owned home telephone systems are over. You may have noticed that little new industry-supplied telephone-interference information has become available to amateurs since deregulation. It's a good bet that your local telephone company service department hasn't read anything new either! When repair personnel use outdated procedures on contemporary interference problems, RFI problems are usually not cured.

Customer service policies vary widely, so you need to ask your local telephone company what they can and will do to help you with an RFI problem. If assistance is available, be sure to ask what it will cost. Telephone companies usually charge for service to customer-owned wiring or telephones. Interference investigations can fall into the same billing category as repair service. If you and your neighbor are discussing calling in your telephone company, make sure you and your neighbor understand how the telephone company is going to be paid.

Consumer Responsibilities

Consumers have responsibilities, too. The FCC has put a lot of effort into the factual information that is designed to help consumers understand the issues surrounding telephone RFI and to proceed correctly toward locating a cure. Consumers should educate themselves by reading the FCC's information and following the FCC's advice. A consumer who refuses to discuss the problem or who demands that the transmitter operator correct the problem at the radio station is not doing his or her part toward finding a solution. Although some types of interference can be the direct fault (and responsibility!) of the radio operator, interference to non-radio devices such as telephones is caused by inadequacies in the design of the telephone equipment. Telephone RFI can generally *only* be fixed by applying the appropriate cures at the telephone end.

Costs

Telephone companies often charge to come out and repair wiring. Repairing consumer electronic equipment usually isn't free, and telephone RFI filters require a bit of wallet digging, too! These issues should be discussed up front, before filters are ordered or that call is made to the service department. You are *not* responsible for purchasing filters or repairing defects in your neighbor's house wiring. If you want to be neighborly and buy a filter for a neighbor, that's your choice. However, doing so may set a precedent. If you live in an apartment building where there are hundreds of telephones to contend with, you may have to re-evaluate your generosity!

A TYPICAL TELEPHONE INSTALLATION

Figure 9.3 shows a typical residential telephone wiring system. Since deregulation of the telephone companies, telephone wiring is often installed by contractors. While there are many competent qualified contractors, they do not have the same level of support of dedicated labs and engineers as before deregulation.

The Service Entrance

The service entrance is also referred to as an interface or connector block. From this point, telephone wiring is distributed outward to the phone jacks, usually in one of two wiring configurations: straight cable runs (parallel wiring) or loop series.

Wiring Styles—Straight Cable Runs

Parallel wiring (Figure 9.3A) uses a separate cable run to feed each jack in the house. The wiring style can be identified by looking at the service-entry connections. When the number of used wire pairs leaving the service entrance equals the number of phone jacks in the house, the house is parallel wired.

Wiring Styles—Loop Series

Loop series wiring (Figure 9.3B) is the method usually used to wire residential homes. A single wire pair connects the service entrance to the nearest phone jack and continues to each of the system jacks. Except for the most distant jack (where the wires terminate), the cable enters and leaves each jack location.

Some homes, especially where customers have done their own wiring, will use a combination of wiring methods. In general, there is little difference in RFI potential between the two methods.

Resonance

No matter which wiring style is used, the house and street telephone wiring is a large long-wire antenna. When a local transmitter is in use, the wiring *will* pick up RF energy (usually common-mode—see the EMC fundamentals chapter). The telephone wiring will then conduct the RF signal to the telephones or telephone devices, which, if they are susceptible, will create telephone interference. The wiring resonance is usually more pronounced on the lower HF bands, so it is possible to have interference only on 160 through 40 meters, with the higher bands being less troublesome.

You may need to break up the resonances in the telephone wiring. Several manufacturers sell "wired" telephone RFI filters that can be installed at strategic points to detune the resonances. This is similar to the effect of using insulators to break up the guy wires on a tower to prevent them from resonating at amateur frequencies. Figure 9.8B shows an example of a filter that can be easily attached to the telephone wiring system.

There may be several reasons why your sophisticated telephone is particularly susceptible to interference from one particular band. Internal resonances in the telephone circuitry may allow more energy on that one band to reach the susceptible component inside the telephone. In this case, however, it's more likely that the telephone wiring in your home is resonant on that band. The telephone wiring, its associated grounds and any other equipment connected to the system form a large antenna. Like any antenna, this system has high and low impedance points. If the telephone happens to be located at one of these nodes (either high or low impedance) it may be subject to quite a bit of RF energy.

Wiring Types and Color Codes

Much of the telephone wiring found in homes is not very good from an RFI point of view. Flat ribbon cable is commonly being used in new construction; it is not twisted and is not very well balanced. RFI and crosstalk are common problems.

Most telephone wiring found in residential homes consists of two pairs of wires run to each telephone jack. In most cases, the red and green are used for single-line telephones. The yellow and black wires are reserved for a second line or for auxiliary uses. Rarely, these wires are actual "twisted pairs"; untwisted wiring is much more common, even in installations done by the telephone company. Untwisted wires can be found in either a flat ribbon cable or the more or less round wiring shown in Figure 9.4B. Such cable is not as good as a twisted pair for RFI control. It is prone to interference when subjected to radio signals.

The telephone companies generally have access to better telephone wiring. Category 3, or better yet, Category 5 telephone wiring can make a *big* difference in RFI cases. If you have a telephone interference problem that seems to be stemming from RF being *everywhere,* and it doesn't respond well to the usual cures, consider having the telephone company supply Category 5 wiring and rewire the house.

Shielded telephone wiring can help reduce RFI problems by minimizing the amount of RF energy conducted to telephones. The wiring will pick up most of the energy on the shield, which is grounded to help shunt the unwanted energy to ground instead of to

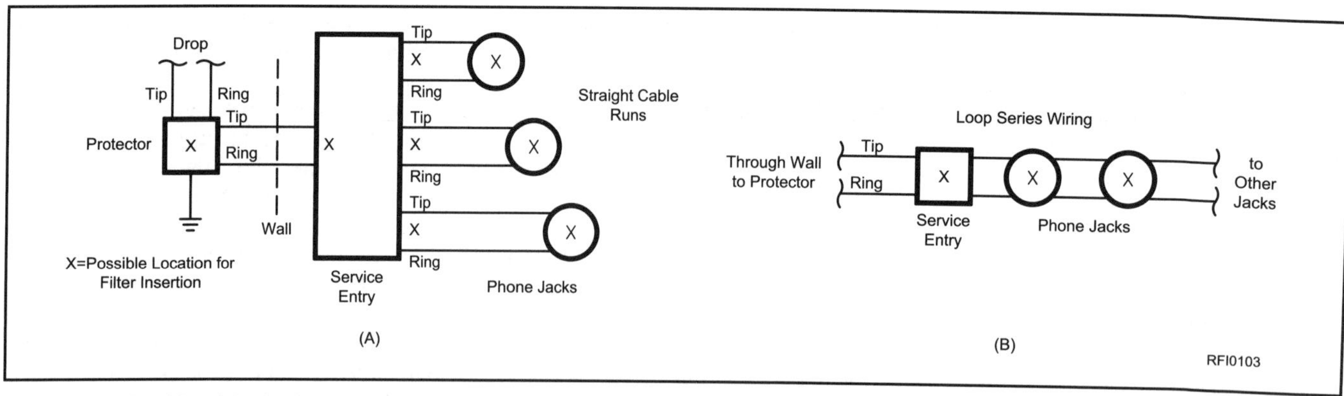

Figure 9.3—Residential telephone systems usually depend on straight cable runs (A) or loop series wiring (B) for phone interconnection. The owner's system connects to the telephone company via the pole-to-house drop (which may run underground in some installations); a ground, fused protector minimized lightning damage to the house wiring and telephones. The terminal block where the wires actually enter the house is called the "service entry." This drawing shows possible locations for the RF filters discussed in the text. Tip and ring are telephone-industry nomenclature for the talk-circuit wires, which date back to the days when telephone operators used phone-plug patch cords for call routing.

the telephone. Unfortunately, this wiring is expensive so it is rarely found in residential homes. It can usually be installed by the telephone companies, but it can be expensive to retrofit an existing home.

The Twisted Pair—Built-in Interference Rejection

Aside from technically advanced telephones and the switching equipment, the rest of the system might be described as "low-tech." Telephone engineers, nonetheless, have established some brilliant design criteria and industry standards for basic telephone systems. Many of these design conventions have been incorporated to minimize noise pickup. Naturally, this also helps with respect to interference rejection.

Consider the telephone cable itself—commonly referred to as the "pair" or "twisted pair." Twisting the wires results in nearly complete cancellation of induced differential-mode signals. (A common-mode signal may be induced on the pair, but such signals are easily removed by common-mode chokes.) When a twisted pair is combined with the common-mode rejection inherent in old-style telephones (because all of their circuits are balanced), external signals are effectively rejected. This rejection of external electromagnetic fields is significant because telephone wiring is often installed near electrical wiring, motors and so on.

Twisted-pair phone systems with old telephones can pass through noisy environments and still remain quiet and interference free. This benefit is lost if the twist in a phone wire is physically or electrically eliminated by damage, unbalance or installation of non-twisted wire. The benefit is also lost if the telephone responds significantly to common-mode signals, as do many modern devices.

System Unbalance

Telephone systems use balanced transmission lines to transfer audio and control signals efficiently over long distances. When a condition in a telephone installation upsets critical system balance, the unbalanced system is more susceptible to RFI problems.

Many different things can result in system unbalance. Defects in the telephone devices connected to the line can unbalance the entire system. Other causes of unbalance include permanent or intermittent connection of a ground or unused cable wire to one side of the pair. Such connections can be caused by wiring errors, wire-insulation breakdown, staples driven through a telephone cable or poor connections at joints or terminations. All of these problems can be found in customer-installed wiring, although it is not impossible to find them in work done by the telephone installers.

RF Pickup

As mentioned earlier, telephone wiring is a longwire-type antenna. This pickup is usually the root cause of telephone interference—most telephone devices do *not* pick up much RF energy directly.

When a nearby transmitter is on the air, *all* nearby conductors will pick up the transmitted RF. In the case of telephone wiring, this signal is usually picked up as a common-mode signal. This means the signal will be picked up by all of the wires on the telephone line in phase, with all the wires acting as if they were one wire. A good way to look at this is that the common-mode signal is the signal picked up *on* the phone wiring, as opposed to a differential-mode signal, which comes down on one wire and returns on the other. A differential-mode signal can be thought of as being *inside* the phone wire, just like the desired telephone signal. Both

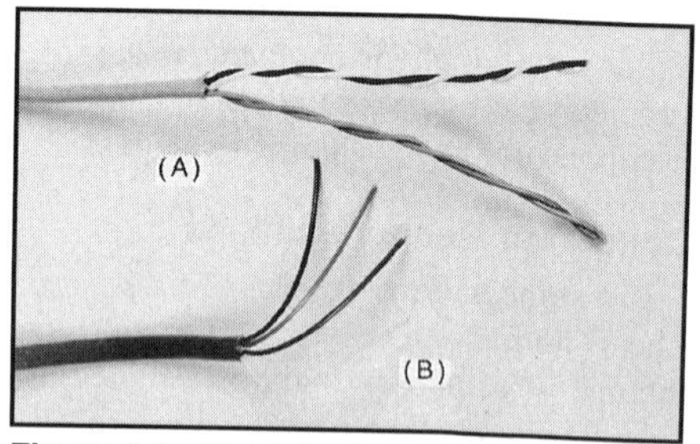

Figure 9.4—Flat telephone cable (B) is rapidly replacing the traditional twisted pair (A) in residential installations.

types of signals are shown in Figure 9.5, and are discussed in detail in the EMC Fundamentals chapter.

If the phone wiring and system is well balanced, the pickup will be almost entirely common mode. If, however, there is any unbalance, the pickup on each wire on the pair will *not* be identical in amplitude and phase. This will result in a differential-mode RF signal being present on that pair, in addition to the common-mode signal. Most telephone interference is caused by a common-mode signal and can be cured with a common-mode choke. Most telephone filters are common-mode chokes, consisting of a number of turns of wire on a small piece of suitable ferrite material.

Unused Pairs Should be Grounded

Good telephone wiring practice calls for all unused conductors inside active cables to be grounded. This is done at the service entrance. Simply attach all unterminated wires to the system ground at the service entrance. This can often reduce the RF pickup substantially.

The Lightning Arrestor

Telephone service enters a house at a

grounded, fused lightning arrestor located outside at the house end of the telephone company drop. Years of exposure to weather or moisture can cause corrosion or discoloration (onset of corrosion) of wires, junction boxes or components inside the lightning protector housing. If the arrestor is accessible, a good visual inspection may reveal potential problems. Lightning arrestors, especially those that have done their job a few times, can become nonlinear, acting like a diode, rectifying any RF energy present on the phone lines (just like the crystal radios many hams built in their early years). If you discover that a lightning arrestor is creating RFI, the fact that it's rectifying RF is one indication that it needed to be replaced anyway! Modern arrestors are less prone to RFI problems than older ones.

The Telephone Ground

Correctly installed telephone systems use their own ground rod (which should be tied to the power company safety ground) or are tied directly to the power safety ground rod. In the past, it was common practice to tie the telephone ground to a cold water pipe, either at the protector or inside the service entry. In fact, the telephone ground is usually a safety ground for the lightning arrestor only.

There have been some installations where the telephone ground is tied to a cold water pipe at one end of the home, and the water pipe is then tied to the power safety ground at the other end of the house. This very bad practice creates a large ground loop. Such ground loops may be in violation of local electrical codes, may pose a lighting hazard and are more susceptible to interference than a proper installation. (Not all copper water pipes in homes are connected to earth ground. Some plumbing installations mix copper and various plastic pipes.)

If you see evidence of an improper ground installation or a problem at the protector, contact the telephone company repair department. Local phone companies generally correct these installation problems without charge. If you're unsure who pays for a particular phone-system repair, ask first.

Correcting Wiring Problems

Most of the time, telephone wiring does not contribute to telephone RFI problems directly. But don't overlook the installation as a possible source of trouble, especially in cases that seem difficult to solve.

In some cases, you will be lucky—the telephone wiring is easily accessible and can be easily inspected. In other cases, the wiring may be hidden behind walls. In either case, do the best reasonable job inspecting the wiring for problems.

Look for obvious problems. A careful inspection of the wiring may reveal the problem. Years of exposure and moisture in damp basements, walls or crawl spaces takes its toll. Splices are rarely soldered in residential wiring and it is quite possible that twisted connections can become corroded. Be suspicious of splices that are corroded or badly discolored. Corroded copper is most often greenish or bluish. Corrosion can often be nonlinear, resulting in diode detection and audio rectification. Metal corrosion can form a fairly efficient diode, especially if dissimilar metals make contact. Loose or corroded splices can also be noisy, creating annoyance that is sometimes confused with RFI.

In many cases, homeowners have installed their own telephone wiring, often using substandard wiring. If you find sections of telephone wiring made from two-conductor zip cord, 300-ohm television twin lead, or some other mystery cable, replace it with standard twisted-pair wire. RadioShack, among others, sells several types of telephone wire. The best telephone wiring systems use twisted-pair, balanced wiring to minimize pickup of external electromagnetic fields.

Installation problems can also cause system unbalance. If one of the two wires in a pair is shorted to another conductor, such as a plumbing pipe, the system can become badly unbalanced. The phones may continue to work, but the line may be noisy, subject to crosstalk and extraordinary amounts of RF pickup.

If the telephone wiring looks jury-rigged, held together with baling wire and chewing gum, chances are good that installation problems are contributing to the RFI problem. Correct any installation problems found during this overall inspection.

If there are problems in a neighbor's telephone wiring system, however, you'll probably want to call a professional (often your local telephone company) to do the actual repairs. By performing repairs on a neighbor's wiring, you may be held liable for any problems that occur—even if the problem is with his toaster. If you lack experience, you may want to hire local experts for your own wiring, too. Before you start troubleshooting your system, check local regulations. There may be state or local requirements that must be met.

TELEPHONE DEVICES

Today's new generation of electronic telephone equipment is remarkably sophisticated. With a host of special features, new phones have made obsolete the less glamorous (but much less interference-prone) instruments made in the past. (These simple telephones are often called POTs in the industry—"plain old telephones.") Unfortunately, as is all too common in consumer electronics, these new devices are a recipe for problems—lots of sophisticated electronics to run all the bells and whistles, but little or no internal protection against RFI.

Two types of telephones seem to be most susceptible: the inexpensive ones that have virtually no shielding or filtering, and the expensive ones that have many solid-state devices rectifying RF signals. Of course, telephones that fall between these two categories can have problems, too.

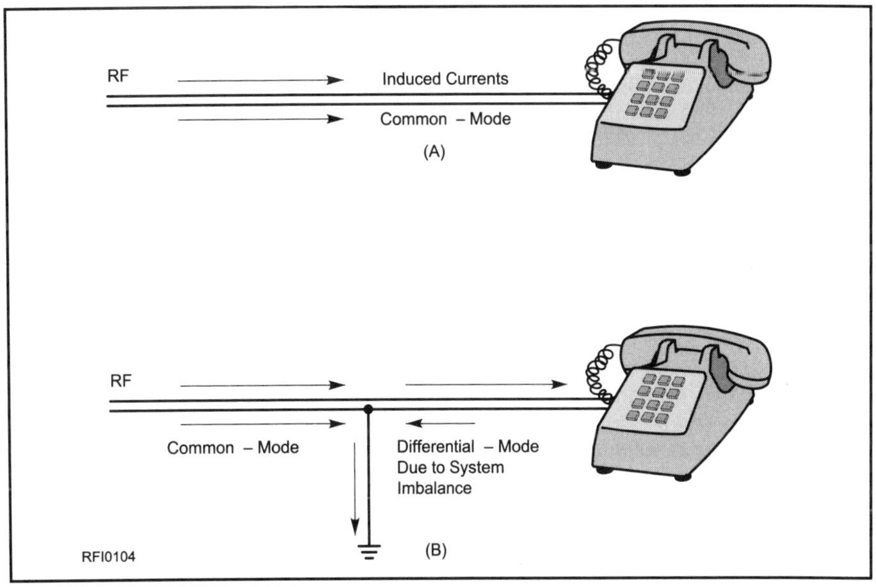

Figure 9.5—Telephone wiring usually picks up nearby RF in the common mode (A). System unbalances, however, can result in a strong differential-mode pickup (B).

Cordless Telephones

Cordless telephones use radio frequencies. They are actually small two-way radio systems. There is often little front-end filtering, making them very susceptible to overload by nearby radio signals. As with any two-way radio, they are susceptible to interference from nearby radio transmitters, including other cordless phones.

Cordless telephones are operated under Part 15 of the FCC regulations. A cordless telephone is an unlicensed transceiver that shares the spectrum with other users and services on a non-interference basis. The FCC does not intend Part 15 devices to be protected from interference.

Part 15 devices are required to carry a label indicating that: Part 15 devices must not cause any interference to other services and Part 15 devices must accept any interference that is caused to them.

The owner of a cordless telephone may be unhappy about any interference problems that may occur, but the law is clear about the whole matter. The owner should contact the telephone manufacturer for assistance. Some manufacturers may help.

Cordless telephones come in many styles and have several different features. The oldest cordless telephones, operating under an FCC waiver, use frequencies between the upper end of the AM broadcast band and the 160-meter amateur band. They are very susceptible to RFI from HF operations and also cause harmonic interference to amateurs.

Many older cordless phones use frequencies near 49 MHz. These devices are somewhat immune to interference from HF operation, especially from the lower HF range. They are only moderately immune to interference from the upper HF range. They provide poor immunity to interference from 6-meter operation! It is virtually impossible to eliminate 6-meter interference to a 49-MHz cordless phone; the frequencies are closer together.

The newest cordless phones are digital and operate in the 900 MHz and higher bands. These devices are often immune to most interference, although a UHF amateur transmitter is apt to cause them some problems.

Cordless telephones often include other features. For example, some are two-line phones, some include automatic answering machines, intercom systems, hands-free speakerphones, three-party conference features, memory features and so on. Such additional features are subject to RFI.

Cordless telephones can interfere with other cordless telephones when operated on the same frequency.

The illustration shows a block diagram of a typical cordless telephone system.

CORDLESS RFI CURES

Other than 50-MHz overload of a 49-MHz cordless phone, the most common source of interference to cordless phones is through the telephone wiring. The unwanted RF is almost always a common-mode signal. It doesn't always work, but you can try to solve this problem in the usual way—install a telephone RFI filter in the telephone line that connects to the modular jack on the master station.

This often solved the interference problem. If it does not help, leave the filter installed and proceed to the next step.

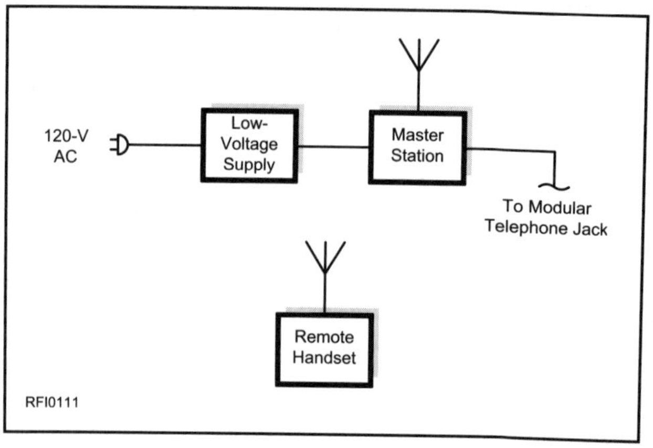

The Power Supply

Master stations operate from the ac line, and common-mode RF signals can enter via that route. The power supplies contain rectifiers, which are possible sources of audio-rectification interference.

The power-supply lead is often about 7 feet long, and it may be resonant on 10 meters. If there is any excess length, wrap it in a small loop, which is less likely to act as a good antenna. If the trouble persists, install a common-mode choke where the power cord connects to the master station. Wind 10 to 20 turns of the power-lead wire on a (no. 43 material) ferrite toroid or rod. This should solve the problem. Some people have eliminated RFI by installing a brute-force ac-line filter at the ac outlet and a telephone RFI filter at the device. The chapter on electrical interference shows an appropriate ac-line filter.

Direct Radiation Pickup

Both the master station and the cordless unit may be subject to direct radiation pickup (especially if the unit is close to a transmitting antenna other than its mate). If the unit contains an audio amplifier, that can also pick up RF. Direct radiation pickup is difficult to solve, but sometimes the master station can be relocated to another area of the home (as far away from the other transmitter as possible). If this does not solve the pickup problem, contact the manufacturer for a remedy.

The Cordless Handset

The cordless-telephone handset of 49-MHz units is not often a problem. Older units (on 1.8 MHz) may have more problems. Problematic older systems should be replaced with newer equipment.

Other

Cordless telephones can have other problems that might be confused with actual interference. When the batteries are low range is reduced, audio sounds distorted, and the phone is more susceptible to interference. A broken antenna reduces the range and produces "scratchy" audio. This also occurs when the handset is operated too far from the base station.—*John Norback, W6KFV*

Electronic telephones offer many convenience features like DTMF key pads, displays, memories, clocks, calendars, electronic ringers, audio amplifiers and so on. Such added features mean additional circuitry. Modern phones contain ample components to build several basic radio receivers! Ironically, a phone that costs more may be more difficult to provide with RFI immunity. Higher cost often indicates a greater number of special features using numerous diodes, transistors and integrated circuits. Figure 9.6 shows the ultimate result—each semiconductor junction in a telephone can function as a simple diode detector and turn RF into audio.

Most telephones for the US market are now built overseas. A few years ago, ARRL Lab Supervisor Ed Hare, W1RFI, purchased an inexpensive desk telephone that had only a manufacturer name and an FCC identification number. If there were an RFI problem with this telephone, there would be no easy way to contact the manufacturer for assistance. The FCC can supply manufacturer information from their files about registered ID numbers. If you supply the FCC with the ID number, they can give you information about the manufacturer or US distributor. Write to Federal Communications Commission, Manager Part 68 Rules, Room 235, 2000 M St, NW, Washington, DC 20554, tel 202-418-2343.

Modern telephone equipment can be quite sensitive to RF signals or fields. When a telephone that is not affected by RF energy is replaced with a different telephone that is susceptible to RF, telephone RFI can result where none existed. Or, previously insignificant interference levels may become significant.

Faulty Telephone Equipment or Installation

Telephone systems can continue to operate despite defects in the telephone wiring—or installation errors made recently or decades ago. (Telephones with internal component failures can also continue to operate.) If the twisted pair is not broken, the telephone will usually function normally, except for some noise. A telephone user may never notice deficiencies that increase interference susceptibility.

As a result, interference can be more severe at one location than another when both are exposed to the same RF field. For example, RFI might be worse at a neighbor's house than at a ham's—even though the antenna is closer to the ham! The neighbor could have a number of wiring and installation problems, and the phones still work, but still have telephone RFI when the ham transmits. Unfortunately, this sometimes leads to the erroneous conclusion that the amateur is at fault.

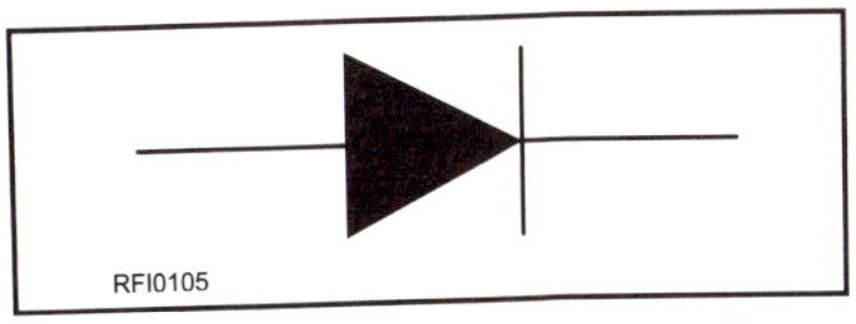

Figure 9.6—This is the equivalent schematic of the part of a telephone that is the root cause of interference problems. Every semiconductor junction inside a telephone can act as a diode detector, turning RF into audio. A single IC can have millions of semiconductors, each of which can act like a diode detector. It only takes one diode to cause interference!

Direct Pickup

Although most of the time it is the telephone wiring that picks up the RF energy, it is possible for components in the telephone system to demodulate the RF energy produced by a radio transmitter. Some telephones and accessories are so sensitive that they produce audible interference even when on-hook or disconnected from the telephone lines. If equipment demodulates RF (audio is heard from the earpiece or speaker) when hung up, it may be:

- *Defective.* The voice circuits are not disconnected from the line when on hook. Repair or replace the instrument.
- *Directly picking up RF* (see Figure 9.1). Little can be done to remedy this. A shielded enclosure and telephone RFI filters should help, but an enclosure may be impractical. Replace the instrument with one that does not detect RF fields directly.
- *Picking up RF from the ac line* (if the instrument is ac powered). An ac-line filter should help.

Some consumer devices are so susceptible they can't be filtered externally. If the manufacturer is unable to help, the best solution might be to give the telephone to someone who lives far away from your transmitter and purchase an RFI-resistant telephone to replace it!

Bad Hookswitch

A hookswitch is located inside of all telephones. If it is working properly, it disconnects the telephone from the line when the user hangs up the phone. When the handset is on-hook (hung up), switches are supposed to disconnect the voice circuits from *both* sides of the wire pair. This *completely* removes the telephone transmission network and handset from the system. By design, a telephone that is hung up leaves only the ringer across the line. If one of the cradle switches fails and the telephone remains connected to *one* side of the pair, watch out! (This is not common, but it can happen.) If the failure is such that the hookswitch disconnects only one of the two wires in the pair, this can unbalance the system by leaving many feet of wire connected to one side of the pair. This can cause RFI without disabling the telephone.

It's a coincidence, but Murphy's law is always working in Amateur Radio: Popular telephone cord lengths come very close to resonant antennas for ham bands between 10 and 20 meters! The extra wire will be an antenna, picking up a strong signal, in both the common and differential mode. Most telephones use a 12-foot coiled cord for the handset. If one of the switches fails to open, this leaves 24 feet of wire connected to one side of the line. This unbalances the system and leaves RF-sensitive components in the telephone connected to the line! The telephone may detect RF when it is hung up and put the rectified audio on the line. The audio then appears throughout the entire installation.

Further, the telephone could operate properly when the handset is lifted off the cradle (off-hook). Therefore, it is of no value to pick up the phone and listen for this source of interference while troubleshooting.

Dirty Contacts

Dirty hookswitches or dirty microphone contacts cause noise that is often interpreted as interference. The noise sounds like static or a hissing/frying sound. It is most bothersome in humid areas. The cure is to clean the contacts.

Magnetic Induction

Audio-frequency electromagnetic fields can couple into telephone lines or equipment by magnetic induction (see Figure 9.7). The most common magnetic-field sources are the house ac wiring or transformers inside electrical equipment. The best cure is to physically separate the phone line and field source.

What To Listen For

Anyone with a "trained ear" (anyone who can pull a weak DX signal out of the "mud" on an Amateur Radio receiver) can hear symptoms of straight-line induction in telephone circuits. Induced power signals, for example, are often heard when the amateur station is not even switched on. Such signals in the telephone system will sound like a hum or sometimes a "pop" accompanying current inrush to an appliance.

If the ac circuit feeding a transmitter induces current on the pair, load variations can modulate the induced signal. This can make it appear that the interference is a result of the transmitted signal.

Determine whether the interference is a result of the transmitted signal by connecting the transmitter to a well shielded dummy load and transmitting. If the interference remains, it's not associated with the presence of RF, and must be caused by something else. The cause could be either direct induction between the transmitter or its power supply and the telephone or its wiring, or, less likely, the transmitter could be conducting RF onto the power lines and it is getting into the telephone through that route. You can test for direct induction by moving the telephone; test for RF on the power lines by installing a brute-force power-line filter on the transmitter. (See the References chapter for sources of power-line filters.)

An Example—Telephones Near Amateur Equipment

Fully outfitted amateur stations often include a telephone. One operator experienced telephone interference for years—always assuming his transmitted signal was getting into the phones. Upon investigation, the problem turned out to be magnetic induction (affecting the whole telephone system) through one telephone on the operating desk!

The phone was physically close to a high-power amplifier. Hefty magnetic fields are generated by transformers in high-voltage power supplies, and some models of telephones are easy targets for induction! (Older telephones that use induction coils for electromechanical bell ringers are especially prone to pick up nearby fields.) The telephone was picking up the 60 Hz electromagnetic field from the high voltage transformer. The phone was modulated as the operator spoke, thus disguising the true nature of the problem. The simple solution to this sneaky problem is just to move the phone away from the transformer.

Don't Work On Telephones

Do not attempt to modify a telephone instrument internally or alter telephone wiring owned by the telephone company. Only telephone company personnel may service telephone-company owned equipment. Believe it or not, one must be federally licensed to work on telephone equipment. Your amateur license is *not* enough. This is especially true for your neighbors' equipment; working on their telephones could create liability issues adding to an already difficult situation.

FILTERS

Filters are the number one cure for telephone interference problems—if you can keep the RF energy away from the detector, you will have no interference. The easiest way to do this is to select and use the correct telephone filter. Filters come in quite a few "flavors," each designed to do a specific job. Some filter manufacturers sell filters designed for different frequency ranges. Others sell filters for single-line telephones, two-line telephones or for the handset cord. In some cases, a separate filter may be needed for any power connections to the telephone device, usually through one of those "wall cube" type power supplies.

Telephone RFI filters are made by several different manufacturers. The ARRL Laboratory Engineers have received reports from the field that the effectiveness of these filters varies from installation to installation, possibly dependent on the exact nature of resonances, impedances and system reactances. Many of these factors are hard to predict. Although most filters work well in routine RFI cases, be prepared to experiment with several different manufacturers' filters in difficult cases.

In most cases, you are better off to select and use a commercial telephone interference filter, especially where a neighbor is involved. If you supply a neighbor with a homebrew filter, this does strengthen the belief that *you* are responsible for curing the interference problem. By identifying a commercial filter for your neighbor, and involving him or her in the installation, you are emphasizing that you are locating, not providing, a solution for an interference problem that is not your doing. (You didn't design or build the susceptible telephone!) You are also shifting some of the burden of success to both your neighbor and the manufacturer of the filter.

Figure 9.8 shows some examples of commercially available telephone interference filters. These are made by a number of manufacturers. RadioShack sells a single-line filter optimized for the HF range.

Common-Mode RF Chokes

Most of the time, RF is picked up on telephone wiring in the common mode. Most commercially available telephone-interference filters are common-mode chokes. A reasonable common-mode choke could be made by wrapping about 10 turns of the telephone wiring onto a suitable ferrite core or rod. (Use #43 for upper HF and VHF, #73, #75, #77 or -J type ferrite material for the lower HF range. A good all-around material is #43.) An example of a home-brew ferrite common-mode choke is shown in Figure 9.9. This could actually form a reasonable solution for a desk-type telephone. Most people would not find it suitable for use with smaller telephones designed to be carried around the house. It would make a downright objectionable handset-cord filter!

If you wrap your own ferrite common-mode chokes, be sure to select the correct material. The wrong material may not work, leading you to believe that you should pursue a different cure. Figure 9.10 shows what will *not* work on most of the HF range.

Chokes made with ferrite materials are the most effective devices for reducing common-mode RF current on telephone wiring. Chokes accomplish this by presenting a high impedance to RF in series with the telephone wiring. In order for a choke to act as a filter, there must be a dissipative RF path available with lower impedance than the path through the choke and telephone equipment. (Often the dissipative path is the phone line.) If the RFI is not severe, simply wrap the telephone cord around a ferrite rod or toroid core (about 10-15 turns in a single layer). This forms a low-performance RF choke.

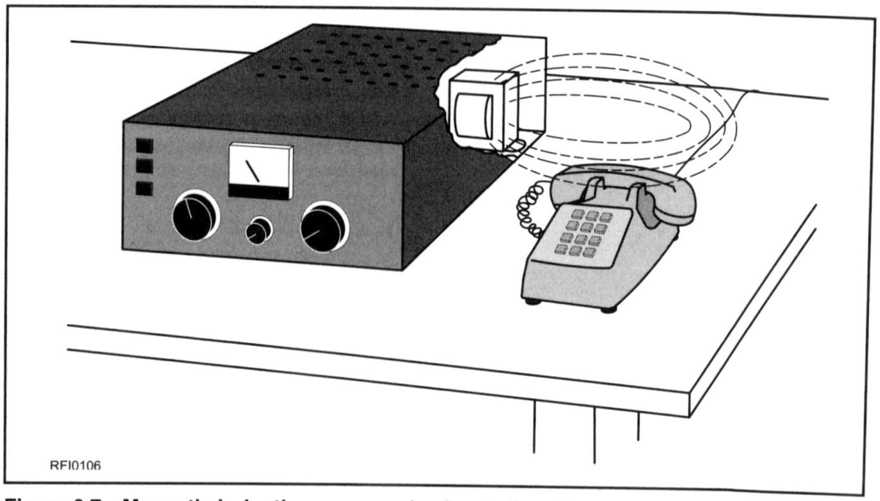

Figure 9.7—Magnetic induction can couple signals into the telephone line. Separate the phone and the magnetic-field source to eliminate the problem.

Bifilar Chokes

Bifilar chokes are preferable to separate chokes for each side of the pair because bifilar chokes maintain phone-circuit balance. Core permeability variations are of no consequence in bifilar chokes because both conductors are wrapped around the same core. Hence, circuit balance is ensured.

A bifilar choke has no adverse effect on telephone operation. This is an important consideration in moderate and severe cases of RFI, where several chokes may be required (spaced several feet apart) to eliminate resonances in the telephone wiring. Figure 9.11 shows a broadband bifilar choke with inductive characteristics appropriate for phone-line filtering in the 3 to 30-MHz range.

Differential-Mode Telephone Filters

Although *most* telephone interference is caused by common-mode pickup, even twisted-pair wiring can pick up *some* differential-mode RF signal. If there is any imbalance in the phone system (likely at upper HF and VHF), differential-mode pickup can be worsened. Although most telephones are somewhat immune to differential-mode pickup, and most common-mode filters do offer some rejection of differential-mode RF signals, in some difficult cases, it may be helpful to use a differential-mode filter, often in conjunction with a common-mode filter.

If a differential-mode filter *is* required, one like the one shown in Figure 9.12 will generally outperform a capacitor. It will offer more rejection of the differential-mode signal than a simple capacitor. The two inductors will also offer some degree of rejection of any common-mode signal present on the line, although it is not likely that the balance of the line will be well preserved. It may be possible to use this type of filter on a telephone connected to a line with a conventional high-speed modem, although this would have to be determined experimentally. In most cases, however, if there is *not* any differential-mode RF pickup, this type of filter will not work as well as a common-mode telephone filter.

Capacitors

The simplest differential-mode filter is the old "capacitor across the line." This has a few significant advantages—simplicity and cost. A 0.001 µF or 0.01 µF 1 kV ceramic capacitor placed across the telephone line, usually at a phone jack, can cost less than a dime, and can solve some interference problems nicely. AC ring voltage can exceed 100 volts, so use ac-rated capacitors, such as those sold by Digi-Key.

This simple approach won't always work, however, and it can create some unexpected and unwanted complications. At 2 kHz, a single 0.01 µF capacitor has an impedance of about 8000 ohms. This will usually not affect a *voice* telephone. If you had 10 of these installed on a line—one across each telephone jack in a large house, the total impedance of the resulting 0.1 µF capacitor would be 800 ohms, apt to have a significant effect in a telephone system with a nominal audio impedance of 600 ohms!

The effect of capacitance across the line on a high speed modem could be even worse. It is not likely that a 56 kbaud modem would function at all with too much capacitance across the line; the phase variations vs frequency caused by the capacitance will wreak havoc with the performance of high speed modems. The telephone industry is now deploying very high speed DSL modems that can manage multi *mega*bit/second data rates. These just will *not* work with even low-value capacitors across the phone lines at various places.

It should be noted, too, that a capacitor across the line will not have *any* effect on a common-mode signal! In the likely event that both common-mode and differential-mode signals are present, you will have to consider other methods, usually multiple filters of each type.

WHAT CAUSES TELEPHONE RFI?

Now that you have learned about phone systems and devices, you need to understand just how they work together to cause RFI problems. The following are necessary components to a telephone RFI problem:

- *A source of RF energy.* This is usually an authorized radio transmitter, operating legally under FCC rules.
- *Pickup of that RF energy by conductors that*

Figure 9.9—This shows a home-brew ferrite toroidal common-mode choke suitable for use as a telephone RFI common-mode filter. Choose the correct ferrite material for the frequency range being filtered.

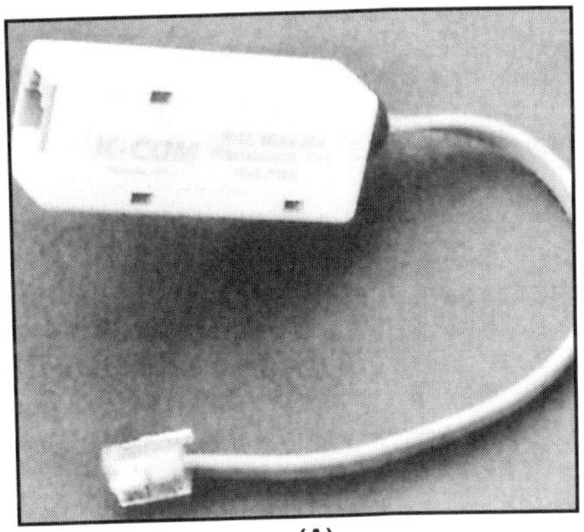

(A)

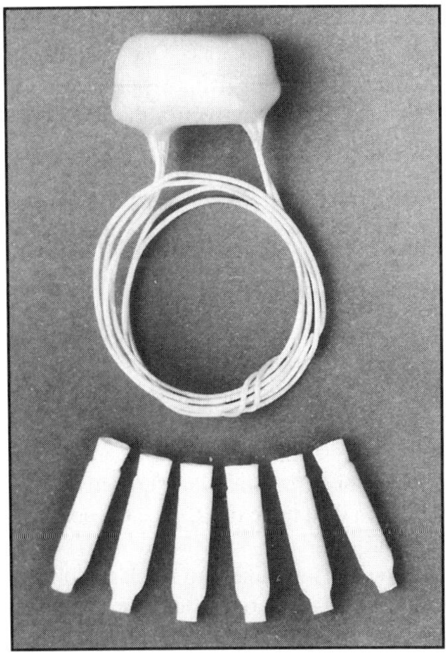

(B)

Figure 9.8—Typical telephone RFI filters. The filter at A has modular connectors for installation at a telephone. B shows a filter used behind wall jacks, at the service entrance and other places where modular connectors are inappropriate.

Figure 9.10—This "split-bead" type ferrite just doesn't have enough inductance to be a good common-mode choke, especially for HF.

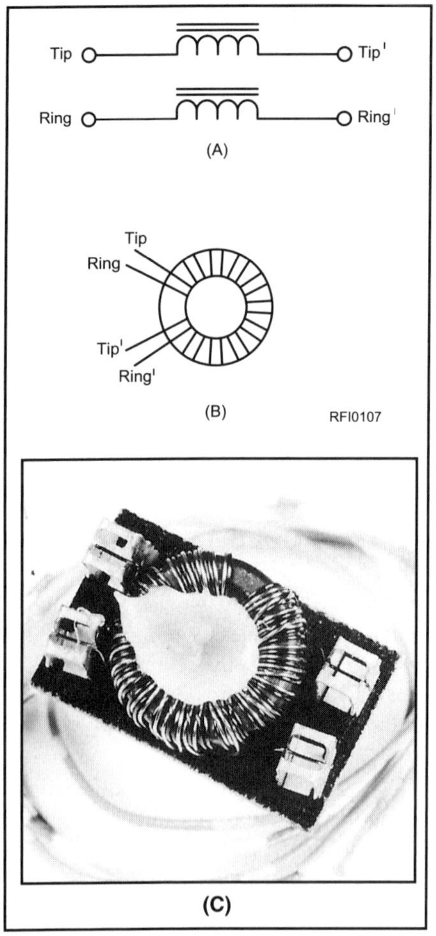

Figure 9.11—Simple RF chokes like this can help stop RF flow on telephone wires. Schematically (A), the choke consists of two 1.7-mH windings (25 turns of no. 30 enameled wire wound over 80% of the circumference of a ½-inch OD, mix-75 ferrite core). This choke is effective against RFI from 3 to 30 MHz.

B shows how the two windings are wound together (*bifilar*) on a single ferrite core. To make construction easier, twist the wires together (five to ten twists per inch) and wind them as a single wire. Once the choke is wound, use an ohmmeter to check for continuity through each winding and label the ends.

C shows a commercial bifilar choke mounted on a board for potting. A dab of silicone adhesive (such as GE RTV) holds the completed choke to its carrier board.

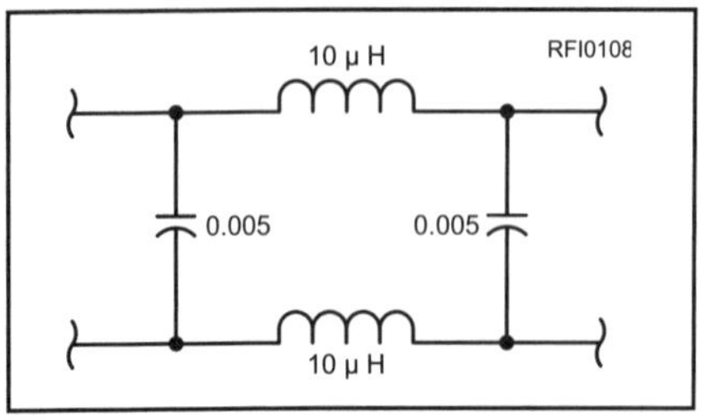

Figure 9.12—The inductors and capacitors effectively filter differential-mode RF signals that can be picked up on telephone lines.

are part of the telephone system. This is usually the telephone wiring, but RF can also be picked up directly by the circuitry and wiring inside a telephone device.

- *Conduction of the RF energy* into the telephone device or other interference-causing mechanism.
- *Detection.* Some nonlinearity (diode) in the telephone system must AM detect the RF energy and turn it into an audio signal. This usually occurs in one or more telephone devices, but rarely can occur in defects in the telephone wiring or telephone company equipment.

The Diode

The detector is the key element in most cases of telephone RFI! This is shown in Figure 9.6. Modern telephones can have millions of transistors (located inside the various ICs inside the telephone), any of which can be the detecting diode. Even in this area, once again the deregulation of the telephone system has had an impact. Many customers are now responsible for their own wiring and many of them have managed to hook up some of the wiring wrong. With two wires on each pair, if the pair gets connected wrong, with the tip wire connected to the ring wire of a splice, the polarity of the wiring would be incorrect. To overcome this, telephone manufacturers have designed "steering" circuitry that ensures that telephones work even if they are hooked up to the line backwards. The most common steering circuitry is—you guessed it, a bridge rectifier! Other diodes are found inside the telephone, often in some of the audio amplifiers used to amplify the sensitive microphone or to amplify the voice or signal on the line.

Once the RF energy is picked up on the telephone wiring, it is usually conducted efficiently to the telephone. Once it hits the detector, the result is RFI. In some cases, a telephone device can rectify the RF signal and put it back on the line as audio, making it appear as if every device on the line is susceptible. In any case, once the RF energy hits the diode, the result is audio rectification.

Audio Rectification

The most common form of telephone RFI, audio rectification occurs when some component attached to the phone line rectifies (detects) RF and places the detected audio on the phone line. Filters are usually quite effective at eliminating interference caused by RF on the phone lines, once the detector is found.

RF signals are likely to reach the phone through the telephone wiring because the telephone wires form a large radio antenna (see Figure 9.13). Induced RF signals usually travel along telephone wires in the common-mode (with all the pairs in the cable acting as if they were one wire), with the return path through earth ground. The radio signal is traveling on the telephone wiring but it is not part of the balanced circuit. However, some interfering signals can be differential-mode, so an effective filter should attenuate both modes.

The detection process that results in telephone RFI can occur anywhere in the telephone system. A poor or corroded connection can actually convert an RF signal to audio with no help from a telephone instrument. If detection occurs outside the telephone (on the telephone-company lines or in a line amplifier, for example) "foreign" differential-mode audio signals are heard along with the desired audio. It is impossible to filter such foreign signals because they are at the same frequency as the desired signals.

"Talking" Telephones

One form of audio rectification can manifest itself in a strange way. RFI may be heard on one or more telephones that are hung up! A phone should not produce audio when on hook because the handset should be completely disconnected from the lines. Such phones are either subject to direct pickup or still connected to the lines.

Simplify the system to one phone, and then reconnect one instrument at a time. Listen for interference as each instrument is added, and pay particular attention to minor variations in signal levels. Telephones with ac-line connections are prime suspects, followed by those phones with the most electronics. Some electronic telephones keep talking even when disconnected from the telephone line!

Eliminate the Diode or Use a Filter?

In most cases, the detector is located *inside* one or more of the telephone devices. In this case, the only practical cure is to use a filter to prevent the RF energy from reaching the diode. If you find that corroded wiring or a defective lightning arrestor is acting as a

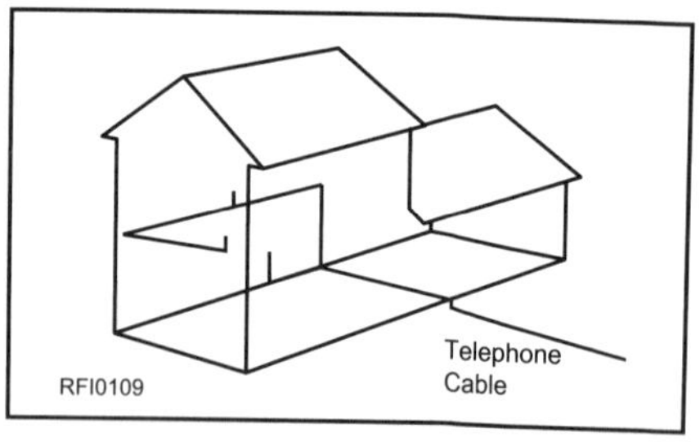

Figure 9.13—The wiring of a typical home installation forms a large random-length antenna.

diode, it is often easy to cure the problem by correcting the defect, although filtering would be an option.

Installing Filters

The most convenient and effective filter location is usually the modular female connector at the telephone or phone jack. Some telephones have most of the circuitry (including the keypad) built into the handset. Locate a filter close to the handset with phones of this type.

If interference is partially reduced by installing a filter at the telephone, additional filters placed in series may help. If interference is only slightly reduced by installing a filter at the telephone, this is probably because the telephone (and filter) are at an RF high-impedance point in the telephone-wiring "antenna." Additional filters will be required to reduce the interference, and they should be located several feet (ideally ¼ λ) away. Additional filters placed at an RF low-impedance point are more effective. Find effective filter locations by trial and error.

Don't Give Up!

When the first filter is not effective, some people conclude that the problem can't be solved and give up—*Don't!* Filters that are properly designed and built do offer RF attenuation, and enough attenuation will resolve an induced-RF problem. There must be some other reason that the filters do not reduce interference.

There are two likely explanations. The filter might be (by chance) at an RF high-impedance point in the circuit. The filter impedance is in series with the system impedance at the point of installation. For a filter to attenuate unwanted signals, the filter must have an impedance much greater than its termination impedance. This is not the case at a high-impedance point in the circuit.

There is another possible explanation. The RF-current level at the filter could be so great that the filter core is saturated. When saturation occurs, any current increase diminishes the inductance. For example, a filter that works quite well at low power levels may not work when the RF level at the filter exceeds a few watts. If you need to filter a telephone line in a high-RF-power environment, ask the filter manufacturer for information about the power-handling capability of the filter.

It is possible to build filters for high-current applications, but the required core size can be prohibitive. Another approach is to reduce the current-handling demand placed on the filters by reducing current level in the telephone wiring. As you recall, telephone wiring acts as a large antenna. Our goal is to find a way to make this "antenna" absorb as little RF energy as possible. There is a way to do this.

Eliminating Resonance in the Telephone Wiring

The task is to make the phone line less responsive to RF. Let's take an approach often used with conductive guy wires. Tower-guy resonance at amateur frequencies is eliminated by breaking the wires into electrically short, nonresonant lengths. A telephone system must be continuous at AF, so we'll discourage RF-current flow in the system with high-impedance RF filters.

A Shotgun Approach

In a parallel-wired telephone system, a "shotgun" approach places filters at the beginning and end of each cable run (the fast way). Another approach places a filter at each jack in a loop-series system. (Pay attention to the wiring at the jack to make sure your filter is installed in the series loop, not just in the wires terminating at the jack.) This approach solves nearly all induced-RF problems

Telephone Modem/Fax Installations

Telephone modem and fax installations are becoming commonplace in the home. (Modern fax machines exchange digital data, so their RFI problems can be treated just as those of modems.) Modems are most often associated with computers. RFI to computers is covered in another chapter of this book. Here we will deal with modem RFI only as a telephone accessory. Modem installations fall into two categories: internal and external.

Internal modems are located inside the computer cabinet. They draw power from the computer supply and their data is placed directly on the computer bus. So long as the modem was properly designed, it should suffer no RFI from its computer. An internal modem is connected to the outside world only through the phone line, so that is the only place where RFI can enter.

External modems are located outside the computer cabinet. They are usually powered from the ac line. The phone line carries audio to the modem, and a serial-communications cable carries data from the modem to the computer. Thus, most external modems can receive interference over three paths.

"Pocket" modems are a special class of external modems. They are enclosed in a case that is nearly as small as the data connector (small enough to carry in a pocket). Although external to the computer, pocket modems are battery powered, and they connect directly to the computer serial port with no cable. From an RFI viewpoint, they are equivalent to an internal modem.

MODEM RFI CURES

The most common source of interference to modems is through the telephone wiring. The unwanted RF is almost always a common-mode signal. To solve this problem, install a telephone RFI filter where the telephone line connects to the modular jack on the modem. (Modems sometimes use two telephone lines to send and receive data simultaneously. In such cases, filter both lines.) This normally solves the interference problem. If it does not help, contact the manufacturer.

The Power Supply

Many external modems operate from the ac line, and common-mode RF signals can enter via that route. The power supplies contain rectifiers, which are possible sources of audio-rectification interference.

The power-supply lead is often about 7 feet long, and it may be resonant on 10 meters. If there is any excess length, wrap it in a small loop, which is less likely to act as a good antenna. If the trouble persists, install a common-mode choke where the power cord connects to the machine. Wind 10 to 20 turns of the power-lead wire on a (no. 43 material) ferrite toroid or rod. This should solve the problem. Some people have eliminated RFI by installing an ac-line filter at the ac outlet and a telephone RFI filter at the device. The chapter on electrical interference shows an appropriate ac-line filter. If interference persists, consider switching to an internal modem.

The Serial Communications Cable

RFI can enter external modems through the data-communication cable. Since the cable carries data that could be affected by filters, a shield is the only option. Most quality cables are shielded. If the cable is not already shielded, replace it with a shielded cable.—*John Norback, W6KFV, and Bob Schetgen, KU7G (SK)* quickly, but it is costly: chances are that not all of the filters are necessary.

A Systematic Approach

A systematic approach identifies and treats specific RF-entry points; it is relatively slow, but more precise. This approach is better because it draws attention to sections of wiring (and/or connected telephones) that contribute to, or cause, interference.

If there is a telephone RFI filter at the telephone or telephone jack, leave it for now. Install the first additional filter at the telephone service entrance.

Check the interference level on all bands. If the interference is gone, congratulations! If not, don't give up! Even if a service-entry filter doesn't reduce or cure the RFI, that filter narrows the search by eliminating RF (but not AF) signals that might enter via the service drop. Continue by grounding all other cables (except the one already filtered) at the service entrance.

The filters already installed should solve all but the most severe telephone RFI cases. If interference remains, physically break any long cable runs at one or more locations and insert additional filters. (When cutting a long cable to insert a filter, remember to maintain the dc continuity of all cable wires through the filter.)

By this time, there are several high-impedance filters between the RF signal and the telephone. Even severe interference should be eliminated. If not, suspect direct radiation pickup or audio rectification.

When interference is removed from the single cable, it's time to reconnect, one at a time, the cables removed earlier at the service entrance. Continue filter installation as described above, listening for interference on the reference phone and other phones as they become active. Work your way through to the final cable and the job's done!

Automatic Telephone Answering Machines

Answering machines come in many styles and often include many features. Some are standalone devices that are used with an external telephone; others have a telephone instrument built into them. The illustration is a block diagram of a typical installation.

ANSWERING MACHINE CURES

The Power Supply

Answering machines operate from the ac line, and common-mode RF signals can enter via that route. The power supplies contain rectifiers, which are possible sources of audio-rectification interference.

The power-supply lead is often about 7 ft long, and it may be resonant on 10 meters. If there is any excess length, wrap it in a small loop, which is less likely to act as a good antenna. If the trouble persists, install a common-mode choke where the power cord connects to the machine. Wind 10 to 20 turns of the power-lead wire on a (no. 43 material) ferrite toroid or rod. This should solve the problem. Some people have eliminated RFI by installing an ac-line filter at the ac outlet and a telephone RFI filter at the device. The chapter on electrical interference shows an appropriate ac-line filter.

Direct Radiation Pickup

Radiation pickup may affect the answering machine or an attached telephone. A shielded enclosure should help, but an enclosure is inconvenient when the machine includes an integral phone. Shielded enclosures are often unacceptable for convenience and appearance reasons.

Other

RF signals can be picked up by the line that connects the telephone to the answering machine (watch for resonance). Install a telephone RFI filter in the line. It may be necessary to install one near the telephone and another near the answering machine.

Combination telephone/answering machines are very difficult to cure with external filters. Most are impossible. The best way to attack RFI problems with a combo unit is to replace the unit with a separate telephone and answering machine. Then filter each one independently. If the RFI problem involves a telephone and a digital answering machine as one integral unit, the approach for RFI resolution is to consider the unit to be a telephone.—*John Norback, W6KFV*

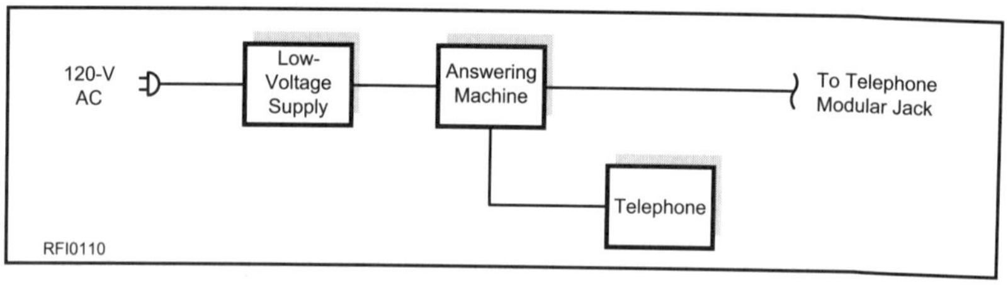

Locate the Detector

It can be difficult to discern the difference between externally rectified signals and those generated in a telephone in the building. In both cases, the audio varies with the radio transmitter modulation. The major difference is that externally rectified audio is present in the telephone circuit even when there are no telephones or accessories connected to the line. Keep in mind that any telephone or accessory can couple detected audio onto the telephone lines, resulting in interference to all other devices connected to that line. External audio-rectification interference can still be heard on a completely RF-proof telephone!

If audio rectification is present, the detector must be located to effect a cure. The simplification procedure should indicate defective telephones and accessories.

If no instrument or accessory is indicated by the simplification procedure, visually check the physical integrity of as much of the system as you can. Start at the lightning protector, move to the service entry and continue to the connector block. Then check all accessible phone-cable runs and the conditions inside each phone jack.

Look carefully for physically or electrically poor connections, corrosion, moisture, wire-insulation breakdown, staples driven through cables, stretched or otherwise damaged cables. You may also find wire other than telephone wire. Unqualified personnel may have installed speaker or thermostat wire, or other not-for-telephone conductors. Improper wire will pick up more RF energy than a properly balanced twisted pair. Replace such incorrect wiring if possible.

Devices that are not FCC approved should not be connected to the telephone system wiring. They may cause interference by unbalancing the line or acting as an RF detector. If practical, eliminate the defect that causes the problem; if not, install a telephone RFI filter to prevent RF from reaching the detector.

If the telephone installation uses parallel wiring, it might help to disconnect cables to unneeded telephone jacks. Disconnect unneeded pairs from the service entry and connect them to the telephone ground. This practice reduces the size of the telephone-wire antenna and usually the amount of

F picked up (especially if one or more of he unused wires is resonant on the transmit requency). This may result in an immediate olution to the problem.

STEP BY STEP

Now that all of the bases have been covered, it is time to pull it all together. Here is a ummary of how to solve telephone interference, step by step. This summary draws on he material in the First Steps, Troubleshooting and Telephone RFI chapters.

If you are involved with the interference problem, organize yourself. Obtain local help, if possible, possibly with you and your neighbor eliciting the help of trusted, diplomatic friends. Obtain an RFI-proof telephone and an assortment of ferrites and filters.

Understanding

First, if you have an involved neighbor (or family member), ensure that your neighbor understands the issues. When you discuss it with your neighbor, be diplomatic. RFI can be a stressful situation and tempers *can* flare. First, help your neighbor understand the issues behind telephone interference. The FCC material will help explain things authoritatively.

Keep a Log

You and your neighbor should keep written records. Your station log will show when you are on the air, and on which bands and modes. If your neighbor writes down information every time interference occurs (dates, times, severity, and so on) you can quickly determine if you are involved in the interference problem, and what bands, modes, power and antenna combinations cause the problem.

"It Ain't Me!"

After discussing things calmly with your neighbor, one of the first troubleshooting steps is to determine if the interference is really related to your transmissions. Conduct on-the-air tests to determine if the interference is coming from your station and if so, what bands, modes, antenna and power levels are causing the problems.

Sometimes telephone RFI problems are related to power lines or crosstalk within the telephone-company circuitry. Most of us have heard others talking faintly in the background on the phone. If the RFI does *not* follow transmission patterns, and you can demonstrate that to the complainant, relax. At this point, the matter is the telephone company's problem. [This is an example of my favorite interference problem—"It ain't me!"—*Ed.*]

When interference begins and ends with RF transmissions, RF is somehow involved in the problem. This does not mean that the transmitter is *causing* the problem. Since telephone equipment is not intended to receive radio signals, the affected equipment is not functioning properly. Continue troubleshooting to discover the exact telephone or device that causes the problem and apply a remedy there.

Telephone RFI Troubleshooting

If you do determine that there is a real

Home Security and Medical Alert Systems

Automatic Dialers

Home security systems are often connected to the telephone network by means of an automatic dialer, which can be a two-way source of RFI. The automatic dialer and alarm system master station are normally located in a hidden, secure area of the home. Alarm cabinets are usually locked. Security-system details are beyond the scope of this book, so we shall consider them as single-port systems. The illustration shows a typical alarm system connected to an automatic dialer.

Alarm/Telephone RFI Cures

The figure shows great potential for two-way RFI. These systems can *cause* TVI and RFI to broadcast radios and telephones. The main sources of TVI/RFI are the rectifiers and (in the case of the automatic dialer) a polarity-guard circuit. It is best to solve the telephone RFI problem first because of effects from the polarity-guard circuitry in the automatic dialer.

The telephone circuit to the automatic dialer is normally hard wired to terminals inside the unit (it does not use modular jacks), so modular telephone RFI filters cannot be used. Fortunately, appropriate filters are available for use with wall telephones and installation behind telephone jacks. Since the RFI-causing electronic equipment is located inside the automatic dialer, install filters as close to the dialer as possible.

Next, filter the two-wire circuit between the alarm master station and the automatic dialer. Use a ferrite choke (43 material), and install it as close to the dialer as possible. These two steps solve most alarm/telephone RFI problems.

If RFI remains, apply common-mode chokes or ac-line filters at the line cord. Where separate low-voltage supplies are used, common-mode chokes may be needed on the low-voltage leads as well (see the sidebars about cordless phones or answering machines for details of low-voltage supply filtering). Since alarms are constructed to operate during ac power failures, there may be complex and tenuous battery charging and switching circuits. The chief RFI threat in such circuits comes from audio rectification in switching or rectifying diodes. Read the discussion of audio rectification in this chapter and the chapter about external causes of RFI for information about locating and treating audio rectification.

Alarm-TVI/BCI Cures

To solve alarm-TVI/BCI problems, install a ferrite choke (43 material) at each alarm sensor (as close to the sensor as possible). This also prevents common-mode RF signals from falsely triggering the various alarm sensors. Be sure to treat *all* alarm sensors in the home.

If the lead from the external audible alarm to the alarm master station suffers from common-mode RFI, install a ferrite choke (43 material) at the master-station end of that line. This prevents RFI from reaching the alarm master station by that route.—*John Norback, W6KFV*

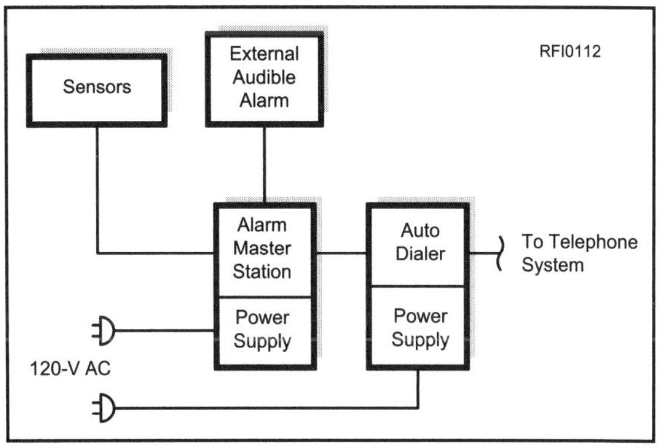

RFI problem, you need to narrow the focus a bit. Naturally, the Troubleshooting chapter covers this complex subject in some detail. Take the time to review that chapter before you continue. Some of the specifics that apply to telephone RFI are worth repeating, however. Many generalized troubleshooting principles are not repeated here.

Simplify the Problem

Most cases of interference stem from more than one cause. If you start with a systematic approach to troubleshooting —debugging and fixing the problems one at a time—you can usually resolve the interference quickly.

Rule number one: as outlined in the Troubleshooting chapter, *simplify the problem!* If your home telephone system consists of eight two-line outlets with a telephone in each room, two answering machines, a modem and a fax machine, it may take quite a while to get it all straightened out. Simplifying the problem saves lots of time and headaches!

Start by disconnecting (unplugging) all telephones and telephone devices. Don't forget to consider hidden, unexpected connections to the telephone line. If your telephone line is hard-wired to your automated alarm system, for example, you really haven't simplified the problem completely—there's still something hooked up to the line!

If you have an RFI-proof telephone, this will be an excellent troubleshooting tool! Some manufacturers have made guaranteed RFI proof phones, but many of the older rotary-dial telephones are almost as immune. If you have a telephone known to work well in your own home, this can also serve the purpose, although you may want to add some telephone filters to it as a precaution.

Once all of the telephone devices have been removed from the system, you can proceed to see if the telephone wiring or telephone company equipment is directly responsible. In many cases, the telephone company has installed a troubleshooting telephone jack right near the service entrance. It is sometimes possible to disconnect the entire house wiring easily at this point; if so, doing so will simplify things even further. If there is no jack near the entrance panel, use a convenient jack inside the residence. Plug your troubleshooting telephone into this jack. If it is a known-good phone and you *do* hear interference, the audio detection is occurring in the wiring or telephone company equipment. By plugging your RFI-resistant telephone in at the service entrance and finding no problems, you've proven that the RFI-resistant telephone and the phone system are clean. When the problem reappears after you reconnect the house wiring, the house wiring itself is the prime suspect.

If you don't hear interference—the likely case—you have cut the size of your troubleshooting problem by quite a bit. You know that the wiring and telephone company equipment are okay. If you *do* hear RFI using the test telephone, the RFI-resistant telephone is probably responding to the audio that's present after the "diode" as a result of RF rectification. Once the RF has been detected and turned into audio, it cannot be filtered out because the interference is at the same frequency as the desired audio signal. This principle applies whether rectification occurs in the telephone company system, your house wiring or a defective telephone or answering machine in your home.

One at a Time

Start plugging in telephone devices, one at a time. Verify whether each has RFI problems. If so, cure the RFI bugs (usually with the proper application of filters), then begin adding other devices one at a time, curing each RFI problem as you go. Filters are the usual cure. Remember that in some cases, it is necessary to use multiple filters, either to cover multiple RF entry points or to cover multiple frequency ranges. For example, at ARRL HQ, many of the phones required a line filter, a handset-cord filter and a common-mode choke on the dc power supply lead powering the digital electronics in the telephone. With any luck you'll get all the RFI bugs out of all of the telephones. If not, at least you'll identify the specific equipment that's not immune. See Figure 9.14.

If there are different models of phones in the house, make comparisons to determine which are the least sensitive to RF energy on the telephone line. Modular connections make this easy. Test each of the phones at the same phone jack. By the process of elimination, you may identify a single susceptible telephone that is more trouble than it is worth—a result of either inferior design or component malfunction.

If RFI is present with each phone or accessory, suspect a wiring fault or external device. Once the simple system is RFI free, add devices back one at a time, eliminating problems as they appear. At each step, listen for improvements on all telephones in the building. If, for instance, RFI is eliminated by reducing the RF field strength, no further work or expense is necessary. In some cases, one bad telephone (or other device such as an answering machine) may feed detected audio to all other telephones in the house.

There are a great number of variables from one site to another. You may need to spend time troubleshooting the fundamental cause of the interference problem. The good news is that solutions are possible and seldom technically complex. You need not be an electrical engineer or professional telephone troubleshooter to understand telephone RFI and apply the appropriate cures. With good information (found in this book) and a healthy dose of patience (you must provide that yourself!), you can solve an interference problem.

Even Though It's Not Your Fault. . .

In no case is telephone RFI the fault or direct responsibility of the amateur. Although it is not required by FCC regulations, however, sometimes expedience does rule the day and amateurs may need to make some operating changes at their stations to reduce telephone RFI. (The ham who lives in an apartment building may be told to fix the problem or stop operating.) A station licensee may *elect* to make changes to improve the situation, but that decision does not establish cause or responsibility.

One obvious solution is to reduce the transmitter's output power. Many a QRP operator got started with QRP to eliminate problems caused by apartment living. Different operating modes also have different RFI potential, especially in cases involving audio rectification. SSB might cause objectionable interference, while CW results in only a few minor pops and thumps. RTTY might create almost no telephone interference at all!

One of the most effective cures can also improve station effectiveness! Moving that antenna high and in the clear will maximize your signal strength at that coveted DX station while minimizing your signal at neighboring electronic devices. Not a bad trade-

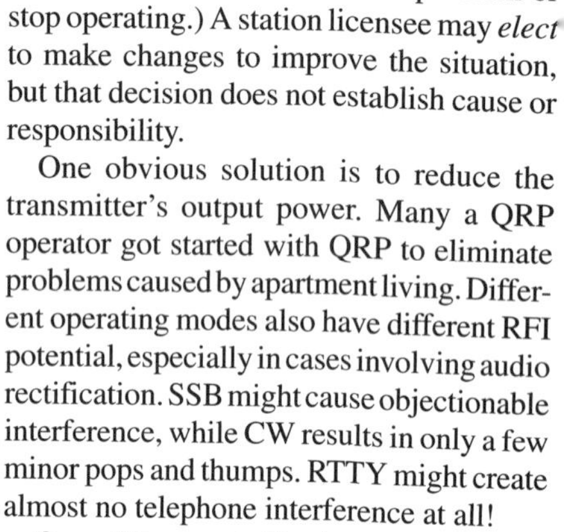

Figure 9.14—To cure interference to the telephones at ARRL Headquarters, it was necessary to install a phone filter on the telephone line, a separate filter on the handset cord and a home-brew common-mode choke on the dc power supply lead. This phone now works at W1AW when the bulletins are running on all seven bands!

off! If moving an antenna a few feet can solve a neighborhood (or family) dispute, that may be the best solution.

We amateurs can create our own devil, so to speak, with antennas near the house, on the house and sometimes in the house. This results in strong RF fields at the telephone installation. It is much better to send the signal skyward (to the other end of the QSO) than to a neighboring telephone!

Even a slight increase in the distance between antenna and house could make a big difference. It is not practical to move a tower to solve an RFI problem, but moving one leg of an inverted-V is an easy way out. This is not an ideal solution because it does not cure the fundamental problem, which is a defective telephone or installation. It is mentioned here so a ham can consider all alternatives.

Telephone wiring picks up RF energy from nearby transmitters. In some cases, telephone wiring acts as part of the transmitting antenna! Telephone wiring that is less than about 0.3 λ from an antenna and nearly parallel to it may act as a parasitic element in the antenna system. The telephone wiring would then contain a *lot* of RF energy.

Cable runs in the upper levels of a home are particularly susceptible to RF pickup. Place telephone wiring near ground level (in the basement or crawl space) to reduce RF on the lines.

Transmission-line radiation can contribute to an RFI problem. Do not place transmission lines close to telephone wiring. Feed-line radiation is more of a problem with dipole antennas when the feed line does not leave the feed point at 90° to the antenna.

If the feed line must pass close to telephone wiring, all of these practices may help reduce feed-line radiation:

- Use high-quality coaxial feed line
- Use a balanced antenna with a balun at the feed point
- Ground the feed-line shield where it enters the house
- Effectively ground all station equipment.

Summary

Past troubles in the treatment of modern telephone RFI resulted from a lack of understanding. Solutions designed for carbon-mike rotary-dial telephones (such as bypass capacitors) do not usually apply to modern telephones, which are more sensitive to RF signals.

Modern phones exhibit RFI for three reasons:

- Direct radiation pickup
- Audio rectification
- Magnetic induction.

Direct radiation pickup is either a design defect or malfunction of the telephone instrument, which should be taken to the manufacturer. Audio rectification results from an AM detection of the RF signal picked up on the telephone line; it can be eliminated by correcting the defect or proper application of telephone RFI filters. Magnetic induction is a problem of proximity that can be eliminated by moving the magnetic source away from the phone line.

Most of the time, telephone interference can be cured. Use your best diplomacy skills, diagnose the root cause, select and use the right filters, and you and neighboring telephones can coexist in peace.

Work Safely!

The risk of electrical shock from working with telephone wiring is minimal, but use good judgment when coming in contact with any active circuit. In the US, 24 to 48-V dc appears across the telephone pair at all times, and incoming ring signals are 20-Hz ac at 90 V or more. The ringer can give you quite a start!

Telephone wiring is often hidden from view as it travels inside walls, ceilings and so on. Although chances are slight, hidden wires could come in contact with live ac wiring. As a precaution, use a voltmeter to verify that the talk-circuit voltage is not abnormally high before working with a circuit. Protect yourself from any incoming rings by taking all phones off-hook.

Don't work on telephone wiring when lightning storms are in your area. All of these types of electrical shocks can be especially dangerous to those using pacemakers.

Chapter 10

Stereos and Other Audio Equipment

This chapter covers a lot of territory, ranging from high-end audio equipment to the low-fidelity amplifiers used in a house intercom. In all of these cases, some device in the audio equipment is acting like a diode and rectifying RF signals. The result is interference. The cure? Prevent RF energy from getting to the susceptible component in the first place.

By James Lee, W6VAT
PO Box 357
Cupertino, CA 95015-0357

OVERVIEW

If you have an audio interference problem, it is likely that you have skipped immediately to this chapter. The cures for audio interference are usually easy to understand and implement, but there are often other issues that need to be addressed first. Technical issues are really the *second* half of an interference problem. First you must deal with personal and political matters. After all, knowing all of the technical solutions will do no good if your neighbor won't let you into the house to try them!

The First Steps chapter helps you deal with your neighbor, the EMC Fundamentals chapter explains technical principles and terms, and the Troubleshooting chapter explains how to track down interference. This chapter covers specific information about audio devices, followed by the steps to take to diagnose and cure audio interference.

Patience and understanding are important tools for solving audio/stereo problems. Use understanding to keep on good terms with an impatient equipment owner. Strife causes more problems than it solves, so keep your cool. Try to see the scene from the other viewpoint, and do your best to build confidence in the minds of others involved in the problem.

RESPONSIBILITY

Each Amateur Radio operator is responsible for the proper operation of his or her station. If spurious or harmonic emissions generated at your station cause interference to another radio service (such as broadcast AM or FM radio), the FCC requires you to reduce those emissions as much as possible. If your station transmitter is clean by FCC standards, however, the problem lies solely with the affected product. In the case of audio equipment, most cases of interference are *not* caused by the improper operation of the radio transmitter!

The Role of the FCC

The FCC regulates radio transmitters. The Federal Communications Commission (**www.fcc.gov**) doesn't have any specific rules that govern interference to audio devices. This is because audio devices are *not* radio receivers. The FCC doesn't require that audio devices have any interference protection. Radio transmitters do not transmit audio signals so there are no applicable rules that govern radio transmitters interfering with audio devices.

This means that when audio RFI occurs, there is nothing for the FCC to enforce! The FCC rightfully believes that most interference is a matter to be resolved by the consumer equipment manufacturers as a customer service issue.

Nonetheless, the FCC realizes it will continue to receive inquiries from the public about audio RFI. They offer help to transmitter operators and consumers, primarily as an authoritative information source on self-help cures. The former FCC Compliance and Information Bureau (CIB) had created *Interference to Home Electronic Entertainment Equipment Handbook—May 1995 edition*. This was released as *Bulletin CIB-2*. No longer available from the FCC, a copy can be downloaded from the ARRL Web site at **www.arrl.org/fcc/tvibook.html.** This bulletin has been reproduced in the appendices of this book. A page from *CIB-2* is shown in Figure 10.1.

Over the past several years, the FCC has made tremendous strides forward in collecting and distributing unbiased, factual information about all types of interference problems, especially interference to non-radio devices such as audio devices. The FCC clearly explains that the burden of resolution of audio interference problems is *not* on the transmitter operator, but lies with the manufacturer of the affected equipment. Hams knew this all along, but in the past, the FCC offered scant material on the topic. Thanks to a major effort a few years ago by a national team of FCC staff, the information the FCC now provides on interference is useful and complete. The ARRL was pleased to serve as a consultant to the FCC as this information was being developed.

"Oh no, here it goes again!"

The Manufacturer and Dealer

If there is an interference problem with audio equipment, the FCC suggests that the equipment owner should have the equipment repaired by a qualified technician. The dealer who sold the unit may help, *if he understands the problem.* (After all, he wants to keep his customers happy.) Manufacturers can offer information about problems unique to their products. There may be a manufacturer's representative nearby who can help with the problem. A number of manufacturers' contacts are contained in the FCC material printed in the appendices. The owner should look in the owner's manual or contact the dealer for the manufacturer's phone number. The Consumer Electronics Association may also be able to help you contact the manufacturer.

The Ham

In those cases where your neighbor can't get satisfaction from the dealer, manufacturer or repair person, you may choose to help. Most of the time help is appreciated, and an excellent relationship with the neighbor may result. Occasionally there are pitfalls, which can be *very* deep and nasty. Remember that an Amateur Radio operator is *not* responsible for the cost of modifications or filters required to restore operations of a neighbor's equipment. You may choose to buy a few filters as a neighborly gesture, but the choice about helping or purchasing filters is up to you. If you do choose to help, do consider the possible extent of the problem. A ham who lives in an apartment building can spend more time helping than operating! If you buy a filter for a neighbor, you may find yourself asked to supply a lot of filters.

HOW AUDIO RFI HAPPENS

RFI occurs in audio equipment in two basic ways: audio rectification and direct radiation pickup. These almost always result from the AM detection of the transmitters *fundamental* signal, so they are all forms of fundamental overload.

Fundamental Overload

Proximity is the usual cause of high RF levels at audio and stereo systems. The normal audio-frequency input levels of these systems range from millivolts up to a few volts. RF signals near transmitters can sometimes result in several volts of RF energy being coupled to the affected equipment. The audio input path is not, however, the only way an undesired signal can enter the system. Any stage in the signal path can be affected by fundamental overload, so a process of elimination is used to isolate the affected stages and apply corrective measures.

Audio Rectification

Audio rectification occurs when an RF signal is strong enough to saturate the amplifier stages in the audio equipment, making it respond in a nonlinear fashion like a diode and AM detect the RF signal. It may seem puzzling that a transmitter at 1.8 MHz can affect audio equipment with an upper frequency limit of about 100 kHz. However, audio rectification can occur at essentially all radio frequencies.

INTRODUCTION

Interference is any unwanted signal which precludes reception of the best possible signal from the source that you want to receive. Interference may prevent reception altogether, may cause only a temporary loss of the desired signal, or may affect the quality of the sound or picture produced by your equipment.

Interference to home electronic equipment is a frustrating problem; but, fortunately, there are several ways to deal with it. This handbook provides a step-by-step process for eliminating the interference.

If your problem is not eliminated by following the steps in the handbook, you should follow the instructions in the owner's manual of your equipment for contacting the manufacturer. We have provided a list of manufacturers. If your equipment manufacturer is not listed, there are additional resources at the end of this book from which to seek help.

1. Check the Installation of Your Equipment

Many interference problems are the direct result of poor equipment installation. Cost-cutting manufacturing techniques, such as insufficient shielding or inadequate filtering, may also cause your equipment to react to a nearby radio transmitter. This is not the fault of the transmitter and little can be done to the transmitter to correct the problem. If a correction cannot be made at the transmitter, actions must be taken to stop your equipment from reacting to the transmitter. These methods may be as easy as adjusting your equipment or replacing a broken wire. These and other simple corrections may be accomplished without the help of a service technician.

Figure 10.1—This section of the FCC *Interference Handbook* explains interference to audio systems. Interference to non-radio systems is not caused by a violation of FCC rules.

Audio rectification occurs because a transistor can act as a diode when power is removed or it is driven into nonlinear operation (saturation) by a strong RF signal, as shown in Figure 10.2. It occurs because transistors (and ICs that contain them) are essentially combinations of diodes arranged according to a specific internal metallurgical scheme. The "diode" envelope detects RF, and the AM-detected result appears in the affected device.

Audio rectification is a form of fundamental overload. It is not the fault of the amateur station that the affected equipment was inadequately designed to reject unwanted signals. Most audio rectification results from RF pickup on wires connecting stereo components. It can also occur in telephones, answering machines and even things like alarm systems that produce no audio.

You cannot "make" a transistor from individual diodes, but a transistor contains "forward biased" and "reversed biased" diode junctions that act like individual diodes. Since diodes are good "envelope" detectors (or rectifiers), they detect any unwanted ac signal of sufficient strength. This occurs in spite of any desired audio signals that may be present.

Audio rectification occurs even though the stronger-signal frequency may be well outside the audio range. This is a form of fundamental overload (a very strong signal at the transistor or vacuum tube) that disrupts the normal operation of the device. The unwanted signal can arrive at the transistor or tube via several paths. Once an RF signal gets detected and turned into audio, it will follow the path through the amplifier to the speakers, with the result being interference. Once it is turned into audio, it cannot be filtered without affecting the desired signal, too.

PATHS

The energy that causes these problems may arrive at the affected audio equipment by three essential paths: radiation, conduction and magnetic induction. These basic mechanisms are explained in detail in Chapter 2; they won't be repeated here. (Remember the admonition to read the basics chapters first!) This chapter touches only the highlights as they relate to audio interference.

Radiated

Radiated interference arrives at its victim in the same fashion as any other radio transmission. There is a radiating antenna, a transmission path and a receiving "antenna." The antenna of the victim device is often the interconnecting cables between various elements of the audio/stereo system. Figure 10.3 shows a typical layout of an audio/stereo system that consists of at least two input devices (a tape player and a CD player), a main stereo amplifier and at least two speakers.

Cables from input devices are often shielded, but sometimes cause difficulty anyway.

Speaker cables are generally unshielded and often much longer than necessary. The excess length is usually just stuffed in any convenient nook. Long cables make good receiving antennas for any RF fields that may be present. In most cases of interference to audio amplifiers, the RF energy is being picked up on the speaker leads. Many speaker leads just happen to be cut to lengths that are resonant on one or more HF bands, adding to the problem.

Direct radiation pickup is a special case of radiated interference. The term describes cases where a radiated signal is received by conductors inside the affected equipment. RFI that arrives via radiation can be blocked by shielding the affected device (if that is possible; antennas, for example cannot be shielded). When shielding is not possible, direct radiation pickup can be treated with the proper filtering components applied at the component level, inside the affected device. Such modifications should be left to the manufacturer or a qualified service technician.

Wires Are Good Antennas!

To summarize one important point—components don't pick up RF; wires do! In most cases, the RFI problem is caused by radiated RF signals. Most audio devices are connected to three antennas—the input leads, the

Interception of Radio Signals

This discussion applies to interception of radio signals where the condition is observed on all channels or all across the dial. These units, and similar units, are referred to as "audio devices":

phonograph
CD player
hi-fi or stereo amplifier
tape recorder
public-address system
home-music system
telephone
the audio section of a television or radio receiver
computer speakers

Audio devices are designed to amplify audio signals such as music or speech and are not intended to function as radio receivers. The FCC does not give any protection to audio devices that respond to signals from a nearby radio transmitter. The problem is not caused by the improper operation or technical deficiencies of the radio transmitter. Strong radio energy gains entry to the audio circuitry, "overloads" the amplifier, is "rectified" and amplified, and appears at the loudspeaker as undesired sound. The only real cure is achieved by treatment of the audio device. Owners should contact a qualified technician, the seller or the manufacturer of the audio device for assistance.

Why the Owner?

Owners may reasonably ask, "Why should I do something to my audio device? It works fine except when the radio station is transmitting. Why is it my problem and not the responsibility of the radio operator?" The answers lie in policies concerning the economics, design and sale of these devices in a highly competitive market. The audio device has two objectives: (1) to reproduce a desired audio signal, and (2) to reject unwanted signals that may degrade the overall performance of the device (at a reasonable cost).

In their *Interference Handbook*, the FCC encourages consumers to be sure that their TV, stereo or other electronic equipment can sufficiently reject undesired radio frequency signals. The consumer should also be sure that any newly purchased electronic equipment meets contemporary voluntary immunity standards.

Relatively few audio devices are located near strong RF fields and respond to unwanted radio signals. Those that exhibit interference require added filters or shields or both.

Manufacturers believe it is unfair and unnecessary to burden all consumers with the added costs of special circuits and designs, inasmuch as the number of devices affected is relatively small. Many manufacturers, dealers and technicians have devised procedures to improve the RF rejection capability of audio devices.

The conditions at a particular location are not necessarily stable. A consumer residing in an area with no strong unwanted signals may begin experiencing RFI in three ways:

1) Local unwanted signals increase because a new transmitter is installed or an existing transmitter increases power. (Remember that "transmitters" may be consumer devices such as computers, touch-controlled lamps and light dimmers.)

2) The consumer moves into an area where strong unwanted signals are prevalent.

3) The affected audio device malfunctions.

In cases one and two, the situation has changed from the majority case (where special treatment is not necessary) to the minority case, where special treatment is needed. In case three, the equipment should be repaired or replaced.

The foregoing information is not widely distributed or fully recognized by all manufacturers, dealers and technicians. *The statement of a sales person or dealer that he sells a "good quality" device, or that there is "nothing wrong with the unit," is insufficient; it avoids the issue.*

Those who possess an audio device incapable of rejecting unwanted signals must fully understand the situation: *An audio device may perform the task for which it was designed with excellence and still require special treatment to improve its capability to reject strong unwanted signals in some locations.*

An audio device that responds to strong unwanted signals requires added filters, shields or both. The consumer is urged to bring this information to the attention of the electronics technician, dealer or manufacturer.—*John Norback, W6KFV, SB Section Technical Coordinator*

output leads and the power-supply leads. These conductors are usually the way that unwanted RF gets into audio equipment.

A quick fix: Shorten the wires! This is especially effective with speaker leads. Many times speakers are located next to the stereo, with several feet of speaker wire coiled into a large loop behind the equipment. Shorten the wires or wrap the excess into a small loop that fits around your hand and secure it with a wire tie. This technique makes the wire a much poorer antenna and also forms a simple common-mode choke.

Conducted

Conducted interference arrives by wire (or other conductor). Conducted interference may be controlled by filtering the conductor. Conducted interference may appear as either common-mode (all conductors act as though they were a single wire) or differential-mode signals (the signal arrives on a pair of conductors, with a 180° phase difference between the pair). Most of the time, when RF is picked up on conductors, it will be in the common mode, but if the conductors are not well balanced at RF, there will be a fair amount of differential-mode pickup, too. You should have both common-mode and differential-mode filters on hand for testing because some cases require both kinds.

Very seldom is conduction the sole path of an interfering signal. The incoming signal is usually radiated to a conductor near the affected device, which then passes the signal to the affected device. In those cases, interference may be reduced by shielding, or somehow detuning, the conductor. (Detuning a conductor might include changing the length, or inserting RF chokes to break its apparent length at RF.)

Magnetic Induction

Magnetic-induction interference requires a very close proximity of the interfering device (a coil or transformer) to its "victim." Hams are not normally confronted with this situation, even in their own homes (except in the case of telephones; that chapter has more about "magnetic induction"). Magnetic induction can be reduced by physically separating or magnetically shielding (as by an enclosure of soft iron) the interacting devices.

TROUBLESHOOTING AUDIO RFI

The basic principles of RFI troubleshooting are covered in the troubleshooting chapter. This chapter offers a brief overview and a few specifics related to audio interference.

Most audio interference is easy to identify: voices, often garbled from amateur SSB, thumps and bumps from CW or a hum can be heard in the device speaker when a nearby transmitter operates. Amateurs should be familiar with the sound of AM-detected SSB. If your receiver has an AM mode, listen to SSB signals in that mode. The resultant audio will sound muffled and distorted. If interference to audio equipment from your SSB transmitter sounds similar, the problem is audio rectification. If the interference is not affected by the volume control, the problem is almost certainly rectification in one or more audio stages.

In some cases, the interference can be more subtle. A high-end audio device that normally has a very low distortion level may still continue to function in the presence of RF, but the distortion level may rise significantly higher than the manufacturer intended in the presence of RF.

Identifying the Problem

It is not difficult to diagnose audio rectification in a device that contains no RF components. The FCC RFI pamphlet clearly states that audio devices that pick up RF signals are functioning improperly as radio receivers. When this occurs, it is clearly a case of audio rectification. Steps that reduce the strength of the amateur's fundamental signal may help the situation, but the fault and solution both lie at the audio-equipment end.

Determine, if possible, how RF energy is entering the affected equipment. The basic step of simplifying the problem is a good start. For example, if dealing with a multi-component audio system, disconnect all of the input cables from the audio amplifier or tuner and see if the interference is present with only the amplifier and speakers. If so, that leaves only direct pickup, the ac line or the speaker leads as suspects.

Troubleshooting Steps

The first step in identifying a problem is to characterize it as completely as possible. This is covered in detail in the First Steps and Troubleshooting chapters. Ask

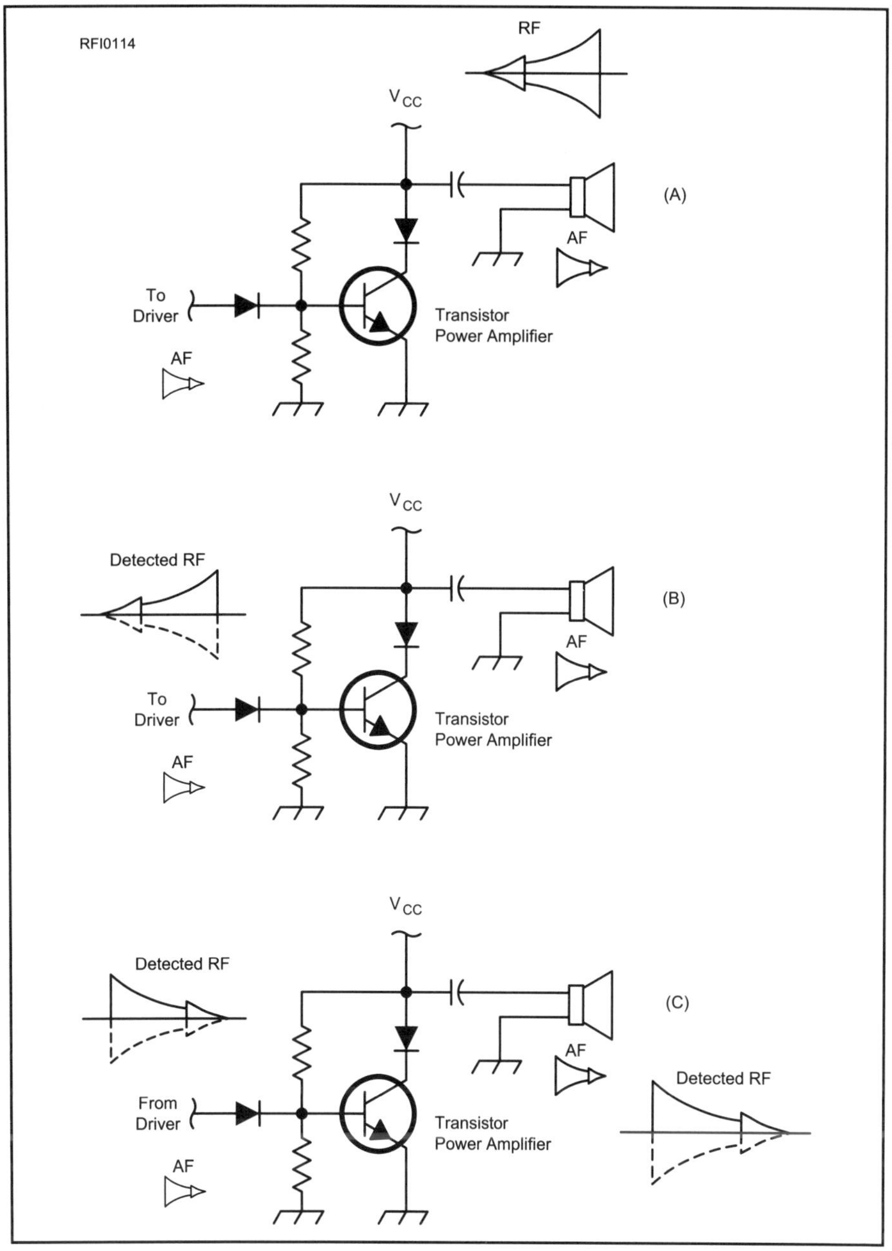

Figure 10.2—In audio-rectification interference, some device that is not designed to act as an RF detector does so. RF on the output lines (A) is detected by the output device (B). The detected signal is then amplified and applied to speaker (C).

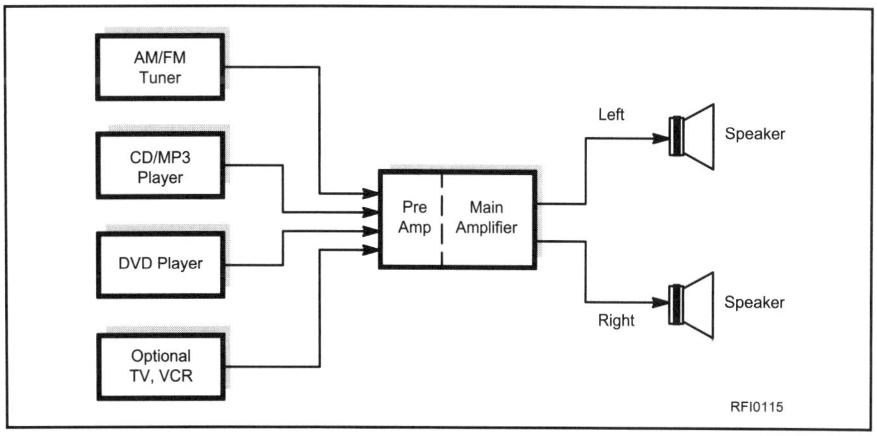

Figure 10.3—A typical modern stereo system.

the question, "What does the interference sound like?" Get the best description possible from nontechnical people. Certain kinds of interference can be identified by characteristic sounds. Amplitude modulation, for example, is readily detectable and audible on an AM radio or audio/stereo system. Single-sideband is also an AM system, but it is usually not intelligible. Instead, it produces a "muffled" sound on an AM radio or audio/stereo system. See Table 10.1 for a list of characteristic interference sounds.

You may not have many options on small or simple systems. A tabletop radio contains its own antenna. Battery-powered equipment has no line cord to filter. It is not possible for the typical owner to add external filters to such self-contained elements. Fortunately, systems with few options also usually have few RFI paths. Any external connections, such as the power cord on a tabletop radio, can be treated in the same manner as the corresponding part of a large, complex system. Therefore, let's discuss a large system with the understanding that the techniques can be applied to smaller systems as well.

Simplify the System!

When you begin troubleshooting, keep an open mind and don't be afraid to experiment. Make your approach to the problem a logical one, rather than a hit or miss proposition. Begin troubleshooting by simplifying the system. For example, in a complex stereo system, turn the system power off and disconnect all separate components from the main amplifier. Disconnect the speakers as well, and listen with headphones on a short cord. If a radio tuner is part of the main unit, disconnect the radio antenna. If the interference persists, it is entering the main amplifier through either the power cord or by direct radiation pickup.

Reconnect Step by Step

Once the minimum system is free of interference, begin adding the components that were removed. Begin with the speakers. If interference returns, then the speakers and their cables must play a part.

Continue troubleshooting: reconnect a component, test for interference, consider likely paths and apply likely cures until the interference is gone. Repeat the process for each component by working back to the signal inputs, until the system is again complete and all interference is gone. If the system contains several pieces of equipment, such as CD players, turntables, and FM tuners, reconnect only one source at a time. If you are working with a neighbor, have your neighbor do the connecting and disconnecting. This will minimize the likelihood that you will be accused of damaging the equipment. If you must do the job, first establish that you will not do anything that affects the "inside" of the equipment, or its operation.

When the amplifier is clear of all RFI, add the next component upstream. Repeat the same troubleshooting techniques as you reinstall system components. Each system component may need its own RFI cure as part of the overall fix. Turntables, tape decks, CD/cassette players, microphones and AM/FM tuners may all be connected as inputs to an audio system and each may need treatment.

For example, if you are able to get the system cleared of RFI with a turntable as the input, but the RFI reappears when an AM/FM tuner is reconnected to the amplifier, then the troubleshooting must continue to solve the new problem. At least when you have one complete RFI-free path through the system, it demonstrates to the owner that you can solve the problem, even though other components may still need work. The equipment owner can then see that—with his cooperation and understanding—there may be a satisfactory solution for all concerned.

Individual components such as turntables, tape decks, and CD/cassette players can have the same basic problems as the main amplifier. Just as speaker cable can be resonant at amateur frequencies in the HF range, cables from the pick-up stylus in the turntable tone arm can be resonant at VHF/UHF. A phono preamplifier has very high gain, and it is just as susceptible to rectification as the input to the main amplifier.

As you continue adding units to the system, mentally review other chapters in this book for potential solutions. For instance, if interference reappears when an AM/FM tuner is connected back into an otherwise clean system, other techniques may be required to solve this problem. This is because the problem may move from the AF range to the RF and IF ranges of the tuner.

The Volume Control

If there is a volume level control in the affected equipment, it can provide valuable clues about where the audio detection causing the RFI may be occurring. When the control has no effect, interference is rectified after the control. If the control does affect the level of RFI, the overloaded circuits are ahead of the control.

RFI being generated before the volume control is usually direct radiation pickup (in tape recorder heads and preamplifier circuits, for example). Circuit modifications could cure the problem, but leave such action to the manufacturer or a qualified technician, unless you own the equipment being modified and you are qualified to do circuit work.

When the Volume Control Has No Affect

If the volume control has no affect on the level of the interference, the interference is occurring after the volume control. Interference after the volume control is usually audio rectification in the amplifier output stage. While such interference should have been cured in the early stages of troubleshooting, the last piece of reconnected equipment may provide an interference return path or ground loop that affects the previously cured amplifier problem.

For example, a CD player that is grounded through its power cord, uses a shielded connecting cable and also provides a ground connection to the system amplifier adds ground loops (alternate ground paths) to the system. The problem might be cured by disconnecting the CD player ground from the amplifier, by

Table 10.1

Characteristic Sounds Produced by Audio Rectification of Various Signals

Source	*Sound*
AM	The voice or music is heard as any normal signal applied to the amplifier. The interfering signal may be extremely loud and slightly distorted.
SSB	Voices are garbled and unintelligible.
FM	Usually no sound is heard, but the amplifier volume decreases during transmissions. Clicks may be heard as a transmitter is keyed.
TV	The audio rectification of a TV signal makes a loud buzzing noise. The buzz changes as the TV picture changes.
Data	Data transmissions may affect the receiver in the same way as FM. There may be an audible "tweedling" sound, constant or intermittent or constant-level static.
CW	Usually a rhythmic clicking with keying.
Motor	Constant-level static that varies with motor speed.
Atmospheric	Uneven crashing sounds.

placing an RF choke in that lead or by electrically breaking the shield connection of the CD player signal cable.

When the Volume Control Has an Affect

If the interference is affected by the volume control, then it is either being picked up directly by high-gain stages in the amplifier or it is entering on the signal lines and not the ac cord or ground loops. If this is not the case, the problem can be difficult to diagnose. The interference may be acting in the amplifier between the signal input connector and the volume control, or it may be rectified in the audio source and passed to the amplifier, as AF, along with the desired signals.

Check the interconnect cable between the two units to see if it is shielded. If it is not, and you have a shielded cable you can substitute for it, do so. If the interference disappears with a shielded cable, then the solution is obvious. If you don't have a shielded cable to substitute, then apply the same ferrite core treatment to this cable as you did to the speaker cables. Even if the existing cable is shielded, try a common-mode choke; it should protect the amplifier against RF on the cable shield (within the limitations of chokes, covered later in the chapter). If a shielded cable and chokes do not help, suspect that the auxiliary equipment connected to the amplifier may be the source of the RFI.

When it is Not Even Turned On!

Another simple audio-rectification test is performed by turning off the affected device and listening for detected RF (clicks from CW or distorted audio from sideband) from the speakers while operating the amateur transmitter. If clicks are heard with the audio equipment turned off, audio rectification is taking place in the final audio transistors.

Good Engineering Practice

Stereo installations often have a rat's nest of wires, cables and antennas forming large loops behind the equipment. Clean up this mess by using minimum-length cables. Bundle signal-input cables, speaker wires and antenna leads away from each other as well. Things can get a lot worse than the example shown in Figure 10.4!

AUDIO RFI CURES AND APPLICATION

Historically, amateur treatment of RFI problems has included detailed information about circuit modifications to affected equipment. These days, however, such measures are seldom appropriate. Nearly all RFI problems can be solved by means of filters and external shields. Also, some states require that anyone who works on electronic equipment must hold a state-issued technician's license. (Your FCC license will *not* be sufficient.) Many of the cures shown here can be applied externally, but in-circuit modifications are shown here as guidance for licensed technicians who are unfamiliar with RFI cures.

Ferrite Common-Mode Chokes are Your Friends

In most cases, ferrite material is used to form a common-mode choke. Common-mode toroid chokes are often effective against RF in cable shields. Wrap any excess interconnect cable around a ferrite core (such as FT-140 or FT-240), made from a material suitable for the frequency being filtered, and see what affect it has on the interference.

Chokes work by presenting a high impedance to common-mode RF in series with the conductor. This impedance is a combination of reactive, which tends to reflect the RF energy and resistive, which absorbs it. If using one choke does not cure the RFI, add a second choke at the other end of the cable. In long cables, add another choke 8 to 10 feet from the first. Most of the time, a common-mode choke has no affect on the desired, differential-mode signal *inside* the cable or wire being filtered.

Simply wrapping signal cables (or an ac cord) around a ferrite rod often eliminates RFI. Figure 10.5 shows another way to make a ferrite choke. As a rule, more turns yield more rejection of unwanted energy. For severe RFI cases, more than one core can be used (the effect is additive). Use ferrite with a permeability (μ) of at least 850. Higher permeability is okay, as long as it does not lessen at the frequency to be rejected.

Place the choke as close to the affected chassis as possible. A choke at a speaker is not usually as effective as one at the chassis terminals. Excess speaker cable is generally sufficient to wind a choke. Power cords may not be long enough unless the power outlet is nearby.

Keep a variety of chokes and choke/cable combinations on hand with different connectors on the ends to match the usual audio/stereo component connectors and terminals. If you have had TVI or other RFI problems in your own house, you may already have pretested chokes on hand. Note that if you need ferrite chokes to tame RFI in one unit, you will likely need more for others. (You won't be able to remove any such chokes from previously cleared units because their RFI will return.) Therefore you may need two or three ferrite chokes for each piece of equipment as you continue clearing the system. If you are dealing with a neighbor's interference problem, you will look more creditable if you come prepared with effective and sufficient cures to solve the problem quickly.

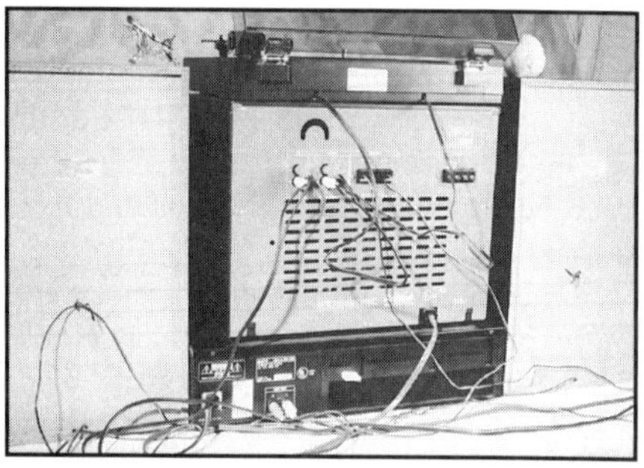

(A)

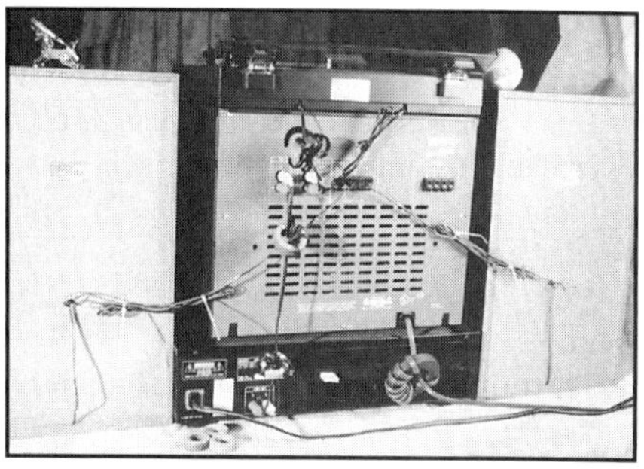

(B)

Figure 10.4—A typical stereo installation at A and its EMI-resistant counterpart at B. (*Photos by Suli Sullivan*)

A good audio common-mode filter suitable for speaker or input leads can be made from an FT-240 ferrite core (see Figure 10.6). Use a ferrite material suitable for the frequencies involved. Wrap approximately 10 turns (a single layer) of shielded speaker or interconnect cable. In most cases this filter greatly attenuates the level of audio rectification.

Ferrite Beads—Their Little Brothers

Ferrite material can also be shaped into beads. These dowel-like ferrite cores have one or more holes through them. When placed on wires, they act as RF chokes. The beads are made from several different ferromagnetic materials, in order to vary their permeability. Beads work where their larger brothers won't fit. In most cases, ferrite beads do not provide much inductance, so they do not work as well as ferrite chokes as do toroids with multiple turns of wire. If possible, using a few turns of wire in conjunction with a ferrite bead will help improve its effectiveness.

Figure 10.7 shows four different bead configurations. As the figure shows, one need only thread one or more beads on a lead, or wind one or more turns through them

(Type 2 and 3 beads) to impede unwanted RF signals. The impedance will rise linearly with the number of beads; it will rise as the square of the number of turns of wire on each bead. Passing a wire through a bead is one turn.

To install a ferrite bead in a transistor amplifier built on a PC board, unsolder the base lead, lift the lead clear of the board, place a ferrite bead over it, reinsert the lead and solder it. This places an inductively coupled impedance in the signal path without adding resistance. Remember you are trying to eliminate RF, and the desired signal is AF.

Bead type 4 is a split bead that can be used over cable assemblies (¼ to ½ inch diam) without disassembly of the cable. The split bead is mounted around the cable to enclose it. Flat, split beads are available for use on multi-conductor flat ribbon cable.

The impedance coupled into a circuit by a bead is a function of the bead material, the signal frequency, and the bead size. As frequency increases, permeability decreases, resulting in a band-reject response. Mix 43, 73 and 75 ferrite beads are most commonly used for RFI suppression.

Impedance is directly proportional to the length of the bead, and it increases as more beads are used. The magnetic field is totally contained within the beads, and it does not matter whether they touch one another or not. Ferrite beads need not be grounded; in fact, take care with these high-permeability materials because they are semi-conductive. They should not touch uninsulated wires or ground.

Beads come in various sizes and permeabilities, so manufacturers' literature should be consulted to determine their size, permeability and optimum frequency range. Manufacturers usually publish this data, along with an "impedance factor" (related to the permeability), which relates impedance to bead size and the number of turns.

Bypassing

Conventional bypassing techniques (as discussed in the RFI Fundamentals chapter) can work against audio interference problems. Over the years, *QST*'s Hints and Kinks column has featured audio-RFI cures that involve placing capacitors across audio inputs or outputs. In many cases, these can also work together with inductors to form a very effective filter. Several examples are shown in Figure 10.8.

Bypassing does have its side effects, however. A 150 pF capacitor has an impedance of about 100 ohms at 10 MHz, making it an effective RFI filter at that frequency if used in a high-impedance audio circuit. It has an impedance of about 50,000 ohms at 20 kHz, however, and if used in a typical audio high-impedance circuit, it will roll off the 20 kHz response by about 3 dB. This is tolerable in some applications, such as a house intercom, but will probably be unacceptable to most high-end audio users.

When Not to Bypass

Do not apply bypass capacitors to the output of transistor amplifiers. See Figure 10.9. Contrary to previous literature, it is not a good idea to bypass speaker leads with a capacitor. This works just fine for tube amplifiers, but solid-state amplifiers are apt to break into full-power oscillation with capacitive loads. Damage to the amplifier is the probable result.

Transistor amplifiers can provide low-impedance outputs. This lets them drive speakers without matching transformers that would attenuate frequencies above the audio range. Transistors are extremely sensitive to bias variations. Their gain typically decreases as frequency rises and increases with temperature. Feedback circuits are normally used to compensate for these traits. Because of the gain-frequency relationship, the feedback circuits have a high-pass characteristic.

A high-pass feedback network (with no output transformer to limit high-end response) permits potentially dangerous oscillations. The feedback network is adjusted to prevent such oscillations with a resistive load. When a capacitor is added across the amplifier output, however, the phase relationships in the output stage can change. The feedback circuit that helped keep the amplifier linear can now result in oscillations. Since the oscillations are often above AF (ultrasonic), the only noticeable effect may be overheating or destruction of the amplifier. If you cure the RFI, but the amplifier fails as a result, you will have no credibility. (Imagine the effect if you do this to your teenager's stereo!) Fortunately, there is a way out of this dilemma.

The outputs of transistor amplifiers can be safely treated in two ways: Ferrite toroid common-mode chokes do not place a capaci-

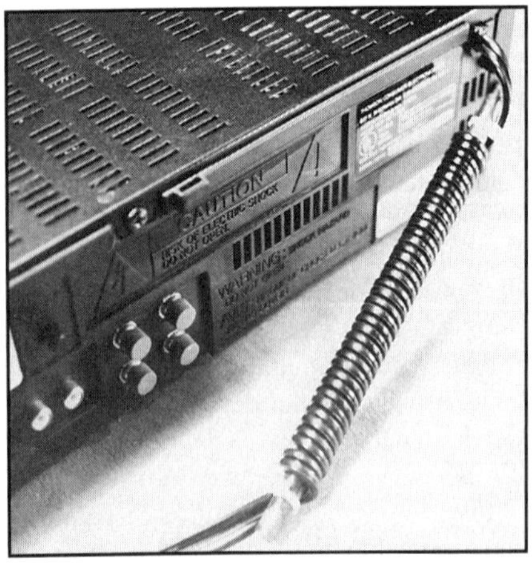

Figure 10.5—Wrap the ac-line cord around a ferrite rod to break ground loops that include the ac line.

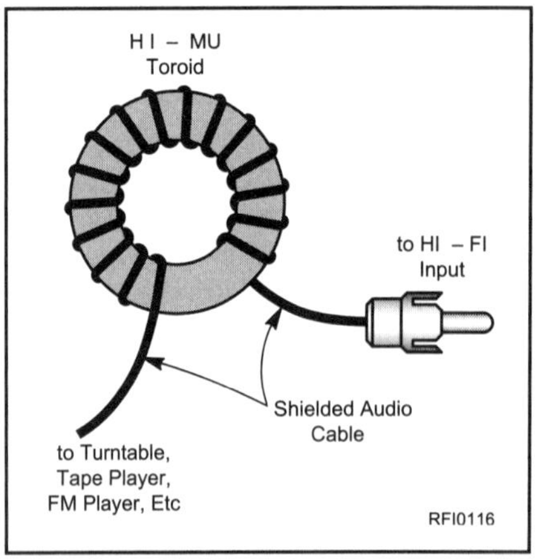

Figure 10.6—A ferrite toroid common-mode choke for use on speaker leads and interconnect cables.

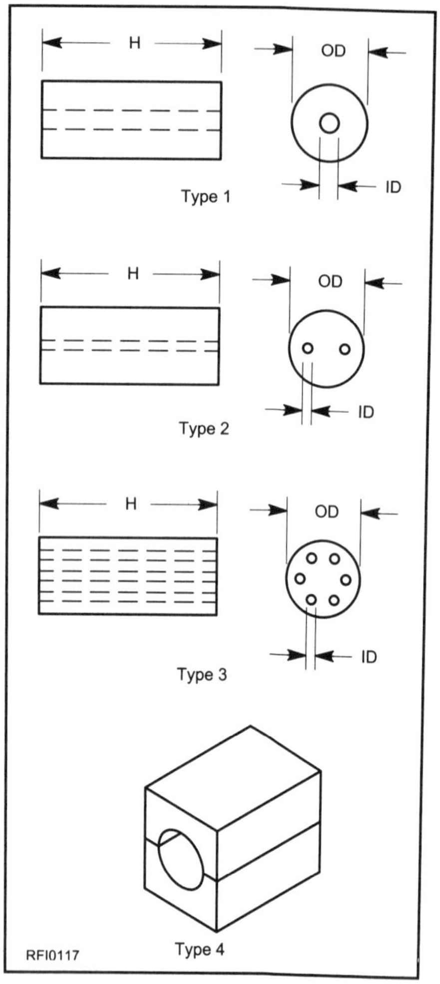

Figure 10.7—Typical ferrite bead configurations. Types 1 through 3 are one-piece beads. Type 4 is a split bead for assembly around cables or wire bundles.

tor across the output. A bypass capacitor may be placed across the output if it is preceded by an RF choke. The choke and capacitor form a low-pass filter and present an inductive, not capacitive load to the amplifier. See Figure 10.8D for an example of a filter that can work on most speaker leads. If you have any questions about what cures are appropriate, ask the equipment manufacturer.

Shields and Grounds

In rare cases, the previously discussed cures will not completely remove RFI. Such cases may require more stringent techniques to eliminate the last traces of RFI. Grounds and shields should be investigated. Grounding techniques are important to eliminate ground-loop currents and help eliminate conducted interference. Shields serve to protect circuits from direct radiation pickup.

Shields

Amateur equipment has shielded cases to prevent chassis radiation. Transmit filters permit only the fundamental to reach the antenna. Most consumer devices do not have shielded cabinets, nor do they have any internal shielding. Some system components are poorly shielded. When they suffer from direct radiation pickup, the owner should contact the manufacturer to request shielding modifications. The owner's only remaining options are a shielded enclosure or replacement of the affected equipment.

It is possible to shield an affected device, its enclosure or cabinet. Although it is not often a practical solution, direct radiation pickup may be treated by placing the amplifier inside a home-built shielded enclosure. (Many equipment owners would not like this solution!) The owner may make (or have made) a shielded enclosure to hold the affected device. A suitable enclosure is reasonably easy to find or construct. For example, a stereo system might be placed inside a metal entertainment-center cabinet. The main drawbacks are appearance and convenient access to the equipment inside the enclosure.

Perforated aluminum and metal screen are suitable materials. Be sure that the enclosure provides enough ventilation to prevent overheating of power amplifiers. Alternatively, the manufacturer or a qualified technician can add shielding or a conductive spray inside the amplifier cabinet or make circuit modifications to reduce RF susceptibility. Spray shielding is discussed in the EMC Fundamentals chapter.

A suitable enclosure may be made of metal, or a nonconductive enclosure may be lined with metal or a conductive spray. A spray EMI-RFI coating is available from GC/Thorsen. Copper and brass offer ease of connection, aluminum is cheaper (use lock-washer solder lugs for electrical connections to aluminum). A small metal case should be perforated (such as decorative screening) to allow adequate ventilation. Use piano hinge for good shielding at openings. Opening covers should fit closely and overlap the opening edge. If the enclosure cures the RFI without a ground connection, leave it that way. A good earth ground helps in some situations.

RF easily gets through long, narrow slots. Fasteners used to secure a cover should be spaced about 0.05 λ or less to maintain good shielding integrity.

The owner or ham should not attempt to install shields *inside* an affected device. There are safety, regulatory and other issues involved. For example, spray shielding on transformerless consumer equipment could create a shock hazard. It may also change the thermal balance inside the equipment, making it more prone to long or short-term failure. The manufacturer or a qualified technician may be able to install rigid (often foil bonded to cardboard) or spray shielding inside the case.

Grounds

A ground is supposedly a "zero-potential" surface or point, but most float at some small voltage. In reality, our best hope is that there is little circulating current (particularly RF). Some households have multiple ground reference points: the ac power ground, the plumbing system and sometimes separate grounds for telephone and cable. Multiple grounds, however, are not a good practice; the National Electrical Code requires that all grounds be connected to a single point.

Water pipes are buried, but their ground resistivity can be quite high, and there are no precautions to ensure electrical continuity. As a result, the plumbing is usually tied to the ac ground return at some point in the house. This point is often at an outside faucet, where a wire is attached from the ac ground to the water pipe with a clamp.

Improper grounding can introduce unwanted audio hum and may violate the National Electrical Code or local regulations. An added ground for an audio system could create a big ground loop, which could act as an antenna and worsen the RFI.

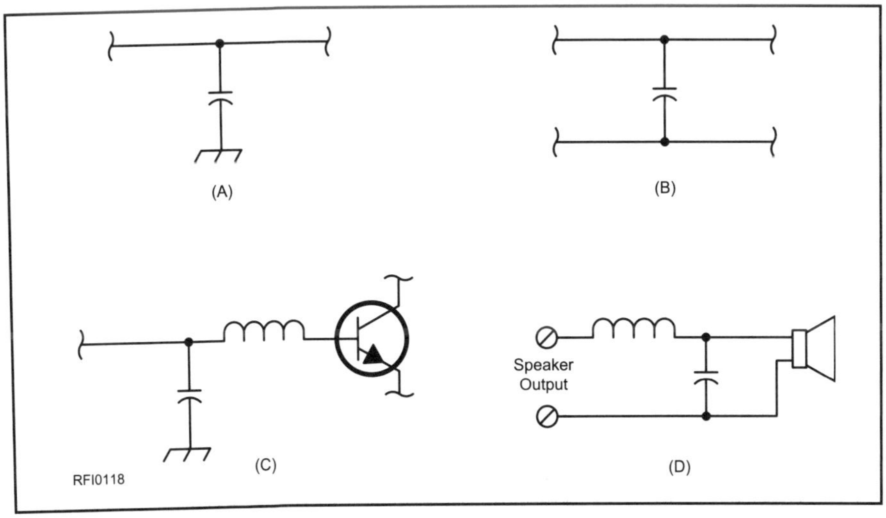

Figure 10.8—There are a number of ways to bypass audio signals. The simple capacitor in A is connected from a signal line to ground. The capacitor in B is connected across two ungrounded conductors. A good way to bypass the input of an audio amplifier is shown in C. The unwanted RF would prefer to flow to ground through the capacitor rather than pass through the inductor. A way to bypass speakers is shown in D. The inductor should be rated to carry several times the current that will be present in the speaker system. In an 8-Ω, a 100-watt signal would require 3.5 A. Commercial inductors are not available economically in this current range. Make a suitable inductor using ferrite cores.

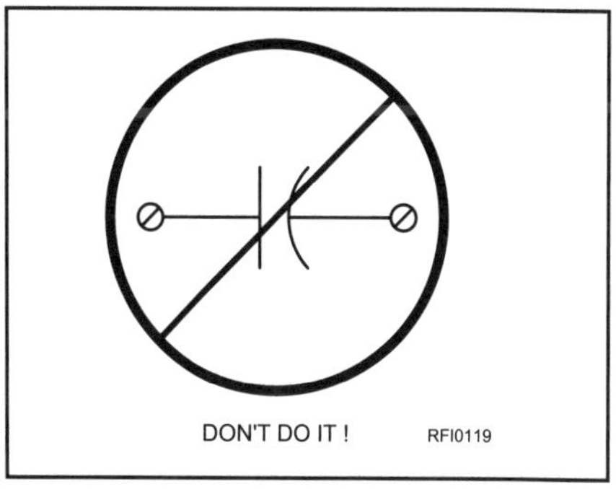

Figure 10.9—Modern solid-state amplifiers do not usually operate properly into a capacitive load. *Do not* follow the old advice to install 0.01 μF capacitors across the speaker terminals of a solid-state amplifier.

Ground Loops

In nature, it is very unlikely that any two grounds are at exactly the same potential. Therefore, when a conductor is connected to two different ground points, a "circulating" current flows through the (supposedly grounded) wire. By connecting the wire to a second ground, a ground loop was created. The currents circulating in ground loops can cause or worsen RFI problems.

A "single point" grounding system is usually necessary to avoid circulating currents between components of a system. Figure 10.10 shows how to avoid ground loops between system components. In the figure, a turntable that has a shielded interconnect cable should not have a separate ground lead to the amplifier. Other components such as AM/FM tuners and CD or cassette players should be treated likewise.

To avoid ground loops, there should be one (and only one) path from each point in the system to earth ground. (Visualize the ground system as a tree, with one trunk and many branches.)

When each component has its own three-wire ac cord, it may be difficult to avoid ground loops. It may be satisfactory to connect them all to the same ac outlet. If not, try a ferrite choke on each ac cord. Then add ferrite chokes on any shielded interconnect cables. In some cases it may help to use unshielded interconnect cables. There are no hard and fast rules, because each audio/stereo system is unique. Try all combinations of potential cures; some are bound to work. Don't give up! When the final solution is achieved, it will hold lessons that can help in future cases.

What does this mean for an audio/stereo system that may be located on a second story or in a far corner of the house? Establish a reference ground at the system itself. Then connect the reference ground to the main ac ground with as short a lead as possible. This may not be easily done, but it may be the only solution in severe RFI cases.

AC-Outlet Grounding Practices

All this may sound difficult, but often it is not. Many homes have three-wire grounded outlets near water sources, while the rest of the home has only two-wire outlets. In such cases the home is usually wired with three-wire cable, but the ground wire is not connected at the two-wire outlets. In this case, outlets can be changed to a three-wire style, and the ground wire connected to its proper terminal. Of course, the ground wire must be continuous to the ac ground point. For safety reasons, the work should be done by a licensed electrician. See the electrical RFI chapter for more information.

AUDIO/STEREO SYSTEM DESIGN

While today's audio and stereo equipment is mostly solid state, the occasional vacuum-tube equipment can still be found. This chapter covers both technologies with, of course, much more emphasis on solid state. RFI-reduction techniques apply to both semiconductor and vacuum-tube audio devices, but semiconductor circuits have some unique characteristics that preclude the use of some traditional RFI cures.

Transistors, ICs and Diodes

The size and economy of integrated circuits (ICs) and other semiconductors have brought about a revolution in consumer electronics. (Almost all semiconductor devices contain transistors, so we can talk about "transistors" and apply that term to ICs as well.) However, several characteristics of solid-state devices make modern consumer equipment more susceptible to RFI than some older equipment.

Transistors are generally less linear than vacuum tubes. They are also more sensitive to operating conditions than vacuum tubes. Transistor bias conditions are critical for linear operation. These conditions combine to make transistors particularly susceptible to RF-signal detection. Non-linear devices act like diodes to one degree or another, turning RF into unwanted audio energy to interfere with the desired audio signal.

The trend toward small size and low power consumption has made it difficult (or impossible) to apply some RFI-reduction techniques used on vacuum-tube circuits. An IC is a hermetically sealed component. There is no access to the internal circuitry. The basic RFI solution for an IC: *isolate it from RF.*

As ICs get smaller, so do the consumer products in which they are used. Reductions in the size and cost of consumer electronics have greatly increased the number of such devices near transmitters. The small size of some equipment makes it difficult to install RFI cures. For example, personal AM/FM cassette players usually have built-in antennas and are used in such a way that it is impossible to apply RFI cures.

Amplifiers

Almost all modern audio devices have one thing in common—an audio amplifier. Amplifiers are the core of audio devices. In modern equipment, transistors or ICs are usually the active devices, but there are still

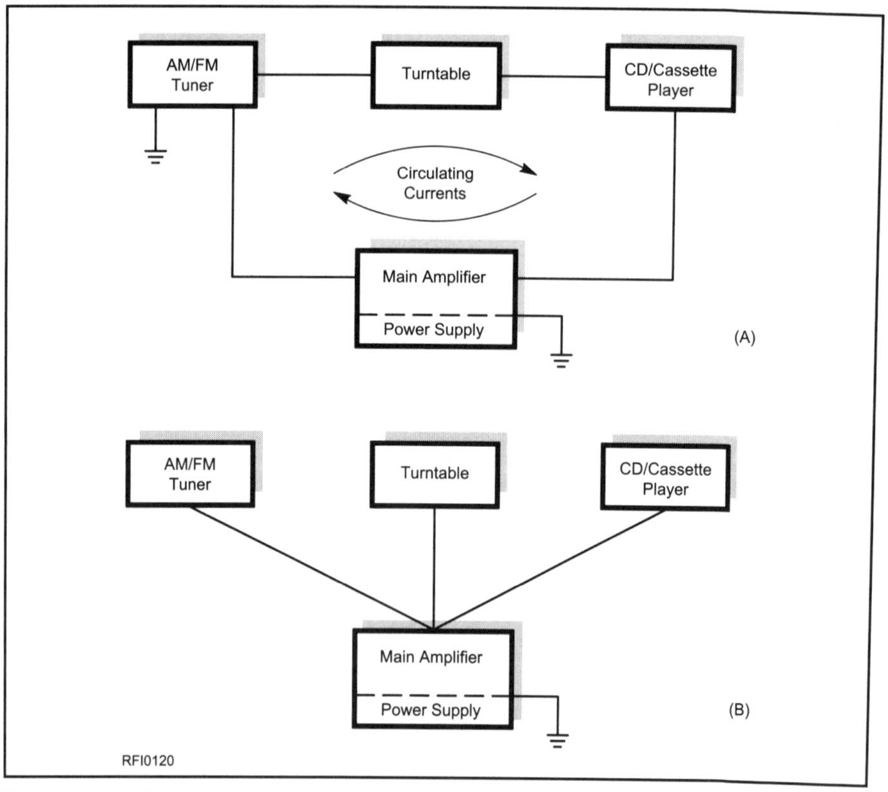

Figure 10.10—Ground loops in stereo systems. A shows a system with multiple grounds. Circulating currents may flow through the loop formed by the ground connections between the tuner and phonograph, phonograph and CD player. There is also a loop between the tuner's separate earth ground and the main amplifier earth ground. Ground loops are eliminated at B, where each component has only one path to a single system ground via the main amplifier.

Intercom Feedback and Ground Loops

Some years ago, I was the project engineer on part of a large data-processing system. It required an intercom system and intercom access to a group of audio recorders. During the preliminary tests, there were severe hum and multiple feedback problems when more than two intercom stations were connected and when intercom stations accessed the tape-recorder system.

All audio cables were twisted-pair shielded lines, which should have been clear of both hum and feedback. Since the system was a prototype, no special care had been taken to keep leads short or provide optimum grounding. While troubleshooting the feedback and hum problem, I noticed that the number of recorders installed directly affected the hum level; more recorders increased hum.

Ring-Around Feedback

When I disconnected the recorders completely, the hum problem was drastically reduced, but the feedback was still there when more than two intercom stations were connected. The feedback problem was determined to be a familiar problem known as "ring around." (This problem first showed up in cross-country microwave relay circuits, where a complete loop was generated by several site operators using the service line simultaneously.) The effect is similar to the feedback in PA systems when the gain is too high. In my case, the connection of the intercom stations was provided "through" each station, rather than by a common node. Once this was determined, phase-delay networks (with slightly different delays) were installed at each station. This interrupted the ring and solved the problem.

Ground Loops

Next the hum problem was examined (it occurred when the stations were connected to any of the tape recorders). A quick look at the recorder schematic revealed the cause. There were eight separate grounds at the output connector of each recorder. The solution to the problem was obvious. Each recorder was internally modified to provide only one ground at each connector. The hum disappeared, and the intercom system operated normally.

The moral to this story is: Too many grounds can spoil an otherwise good system. With eight grounds exiting the recorders the potential for ground loops was enormous; any hum or noise on those grounds could enter the system. A single ground from each unit (connected to a common system ground) completely solved the problem.—*James G. Lee, W6VAT, San Jose, California*

some vacuum-tube amplifiers in existence. (No old-time guitar player will *ever* stop using an old Fender tube amp!) The term "amplifier" covers a lot of territory, from low-level preamps to high-power audio output amplifiers. The cures discussed in this section apply to a wide range of audio devices.

Input Stages

Amplifiers can be subject to interference in two fundamental ways. Refer back to Figure 10.2. This shows the input circuit of an audio preamplifier. The diode shown in series with the input is to demonstrate a point. If RF signals are picked up on the input leads, or get to the input transistor by any means, the circuit can function as a diode detector.

Output Stages

Figure 10.11 shows how the audio output stages of a power amplifier can also create RFI. In this case, the RF is usually picked up on the speaker leads. (Notice how often speaker lead lengths are resonant on HF amateur bands! As shown by the diode, the output transistor can detect the RF signal, turning it into audio. The resistor and capacitor feedback components are designed to help make the audio amplifier linear. They can, however, conduct that audio RFI signal back into the high-gain stages earlier in the unit, resulting in very strong interference from a small amount of RF energy.

Other output taps provide various levels of feedback to preamplifiers or drivers (to control the frequency response of the amplifier). As a result, any RF picked up by the speaker cables is fed directly back to low-level stages of the amplifier. Those stages amplify any unwanted signal along with the desired signals, until one or more stages are overloaded. The placement of the feedback tap on the secondary winding is the only restriction on the unwanted signal fed back to the early stages.

Transistor Power Amplifiers

Figure 10.12 shows a typical transformerless audio amplifier with a complementary-symmetry output stage driving a speaker. (Complementary symmetry uses of a combination of PNP-NPN transistors, Q4 and Q5, to match the speaker.) For clarity, many of the bias-stabilization networks have been omitted. The feedback path (from the output through R3 and R2) is shown, along with R1 (the collector load for Q1). The schematic shows how the circuit appears to signals, or ac conditions.

Without going into great detail, the figure also shows three phase lags that contribute to the overall frequency response of the amplifier. The amount of lag in each section depends upon both the alpha and beta cutoff frequencies of the individual transistors. The feedback path (R2 and R3) from the output to Q1 is adjusted to provide the desired frequency response characteristics.

Vacuum-Tube Power Amplifiers

It is possible to design audio amplifiers with bandwidths exceeding 100 kHz without much difficulty. Vacuum-tube amplifiers are high-impedance devices though, and they need a transformer to match the tube impedance (usually several thousand ohms) to the speaker (4-16 Ω). Most use large iron-core transformers, which are not broadband devices. The transformers have an upper frequency limit of about 20 kHz.

Figure 10.13 shows a simplified diagram of a typical vacuum-tube audio amplifier. Note the multiple output taps on the transformer and also the feedback path to a driver or preamplifier stage. The multiple taps are there to match a range of speaker impedances, which can vary from 3.2 Ω to 32 Ω on individual speakers. One of the output terminals is shown grounded, but this may not be so in every amplifier.

RFI entering vacuum-tube amplifiers via the speaker cables can be controlled with bypass capacitors (C1 and C2) connected from speaker output to ground. Assuming that the ground terminal offers a good ground, the two capacitors should prevent any unwanted RF from reaching early stages in the amplifier. In general, the capacitance should be as little as possible, but large enough to cure the RFI. Start with small capacitors, say 0.001 µF, and increase the value if they do not work. The output impedance of power amplifiers is usually low enough so the capacitors can be relatively large without affecting the high-frequency response of the amplifier. *Don't do this with solid-state amplifiers!*

Multiple Antennas

It is not uncommon for more than one speaker to be connected to one channel of an amplifier in a stereo system. Even monophonic systems have been wired to drive two speakers. Two 16-Ω speakers in parallel are an 8-Ω load, and as long as the amplifier has the power capability to drive them, it is a very practical arrangement.

The Electrical System

Interference that enters by the ac cord is conducted interference. Strong RF fields are inducing currents in the ac wiring, and the wiring then conducts RF to the amplifier. Since nothing useful can really be done with the suspect house wiring, action must be taken at the power supply cord. If the equipment has an ac cord, apply the same technique to the cord, particularly if a choke on the interconnect cable reduced any problems in the main amplifier. In severe cases, wind both the interconnect cable and the ac cord around ferrite cores. The interference is probably a common-mode signal, so apply a common-mode filter as close to the amplifier as possible. You can also try a "brute-force" ac-line filter, such as those sold by Industrial Communications Engineers.

Where to Place Internal Components

Be warned! It is unwise to add components to internal circuitry. If a qualified technician has the understanding and willingness to solve the problem, internal modifications may be a viable solution. (Even then, consult the manufacturer first.) In all cases, strive to protect the equipment owner's equipment and interests. You and he (or she) are on the same side.

Make sure that cures clear the interference, but have no chance of harming the affected unit. Remember that transistor power amplifiers may not tolerate the same cure that vacuum-tube amplifiers accept. See "Do Not Apply Bypass Capacitors to the Output of Transistor Amplifiers" earlier in this chapter.

There is another reason to avoid internal changes. They can void any remaining warranty or service contract. Point this out to the owner; it shows your concern for his/her protection. Volunteer your services as a consultant to the dealer or his technician if they are unfamiliar with RFI problems.

If a technician is to add components, placement may be critical. Figure 10.14 shows components added to the phonograph, tape deck or auxiliary input connectors of an amplifier to remove unwanted RF. Since this occurs at a connector, there is usually room to attach these components without difficulty. As before, these components should be placed as far "upstream" (closer to the input of the system) in the signal path as possible to prevent amplification of unwanted signals by affected circuits.

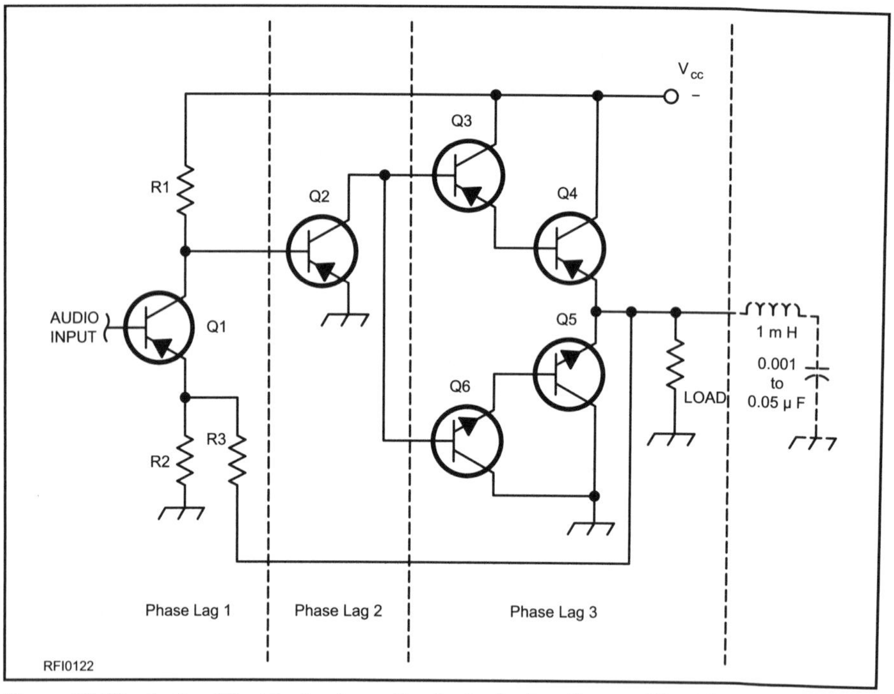

Figure 10.12—A simplified (ac) schematic of a typical audio amplifier. Phase lags occur between the dashed lines that determine the frequency response of the amplifier. The choke and capacitor (shown dashed) may be added to suppress RFI entering the amplifier via the speaker cable.

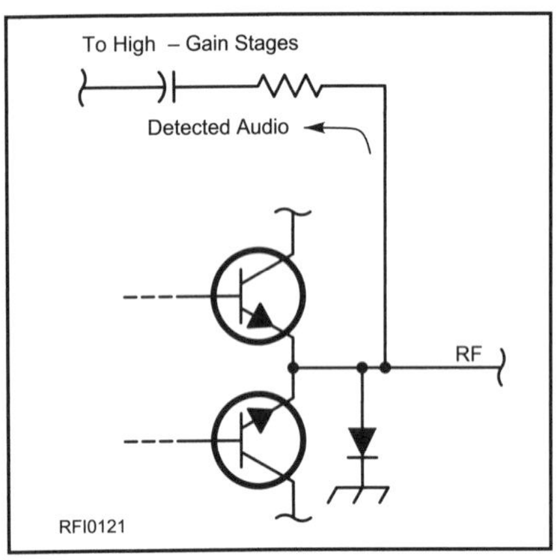

Figure 10.11—RF picked up on speaker leads can be detected. The resultant audio can be coupled back to the high-gain stages by the amplifier's negative feedback circuitry.

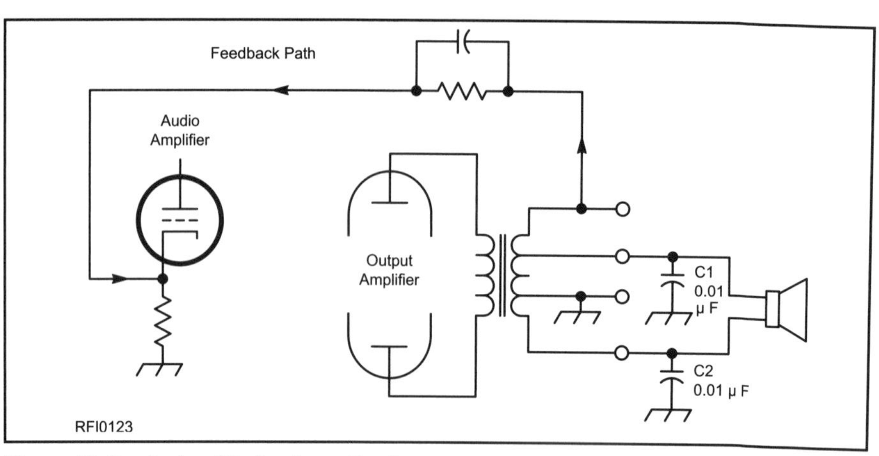

Figure 10.13—A simplified schematic of a vacuum-tube audio amplifier. C1 and C2 bypass RFI entering from the speaker cables.

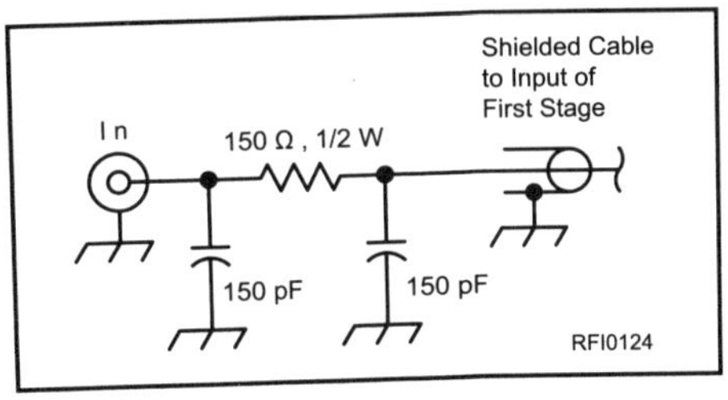

Figure 10.14—A filter for use at the input of audio equipment. The components should be installed by a qualified technician inside the affected equipment as close as possible to the input of the printed circuit board. In some cases, it may be possible to install them externally, at the input connector of the audio equipment. Use the shortest leads possible. Use capacitors with a minimum voltage rating of 1500 V, in case "ground" is inadvertently connected to 120 V ac.

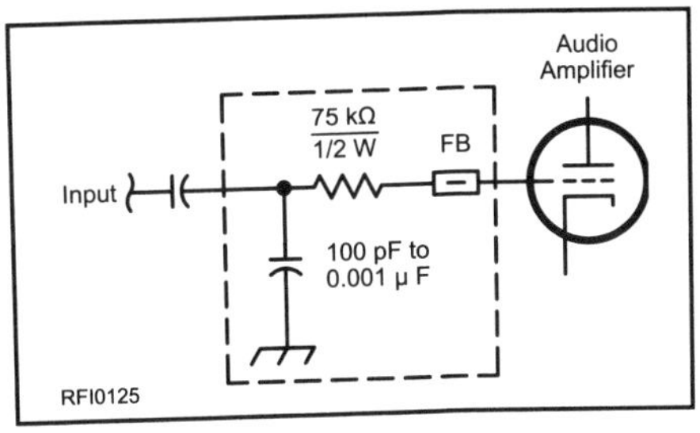

Figure 10.15—Partial schematic of a vacuum-tube audio amplifier with RFI-corrective components installed (inside dotted line). The added capacitor should be suitable for RF, with a dc working voltage appropriate for the circuit. FB is typically an FB-43-101 ferrite bead. This method is sometimes called "resistor-capacitor bypassing."

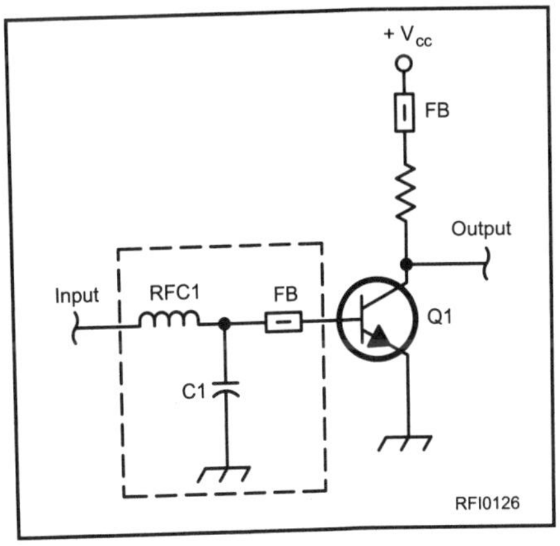

Figure 10.16—Partial schematic of a transistor audio amplifier with RFI-corrective components installed (inside dotted line). RFC1 is a 1-μH RF choke. The capacitor should be suitable for RF, with a dc working voltage appropriate for the circuit. FB is typically an FB-43-101 ferrite bead. A second bead was added on the positive-supply line.

Vacuum-Tube Circuits

Figures 10.15 and 10.16 show ferrite beads applied to vacuum-tube and transistor amplifiers. Figure 10.16 shows two different types of RF chokes (RFC) used as filters. The small square box containing a hyphen (-) is the symbol for a ferrite "bead" choke. (Ferrite beads are discussed in more detail later.) Vacuum-tube and transistor circuits require somewhat different approaches to install the components.

Most vacuum-tube amplifiers are constructed with "point-to-point" wiring. That is, wires are hand installed between various points of the circuit. It is relatively easy to add components to such circuits. In Figure 10.15, for example, the lead to the tube grid is simply unsoldered from the socket terminal, the components added, and the end of resistor R1 is resoldered to the socket pin. A small terminal strip may be added to support the extra components, but often this is not needed. Transistor amplifiers are not as easy to modify.

Solid-State Circuits

Essentially all transistor and IC circuitry is mounted on PC boards. The only leads that can be removed are those holding the components to the circuit board. On densely populated boards, it may be necessary to break a foil trace on the PC board to insert a component. This is not a desirable solution.

There is no problem installing C1 in Figure 10.16 at the base of Q1, but RFC1 cannot be easily inserted unless the foil trace is cut. It is doubtful that dealer's service technician would do this, and the manufacturer would certainly dislike this solution. Fortunately, the little "square" RFCs with the hyphen shown on Figure 10.16 can be slipped onto the lead of existing components. This usually cures RFI and can save the day.

SPECIFIC AUDIO/STEREO SYSTEM COMPONENTS

Tape Decks and CD Players

Tape decks and CD players do not have the long input cable associated with turntables. Normally, head assemblies in these components are shielded to minimize ac hum and other noise pickup. So any RFI from these components usually comes through the signal cables, or possibly through their ac power supply. The same cures apply to these units as to the main amplifier. The same troubleshooting techniques should give you the evidence you need to solve the problem.

Microphones

Many hams have experienced feedback from their transmitters to their microphones. If a microphone is used with a victim audio/stereo system, the same cures apply here as well. Microphones and their cables should be shielded and properly grounded to the main system.

Public Address Systems, Intercoms and Organs

These systems are lumped together because they share the RFI problems of audio/stereo systems, but the fixes are slightly different. Public address (PA) systems are found in churches, meeting rooms and assembly halls everywhere. The lengths of their cables and interconnecting wires set them apart from home hi-fi systems.

Public Address

Public-address systems can be large, with multiple speakers and high-power amplifiers. Organs may feed church PA systems through long cables. Most older systems use vacuum-tube amplifiers. Start by directly bypassing the speaker cables at the *vacuum-tube* amplifier. Common-mode chokes may be needed at the speakers as well (there is usually no ground available for bypassing at the speakers). If the equipment is solid state, use common-mode chokes or a capacitor-choke low-pass filter as described for transistor amplifiers.

Intercoms

Intercom units are seldom well-shielded, often use no shielded wiring and may be designed by engineers who do not fully understand the EMI/EMC field. The nature of the connectors used makes lead filtering almost impossible (because of mechanical considerations). External common-mode filters formed from ferrite cores may give some degree of suppression, but unfortunately they don't work well in this application.

If you do have a problem with an intercom, it may be helpful to contact the manufacturer. This problem is potentially severe because intercoms are never switched off, and they aggravate pets as well as people. Affected systems can awaken people. They can also contain electronic door bells that tend to go off with amateur transmissions.

Intercoms can suffer the same problems as large PA systems. Unfortunately, there are three different kinds of intercoms to treat:

"Wired" systems carry audio signals over dedicated wires. They may operate from ac or battery power.

"Wireless" intercoms place audio signals on the building ac wiring. (They are only called "wireless" because the owner need not install wires.) These intercoms are es-

Grounding Microphone Shields

With so many 'gizmos' generating RF like crazy today, it becomes very important to make certain the shield of your microphone cable is properly grounded. One of the biggest problems in doing that today is many transceivers use a floating ground system for their audio input circuitry and that ground is *not* at the chassis ground potential. Observe the pin out that is in your radio's operations manual and you will note that the shield is perhaps on Pin 7 and the control (PTT) or chassis ground is pin 5. If you experience any RFI or hum problems, you may try connecting those two pins together, thus actually connecting the shield to chassis ground. There are times that this could create a ground loop situation, so you may not be able to ground the shield at that ground point.

One of the main culprits to any noise or RFI pick up on the mic line is that all amateur radio transceivers use an unbalanced microphone input circuit. Only the Yaesu FT-DX9000 uses a true balanced microphone input.

With an unbalanced system, the microphone positive is fed down the inside wire and the negative is fed down the shield. The shield carries *two* potentials. The audio minus and the dc system ground. In many cases, this is no big problem, but certainly leaves the station wide open for picking up outside interference and laying it on that shield —which is also the microphone return signal.

With a balanced system, the microphone minus and plus signals are fed through the two wires that are inside a "conduit" of heavy shield that is grounded to the chassis. The shield has one job to do—keep interference away from those sensitive input wires. The shield does *not* carry any voltage or signal potential.

It is very simple to use a balanced 600 ohm to 600 ohm transformer at the input of the transmitter and use a *balanced*—not balanced—microphone, something the commercial broadcast and serious recording studios have done for decades. This will keep your audio line balanced and free from all types of interference, thus resulting in much cleaner and clearer audio fed into the transmitter.

You also may want to visit your station grounding system and make certain that there is only one path to ground. So many stations have the tower grounded, the coax from the transmitter grounded to the antenna mast, the amplifier chassis grounded to a water pipe as well as to the ground lead on the ac plug, and then the amp and the transmitter are grounded together—a total disaster with four, five or more paths to ground. Additional information on station grounding is contained in the EMC Fundamentals chapter in this book, the *ARRL Antenna Book* and the *ARRL Handbook.—Bob Heil, K9EID,* **www.heilsound.com**

sentially low-frequency FM transmitters and receivers that operate in the 100 to 500-kHz range. They take their power from the ac wiring, and simply couple the FM signals into and out of that same wiring.

The third type of intercom uses normal telephones and their associated wiring to carry signals. Although this system is most often found in business operations, it is sometimes found in the home as well. Any RFI affecting this type of intercom should respond to the same cures as ordinary telephone systems. For more information look at the telephone-RFI chapter.

Wired Intercoms

Some of the most difficult audio-rectification problems involve interference to wired intercom units. Susceptible audio equipment is often connected to hundreds of feet of wire that acts as an immense antenna.

The interference magnitude is affected by many factors: the length of wires connecting components, component shield quality and component lead filtering. RF pickup of long wires may be reduced by using shielded cable. It may help to bypass the base lead of transistor amplifiers with 0.001-µF capacitors, but it is better to apply filtering techniques, usually common-mode, as described earlier in the chapter, and earlier in this book. It is usually not necessary to apply internal modifications.

Solid-state systems may have an output transformer to drive the interconnecting wires. Many systems consist of a central main amplifier, and outstations that simply function as additional microphone/speakers. These systems often use low-impedance lines that couple back into a matching transformer to complete the communications circuit. Obtain a schematic diagram to find out exactly how the intercom functions.

Wireless Intercoms

Wireless intercoms may require a combination of techniques to correct RFI problems. Use the same filter and ground techniques used with other audio systems. Remember, however, that you are working with circuitry that is directly connected to an ac power source. Be careful while troubleshooting, and make sure that any cure applied does not increase the likelihood of shock.

Harmonics from amateur transmitters should not be a problem in wireless inter-

Radio Receivers and Audio Devices

AM, FM and FM-stereo RFI manifests itself as buzzing or voice sounds superimposed on the desired audio. The unwanted signal may enter through the antenna, feed line, power or control leads, earphone or speaker leads, receiver enclosure or joints and openings in the enclosure. Six-meter (50 MHz) operation is most likely to affect FM receivers (88-108 MHz) because the second harmonic from any transmitter is usually the strongest. Interfering RF is commonly conducted via the speaker leads, which act as antennas. With this kind of RFI, the receiver audio-gain control has no effect on the interference. (The RF enters the set after the audio control, in the output stages.)

All kinds of audio appliances—from alarm systems to computers—are subject to overload by strong radio signals. The radio signal is usually amplitude detected then amplified just as if it were the desired signal. The FCC *Interference Handbook* clearly states that any audio device that receives interference from a radio transmitter "improperly functions as a radio receiver." The FCC does not offer protection to audio devices such as amplifiers or telephones.

coms. Fundamental overload is the most likely RFI cause for wireless intercoms, but audio rectification can also occur. The RFI may desensitize the intercom receivers and prevent desired signals from coming through. The standard techniques for trouble shooting and solving the problems for both radio receivers and audio apply here.

Shielded Cables

Shielded conductors may be mandatory for the long cables found throughout churches and assembly halls. However, access to these cables may be a problem. In severe cases it may be necessary to replace one or more of them. Replacing them in existing structures may be difficult, and any such cures should be done only as a last resort. (This is another place where professional help is required, if it comes to such replacement.) The telephone RFI chapter has some good discussion of techniques that can be used to eliminate resonances in long wiring runs. Refer to that chapter for some ideas.

In general, the length of audio cables does not appreciably affect the normal operation of audio equipment. You can usually add wire to add chokes. An extra 10 feet of speaker cable wound around a ferrite core has no effect on the sound when the speakers may be mounted many times that distance from the main amplifier. Home audio/stereo system cures should apply here. You may just need a few more of them. If one common-mode choke does not remove interference at a speaker or amplifier, add about 10-20 feet of cable and a second choke. (This ensures that both chokes are not mounted at an RF high-impedance point.)

Phonograph Tone Arms

If the cables to the turntable tone arm are not shielded, it may be necessary to replace them with shielded wiring. If you do so, be sure that you use wire designed for use in tone arms. Wire that is too stiff interferes with the proper travel of the tone arm, resulting in skips and ruined records.

A tone arm must swing back and forth easily, so it cannot be well grounded through the mechanism. The shield of the stylus wiring should be grounded to the base at the nearest point that allows free tone-arm movement. Use lock washers under ground lugs, unless it is possible to solder the shield. Lock washers help avoid problems with oxidation or lubrication buildup, which occurs when drive motors run for long periods of time. The average user does very little maintenance to keep the inside of the system clean. Any RFI corrective measures must consider this and withstand similar neglect.

Organs and Other Electronic Instruments

Electronic musical instruments, such as organs, vary in size, shape and complexity. Fortunately, many organ manufacturers have recognized RFI problems for a number of years, and have developed "TV chokes" (as they are known) to help solve RFI problems. Tests can help isolate the cause of any RFI. Each function on an organ such as the "swell" pedal, the "band-box" volume or "draw bars" often have their own amplifiers. Each function can be adjusted individually, and the effect on the RFI noted.

When the RFI changes as a pedal or effects control is manipulated, RFI is entering before the control. If a control has no effect, then either there is no RFI to the function, or the RFI is introduced after the control. Once you have determined where the interference is entering the organ circuitry, filters usually solve the problem. Once the RFI has been identified, request "TV chokes" from the manufacturer. If the manufacturer is not acquainted with RFI, or how to cure it, other manufacturers may help by supplying information on how to cure the problem in their product.

Chapter 11

Part 1—Electrical and Power Line Interference

Part 2—How to Resolve a Power Line Noise Complaint

This chapter was written to help hams identify various types of electrical interference. Also see the section later in this chapter on how to resolve a power line noise complaint. Hams can show information in this chapter to power company personnel to help them identify power line problems.

One stern warning:

Leave the power lines and equipment to the pros!

Part 1 by Jody Boucher, WA1ZBL
Northeast Utilities
300 Cadwell Dr, Springfield, MA 01004

Part 2 by Mike Gruber, W1MG, ARRL Lab
and Mike Martin, K3RFI
RFI Services Interference Investigator

Part 1—Electrical and Power Line Interference

INTRODUCTION

The world population of electronic and electrical devices is continually growing and becoming more sophisticated. Equipment is becoming more sensitive and devices that are susceptible to RFI are found in every home and office. Not only are there millions of devices that can be affected by RFI, these same devices can be sources of RFI, too! Any appliance that switches current or oscillates is a potential source. In addition, any appliance that senses, amplifies or reproduces audio may suffer from RFI. Most such devices (telephones, televisions, computers, and so on) are covered in other chapters in this book. Nonetheless, a few notable exceptions related to electrical interference are discussed later in this chapter.

This chapter provides a working knowledge of how sparks and corona discharges can generate RFI. It also tells you how to troubleshoot problems caused by electrical noise and describes effective measures for reducing electrical RFI to an acceptable level. Some of the more common offenders and appropriate cures are explained. With a good understanding of how interference is created and an idea of where to look, you should be able to locate the source and apply an appropriate cure. Although this section is not all inclusive, the principles it teaches you will apply broadly to a wide range of electrical interference problems.

Since many RFI complaints allegedly caused by amateur transmitters are really caused by something else, this chapter is worthwhile reading for most amateurs. Unfortunately, some of the general public continue to blame all RFI on the nearest visible radio tower. This may mean that you will be called if a nearby electrical-noise problem interferes with neighboring equipment. Of course, your sensitive amateur receiver will usually be the first to feel the effects of electrical interference, so you are probably already on the job!

If you receive a complaint and you know that your transmitter or other equipment is not at fault, it might be neighborly to offer to check the neighbors' RFI problem. This may help convince your neighbors that your equipment is not at fault, since you cannot be at their house and operating the transmitter at the same time. It also helps show that amateurs are willing to help track down the offending noise. It will, of course, eliminate a possible source of RFI to your own equipment (which may be coming from your neighbor's house). Most important, it will point your neighbors in the right direction toward the correct solution for the problem at hand and help all involved understand how to achieve it.

SAFETY FIRST!

A better heading might be "safety always!" All areas of electronics can pose safety hazards, but the entire area of electrical interference can involve the electrical and power-distribution systems. Every year, there are fatalities from electrocution. Don't be among them!

RFI work must be done safely. There are many hazards to you and your equipment that can be avoided with a little common sense during the search for a cause and a cure of your RFI woes. There is nothing in this world that is more important than your safety. When it comes to safety rules, *there are no compromises*! The only "dumb questions" are the ones not asked. No one is immune to the possibility of an accident.

Make Sure You Are Covered

First, if you are not sure of what you are doing, find out how to do it right or don't do it! In commercial and residential environments, outside electrical power is brought into a distribution panel. These are often called "breaker boxes" or "fuse boxes." From there, it is distributed throughout the structure. Most electrical outlets are also protected with a cover, to prevent accidental contact with the wiring behind them. Never remove a protective cover plate from a distribution panel or outlet unless the power is off and you are *qualified* to do so. The voltages behind these covers can be deadly. Many state and municipal regulations require that all work on electrical wiring is only performed by licensed electricians. The available fault current (the transient short circuit current available before the circuit breaker trips) in a home distribution panel is in the thousands of amps. If a screwdriver were shorted to ground in it, it would vaporize it in milliseconds. The resultant sparks would be explosive in nature.

Utility Poles and Wiring

The distribution poles and wires outside of your home are the property of the utility company. They are *not* public property! This means you have no right to do anything to the pole and its conductors except look at them! The high voltages present are lethal. They can range from 120 V on the secondary conductors to 34,000 V on the primary conductors (and up to 765,000 V on a transmission system). Should one of these primary conductors fall to the ground a person need not contact it to be electrocuted. The voltage gradient present in the ground around the conductor could injure or kill a person nearby.

See the sidebar "Power Poles and Utility Wiring" for important information about what *not* to do!

Circuits

Do not work on a circuit unless the power

Power Poles and Utility Wiring

There are a number of important safety rules when it comes to power poles and utility wiring and your search for the cause of an electrical RFI problem. They can all be summarized in one important sentence: DON'T DO IT!

Never hit a pole with a hammer to "check for loose hardware." In many cases, electrical noise *is* generated by loose hardware on the poles but it could be so loose it could fall off the pole at the slightest provocation. If there is a cracked insulator, any physical stress on the pole could cause the insulator to shatter thus creating a ground fault to the high voltage conductor. Don't diagnose or cure your noise problem by dropping the defective parts onto passing traffic or a pedestrian.

The same goes for yanking on guy wires. Poles can rot at the base near the ground line. A good tug on a guy wire could send a pole crashing to the ground!

Do not climb a pole or put a ladder up against it to "get a better look" at the hardware. Another dangerous practice is to climb a tree adjacent to a pole. Power lines could be hidden in the branches or a conductor could be touching a branch and any contact could result in electrocution or serious harm.

All these situations might seem unlikely, but they have all actually occurred and have sometimes resulted in injury. Leave problems with poles or utility wiring to the power company. The power company staff is trained and equipped to handle any situation that could arise, so let them do their job.—*Jody Boucher, WA1ZBL* has been removed and locked off. Make sure any capacitors in the circuit have been discharged through a load resistor. Do not open any appliance case unless you are qualified to work on that appliance. Even if you are qualified, it is unwise to modify equipment that belongs to someone else. The FCC places the burden of RFI correction on the owner of the affected equipment. To ensure that you are not held liable for any future problems, equipment modifications should be left up to the manufacturer or an authorized service technician.

DEFINITION OF INTERFERENCE

Electrical noise interference fits the same general definition of interference discussed in earlier chapters—it is the undesired interaction between two electrical or electronic circuits. Noise interference can be caused by natural static such as a lightning storm or other atmospherics heard in a radio receiver. On the other hand, unwanted signals may come from light dimmers, motors, electric fences or the power lines outside our home. These forms of interference are "man-made" and can often be eliminated or reduced to an acceptable level.

Man-made interference has two elementary causes. The first cause is equipment that produces electric arcs. Depending on the equipment and circumstances, an arc may or may not be a desired end product. For example, undesired arcs occur in power lines, bimetallic thermostats, switching circuits and brush-commutator type motors. Desired arcs are found in welding equipment and fluorescent lamps.

The second cause of man-made interference is equipment that contains electronic oscillators. Transmitters, radios, TV receivers, scanners, microprocessor controlled equipment and microcomputers all contain oscillators that can cause severe interference problems. This type of interference is addressed in other chapters in this book.

Evolution of Electrical Interference

In the early days of radio, there weren't many transmitters and receivers competing for airspace. Radios were few and far between and the electrification of America was in its infancy. As time passed, radio flourished and became America's favorite pastime. The demand for electricity grew with equal enthusiasm. More power was needed, since every home was soon to be outfitted with electric light bulbs. Soon everything in the home, from washing machines to toasters, was running on electricity. Parallel to this, the industrial revolution was equally in gear. Factories were getting away from unreliable hydro-power that was dependent on river flow and instead using electric motors to replace the old water wheels. This put a huge demand on the local power companies, which were often tiny municipal cooperatives or local utility companies. The need to supply more power and run larger conductors at higher voltages was a financial burden to many of these small groups. They found it beneficial to merge into larger companies so they could raise more capital to build new equipment. This allowed them to supply ample amounts of power, which then attracted more industry to come into their towns and therefore bring more jobs and prosperity to the local community.

And run wires they did! No longer is the power distribution at 2400 V with #6 wire. Today the distribution voltages can be as high as 34000 V, run both over and underground. The conductors can be anywhere from #6 AWG to much thicker cable. Along with these advances came RFI. Many times the power companies would upgrade the voltage on their lines without changing conductor size or insulators. A higher voltage and a transformer change meant more watts available to the customer and less of a voltage drop on the power line. Raising the voltage on the lines put more voltage stress on the power equipment and often resulted in RFI generated from this. Some of these lines still exist today in many parts of the country.

The radio and electronics field grew at an astonishing rate. The regenerative receiver is gone; now there are cable television and satellite dishes, microprocessor-based electronics and receivers capable of 0.1 μV sensitivity. All are susceptible to interference from power lines and other sources more than ever before. Many parts of the country have power equipment still in service that is over 50 years old. Much of this is in relatively good condition for its age due to the high quality of workmanship when it was manufactured. Some has deteriorated to a point where it has become a considerable source of RFI, however. This is where a lot of RFI-locating resources go to. In some parts of the country, much of the equipment is relatively new and was designed with RFI prevention in mind using new materials and techniques in line construction.—*Jody Boucher, WA1ZBL*

WHO IS RESPONSIBLE?

The Law

FCC regulations limit the amount of RF noise that can be emitted by different radio and nonradio devices. These regulations are found in Part 15 of the FCC regulations. Under these FCC rules, power lines and other electrical equipment are classified as "incidental radiators"—devices that are not designed to use RF energy, but do radiate RF energy as an incidental, unintended consequence of their primary function and operation. In the case of power lines, this function is transmitting 60 Hz electrical energy; other electrical devices, such as motors, have a different intended function. Many electrical devices, however, also act as incidental radiators.

No Absolute Limits

In the United States there are no *specific* or absolute interference radiation limits for incidental radiators—overhead power lines and most electrical devices. FCC regulations require only that incidental radiators not cause harmful interference to authorized radio services

Harmful Interference

The term "harmful interference" can be subject to interpretation. What may be harmful to one party or organization may not be considered harmful by another. In most cases, people can work out their differences of opinion between themselves, but in others, the FCC can also help decide whether an incidental radiator is creating RFI, and to what extent an RFI source needs to be "cleaned up." The FCC is generally reluctant to do this, feeling that imposed solutions to local problems are rarely ideal.

The FCC uses a number of factors to decide to what degree a utility or other generator of interference needs to go to, to eliminate an RFI source. These are cost, number of individuals involved, severity of interference, the number of stations or frequencies and the distance from the originating station (fringe area). For example, the FCC is not likely to rule that a utility spend large sums of money on transmission line repair to resolve a complaint from a few individuals in a fringe area of TV reception. In many cases, the FCC staff will make a determination on the basis of what is in the public interest, balancing many factors. See the sidebar, "Where to Draw the Line." More information on FCC rules about

Where to Draw the Line

Power lines are classified as "incidental radiators" operating under Part 15 of the FCC rules. This means that these devices are not designed to use radio frequencies as part of their operation, but "incidentally" generate and radiate some RF energy anyway.

FCC Regulations

FCC regulations about incidental radiators and power line noise are deliberately not specific. Incidental radiators come in many different configurations and can be used in many different circumstances and locations. Each configuration, such as the difference between a small electric motor and a country-wide power-distribution system, could require a different set of test conditions and rules. Many incidental radiators could be found in locations where it is not at all likely that interference to radio services could occur. For all of these reasons, there are no *specific* limits placed on incidental radiators. To offer general protection about interference, FCC rules state:

•Title 47, CFR Section 15.5 General Conditions of Operation

(b) Operation of an intentional, unintentional, or incidental radiator is subject to the conditions that no harmful interference is caused and that interference must be accepted that may be caused by the operation of an authorized radio station, by another intentional or unintentional radiator, by industrial, scientific and medical (ISM) equipment, or by an incidental radiator.

(c) The operator of the radio frequency device shall be required to cease operating the device upon notification by a Commission representative that the device is causing harmful interference. Operation shall not resume until the condition causing the harmful interference has been corrected.

•Title 47, CFR Section 15.13 Incidental radiators:

Manufacturers of these devices shall employ good engineering practices to minimize the risk of harmful interference.

•Title 47, CFR Section 15.15 General technical requirements

(c) Parties responsible for equipment compliance should note that the limits specified in this part will not prevent harmful interference under all circumstances. Since the operators of Part 15 devices are required to cease operation should harmful interference occur to authorized users of the radio frequency spectrum, the parties responsible for equipment compliance are encouraged to employ the minimum field strength necessary for communications, to provide greater attention of unwanted emissions than required by these regulations, and to advise the user as how to resolve harmful interference problems (for example, see Sec. 15.105(b)).

Harmful Interference

The crux of the regulations centers on "harmful interference." Amateurs, utility companies and the FCC may not always agree on what constitutes interference. Clearly, if electrical noise is causing interference to over-the-air television within the service area of a TV station, and causing an S9 noise level across the HF range, this would be seen as harmful interference by all parties concerned.

The FCC provides a very specific definition of harmful interference:

"Section 15.3 (m) Harmful interference. Any emission, radiation or induction that endangers the functioning of a radio navigation service or other safety services or seriously degrades, obstructs or repeatedly interrupts a radio-communications service operating in accordance with this chapter."

Amateur Radio, however, sometimes pushes radio to its very limits. This has been one of the strengths of the service, as hams have shown the world that the impossible could sometimes be done. This, however, can also push the concept of "harmful interference" to its limits—often much further than it was really intended to go. Hams need to be realistic in their expectations of what can be reasonably accomplished by a utility company that is trying to meet the spirit of the FCC rules.—*Ed Hare, W1RFI, ARRL Lab Supervisor*

all types of interference can be found in the Regulations chapter of this book.

Power Company Responsibility

Under Part 15 rules, the operator of an incidental radiator is responsible for its legal operation. If harmful interference occurs, the operator of the incidental radiator must take whatever steps are necessary to correct the interference, including terminating its use, if necessary. In the case of electrical interference, the power company is clearly the operator of the power lines, and is thus responsible for interference that may result from their use. The power company, however, is not held responsible for RF noise generated by devices hooked up to their lines, such as motors, switches, arc welders or even entire factories full of noisy devices. In these cases, the power companies are usually willing to act as consultants and troubleshooters, as a customer service issue.

When it comes to the power company and their responsibility to repair equipment owned by them, the jurisdictional areas of ownership can vary from one company to another. Obviously the utility pole and the primary and secondary wires attached to it belong to the power distribution company. But what about the wires that feed a home or factory? Whose responsibility are they if they should fail and need repair, or they are suspect in the generation of RFI?

There is a well defined point at which the change of ownership occurs between the utility and the homeowner, and it is important to know who is responsible for what in the electric supply system. Nearly all homes in the United States are fed with a 120/240-V 60-Hz supply at typically 60 to 200 A and all have a meter that measures usage. How it enters the home and where the meter is located can vary widely. You can determine the type of service entrance that feeds your home by first locating the electric meter (most often located outdoors attached to the side of the house, but sometimes located indoors or in a cellar). There are two cables leaving the meter. One goes to the main service panel of the home where the circuit breakers or fuses are located, and the other is the source from the utility. This can be either an underground feed from the street or an overhead feed from a utility pole.

Underground Electric Service

One set of wires connected to the meter box is connected to the Utility Company source. This cable can either be directly buried in the ground or run underground in conduit to a utility pole on the street (or in the case of all underground utilities to a pad mount type transformer or a splice box located in the ground near the street). Ownership of this type of service entrance varies widely; some utilities own the entire cable to the house and the meter box, and some make the homeowner purchase it and have it installed, leaving the final hookup to be done by the utility linemen.

Overhead Electric Service

With this type of service, the cables leave the meter box and run up the side of the dwelling usually about 16 feet above ground level and connect to an overhead cable that comes from the utility line on the street. Again, ownership of this type of service entrance varies widely. Many utilities prefer to own the overhead portion of the cable making the change of ownership at the splice to the cable on the house that goes to the meter. Some utilities even own the cable and meter box, and as mentioned before some companies make the homeowner purchase the entire run to the pole and have it installed by a qualified electrician.

If after all this you are uncertain as to the type of service, location or ownership responsibilities call your local electric utility. They will identify the point of ownership change and help you with any questions. The utility is responsible for locating and repairing any interference sources on their side of the ownership change. The customer is responsible for the rest of the service entrance into the house. If it becomes necessary for the homeowner to have repairs made, they should be done only by a licensed qualified electrician. This is no place for a homeowner to try to make his own repairs.

The Good Guys

Although the power companies are responsible for solving interference that comes from their power lines and equipment, many power companies go over and above the requirements of the law, solving or locating other interference problems voluntarily. This is a customer service issue. These companies do more than their share by locating and correcting interference from many sources other than directly from power lines, perhaps by locating a nearby noisy motor or a defective thermostat in a residential building. These progressive companies may have formal programs and trained staff to help locate electrical-noise interference for their customers, whether the noise is created by problems in the power lines or by devices in the customers own home.

Even though the power utilities are technically responsible only for their own equipment, many do take the complaints seriously. An investigation into an interference complaint has often led the investigators to discover potentially dangerous situations, either in a customer's home or in their own power equipment. RFI is often the first warning sign of a failing piece of power equipment on a pole or an appliance in a customer's home. It could be a heating pad that was left turned on and stuck behind an elderly person's couch, ready to ignite, or the distribution panel with a red hot bus connection. Many things have been brought to a customer's attention that could have led to disastrous results if left unattended.

The utilities did not intentionally try to make themselves the so called "experts" in the field of RFI investigation, nor did they want this role, nevertheless, by default they are often the first source of information and help with RFI problems. Some power companies dedicate many man hours each year to interference resolution.

The Bad Guys

The flip side of this coin, however, is that there are some companies that are not meeting their legal or customer service responsibilities. In some cases, real power line noise problems are left unsolved, often with an excuse that "it is not our problem," or, "we don't have any way to locate the source of the noise."

If this happens to you, the information in this book can be used to help explain the appropriate FCC rules and responsibilities to the power company. Start with the customer service office in your utility company and explain that you have a problem with power line interference. In some cases, it may be helpful to deal directly with the service manager, who is more likely to have some experience with RFI problems. If the power company is not willing or able to help, see the section on how to resolve a power line complaint, later in this chapter.

In addition to providing information on how to best handle a power line noise problem, this chapter provides details on the ARRL cooperative agreement with the FCC. The ARRL RFI Help Desk can offer a wide range of help, including prepared information packages, answers to individual questions, and if necessary, refer you to FCC staff or local attorneys, who may be able to persuade the utility company to honor its responsibilities. Contacting an attorney is generally not a very efficient process; this drastic step should be considered only as a last resort, after all other attempts at resolving the problems have failed.

It is in the utility's best interest to take a proactive approach in solving customer complaints because of the high visibility to the general public. Cooperation and patience by all concerned, whether it be radio, television or other electrical interference problems, are the key words.

TYPES OF ELECTRICAL INTERFERENCE

Most electrical interference is generated in one of two ways, corona discharge and spark-gap type noise. Most devices that generate electrical interference in the home will most likely fall into the spark-gap category, such as a thermostat contact, brush-commutator motor or electric-arc welder.

Figure 11.1 shows a spark gap or an electrical arc. An electrical arc generates varying amounts of RF energy across the radio spectrum. These RF signals are completely random, and can appear as a raspy buzz in a communications receiver, or dots across a television screen.

Corona Discharge

Corona discharge is defined as the partial breakdown of the air that surrounds an electrical element such as a conductor, hardware or insulator. In order for corona to occur there must be a voltage gradient present. Voltage gradient refers to the voltage difference between two points divided by the distance between them. As can be seen in Figure 11.2, the voltage drop to ground is a nonlinear function. In other words, the highest amount of voltage stress occurs closest to the conductor. The smaller the conductor, the higher the voltage stress around it due to the tighter lines of EMF. Voltage gradient also depends on system voltage, the distance to earth or ground, the distance to other con-

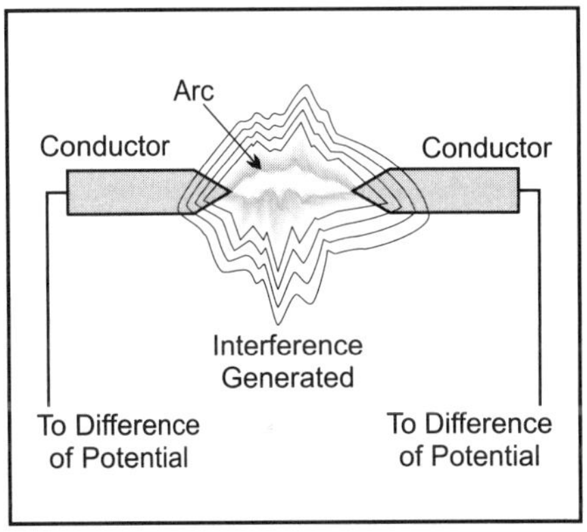

Figure 11.1—Representation of an electrical arc.

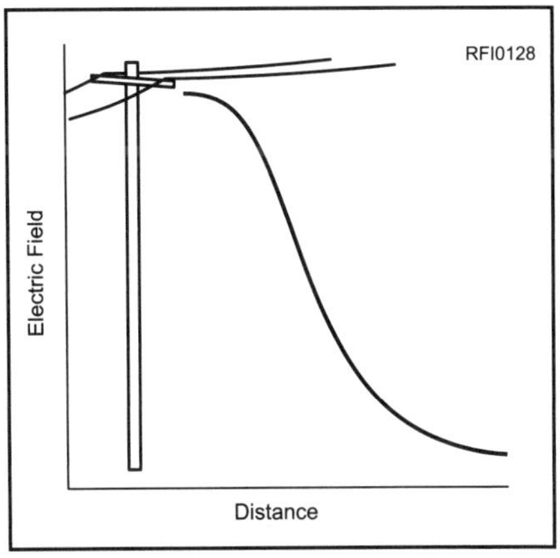

Figure 11.2—The electric field is very intense near this high-voltage conductor, but falls off rapidly with distance.

ductors, the types and size of conductors, and how many conductors are bundled per phase. The voltage gradient near a conductor carrying 230 kV may be as high as 7 kV per centimeter.

The air dielectric around a conductor carrying high voltage is going to have some high voltage stresses imposed upon it. At some point the stress exceeds a critical value of voltage gradient known as the corona-onset voltage. The air partially breaks down electrically by losing or gaining electrons. This produces ions, which now form a good conductive path, extending the range of the breakdown. Energy is released in the form of sound, electromagnetic radiation (RFI) and a pale violet light that surrounds the ionized-air conducting path.

Increases in humidity and precipitation can lower the onset voltage at which corona occurs. Any surface irregularities on the conductor can further result in a distortion of the lines of the electric field. At this point of the irregularity, a *corona plume* can form. Raindrops dripping off the bottom of the conductor can form plumes; these are known as precipitation discharges. These can be up to several inches long. All types of corona generate RFI noise.

The RFI generated from these corona discharges in dry weather tends to fall off sharply at 30 MHz (see Figure 11.3). When the humidity is very high or the conductors are wet, however, the noise can be easily heard into the VHF range, tapering off as the frequency is increased. The way corona behaves in humid weather is characteristically diagnostic. If the interference is caused by an actual arc, the noise may go away entirely when it rains. Corona discharge can affect VHF television, up to about channel 6, and 6 and 2-meter amateur bands. The distance corona-generated RFI travels laterally from a transmission line is dependent on frequency, weather and the severity of the corona discharge. It is, however, usually confined to within 1000 feet of the power line.

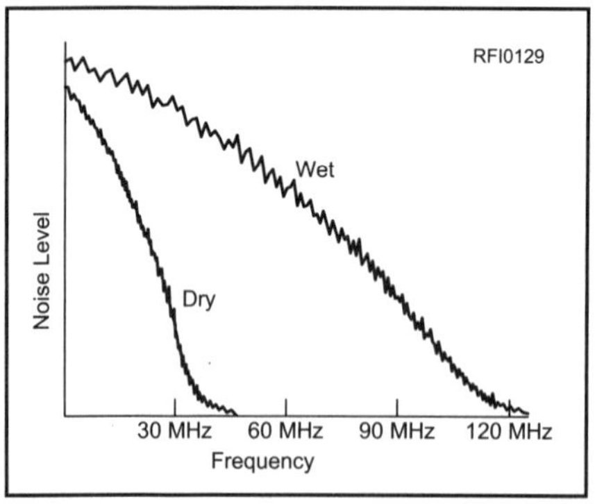

Figure 11.3— In dry weather, the noise from corona extends to about 30 MHz. In wet weather, however, it can extend to about 100 MHz or so.

What is in the Public Interest

As explained in §15.5(c), it is sometimes the FCC that is the final judge as to what constitutes harmful interference. The FCC normally doesn't do this until all those affected by an electrical-noise problem have tried to work out any disagreements between themselves.

The FCC exists to regulate radio transmission and reception. To do so, they often need to consider a number of different factors. The criterion the FCC often applies to their decision-making process is the "public interest." In the case of power line noise, the FCC has to consider both the public interest of interference-free radio reception (broadcast and amateur, for examples) and the maintenance costs involved. Keeping the lights turned on is in the public interest, too.

You Be the Judge

As amateurs, we want to see electric utility companies "do the right thing" and fix our interference problems. We have all heard the horror stories about utility companies not fixing serious electrical-noise problems. As an organization, the ARRL is working with utilities to raise their level of awareness about the causes and cures of electrical RFI. To do so effectively, however, we have to be reasonable in our demands. The utilities have shared some horror stories, too, such as the ham who complained about interference heard on 2 meters on a large Yagi antenna. The utility spent tens of thousands of dollars to locate a number of leaks in the neighborhood. When the utility representative revisited the ham to verify that things were better, the ham responded, "Well, it is quiet now, but look what happens when I turn on the preamp." The noise was gone across most of the rotation of the antenna, but in two directions, it increased by a few dB.

You be the judge. What do you think a local FCC engineer might decide is in "the public interest." The FCC knows that imposed solutions for local problems are not always the best solutions, and often do not please *anyone* involved in the problem. If you are not sure how the FCC might decide a case you are involved with, perhaps it is a good time to redouble your efforts to reach an equitable understanding with the utility.

If the ARRL is going to be effective in helping utility companies improve the way they handle RFI problems, we have to be certain that what we are asking for is appropriate and reasonable. Hams who live in cities, near industrial areas or near high-voltage transmission lines just cannot expect that the lines will be as quiet as that isolated, rural location. Perhaps finding the appropriate middle ground will let us make some real headway with that S9 noise level that really *does* need to be fixed.—*Ed Hare, W1RFI, ARRL Laboratory Supervisor*

Assuming the equipment is of good design and construction, little can be done to eliminate the RFI generated from corona discharge. *Grading rings* are used around high voltage stress areas in substations and at transmission towers to equalize the voltage gradient and prevent corona at critical points. Even the best designed lines and facilities can still have some corona present. These are problems for the power company personnel to solve.

Amateurs can sometimes make some changes at their station that might make corona and other types of electrical noise somewhat less bothersome. One approach is to increase the signal-to-noise ratio of the radio or television by the use of better, more directional antennas, or a preamplifier (although the latter can amplify the noise as much as the desired signal). With satellite dishes now available to the consumer, television interference can be eliminated completely at a reasonable cost.

Spark Gap

Some interference problems are caused by an actual spark. In a spark, the same breakdown and ionization of air occurs, but current flows between two conductors. A sustained spark is often called an "arc." In order for electric current to jump a gap, a *difference in potential* great enough to ionize air must exist in the space between conductors. For an arc to occur, there must be sufficient voltage to ionize (break down) the air in the gap. This voltage can come from a current carrying conductor or it can be induced into pole hardware from the strong electric field associated with high voltage power lines. These induced voltages can be in the hundreds of volts.

There is a definite relationship between the length of the arc, the voltage needed to sustain it and the amount of interference it produces. Sparking usually occurs across gaps of 0.05 inch or less, and the rise and

Who Is Responsible For Finding and Fixing Power Line Noise?

The electric utility is responsible for correcting only that noise generated by the equipment and hardware that it actually owns. In cases where a utility customer uses an appliance or device that generates noise, the operator of the device is responsible for fixing it—even if the noise is conducted and radiated by the power company's power lines.

Electric utility companies are often blamed for and even victimized by noise they do not cause and are not responsible to fix. This can be especially true when the customer-owned noise source generates noise similar in sound to true power line noise. Light dimmers, for example, are often mistaken for power line noise, especially by an untrained ear. Customer owned doorbell transformers are also notorious and often found to be the source of an RFI problem. The latter is an example of a serious defect that should be repaired. *In many cases, power line or electrical noise is the first indication of an electrical failure about to occur.*

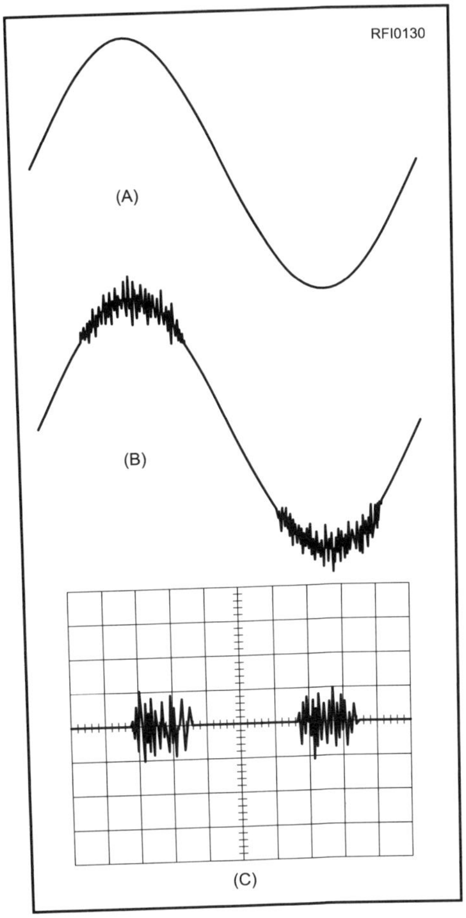

Figure 11.4—The 60-Hz signal found on quiet power lines is almost a pure sine wave, as shown in A. If the line, or a device connected to it, is noisy, this will often put visible noise onto the power line signal, as shown in B. This noise is usually strongest at the positive and negative peaks of the sine wave. If the radiated noise is observed on a 'scope, the noise will be present during the peaks, as shown in C.

decay time is extremely rapid. Under some conditions, sparks can trigger trains of successively weaker pulses.

Once an ionized path is established, current flows at the positive and negative voltage peaks—the times of highest instantaneous voltage throughout the cycle. (See Figure 11.4) Because power lines carry 60 Hz ac, the voltage on them passes through two peaks each cycle (one positive and one negative) and pass through zero twice each cycle. This gives 120 peaks and 120 zero crossings in each second. Corona and spark gap noise follow this pattern, generally starting and stopping 120 times per second. This gives power line noise its characteristic hum or buzz.

The flow of electrons (or arc) is not smooth because the resistance and capacitance of the ionized path changes constantly. The instantaneous current varies widely with the resistance, and the rate of charging of the electrodes depends on the voltage and the capacitance across the gap.

The resulting damped waves contain strong harmonics and can cause severe interference. These gaps can fire extremely fast and can generate radio frequencies as high as 1 GHz! The noise tends to be very broadband due to the fact that the characteristics of the gap, i.e. the resistance and capacitance, are constantly changing. Spark gap noise generally weakens with frequency (see Figure 11.5A), a characteristic that can be very useful as you track interference. One exception to this rule occurs when the power lines connected to the noise source resonate at a particular frequency or frequencies. The noise may peak at these frequencies. (See Figure 11.5B.)

Unlike corona, spark gap noise is usually a fair weather phenomenon; it may disappear in wet weather because precipitation short circuits the inter-hardware gaps. Any interference that changes with the weather is a dead giveaway: the noise source is almost certainly outdoors.

Sources of Spark-Type Interference

In the home environment, electrical noise is created in a variety of everyday appliances, especially those with brush-commutator motors. Electric razors, vacuum cleaners, sewing machines and air conditioners are just a few. Induction-type motors, found in record players, clocks and refrigerators, do not normally cause interference. Figure 11.6 illustrates the difference between the two. Induction motors do not use a brush-commutator system. Since no arcs are normally produced, there is no interference.

Very small (and harmless) arcs are also common in many thermostatically controlled heating systems. Some use a heat-sensing, bimetallic strip to open and close a set of contacts or operate relays to start zone

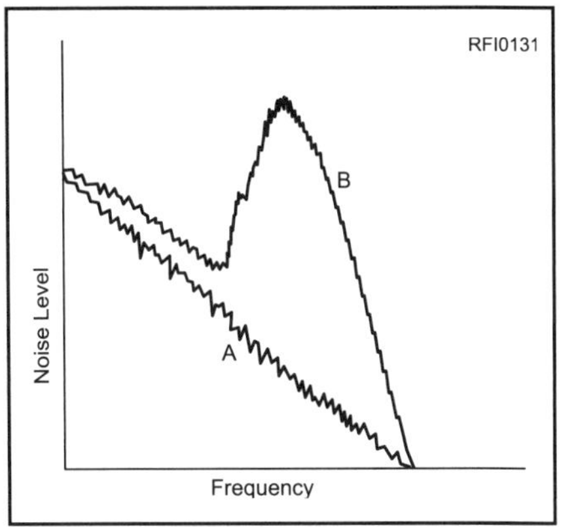

Figure 11.5—In most cases, the noise varies in inverse proportion to frequency, as shown in A. In B, however, a prominent peak can be seen as the noise intensity is modified by a resonant line connected to the noise source.

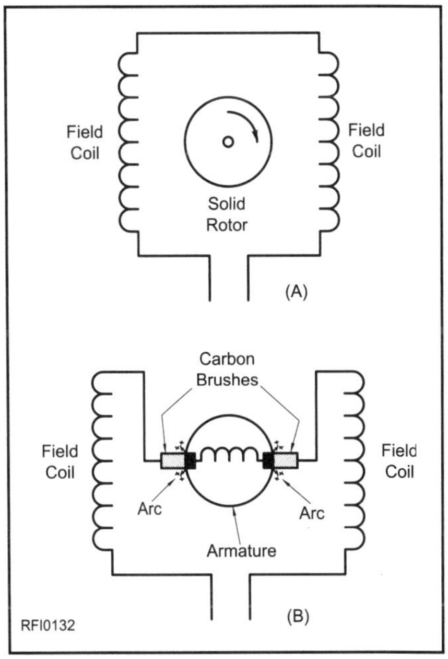

Figure 11.6—(A) Diagram of an induction motor. (B) Diagram of a brush-commutator motor.

pumps or circulation motors. As the contacts age they become pitted and prone to arcing. Arcs, sometimes more dangerous, can also occur in light switches, circuit breakers, or loose ac outlets.

Regardless of the severity, the possibility for an arc exists in almost any piece of electrical or electronic equipment. A common problem with homes built in the 1970s was the use of aluminum wiring. The constant expansion and contraction of the conductors causes the electrical connections to become loose and arc resulting in RFI. House fires have been blamed on this problem.

"Tracking"

Tracking can be defined as a leakage current from the primary source to ground, such as might be found across a high-voltage insulator on a power line tower. It is generally a surface phenomenon, but the term can also apply to discharge current paths in free air. Tracking is a phenomena that occurs mostly with high-voltage equipment. Power company substations and power lines are the most common places to find it. The insulator can become contaminated with dirt or other debris and form a resistive coating on its outer surface. The electrical current flows through this resistance to ground further eroding the insulation. In the long run, the insulator usually fails, diagnosing the problem.

Tracking is usually made worse by wet or humid weather. It can be particularly troublesome near the ocean where salt spray is blown onto equipment. Tracking itself does not directly cause much RFI. It can, however, lead to the degradation of an insulating device. This could lead to an arcing type problem and RFI, or a complete flashover, causing a major fault.

TRANSMISSION OF RFI

RFI noise generated by power company equipment or devices connected to it can be transmitted effectively over a wide range of frequencies over great distances. Power lines contain miles of wires that can act as a multiple-wavelength antenna. RF signals present on the lines can travel miles down a transmission or distribution line. At RF, the spacing of most power lines is a significant part of a wavelength, so RFI noise is also radiated fairly efficiently by the line.

There are three ways an RFI signal can be transmitted from a noise source. The first is by direct *radiation*. This happens if the interference source is strong enough to transmit directly into the free air from the spark gap source. In this case, the ionized air inside the spark is acting as a small antenna. Because the "antenna" is small, the arc is broadly resonant at VHF and higher, so this type of noise is stronger at those frequencies.

In addition to the noise radiated directly by the source, the source is also generating broadband noise across most of the radio range. The lower frequency component of this noise is generally *conducted* down the power line, whether it is a current-carrying conductor or the neutral wire. This can travel many miles on the conductor especially at AM broadcast band frequencies and the lower part of the HF range.

This conducted noise can be coupled to the nearby primary and secondary conductors by *induction*. Thus a noise from an overhead transmission line can induce noise in a nearby distribution line while it in turn can induce the noise into the 120/240 V secondary to your house. This again is usually an HF phenomenon.

If the noise is radiating from the power line conductors, it is usually heard for quite some distance from the source. On HF, the noise may propagate for miles. Even on VHF, the noise may be present for as many as 10 or more poles away from the noise source. In both cases, standing waves on the line can complicate this otherwise simple picture, with peaks and nulls in the radiated noise that vary with distance from the source.

In cases where the arc itself radiates, or if the noise is radiating directly from a piece of hardware not directly connected to the power line conductor (such as a cross-arm bracket), then the noise will seem to be coming from one structure.

In most cases, the noise source is actually a combination of these effects. This can really complicate the troubleshooting process, as different propagation mechanisms set up various standing waves and add and subtract from each other at various points.

The result of the combination of all of these factors is that power line interference tends to be inversely proportional to the frequency, the higher the frequency the shorter distance it is from the origination point. Interference in the upper VHF or UHF region of the spectrum is almost always a result of direct radiation from the source. In a realistic situation, however, the RFI is generally transmitted by all three methods due to the broadband nature of the noise. This can be used to a troubleshooting advantage. You can use the AM broadcast band or lower HF to get to the neighborhood of the noise source; you generally won't be able to hear it on VHF unless you are very close. This often helps you identify the exact pole or house with the noisy device.

RF on Power lines

Until now, this chapter has talked about RFI noise generated by power lines or devices connected to them. This is generally a differential-mode problem. Other than eliminating the source, the cure for this type of problem, either at the source or at the affected equipment, is generally a differential-mode type of filter. See the RFI Fundamentals chapter and the section on filters later in this chapter.

Transmitted RF can also be picked up by power lines, house electrical wiring or on appliances connected to house wiring. Although related to the general topic of this chapter, because electrical wiring is involved, this is actually a different phenomenon than the noise and electrical arc interference discussed earlier. In most cases, the RF energy picked up on electrical wiring will be a common-mode signal. This common-mode signal can often be involved in fundamental-overload problems. The cure, as it is for any common-mode problems, is generally a common-mode choke.

If you are considering cures for electrical interference caused by a local noise source, think of either correcting the problem at the source or using differential-mode filtering techniques, if appropriate and practical. If you are considering cures for interference *to* electrical devices from nearby transmitters, think of common-mode choke techniques.

Filters

The most effective electrical-RFI filters are those installed right at the *source* of the RFI. For example, it is much more effective to filter a noisy motor at the source than it is to filter it at a distant receiver. If the RFI noise were *entirely* conducted between the source and affected receiver, a differential-mode filter could be used at *either* end equally effectively. Unfortunately, however, in most cases the RFI noise is radiated to one degree or another by the electrical wiring. Once noise is radiated, it can't easily be filtered.

If you have done your best to eliminate the offending interference, or you were unable to convince the owner of the noise source to fix it, it may be your only recourse to try to filter the RFI at *your* equipment. There are several methods of filtering the signal or power source to make the signal usable. This section will only deal with filters for arcing type noise and power line filtering. Several other methods and types of filters are covered in other parts of this book and related articles.

Filtering the ac line at the source of the noise, such as a motor or neon sign, can effectively reduce the generation of RFI by preventing the noise from being conducted onto the power line where it can be radiated. An ac line filter can also help prevent noise from entering an electronic device via the ac line. As previously mentioned, simply

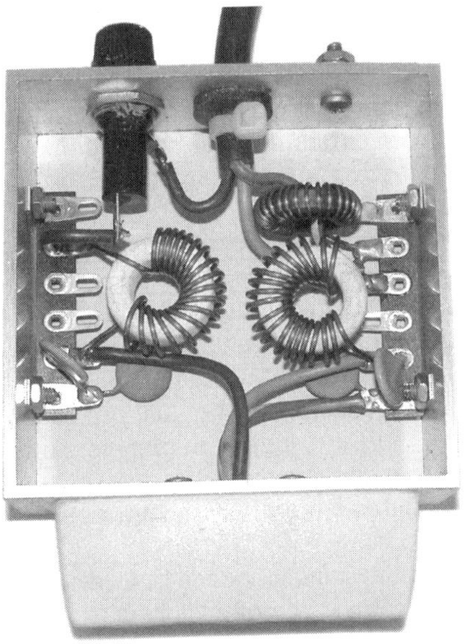

Figure 11.7—An inside view of the ICE 475-3 ac line filter.

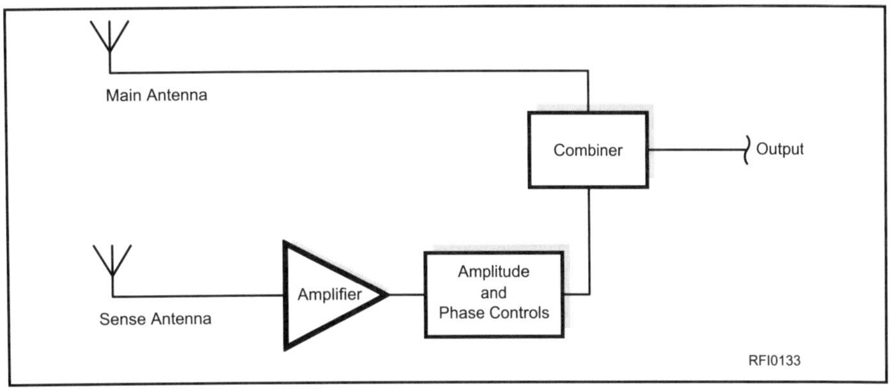

Figure 11.8—A noise-canceller works by combining the desired, but noisy signal from a receiver's main antenna with the signal from a sense antenna that is positioned to pick up more noise than signal. These devices can give up to 50 dB of attenuation. Their limitations are that they work well for only single point sources of noise and have to be carefully adjusted for frequency changes of more than several kHz.

adding bypass capacitors between a signal or power lead and circuit ground can provide a low impedance path to ground for RF signals. Bypass capacitors for HF signals are usually 0.01 μF, while VHF bypass capacitors are 0.001 μF. Any capacitors placed across the ac line should be rated for use in ac circuits.

AC line filters, sometimes called "brute force" filters, are used to filter RF energy from the power lines (see Figure 11.7.) We generally recommend using a UL-listed commercially made filter. The type of filter used depends on how the signal is coupled into the susceptible equipment. The signal can be coupled in the differential mode or the common mode and each takes a different type of filter to solve the problem. More information on all types of filters can be found in the RFI Fundamentals chapter.

Noise Canceller Devices

In some cases, however, it is possible to use techniques to eliminate the noise at your receiver. Devices that actually cancel various noise sources are available on the market. A block diagram of one of these devices is shown in Figure 11.8. The *noise antenna* is generally positioned to pick up more noise than the desired signal. The *signal antenna* is your station's normal antenna. The device has phase and amplitude adjustments that will combine the noise from the noise antenna out of phase with the signal from your main antenna. By carefully adjusting the controls, you can usually find a combination that will almost completely cancel the noise heard in your receiver. It sounds like magic, and to someone who has an electrical noise problem that simply cannot be fixed any other way, it is.

These devices do have some disadvantages, however. They work well only if there is a single point source of noise. They will not work well if there are multiple noisy insulators or multiple noisy electrical appliances nearby. They are also tricky to adjust and will require readjustment for frequency changes of more than about 10 kHz or so. Still, in some cases, the limitations are worth the gain. As shown in Figure 11.9, these devices won't take up much room on your station table.

Audio Filtering

Filtering can also be done at the audio output of the receiver. Conventional audio filters limit the bandwidth at audio frequencies. This can help eliminate noise because most noise is broadband. A reduction in bandwidth means there is less noise power. CW and digital signals are very narrowband, so you can reduce the bandwidth and noise quite a bit without affecting the desired signal. This isn't very practical with voice signals because human voice reception needs to include the range of 200 to 2500 Hz to remain intelligible.

The effect of using narrow-bandwidth audio filters is essentially the same whether analog or DSP type filtering is employed. There are, however, DSP techniques that are able to differentiate between noise and coherent signals, such as CW, RTTY, etc. These DSP filters don't help much if the signal is at the same approximate level as the noise, but can help make a somewhat noisy signal sound much better.

Noise blankers, whether analog or DSP-based, can help with certain types of impulse noise. They generally don't help much with broadband noise created by noisy electrical devices.

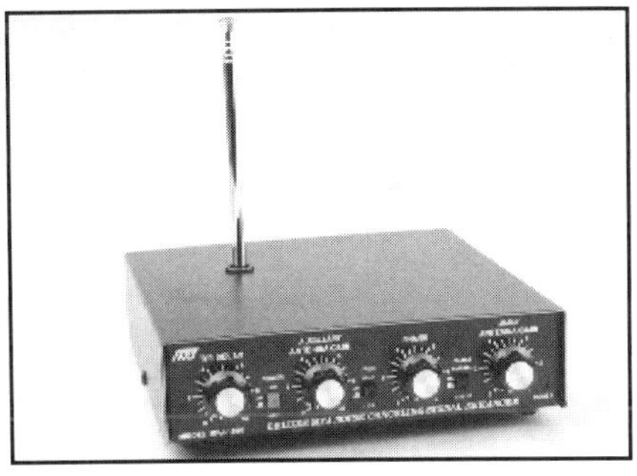

Figure 11.9—This noise canceling device was featured in the April 1998 *QST* Product Review column.

LOCATING RFI

Before an interference problem can be cured, it must be found. In many cases, the power company will be able to find the source of the problem for you. In other cases, however, the power company may not know much about interference, so they might need your help to pinpoint the source. In most cases, if you are able to identify the correct pole or house, the power company personnel can take over from there. The RF Direction Finding chapter, and the section in this chapter on how to resolve a power line noise complaint, offer a lot of guidance about how to find noise sources.

Assistance from the Power Company

In many cases, the power company can help you locate the source of the noise, even if it is not being caused by their equipment. Many power companies have formal, or informal, RFI investigators.

An RFI investigator is an engineer or technician employed by the power company to investigate RFI complaints. The best investigators are well versed in power distribution systems, RF systems, CATV systems, microwave systems and so on. They have access to the equipment needed to conduct a thorough investigation. The power company relies on the investigator to solve power-system RFI problems. If you do not feel comfortable finding it yourself, call the power company and ask to talk to an investigator.

Before Calling for Help

Before you do anything else, analyze what is happening. Answering a few questions will often be a good start toward identifying the problem. What does the noise sound like? Is it a broadband hiss or an intermittent raspy buzz? Does it have a high pitch or a low one? What frequency does the noise appear on and how high in frequency can you go before it disappears? Does it affect both AM and FM receivers? Does it affect your TV and telephone? Each of the answers may provide valuable clues to you or the power company RFI investigator.

What about the time of day, is it morning afternoon or evening? Does the noise have a definite cyclic rhythm to it? What about the weather? Does the noise go away in the rain or is it worse in the wet weather? Is it on only when your neighbors are home? It may help to keep a log when the noise happens, time of day, weather and any other pertinent information that might help. All this will help identify the problem and point you right to the source.

Most of the time, if the noise varies with the weather, it is outside, often on the power company distribution equipment. If it varies with the time of day, it is caused by something people are doing. For example, you may note that the noise occurs every morning for 10 minutes at about 7:15. This could be the approximate time that a neighbor uses a hair dryer or electric shaver.

Another thing to consider is that, in the real world, every device adds a small amount of ambient noise, so every location has a definitely established ambient noise level. An urban environment surrounded by homes and factories simply will not be as quiet as that country mountaintop location. Any attempt to use a radio or television for reception of signals which are below the ambient noise level of a given location is doomed to provide unsatisfactory reception.

The interference experienced in the operation of a receiver depends on the character of the offending radio frequency or random noise, the coupling between the source and the receiver, and the strength of the desired signal. In other words, what is the signal to noise ratio? Today the S/N ratio of the equipment is excellent but, therefore, more susceptible to the noise in the area, and today there are more devices to raise the ambient noise level. So we need to have realistic expectations of what we can expect to receive in an electrically congested area.

If you suspect power line RFI, first make sure the ham station is not the RFI source. Ensure that all connections are sound and cables are properly connected or terminated. For example, a loose antenna feed line can cause noise problems in a receiver. An arcing antenna trap could cause area-wide noise that might sound like a power line problem. Make certain that all individual equipment grounds are connected to a single point in the station, which is then connected to the station ground rod.

Next, check the area outside the house using a portable radio tuned to a frequency where the interference is prevalent. Listen for an increase or decrease of the interfering noise. If there is a power pole nearby and the noise increases as you get near it, note the pole number.

Continue the investigation by walking around the block while listening to the portable receiver. Note any changes in the RFI and any other pertinent information such as the location of an operating arc welder. The information you collect will help the RFI investigator. It is a good idea to tape record the RFI, or videotape any resultant TVI, just in case the noise is not present when the RFI investigator arrives.

When you are reasonably sure that the RFI is coming from the power line system or you are unable to make any definite determination, call the power company. The person who answers the phone will direct you to a power company employee who is knowledgeable in power line RFI problems.

From this point on, let the RFI investigator control the investigation. If you think of additional information that might help the investigator, by all means, speak right up. You might think that a particular trait is trivial, such as the fact that the noise goes away in the rain or that it only happens at a certain time of day, but it just might be the lead that resolves the problem.

There are many possible ways that a power line can create interference. Some of these sources have characteristics that allow an experienced RFI investigator to locate the source of the problem quickly. In all cases of power line RFI, turn to power company personnel for help. Your Section Technical Coordinator may have already established contact with the technical people at the power company who can help you.

TROUBLESHOOTING TECHNIQUES

There are many different tools used to locate RFI, some as simple as a homebrew Yagi, to the hand-held commercially made all in one receiver with a complete array of antennas. A simple 2 meter fox hunting setup could work just fine with an AM receiver, or a portable receiver with shortwave bands could work too. Some of the more elusive RFI sources require some better equipment. Commercially made equipment is expensive but well worth it.

Ultrasonic audio emissions are another by-product of an RFI source and can be used to locate it, if there is a true line of sight path between the source and the probe. Several companies manufacture ultrasonic detectors with parabolic dishes that can be used by the RFI investigator to pinpoint an arc. (Commercial ultrasonic detectors cost a few thousand dollars each, so they are not usually used by hams.)

The ultrasonic probe can be a device that mounts on the end of a "hot stick" (an insulated pole a lineman uses), or it can be in the form of a parabolic dish that can be used from the ground (see April 2006 *QST*, page 41 for a home-brew article).

RF detectors can also be used. For close-in work the Micro-Tech *Little Snoop* is a broadband HF detector that can be used in the home on residential wiring or on the end of a hot stick for a lineman to use at a pole. Don't do this at home! The only people who should put any device this close to a power line are the trained power company personnel.

Other Detectors

Another useful tool is an oscilloscope attached to the output of a receiver's audio. Looking at the noise pattern on the 'scope will help ensure that the same noise is being tracked and identified as the receiver's search frequency is changed. Infrared detectors are useful in finding areas of high heat, which could indicate a poor electrical connection that can result in circuit failure.

Sensory Clues

A light amplifier is another high-cost tool that may be found in the RFI investigator's tool kit. It can be used at night to observe high voltage substations or power lines for any offending problems caused by corona. (At night, corona can sometimes be seen with a set of good binoculars, although some corona may not be visible. Coordinate this with your neighbors to prevent any misunderstandings!) You may be able to hear arcs from the ground if they are particularly bad. Another obvious clue is a red-hot line connector on a primary conductor, indicating a high resistance connection. This should be corrected as soon as possible!

Tracking the Interference from Sources Inside Your Home

You may choose or find it necessary to search for an RFI noise source on your own.

Before attempting this please read the safety section of this chapter carefully, your life may depend on it! Remember, only do the following if you feel qualified to do so. If you suspect that the problem is caused by electrical noise, check for overloaded circuits, frayed wires, loose sockets, etc. These types of problems should be fixed no matter what! First, make sure the interference isn't coming from *your own home* (quite a few power line noise complaints stem from sources within the complainants' homes!). The easiest way to check is to get a battery powered AM radio (noise does not usually affect FM) and tune it to the interference. Then turn off the power at the entrance breaker panel. (A typical unit is shown in Figure 11.10.) If the noise goes away as you cut the power, the noise source is located in your own house and is not a power company problem.

If cutting the power does not stop the noise, turn the power back on and skip to the section on "Tracking Interference from Sources Outside Your Home." If cutting the power reveals that the noise comes from a source in your own home, turn the power back on and confirm that you can once again hear the noise from the monitor receiver. Next turn off all the individual house circuits and the noise should again go away. Isolate the noise generating circuits by turning the individual circuits back on one at a time. When you activate one that brings back the noise, turn it off again and note that circuit. Work through the remaining circuits to determine if any more are noisy. (There may be multiple sources!) Once you have worked through the entire service panel, turn everything back on except the noisy circuit(s). You should hear no line noise in the receiver. After you have determined what circuits are involved, identify which rooms or devices that circuit powers and turn it back on.

Once you have identified the room(s) involved, unplug—don't just turn off—all appliances and switch off all the lights in the suspect room(s). Plug them back in and switch them on one at a time. When you activate the one that brings back the noise, turn it off and note that appliance or light. Work through the remaining appliances and lights to determine if any more are noisy. Powering up the suspected appliance should bring back the noise. It should either be repaired or discarded because it may have malfunctioned internally.

If you can't find the faulty appliance, track down the interference by carrying the receiver around the house to find where the noise is the strongest. An RF sniffer can be built or purchased (see the Troubleshooting chapter). This type of tool is very handy for in-house RFI investigation, and it can also be used by a lineman for power pole troubles. You may be able to use the receivers medium wave antenna directivity to get an idea of the noise source's direction. Even with everything in a circuit unplugged and shut off it is possible that faulty wiring may continue to power the noise source, as in the case of aluminum wiring. If this is the case, the circuit should be shut off and an electrician should be called to investigate the house wiring.

Tracking Interference from Sources Outside Your Home

If your investigation suggests that the line noise is coming from a source from outside your home, it may be time to brush up on your DFing skills. (See the chapter on RFI Direction Finding.) Using certain specialized equipment might substantially improve the RFI investigator's success rate. A portable HF short-wave receiver with an AM VHF band and signal-strength meter would be a necessity, along with some type of directional antenna (see September 1992 *QST*, page 52, for an antenna project), such as a small portable Yagi. It is important that the test equipment be more sensitive than the equipment experiencing interference. In many cases, hams who are doing sensitive, weak-signal work may hear noises that a hand-held sniffer simply can't detect. There are several commercially made receivers, 'scopes, directional antennas and ultrasonic devices designed for the RFI investigator that are usually only found at utility companies due to their substantial cost.

When you are searching for interference over a large area (a neighborhood), interference heard is the sum of *all* interference sources in the area. Ideally, you can use triangulation to locate the source. This is done by obtaining a bearing from several different directions to narrow it down to one area. This technique is only marginally effective on VHF, however. The problem is that the noise can reradiate from several (or many) different poles or be conducted for a long distance depending on the search frequency. And it may be physically impossible to position yourself on three sides of a noise source due to a lack of roads or private property. Signals can also reflect off ground or buildings, or be reradiated by nearby conductors. An experienced DFer can sometimes compensate for these variables, however.

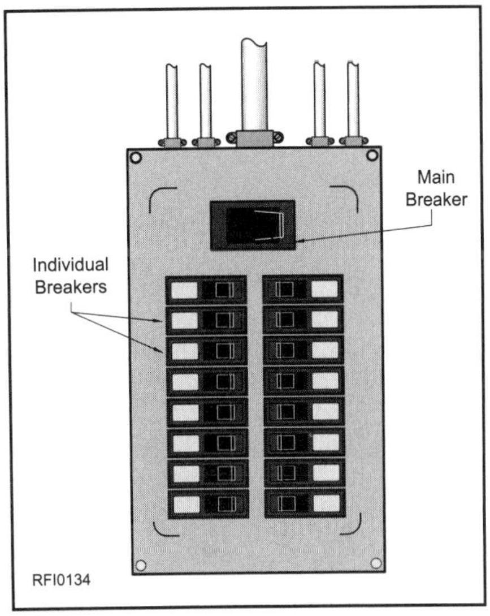

Figure 11.10—A typical home circuit-breaker box. The main circuit breaker is at the top center with the branch circuits below.

If the power lines are on a roadway that is accessible to a vehicle, you may be able to "drive the line" in a vehicle equipped with the RFI location equipment to cover a large enough area necessary to do a good overall assessment of the noise problem. There may be several noise sources, each contributing to the problem. A less severe source may be masked by a more severe one, so each source must be identified and repaired in order of severity. It may take several tries by the power company to reduce the noise to an acceptable level. If the power lines are in a right of way not passable to vehicles, as many transmission lines are, the search will have to be on foot. This will change the amount and type of equipment that can be used due to the fact it will be battery operated and carried by one or two people. Many right of way areas are posted against trespassing, so you may not be able to locate the noise without involving power company personnel.

To track down power line noise, start out at about the same frequency as the noise is on the offended equipment. Having the RFI equipment setup in a car would be the most convenient way to travel especially if an oscilloscope is used on the audio output. A copilot is a necessity, so one person can drive and one person can work the equipment.

Frequency is a key factor when searching for the noise. If the frequency you are monitoring is too low, the noise may seem to come from "everywhere." At low frequencies, such as the AM broadcast band, the noise can travel for miles down a line with no obvious peaks or nulls in magnitude. It may be necessary to move up in frequency going as high as possible where the interference can still be detected. If you change frequency, ensure that you are listening to the same noise that was heard on the frequency you were monitoring earlier. As you move along toward the source you will notice that the magnitude will increase and decrease in intensity in "waves," getting stronger as the source is approached and at a more rapid

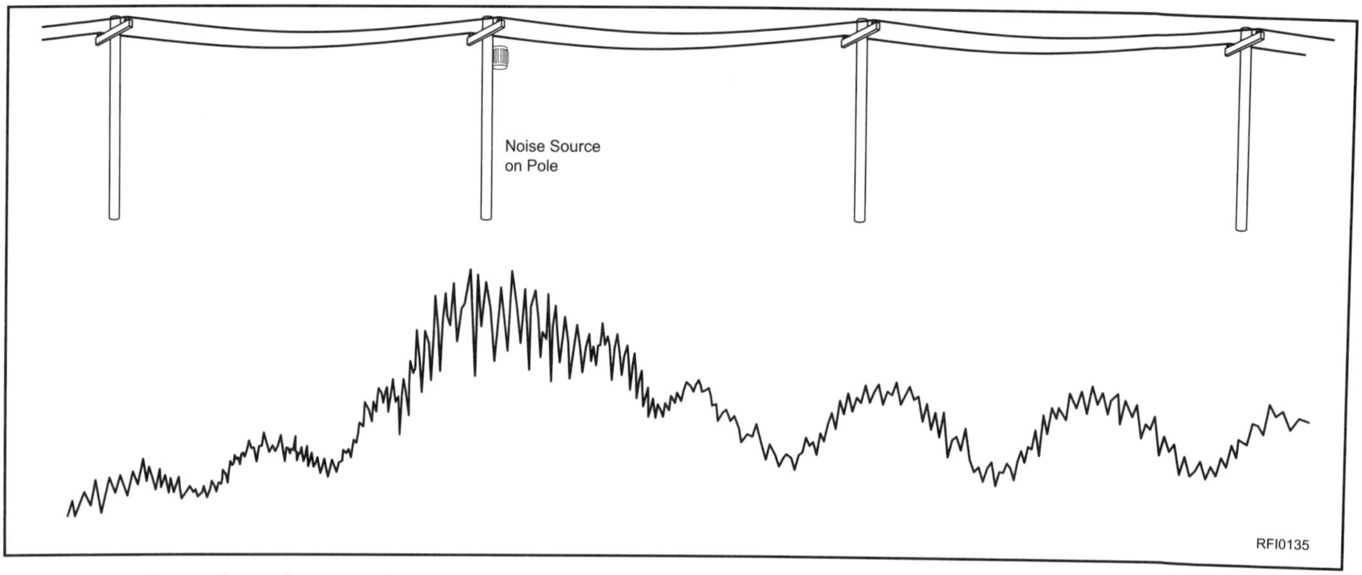

Figure 11.11—Power line noise can exhibit peaks and nulls that can fool the unwary RFI investigator. These peaks generally get closer together close to the source of noise.

rate until the peak is reached, then the magnitude of the waves will fade as you move away from the source. (See Figure 11.11, a diagram showing how noise peaks relate to power line structures.)

Be wary of multiple reflections that can occur where various lines intersect, or where there are other conductors near power lines. These reflections and impedance irregularities may give false peak readings in the conducted interference paths at these points. Another thing to watch for is more than one noise source. This can easily confuse the investigator, especially if lower frequencies are being used for direction finding.

If possible, narrow the highest magnitude noise level to a few structures using VHF or UHF and a Yagi antenna, taking advantage of the Yagi's directivity to determine exactly which pole generates the noise. The power company RFI investigator may have access to equipment such as that shown in Figure 11.12. A homebrew version of one of these directional devices was featured in September 1992 *QST,* "A Line-Noise Sniffer That Works."

Most power line arcs make some noise, so if the arc is particularly bad, it may be possible to hear it. Most arcs, however, make noise in the ultrasonic range. Under these circumstances, an ultrasonic dish would be useful in locating the exact piece of hardware that is sparking. Figure 11.13 shows a commercially available ultrasonic dish and amplifier being used by a power company RFI investigator.

If you rely solely on a whip antenna to locate the source, then going as high in frequency as possible (around 450 MHz would be best) will probably narrow the source down to a single structure or two. The utility lineman should then be able to find the problem with a hand-held RF snooper or with a little luck, just by ear. Figure 11.14 shows a commercially available RF sniffer being used by power company personnel. A hand-held sniffer such as that shown in Figure 11.15 can be used to pinpoint local sources in and near homes.

Residential Noise Sources

Relatively little electrical noise comes from power company wiring and facilities. In many cases, the noise may come from your own home, or a neighboring house or business. If it is your own home, you, of course, can usually correct the problem there. If, however, it is a neighborhood problem, you may need to rely on the RFI investigator's "personal diplomacy" skills to help the owner understand the issues.

It may be helpful to explain that the noise could indicate a serious problem that needs to be corrected. Your neighbor will probably not understand the FCC regulations and responsibilities. Let's face it, the average citizen doesn't know much at all about RFI. If the other party does agree to look for the problem, help them understand the steps of locating the source, and leave it up to them to actually shut off and unplug appliances, etc. This will help insulate you from any liability in case some appliance fails in the process

Figure 11.12—AN RFI investigator uses an AM VHF receiver and directional array to identify a structure that is radiating RFI. Noise normally would only be heard on VHF if the investigator is close to the source.

Figure 11.13—An ultrasonic detector made by UE Systems of Elmsford, New York, is used to pinpoint the source, often to a single device on the structure.

Figure 11.14—These power company personnel are using a hot stick and an RF sniffer made by Radar Engineers to locate a source of noise on a power pole. This should be done only by trained personnel; most power companies will not allow their RFI investigator to get this close to high-tension lines.

Figure 11.15—This handheld RF sniffer made by Micro-Tech, Inc can be used to find noise sources in or near residences. If the building is the source of RFI noise, the noise will be very strong near the power meter. A hand-held sniffer should never be used near high-tension lines and should never be allowed to contact any source of ac-line voltage in the home.

of the search. Once the problem is found, it would be entirely up to the other party to repair it. We can only ask them to correct the offending device because it's creating interference for himself and his neighbors. (Some desperate hams have been known to purchase a new appliance for a neighbor, but this is a "neighborly gesture" that extends past what most hams would be willing to do.)

NOISE SOURCES

Utility Noise Sources

There are a number of defects that can cause RFI on power company equipment. In fact, the list of possible problems seems endless with all the different varieties of insulators and pole hardware in use. In most cases, the power companies understand causes such as defective insulators, loose wire ties, etc. What they may not understand, however, is that near power lines, every part is suspect. Power companies have been known to check out a noisy pole and say something like, "Everything is okay. A few of the cross braces were loose, but that isn't the problem." Actually, there is a very strong field near high-voltage lines. This field will induce currents and voltages in all nearby conductors. If those conductors are not firmly bonded together, or insulated, the resultant sparks and arcs can radiate RFI noise over a large area.

Every year new products are developed and put into use to replace the aging equipment. Sometimes they bring solutions and sometimes more sources of interference.

Here are some of the more common problems:

Power Pole Transformers

Transformers do not usually cause RFI problems unless they have been damaged. The power company crews check for external problems with insulators, loose connections or lightning arrestors. "Squirrel guards" that are used to cover the primary bushings have been known to create tracking or arcing problems on their inside surface. These parts are easily replaced.

If those parts are not defective, the crew may remove and replace the defective transformer. There have been cases of certain brands of transformers being recalled due to internal RFI generation. In another brand, the transformer was equipped with a red overload warning light. This light was found to be generating RFI noise! Usually an internal failure will not radiate above 150 MHz and a UHF detector will be of no use.

Pad-Mounted and Underground Transformers

These transformers, thus far, have not been a great source of interference. If they develop a problem, the frequencies involved are usually in the MF and low-HF range. It may be difficult to pinpoint the noise source and be sure that the transformer is indeed the culprit. These transformers are dangerous and should only be checked by utility personnel. These transformers are subject to the same kinds of damage as are the pole transformers discussed earlier. They are much more dangerous to the public, however, because they are at ground level and are susceptible to vandalism or physical damage. Some of the older type pad-mounts were known as *live front* types. This type of transformer has the primary voltage terminal exposed when the enclosure hood is open. These are very dangerous for utility personnel to work on and should be treated with caution.

Primary Conductor Problems

Most of the problems associated with the primary conductors and the creation of noise result from the method used to fasten the conductor to the porcelain insulator. Many of the power line primary conductors are made from bare wire. This requires that the wire be held onto the insulator by a *tie wire*. This wire was wrapped around the conductor and the insulator to mechanically hold it in place. (See Figure 11.16.) As the tie wire ages it can become loose and corroded. It will then begin to arc to the primary conductor and to the insulator. If the insulator is dirty or cracked there can be leakage current to ground which can further exaggerate any loose hardware problems.

One cure is a type of insulator with a black conductive coating on it to equalize the voltage gradient. These are known in the utility trade as "radio free" insulators. These work well when new, but as the insulator ages the tie wire can again become loose and the arcing process will start over.

Another common problem is the use of a wire tie over *poly-insulated* wire. The gap between the tie wire and the conductor will arc and burn through the poly coating and create severe RFI. There have been cases where the arcing actually burned through the conductor, causing it to fall! (Poly covered wire is used heavily in the northeast part of the US.) Both these scenarios can be cured by the use of a *vise top* style of insulator made of an ultraviolet resistant plastic. This method secures the conductor in a vise type clamp and can be used for bare or poly-covered wire. (See Figure 11.17.) Other parts of the country use post type insulators. This solves the tie wire problem but still can have metal to metal arcing in the hardware if it becomes loose.

Still another common problem is known as a *slack dead end insulator*. A dead end is a string of bell-shaped insulators strung in series, with the number of bells depending on the primary voltage. (See Figure 11.18.) These work very well, and are generally radio quiet. The span of wire connected to them must be kept tight, however, or the clevis and pin arrangement used to tie them together can become loose and corrode causing a gap. If an open circuit exists between two clean, dry insulators the voltage across the gap can be as much as 3 kV. It is easy for an arc to start across the small gap. This is instantly snuffed out due to the instantaneous drop in voltage as current tries to flow through the high impedance of the insulators. The voltage then begins to build up again on the next cycle peak, and the process begins again. There are special clips used on slack dead ends to make the connections tight and eliminate arcing or they can be replaced with a solid polymer unit. There still is a chance of arcing at either end connection, however.

Any type of insulator can develop defects that can cause RFI. Figure 11.19 shows a tiny crack in an insulator that generated a *lot* of RFI noise. Other "bugs" can crop up, too. The insects that built nests in the old fuse shown in Figure 11.20 were also creating a problem for a ham and his neighbors. These problems can only be corrected by trained power company personnel.

Ground Wires

A ground wire is usually run from a ground rod at the base of the pole at least up to the secondary conductors bonding to the neutral or a transformer. Some utilities near the sea coast or areas with high airborne contamination also run a ground wire across the cross arms to prevent tracking and burning. The ground wire eliminates a difference of potential across the cross arm. These ground wires are held in with staples. The staples can become loose and arc to the ground wire. They must be banged into the pole to tighten them or use a plastic insert on them to insulate the staple from the wire.

Loose Hardware

The arcing source does not have to be connected to the primary or secondary power line to cause a problem. With the strong electric fields surrounding the wires, metal hardware can have high voltages induced in them and be at a potential difference to other metal parts or the wood pole itself. Virtually any piece of hardware on a pole can be a noise source. There can be dozens of nuts, bolts, washers and other related items on a single pole. Every one of these is a potential source of arcing if it becomes loose or corroded. The arcing can be metal to metal or even metal to wood!

All connections must be kept clean and tight. This is no easy task considering the exposure to heat, cold, acid rain and wind. The wood poles constantly swell and contract and the hardware inevitably loosens.

A *helical spring washer* can be used to keep the hardware tight. It is expensive and time-consuming to install on pre-existing construction, however. The utility company can use a variety of plastic washers and miscellaneous hardware to prevent RFI. Each is

Figure 11.16—This photo shows the correct way to tie a conductor to an insulator with a tie wire.

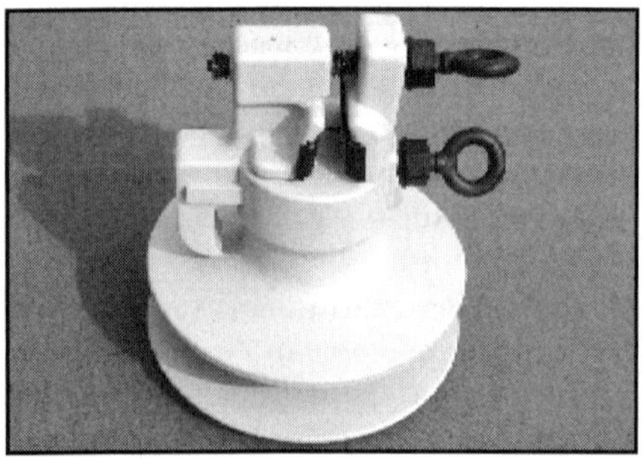

Figure 11.17—This "vise-top" insulator is less prone to RFI problems than some other types.

Figure 11.18—These "bell" insulators are rarely the direct cause of RFI noise. If the line that is connected to them is too slack, however, the internal components can corrode and arc.

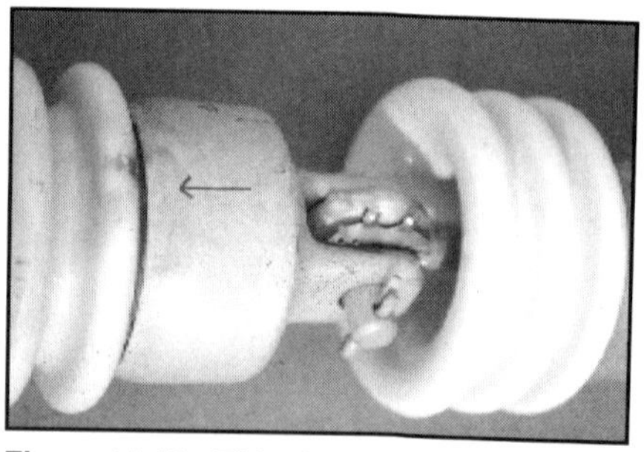

Figure 11.19—This tiny crack in a power company insulator generated a lot of RF noise.

Figure 11.20—The bugs in this fuse were hard to track down. Once this fuse was replaced by power company personnel, a local noise source went away.

designed to cure a specific problem.

External Rectification

A multitude of devices that are not considered RF radiators can be RFI sources because of corroded joints. Corrosion may form a semiconductor junction (rectifier), and produce harmonics of any strong radio signals in the area. Rusty joints in pipes, ducts, fixtures, poor electrical connections, corroded antenna joints and even fences can create and radiate harmonic energy. For example, a fence of metal posts and wire can radiate harmonics when exposed to a strong signal from an HF antenna. Other common suspects are:

Electrical conduit
Heaters
Sheet-metal roof
Electrical wiring
Stove pipes
Furnaces
Water and sewage pipes
Gutters and drain pipes
Lightning arrestors
Guy wires
Radio and TV antennas

Anything that resembles the items on this list has a potential of producing RF harmonics by rectification and cross modulation (IMD). The cure is to locate the "rectifier" and eliminate it by electrically bonding or separating the corroded parts. This topic is covered fully in the External Rectification chapter.

Electrostatic Discharge (ESD)

Static electricity is the accumulation of electric charges on a surface. These charges of electrical energy can build up on insulators as well as conductors. A discharge, or small spark, occurs when two materials at different potential are brought together. Static electricity is more common in cool, dry weather. Static build-up usually occurs on windy days, during periods of low humidity, during temperature changes and when objects are moved or vibrated.

An electric discharge can be induced in a metallic object that is in the field of a strong power source. In a receiver, this discharge sounds like lightning static, similar to a spark discharge. The nature of ESD makes it very difficult to locate. Two pieces of metal rubbing together, or a piece of sheet metal lying on the ground under a power line on a windy day can cause ESD RFI. The only way to cure this kind of ESD is by electrically separating or bonding the offending objects.

Some Common Interference Problems and Cures around the Home and Shop

Here are cures for an array of specific RFI sources that can occur in your home or business.

Brush-Commutator Motors

Figure 11.21 shows a typical brush-commutator motor armature. For clarity, only one winding is shown. The coil is wound in a slot in the armature, and the ends of the coil are attached to commutator contacts 180° apart. There are usually 10 to 15 separate coils. With two connections per coil, there are twice as many contacts as windings. Contact between the field coils and the armature windings is made through soft carbon "brushes." If the brushes are worn, fit loosely, or if the commutator is uneven or dirty, the brushes bounce as the armature turns and create electrical noise (interference). The constant making and breaking of contacts produces arcs. With properly fitting brushes and a clean commutator, however, arcing is minimized (and so is RFI).

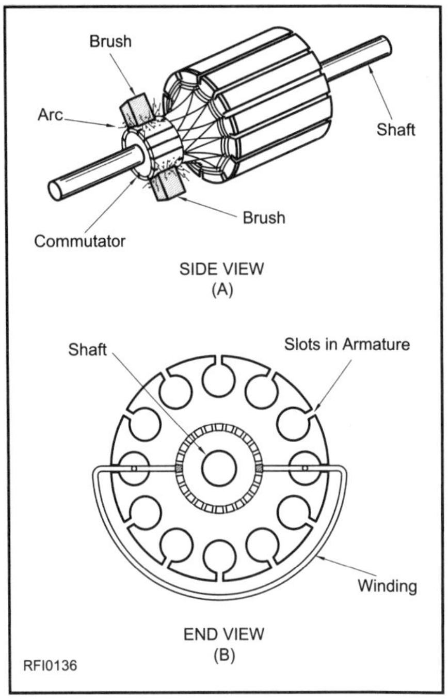

Figure 11.21—(A) A side view of a typical motor armature. Arcing occurs between the commutator and each brush. (B) An end view of a motor armature shows how the individual armature windings are connected.

Most small motors of this sort can be disassembled with ease if they are suspected of generating RFI. Check the brushes for correct fit, and replace them if necessary. Clean the commutator; there may be carbon buildup on the copper bars. Emery cloth, steel wool or an ordinary pencil eraser removes the deposits.

If a brush-commutator motor is an RFI source, and it is in proper working order, install bypass capacitors from each brush to the motor frame. The capacitors should be installed as close to each brush as possible (Figure 11.22). There may be little room inside the motor housing so use physically small capacitors. A 0.01-µF (or greater) capacitor rated for 1.4-kV should be sufficient. These capacitors should be rated for use across ac lines. Disc-ceramic or mylar capacitors are the best choices. Tantalum capacitors are physically small, but they are not designed for such ac applications.

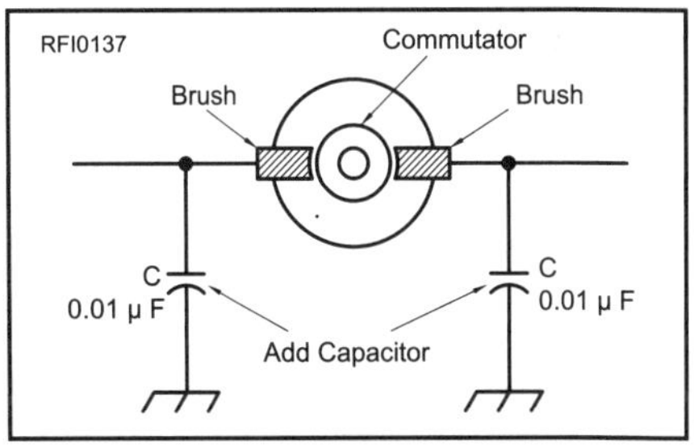

Figure 11.22—Install bypass capacitors inside a brush-commutator motor as shown. Use *only* ac-rated capacitors across household ac lines.

When it is not possible to install capacitors inside a motor, add them (or a line filter) at the point where the power cord enters the motor frame. RadioShack sells a power line filter, catalog #15-1111, for lower-power applications.

Industrial Communications Engineers also sells power line filters rated at full amateur power. See page 48 of the March 2005 issue of *QST* for a Short Takes review of the ICE Model 475-3 AC Line Filter.

If it is not feasible to mount a filter directly at the motor, a plug-in line filter can be used. The filter plugs into the wall outlet, and the offending appliance simply plugs into the filter. Consider the current requirements of the RFI source, and choose a filter rated for that current or more. A filter installed at the wall outlet is less effective than one at the motor because the cord connecting the appliance to the wall outlet can radiate.

Sewing-Machine Motor

The likelihood of RFI from a sewing-machine motor depends on the age of the motor. Because sewing machines are fairly rugged, there are quite a few old models still in use. Most new machines have been treated for RFI, although many newer machines also have computer-controlled features. If the noise coming from the machine is related to the digital electronics, see the Computer chapter for more information.

Many models use a rheostat to control motor speed. Start by connecting a 0.05-µF, 1.4-kV capacitor across the control. These capacitors should be rated for use across ac lines. If space permits, install two bypass capacitors, one from each brush to the frame, inside the motor. If there is not room for such an installation, mount a line filter as close to the motor body as possible. See Figure 11.23.

Electric Lawn Mowers

The motors used on electric lawn mowers are of the high-torque (high-current) variety. Mowers are usually powered by lengthy extension cords (antennas!), and severe RFI is possible. The motors are quite compact, so bypass capacitors cannot be installed directly at the brushes. A line filter mounted directly at the motor is the next best solution. Place the filter as close as possible to the point where the wires enter the housing, and make sure there is a good bond between the lawn mower chassis and the filter case.

Vacuum Cleaners

Most modern vacuum cleaners are fairly well protected against RFI. The inside of the motor gets fairly dirty, however, because not all dust is stopped by the collector bag and filters. The dirt can lead to excessive arcing. Treat the motor as described under "Brush-Commutator Motors." There should be ample room inside most units for bypass capacitors at the brushes. If there is not sufficient room, use a line filter.

Many vacuum cleaners have an electric rug-beater attachment (power head). This is easily spotted because it requires an electrical connection to the cleaner when the attachment is changed. Most have a small motor in the attachment, which drives a system of bars and brushes. Use the same methods of RFI suppression with this motor as with the main motor.

Electric Shavers

Most currently manufactured electric shavers have built in RFI-suppression capacitors. If the shaver is causing more-than-normal interference, check the capacitors.

Some residual RFI is likely, even when the capacitors are in good shape. It can be reduced to an acceptable level with an external line filter. A plug-in style filter works nicely.

Electric Knives, Mixers, Hair Dryers and So On

Most small household appliances use compact, brush-commutator motors. Since many of these items are operated only for short periods, only a fanatic would tackle them all. "Brush-Commutator Motors" treatments apply here as well. If there is no metal motor frame (as is common in small appliances), use a line filter at the motor or a plug-in style line filter.

Office Machines

Some old office machines produce a fair amount of interference. Most new models are well shielded against RFI. Where motors are used, apply "Brush-Commutator Motors" methods. For digital machines (computer cash registers, for example) look at the Computers chapter of this book.

Power Tools

Shop tools are a frequent cause of interference to consumer devices and amateurs as well. Most power tools contain a series-wound (universal) motor. These motors contain brushes that make and break connections as the motor turns. This hot-switching causes sparking. The spark, like any other arc, creates a signal that can interfere with amateur receivers or consumer equipment.

In many cases, an ac-line filter installed close to the motor eliminates the problem. In extreme cases install 0.01 to 0.05-µF capacitors across the power line inside the tool. The capacitors should be rated for ac-line use (at least 1.4 kV)! In most cases a motor generates differential-mode interference, but there may be some common-mode problems as well. Supplement a commercial ac-line filter with a common-mode ferrite choke when noise persists. Electric-motor interference appears on a TV as a large number of black dots or streaks.

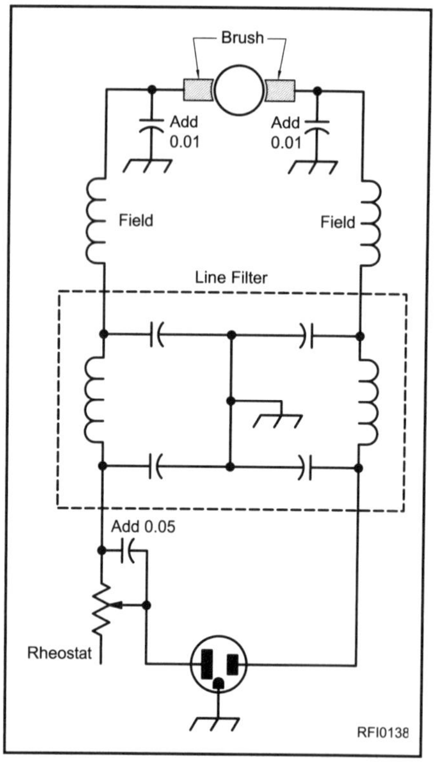

Figure 11.23—Schematic diagram for a typical sewing machine. Bypass the rheostat and brushes, and add a line filter in series with the power cord.

Electric Water Heaters

The typical electric water heater has two heating coils controlled by two separate thermostats. Since the coils draw large amounts of current, the thermostat switch contacts may become pitted. Pitted contacts are prone to excessive arcing; they should be replaced. In some cases, it might be necessary to replace the entire thermostat-switch assembly. The assemblies are not very expensive or difficult to replace. When the contacts are in good shape, a 0.01-µF, 1.4-kV capacitor (installed across them) may cure the problem. This capacitor should be rated for use across ac lines.

Heating, Ventilation and Air-Conditioning Systems

There are several potential RFI sources in "HVAC" systems. (See Figure 11.24.) "Brush-Commutator Motors" are used in many roles, pumping air, water, coolant or fuel. Check each motor for RFI, and treat them as described earlier.

Gas and oil-fired furnaces may use an ignition system (rather than a continuous pilot light) to light the air-fuel mixture. The igniter draws an arc across a spark gap. The gap is fed with a high voltage (sometimes RF, it strikes an arc more easily than 60-Hz ac) from an ac-powered step-up transformer. If the leads on the primary side of the transformer are not properly bypassed or filtered, RFI can reach house wiring. In high-RFI situations, resistors may

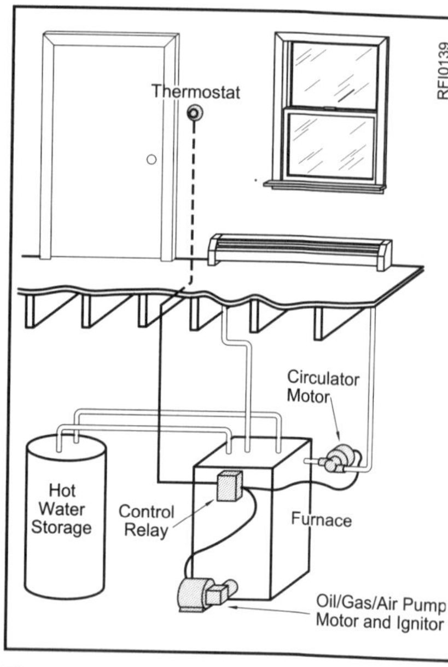

Figure 11.24—A typical home heating system.

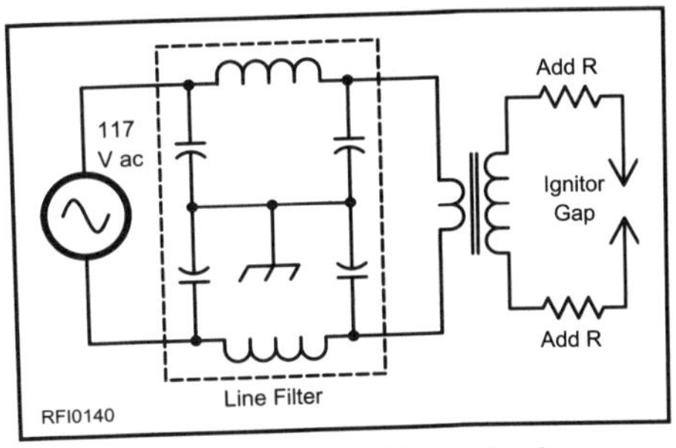

Figure 11.25—A method for reducing interference from the ignition system used in many furnaces. Use a commercial electrical-noise filter, installed by qualified personnel.

Figure 11.26—A mercury-switch thermostat.

be required in each of the high-voltage leads (see Figure 11.25).

HVAC Thermostats

Thermostats sense temperature and automatically control HVAC systems. Some old thermostats use a set of contacts attached to a bimetallic strip. The bimetallic strip bends slowly with temperature, and contact is made or broken very slowly, so the chances for pitting and arcing are great. Newer, mercury-switch thermostats (Figure 11.26) are much quieter from an RFI viewpoint. Programmable, solid-state thermostats are quieter as well. Treat bimetallic thermostats by cleaning the contacts and installing a bypass capacitor (0.01-µF, 1.4-kV) across them. These capacitors should be rated for use across ac lines. Alternatively, install a mercury-switch or solid-state thermostat.

Many heating systems use several relays to control different zones. Because the relays open and close many times they may become dirty and pitted. Replace them as needed, and install a 0.01-µF, 1.4-kV capacitor across each contact pair. These capacitors should be rated for use across ac lines.

Most electric heating systems use a combination thermostat-switch assembly to control the radiators. There is often a separate system (and thermostat) for each room. A typical thermostat-switch assembly is shown in Figure 11.27A. Since the contacts switch high currents, there is a good chance they will become pitted. When the contact is opened arcing is possible. Some heating systems utilize single-pole switching; others use double-pole switches. Bypass each set of contacts in the switch assembly as shown in Figure 11.27B.

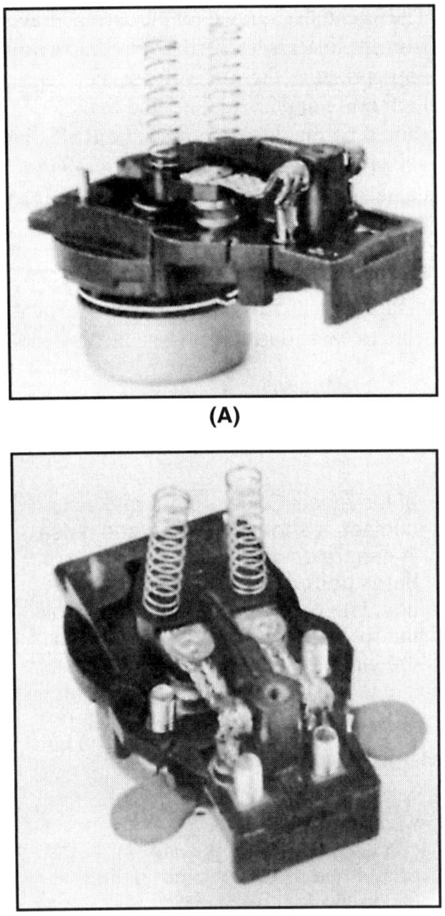

(A)

(B)

Figure 11.27—(A) Typical electric-heat thermostat-switch assembly. For RFI suppression, add capacitors as shown at B.

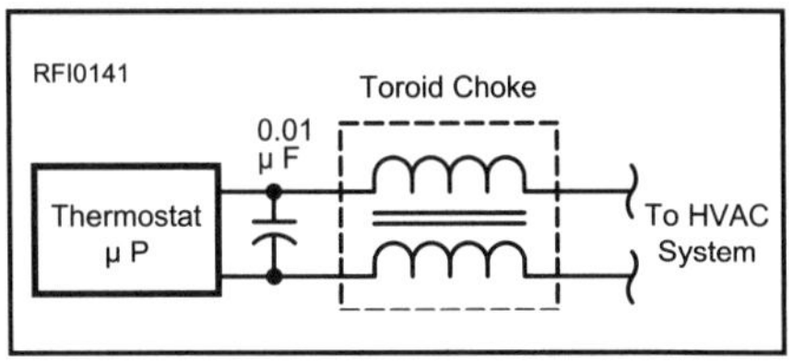

Figure 11.28—A choke-capacitor combination for use with thermostats and alarm systems.

Setback Thermostats

Microprocessor-controlled "setback" thermostats are a problem as well. They respond to RF energy picked up on their wires. In most cases, the interference shows up as a random reset of the thermostat program.

Appliance Thermostats

These devices control temperatures in electric blankets, automatic coffee pots, toaster ovens, water-bed heaters and just about anything else that requires temperature control. (The thermostats in old aquarium heaters are real troublemakers!) It is fairly easy to troubleshoot thermostat RFI: Switch off or unplug each thermostatically controlled device until the interference stops. Then remove and replace the thermostat with one that does not cause interference. An ac-line filter may help in some cases.

All types of thermostats respond well to common-mode ferrite chokes. A single-layer winding on a 3-inch core is usually sufficient. In difficult cases, wire a 0.01-µF capacitor across the leads to suppress differential-mode signals. These capacitors should be rated for use across ac lines. Cut a hole in the wall behind the thermostat to hide the choke. Make the hole big enough for the wound core, and pass the core through the hole to hide it. Then remount the thermostat. Figure 11.28 shows a general-purpose filter that can be used on most thermostats.

Doorbell Transformers

Some doorbell systems have transformers with a temperature-sensing, shut-down mechanism. If there is a malfunction in the

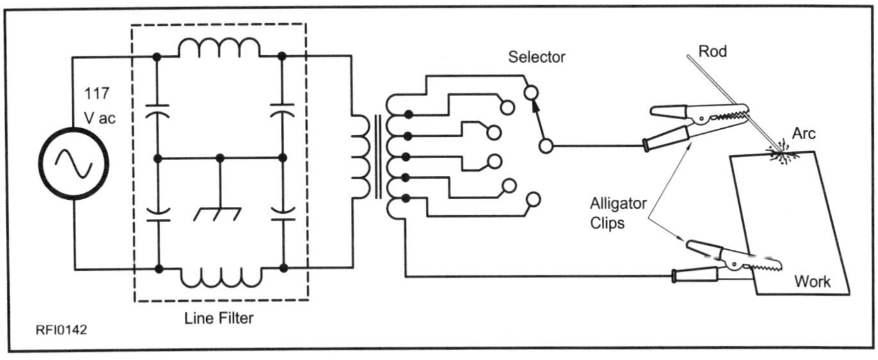

Figure 11.29—An arc-welding setup with a heavy-duty line filter installed.

system, the transformer circuit is opened by thermal overload to prevent further damage. If the contact becomes intermittent, RFI can be generated. Repair or replace the system to eliminate the RFI.

Arc Welders

Arc welders can cause severe interference. Unfortunately, there is no way to suppress the arc and leave the machine functional. Install a heavy-duty line filter directly at the transformer primary as shown in Figure 11.29. The filter must be rated for the current drawn from the line. Use 0.01-µF bypass capacitors appropriate to the line voltage. For such a high-noise application, ac-line bypass caps should be rated for 14 times the RMS line voltage. (This allows for spikes at 10 times the peak line voltage.) Wind the coils from at least no. 8 wire.

Pipe Heaters

Pipe heaters are used to prevent water pipes from freezing during cold weather. Normally, an electric-heat element is wrapped along the entire length of the pipe to be protected, and powered from a wall outlet. Most units do not use thermostats. If a thermostat is used, treat it for RFI in the manner already described. Replace it if necessary.

Electric Fences

Electric fences are still common in rural areas. A transformer steps up the line voltage, and an automatic switch sends a high-voltage, low-current pulse down the fence once every

Electric Fence Noise

A Step-By-Step Procedure

An electric fence can generate radio noise, even if it is in otherwise good working order. This noise is not normal, however, and it almost always can be corrected. In many cases, this noise is caused by a portion of the fence that may fail as the spark causing the noise weakens the wire. Whenever noise from an electric fence causes harmful interference to a licensed radio service, Part 15 of the FCC rules require the fence operator to correct the problem or cease operation of the fence. Fortunately, in most cases, a little maintenance is all that is required. Let's take a closer look at the problem of unwanted radio noise from an electric fence, and ways to find and fix it.

Virtually all radio interference originating from an electric fence is caused by a spark or arcing across some fence related hardware. The noise can interfere with radio and television reception and propagate for a considerable distance. In some cases, the noise can disrupt radio reception for a radius of over a mile from the fence. The interference is most noticeable on an AM radio and typically heard as a "tick-tick-tick" sound. This is a unique characteristic of electric fence noise.

Fortunately, correcting most of these problems is a relatively easy and simple process. Many cases can also be corrected at no cost. For example, it is unlikely for the fence charger to be the culprit and require replacement. Troubleshooting electric fence noise typically involves locating the offending spark gap and correcting it. **Bad splices in the fence wire and gate hooks are two of the more common problems associated with electric fence noise.**

Vegetation can also be a problem. A typical scenario involves noise that will cyclically come and go. First, the weeds grow until they reach the height of the fence. Once a plant makes contact, a short can occur and noise is generated. After a while, the plant burns back and breaks the connection. The process doesn't repeat until the weed recovers and grows up to the fence again.

It is also possible for an insulator to go bad, thereby allowing the fence to arc to one of the fence posts. This is more likely if the problem changes with weather, either getting better or worse when it rains.

It is unlikely (but possible) that the problem is an arc or other defect inside the fence controller. Since most problems occur along the fence wire and related hardware, the fence wire can act as an antenna and radiate the radio noise generated by the arc. A filter, such as a brute-force ac line filter, will only help filter noise being conducted into and radiated by the ac power lines. In the case of the fence however, such a line filter will unlikely be of much if any help. **The only solution in most cases is to find the source of the arc and correct the defect causing it.**

Here is a step-by step approach to troubleshooting a noisy electric fence:

1) Visually inspect the fence for obvious defects. Remove or cut back any problem vegetation and replace any broken hardware. Look for and take note of potential problem areas such as splices, gate hooks, turnbuckles and similar hardware. Rust or corrosion at these points is often an indicator that the splice or gate hook is making radio noise. In some cases you may be able to hear the spark by ear.

2) Confirm the presence of the noise with an AM battery powered portable radio. If you have one, a radio capable of receiving the aircraft band can also be used. Because of their shorter wavelengths, aircraft band frequencies can in some cases be used for troubleshooting purposes. The noise occurs in short bursts in tempo with the fence charger.

3) Unplug the fence charger to verify that the noise goes away. Also confirm that the noise also goes away at the affected radio or television receivers—*especially if your neighbor is involved.* If it does not, there may be additional sources of noise causing the problem.

(Note: There are electric blankets and heated mattress pads that cause pulsed noise, similar to an electric fence. Products made by Perfect Fit generate noise even when turned off. These devices must be unplugged from ac power in order to eliminate the noise.)

4) With the fence controller disconnected from ac power, remove the fence connection to it. Confirm that the noise goes away. If it does not, you may have a bad charger. A brute-force ac line-filter[1] may help in this case. If not, try replacing or using a different charger.

5) Again remove power to the fence charger. Add a short length of fence wire to the charger. Several feet should be adequate. Insulate the wire as appropriate to ensure that arcing cannot occur. It must not come it contact or be near anything that could result in an arc. Spare antenna insulators may serve temporary duty for this test. Turn on the charger and confirm the noise does not return. If the short "fence" wire appears to radiate noise, you may need to replace the fence charger. Fortunately, this is not a typical case.

6) Remove power to the charger and reconnect the fence wire. If there are multiple sections, connect one section at a time and turn on the charger. Make note of which fence sections generate—and do not generate—noise. This will help narrow down the search. Reconnect all portions of the fence and turn on the power.

7) Walk the perimeter of the fence

few seconds. The voltage is pulsed so that it is not a serious shock hazard to humans. A single conductor is used for the fence, and it is connected directly to the switch mechanism. The basic system is shown in Figure 11.30. Since the wire used for the fence can be very long, it may act as a good antenna. Electric-fence interference will consist of a series of pulses, each lasting 0.25 to 1.0 second or so, spaced a second or two apart.

In most cases, the electric-fence controller does not create electrical noise and does not radiate it well if it does. The fence wire itself forms a large, long-wire antenna. Most problems with electric fences come from problems in the fence wire.

Splices are a common source of problems. In time, splices can become loose or corroded, resulting in a strong arc at the splice. Another common source of noise is found in the *gate hooks* used to gain access to various areas of the protected area. These can become corroded or stretch a bit, with the resultant poor contact creating an arc. In other cases, weeds can grow up to the fence. The latter often causes a strong arc that suddenly appears, then disappears for a few days as the unfortunate weed burns off, ending the short to ground. Also, look for broken or dirty insulators. Clean or replace them as needed. See "Electronic Fense Noise" sidebar.

In extreme cases, a resistor placed in series with the fence wire damps the pulse. A capacitor across the switch mechanism and a line filter in the ac line should finish the job.

Figure 1—Electric-fence controllers are not often a source of noise. It is more likely that electric-fence interference is caused by a problem on the wire itself.

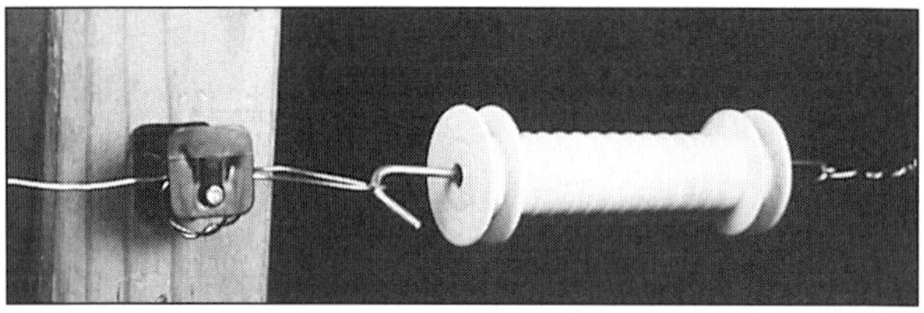

Figure 2—Gate hooks are a potential problem. They often become corroded where they contact the wire, resulting in an arc or spark. The result is a "spark" transmitter, hooked up to the longwire "antenna" fence. The arc creates radio noise and ultimately wears the wire—a problem for the fence owner as well as anyone affected by the radio noise.

Figure 3—Splices can also create problems, especially if they are between two different types of wire. Splices should be mechanically secure, lessening the likelihood of poor contact, sparking and wire failure.

while listening with the battery portable radio. If you have an aircraft band receiver with a telescoping antenna, use the minimum length of antenna necessary to hear the noise. You may be able to further isolate the problem by carefully observing the signal strength of the noise. It will increase as you approach the source. This is not always a reliable test with an AM broadcast receiver, however.

8) Turn the radio level down and listen for faint audible sizzling at any and all suspect areas of the fence. Arcing may also be visible. (This may be especially noticeable at night.) Splices and hardware (such as gate hooks) in fence sections known to be causing the noise require particular attention. Any areas that look corroded, pitted or frequently fail are suspect. (The spark can eat in to a conductor.) See Figures 2 and 3.

9) Remove power from the charger. Clean and resplice all areas of the fence identified in step 8. It is important to ensure good electrical contact in all fence circuit connections. Gate hooks may require replacement but may be jumpered as a test or temporary solution to the problem.

10) Restore power to the fence. If the noise does not go away, repeat steps 8 and 9 as required. As a last resort, try placing a jumper across all connections with short clip leads. Identify problem areas by removing the jumpers one at a time until the nose returns. Turn off the fence each time you handle a jumper in order to avoid risk of shock. Correct each problem area as soon as you find it. (Clip leads are short wires with alligator clips at each end. They are available at RadioShack.) Alternatively, you can redo all splices after careful cleaning of the wire. Steel wool or a wire brush may be helpful for cleaning connections.

11) Restore power to the fence charger and verify the noise is no longer present. If your neighbor is involved, be sure to have him or her check the affected radio or television receiver.

12) Additional information may be obtained from the ARRL's Web page on electric fence noise at: **www.arrl.org/tis/info/fence.html**.

Notes:

[1]A "brute-force" ac line filter can help eliminate a radio signal from getting to and being radiated by power lines. While this is rarely the problem, it may help in some cases.

If the device draws less than 300 watts (about 2.5 A), try using a RadioShack catalog #15-1111. If not, some of the filters sold by Industrial Communications Engineers can handle higher current. More information is available on line at: **www.arrl.org/cgi-bin/tisfind?patt=Industrial+Comm**.

[2]This information is taken from the ARRL Web Site at: **www.arrl.org/tis/info/HTML/elec-fence%20noise/**.

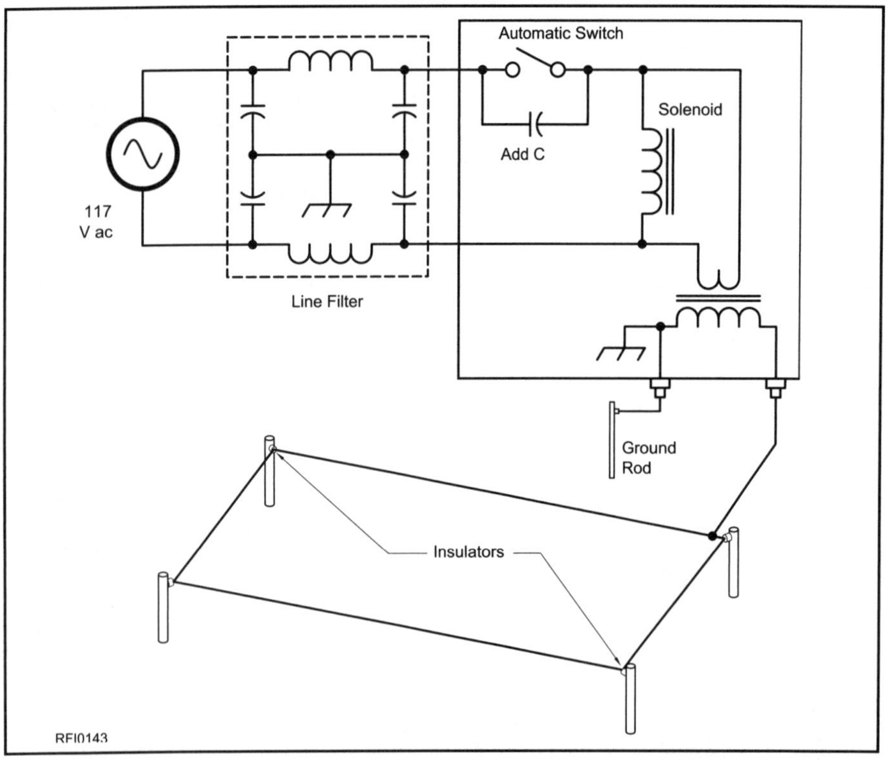

Figure 11.30—Electric-fence RFI cures: You should first cure any problems caused by bad splices, poor connections at gate hooks or from weeds arcing to the line. If those are not effective, a capacitor is used across the automatic switch, a resistor is placed in series with the high-voltage lead, and a line filter is inserted in the ac line.

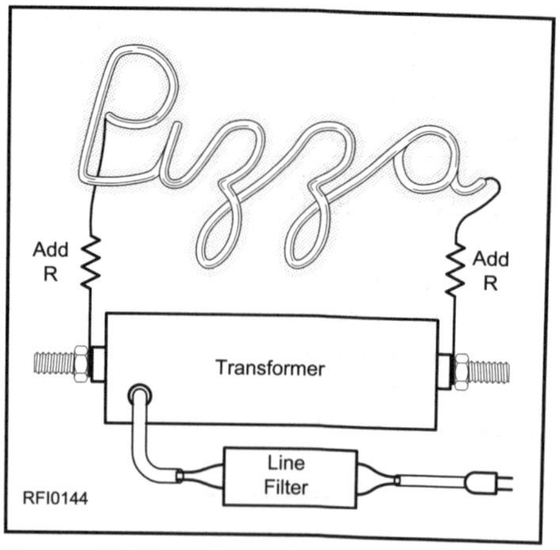

Figure 11.31—A typical neon-sign setup. Add a 10-kΩ resistor in series with each high-voltage lead and a commercial line filter at the transformer. (The transformer should be well grounded.)

Fluorescent Lamps

While these lamps operate from 60-Hz current, they are switched on and off 120 times per second. This rapid switching, along with the stream of electrons (arc) going back and forth in the tube, creates rough radio waves that can cause severe RFI. Most of the RFI is in the broadcast band with secondary peaks in the 7 to 9-MHz region. The noise is coarse and continuous while the lamp is lit.

If the cure cannot be accomplished by reorienting the receiver antenna, RF filters for fluorescent lamps are available by special order from distributors. The filter should be installed on the lamp fixture as close to the terminals as practical.

Neon Signs

The basic circuit of a neon sign is shown in Figure 11.31. It is very similar to a fluorescent light in that a long arc is drawn through the tubing to excite the gas. A transformer steps up the line voltage to a level suitable for excitation of the gas. Relatively high voltage (approximately 1000 V/ft of tubing) is required for proper excitation.

The cure for neon-sign RFI is a line filter mounted directly at the transformer and resistors (10-kΩ, 1-W) in series with each high-voltage lead. Also, make sure that the transformer case is attached to a good ground. If these measures do not reduce the interference to an acceptable level, try winding thin magnet wire around the tube along its entire length. Six or seven turns per foot should be sufficient. Ground each end of the wire.

Light Dimmers and Speed Controls

Most such controls use SCRs or TRIACs. The controls function by conducting on for a short, variable part of each 60 Hz cycle (with a rise time of about 1 µs). The resultant waveform is rich in harmonics and creates severe RFI well into the HF range. A loud buzzing that covers the lower frequencies is a characteristic feature of SCR RFI. The best cure is to replace the control with a better one that has a built-in RFI filter. (Beware of dimmers in plastic cases!)

A combination of an ac-line filter and a common-mode choke, formed by wrapping about 15 turns of ac wiring through an F(T)-240 or F(T)-140-size ferrite core made from the correct ferrite material, placed close to the control, often helps. It is usually necessary to use two cores, one at the control input and the other at the output, to eliminate the interference. (The cores are rather large; they may not be practical in many cases.)

Light Bulbs

RF energy is starting to be used in light bulbs. Industrial devices marketed by GE use signals in the HF industrial bands to power efficient light bulbs. The ARRL Lab has tested a few of these bulbs and found that they do not emit harmful interference to nearby ham bands.

The ARRL has received reports that some 25-W light bulbs emit an RF signal in the 60 to 70-MHz range. Not many of these offending light bulbs exist, but occasionally one is found. The problem created by these bulbs is usually found on TV channel 2. It appears as two horizontal black lines across the picture.

TV "Booster" Amplifiers

In fringe TV-reception areas it is common to employ a booster (preamplifier) mounted at the antenna. These amplifiers may malfunction and oscillate in the upper HF, VHF or UHF frequencies. This can be caused by loose cables, corroded contacts, poor alignment or a defective component. If tightening the connections does not help, replacement is probably necessary.

Touch-Controlled Lamps

Lamps that switch by sensing human contact can be both sources and victims of RFI. Touch lamps use a free-running oscillator usually running at 50-200 kHz or so, a touch plate, and a simple circuit that senses a change in frequency. The oscillator is connected directly to the lamp chassis, so that hand contact changes the oscillator frequency. When the oscillator frequency changes, the lamp switches. In order to sense a touch, the oscillator runs even when the lamp is off. Naturally, the touch plate can also pick up nearby RF.

When it does, the sense circuit thinks the frequency has changed and the lamp changes state. The oscillators are easily "pulled" by nearby transmitters, and the lamp responds by switching with each transmission. In fact, a 2-meter H-T makes a good remote control. RF can also be picked up on the power lines and conducted into the lamp circuitry.

The oscillator is also rich in harmonic energy. These lamps can radiate a fair amount of RF noise. Touch-lamp noise will appear as groups of raucous buzz, spaced at the same frequency as the lamp oscillator. In addition, most such lamps use a dimmer circuit to select several brightness levels. The lamps can source RFI from the oscillator (usually 50 kHz - 2 MHz) or the dimmer circuit. (The dimmer should be quiet when the lamp is off.) Once on the ac line, the interference can affect receivers many blocks away.

Several of the cures that have been featured in past *QST* articles are included in the Hints and Kinks at the end of this chapter. These cures generally consist of inserting a resistor or inductor in series with the touch plate. This seems to work in about half of the cases, especially if the RF level is not significantly above the threshold of interference. In some other cases, using a ferrite core made from #43 material (upper HF, VHF) or #75 or -J material for lower HF, with about 10 turns of the lamp's ac cord wrapped onto the core, may also effect a cure. In rare cases, a "brute-force" type ac-line filter may be helpful.

Use the same cures whether the lamp is acting as a source or victim. First, install an ac-line filter. It should block RF entering on the ac line, and that may cure the problem. Next, add a common-mode choke (7 turns on an FT-240-43 core) at the lamp base. Some hams have cured these lamps with a 1 to 4-kΩ resistor and a 10-µH choke in series with the oscillator "sense" line. Have the work done by a qualified technician. Even if you are qualified, it may be unwise to work on equipment you do not own.

Some hams have offered that the best advice is to throw away the lamp. This is probably a good idea, for a number of reasons. In most cases, it is *not* a good idea to modify your neighbors' equipment. The potential liability is not worth the risk. There are dangerous voltages inside the lamp, and you may be electrocuted trying to fix it, or your cure may inadvertently put 110 V onto metal parts of the lamp. In some cases, the cure may cost more than the lamp is worth.

The FCC is pretty clear about touch lamps and other "non-radio" devices. If non-radio devices pick up RF signals, they are improperly functioning as radio receivers. The FCC suggests that the owner of the device should contact the manufacturer for help, although this rarely works out.

Garage-Door Openers

Openers use both radio and dc control. Strong RF may enter via the line cord, dc control lines or antenna. Unwanted opening and closing is the usual symptom. Apply a line filter at the line cord and a common-mode ferrite choke at the opener end of the dc control lines. If interference remains, a band-pass filter (in the antenna lead) for the frequency of the opener should cure it. Contact the manufacturer or a repair facility for help.

Ground-Fault Current Interrupters (GFCIs)

GFCIs are special circuit breakers or outlets that quickly interrupt the current flow when a power circuit is shorted to earth ground (a "ground fault"). While normal current flows in the differential mode, ground-fault current flows in the common mode. Thus, a GFCI is constructed to sense common-mode signals by determining that all of the current that is present on the hot wire is also present on the neutral wire. If it detects that some of the current is not flowing back on the neutral, it is quite likely that there is a fault causing some of it to flow to ground. This often indicates a short circuit, or that someone or something is conducting some of that current to ground, as would occur if someone is being shocked. If it detects a fault, it opens the circuit. They may be very sensitive to common-mode RF on the ac line. RFI appears as false triggering.

Any RFI-suppression components may prevent the GFCI from functioning properly. Report any problems to the manufacturer. Any internal corrective measures should be performed by the manufacturer or the manufacturer's representative. It is possible to add some external components to a GFCI protected circuit, but this usually must be done by a qualified electrician. A common-mode choke may be helpful, but in most cases, the wiring is behind a wall or inside the breaker panel or fuse box. In most cases, house wiring is too fat to comfortably fit through a ferrite core. There is not usually any slack wire to use to form a number of turns of wire to make an effective common-mode choke.

Some manufacturers publish RFI immunity specifications for GFCIs. Replacing a GFCI with one that is particularly immune to RFI, or even one made by a different manifacturer is often a solution.

Smoke Detectors

Smoke detectors suffered some RFI problems in the early 1980s. According to a letter from Charles E. Zimmerman, PE, a Fire Protection Engineer with the National Fire Protection Association (July 1981 *QST*, p 46), Underwriter's Laboratories includes an RFI test on all smoke detectors manufactured since 1981. If you experience smoke-detector RFI, *do not attempt any modification.* Replace detectors from the early 1980s with newer models. Report any problems with current models to the manufacturer.

Burglar Alarms—A Two-Way Street

Many modern burglar alarms are full of logic gates and microprocessors. Because of this marvelous technology, the burglar alarm can detect RF fields. Alarms have

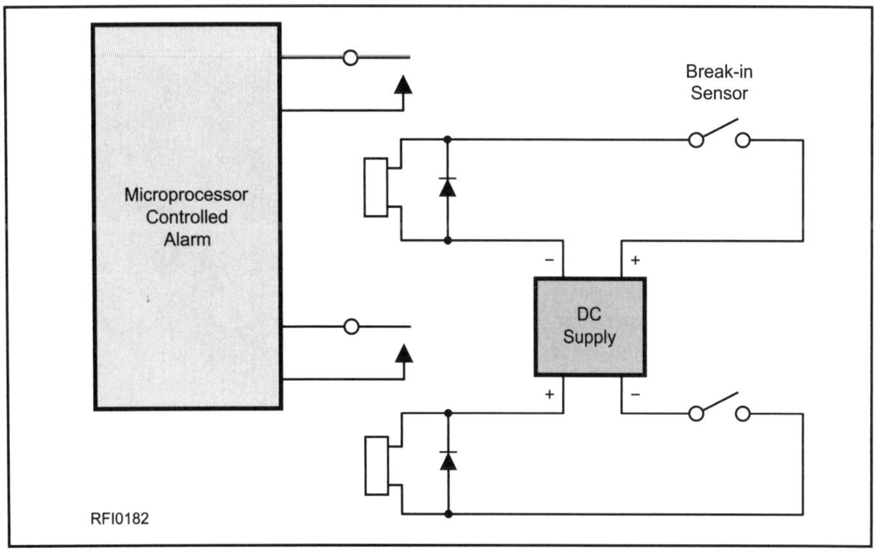

Figure 11.32—Schematic of dc loops used to harden a microprocessor-controlled alarm system.

generated serious harmonics when used in the vicinity of amateur stations. Problems include false alarms when mobile radio is used near the alarm. Police radios as well as amateur radios frequently generate fields strong enough to "false" these electronic wonders.

Microprocessor-controlled alarms may emit interference at the clock frequency and its harmonics. Some electronic alarms fill the ham bands with birdies as well. The most effective way to reduce these problems is to use dc loops and drive relays from the perimeter sensors. (See Figure 11.32.) If the relays are mounted close to the burglar alarm, little wire is present to act as an antenna. Burglar alarms hardened in this fashion also are much more resistant to lightning damage—isn't technology wonderful?

Switching Power Supplies

Switching power supplies are more efficient than linearly regulated supplies, and they certainly save weight. The "down side," however, is that they can generate substantial amounts of noise. (See Figure 11.33.) This noise is generated by their fast switching characteristics. Noise from switch-mode power supplies is generally created from harmonics of the internal switching frequency. This can interfere with HF reception or nearby televisions. The resulting herringbone TVI pattern can be mistaken for amateur transmitter harmonics. A review of the Yaesu FT-757GX (in December 1984 *QST*) noted that the transceiver generated TVI from the switching power supply.

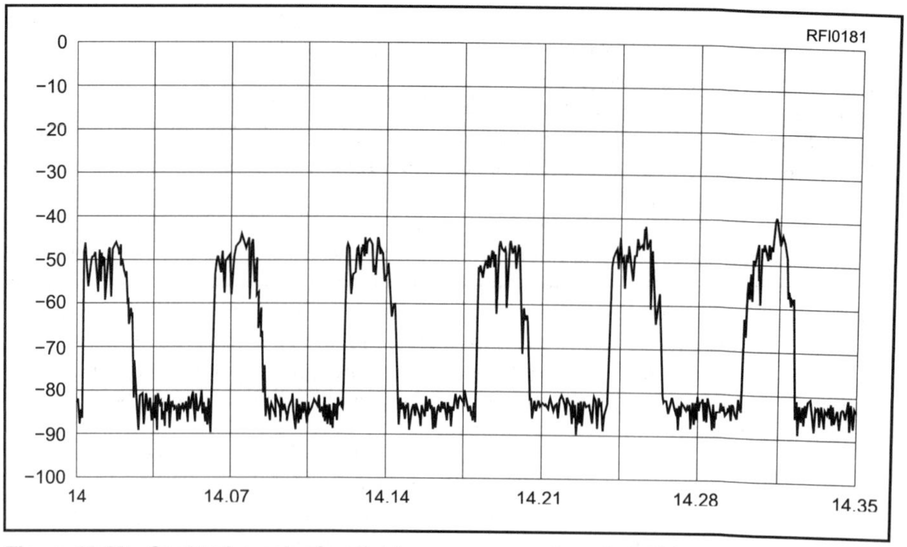

Figure 11.33—Spectral graph of switching-power-supply noise. This can consist of discrete spurs and broadband noise. This example shows spurs across the 20 meter Amateur band at approximately 50 kHz intervals.

Some high-power battery chargers use switching regulation. They can certainly create TVI and cause interference on the ham bands as well. See the Part 15 Devices chapter for more information on switching mode power supplies.

Microprocessors and Microcomputers

As mentioned earlier, the population of microprocessors and microcomputers is increasing daily. Although a microprocessor may only control a coffee pot, its RFI treatment is similar to that for computers. In appliances, RFI usually exits through a plastic case or the line cord. Use shielding to cure case radiation, a ferrite choke for the line cord. Application details appear in the Computers chapter.

Part 2—How to Resolve a Power Line Noise Complaint

INTRODUCTION

Although the problem has been around since the dawn of radio communications and broadcasting, power line noise is on the rise. The proliferation of electrical and electronic devices that are potential victims of power line noise, coupled with today's increased dependence on mobile and wireless communications, have each contributed to this increase.

Unfortunately, dealing with a power line noise complaint can sometimes seem like an exercise in futility and frustration. Power line noise continues to be the single most reported type of interference to the ARRL. Some cases have dragged on for an extended period of time. A little knowledge, however, can go a long way toward a more timely resolution. Fortunately, as we shall see, the FCC rules favor amateurs when it comes to this type of interference!

Power line noise can interfere with radio communications and broadcasting. Essen-

FCC Part 15 rules require that utilities and other operators of "incidental radiators," such as power lines, cause no harmful interference to licensed operations.

tially, the power lines or associated hardware improperly generate unwanted radio signals that override or compete with desired radio signals. Power line noise can impact radio and television reception—including cable TV head-end pick-up and Internet service. Disruption of radio communications, such as amateur radio, can also occur. Loss of critical communications, such as police, fire, military and other similar users of the radio spectrum can result in even more serious consequences.

Let's now take a look at power line noise, and how to best handle a complaint with your local utility.

The Cause of Power Line Noise

Virtually all power line noise, originating from utility company equipment, is caused by a spark or arcing across some power line related hardware. A breakdown and ionization of air occurs, and current flows between two conductors in a gap. The gap may be caused by broken, improperly installed or loose hardware. Typical culprits include insufficient and inadequate hardware spacing such as a gap between a ground wire and a staple. See Figures 11.34 and 11.35. We'll be discussing more specific causes of power line noise later in this section.

Note: The terms "gap" and "conductors" should be interpreted broadly in this case.

While not a source of power line noise, a gap can exist in the commutator of a motor. A gap can also exist between insulator units and other parts of a utility structure. In some cases, the "conductor" can be the wood on the utility pole.

A brief mention should be made concerning corona. Contrary to a common misconception, corona discharge is rarely, if ever, a source of power line noise. Corona discharge is defined as the partial breakdown of the air that surrounds an electrical element such as a conductor, hardware or insulator. In reality it is typically nothing more than a minor annoyance, as the noise caused by it is usually confined to lower frequencies. This noise does not propagate very far from the source because it is a low-current phenomenon that does not couple into the adjacent wires.

May Not Be Your Typical Problem!

Power line noise is not your typical problem in several respects. First and foremost, the offending source is never under your direct control (unless you work for the utility). You can't just simply "turn it off" or unplug the offending device. Nor will the source be under the direct control of a neighbor, or someone that you are likely to know. In the case of power line noise, the source is usually operated by a company, municipality and in some cases, a coop. Furthermore, shutting down a power line is obviously not a practical option either.

Figure 11.34—"Bell" insulators are often the cause of RFI noise. The problem is most often associated with corrosion in a joint, as shown by this photo.

Figure 11.35—Arcing occurred between two of the three components in this insulator. It was a significant source of noise until it was replaced by utility personnel.

FCC rules specify that the operator of a device causing interference is responsible for fixing it. Whenever encountering a power line noise problem, you will be dealing with a utility—and won't have the option of applying a relatively simple technical solution to facilitate a cure, as you would if the device were located in your home. Utilities are a mixed bag when it comes to dealing with power line noise complaints. In some cases, a utility will have a budget, well trained personnel, and equipment to quickly locate a problem. In other cases, however, the involved utility is simply unable to effectively deal with power line noise complaints. Some utilities have even been known to deny their equipment can cause radio interference, claiming they only operate at 60 Hz.

Another unique aspect of power line noise is that it almost always involves a defect of some sort. The cure for power line noise is to fix the defect. Furthermore, it's a repair over which you do not have any direct control. And, since it's a broadband type of noise, you simply can't tune off frequency or QSY to eliminate it. It can plague everything from AM broadcast band up through UHF and television. There is seldom anything you can do to help reduce or eliminate power line noise, especially if multiple sources are involved. The best cure is almost always a utility implemented repair.

What does this mean for an Amateur with a power line noise complaint? Many utilities are large corporations. Others are relatively small companies or coops. Some are owned and operated by municipalities. Regardless of the category in which your utility may fall, it must follow Part 15 of the FCC rules. These are the rules that apply to interference to licensed radio services, including Amateur Radio, from power lines and associated equipment. Dealing with a company, coop or municipality, however, as opposed to a device in your home, or a nearby neighbor that you know personally, can present its own set of unique challenges. Multiple individuals are often involved, including an RFI investigator, a line crew and associated management. In some cases, the utility may not have ever experienced an RFI complaint before yours.

Before Filing a Complaint...

Obviously, before filing a complaint with your local utility, it is important to verify the cause of the problem as best as possible. Other sources, such as lighting devices, can often mimic power line noise, especially to an untrained ear. Utilities are not responsible for noise generated by customer operated Part 15 devices—*even if the noise is conducted onto and being radiated by the power lines.* They are responsible for fixing only that noise being generated by their equipment.

Generally, it's best to try the easy things first. A good first step is to eliminate some obvious sources. Although not usually necessary, you can quickly determine if the problem is being generated within your radio (or other affected device) by a simple test. Remove the antenna connection to the radio in order to see whether the noise goes away. If no change or little change in the noise results, the problem with the receiver or its power supply, or the RFI source may be located near the receiver or connected to the same ac circuit. Typically, this is not the case, however. This is mentioned only for situations where you suspect a problem with the affected device, such as your radio or television.

Assuming you don't feel this test is necessary, or observe a significant noise drop while performing it, proceed to verify the source is not in your house. The proliferation of electronic devices and electrical appliances can often result in a plethora of confusing and hard to identify sources. Many of these sources are actually the cause of harmful interference. See the procedure in the "'How to' Locate Tips" sidebar for locating sources in your home.

Here are some household items commonly found to cause interference:

- Door Bell Transformers
- Electric Blankets
- Heating Pads (of all kinds)
- Recessed Ceiling Light Fixtures
- Furnace Control Circuits
- Refrigerators (Becoming a frequent problem)
- TV Top & Stereo, Amplified Antennas
- Light Dimmers
- Aquarium Heaters
- Screw-in Photocells
- Low Energy Compact (screw-in) Florescent Lights
- Touch Control Lamps
- Clean Air Machines (table-top and furnace type)

These devices, when causing harmful interference, are in violation of Federal Communications Commission rules and regulations and can be a nuisance to you and possibly your neighbors. The rules however specify the operator of the device is responsible for correcting the interference. Not surprisingly, the Commission has never shown much sympathy for anyone causing interference to him or herself. Perhaps more importantly however, the interference may signal a potentially hazardous situation. It is important to have the offending device repaired or replaced to ensure normal safe operation. Many sources of radio and television interference are also caused by arcing. The arcing will generate heat and may signal a fire hazard.

"How to" Locate Tips

"How to" Locate Radio Frequency Interference (RFI) or Television Interference (TVI) in a Home

All steps should be performed while interference is active

1. Go to your main circuit breaker or fuse panel with a battery powered AM radio tuned between stations so all you hear is the offending noise. (*If at night or in an enclosed room, be sure to have a working flashlight.)*

2. If the noise is present and corresponds to the disturbing interference, shut off all power to your premises by turning off the MAIN circuit breaker or equivalent enclosed device. (**DO NOT ATTEMPT TO REMOVE CARTRIDGE FUSES or OPERATE EXPOSED or OPEN-TYPE DISCONNECTS IF PHYSICAL CONTACT WITH ELECTRICAL CIRCUITS IS AT ALL POSSIBLE.**) If the noise on the AM radio stops while the power is off, the source of the interference is within your own residence.

3. Restore the main circuit breaker. Don't forget to reset clocks after the interference source is located.

4. Assuming the noise stopped while the electricity was off, you can now locate the circuit supplying the power to the noise source. While monitoring the battery powered AM radio as before with the offending noise present, turn off and on the individual circuit breakers one at a time until the noise stops. Leave off the breaker that stops the noise.

5. You must now determine what has been turned off by going from room to room, if necessary, checking outlets, appliances, and lights for the absence of electricity. The offending noise will be something on this circuit. Turn the breaker back on and wait for the noise to return.

6. With the noise back on and using the AM radio to monitor it, return to the area of the noisy circuit and unplug everything on this circuit one at a time until the offending device is found.

Things most commonly found to cause interference are:

Door Bell Transformers
Electric Blankets
Heating Pads (of all kinds)
Recessed Ceiling Lights
Furnace Control Circuits
Refrigerators
TV Top Amplified Antennas
Aquarium Heaters
*Screw In Photocells
Low Energy (screw in) Florescent Lights
*Touch Control Lamps
Clean Air Machines (table top and furnace type)
*Light Dimmer Controls

(*) Radio Interference Only

These devices, when causing interference, are in violation of Federal Communications Commission rules, and are a nuisance to you and your neighbors. It is important for your own benefit to have the offending device repaired or replaced to insure normal safe operation. Most radio and television interference sources are the result of an arc. This arcing will get hot and could create a fire hazard. Your interest and assistance is sincerely appreciated.—*Michael C. Martin, K3RFI, RFI Services' Interference Investigator*

If the noise source is not in your home, check with your closest neighbors. The place where the interference is the most intense may indicate the source of the disturbance. See "How to 'Locate the Residence' Containing the RFI / TVI Source" sidebar for a procedure to isolate the residence containing the offending source. If this procedure points to a nearby residence, or if one of your neighbors has a similar problem, ask him, or her, to run the breaker test to try to locate the faulty equipment. A household appliance or electrical device rarely causes interference that extends beyond a few houses on a secondary system.

Note that, if the source is not in your or a neighbor's home, the noise is originating from a source that is beyond your control. More advanced techniques, such as Direction finding, may then be used isolate the noise to a particular residence or an area of the utility's power line system. We'll be exploring some of these techniques in a later section.

How to ID Power Line Noise

Many electrical devices, such as electrical motors, tools and appliances, can cause interference. The types of interference differ greatly from one electrical device to another. Interference caused by a computer, for example, is not the same as that produced by a household appliance.

Noise that varies with the time of day is related to what people are doing, usually pointing to some electrical device or appliance. Noise from consumer type devices, as opposed to power line noise, will often come and go with periods of human activity. It will frequently correlate with evenings and weekends. Unless it is associated with climate control or HVAC system, an indoor RFI source is less likely to be affected by weather than power line noise. The importance of maintaining a good and accurate interference log cannot be overstated. Be sure to record date, time and weather conditions. Correlating the presence of the noise with periods of human activity and/or weather often provides very important clues when trying to identify power line noise.

Often Weather Related

If the interference appears and varies in intensity depending on weather conditions (dry or damp weather, or wind), and if the breaker test excludes a source inside your home, the interference may be caused by faulty components associated with the electrical power lines near the home. Wet weather may temporarily reduce or eliminate the noise by shorting out spark gaps on the power line. Windy weather may cause the noise to vary or even stop for a while, as loose hardware is affected.

Is There A Smoking Gun?

While there may not be a smoking gun, power line noise often reveals itself with some important clues. As previously discussed in this chapter, virtually all radio noise originating from utility company equipment is caused by a spark or arcing. The radio noise is only generated during the times when a breakdown and ionization of air occurs, and current flows between two conductors in a gap.

Once an ionized path is established in the gap, current flows at all parts of the cycle where the voltage is higher than the breakdown voltage of the gap. This typically occurs only near the positive and negative voltage peaks—the times of highest instantaneous voltage. Sometimes, the gap may break down only on one polarity of the waveform.

Because power lines carry 60 Hz ac, the voltage on them passes through two peaks each cycle (one positive and one negative) and pass through zero twice each cycle. This gives 120 peaks and 120 zero crossings in each second. Power line noise follows this pattern, generally occurring in bursts at a rate of 120 (sometimes 60) bursts per second. This gives power line noise a characteristic sound that is often described as a harsh and raspy hum or buzz. Because the peaks can occur twice per cycle, true power line noise usually has a strong 120-Hz modulation.

Noise occurring in bursts at a rate of 120 bursts per second, and the resulting characteristic raspy buzz or frying sound, is often the first and most obvious clue of power line interference. It is typically a broad banded type of noise starting at the low end of the radio spectrum. Power line noise is usually stronger on lower frequencies. It occurs continuously across each band, up through the spectrum to some upper frequency where it will taper off.

A good test for the 120 Hz burst rate for both indoor and power line noise sources involves an oscilloscope. The oscilloscope should show the bursts occurring every $\frac{1}{120}$ seconds, or $8\frac{1}{3}$ ms. Look at the suspect noise from a radio's audio output using the AM mode. Use the wide filter settings and tune to a frequency without a station. Power line noise bursts should repeat every 8.33 ms. If this is not the case, you probably don't have power line noise. In some cases, rectification in the gap can occur, and the noise bursts occur ever 16 2/3 ms. See Figure 11.4.

Alternatively, you can perform a similar test if the noise pattern is visible on a TV set. The noise occurs in two horizontal groups or bands. Typically, these two bands drift slowly upward on the screen. One group is a result of arcing during the positive half of the 60 Hz sine wave. The other group results from the negative half of the sine wave. If rectification is taking place, you may only see one bar drifting upward, or you may see alternating bar widths as they form at the bottom of the screen and roll upward.

Note: The slow drift upward is caused by a slight difference in the power line noise burst rate and the rate at which the TV images are transmitted. The TV images are transmitted at a rate of 59.94 Hz. This is because when television was first developed, a 60 Hz vertical scan frequency was selected so that any power line noise would remain stationary on the screen, and be less annoying. When the color burst signal was added some years later, the scan frequency had to change to accommodate the color burst signal. Power line noise occurs at 120 bursts per second. Since the power line noise burst rate is almost twice the TV rate, two synchronized bands of noise appear on the screen. The slight difference in frequency causes these two bands to slowly drift upward on a TV screen.

It is usually best to perform this test at the lower VHF TV channels and with an antenna (as opposed to a cable hook-up). In addition, as previously discussed, the positive and negative power line noise burst may also have slightly different characteristics. This can cause each half of the cycle to have a slightly different pattern on the screen. As you turn the channel selector to higher frequency channels, the interference should diminish. If the interference can be observed on UHF channels, the source is probably relatively close by. See Figure 11.36.

(A)

(B)

(C)

Figure 11.36—The power line interference shown in A is not typical. In this case, the noise occurred as a result of two sources that were out of phase from each other. The interference from each source overlapped on the screen. Once one phase was corrected, the more typical upward drifting horizontal snowy bars can be seen. Both noise sources were corrected in C and the noise was gone!

What to Look For

As previously discussed in this chapter, corona typically does not cause radio noise. Radio noise is almost always caused by a spark or arc across an air gap. *(There are also many other non-arcing sources, such as lights.)* Any voltage across an air gap can cause radio noise — even ground wires, neutral wires and wires not directly connected to a power line.

Typical culprits include broken or loose hardware such as bolts on wood cross arm brackets, a broken lightning arrestor lead wire, inadequate hardware spacing such as a gap between a ground wire and a metal staple, metal tags left on hardware, or metal objects thrown on the power line. Any metal parts that are not well insulated from, or well connected to, one another may form a spark gap. See Figure 11.37. We'll discuss more of the specifics later this chapter.

Locating the Source

It is the utility's responsibility to locate a source of noise emanating from their equipment. Utilities, however, do not always possess the necessary expertise or equipment to locate sources of radio noise. As a practical matter, many hams have assisted their utilities in locating noise sources. In some cases, this can help expedite a speedy resolution. This section is, therefore, optional and only intended if you wish to pursue locating the source in an effort to assist your utility. If not, you may proceed directly to the next section, "Filing a Complaint." If you wish to attempt locating the source, read on…

First, it is important to understand that there is a significant caveat to this approach. Should you mislead the power company into making unnecessary repairs, the more frustrated they will become. This expense and time will be added to their repair list. Do not make a guess if you don't know the cause. And don't suggest they replace insulators and transformers. Utilities will often make these "repairs" but they rarely help. While some power companies might know less about the locating process than the affected ham, indiscriminate replacement of hardware almost always will ultimately make the problem worse. Nonetheless, depending on your level of expertise and the specifics of your situation, you may be able to facilitate a speedy resolution by locating the RFI source for the utility.

Here are a few pointers and suggestions should you wish to attempt this approach. Although not intended to be complete or comprehensive, they should give many amateurs an understanding of the procedures involved and an idea of where to start. See the *AC Power Interference Handbook* for a more in-depth treatment of this subject.

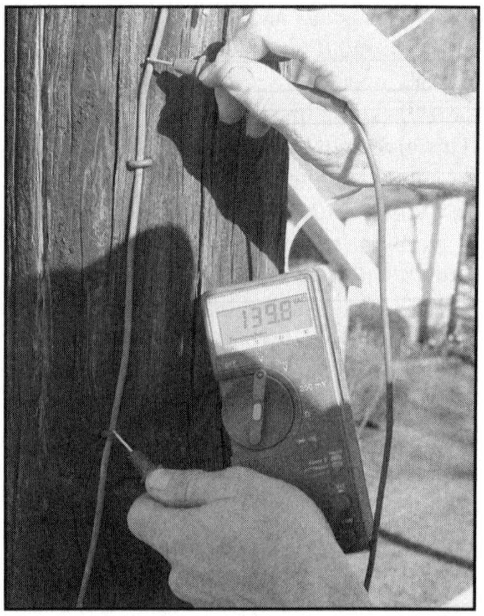

Figure 11.37—Power line noise can be generated by a variety of sources, many of which are not intuitively obvious to a beginning RFI investigator. In this case, arcing between a staple and a ground wire was the source. This is a frequent cause of power line interference.

A Few Preliminaries

By now, you should have determined to the best of your ability that your interference problem is indeed power line noise. Before getting started, here are a few preliminary steps to help you get started on the right foot:

- It's usually best to file your complaint with the utility *before* starting your RFI investigation. It can take a few days for your utility to respond. Why not start the clock ticking as soon as possible?
- Listen to the noise with a radio capable of receiving AM at VHF and UHF frequencies. Make note of the frequency at which the noise starts to diminish. As indicated previously, this can provide an important clue as to the proximity of the source. The closer the source, the higher in frequency you can receive it. If the noise can be heard at 440 MHz, you can expect it to be relatively close by—perhaps within less than a quarter mile radius. If it diminishes around 80 meters, however, the source can be over a mile away.
- Be sure to use any directional antennas you may have at your station to get an initial heading on the noise.
- Most power line noise sources are not readily obvious by a visual inspection. Nonetheless, it may be worthwhile to look for any telltale clues by an inspection of the utility poles in any suspect areas. A pair of binoculars can help. Arcing can also sometimes be visible on a dark night.

Radio Direction Finding (RDF) Techniques

Radio Direction Finding (RDF) techniques typically offer the best and most efficient approach to locating most power line noise sources. It is the primary method of choice used by professionals. While it is usually the most effective method, it also requires some specialized equipment, such as a handheld beam antenna.

Let's now take a look at some procedures used by Mike Martin, K3RFI a professional RFI investigator for over 25 years.

Mike, you're on…

The First Step

As previously discussed, a good first step for all RFI investigations is to eliminate the possibility that the interference source is located in your home. Since the majority of power company Radio TV Interference (RTVI) complaints that I investigate are actually caused by equipment inside a customer's residence, this step bears repeating. Think about it! Since interference gets worse as you get closer to the source, it makes sense to start your investigation in your own home. (If the source was closer to someone else they would most likely be complaining.)

Fortunately, locating and eliminating an inside source or a particular residence is a very easy process. It's simple enough that you can actually perform this test as a prelude to your RFI investigation or formal complaint without specialized equipment or your utility's involvement. Be sure to review the sidebars, "'How to' Locate Tips" and "How to 'Locate the Residence' Containing the RFI / TVI Source", if you have not already done so.

Finding the Source

It is difficult, if not impossible to locate an RTVI source that doesn't exist. Remember —always confirm the noise is active before you start looking for it. Once you have done so, you are ready for the next step—locating the source. You have several options, depending on equipment availability and other factors.

Although it requires some extra equipment and expertise, RDFing is usually the choice of professionals. While the pros often use more advanced techniques and equipment to locate power line noise, hams can often improvise with available equipment in order to achieve similar results.

A professional RFI investigator would first connect a Defect Direction Finder (DDF) receiver to the customer's antenna. See Figure 11.38. This specialized equipment enables the investigator to monitor the symptoms as received by the customer's

How to "Locate the Residence" Containing the RFI / TVI Source.

Please read all instructions thoroughly before using this procedure.

1. **Safety**: The following procedure should only be performed by qualified technicians familiar with the potential hazards of working around energized electrical equipment.

2. To locate the home (town house, apartment or condominium) containing the source, always start at the complainant's location following the steps in the **How to Locate Radio & Television Interference Sources in Your Home** sidebar. All instructions are to be followed while the interference is active. Remember, always start with the complainant. This person is complaining due to the severity of the problem. The closer to the source of the interference you get, the more severe the symptoms will be. Meaning, if the source were elsewhere the symptoms there would be worse and *that* person would probably be complaining.

3. **Locate the home containing the strongest noise signal.** The noise will be strongest at an electrical device connected to the home containing the source. You want to find an electrical device on the exterior of the homes, common to all the homes attached to the same secondary system as the complainant. These devices, such as the electric meters, main service breakers (whether outside or in a utility room), front porch lights, electric lamp posts, outside air conditioner units or doorbell buttons, will be the radiators for the offending noise. Whatever radiator you choose, it should be the most accessible at each home. The device you select to use as the noise radiator will be referred to in these instructions as the "radiator".

4. **Start at the complainant's home.** Start with your detector on and the gain turned up high enough to hear the offending noise. With your detector about 2 inches from the radiator you've chosen, turn the detector's gain control down to the point where you can barely hear the noise.

5. Start on your way to the next home. The next home may be next door or across the street, but almost always will be connected to the same transformer/secondary circuit as the complainant. As you move away from the starting point, the noise level on the detector will diminish. This is because the radiated noise signal is getting weaker as you move away. In order for the detector to hear the noise at the next house, the noise level will have to be the same or higher than the previous location.

6. When you reach the next location it's important that your detector be held in the same position as the previous location. The noise level should change. If the level is lower or not heard, you're moving further from the source and need to continue your search to the house in the other direction or across the street. If the level is higher, then you're headed in the right direction and again must turn the gain control down to the point of just barely being able to hear the noise.

7. Continue on to the next house repeating steps 5 and 6 if necessary. If the detector picks up no noise at the next home, then you're moving away from the home containing the noise source. Return to the previous home to hear the noise again.

9. Now that you've located the residence with the source, repeat step 2. **However, don't forget step 1.**—*Michael C. Martin, K3RFI, RFI Services' Interference Investigator*

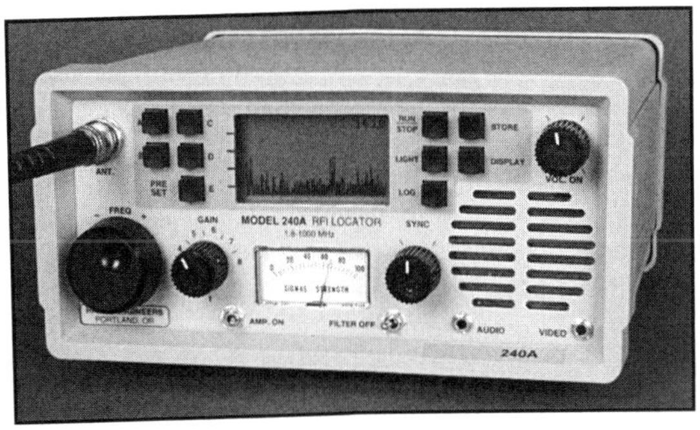

Figure 11.38—The Radar Engineers Model 240 HF-UHF RFI Locator is a professional grade receiver for locating RFI sources. It has a built-in oscilloscope display for recording and observing noise signatures. This receiver operates from 1.8 to 1000 MHz.

antenna. The DDF setup should include a broadband AM receiver that covers the frequency range affected by the problem, an oscilloscope (scope) and an attenuator or RF gain control to adjust the RF signal level. With these tools, the investigator can monitor the sound and pattern produced by the RTVI source(s).

Scope patterns show many important facts about the source(s) plaguing the affected equipment. They can reveal the number of simultaneous sources, determine which source is the strongest, and even provide an indication as to the size of gap across which the spark is occurring. When working with TVI complaints, the scope can show which source is having the most impact on the TV picture.

Television Interference

For starters, let's consider a TVI problem on television channel 4 as an example. While not an amateur radio issue, the concepts are the same, and the video aspects of the interference can be used to your advantage during an investigation.

While viewing channel 4 on the affected TV, we should see dots and lines in two horizontal groups slowly moving upward on the screen. As one group goes out of view at the top of the screen a second group re-appears at the bottom. (This characteristic of power line noise has already been previously discussed in this section.) Tune to 67.25 MHz, the frequency of channel 4, with our DDF receiver. We should now be able to view the noise as displayed on our scope and observe its unique noise signature.

Don't be intimidated by this instrument. It is a very powerful tool and a simple process to use. Once it is set up, a scope rarely needs adjustment. You can also use an amateur receiver with an oscilloscope to observe noise signatures. Connect the oscilloscope to the audio output of the receiver, set the receiver to the AM mode, and use the widest available filter option. Adjust the receiver and oscilloscope for proper gain, level and sweep settings. Be sure to observe the noise with the receiver using a frequency that is otherwise clear of any intentional stations. You want to be able to see the noise—and just the noise—with the oscilloscope.

The Signature or Fingerprint Method

Each sparking interference source exhibits a unique pattern. By comparing the characteristics between the pattern taken at the affected station with those observed in the field, we can now determine which is an offending source from the many sources we might encounter. It therefore isn't surprising

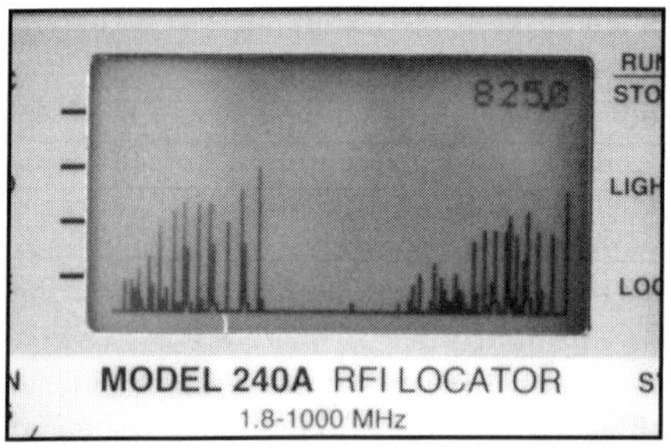

(A)

(B)

Figures 11.39—An unknown noise source at the complainant's antenna is shown in A. During the RFI investigation, noise signatures not matching this pattern can be ignored, such as shown in B. Once the matching signature originally shown in A is found, the offending noise source has been located.

that a pattern's unique characteristic is often called its "fingerprint" or "signature." See Figure 11.39 for an example.

This is a very powerful technique and a real money saver for the utility. Even though you may encounter several different noise sources in the field, this method helps identify the sources that are actually causing the interference problem. The utility need only correct the problem(s) matching the pattern affecting your equipment. You can accomplish this method one of three ways:

1) Record the pattern on the scope by drawing it on a note pad.

2) Take a photograph of the pattern. You can then compare it to patterns found in the field.

3) Or, use a receiver with a built in scope and the ability to store the pattern. This is the most modern method.

Professional interference locating receivers, such as the Radar Engineers Model 240 shown in Figure 11.38, have a built in oscilloscope display and waveform memory. They are ideal for the third method described in the previous paragraph. This is now the preferred method used by professional interference investigators. It provides the ability to toggle between the pattern saved at the customer's house and those from sources located in the field.

Once armed with the noise fingerprint taken at the affected station, you are ready to begin the hunt. If you have a directional beam, use it to get a heading on the noise. If you do not have a beam heading, start your search in front of your home or residence with the affected equipment. Next, travel in a circular pattern around the house, block-by-block, street-by-street, until you find the noise pattern matching the one recorded at the affected house. Use VHF or UHF if you can hear the RFI at these frequencies. The longer wavelengths associated with the AM Broadcast Band, and even HF, can create misleading "hotspots" along a line when searching for a noise source. At these frequencies, you may find that the noise peaks at certain poles with different types of hardware mounted on them. As a general rule, only use the lower frequencies when you are too far away from the source to hear the offending RFI at VHF or UHF. Work at the highest frequency on which the noise can be heard. As you approach the source, keep increasing the frequency. See Figure 11.11.

By now you're probably beginning to see the value of having the correct equipment. Once you've matched the pattern obtained at the affected equipment's antenna with one in the field, you're well on your way to locating the structure containing the source. The process now gets a bit interesting, but let's first go back for a moment and change the scenario a bit.

Interference to Amateur Radio

Recall that the original complaint was TVI on channel 4. In this case we tuned our DDF locator to the frequency of channel 4. Let's now consider an Amateur Radio problem with RFI at 21.4 MHz. The rules are still pretty much the same as with the TVI problem:

• The source must be active at the time of your investigation.

• As always, observe the symptoms on the affected equipment. (Obviously, this only applies if you are conducting the investigation for someone else.)

• Start the investigation by verifying that the source is not located in your home or residence.

• Connect the DDF receiver to the affected equipment's antenna before investigating the area outside the house.

In this example however, tune the DDF receiver (while connected to the affected equipment's antenna) to a frequency of 21.4 MHz. Again, observe and record the noise pattern for future viewing. Once ready to begin the hunt, start traveling in a circular pattern away from the house until you find the matching noise fingerprint. If, however, you have a rotating antenna, use it to your advantage. Determine the direction of the noise source from the station and reduce travel to a minimum.

Whether the interference problem is TVI or RFI, a rotating antenna is always helpful. Instead of traveling in spiral away from the house to find the noise, you can reduce your search to only one direction. You need now travel only in the direction toward the source. Obviously, you can ignore any noise patterns that don't match your recorded fingerprint and concern yourself only with the offending source(s).

Another important clue can be obtained by tuning the DDF receiver up in frequency. Listen to the noise at VHF and UHF and make note of the frequency at which it starts to diminish. This frequency can provide an important clue as to the proximity of the source. The closer the source, the higher in frequency you can receive it. If the noise can be heard at 440 MHz, you can expect it to be relatively close by—perhaps within less than a quarter mile radius. If it diminishes around 4 MHz, however, the source can be over a mile away.

An Important Rule

By now I hope you can easily see a tremendous improvement in our noise locating efficiency. We can now quickly locate the direction of an interfering signal and match its pattern with any number of suspect noise sources. Using this fingerprint technique, we can more easily locate the structure containing the source.

Perhaps the most difficult hurdle to overcome in this process is to ignore those noises not affecting your equipment. Whenever a suspect noise pattern doesn't match the recorded one, you must ignore it. Whenever attempting to locate the source of an interference problem, you may encounter many power line sources and other noise signals. This is normal and to be expected. If, however, the utility were to repair all of them, the task of locating and solving RTVI complaints would become more difficult. As a result, the repair cost would quickly become unacceptable. An important rule for efficient and economic RFI troubleshooting is to locate and repair only the source causing the problem.

During the Hunt

Let's now pick up where we left off a few paragraphs ago—on the trail of an interference source. Thanks to your beam antenna, we had a good idea of the direction to start the hunt. We started out heading in the direction from which the antenna indicated the noise was the strongest. This cut our travel distance down considerably since we didn't need to travel in a spiraling path away from your house. After a few blocks, we start receiving a noise with the same exact pattern as the one we recorded at your house. Let's now determine the actual structure containing the source.

At this point we want to reduce our signal level on our DDF receiver. We can do this in one of two ways. Typically, in most cases with a modern DDF receiver, simply turn the RF gain control down to the point we can achieve a minimum signal level (as indicated by the receiver's signal strength meter) and still have a clear noise pattern on the scope. If the receiver does not have an RF gain control, a step attenuator between the antenna and receiver can be used to reduce the signal level at the receiver's input. (Since step attenuators do not need to be calibrated, hams can often find them at flea markets at bargain prices.)

We'll know if and when we are approaching the noise source by observing its signal strength during the hunt. The pattern amplitude increases as the signal gets stronger and we get closer. If we find the signal getting weaker, we'll know we are going the wrong way and may lose the signal.

One secret to effective Defect Direction Finding is to maintain proper signal levels at the receiver. As indicated previously, we control this level by either an RF Gain control or an attenuator. Always maintain the minimum signal level necessary to observe the signal. As we approach the source, the signal level will increase. We must continuously adjust the gain to accommodate changes in the signal level. The importance of this rule cannot be overstated. Improper gain settings can make it extremely difficult to determine the direction of the source. In extreme cases, weaker signals may no longer be detected and stronger signals can produce abnormally high noise levels.

As we previously discussed, always ignore those patterns not consistent with the interference as observed at the affected equipment. As you approach the source and reduce your gain, the number of sources you want to ignore will decrease due to the receiver's reduced sensitivity. The receiver will no longer be able to hear weaker signals. Always maintain the lowest level possible when viewing noise signal patterns.

Directional Antennas

The hunting process is much easier when starting with a directional antenna at your house. You can also use this same method while on the street. With an omni directional or whip antenna, you must move the vehicle to determine the direction of the higher signal level. If we use a handheld or vehicle-mounted Yagi (directional) antenna, we can follow the direction of the strongest signal to the noise source. This will greatly reduce the amount of time and travel distance required during the hunt.

Radio Direction Finding (RDF) techniques typically offer the best and most efficient approach to locating most power line noise sources. It is often the primary method of choice used by professionals. A handheld Yagi works at VHF and UHF, but must be used within its specified frequency range. Not only are VHF and UHF antennas typically smaller, but also direction headings are more reliable. An attenuator is required between the antenna and the receiver if the receiver does not have one. Use as much attenuation as you can in order to minimize the area of search. You'll need to add more and more attenuation as you approach the source. See Figure 11.40.

Pinpointing the Source

Once you have located the source pole, you can consider your role as an interference investigator done. Some hams, however, like to go one additional step—pinpointing the offending hardware on the pole. A pair of binoculars on a dark night may reveal visible signs of arcing, and in some cases you may be able to see other evidence of the problem from the ground. These cases are rare, however. More than likely, you'll need to consider a better approach. Once again, the professional interference investigator has specialized equipment for this purpose. You, as a ham, have a more limited set of options. Equipment, however, can be homebrewed! We'll talk more about this later.

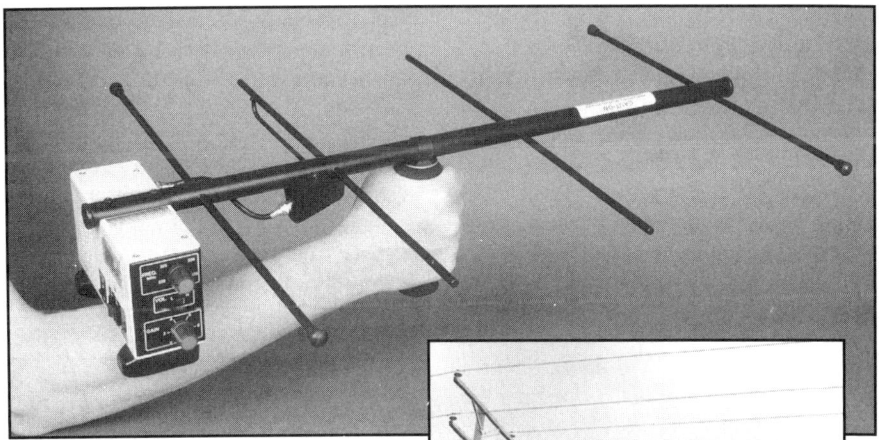

(A)

(B)

Figures 11.40—The Radar Engineers Model M330 is a professional grade Mini RFI Locator. This photo was taken during an RFI investigation at an Amateur Radio station. The source of the harmful interference was traced to this pole. The Model M330 Mini Locator operates from 320 to 340 MHz. It's ideal for locating the source pole once the search has been narrowed down to several poles.

Once a utility or professional RFI investigator knows the structure containing the offending noise source, the next step is obvious. He or she must find the source on that structure. The RFI investigator, even if not a lineman, must be able to pinpoint the source on the structure down to a component level from the ground. Alternatively, an investigator can instruct the lineman on the use of a "hot stick" mounted device used to find the source. Both methods are similar, but as a ham you have only one option. A hot stick mounted device can only be used by utility personnel. See Figures 11.42 and 11.47 later in this chapter. **Hot sticks are not for hams! Proper and safe use of a hot stick requires specialized training. Do not use them.**

Note: In most localities, it is generally unlawful for anyone unqualified by a utility to come within 10 feet of an energized line or hardware. This includes hot sticks.

An ultrasonic dish is the tool of choice for pinpointing the source of an arc from the ground. While no hot stick is required, an unobstructed direct line-of-sight path is required between the arc and the dish. This is not a suitable tool for locating the structure containing the source. It is only useful for pinpointing a source once it has been highly localized. An ultrasonic dish, for example, is not useful for locating the pole on which a noise source is located. It is, however, ideally suited for pinpointing the arcing hardware once the offending pole has been isolated. See Figure 11.41.

Caveat: Corona discharge, while typically not a source of RF power line noise, can and often is a significant source of ultrasonic sound. It can often be difficult to distinguish between the sound created by an arc and corona discharge. This can lead to mistakes when trying to pinpoint the source of an RFI problem with an ultrasonic device.

The key to success is the gain control. Ultrasonic locators used from the ground can work very well provided you maintain minimum gain after initially detecting the noise. If the source appears to be at more than one location on the structure, reduce the gain. In part, this will eliminate any weaker noise signals from hardware not causing the problem. Incidentally, proper gain control is just as important for hot stick mounted locators.

Professional grade ultrasonic locators are beyond the budget of the average ham. Home brewing options, however, can make a practical ultrasonic locator affordable in most situations—and can make a great weekend project, too. See "A Home-made Ultrasonic Power line Arc Detector" in the April 2006 issue of *QST* on page 41.

Figure 11.41—This RFI investigator is using a Radar Engineers Model 250 Parabolic Pinpointer to find an RFI source on a utility pole.

Common Sources

The following list contains some of the more common power line noise sources I've encountered over the years. They're listed in order from most common to least common. Note that some of the most common sources are not connected to a primary conductor. This in part is due to the care most utilities take to insure sufficient primary conductor clearance from surrounding hardware:

- Loose staples on ground conductor
- Loose pole top pin
- Ground conductor touching nearby hardware
- Corroded slack span insulators
- Guy touching neutral
- Loose hardware
- Bare tie wire used with insulated conductor
- Insulated tie wire on bare conductor
- Loose cross arm braces
- Lightning arrestors

A Common "Non-Source"

Note that transformers don't even receive honorable mention in the list of most common power line noise culprits. Despite their reputation, only a very small percentage of transformers are actually found to be the cause of an RTVI complaint.

Why are they blamed so often for noise they do not actually cause? Let's take a closer look at a typical scenario for some insights:

A utility customer calls with an RTVI complaint. He'll typically say he has looked long and hard for the cause of the problem. He'll also add that he found the source on a transformer pole and that he believes the cause is the transformer. When power company investigator comes to start his investigation, like the customer, he finds the highest level to be at that pole. He too may then conclude the problem is the transformer. The transformer is changed and the problem is gone. Problem solved!

You may now ask the obvious question, "If the transformer wasn't the source, why did the noise disappear? The actual reason may be that the source was only loose hardware. The hardware was tightened when the transformer is replaced. Obviously, it is far more economical to only tighten the loose hardware and not change the transformer. There is also added hardware associated with the

Figure 11.42—This RFI Services lineman is using a hot stick and an RF sniffer made by Radar Engineers to pinpoint a source of noise on a power pole. In this case, the noise source was found to be a pole-top pin sparking to a bolt.

transformer pole. Remember, the pole will have a driven ground conductor, lightning arrestor, often a down guy and other hardware that can act as an antenna and radiate noise. This can cause a high level of noise that fools the investigator into believing he has found the source structure. He hasn't found the source of the noise, only a better antenna to radiate it.

The Final Step

Once a professional RFI investigator has successfully located and repaired the source, he or she should always check back with the customer to verify the complaint has been entirely solved. As a ham and utility customer, however, you should not rely on your power company to do this. Be sure to ask when you talk to your utility. The RFI investigator may need to check for additional sources. In some cases, it may be necessary to repeat this process until all the sources are corrected.

Some Final Tips and Comments

Let's now review and highlight some of the key points we've discussed in this section.

• Use a VHF or UHF receiver in the AM mode, if you can hear the RFI at these frequencies. The longer wavelengths associated with the AM Broadcast Band, and even HF, can create misleading "hotspots" along a line when searching for a noise source. Not only are VHF and UHF antennas typically smaller, but also direction headings are more reliable. As a general rule, only use the lower frequencies when you are too far away from the source to hear the offending RFI at VHF or UHF.

A ferrite loop works at HF, and a handheld Yagi works at VHF and UHF. An attenuator is necessary. Use as much attenuation as you can in order to minimize the area of search. You'll need to add more and more attenuation as you approach the source. An aircraft band receiver is ideal since it is both AM and VHF. Even a battery powered portable with a loop stick antenna will have some directional capability, and it will get stronger in intensity as you approach the source. The MFJ-857 is also relatively small and inexpensive power line noise receiver.

Note: See *The ARRL Antenna Book* for RDF antenna construction plans, including antennas for HF. Look for them in Chapter 14 of the 20th edition.

And now for some final tips:

• If you are not the complaintant, always view the problem as seen by the complaintant *before beginning the RFI investigation.*

• Always eliminate your residence first.

• Maintain minimum receiver gain once you've detected the offending noise pattern.

• If the noise appears to be coming from more than one source or direction, reduce the receiver's RF gain.

• Ignore any noise patterns not seen by the affected equipment's antenna.

• Locate and report only the source causing the problem.

• Every antenna works best at one frequency. Use that frequency when using that antenna.

Equipment

By now you may be wondering about some of the DDF equipment we've discussed. Let's take a look at some professional grade instruments that might be used in a typical noise-locating arsenal.

Receiver

A good DDF receiver should be broadband and cover the AM broadcast band through at least 250 MHz and operate in the AM mode throughout this entire range. In addition, it must have an attenuator or RF gain control, and signal strength meter or indicator and preferably the option of battery power. The Radar Engineers Model 240 receiver was previously shown in Figure 11.38.

Portable ham and shortwave receivers can always be substituted. Of course, it's imperative that whatever receiver you use, it should be capable of receiving AM. It should, preferably, have an RF gain control, but an external step attenuator can be substituted. It should also be capable of receiving the affected ham bands, and tune high enough to be used with your VHF or UHF Yagi antenna.

Antennas

A professional RFI investigator will typically have several antennas to complete his DDF arsenal.

A 7 MHz or lower frequency amateur radio antenna. This will help narrow down the area of sources that affect the AM Broadcast band as well as the lower amateur radio band frequencies. Such antennas are readily available from Amateur Radio suppliers. They are used for mobile HF operation and can be easily mounted on a vehicle with a multi-magnet base. See Figure 11.43.

A 140 MHz-150 MHz or higher whip antenna used to locate the source structure. This antenna can be easily installed on a vehicle with a magnetic base for this purpose. While this antenna does not have directional capabilities, you can use it for monitoring relative signal strength as you approach the source. See Figure 11.44.

A directional Yagi type antenna. This antenna provides the capability to determine the direction of a noise source. Because it is larger and must be rotated, this is not an antenna that can easily be mounted on a vehicle. It can, however, be hand carried. Look for a suitable directional Yagi antenna that covers a frequency or frequency range

Figure 11.43—Mobile Amateur Radio antennas such as this are typically available for 4 and 7 MHz, depending on the particular design. This antenna is shown with a four-magnet mount. Most Amateur Radio dealers carry these or similar products.

Figure 11.44—VHF and UHF Amateur Radio antennas are ideal for RFI locating. This particular antenna is shown with a single magnet base. They are readily available at almost all Amateur Radio dealers.

Figure 11.45—This antenna on a van specially equipped by the author makes tracking down power line noise a bit easier. Similar directional handheld antennas are also available.

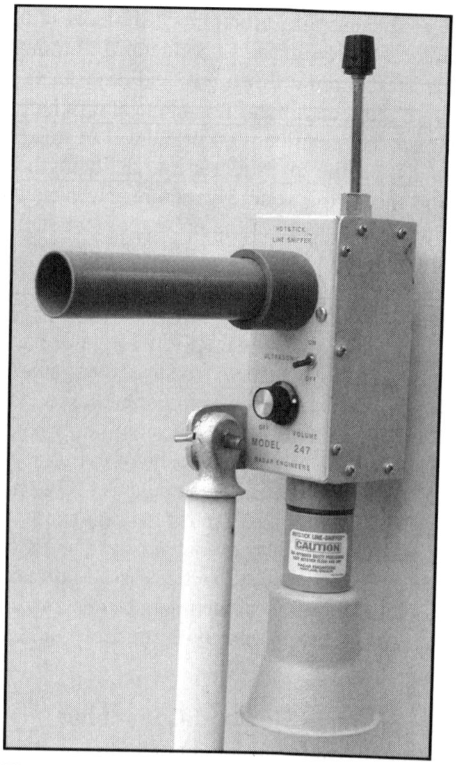

Figure 11.47—The Radar Engineers Model 247 Hotstick Line Sniffer is an RF & Ultrasonic Locator. It's ideal for pinpointing a noise source once the offending pole is known. It requires a hotstick, and is typically used from the pole or a bucket.

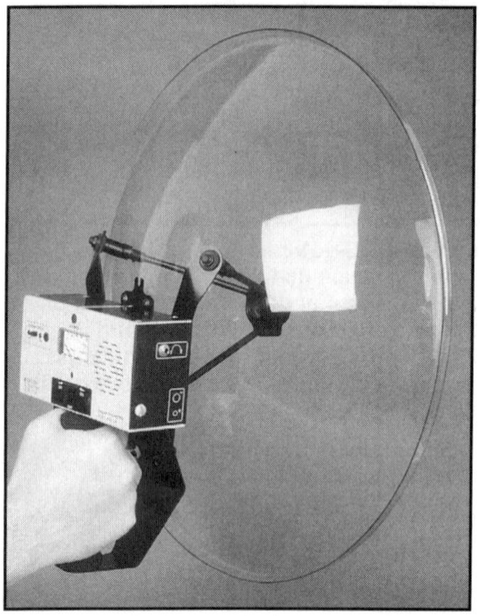

Figure 11.46—The clear-plastic parabola is an "ear" connected to an ultrasonic detector that lets utility personnel listen to a power line.

somewhere between 140 and 500 MHz. See Figure 11.45.

Ultrasonic Dish

My professional grade ultrasonic receiver/pin pointer is great for pinpointing a noise source from the ground, once the structure it's on is located. Even though it takes some experience to discriminate between corona and sparking sources, I wouldn't leave the office without one. See Figure 11.46. As previously discussed in this section, these tools are beyond most ham budgets. If you wish to go this extra step, you may want to home brew an ultrasonic pin pointer.

Hot Stick Sniffer

A hot stick mounted sniffer is not suitable for ham use. Leave this one to the professionals! See Figure 11.47.

Equipment of Minimal Direction Finding Value

Thermal/Infrared detectors and corona cameras are not recommended for the sole purpose of locating RTVI sources. It is rare that an RTVI source is detectable using infrared. In some cases, thermal vision and corona camera techniques can actually *cause* RTVI. These are not useful tools for locating power line noise. Nonetheless, it seems many utilities still use them for such purposes with minimal or no results.

Final Suggestion—Form an RFI Committee

Many Amateur Radio clubs feature fox-hunts and similar club activities. If your club is one—you may wish to consider or suggest the possibility of forming an RFI committee. Many of the skills and equipment used in fox hunting are ideally suited for RDFing sources of RFI. Sharing equipment, knowledge and expertise can work to everyone's advantage!—*Mike Martin, K3RFI*

FILING A COMPLAINT

Once you have verified the problem to be true power line noise, and that it is not coming from a source internal to your home, contact your utility's customer service department. In addition to your local phone book, customer service phone numbers are included on most power company websites.

It is important to maintain a log during this part of the process. Be sure to record any "help ticket numbers" that may be assigned to your complaint as well as names, dates and a brief description of each conversation you have with electric company personnel.

If The Power Company is Unresponsive

Unfortunately, the reality is that power line noise cases can sometimes drag on for a considerable period of time. Power companies in these cases will typically try to fix the problem at first. If their initial attempts are unsuccessful, they often become frustrated; especially as the costs associated with this "repair" begin to escalate. In other cases, the utility is simply unequipped or unwilling to deal with a power line noise complaint. They may simply decide to ignore the problem altogether.

Fortunately, there is a source of help in these cases. It's called the FCC/ARRL Cooperative Agreement. While the program is not a quick or easy solution, it does offer an opportunity and step-by-step course of action for relief. It emphasizes and provides for voluntary cooperation without FCC intervention. There are several built in waiting periods and a number of requirements that a ham must follow precisely.

The Role of The ARRL

It must be emphasized that the the ARRL is not in the enforcement business. The role of the ARRL is primarily a technical one. As such, we serve as a source for help rather than enforcement. Our role is to help solve a power line noise case *before* it ever gets to the FCC. Only those cases that require some sort of enforcement are therefore handled by the FCC. It is an attempt to focus the FCC's resources where they are needed the most.

Riley Hollingsworth, Special Counsel at the FCC's Enforcement Bureau, offers the following explanation:

"We do not have the staff to deal with

all of these issues directly. We therefore depend upon the technical folks at ARRL to take first cut at resolving the issue and in helping the complainant define and isolate the problem. Many times what is thought to be power line interference is actually originating from within the household or a neighbor's property."

The ARRL / FCC Cooperative Agreement Program Process

Let's take a step-by-step look at the entire procedure starting from the very beginning:

1) Eliminate the RFI source as being internal to your home and verify it as power line noise to the best of your ability. Maintain a log of frequencies, noise levels in S Units, the date, weather and time of day. It can be just as important to log the presence of noise as well as the absence of it.

2) Get the ball rolling as soon as possible! File a complaint with your local utility's customer service department immediately after performing step 1, above. And as before, don't forget to maintain an accurate log. Be sure to record each contact and any associated help ticket numbers with your complaint.

3) If, after a reasonable period of time, your utility remains unresponsive to your complaint, contact the ARRL RFI Desk. Sixty days is probably a good approximation, but this can be adjusted on a case-by-case basis. You'll need to provide the name of your utility, plus the name and address of your utility's CEO.

4) The ARRL will send your utility a letter. Typically, this letter is addressed to your utility's CEO. It describes the FCC/ARRL Cooperative Agreement, pertinent FCC Rules, and and includes pamphlets on how to fix the problem. It also urges voluntary cooperation with an offer to help from the ARRL. You will automatically receive a copy of this letter.

5) The ARRL letter allows 60 days before proceeding to the next step. An extra week or more is also added for mail delivery. During this time, you must cooperate in any reasonable way that you can with the utility's RFI Investigator. This includes allowing the investigator to hear the noise and take samples of it at your station. You must also inform the ARRL of any activity or correspondence from the utility regarding your case.

6) If, after 60 days, the utility has failed to make a good faith effort to fix the problem, you need to once again contact the ARRL. This is an important step, and the only way the ARRL has of knowing the problem has not been resolved. Should you fail to contact the ARRL at any time during or after this step, your case may incorrectly be assumed to have been resolved.

7) The ARRL prepares your case for FCC consideration. Assuming the Commission does not reject your case, the FCC sends an official advisory notice to your utility. This step typically takes up to two weeks.

8) You do not receive a copy of the FCC letter. It will, however, be released into the Enforcement Log approximately two to four weeks after it is issued. Review the FCC Enforcement Log for your letter. (Alternatively, a search for your call sign on the ARRL Web site is often another convenient approach.) If your letter does not appear in the log for a month after it is issued, you need to inform the ARRL RFI Desk. Here is a link to the FCC Enforcement Log: **www.arrl.org/news/enforcement_logs/**

The FCC letter allows another 60 days for utility to fix or demonstrate a good faith effort to fix the problem. The same rules apply during this time period as with the ARRL letter. An extra week or more is added for mail delivery, and you must cooperate in any reasonable way with the utility's RFI Investigator. You must also inform the ARRL of any activity or correspondence from the utility regarding your case.

9) If, after the 60 days allowed by the FCC letter, the utility has failed to make a good faith effort to fix the problem; you need to once again contact the ARRL RFI Help Desk. As with the previous ARRL letter, this is an important step. It is the only way the ARRL and the FCC has of knowing the problem has not been resolved. Should you fail to do this, the ARRL and FCC may incorrectly assume your case is closed.

10) The FCC issues a second notice to the utility. Depending on the specific circumstances of your case, this notice typically allows 15 to 30 days for a response. As with the first FCC letter, you do not get a copy. You should again look for your letter to appear in the enforcement log. If it doesn't appear after four weeks, you should notify the ARRL RFI Desk.

11) If after the second FCC Notice, the utility fails to demonstrate a good faith effort to resolve the problem, again contact the ARRL Help Desk. The next step is to form a committee to provide an independent assessment of the situation. The committee will generate a report, which the ARRL's RFI Desk will present to the FCC.

The FCC will then make a decision as to whether or not a formal field investigation is in order. Should such an investigation be requested, it is important to understand that the outcome will be difficult to reverse. The FCC may or may not decline to take any further action in your case. Monetary forfeitures may also be levied against the utility. This is a risk that most amateurs would obviously prefer to avoid. It is in everyone's best interest to achieve a voluntary resolution with which everyone will be satisfied.

Your Role In The Process

While every case is different, and therefore handled differently, there are some general expectations and requirements to which everyone must adhere. Most are common sense and good manners. The following summary should be helpful for anyone that wishes to be included in the program.

1) Depending on the specifics of a particular case, other ARRL volunteers and staff may be involved. ARRL Section Managers and local Technical Coordinators often provide help at the local level. Don't be surprised if you are contacted by other individuals from the ARRL attempting to help in your case.

2) Individuals participating in this program are expected to treat all utility personnel, ARRL staff and volunteers with respect. Anyone that exhibits abusive or hostile behavior may be dismissed from the program.

3) Individuals participating in this program are expected to cooperate in any reasonable way with RFI investigators from their utility. This also includes providing access to signals from their station's antenna for the purpose of assessing the problem and recording a noise signature. In cases where the amateur will not be home during the RFI investigation, a cable from the station can be run out a window for the RFI investigator.

4) During the mandatory waiting periods, i.e., the time periods allowed after each letter is sent to the utility, it is not necessary to flood the ARRL RFI Desk with daily or excessive reports that nothing has happened. Be reasonable about updating the ARRL RFI Desk when nothing has taken place. Unless circumstances unique to your particular case that require more frequent updates, reports every two to four weeks should be adequate. It is important, however, to report any contact or action taken by the utility during a waiting period.

5) It is imperative to update the ARRL RFI Help Desk of your RFI status *after or just prior to* the expiration of a waiting period. Should you fail to do this, your case may be incorrectly assumed to have been resolved during the waiting period.

6) Any and all formal contact with your utility must be through the normal complaint process in order to be properly logged. Informal contact with utility personnel, such as a lineman that might be working in your neighborhood, may not be properly recorded in the company's system. It is important to recognize the difference. While utility personnel

may be able to help, the complaint will, most likely, not be recorded in the utility's system and be available should future ARRL or FCC involvement be required. Always follow-up with a formal complaint whenever an informal request for help does not resolve the problem. And always remember to request the help ticket or complaint number associated with your case.

7) The last and probably most important requirement is patience. While it is true that some cases of power line noise are resolved in a timely fashion, the reality is that many cases can linger for an extended period of time. There are often no quick solutions. It is important to understand that the ARRL Cooperative Agreement Program does not offer a quick fix. It does, however, provide a step-by-step and systematic course of action under the auspices of the FCC in cases where a utility does not comply with Part 15.

ARRL Membership and The Cooperative Agreement

ARRL membership is not required to participate in the Cooperative Agreement program. Most amateurs, however, recognize the value of this service and wish to support this effort with their membership. A typical case can take many hours of staff time, plus other ARRL resources. Furthermore, the price of a single membership is typically only a small fraction of the costs associated with a single power line noise case under this program. Membership in this case is a real bargain.

Filing A Complaint

Assuming a power line noise problem has been ongoing for a considerable period of time, and you have exhausted every reasonable possibility to resolve it with your utility, it may be time for the next step. The procedure is relatively straightforward.

All complaints filed with the ARRL under the Cooperative Agreement with the FCC start and are initiated by the ARRL RFI Desk. If you contact the RFI Desk, be prepared to explain *briefly* the steps you have taken to try to resolve this with your power company. If you and the person handling your complaint agree that an "ARRL Power Line Noise Letter" is appropriate, be prepared to supply them with the utility's name, address and an individual to receive the letter. This individual should be someone at the executive level, typically the corporate CEO, if possible.

Here is the contact information for the ARRL RFI Desk:

American Radio Relay League
RFI Desk
225 Main Street
Newington, CT 06111
E-mail: rfi@arrl.org
Tel: (860) 594-0392

Filing A Complaint With The FCC

In some cases, for one reason or another, an amateur may prefer to directly contact the FCC, or if the interference is to a broadcast service, there is a procedure to file a complaint with the FCC. The Consumer & Governmental Affairs Bureau Call Center can help with these RFI problems.

Before contacting the FCC, however, people should make reasonable efforts to resolve power line interference through the normal customer-service procedures at their power-utility company. In cases that involve the Amateur Radio Service, the FCC often directly asks the ARRL to help resolve cases before they consider other steps.

Once someone has exhausted every reasonable possibility at resolving a power line or other interference problem with a utility company (or operator of any other Part 15 device), they can contact the Consumer & Governmental Affairs Bureau Call Center and discuss their problem with one of the FCC personnel.

If you do contact the Call Center, be prepared to explain *briefly* the steps you have taken to try to resolve this with your power company. If you and the FCC staff agree that having the FCC send the "RFI-Power-Utility Letter" is appropriate, be prepared to supply them with the utility name, address and, if possible, an individual to receive the letter. This individual should be an upper manager or vice president, if possible.

Here is the contact information for the Consumer & Governmental Affairs Bureau Call Center:

Federal Communications Commission
Consumer & Governmental Affairs
Bureau Call Center
1270 Fairfield Rd
Gettysburg, PA 17325
www.fcc.gov/cib/ncc/Welcome.html
E-mail: fccham@fcc.gov
Tel: 1-888-CALL-FCC

—*Mike Gruber, W1MG*

CONCLUSION

The information contained in this chapter is a summary description of the most common RFI problems caused by power lines and electrical devices. It provides a foundation for investigation, analysis and cure of specific cases not covered here. The entire second half of this chapter is devoted to the resolution of a power line noise problem, which usually requires cooperation between the affected ham and the utility.

ACKNOWLEDGEMENTS

Jody Boucher would like to acknowledge that much of the information in the first part of this chapter was learned from *Interference Handbook* by William R. Nelson, WA6FQG, and from the many articles published over the years by Marvin Loftness, who spent decades in the field of RFI research and has written many books and publications on the subject. Mike Gruber would also like to acknowledge and thank Michael C. Martin, K3RFI, RFI Services' Interference Investigator for his contributions to the second part of this chapter.

Chapter 12

External Rectification—"The Rusty Bolt Effect"

In rare cases, nonlinearities near transmitters can create interfering signals that would not otherwise be there. It is hard to think that corrosion in your transmitter antenna system could cause interference, but it can happen. When it does, this can be the hardest type of interference to track down.

By Mitchell Lee, KB6FPW
172 N 24th St
San Jose, CA 95116

Low-pass filters on the transmitter, high-pass filters on nearby TVs and careful bypassing at key consumer items such as stereos and telephones are touted as the "righteous" path to RFI-free hamming. But, even the most thorough application of these techniques can be foiled by a single rectifying joint in the vicinity of the transmitter or receiver. Nature is a prolific creator of diodes. Weathered joints between pieces of metal (such as TV mast sections, barbed wire and fence stakes or sections of rain gutter) form crude diodes that are efficient generators of spurious signals. The associated lengths of metal on either side of the joint act as antenna elements. They feed energy to the rectifier. The resulting nonlinear current flow is rich in harmonics, which are reradiated to wreak havoc in nearby receivers.

The effects of these harmonics are identical to those of harmonics produced in a transmitter. Two forms of "external" rectification are troublesome to amateur operators. First (and most familiar to amateurs), rectification of amateur signals causes interference that is associated with amateur station activity. Signals from other sources may contribute to intermodulation distortion (IMD). This obscures timing relationships between source activity and interference (because both sources must be active to produce the interference) and adds to the insidious nature of this interference.

The second (and often most frustrating) form results from nonamateur transmitters, especially commercial broadcast transmitters. In congested areas (where one or more AM broadcast stations operate in close proximity to buildings and power lines) the resultant interference can be quite strong in the lower HF bands.

This chapter addresses three important points:

- how (and where) rectifying joints are formed.
- how to track them down.
- how to disable them.

In general, any conducting structure (such as electrical, plumbing and antenna systems) can harbor a rectifying joint. Oxides and other corrosion products form crude, but effective, interference generators. Actually, some of these compounds are remarkably efficient. Lead sulfide is the galena crystal of early radio fame. In a crystal radio, small impurities of lead sulfide occurring at the surface of a lead crystal form diodes that are contacted through a "cat's whisker." Copper oxide, a common substance in many antenna and ground systems, was once commercially exploited in power rectifiers.

Any place where two pieces of metal touch is a candidate for corrosion. The process is accelerated if the joint is subject to humidity or weathering.

Rectification may also take place in "real" diodes—diodes contained in equipment around the shack. This includes amateur equipment such as antenna rotator controllers and VHF transceivers (even when they are unpowered, but connected to an antenna). Also include such nonamateur devices as alarm systems, power supplies, telephone automatic dialers and so on.

The task of locating a rectifying joint is much simpler if you know where to look. Classic rectifying-joint RFI generators are described below. While by no means exhaustive, these examples should give RFI sleuths a good frame of reference.

THE USUAL SUSPECTS

Guy wires: Metallic guys are used in some antenna and tower installations. The connections at the tower or mast can rust and form rectifying joints. Wire guys are normally broken at regular intervals by insulators, but there is always a piece at the tower end that connects directly to the tower. Even relatively short lengths can cause problems in a strong field; the short length enhances the radiating properties on harmonic frequencies. Egg insulators sometimes break, bringing the otherwise separate wires together and possibly forming a diode. Beware of continuous guys (as used on TV masts); they may be long enough to form a resonant element. Beware of spliced guy wires, guy rings and the tie point; these are likely to harbor rectifying joints.

Occasional broken strands can cause problems described later under "Stranded Copperweld." If guys are properly tensioned, however, broken strands are obvious. They should be replaced immediately.

Jointed antenna elements: The joints between telescoping sections of aluminum antennas must be cleaned thoroughly, coated with conductive grease (available from electrical supply companies under several different brand names), mechanically secured and weatherproofed with a nonacidic caulking product. The ends of the finished element must also be sealed. A diode can easily form if moisture penetrates the joint.

Corroded TV antennas: An old TV antenna, immersed in a strong field, is a likely rectifier. The usual trouble spots are element-to-boom joints and feed-line connections. Replacement is the best option for a deteriorated TV antenna or antenna system.

Towers and masts: The joints between tower and mast sections are subject to the same problems that beset antenna sections. Joints should be mechanically bonded, or short lengths of solid grounding strap should be used to ensure electrical continuity. If the mast or tower is of the crank-up variety, this is impossible.

TV mast sections: The actions for towers apply, but with some interesting twists. TV mast sections are normally erected in multiples of 5 or 10 ft. A 5-ft mast section attached to a chimney renders useless the best efforts at horizontally polarizing a 6-meter beam to suppress RFI. A 15-ft mast attracts the nearest 10-meter signal. The 30-ft masts used in fringe areas resonate beautifully on 20 meters. Even though a mast is not an exact resonant length, it can still gather enough current to excite an otherwise marginal diode.

"Stranded Copperweld" antenna wire: This product is available through many retail outlets. Some stranded Copperweld is of marginal quality. It becomes an unequaled TVI generator when aged. The strands gradually corrode over time; the steel core is eventually exposed and quickly rusts through. Then, current in a broken strand must bypass the break through adjacent conductors. But, the conductors are covered with copper oxide—once used in commercial power rectifiers! Single-conductor wires are not subject to these effects.

Front-end switching diodes and RF stages: Auxiliary receivers and transmitters are often left permanently connected to their antennas. Strong fields can induce enough voltage in the front end to forward bias switching diodes or an active device in the first RF stage—even when the radio is off. TVI travels right back to the antenna. One of the most insidious sources of trouble is the infamous masthead TV preamp. Strong fields can send these units into oscillation, wreaking havoc across VHF/UHF TV and amateur bands.

Antenna-rotator control boxes: These often contain circuits that rectify signals picked up by the control cable. Ferrite chokes, ferrite beads and bypass capacitors are the solution.

Remote coax-switch control boxes: Treat as rotator controls. Here RF energy travels in close proximity to steering diodes and control circuitry, compounding the problems.

Ground radials: Nothing beats a copper plate covered with sheet-metal screws as a ground "hub" at the base of a vertical antenna. But, each ground wire must be attached and weatherproofed with care or diodes may form. Ground wires can corrode clear through with time, possibly forming a diode at the break. Even if a radial is buried, harmonic currents are injected back into the system, where they radiate.

RF "probes" and monitors: These invariably contain a diode or transistor that rectifies the received field. A short sensing wire efficiently radiates harmonics on VHF. Disconnect these instruments except when needed.

Power supply diodes: Especially on the lower HF bands, RF traveling up a power cord can enter a power supply for rectification by the diodes. 0.01-μF capacitors are often included both across the ac line and across the diodes for protection against this phenomenon. On higher HF bands, the supply output leads easily receive RF fields and feed energy back into the regulator circuitry and/or diodes.

RF ammeters: Believe it or not, a 2-A thermal RF ammeter in series with a $^{3}/_{8}$-λ 160-meter inverted L can cause interference in a receiver. A shorting switch solves the problem.

Cold solder joints and crimps in RF connectors: This is a tough one to trace, but a bad joint in an RF connector (or connections at the antenna, for that matter) causes interference of large proportions. Beware of crimped TV-grade coax. Coax can develop corroded breaks in the center conductor (especially where the coax is constantly flexed; see Figure 12.1) that conduct well enough to transmit power, yet still rectify transmitted and received signals.

Around the House

Burglar alarms and garage-door openers: Both of these contain solid-state devices and plenty of actual diodes. Both are attached to long lengths of wire. Treat them with common-mode chokes near the rectification source.

Metal roofing: Joints in metal roofing can rectify. When a vertical antenna is located directly overhead, the roofing carries return currents. Bond sections together with short lengths of solid strap and suitable soldering materials. Metal siding on mobile homes, motor homes and travel trailers is equally troublesome. Bond aluminum panels with stainless-steel lock washers and straps.

Duct work: Joints between ducts can rectify. Sheet-metal screws are an instant cure in most cases. Alternatively, connect the sections together with short lengths of solid strap.

The story of "talking ductwork," is often told late at night on 75-meter phone: A voice was heard emanating from a register or vent. The strong RF field from an adjacent amateur station (rectified by the ductwork) was responsible. Another version describes an apartment building laundry chute that receives a local broadcast station.

Pipes touching pipes: Service pipes sometimes touch to form diodes. Ground currents are the usual culprit. While it is possible to interrupt the current flow with ferrites or coupled tuned circuits, those solutions are difficult. It's easier to isolate the two pipes with a shim of cork or acrylic sheet, as in Figure 12.2.

Pipe joints: Where pipes carry RF current, problems similar to those mentioned previously (towers and masts, jointed antenna elements, and TV mast sections) can occur at joints. Renew or bond suspect joints.

Figure 12.1—The conductors of cables that are subject to repetitive load and flexing may fracture to create a rectifying junction inside the jacket.

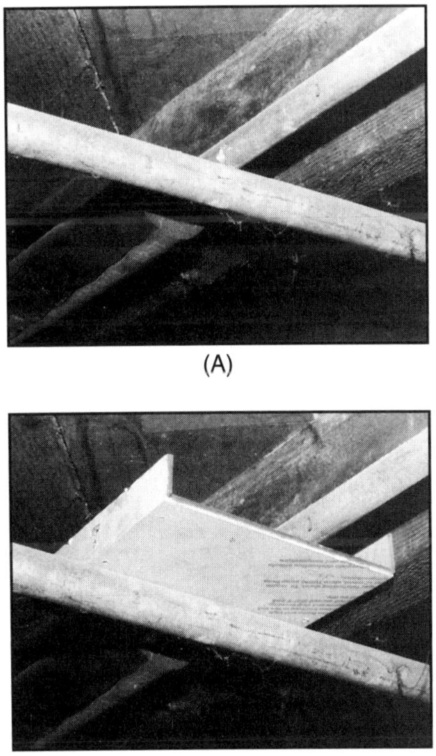

(A)

(B)

Figure 12.2—At A, two pipes touch where they cross, to produce a rectifying junction. An acrylic insulator has been inserted at B to prevent them from touching.

Rain gutters: These are, without question, the most well known of interference generators. In reality, the likelihood of a rain gutter causing external rectification is small, but ham folklore blames the rain gutter for everything. Rain gutters and down spouts are common components that can function as antennas, counterpoises, or grounds in radio systems. Rain gutters and down spouts are guaranteed to have at least a few suspect joints. Bond them with solder and screws. Better yet, convert to plastic rain gutters and lay in wire for any intended radio use.

GFCIs (ground-fault current interrupters): These pick up RF energy from the power lines. A properly installed and operating GFCI should not cause problems, so replace any units that cause trouble. If proper operation is verified, but the GFCI causes RFI, have a qualified electrician install metal conduit and a metal outlet box to shield the GFCI and adjacent conductors.

Metal window frames: Where metal frames are set in stucco walls there is a possibility of rectification between the frame and underlying wire. It may be nearly impossible to reach the connection and effect proper bonding.

Plumbing joints: The faucet-to-sink interface, bathtub drain-link joint and various combinations of plumbing and household fixtures produce rectifying joints. Tighten, clean and bond as necessary.

Conduit joints: EMT conduit is joined by sleeves and set screws. If the joint loosens a diode may be formed. RF is fed to the joint by either the field, ground currents, or RF flowing on the power lines.

Bed springs: This is more popular radio folklore. (Is there anyone who hasn't heard this story?) The ham next door transmits, and the audio is heard coming out of the bed, where RF is rectified between the various springs. Maybe it could happen, especially in apartment buildings where indoor antennas are employed. (This could prove quite interesting if the transmitted signal is full-carrier, double-sideband AM.) For cures, experiment with aluminum foil shields, bonding springs together or grounding the springs.

Around the Neighborhood

Loose power line hardware: Loose hardware often produces line noise, but it can also rectify. There is evidence that arcs are modulated by RF energy present on the line. Sometimes broadcast audio can be heard emanating from the arc. Widespread interference can be the result. The power company must fix this problem.

Utility lines touching other lines, guys or

(A)

(B)

Figure 12.3—Cables that rub against conductive poles or pipes may be sources of external rectification. At A, the second cable from the bottom has a protective sleeve. At B, there is no protection between the metal line-amplifier housing and the traffic-signal support.

poles: Power poles, overburdened with high-tension lines, 220-V distribution lines, telephone lines and CATV lines offer many sites for dissimilar metal-to-metal contact (Figure 12.3). One common source of contact is telephone or CATV lines that cross paths with an adjacent metal pole. Utilities appreciate hearing about damaged insulation—before major damage is done. When this occurs, the utility dresses the affected line with a plastic sheath to prevent damage.

Unfortunately, sheaths can wear through, again exposing the wires inside. Protective sheaths also creep from wind action and rubbing; watch for "misplaced" sheaths that don't cover the point of contact. The same problem can occur where lines cross guy wires or each other. When metal touches metal, a rectifier is formed.

Metal fences: Again, pieces of metal that are not securely joined corrode and develop rectifying joints. (See Figure 12.4.) Bonds at a few strategic points may cure the problem—experiment with a short clip lead.

Elevator shafts: An elevator shaft (probably the vertical beams in the shaft) once caused a severe BCI problem. Not unlike the ductwork phenomenon, the rectification was so severe that audio could be heard in the elevator.

This list should give the RFI sleuth some good ideas of where to start looking for sites of external rectification. It's always easier to find something if you've got a general idea of where to look.

THE SEARCH

Aside from a shotgun approach (tapping every joint in a one block radius with a hammer), the search for external rectification can be reduced to a semi-science. Use a number of experiments and tools to narrow the field of view. In some areas, similar methods are employed for amateur and nonamateur interference. In most areas they differ significantly. We'll treat them separately, considering TVI first.

AMATEUR TRANSMITTERS

Before testing, remove any masthead preamp. Even if the preamp is unpowered, it can easily rectify strong signals when connected to an antenna system. If only one TV is affected, the rectification problem is probably in the antenna system connected to that TV, or the TV has a problem. A random rectifier (downspout, mast, rusty fence and such) usually affects many TV receivers. If the TVI is noted in a CATV system, stop: Either the CATV company has some work to do, or the TV needs some attention. In either case, the problem is probably not the result of external rectification. Look in the Televisions chapter for more help.

Is It External Rectification?

It is difficult to differentiate between TVI resulting from external rectification and other causes. When the TV and transmitter already have filters and other standard precautions, place a step attenuator in line with the affected TV. Gradually increase attenuation while the transmitter runs at full power. If the TVI reduces faster than the legitimate TV signal weakens, the problem is probably fundamental overload (in spite of the high-pass filter). If the TVI level remains constant as attenuation increased, direct injection (common-mode current, via the power cord or feed line) is the path.

In some cases, the TVI level remains constant (relative to the TV signal). If there is no masthead preamp, the interference comes from either the transmitter or a rectifying joint.

TVI that tracks the attenuated signal (with little or no change when the transmitter is

(A)

(B)

Figure 12.4—These two metal fences are likely sources of external rectification interference.

shielded, choked and filtered) indicates external rectification. External rectifiers affect all receivers in the vicinity. (Substitute a second TV as a double check.)

There is another test that is not conclusive: Rotate the TV antenna. If the TVI worsens or peaks in a direction other than that of the transmitting antenna, it may be pointing at the rectifier. The same is true at the transmitting end: TVI should be strongest when the beam is pointed at the rectifier.

Where Is It?

Next test is to determine the power threshold at which rectification begins. Reduce power gradually until the problem disappears. TVI with less than 5 W indicates the transmit system. Give the transmit antenna, connectors and ground system a detailed inspection. 10 or 20 W is a more common threshold (higher on 80 and 160 meters, where the average bit of metal around the house isn't of significant length).

It is difficult to exactly locate a rectifier. Make a "first pass" by rotating the TV antenna and looking for a peak in the interference. Be careful, it is easy to mistake a signal null for an interference peak. Rotate the transmit antenna (where possible), and look for a TVI peak in the direction of the rectifier. (Do this test with the minimum power required to cause interference when the transmitting antenna is pointed in the most sensitive direction.)

A portable TV or scanner is required to "home in" on the actual interference source. Check suspicious metallic objects by proximity, one at a time. Don't confuse a drop in legitimate signals with a "hot spot" of interference. The usual clanging, banging, twisting, torquing and pushing may produce recognizable interference changes that lead to the source. Above all else, check the transmit and receive antenna systems thoroughly before beginning a rectifier hunt.

When searching for TVI always remember that rectification in the TV antenna system generates lots of TVI, but it is very difficult to spot. Many a TVI goose has been chased when the source of interference was the TV antenna system.

NONAMATEUR TRANSMITTERS

Even in cases where the amateur is not directly involved, the ham may choose to help track down the problem. The amateur is often the technical "wizard" of the neighborhood. The ham may also be experiencing interference from the nonlinear junction.

Commercial Broadcast Stations

If the transmitter is a commercial broadcast station, the hunt follows a slightly different course. Since the transmitter is always on, and its antenna can't be rotated, the rectifier must be "DFed" from the receiver end. This isn't too difficult—the signal is coherent, narrow band, and easily identified. After DFing the general direction, set out on foot with a portable receiver and search by proximity. If the interference seems to be farther afield, use a mobile rig or portable receiver in a car to locate the general vicinity of the source.

There are four possible sources of broadcast transmitter interference: (1) receiver overload; (2) IMD in the transmitter (second harmonic of one signal ± the fundamental of the other); (3) rectification at inanimate objects; (4) "active" rectification on power lines.

Fundamental Overload

Especially in contemporary upconverting solid-state radios, broadcast energy can directly overload the receiver. Overload-induced interference is a gross problem that covers whole bands, not just a spot frequency. The interference tends to vary in intensity as antennas, rigs and matching networks are changed.

Differentiate between external rectification and receiver overload by placing a tuned filter in series with the receiver input. External-rectification interference peaks as the network is tuned. Overload-induced interference disappears when the network is installed. Read the Fundamentals chapter for a detailed description of fundamental overload. Cures for receiver overload appear in the Stereos and Televisions chapters.

Intermodulation Distortion (IMD)

Intermodulation DFing usually leads to the transmitter site. DF bearings from various locations with several methods all point to the transmitting site. Often the second and third harmonics are not strong. To effect a cure, the "mixer" must be located and one or both of the driving signals removed. If mixing occurs in electronic equipment, use filters to remove the driving signal. If mixing occurs in an external rectifier, bond or insulate the offending junction. See the Fundamentals and RFI at the Receiver chapters for more discussion of IMD.

External Rectification

Rectification DFing leads to rain gutters, rusty water pipes and so on. Second and third harmonics are usually evident, as well as third and high-order IMD with broadcast stations. More often than not, the audio is remarkably clear, but scrambled by the simultaneous presence of two sources.

The antenna or ground system may be at fault. Significant broadcast-frequency currents flow in low-band antennas and towers. In unbalanced systems, broadcast currents are shunted to ground by radials and water-pipe connections (Figure 12.5). To test: (1) Take an interference reading on a portable receiver some distance from the suspect antenna. (2) Interrupt the possible shunt paths, and (3) take a second reading. When the interference is generated in the antenna/ground system, the second reading is much lower than the first.

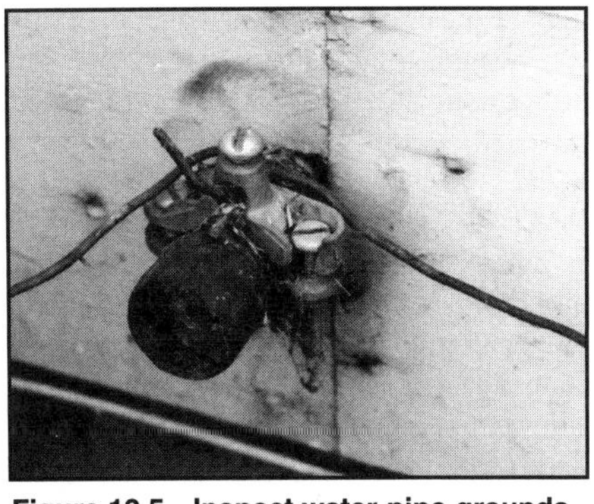

Figure 12.5—Inspect water-pipe grounds to be sure that the connections are tight and clean of corrosion.

Since balanced antennas have no common-mode path to ground, any rectification must take place across the antenna feed point. Perform the above test, but open the antenna feed point between readings.

Power line Rectification

Hum on an interference signal is a possible sign of power line rectification. A mobile rig and a quick drive around the neighborhood should pinpoint one or more poles. Report them to the utility company for repair. (Sometimes you can actually hear the broadcast audio while standing near the offending pole.)

Arcs are notorious for producing a wide variety of RF products. Arcs can generate second and third harmonics of the fundamental, as well as second and third-order IMD of many broadcast signals near and far.

Rectification may occur in passive conductors (grounds, guys and so on), yet a strong hum component may be evident. This is caused by 60-Hz energy in close proximity; it modulates the current flowing in the rectifying junction.

If IMD products are intermittent but seemingly periodic (seconds), suspect a long expanse of wire. Power lines in residential areas swing like pendulums, with a period of 1 to 2 seconds. As the line sways in the breeze, it tightens and loosens or gently rocks the insulators back and forth. This action may be just sufficient to repetitively break or short the offending rectifier, thus switching the BCI on and off.

Any lengthy conductor can be checked with a portable receiver. Couple the internal loop antenna to the conductor under test. The BCI always gets louder (the metal object is an antenna); but in BCI sources it gets disproportionately louder. Test several objects in the area, using an attenuator (or proximity) to control receiver sensitivity. It doesn't take long to get an intuitive feel for levels that are "normal" and those that aren't.

If the interference is generated in residential wiring or plumbing, check for ground-return currents in the water pipe between the street (meter) and dwelling. In temperate climates (where pipes are buried relatively close to the surface) return currents radiate quite well. By simply "sweeping" the sidewalk near the water entrance with a portable receiver, the relative interference level can be detected.

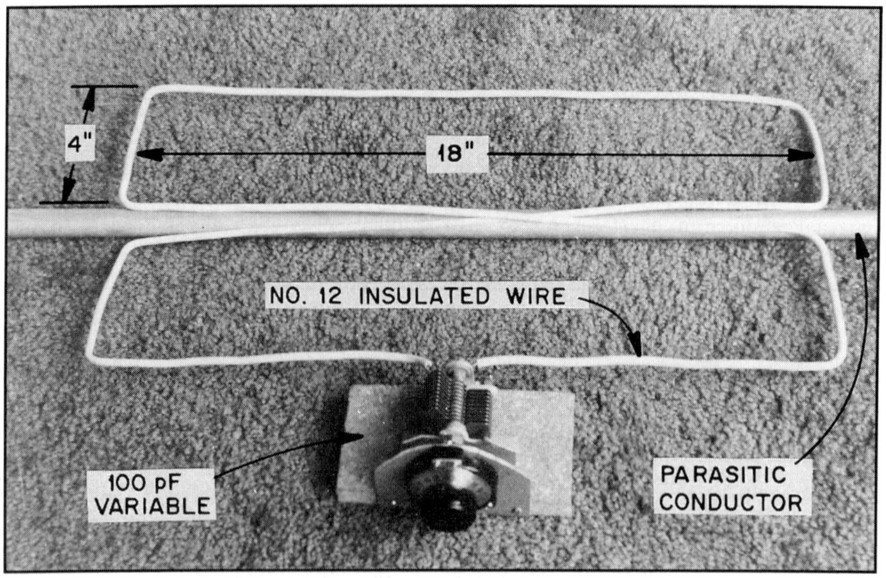

Figure 12.6—Use a "resonance breaker" such as shown here to obstruct RF currents in a conductor without the need to physically break the conductor. Use a vernier dial for the variable capacitor because tuning is quite sharp. The 100-pF capacitor is in series with the loop. This resonant breaker tunes from 14 through 29.7 MHz. Larger models may be constructed for the lower frequency bands.

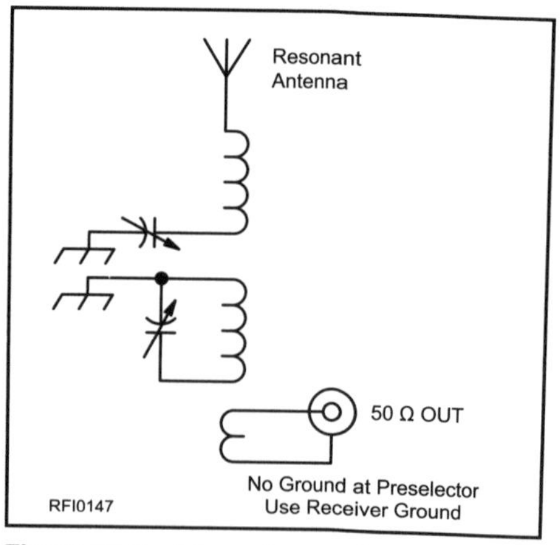

Figure 12.9—A loosely coupled preselector to reject strong broadcast signals or adjacent-band interference.

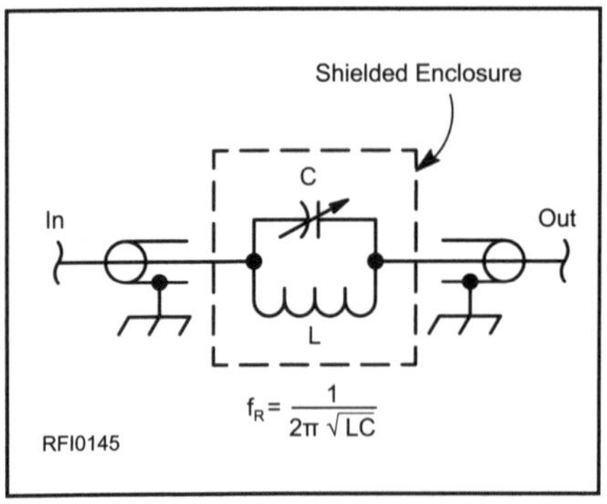

Figure 12.7—A parallel-tuned trap. Select L and C to yield resonance at the frequency to be rejected. Use a similar trap in each line of balanced feed lines.

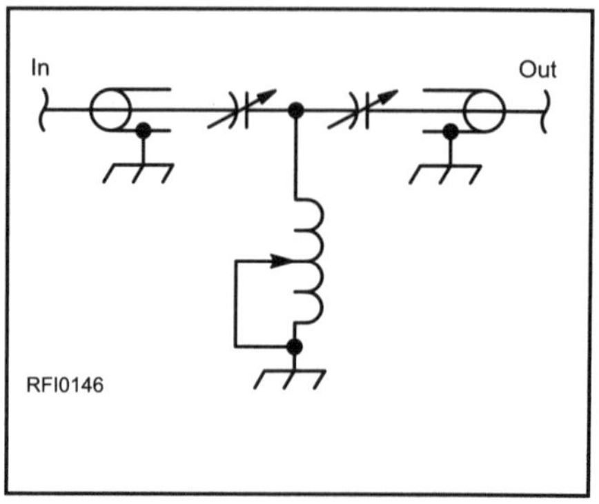

Figure 12.8—A high-pass Transmatch, such at this "T" network, can help reduce broadcast energy at the transceiver input.

Signals are strongest at the house with rectification problems.

A search by foot and by car is usually required to locate IMD sources. If the IMD is weak, it may be difficult to detect on a portable receiver. Check other related frequencies for a stronger IMD product. The second and third harmonics should be strongest, followed by third-order IMD with other local stations. The search for TVI and BCI is more art than science. A few simple experiments, however, and some practical experience with a portable receiver quickly make an expert out of a beginner.

FIXES

Once the rectifier is found, apply cures. There are four ways to fix a rectifying joint: (1) disassemble the joint, clean, reassemble and weatherproof; (2) short circuit the joint with a piece of solid strap, a screw or a bypass capacitor; (3) physically open the circuit with an insulator; or (4) impede the RF current with ferrite or coupled, tuned circuits. If the rectification occurs along a utility line, the utility company must repair the damaged joint.

BYPASS THE RECTIFIER

Where the RF current is intentional (as in grounds and antennas), short or bypass the rectifier. On water pipes, attach ground clamps at either side of the affected joint and connect them with a short piece of heavy wire (no. 10, 12 or 14) or solid strap. If two pipes cross each other, make the connection solid with a grounding clamp. Do not ground gas pipes to adjacent pipes because there is a risk of corrosion. Use an insulator to separate gas pipes (see Figure 12.2).

Where the RF current is unintentional or unnecessary, short the joint or add a component to reduce or null the current. Several methods are common. Install large or multiple ferrite beads, tuned traps coupled to the conductor or some sort of insulator. For example, add egg insulators to guy wires, or use plastic isolators in water pipes.

Clean and reassemble antenna elements and terminals, connectors, radial connections, household fixtures and conduit joints. Replacement may be the best policy with defective antenna wire, guy wires, corroded TV antennas and bad pipe joints (are they leaking?). In cases such as loose tower sections, replace rusted hardware and tighten viable hardware that is loose. Use a shorting strap in difficult situations where it isn't convenient to disassemble the joint. Leave all power line maintenance to utility companies.

Apply a few sheetmetal screws and a coat of RTV to rain gutter joints, TV mast sections, metal roofing and so on. TV masts can be completely sealed at each joint and at both ends with RTV.

If rectification is caused by an electronic appliance (power supplies, antenna relays, rotor servos), bypass diodes with 0.01-µF ceramic-disc capacitors. Similarly bypass wires entering and exiting the product. Use ferrite beads to inhibit the flow of unwanted RF current. Shielded control cable entrances help impede unwanted current and provide a shunt ground capacitance before the signals reach the control box. Shield the power cord and dc-output lines of power supplies with ½ to 1-inch braid. Ground the braid at the supply.

REDUCE RF CURRENTS

A totally inaccessible joint calls for special techniques: Stop the rectification by somehow impeding the flow of current. Current flow can be altered (in hopes of reducing the interference) by changing the electrical length of the conductor. Add a piece of wire at an accessible point in an effort to make the object less resonant at the fundamental frequency.

A tuned trap is a very successful means of breaking current flow without actually breaking a conductor (Figure 12.6). It is often employed on pipes, when rectification occurs at a point that is buried. It may be impossible to keep RF current off the pipe, but a coupled tuned circuit can quickly block fundamental current flow. The network requires a bit of wire on broadcast frequencies, but it works quite effectively. The inductor is folded into a "figure 8" pattern. This reduces the efficiency of the inductance as a loop antenna, without affecting the transformer action that couples the system together. The "resonance breaker" shown in Figure 12.6 was first described by Fred Brown, W6HPH, in the Oct 1979 *QST*.

RECTIFIERS IN ELECTRONICS

Unused (but antenna connected) rigs were mentioned as possible rectification sources. In many cases it is inconvenient or undesirable to disconnect a rig, but another means of blocking the flow of current is possible. A parallel trap (Figure 12.7), tuned to the offending frequency (whether ham or broadcast) is one answer. A high-pass filter or a high-pass T-network Transmatch (two capacitors and one coil) helps reduce broadcast energy (Figure 12.8). High-pass filters protect VHF and UHF radios from HF energy.

Loose-coupled, double-tuned networks (Figure 12.9) block energy extremely well, especially on closely spaced frequencies. The broadcast and 160-meter bands, for example, or any two adjacent ham bands. Unfortunately such networks are tricky to design and adjust. They also have limited power-handling capability. Don't leave them in the line while transmitting.

Chapter 13

"Intermod" —A Modern Urban Problem

Getting squeaks and squawks from your handheld transceiver? Are pagers crunching your mobile rig's front end? Here's how intermod happens— and how to fight back!

By Ed Hare, W1RFI,
ARRL Laboratory Supervisor

Some old-time hams remember when VHF was quite deserted. Nearly all stations found above 50 MHz were TV and FM broadcasters. The 2-meter band was often silent, and there were relatively few VHF transmitters used by the police and businesses. With so few stations on the bands, and those stations using simple equipment (insensitive receivers), interference was unlikely.

If you use VHF today, you know that those days are gone forever. Thousands upon thousands of transmitters operate above 50 MHz for business, public service, amateur, satellite and many other uses. In addition to licensed transmitters, there are countless unlicensed transmitters, operating under FCC Part 15. It is likely that some of these radio systems will interact in undesirable ways.

When a new ham first fires up a VHF receiver and hears a few pagers in addition to the normal signals—or, even worse, hears bleeps, noises and unidentified voices across most of the band—he or she often turns to more experienced hams for an explanation. Most of these explanations blame the problem on intermod (intermodulation distortion, or IMD)—often with little explanation of the term.

So, you think you might have an Intermod problem? First, let's learn about what "Intermod" is.

WHAT IS INTERMOD?

Communicators sometimes use the term "intermod" incorrectly. While intermodulation of two or more signals causes some VHF interference problems, other problems, such as front-end overload, poor image rejection or IF (intermediate frequency) leakage are sometimes the real culprits. Hams tend to call them all "intermod," which complicates the explanation a bit.

In a perfect world, every amplifier would amplify signals without distortion, every mixer would convert RF signals to the IF perfectly, and a radio would hear only the desired signal. In the real world, however, all of these processes are nonlinear to some degree. This results in the creation of interference.

What does "nonlinear" mean? When applied to an amplifier, it means that the output voltage does not follow the input voltage perfectly. Nonlinear circuits can generate harmonics and mix signal frequencies. The RF amplifier or mixer circuits in a receiver can be somewhat nonlinear, creating additional signals from the desired signal—and perhaps others—present at the nonlinear stage.

Harmonics

When a single frequency (the fundamental) passes through a nonlinear circuit, distortion signals appear at integer multiples of the fundamental frequency. These are called harmonics. See Figure 13.1. Each harmonic is identified by its relation to the fundamental: The second harmonic is at two times the original frequency, the third at three times the frequency, and so forth. Unwanted harmonics can cause interference wherever they occur, ranging from HF-transmitter harmonics that interfere with a TV, to a 49-MHz cordless-telephone transmitter's third harmonic that interferes with a 147-MHz 2-meter station (49 MHz * 3 = 147 MHz).

Mixers

Mixers are nonlinear devices by design. In a typical mixer used in a superheterodyne receiver, a desired signal mixes with that from a local oscillator (LO) to produce sum and difference signals. The IF circuitry selects and amplifies either the sum or difference signal. In older radios, the IF is usually the difference frequency. Modern radios, some of which use multiple conversions, sometimes use the sum frequency.

All Mixed Up

IMD, however, is a mixing process gone bad—a form of distortion. Whenever two or more signals are present in a nonlinear circuit at the same time, IMD creates new, unwanted frequencies from them.

The relationships between the two signals and the resultant distortion products can be quite complex. (This discussion will consider only two input frequencies, f1 and f2, to be as simple as possible.) Signals and their harmonics can mix together to form still more new frequencies. Any signals created in the circuit can then mix with each other and the original signals to form a complex spectrum, indeed.

The strongest IMD products are those that involve the sum and difference of the input frequencies, the harmonics of the input frequencies, and the mixing of the harmonics with each of the original input frequencies (harmonic mixing). There are higher-order mixing relationships, but they're more complicated than we want to discuss here.

IMD Product Orders

The term "order" is often used to describe a group of IMD products. Because IMD results from combining frequencies, we can identify each IMD product with an equation describing the sums and differences of the various signals involved. For example, f1 + f2, 2f1—f2 and 3f2—2f1 are three such

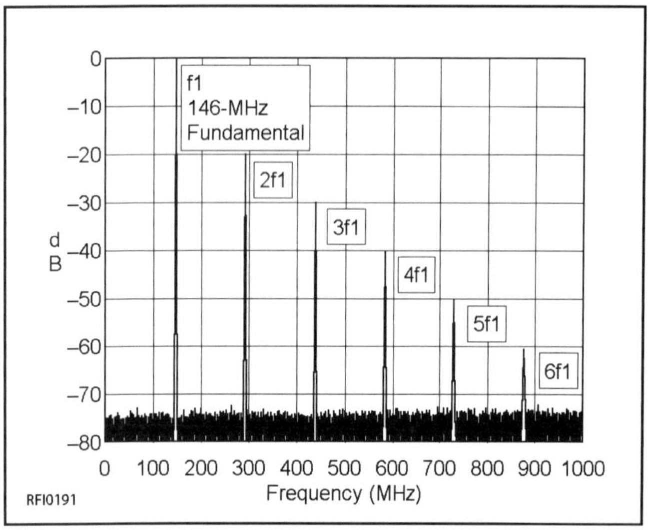

Figure 13.1—This 146-MHz signal contains harmonics that extend well into the UHF range.

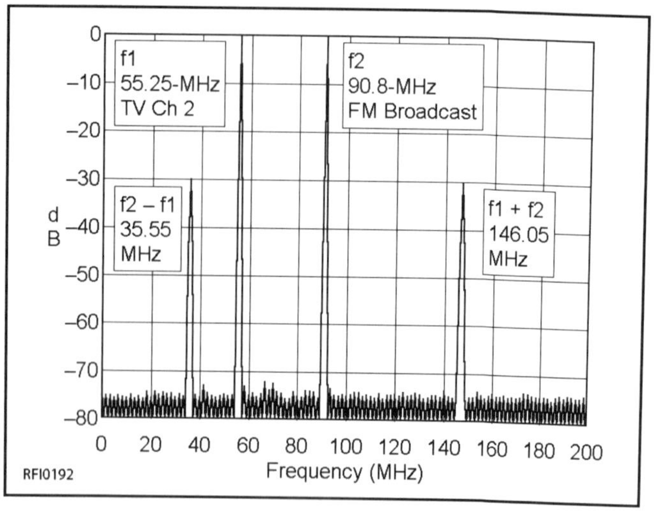

Figure 13.2—Here, f1 (55.25 MHz) and f2 (90.8 MHz) are present at the input of a VHF receiver, they may generate an interfering signal (f1 + f2) on 146.05 MHz.

Table 13.1
Orders of Some IMD Products

IMD Product	*Order*	*IMD Product*	*Order*
f1 + f2	2	2f1 – 2f2	4
f2 – f1		3f2 + f1	
2f1, 2f2			
2f1 – f2	3		
f1 + f2 – f3			
3f2			

equations. An IMD product's order is the sum of the coefficients (in the term 2f1, 2 is f1's coefficient) of the terms in the equation. Remember that, even though you don't see it, a term like "f1" has a coefficient of 1. Table 13.1 shows some example IMD equations and their orders. Figure 13.2 shows the result of a mixer circuit (or very nonlinear amplifier) generating sum and difference products.

IMD is more complex than simple sum and difference second-order relationships. A nonlinear circuit creates harmonics of all input signals, and those harmonics mix with all of the fundamental signals plus those created by the nonlinearity. Third-order IMD between paging systems causes much of the IMD problems that plague VHF operators.

Third-Order IMD

The second harmonic of f1 or f2 can mix with the other fundamental to form a product. These products are especially important because they are low-order products, and therefore relatively strong. Also, when f1 and f2 are relatively close in frequency, these products are close to f1 and f2. These equations characterize third-order products:

2f1—f2	Eq 13.1
2f2—f1	Eq 13.2
2f1 + f2	Eq 13.3
2f2 + f1	Eq 13.4

The first two are probably the most important to the VHF operator because the frequencies involved are so close to the desired frequency. For example, a pager on 153.75 MHz (f1) can mix with a 160-MHz (f2) signal to produce an interfering signal (2 * 153.75 MHz—160 MHz = 147.5 MHz) on the 2-meter band, according to Eq 13.1.

Some of the frequencies in the above example are assigned to paging transmitters, which are very common in urban areas. Figure 13.3 shows the result of third-order IMD in a nonlinear circuit.

While 3rd-order products are often the most problematic, other "orders" can cause problems, too. Even-order IMD [2nd order: f1 + f2 or f3—f4; 4th order: 3f1-f2] are certainly possible, though they usually involve signals well out of the range of the one they are received on. Higher odd-order IMD may be less common than 3rd-order, but it can certainly cause serious interference. Figure 13.3 is an illustration that shows what 3rd, 5th, 7th, 9th and 11th order IM products would look like on a spectrum analyzer. While only the 3rd order products are labeled, the rest are shown, at steadily decreasing levels. This sort of display is often labeled a "Christmas tree" in technician slang.

There is one other mechanism that can also complicate things if a source of IM distortion is present—a strong IM product can in turn mix with another signal just as if it were a transmitter of it's own—which will complicate the process of deducing just what signals are creating the problem. These "multi part" IM products can pose a serious challenge in calculating the source of a problem and they represent a truly high "high-order" mix, as all of the involved terms are added—for example: 2f1-f2+3f3-2f4-f5 is a 5-transmitter mix of 9th order—and very possible on a active site. In modern high-density radio sites with tens or even hundreds of "in band" transmitters and receivers, operators have reason to be concerned about very high-order IM products.

So Is It IM?

In the chapter on troubleshooting, the first step is "Identify the problem"—so now you need to try and figure out if what you're hearing is IM—or something else.

Intermod is *mixing*—by definition; it is composed of more than one signal, so if what you are hearing has more then one modulation (voices, pager tones, etc.) then it is pretty likely to be IM. Remember, though, some of the signals might be unmodulated (like a wayward receiver local oscillator) or their modulation might not be audible in your receiver (a 2 meter FM receiver won't demodulate much from an AM broadcast signal, for instance). If the IM involves harmonics of one of the mixed signals, remember that the apparent deviation of FM signals will multiply with the number of the harmonic—for example the 2nd harmonic will have twice the FM deviation of the original. If one of the signals is noticeably "louder", or is so "hot" it is "chopping out" of the receiver, that is an excellent clue as to where that signal falls in the IM calculation.

Some signals, such a FM broadcast have such high deviation, or are so wide (like television) that IM products involving them only seem to appear when the station's modulation is at a low level. A signal that comes and goes in a rhythmic or syllabic (speech pattern) manner is often a good clue in these cases.

If you are sure you have an Intermod problem, then you are ready to move on to the next step—figuring out where the offending mix or mixes are taking place. Even if you are fairly sure what you have *isn't* IM, don't put it completely out of your mind (see sidebar).

A Case of Intermod?

In the late 1980s our 2 meter repeater was moved from it's old location on a downtown building to the top of an "urban mountain" 10 miles to the North. The old malfunction-prone tube transmitter was replaced with a much more modern solid-state Motorola Micor, and initial signal reports were very favorable.

Soon, however, reports that our new repeater had a "spur" on a Civil Air Patrol frequency (just above the 2 meter band) started coming in. The CAP stations reporting the "spur" were all fairly close to the new repeater site, and they all reported it as weak but steady, and with no other modulation on it. That seemed to rule out IM, so we hurriedly scrounged the use of a spectrum analyzer and made our way up to the site.

When we got there, though, we discovered two things that really had us wondering—first, a handheld transceiver on the CAP frequency didn't hear any "spur" and second, the analyzer confirmed that the repeater transmitter's signal was clean! We adjourned off the hill to a local pizza place (*love* those urban repeater sites!) and sat down with some CAP members to try and figure it out. We decided it must be some sort of Intermod, but all of the frequencies that fit into the 3rd, 5th and 7th order mixes seemed to be unoccupied. The repeater was being heard on a frequency 910 kHz above it's output frequency, but we couldn't figure out why!

Then one of the CAP guys with "local knowledge" said, "910 kHz? That's the frequency of the AM station at the big mall"—and we had our answer. The mix was happening in the CAP member's receivers when they were close enough to that high-powered AM broadcast transmitter to make a simple f1+f2 mix—and the AM stations modulation was inaudible in the FM receivers.

Subsequent study by the CAP ops showed that the problem affected only certain older radios, and the solution was to keep those radios at least a few miles away from the AM station's towers—problem solved and a lesson learned!

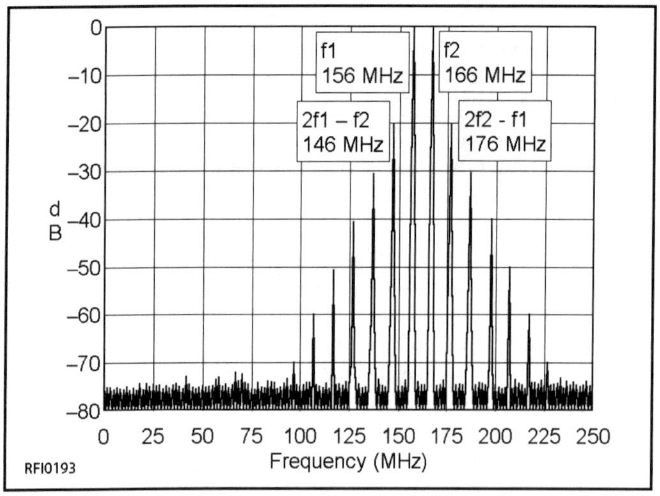

Figure 13.3—These pager frequencies can intermodulate to form a product in the amateur 2-meter band. If there are many pagers on different frequencies in the same area, the result can be interference across the band. This actually occurs in some major cities. This graph shows third- and higher odd-order products.

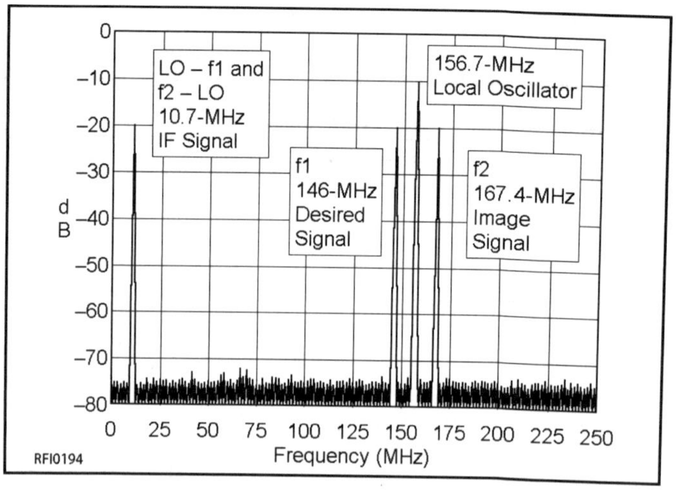

Figure 13.4—A superheterodyne receiver can respond to signals other than the desired signal. If the RF selectivity of the receiver is inadequate (poor image rejection), the receiver could hear the signal on 167.4 MHz in addition to the desired signal on 146 MHz.

Where is it Mixing?

Once you suspect IM, the next question is "where is it mixing?" While IM mixes can occur almost anywhere, the "big three" are 1) the affected receiver itself (about 90% of the time), 2) a transmitter (about 8%, and usually one of the "parties" to the mix, but not always), and 3) a "Rusty Bolt" situation (about 2% of the time). The reason we want to know where the mix is occurring is that *some* of the "cures" can be expensive, difficult, and/or involve compromises of performance, so we would rather not spend a lot of money or reduce our performance only to discover the problem was elsewhere!

For most hams, especially for those who are experiencing problems while mobile or operating a handheld radio, the characteristics of their receiver are the most critical. Hams serving in a Public-Service role in places like a crowded Emergency Operations Center or other radio-active locations will also need to look first to their own rig.

ALL RADIOS ARE NOT CREATED EQUAL!

There are considerable differences in IMD performance among different radios. Many of these differences result from the radio design. A single-band radio that cannot receive frequencies outside that ham band usually offers good rejection of interference from nonham frequencies. Some radios have tracking filters in the front end that automatically pass the frequency a user selects. Other differences in performance can result from the specific design of front- end or mixer circuits.

The ARRL Laboratory performs an extensive battery of tests on all equipment featured in the *QST* Product Review column. One of those tests is a wide-band dynamic-range test on VHF equipment, using two strong signals just outside the amateur band (usually the abode of nearby pager transmitters). This test is a good indicator of relative IMD performance. *QST* lists the result in the Product Review's test-result table as "Two-tone, third-order dynamic range, 10-MHz offset."

The November 1995 Product Review column compared eight dual-band FM handheld transceivers. The 10-MHz spacing dynamic range varied quite a bit from radio to radio, ranging from 73 dB to 91 dB.

Another problem that can occur in some urban areas is a mixing of television channel 2 (55.25 MHz video carrier) with radio stations in the low end of the FM broadcast band (89-93 MHz), to produce second-order IMD products in the 2-meter amateur band. Product Reviews of 2-meter FM equipment from July 2002 onward include this performance figure.

All Product Reviews (and test data) from 1980 to the present are available on the ARRL Members Only Web Site. The URL is **www.arrl.org/members-only/prodrev/bymfg.html**.

The higher the dynamic-range number, the better the radio. High dynamic range, however, is no guarantee that a radio will perform well in the "Intermod Alley" near many large cities.

Why are they different? Many factors combine to affect a radio's IMD performance. Most VHF receivers are designed to be very sensitive because they will be used with inefficient "rubber ducky" or short mobile whip antennas. Unfortunately, this sensitivity can make the receivers prone to overload, especially when used with good outdoor antennas in urban areas.

Band-Pass Versus Broadband Designs

Radios designed to receive only ham-band signals contain a band-pass filter in the RF or "front end" stages, and these filters will reduce or eliminate out-of-band signals, such as paging and public-safety systems. Unfortunately, most current amateur HF, VHF and UHF radios are designed as general-coverage receivers to allow monitoring of police, fire, weather, aircraft and other signals often far removed from the amateur bands. Although this is convenient, it also means that these have very broadband "front ends". While some high-end radios are designed with "tracking" front-end filters, the vast majority do not, and the combination of limited selectivity combined with high gain (for good sensitivity) yields a receiver that can easily produce mixes involving signals far removed from the desired receiving frequency. Handheld radios, which are designed to work with very small antennas and which have very limited space for filtering circuitry are often the most affected, especially if they are connected to outside antennas. In high signal areas, even the best radios can be affected by interference from these "receiver-produced" IM signals.

Commercial radios, which are designed to operate in more congested areas, are often equipped with "tighter" front end filtering, but this often results in slightly lower sensitivity and higher production costs. In areas subject to large numbers of strong signals (like repeater sites and EOCs) *any* receiver, no matter how high of a quality, can be affected and require additional filtering to function properly.

So How Do I Know It's My Receiver?

The next chapter in this book "RFI at the receiver" has some specific tests you can do to isolate your receiver—and also tells you how much improvement you need to make. But there are a couple things you should know that will make many cases easy to figure out.

First, see if your radio is being affected more (or less) than others—borrow another rig (or two) and try it in the same area on the same antenna, trying to swap the various rigs around so you have a good idea of where your rig stands in relation to the others. If you find that other rigs are unaffected, or affected much less than yours, that's a good sign your rig may need help, and it also confirms that the interference is indeed being caused in the receiver, as an external source of interference (mix or "spur") should be the same in all the radios.

Another good test is to add attenuation—reducing the level of RF reaching your receiver. If you have a handheld, try grabbing the antenna—if the interference disappears or drops way down relative to the desired signal, it is a good bet that your receiver is being affected by high RF levels. For bases or mobiles, try to borrow an attenuator (fox hunters almost always have them handy, by the way—but don't transmit into them or they'll be upset!). Mobile ops can try the same trick of grabbing the antenna—but be *sure* you don't transmit while someone is touching (or close to) that antenna! The effect you're looking for is the same—if the interference drops relative to the desired signal, it is a sure sign that the receiver is producing the interference.

One additional note: make all these tests with an "open squelch"—some manufacturers have installed some sneaky circuitry that closes the squelch on interference when it is even slightly off-frequency, which masks it until your desired signal opens the squelch and then you hear all the noise! They do this because they know that masking the problem makes it "sound better"—but it really doesn't solve the problem.

Your receiver is the problem and a "new and improved" model is not in the cards? [Incidentally, "new and improved" it may not be—check out the *QST* reviews and also the various Internet boards for user reports before plunking down your hard-earned cash!]

Here are some filters and suggestions that may help:

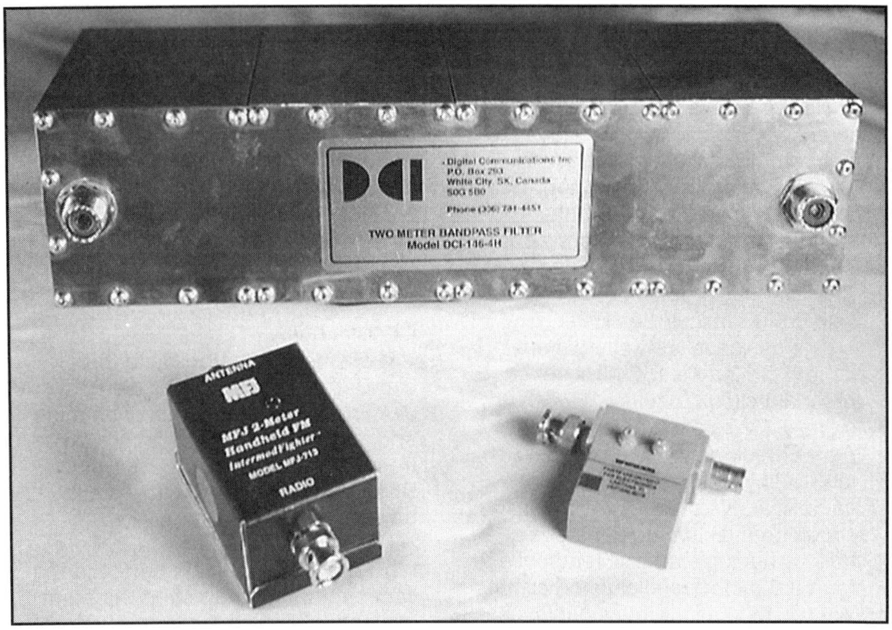

Figure 13.5—Each of these three filters helps with intermod problems.

EXTERNAL FILTERS

External filters can significantly improve the IMD performance of existing radios. These filters are usually either band-pass or band-reject (notch) designs. Several companies sell filters designed to solve IMD problems.

Band-Pass Designs

Band-pass filters do just what the name implies—they pass a band of frequencies and attenuate all others. A well-designed filter can exhibit less than 1 dB of loss inside the passband and up to 60 dB or more attenuation outside the passband. Figure 13.5 shows two commercial band-pass filters. (Any of these filters will help reduce IMD problems, each in its own way.) The large filter is a helical band-pass filter made by DCI.[1] The ARRL Laboratory measured 0.8 dB insertion loss (a fraction of an S unit) and about 55 dB of attenuation at pager frequencies. This filter would make an excellent IMD filter for a base or mobile station. It is much too large to use with a handheld transceiver,

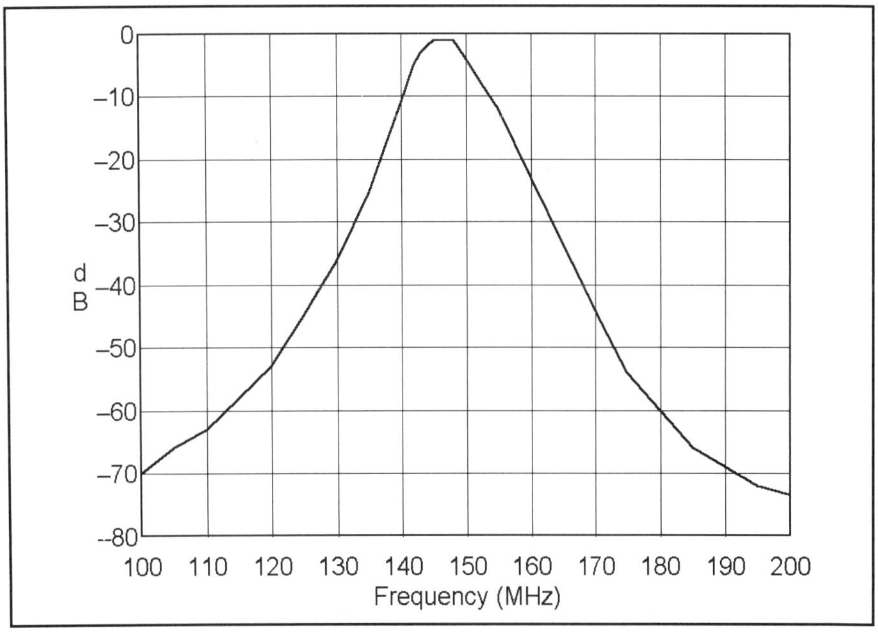

Figure 13.6—The frequency response of a typical band-pass filter. This filter's passband centers near 147 MHz. Its 3-dB bandwidth is about 7 MHz (143 to 150 MHz).

A 2-Meter Notch Filter You Can Build

Having trouble with VHF pager interference? The notch filter in Figure A can remove signals just above the 2-meter amateur band, without significantly degrading 2-meter signals. As a bonus, it notches out some FM broadcast signals as well. The filter has a notch depth of 27 dB, with 0.5 dB of insertion loss.

The design is fairly straightforward. L1 and C2 form a 152-MHz series-resonant circuit to reject 152-MHz signals. C2 sets the notch frequency. As the frequency decreases below 152 MHz, the circuit becomes capacitive. We prevent the capacitive circuit from disturbing 2-meter signals by adding inductive reactance to form a 2-meter parallel-tuned circuit. We could do this with a tunable inductor, but a variable capacitor and a transmission-line section (C1/TRL1) are more practical. As a bonus, the capacitor and transmission line form another series-resonant circuit, which rejects signals in the FM broadcast band.

This design assumes that you want a notch above the 2-meter band. For a notch below the 2-meter band, the unwanted reactance would be inductive, which would require only a capacitor to obtain parallel resonance. This would eliminate TRL1.

The filter is bidirectional—it works both ways. I have assumed that you would attach it directly to a BNC connector on a handheld transceiver.

Construction

If cost isn't a factor, build this circuit in a Pomona 2391 die-cast aluminum box that comes with male and female connectors (available from Newark Electronics).† Male BNC panel-mount connectors can be difficult to locate.

RF grounding is important. To establish good electrical contact to the capacitors, I first removed the paint by lightly drilling the chassis with a Black and Decker 5/16-inch "bullet" drill bit. I then enlarged the holes (with an ordinary 1/4-inch drill bit) to mount the capacitors. Similarly, scrape off the paint underneath the connectors for good electrical contact. I prepared RG-188 Teflon coax for TRL1 as shown in Figure B. RG-174 will work just as well, but it is easy to melt the RG-174 dielectric and short the cable while soldering.

Alignment

C2 controls the notch frequency; adjust it first. Even with an expensive piston trimmer, the notch in tuning is rather sharp—use a high-quality capacitor. Next, tune C1 for best SWR. It should not affect the notch frequency, although C1 may need retuning if you change the notch frequency. Use an H-T and an SWR meter or return-loss bridge†† to tune C1.—*Zack Lau, W1VT, ARRL Senior Lab Engineer*

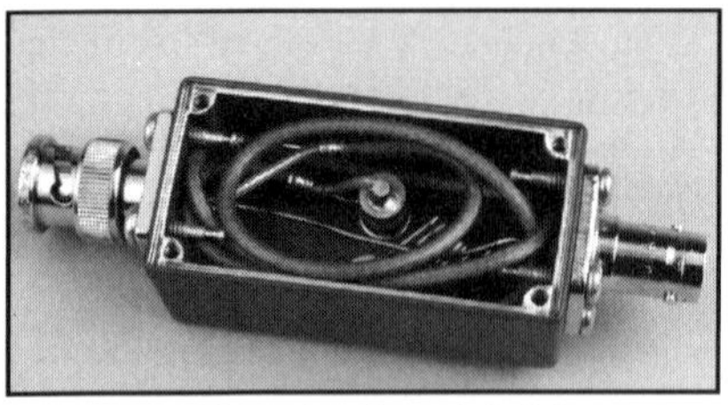

Figure A—This simple notch filter is easy to tune and uses available components.

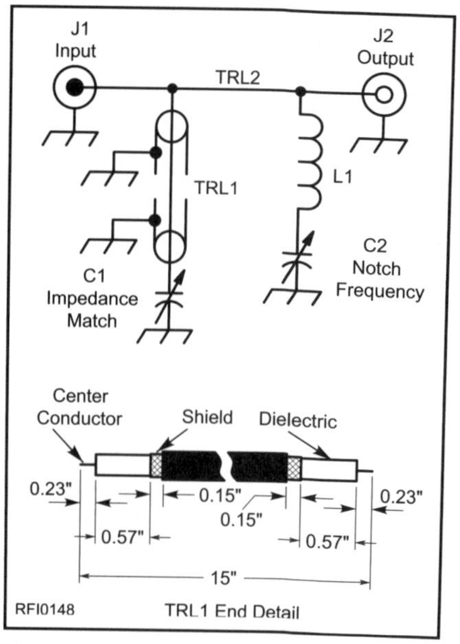

Figure B—Schematic of a pager notch filter that allows undisturbed 2-meter reception.

L1—8 turns #18 diameter, 5/16-inch diameter air wound, space equally.
C1, C2—1 to 14 pF 5402PC Johanson air-dielectric variable capacitors (Newark No. 17F161; minimum Q is 3000 at 100 MHz).
J1—Male BNC connector. See text.
J2—Female BNC connector.
TRL1—15 inches RG-188 Teflon coax with ends prepared as shown.
TRL2—1.75 inches, #18 AWG enameled wire between coaxial jacks (to hook the connectors together).

†Newark Electronics, Inc. For contact information, see the Suppliers List in Appendix A.

††A suitable return-loss bridge circuit appears in April 1996 *QST*, p 76.

however. Its band-pass characteristics are similar to those in Figure 13.6. It filters on transmit as well as receive, offering some additional RFI prevention as a bonus. This filter is left in-line during transmit. It does not need to be switched out.

MFJ Enterprises[2] sells the smaller filter shown in the lower left of Figure 13.5. It requires a small external dc power supply, and you must switch it off line manually or via the built-in RF-sensing circuit before transmitting. It has 6 dB of insertion loss on receive and offers up to 50 dB of attenuation at pager frequencies. It is small, and its BNC connectors mount it to a handheld transceiver easily. It is relatively easy to find the 25 mA of dc current required to operate the RF switch and internal relay.

Although up to 6 dB of insertion loss may seem excessive, most local FM amateur signals are much stronger than 6 dB above the local noise. They will still fully quiet the receiver even when reduced by the insertion loss of the filter. More important, the filter removes third-order IMD from 150 to 160-MHz signals, allowing successful contacts on an otherwise unusable radio!

One disadvantage to band-pass filters is that they will degrade the sensitivity outside the band for which they are designed. They could affect wide-band/scanner coverage on some radios. They are also single-band designs, so they must be removed when using other bands on multi-band transceivers.

Notch Filters

Most 2-meter IMD problems are true IMD cases: third-order distortion from pager transmitters. Pagers near 2 meters operate from 152 to 156 MHz. If these are your only problem, a notch filter on your receiver will eliminate the IMD interference. Figure 13.5 (lower right) shows a notch filter made by Par Electronics[3]. It has 0.2-dB insertion loss on 2 meters and 45-dB attenuation at 152 MHz. It is entirely passive and suited for use with handheld, base or mobile transceivers. It does not protect against interference problems caused by transmitters other than 152 to 156-MHz pagers, though. If you like to build your own accessories, see the sidebar "A 2-Meter Notch Filter You Can Build."

Attenuation

In some cases, a simple attenuator can help reduce IMD problems. Third-order IMD has an unusual property: When you reduce the level of a signal causing a third-order product by 1 dB, you reduce the third-order product by 3 dB. In cases where the desired signal fully quiets the receiver and yet is masked by IMD, a 10-dB attenuator placed between the receiver and antenna reduces the desired signal by 10 dB and the IMD by 30 dB! In many cases, a 10-dB reduction of the desired signal won't matter, but a 30-dB reduction of the IMD product makes the contact possible.

A directional antenna, such as a Yagi, is an "attenuator" solution to IMD problems that is sometimes overlooked. This solution is useful when the IMD source and desired signal come from different directions. Orient the antenna for either maximum desired signal strength or minimum IMD.

A physically small antenna may help reduce IMD problems. They are often lossy, acting as an attenuator. They may also be narrow-band, operating well over a limited frequency range.

But What About the Repeater?

Repeaters (and other "infrastructure" like packet nodes and links) are a special case with Intermod. Since the receiver is generally of good quality with high selectivity, such as converted commercial gear, and additionally protected by a duplexer and often other filters such as bandpass cavities, IMD from sources *outside* the receiver become much more likely.

Before we get into looking for those sources, however, we need to do something first, and that is to measure the strength of the interfering signal. If you have a spectrum analyzer, you could measure it directly (be careful—see below) by connecting the receiver cable to it—but if you don't, that's OK—all you really need is something that will tell you if an adjustment or change you've made has improved things or not. A limiter reading is often enough, though you want to make sure it is really reading the interference when you record it—and you should record it so you can compare later.

The next candidate in our search for "Where is the mix" is a (relatively) nearby transmitter, and the mechanism for producing the mix is usually one of two things. *First,* it is possible for a poorly-shielded transmitter to have enough RF energy coupled into it from a strong signal to allow that signal to be mixed in the stages of a transmitter—

RF Isolators

An RF Isolator is a device that is used to block the flow of RF energy in one direction while allowing it to flow freely in the other—a "trap valve" for RF, to use a plumbing analogy.

How does it work? First, some terms—a "circulator" is a three-port device that, because of a strong magnetic field, induces RF conducted into one port to be "circulated" to the next port in the direction of that field. RF energy entering that port is circulated to the next port in turn, in a circular fashion. An "Isolator" is a circulator with a dummy load connected to one of the ports. The figure below depicts a single-junction isolator, and the arrows denote the direction that RF flows in the device. Such isolators provide between 20 and 35 dB of isolation from energy trying to flow the "wrong way", and if more performance than that is required, packages with two or even three junctions are also available.

Isolators also provide some other services besides protecting a transmitter from invasion by outside signals. They will also absorb reflected power caused by a failed antenna or feedline, and they also offer a good 50-ohm impedance to the output stage of the transmitter—in fact, some early solid-state UHF transmitters were designed with an integral isolator just to keep the output stages from reacting adversely to SWR problems often experienced by mobile antennas.

Because the dummy load may have to be able to absorb the full output of the transmitter in the event of an antenna failure, it is important that it be large enough to safely handle that power level. If the load burns up, the full reflected power will flow back into the transmitter and potentially damage it as well.

An RF isolator is usually installed at the output of the transmitter, between it and the duplexer or other filters, and is often equipped with a harmonic filter on it's output as it can generate harmonics itself. For a transceiver (like a packet node), it must be installed between the transmitter and the T/R relay.

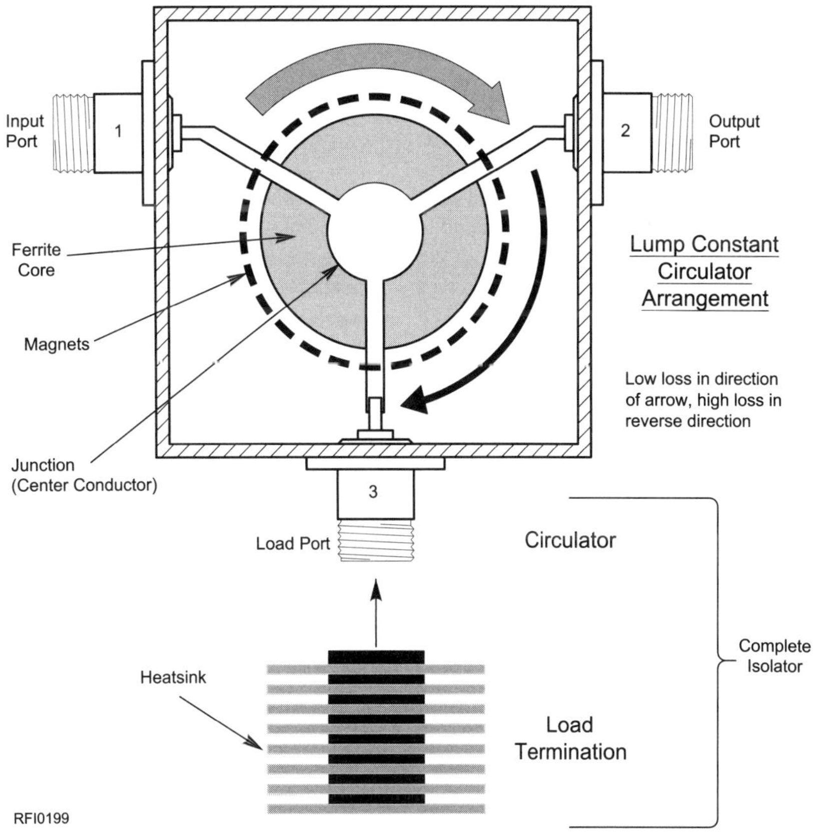

Figure courtesy of Bill Lieske, W7GLT and EMR Corporation

since multiplier stages are excellent mixers with lots of harmonics present, this can be the culprit when out-of-band signals are involved. The resulting mixing product is then amplified along with the transmitter's own signal and sent out the antenna. With solid-state equipment, the power leads are often the source of leakage into a transmitter, especially if the bypass elements have been removed, or have deteriorated with time and heat.

Second, for transmitters that are properly shielded, a signal can be conducted from the antenna, through the filtering and enter the PA stage, where it can mix with the normal signal and be re-radiated back out the same antenna. Because most FM transmitter PAs are Class C, abundant harmonic energy is present in the amplifier where they can mix with outside signals and produce all sorts of undesired outputs. While the duplexer can attenuate these signals (in both directions), relatively few modern duplexers provide serious attenuation to transmitted signals other than those near the repeater's receive frequency.

So how do we tell if a transmitter is the problem? For problems where the extra signals are "sneaking through" the shielding, you can supply extra shielding (even your hands and body will affect the level of signals on the shielding—if doing this effects a change in the interfering signal, then you should continue the approach by augmenting the shielding and bypassing of the suspect transmitter. But what if doing that makes no difference? At this point, you are going to need some specialized hardware—things like an RF *Isolator* (see sidebar) or selective cavities. What we need to do is to cut down the level of signals that are being conducted into the transmitter via it's antenna connection—if you insert a selective cavity tuned to the transmitter's output and the level of the interfering signal drops by a significant amount, then it is a good bet that the transmitter may be the source of the interference. The ultimate tool in combating transmitter-created IM, however, is the RF Isolator—but until you know you need to purchase one (and unfortunately, they are not inexpensive) it is sometimes possible to borrow one for testing. If installing an isolator significantly reduces or eliminates the problem, then you know what you must do. Normally, a single-junction isolator is all that is required, but for cases with particularly serious problems or very numerous transmitters, a dual-junction isolator may be required. In all cases, you need to be sure that the isolator is properly tuned for the frequency of the transmitter you are using it with.

There is another type of transmitter-induced IM that is, unfortunately, becoming more common and that is IM produced in the diodes of a solid-state T/R relay. These diodes are normally biased on and off using dc voltages, and they switch the RF path from transmit to receive by switching those dc voltages around—but if the amount (or, more specifically, the *voltage*) of the RF being conducted into the T/R relay is larger than it was designed to handle, those diodes can begin to conduct (or stop conducting) during part of the RF wave, and an often quite efficient mixer stage is born. These problems are more common with things like packet nodes and remote bases than repeaters, but any time a solid-state T/R relay is encountered, you need to consider carefully how much RF voltage it has to handle. Those with radios near high-power broadcasting transmitters are probably most at risk, but other sites can also experience this problem, especially if they have numerous other transmitters nearby. Remember, it's the *peak* voltage you have to worry about!

Note on using Spectrum Analyzers

A spectrum analyzer is a *wonderful* tool for troubleshooting Intermod and interference of all sorts – but you need to be aware of some of it's limitations, especially if the instrument you are using is not familiar to you. Spectrum analyzers are specially-designed receivers that rapidly sweep across a range of frequencies, and they display the signal levels they detect on a screen or digital display. Because these instruments are built to cover a wide range of frequencies, they have very little or no "front end" selectivity, which unfortunately makes them very susceptible to creating (or "re-creating") the mix you are looking for inside themselves—and they can lead you badly astray unless you are wary. Fortunately, analyzers are usually equipped with accurate step attenuators, and adding attenuation will usually allow you to separate the "real" from the "fake" mixes—if a signal you see on the screen drops by *more* than the amount of attenuation you insert (10 dB, say) then you know that the signal is *not* "real", and should be disregarded. If you have extra filters such as cavity resonators available, you may need to use them to protect the analyzer and allow you to "see" the *real* interference.

Notes

[1]DCI Digital Communications, Inc. For contact information, see the Suppliers List in Appendix A.

[2]MFJ Enterprises , Inc. For contact information, see the Suppliers List in Appendix A.

[3]Par Electronics. For contact information, see the Suppliers List in Appendix A.

[4]*Transmitter Hunting* (Order No. 2701) can be found at your local amateur-equipment dealer or can be ordered from ARRL Headquarters, 225 Main St., Newington, CT 06111, tel 888-277-5289, Internet: **pub-sales@arrl.org**, Web site: **www.arrl.org/catalog/index.html**.

Chapter 14

RFI at the Receiver

Most of this book deals with curing interference complaints that the amateur may receive from others. Here, we will consider the kinds of RFI that could affect Amateur Radio receivers. While interference is more often a mutual interaction than cause-and-effect situation, this chapter looks at the problem from the perspective of the receiver input terminals.

By Ed Hare, W1RFI,
ARRL Laboratory Supervisor,
Bob Schetgen, KU7G (SK) and
Hartley Gardner, W1OG

Most amateurs have encountered signals in their receivers that weren't supposed to be there, sometimes from a noisy power line or a personal computer. Many have encountered interference from Broadband over Powerline (BPL) systems and other elements of modern technology that radiate (intentionally or otherwise) in the amateur bands.

There are countless ways these signals come into being in our receivers — sometimes because they are actually on that radio frequency and sometimes because they are "manufactured" by our receiver (or another device) out of signals on other frequencies. Sometimes these signals are unheard but, by various means, they prevent our receivers from hearing the signals we are looking for.

THE THREE STEPS

The first Step in addressing a problem (or *suspected* problem) is to make an identification of that problem, then troubleshoot the true nature of the problem and its source on our way to (hopefully) a solution that allows us to enjoy the unimpeded use of our receiver. We need to recognize that the Identification and Troubleshooting steps are often linked together as we conduct tests and gather information, with the data derived in the troubleshooting step sometimes making us re-assess the suspicion we had started with.

Once we've determined *what* the problem is and *how* it is occurring, we can then (and *only* then) start taking steps to solve it. Solutions to receiving problems can be as simple and satisfying as discovering a noisy (and easily replaced) power supply or as frustrating and potentially insoluble as discovering that a foreign military is transmitting data on your favorite net frequency. Most problems will fall somewhere in between, with the vast majority being soluble, at least in my experience!

IDENTIFYING THE PROBLEM

The first step in resolving an RFI problem with *your* receiver is to identify the exact nature of what you are hearing (or not hearing). As we said above, you may end up jumping back and forth between the identification and troubleshooting steps as you conduct tests, gather information and are able to eliminate or modify your initial theories as to what might be occuring. You may discover that what seemed a simple situation is actually complex or (hopefully) what seemed an intractable mess suddenly resolves to a single source that is easily eliminated. The point is to try to keep an open mind and not to be reluctant to examine your conclusions in light of new data.

In general, RFI to the receiver takes two distinct forms: 1. where an interfering signal appears, obscuring or distorting the signals you want to receive; and 2. where expected signals are not heard, blocked or obscured by interference that, while you don't detect it, nonetheless prevents your receiver from hearing the desired signals. Note: I use the word "hear" often with respect to the reception of signals, but this can also mean "see" as in the case of a television signal, "read" as one might with a digital signal, or even "act" as with a remote-control receiver. In any case, what the receiver "hears" is what we're concerned with, no matter whether the output of that receiver is audio, video, a data stream or just a relay closure.

The first case is perhaps the most dramatic — we hear (or see) the problem directly though it's source may not be clear at all, while the second case can be much more subtle — but just as real.

One suggestion before we start in — whenever solving a problem of this sort, begin a notebook or at least a note-card where you can record your observations and tests (and even theories) as they occur. You will find this will save you much time and repeated effort, and in some cases reviewing that record will suggest solutions that never occurred to you at the time. It is a sad fact that we humans are *not* particularly good at remembering collections of numbers, sounds and perceptions — but we have the ability to record them when they're fresh and have that information available later — if we want to. No matter how vivid the memory of frequency, time or sound is right now, by next week the essential details will have fled.

So You're Hearing a Signal?

When you can hear the interfering signal directly, there are several possibilities we need to consider, and the primary thing we need to figure out is whether the "source" of that signal you're hearing is inside or outside of your receiver. Here are some factors to consider when trying to resolve this:

First, the signal may be a legitimate, deliberate transmission that is reaching your receiver through an unexpected propagation condition. On the HF bands, there are all sorts of amateur and non-amateur signals "floating around" and oddities of propagation are not unexpected. Some amateur bands are only amateur in this region and are used by the fixed, mobile, military, broadcast and other services in other regions, so the sudden appearance of a non-amateur signal may represent nothing more than unusual propagation. Other possibilities include legitimate transmissions — that *shouldn't* be in the amateur bands. One way to figure out if the signal is really there is your neighboring hams — if two or more receivers several miles apart hear the same signal, it's probably real — the next step is to figure out if it's supposed to be there or not. The ARRL runs a service called *Intruder Watch* that specializes in doing just that — including taking action if that signal really isn't supposed to be there.

For the bands above 6 meters, while far rarer than on the lower bands, propagation "events" can occur that will bring in signals from startling distances — hundreds or even thousands of miles away. If your VHF or UHF receiver suddenly appears to be invaded by "visitors" from another call area, or there is an unexpected "heterodyne" on the local repeater one morning, open your squelch and listen — you may be in for a DX treat! While open water, stable weather and long valley systems are often a part of such phenomena, sometimes these openings "just happen". Amateurs in Southern California have been startled many times over the years by signals from Hawaii, and openings up and down the east coast over water paths are quite common under certain weather conditions.

"Intruders" can also happen in the VHF an UHF bands, and a good source of support in resolving these is often your local OO or OOC, who will have experience and knowledge of "who's who" in your area, and who can also offer help in resolving whether such signals are really there or not.

Spurious Signals

A second possibility is that the signal you're hearing, while actually on the frequency you're listening to, isn't supposed to be there, but is really an "extra" signal generated by a transmitter that is supposed to be on another frequency entirely, a signal that is being deliberately generated but is supposed to have been contained inside of shielding or possibly a signal generated by a device that isn't intended to be a transmitter at all.

For on-the-air transmitters, some of the possibilities include harmonics of lower-frequency transmitters, spurious signals generated by defective or mis-tuned transmitters, signals generated by the mixing together of signals on other frequencies (also called "Intermod") and possibly combinations of these things — Intermod, for example, often involves harmonics as well as mixing.

For signals that are leaking from supposedly shielded systems, cable TV systems, BPL systems, and even home entertainment devices can be the culprits. This book has entire chapters devoted to radiation from electrical devices and other commonly found things like computers and associated equipment.

Internally-Generated Signals

A third possibility is that the signal you're

hearing is being generated inside your receiver. We've already mentioned "Intermod", which is the mixing together of multiple signals, but other possibilities include "birdies" and receiver spurious responses.

Birdies are internally-generated signals inside the receiver that are usually an artifact of the mixing and local-oscillator circuitry. They are found in pretty much all receivers, though of course, the higher grades of receiver have fewer of them and they are at lower levels!

"Receiver Spurious Response" is a fancy way of describing the process where a receiver hears signals that are on frequencies *other* than the one they are tuned to. Depending on details, such as IF frequency, front-end selectivity, local oscillator purity and numerous other design factors, receiver spurious responses can be very troublesome.

I Don't Hear a Signal — and I Should!

A much more subtle situation exists when your receiver *should* be hearing a signal — and doesn't.

Receivers can be rendered "deaf" for a variety of reasons, but most of these reasons are related to some sort of strong signal that inhibits its proper operation.

Desensitization (or "Desense") of a receiver is a process where the receiver is made less sensitive by the presence of the interfering signal, generally by the action of reducing the gain of the RF amplifier stage or affecting the AGC system of the radio. In HF receivers, this is often called "blocking".

Different from, but unfortunately difficult to differentiate from Desense is the condition where the RF noise level at a location is elevated, and all the receivers in the area become effectively less sensitive because the incoming signals have to compete with this local noise.

Making the Case — Troubleshooting

In troubleshooting an interference case, we are trying to determine first, what the source of the problem is, then where it is being created, and lastly, whether there is any way we can control it. All of these things can be interrelated — for example, the signal might be created inside your receiver but as a result of outside signals, and while we can control it, the solution might be more costly than we are willing to invest! So here is the list of decisions we must make:

A. Inside or outside? Is the signal being formed inside my receiver or outside?

B. What thing or process is creating the signal?

C. What are the components of the signal?

For many of these cases, the key is found in carefully listening to the interference and identifying it by what you hear.

In order to decide whether the signal is coming from inside or outside of your receiver there are several tests we can perform:

1. Try a different receiver — if only one (good working) receiver hears the signal, then something about that receiver is probably responsible, and the source may well be inside.

2. Remove the antenna — if the signal is still there, it's a good bet it's coming from inside!

3. Try different antennas or even take the receiver mobile or portable — if the signal is only heard at your shack, or only on one antenna, it may still be "real", but the source is probably close to you.

It is important not to jump to a conclusion regarding whether a signal is really coming from inside or outside — even the "remove the antenna" trick can mislead you if the offending signal is sneaking in via another route, say through a broken interconnection cable or via the power wiring. Keep an open mind as you proceed!

WHAT IS CREATING THE SIGNAL?

Once you have determined whether the signal is internal or external, we then need to figure out what component or process is creating it.

For signals that come in from the outside, it is useful to determine whether the source is close to you or far away and for HF signals, this could be anything up to half a world away, while for VHF and UHF this means up to possibly tens to hundreds of miles. Your best resources for this are your fellow amateurs — if everyone in your area is hearing it, it is probably *not* local to you! Signals like that are usually best handled by referring to outside help, such as the *Intruder Watch* program run by the ARRL, or your local OO or OOC for more local issues such as VHF/UHF signals or things like AM broadcast. These programs offer much expertise and experience, and also can assist you in taking steps to eliminate these problems. They may also be aware of potential propagation events or enhancements that might be causing an unexpected signal.

If the source is relatively close to you, it might be time to learn the art and science of direction-finding — or, as enthusiasts call it: "fox-hunting". There is an entire chapter on the subject in this book, and you may be amazed at what you find radiating in your neighborhood!

The possibilities of devices and systems that can potentially contribute interfering signals is enormous and growing every day, with microprocessors being installed in every conceivable appliance, RF being deliberately created for all sorts of uses and high-speed switching circuits employed for all sorts of electrical and electronic purposes. Gone are the days when the biggest offender was the control on your electric blanket!

One problem that can occur is that an outside signal on a frequency close to yours but just outside of it can interfere with your reception. On the VHF and UHF FM bands, this is usually called *Adjacent Channel* interference. This problem can require some delicate testing in order to determine whether the fault lies with your receiver, or with the offending signal. Again, your fellow amateurs can help here, but caution is warranted — a very strong signal can cause similar "spurious" signals in many different receivers. It takes a thorough knowledge of your receiver or access to sophisticated test equipment to accurately determine whether an interfering signal is "clean" or not. Probably the best way to do a quick check is to see if *all* signals with the same S-meter reading cause similar "off frequency" symptoms in your receiver — if only one signal is doing it, then you have good cause to suspect they have a problem. Conversely, if all signals of a similar strength seem to be "broad", then your receiver should be the suspect.

For FM signals, a check of the offending station's *deviation* is usually sufficient to see whether it is too broad in bandwidth. Remember that the sidebands of an FM signal extend beyond the deviation, however. A normal "5 kHz deviation" signal modulated with speech will be about 16 kHz wide, and more if high-frequency tones are being transmitted, such as with packet radio or SSTV.

If the offending signal is being created inside your receiver, then we need to figure out what the mechanism is. Intermod, or the mixing of two or more signals to create the interfering signal, is a separate subject of its own, and gets a special section (and its chapter in this book!). Other problems might include harmonics, which are sometimes related to Intermod, "birdies" and receiver spurious responses.

Birdies are usually easy to resolve — if removing the antenna doesn't make them go away and they're always there, they are probably birdies. Sometimes birdies manifest themselves in strange ways, particularly if they are caused by corroded shielding or intermittent connections inside. While this delves somewhat into the area of "repair", if shaking or moving the receiver changes the signal, consider having a tech take a look for just such a problem.

Receiver Spurious Responses are present in every receiver made — the better receivers simply do a better job in both reducing their number and the sensitivity of the receiver to them. Some of the possibilities that are often seen are:

1. *Images*. In a superheterodyne receiver (which most receivers are nowadays), the first mixer always has the possibility of mixing both the signals located in the Intermediate Frequency (IF) *above **and** below* the local oscillator frequency into the IF amplifiers. One of these signal frequencies is the desired one; the other is called the *image*. See Figure 14.1 for an example. Only the front-end filtering prevents us from hearing the image frequency, though more modern receivers with very high first-IF frequencies make the problem less serious by reducing the need for a "tight" front end. In VHF receivers with limited or non-existent front ends, such as scanners and inexpensive 2-meter radios, it is very common for the receiver to hear the image as well as the desired frequency. It is not at all uncommon for scanner enthusiasts who are unaware of this phenomenon to accuse amateurs of having "spurs" on the public-safety frequencies they are trying to listen to — which are exactly twice the IF frequency of their scanner above (or below, depending on the receiver) the amateur channel. A common IF frequency for VHF receivers is 10.7 MHz, though many others can be found, and it is a good idea to know the IF frequency of your radio, so you can mathematically check to see if an offending signal is an image.

2. *Local-oscillator (LO) harmonics*. Particularly with a VHF or UHF receiver, it is common to create the LO signal by taking a much lower frequency source, either VFO or crystal, and multiplying it to create the VHF LO frequency. The problem is that the multiples of the source above and below the desired LO frequency can also be present in the first mixer, and these signals are fully capable of mixing with undesired signals and being detected in the receiver. As for images, the only defense is good front-end selectivity, though good receiver design that includes filtering of the LO also helps.

3. *IF feedthrough*. If the receiver isn't well shielded, it is possible for signals at the IF frequency to "sneak" into the receiver and be detected as if they were on the tuned frequency. These signals tend to show up on *every* frequency, regardless of the dial setting, but if the receiver design includes a variable IF design (found in some older but high-end receivers), it can sometimes appear to be tunable. Only a close familiarity with your receiver design will really verify this problem.

4. *Other mixing products*. Some older but very high-end HF receivers employed a number of IFs, and even some modern receivers can employ a number of different oscillators in their design. If the internal shielding of the receiver is compromised, it is possible for these different oscillators to mix and match and generate all sorts of possible spurious components — with older receivers or those that have been subjected to environmental stress, making a general investigation of the internal shielding might prove beneficial if "extra" signals seem to be cropping up.

Intermod

If you are hearing more than one signal mixed together, you are probably dealing with Intermod — but it is still important to figure out whether the Intermod mixing is taking place inside your receiver or outside of it. Intermod does not, however, mean you *always* hear multiple signals — if one of the signals is a "dead carrier" or say, an AM signal in an FM receiver, you might hear only one of its components. This book has an entire chapter on Intermod, but here are a few easy tests you can do to try and resolve whether or not the problem is Intermod and whether or not it is being formed inside your receiver:

1. If you can add attenuation to you receiver's input, note how much the signal is degraded when you add attenuation. If a normal signal drops four S units when you turn on the attenuator, but the interfering Intermod signal drops more, then you can probably conclude the mixing is taking place in your receiver.

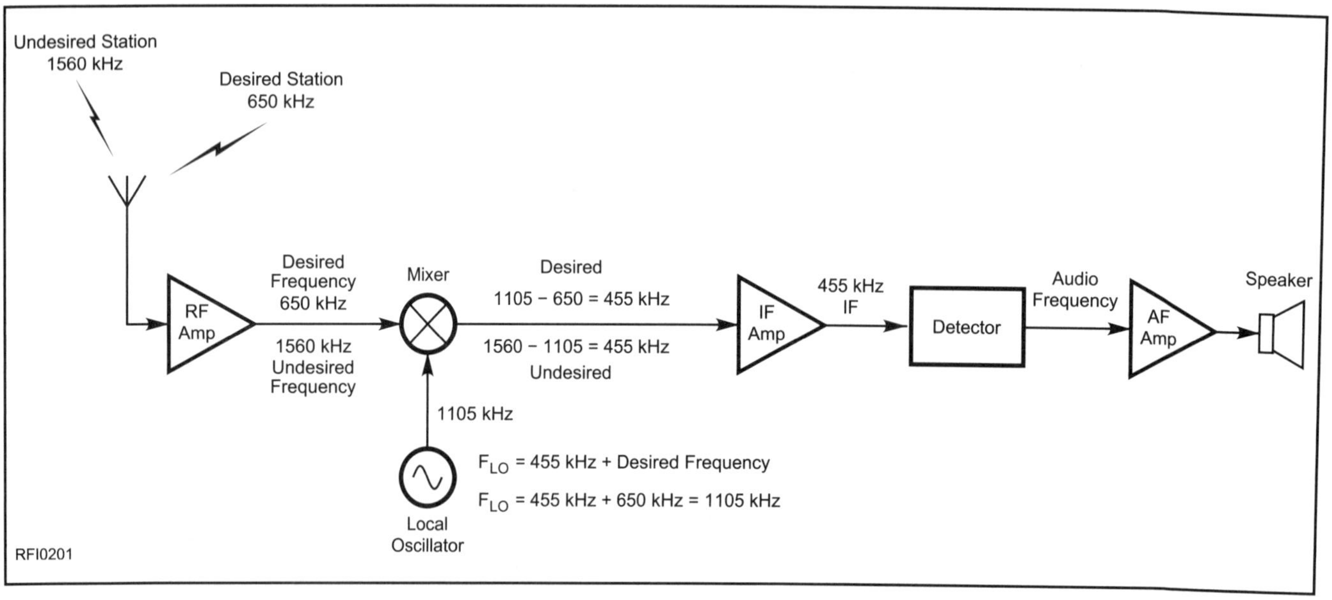

Figure 14.1 — An image can result when the undesired image frequency is not sufficiently reduced before the mixer stage of a superheterodyne receiver. In the example shown, an AM broadcast receiver has an IF of 455 kHz. The receiver is tuned to a station at 650 kHz. A nearby station is also at 1560 kHz, which happens to be the image frequency in the case. This is because the mixer output will be both the sum and difference of the local oscillator and IF:

–Desired Station: 1105 kHz - 455 kHz = 650 kHz

–Image Station: 1105 kHz + 455 kHz = 1560 kHz

Even though the receiver is tuned to 650 kHz, it will simultaneously hear the station at 1560 kHz. A well designed receiver would have a tuned RF stage to reject the image frequency. Modern Amateur superheterodyne receivers also use multiple IF stages to further help reduce images.

Notch Filters

Q Charles, KB3FSX, laments: I listen to a scanner and just put up an outside antenna. Now the intermodulation interference (intermod) from pagers operating on 152.175 MHz from 5 miles away is so bad I can't use my receiver. Is there some way I can block this signal with something I can make or buy?

A Yes, Charles, there are a number of possible solutions. Their effectiveness will depend both on the linearity designed into your receiver's circuitry and the relationship between the frequencies you want to listen to and the offending signal. The two general categories of solutions, other than moving farther away, are as follows:

- A notch filter can be used to severely attenuate the offending signal. If the desired signals are not too close in frequency to the signal you are trying to eliminate, this can be very effective.
- If you are interested in only monitoring a relatively narrow band of frequencies, such as the 144 to 148 MHz 2 meter amateur band, you can use a band-pass filter that will pass 2 meters and attenuate all other frequency ranges. This is particularly useful since intermodulation, by definition, results from signals on multiple frequencies.

For a thorough description of the topic, including a nice notch filter that you can build, see **www.arrl.org/tis/info/intermod/intermod.html**. The filter designer, ARRL Senior Lab Engineer Zack Lau, W1VT, notes that this filter provides a secondary notch at the FM broadcast band, often a contributor to the strong signals your scanner will have to deal with. Zack also cautions that the filter won't work unless you use high quality trimmer capacitors.

The simplest notch filter is a resonant stub across the feed line. It's not as fancy as Zack's filter but it might work and doesn't take much to find out. Start by putting a T connector in the antenna line near the radio. On the extra port, try an open section of coax cable a quarter wavelength ($\lambda/4$) at 152.175 MHz (taking into account the propagation delay of the coax). This will place a short across the antenna input for signals at that frequency.

Unfortunately, it will also place a short at every odd multiple, so it will null 456.525 MHz as well. It will also result in some level of mismatch at every other frequency, but it is a cheap and easy fix that may solve your problem.

For regular polyethylene dielectric coax, the propagation velocity is about 0.67 times that of free space, so a $\lambda/4$ stub should be about $300 \times 39.37 \times 0.67 \times 0.25 / 152.175$ or about 13 inches. Start a bit longer and trim the coax about 1/16 inch at a time until it is resonant and the signal is nulled. You will need to correct for the connectors if you measure it outside the radio. For best results feed a signal at 152.175 MHz into a length of coax before the T and trim for minimum signal strength at the receiver. The advantage of the open stub is that you can just keep cutting until it works, you don't need to do anything to the end except make sure the shield doesn't touch the inner conductor.

Commercial notch filters designed for the purpose are available from PAR Electronics, **www.parelectronics.com/scanner.htm** (see Figure A). MFJ at **www.mfjenterprises.com** offers the MFJ-713 and '714 compact 2 meter band-pass filters (see Figure B) while DCI, **www.dci.ca**, makes a line of larger low loss band-pass filters, suitable for mobile or home station use.

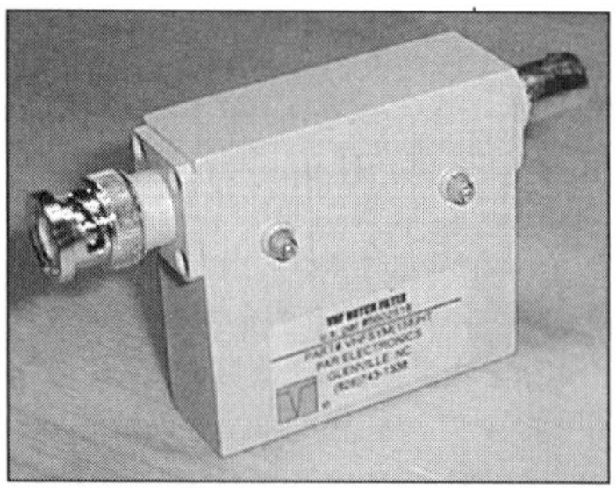

Figure A — One of the PAR Electronics line of notch filters designed to eliminate interference from pager transmitters.

Figure B — The compact MFJ-713 2 meter band-pass filter designed especially for handheld receivers and transceivers. It automatically switches to bypass on transmit to avoid damage. The MFJ-714 is similar but has UHF instead of BNC connectors and uses an external 12 V supply instead of the 9 V battery in the '713.

2. If you add selectivity — perhaps in the form of a pass or notch filter (see sidebar) and the interference drops significantly, you have established that your receiver is doing the mixing.

3. If turning a beam (or any other directional antenna) makes larger-than–normal changes in the signal strength of the interference, you should suspect your receiver.

Note that *all* of these steps can also resolve a case of harmonic interference. While seen most often with nearby AM Broadcast transmitters, the non-linear components of a receiver can "create" harmonics of any signal it hears, particularly if the signals are strong.

If you discover that the problem is Intermod, but your receiver is *not* creating it, then you should go to the chapter on Intermod for some good ideas on where to look. Intermod can form in *any* non-linear joint in a conductor, and even though it gets the most press on the VHF and UHF bands, it can affect any frequency.

TROUBLESHOOTING THE "DEAF" RECEIVER

If you've discovered that sometimes your receiver is not hearing what it should, but performs properly on other frequencies and bands, or you have had it tested and pronounced OK, you should be trying to figure out if some sort of interference is inhibiting your receiver.

This sort of interference takes two basic forms — either a strong signal is disrupting your receiver or a noise problem is causing your receiver to be unable to hear signals it really should be hearing.

When a strong signal — sometimes close to your desired frequency, sometimes not — enters your receiver, it can cause the circuits inside to do things that compromise its performance. If you suspect that such a strong signal is causing the problem, you probably already have a good idea of what the potential problem is, but even if you don't, searching the immediate band for such a signal can sometimes prove valuable, as the closer the interfering signal is to your desired frequency, the more likely it is to be causing a problem. If the suspected interfering signal drops and your receiver returns to proper operation — congratulations — you've found it! If it doesn't, keep looking — if you have a friend with a spectrum analyzer or "band scope", it will help immensely in conducting this search.

The other aspect of the "deaf receiver" problem is when the radio-frequency noise at your site increases and your receiver appears to be becoming "deaf", even though it tests OK on the bench. While this problem is becoming epidemic at some crowded repeater sites, home shacks can be just as affected — though the home shack probably has a better chance of finding a "fix"!

The noise in question can come from a myriad of sources, but the most common are electrical systems, which produce all sorts of noise under some conditions, and other transmitters. All transmitters put out noise — no exceptions — though good circuit design and installed selective filtering can cut an individual transmitter's noise down to immeasurably low levels at frequencies even a few dozen kilohertz away from the main carrier. The problem on many crowded electronic sites is that there are so many transmitters in a small area that these very low individual levels all add up to a measurable level of noise and the sensitivity of all the receivers at the site will be affected.

Noise from power lines and other electrical faults is often distinguished by a very distinct 60 or 120 Hz AM component, and can be detected using an AM or SSB receiver on the affected frequency — the chapter in this book addressing power lines and electrical devices can give you some excellent advice on finding and fixing these problems.

A simple test for "site noise" of the sort created by other transmitters can be conducted using a device called a "line tap", which is a device for inserting (or withdrawing) a sample signal onto (or from) a transmission line. These devices can be purchased, but they can also be easily and cheaply constructed by using a common "T" connector (see sidebar).

Constructing a Line Tap

A line tap is simply a coaxial device with a very small value capacitor installed from the center conductor to a test port. Commercial devices that do this job can cost a significant sum, but we don't need their multi-octave stable performance to do the tests we need to perform. We can even substitute a suitably high value resistor for the capacitor, depending on the specifics of the particular application.

One approach, if you can find one, involves a "UHF"-type "T" connector with three "female" ports, and has the "side" or middle port center conductor *screwed* into the straight-thru center conductor. With a pair of needle-nose pliers, unscrew the female center pin from the side port and set it aside. Then take an adapter that is a male "UHF" to something suitable on the other side – I prefer a BNC myself, but you may prefer something different - and screw it tightly into the port, which should leave the center pin of the adapter close to, but not touching, the pass-through between the other two ports of the "T" connector. We don't want this adapter to move, so make it secure and possibly even use some thread-locker on it so it doesn't work loose. This port is the "test signal" port, while the other two ports are for the antenna line and receiver being tested.

The "load" presented by this line tap to the signal generator is essentially an open-ended cable, and some signal generators have difficulties with this. If your generator is one of these, or if you find it seems to be erratic in output level, put a 50 ohm load on the end by placing another "T" connector on the test signal port and install a 50-ohm load on one side and the generator input on the other. If you've used BNC connectors, one of the small loads used with coaxial Ethernet networking will work well (and they're inexpensive!)

Another approach to line tap construction involves a more readily available UHF "tee" adapter. In our case, an adapter with 3 female ports was selected, and a relatively high value resistor is installed in the PL-259 plug. The resistor used is physically small enough to fit inside the plug's center pin; and is carefully soldered between the coax center conductor and connector's center pin. See Figure A.

Whenever substituting a resistor for a capacitor in a line tap, be certain to consider the impact of dc levels in your application. In both cases, however, component impedance must be high enough to provide adequate isolation for the line tap to work properly. Also, as suggested above, a 50 ohm non-inductive resistor can be added to provide proper load termination for your signal generator, if necessary.

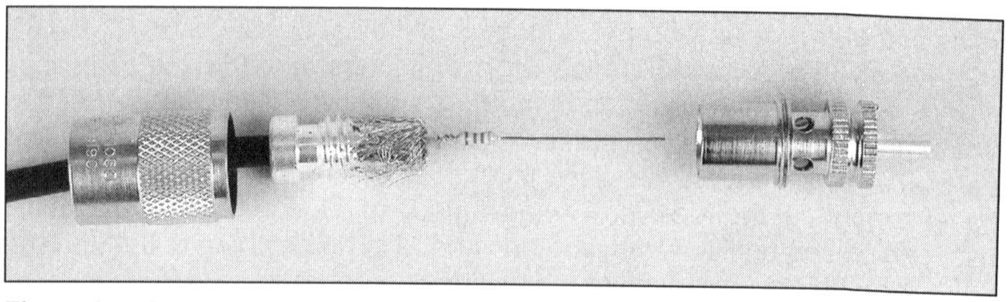

Figure A — For use in a Line Tap using a high value resistor in the center pin. This modified connector, used in conjunction with a standard three port T-connector, makes up the line tap.

Once you have a line tap available, you will need a signal generator that is stable in frequency and level — the absolute level of its output doesn't have to be tightly controlled, but it should have a way to controllably raise and lower the output signal level. You will also need a low-power dummy load (a 50-ohm BNC load that used to be common in computer Ethernet networks is excellent for this).

The line tap will be installed in the antenna line to the receiver with the two connected "in line" connectors going to the receiver and to the antenna, and the "tap" or side connector connected to your signal generator. Don't install the antenna just yet, though — connect the dummy load to that port instead, and adjust the signal generator to give a measurable signal. You can use the S-meter if it has one, a limiter reading, or even a SINAD reading, if your test setup has one — we're not concerned about the absolute sensitivity of the receiver, but rather to establish a baseline sensitivity without the antenna connected. Once you have that level set, remove the dummy load and install the antenna — if the receiver still hears the signal generator at the same level, then you have *no* site noise — but in all probability, you are going to have to increase the output of the generator significantly to get back to the same received signal. The difference in generator output levels is the amount of site noise you have. If your signal generator is calibrated and you know the sensitivity of your receiver, you will know what the best possible sensitivity of any receiver connected to that antenna will be. You may also be startled to discover that a preamplifier actually makes your receiver perform *worse*, rather than better.

On HF, of course, there is *always* some noise, and it will change with the time of day and other factors — on VHF or UHF, it is possible that the level of natural noise is low enough that you can't measure it with ordinary equipment, but few of us are lucky enough to live in areas that are that quiet any more! Moonbounce enthusiasts and others who work with extremely sensitive receivers will not be surprised by this, but many of us will be startled by the level of ambient noise in what we thought was a "quiet" location.

That line tap has other uses, by the way — if you have access to a spectrum analyzer and want a low level, but stable way to look at your transmitter, the line tap makes this easy. It's also a handy tool for checking repeaters for Desense from their own transmitter — insert the signal using the line tap just like you did above, and if the receiver goes deaf as soon as the transmitter is keyed, you've got desense!

CAN WE FIX IT?

So now you've worked your way through identifying the problem you are experiencing, troubleshot it and you know where it's coming from — is there any way to fix it?

The answer is, of course — maybe. Some problems, like the ones caused by 100 other nearby transmitters, or by poor receiver design may just not be soluble at a price we want to pay — but many other problems *do* have solutions, and often quite simple ones.

For external signals, the solution is to find the source of the signal — if it's a deliberate transmitter located out of your immediate area, you are probably going to have to get some help. The ARRL's *Intruder Watch* program can help with non-amateur signals in the HF bands, and your local OO, OOC or Repeater Coordination group can often help with such issues on the VHF and UHF bands.

For more local signals, you should heed the guidance found in the other chapters of this book, which can help you with things like electrical devices, computers, BPL and cable-TV systems.

With problems *inside* your receiver there are fewer outside resources, though there are a number of things worth checking out before you give up and spend big bucks on a new receiver. (If you really want a new rig, though, don't let me stop you just read this to see what you should be looking for!)

These days, it often pays to check out the Internet for ideas and suggestions in solving issues you may be having — there is a "user group" out there for almost every amateur rig or brand ever made, and quite often someone else has had the same problem you are experiencing. If you're lucky, they may have even found a fix for it!

The Internet is valuable in other ways, too. In the old days, you might discuss your problem with the other hams in your local group, or on the 75-meter net you check into, but you had to be lucky to find another ham with the same rig and the knowledge or experience to help. Nowadays, however, you can post your problem on one of the many Amateur Radio bulletin-board and user-group e-mail lists and have literally thousands of active hams looking at your questions, which vastly increases your chances of finding someone who can help.

But, what can you do if all that advice is of no use? Let's go through the possibilities:

1. An *Intermod* problem. If your receiver is experiencing Intermod, the cause is almost always the fact that the signals that are mixing and causing your problem are simply too strong into the input of your receiver. The chapter on Intermod gives many specific ideas on devices and techniques to use, but the overriding need is to reduce the level of those signals. Decreasing their level can involve things like making your antenna smaller, lower, or just moving it farther away; installing filters to reduce the level of those signals, or even begging your neighbor ham to quit using his kW when you are both operating…Be aware that resistance to Intermod generation is a measurable quality in a receiver, and you very likely *can* acquire a better receiver if you want to. Be sure to check the specifications for "IMD Suppression" in both the manufacturer's published specs, and also check and see if the ARRL or someone else has done a review of the model you like and evaluated its performance in that regard. If you commonly operate in "Intermod Alley" (areas of high-level signals on nearby frequencies), make IMD performance one of your "must-haves" when you buy that new rig.

2. A *Receiver Spurious Response*. Fixing these problems can involve many of the same devices and techniques as repairing Intermod problems — if we know, for instance, that a lack of front-end selectivity is causing a problem with image reception, installing either a band-pass filter (tuned to pass the band we want to listen to) or a band-reject filter (also called a "notch" filter, and tuned to reject the problem frequencies) can often make a significant improvement. Spurious responses due to IF feedthrough and internal shielding failures will require serious maintenance to address, and are best left to those with the requisite tools and skills. Receiver Spurious Response is also a specification you can look at when shopping for a new rig!

3. "*Desense*". If your receiver is being affected by strong signals, you have only a few choices — you can attenuate only the strong signals using the filter techniques mentioned above, you can move or change your antenna to reduce those signals, or you could install (or select, if your receiver has one) an attenuator to protect your receiver . You may find that using a smaller or lower gain antenna for receive might just work better than a big one, if it reduces the level of the interference below the point where your receiver starts creating problems. The classic example of this is for a 2-meter handheld radio, which often works fine using the factory "rubber duck", but go berserk or deaf when hooked to an outside or mobile antenna.

If the problem is site noise, you have limited options. If your receiver has a preamp on it, you can try removing it, as sometimes a preamp only makes these problems worse. You can also see if there is a single transmitter that is contributing a disportionate amount of noise to the problem, which you would find by determining that your site noise gets markedly worse whenever this transmitter comes on the air. In that case, you may be able to work with the owner of that transmitter to improve the situation, either by making adjustments or perhaps by installing a filter on the transmitter to reduce the level of noise on your receiver frequency. Keep in mind that a transmitter that is radiating a lot of noise is probably bothering other receivers on the same site, so you may get help from other occupants who would like *their* receivers to work better as well!

CONCLUSION

As you have seen, there are a lot of possibilities to look at when trying to solve a case of interference to your receiver! Here's a rundown of the most important things to do:

1. Stay flexible! Be prepared to change your approach and conclusions if the facts warrant it. Interference problems can be devilishly tricky sometimes, and their true nature is revealed only after much research and thought.

2. Keep Notes. If you keep notes, you will be much better prepared to find the proper solution, especially if you have a

complex problem or one that you can work on only intermittently. Keep the notes after you find the solution, too — it might come back a year from now and your memory might not hold all the essential details!

3. Seek Help. There are lots of sources of help with interference — OOs and OOCs, repeater coordination bodies, ARRL Tech Assistants, HQ itself, and last, but not least, your fellow amateurs, who may have walked down the very same path you are about to and can offer much wisdom.

4. Be nice. Very few people set out to cause interference, and people who buy appliances and devices probably didn't chose them in order to pester your receiver. Be courteous and as helpful as you can if you identify such a source for your problems, and try not to make threats about official action if you can help it — yes, the FCC will probably back you if something is radiating illegally, but this should be used as a last resort!

Good Luck!

Chapter 15

Computers

Every RFI case involves both technical and personal issues. Knowing all of the technical solutions won't do you any good if your neighbor won't let you in the house to try them! Always solve the personal and emotional RFI problems first.

By Ghery S. Pettit, N6TPT

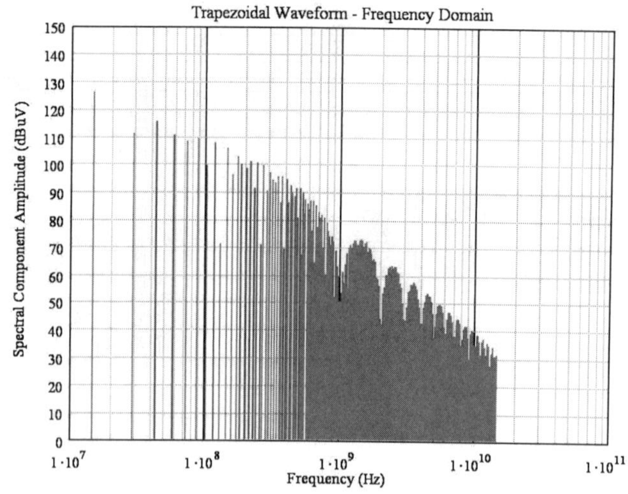

Figure 15.1—Fournier analysis of a 14.318 MHz clock signal.

Computers are used for a variety of tasks in the shack. They may be used for logging, for digital communications modes over the air, or simply for looking up addresses of fellow hams over the Internet. But these handy devices may also be sources of electromagnetic interference (EMI) that prevents us from receiving that weak DX station from the other side of the world.

Computers are digital devices which the FCC refers to as "unintentional radiators". That is, they must generate and use RF energy in order to function, but they do not need to radiate it. The wireless card in your laptop is an "intentional radiator". It must generate and radiate RF energy in order to carry out its intended function. In the case of your PC, there are limits in the FCC Rules and in international standards that limit the magnitude of these interfering signals. These limits are designed to provide a reasonable level of protection against harmful interference — when the computer is in your house and the victim receiver is next door. When you place the computer right next to a receiver or its antenna, you may have problems. We'll get into what to do about this case later in this chapter.

In the early days of personal computers (late 1970s) there were no limits in the United States on the levels of EMI that a product could produce. As a result, no thought to suppressing the generation of RF emissions was given in these early designs. As you might imagine, the interference levels were quite severe. One computer that the author tried in his house over a weekend wiped out reception on any receiver that he had. AM radio, FM radio, television, you name it and you couldn't hear a thing. The joke in the lab where he worked at the time was that the computer was obviously a radio transmitter, so in what service should a license be applied for? The answer was "YES", because it was in all of them. Needless to say, the FCC was receiving large numbers of complaints from all corners of society, both private and public, about the interference. Something had to be done. In concert with industry experts, the FCC investigated limits and methods of measurement of these emissions and issued new rules to limit both power line conducted and radiated emissions from digital devices. These new limits were published as Title 47 of the Code of Federal Regulations, Part 15, subpart J. Any product first placed on the market after October 1, 1981 had to meet these limits. Any existing product had until October 1, 1983 to be brought into compliance or taken off the market. The Rules have been re-written over the years, but the basic requirements are still the same today. This is one government regulatory program that has done what it set out to do. As a result, interference complaints to the FCC have greatly diminished since the rules went into effect.

What does this mean to us today? There are two levels of emissions limits based on where the computer is to be sold and used. Class A limits are intended for products that are not marketed to the public for use in the home. Class B limits are intended for products that are marketed to the public for use in the home. Class B limits are about 10 dB lower than the Class A limits and provide a greater level of protection against interference. As noted above, the Class B limits are designed to protect your neighbor, not yourself. Your neighbor can't turn your computer off. He can't move it. If you interfere with yourself, you can do either of these things. You can turn the computer off when you really want to listen to the radio or watch TV, or you can move it farther away from the antenna, thus decreasing the level of the interfering signal at the antenna. For practical matters, the Class B limits generally offer adequate protection to your own receivers, as well. See, later in this chapter, "Power Line Conducted Emissions limits" and "Testing" for more information on standards and standards test methods.

EMISSIONS—WHERE DO THEY COME FROM?

Digital signals in a computer can be represented by a repetitive trapezoid. This is a square wave with rise and fall times something slower than instantaneous. In the frequency domain this signal can be represented by a series of sine wave signals at integral harmonics of the fundamental frequency of the waveform. To compute the amplitude of each of these harmonics, we can perform a Fourier Transform on the waveform. Personal computers have clocks at a number of frequencies, so emissions from them can occur at harmonics of each of them. A representative motherboard has clocks at 32.7 kHz, 12.288 MHz, 14.318 MHz, 25 MHz, 33 MHz, 48 MHz, 96 MHz, 100 MHz and selectable clocks at 100/133/167/200 MHz and 200/266/333 MHz. These clock signals can be approximately represented as repetitive trapezoidal signals. This signal can be modeled as a succession of sine waves at frequencies that are integral multiples of the fundamental frequency. Thus, a 100 MHz clock would have components at 100 MHz, 200 MHz, 300 MHz, 400 MHz and so on. The amplitudes of these spectral components can be computed by performing a Fourier analysis on the waveform. The example in Figure 15.1 is for a 14.318 MHz clock signal. We have set the amplitude of this signal to 3.3 Volts (typical on a modern PC motherboard) and have arbitrarily set the rise time to 1 nanosecond and the pulse width to 30 nanoseconds (to avoid a 50% duty cycle where the even order harmonics would disappear). The plot was generated by analyzing the equation:

$$G(n) = 2*V*((d + \tau r)/T_0)*(\sin(\pi*n*(d + \tau r)*f0)/(\pi*n*(d + \tau)*f0))*(\sin(\pi*n*\tau r*f0))/(\pi*n*\tau r*f0)$$

where

d = pulse width

τ = rise/fall times

n = harmonic

f_o = fundamental frequency

for the first 1000 harmonics using a *MathCAD* program. This equation gives the amplitude of each harmonic in volts. By taking the array and analyzing it with:

$$Ampl(n) = 20*Log(|G(n)|) + 120 \text{ dB}\mu\text{V}$$

we can plot the harmonics in dBμV. As can readily be seen, this 14.318 MHz clock generates signals at a lot of frequencies. Now consider that each clock on the board does the same thing and the job of the EMC engineer has just gotten larger.

Looking at the graph in Figure 15.1, you can see that the overall amplitude of these signals falls at a rate of 20 dB / decade of frequency up to a point, and then falls at 40 dB / decade of frequency. The breakpoint where this changes is $1/\pi^*\tau_r$, where τ_r is the rise time of the signal. Slowing the rise time of the signal moves this breakpoint to the left, reducing the amplitude of higher order harmonics faster. This can be a valuable tool in reducing emissions – don't use fast rise time circuits if you don't have to.

These signals are moved from one place to another via current flow on the wiring of the computer. These currents flow in loops, from the source to the load and back. Current flowing in a loop creates a magnetic field. This field can and will radiate. The magnitude of the magnetic field is directly proportional to the area of the loop, and this is a key thing that can be controlled in the design of a printed wiring board.

In the past it was common to design printed wiring boards using one or two layers of metallization for circuit traces. Now four layers is typical, with highly complex boards utilizing more layers. In a four layer board one layer typically is used for the power ground, a second layer is used for power distribution and the remaining two layers are used for routing signals. When a signal travels down a trace, the return current in the loop flows on the surface of the plane directly adjacent to the signal trace. It doesn't matter if that plane is power or ground. Current will take the lowest impedance path from the source to the load and back.

Minimum impedance of a loop is minimum inductance, which means minimum loop area. As the length of the path is defined by the distance between the source and load and the path taken between them, the minimum loop area is obtained by the return current flowing directly below the trace on the return plane. Assuming that the plane is unbroken, this results in minimum emissions from the current flowing in the loop. Board designers must insure that the traces carrying signals do not cross breaks in the planes as the return current flowing on the surface of the plane cannot pass over the break. The current will find another path to continue the loop, but at the expense of greatly increased loop area. As the radiation from a loop is directly proportional to the area of the loop, the emissions from that circuit increase in a similar manner. Loop area control is a key concern in laying out a printed wiring board.

In addition to clocks and their harmonics, other imperfections in the design of the computer may result in emissions at other frequencies, not related to the clocks. One key area in a printed wiring board (PWB) that can cause non-harmonically related emissions is improper termination of transmission lines. Just as an improperly terminated transmission line to an antenna has reflected waves that set up standing waves, so it is on a PWB. A wave launched down a transmission line meets a termination of some value different than the characteristic impedance of the line. A reflected wave travels back to the source. If the source impedance is not the same as the characteristic impedance of the line this wave is re-reflected. Ringing results. The frequency of this ringing is dependent upon the propagation velocity on the line and its length. An otherwise quiet product can generate significant emissions at this ringing frequency. Ringing is controlled by properly terminating the line, either by matching the source impedance to the line via a series resistor of the proper value, or by placing a parallel termination at each load to match the load to the line. A single series resistor can stop ringing in an entire clock net at the cost of slowing the net. Parallel resistor termination at each load stops ringing without slowing the net, but requires many more components to accomplish. It also increases power dissipation as resistors are now connected from Vcc to ground at each load. Inserting a capacitor can stop the dc current, but adds yet another component to each termination.

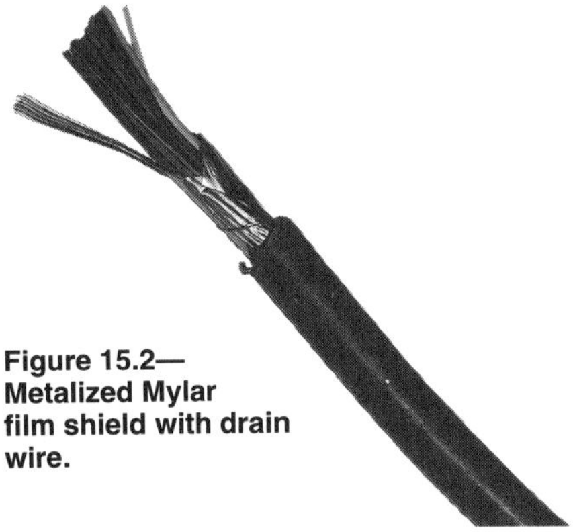

Figure 15.2—Metalized Mylar film shield with drain wire.

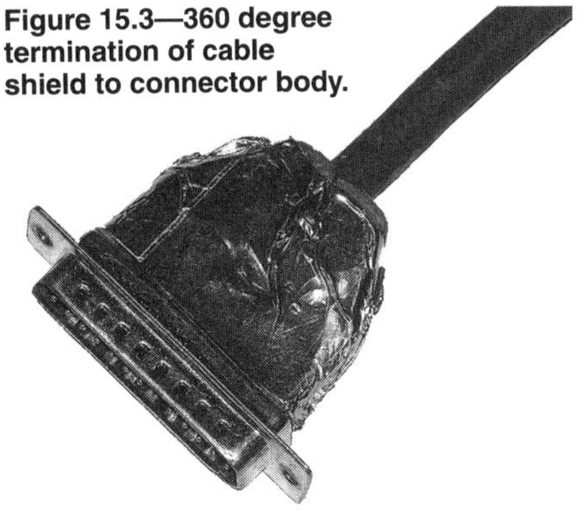

Figure 15.3—360 degree termination of cable shield to connector body.

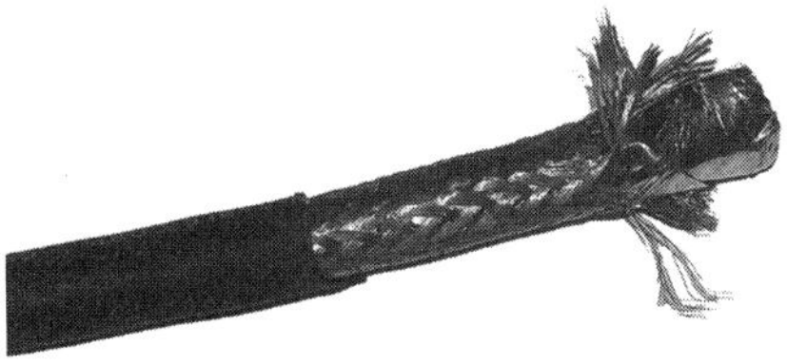

Figure 15.4—Braided shield over metalized Mylar film shield.

Another area to be considered is the selection of bypass capacitors for integrated circuits on the board. As the capacitor is part of a loop, there is inductance associated with its installation. Thus, we wind up with a series LC circuit, which has a resonance at

$$f = 1/(\sqrt{L^*C})$$

Below this frequency the impedance of the circuit falls off with increasing frequency, as would be expected for a capacitor. Above this frequency the inductance from the leads takes over and the circuit appears to be an inductor, with increasing impedance with increasing frequency. It is still an effective bypass circuit as long as the impedance is below what is needed. An interesting aspect of this is that decreasing the size of the capacitor to raise the resonant frequency can actually harm the overall performance of the bypass LC circuit. For common surface mount installations the total inductance of the circuit, consisting of the leads on the PWB and in the lead frame of the integrated circuit, typically is around 4 nH. Typically .1 µF capacitors are used for this bypass function.

Since no conductor is ideal, it will always have some resistance associated with it. Current flowing in a conductor therefore causes a voltage drop. If we have an RF current flowing, generating a voltage difference between different parts of the plane (ground or power) and then connect wires to each end of the board we have what looks a great deal like a dipole antenna, i.e. a pair of wires with a voltage source in the middle. When laying out a board this effect can be minimized in two ways. One is to design the device so that cables only connect to the PWB at one side. The other is to minimize the distance that currents must flow, thus minimizing the voltage drop in the board.

When using shielded cables from one shielded box to another it is important to ensure that the shields are properly terminated. Many common inexpensive computer cables use a metalized Mylar foil shield with a drain wire. Figure 15.2 shows an example of this type of cable. The foil can be an excellent shield, but only if it is terminated correctly. In these cables the drain wire is connected to the body of the connector. This results in a break in the shield, and the drain wire is at a higher impedance. The result is that the RF currents flowing inside the shield couple to the outside of the shield, making it a very effective wire antenna. A proper way to terminate the shield is with a full 360 degree bond around the body of the connector. Figure 15.3 shows an example of such a termination. The illustrated termination could be made better by soldering the aluminum tape surrounding the wire terminations to the body of the connector. With the Mylar foil shield this is very difficult to do. Better cables use a wire mesh over the outside of the Mylar foil, providing a good metallic contact with the metalized surface over the entire length of the cable. Figure 15.4 shows this type of shielded cable. This wire mesh is then more easily terminated to the connector via one of several means, depending on the design of the connector. Remember that if a box is shielded (and many are) then the cable connecting it to another shielded box must be a solid continuation of the overall shield.

When designing a device to use unshielded cables (and at RF that often includes "shielded" audio cables) the connection to these cables should be filtered as close as possible to the point where the cable penetrates the shield. Your personal computer has such filtering components either integral to the connector, or on the board immediately adjacent to it. These filters typically are a capacitor, a ferrite bead or a combination of the two.

When designing a chassis to service as a shield, remember this key concept - as long as the box is made of conductive materials, if it will hold water, it will hold RF. When designing the shield, we negotiate the size of the leaks (perfection typically is not required). The thickness of the metal in a metal chassis is not important. It can be shown that aluminum foil from the kitchen is gross overkill. Leaks come from openings in the shield. These may be deliberate (air vents, places to allow lights to shine through, etc.) or they may be inadvertent (seams where pieces mate together with contact points too far apart). How large an opening that can be tolerated depends on how much shielding is required and the frequency of interest that must be shielded. An opening that is about ½ wavelength (maximum linear dimension) at the frequency of interest has no shielding value. Typically an EMC engineer will ask for no more than 1⁄10 of a wavelength at the highest frequency of interest. Plastic boxes must have a conductive coating which makes proper contact with adjacent pieces. This conductive coating may be accomplished with any of a number of commercially available products, from conductive paint to spray on copper or other metallic coatings. Each has its advantages and disadvantages in commercial applications, which are beyond the scope of this chapter. What is important is that the coating is of adequate thickness throughout and that good conducted contact is made over all joints and seams between adjacent parts.

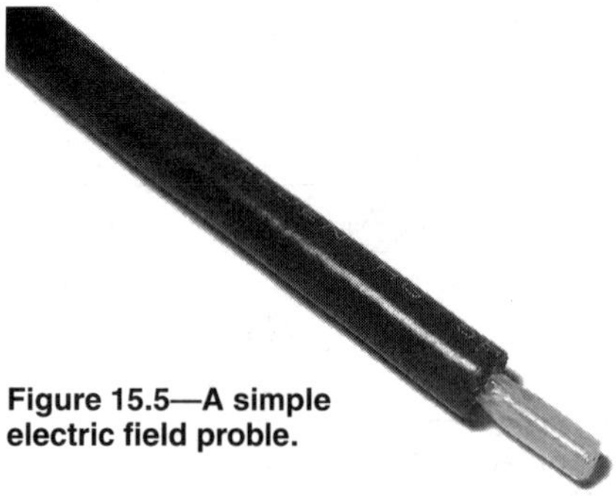

Figure 15.5—A simple electric field proble.

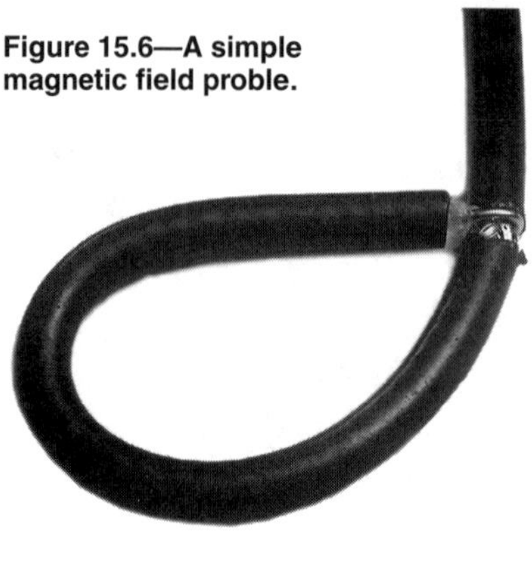

Figure 15.6—A simple magnetic field proble.

TROUBLESHOOTING

Quite often a PC or peripheral that is purchased will not cause interference in a typical Amateur Radio station. This was not always the case, but with the US and international standards for EMC that these products must meet, the incidences of interference have been drastically reduced. In the author's own experience, an XT clone PC (already in use when first licensed in 1988) caused interference in the 20 meter band as the third harmonic of the processor clock frequency fell in that band. It is probable that the system did not meet the FCC limits for radiated emissions. The seller had provided no evidence of having tested the product and no FCC label was placed on the product. The author's next PC, a 33 MHz Intel 486 based machine did not cause interference. But the processor clock had moved from 4.77 MHz to 33 MHz, removing its harmonics from any HF band. His present station is located in the same room as newer machines, interconnected with a 100BaseT Ethernet LAN. No interference on any bands operated is noted from this installation. Sample size is 1, but it has evolved over the years.

So, what do you do if you determine that your computers are interfering with your radio equipment? You have several options, ranging from turning off the offending computer to moving it farther away from the antenna to modifying the design of the computer to make it quieter and reduce the interference.

We'll assume that turning the computer off isn't an option (the kids still want to play their computer games while you're talking on the radio) and that you can't really move it farther away (you're using it for packet radio or PSK 31 or logging or...you get the idea). This leaves fixing it as the only option. Where do you start? Troubleshooting EMC problems is still largely an art, but there are tricks that are used in most labs that can be duplicated in your shack.

First, at what frequency (approximate range) are you suffering the interference? It has been found that in the majority of cases, emissions below about 200 to 300 MHz typically radiate from the cables attached to the system and emissions above these frequencies typically radiate from holes in the shield. While not 100%, it is close enough to be a rule of thumb.

Let's say, for example, that you are suffering a birdie in the 10 meter band and you've determined that it is coming from your PC or something connected to it and you suspect that the offending signal is coming from one of the cables. Please note that the interference may not originate from a desired signal on the cable, it may simply be along for the ride. While watching the S meter on your radio or listening to the interference, disconnect the cables from the PC, one at a time. When the signal goes away, reconnect the other cables, one at a time. If the signal does not return with any of the others, then you know the problem is with that one cable. Otherwise, the one that brought back the noise is also a problem and should be disconnected while reconnecting the others.

Once the offending cable(s) is/are identified, you can fix the problem. Assuming that the connector in the PC (or peripheral) is properly bonded to the chassis (not always a good assumption) you can either try a better quality cable (see the discussion on cable shield termination or you can try ferrites. Use toroids for HF. Beads (blocks, rather than beads, by the time they are large enough to fit over a typical cable) may be adequate for VHF and higher frequencies. Selection of proper ferrite mix can also be important. See the EMC Fundamentals chapter for details on common-mode chokes and ferrite selection.

Fortunately for hams, ferrites will oftentimes fix the problem. This is avoided if possible in commercial applications due to the cost of the large quantities used, but in your shack where you only need a few cables, this is a very viable fix. If the cable is unshielded (audio cables) this may be your only fix.

Let's say the problem was above 200-300 MHz and we don't suspect that cables are the problem. Now what? This is where some troubleshooting tools need to be made and used. Leaks in a chassis can easily be found by building sniffer probes, both electric and magnetic field. There are test equipment manufacturers who make such probes, but effective ones can be made from stuff you already have in your shack—coaxial cable. A simple electric field probe is nothing more than the end of the cable with the shield cut back an inch or so and then wrapped in electrical tape to insulate it from accidental contact with live circuits that could damage the front end of your receiver. Figure 15.5 shows what this looks like before the tape is added.

A simple magnetic field probe takes another minute or so to make. Form a loop at the end of the cable with the center conductor soldered to the shield of the cable. One half to 1 inch in diameter is plenty. Figure 15.6 shows an example of this probe. RG-58 coax works great for both of these applications. Now take the probe and move it along joints in the chassis while listening to the signal or watching the S meter. Areas of higher noise or S meter readings should be investigated to make sure that the adjacent parts of the chassis are properly bonded together. A common problem is paint overspray so that the mating surfaces are insulated. A little sandpaper will cure this (but make sure you don't leave any residue behind before you turn the computer on again). You want good metal to metal contact on all mating surfaces. A few manufacturers make tape out of copper, with conductive adhesives. Copper tape is not inexpensive, but no EMC laboratory can operate without it. Seams can be covered and their contribution (or lack thereof) to the overall emissions from a product shown. Copper tape is not used in production, but can be a valuable tool in fixing a one of a kind problem.

Interference to Computers

Operating computers in close proximity to radio transmitters presents the possibility of interference to the computer or its peripherals. This is less of an issue than it might have been in past years due to EMC regulations in the European Union. Under the EMC Directive in Europe computers must be tested for immunity to various electromagnetic phenomena in addition to meeting limits for radio frequency emissions. CISPR Publication 24 (upon which EN 55024 in the EU is based) provides test levels for various effects. CISPR is the acronym for the French words for the International Special Committee on Radio Interference. As most major manufacturers of computers try to have a single version for a world wide market, computers purchased in the US likely will have been tested to these levels. A fast way to check is to look for the CE Marking on the product. The letters "CE" are the abbreviation of French phrase "Conformité Européene" which literally means "European Conformity". The CE Mark, as seen at right, is placed on the product by the manufacturer to show that the product meets all directives that may apply to it. In the case of a computer without a modem or RF device this would mean the EMC Directive and the Low Voltage Directive (product safety).

If you see the CE Mark on your computer you know that it has been tested to the levels shown in Table 1.

Different pass/fail criteria apply to these tests. The two radio frequency tests (conducted below 80 MHz and radiated above 80 MHz) allow no reaction by the equipment under test. The ESD, surge, fast transient and ½ cycle voltage dip test all allow the system to react, but it must self recover with no operator intervention. An example is the pop you might hear in the speakers when the ESD gun discharges to the equipment. The 25 cycle and 250 cycle voltage dip and dropout tests allow the system to crash, but no data protected by battery backup may be lost and the system must recover when the operator follows the instructions to bring it back up. In other words, there can be no physical damage nor loss of data that was backed up by a battery.

What does this mean to Amateur Radio operators? Generally, computers that meet CISPR 24 fare well in the ham shack. Monitors have been seen to jump when a nearby handheld receivers is keyed or unkeyed, but the computer itself generally is unaffected. The design features that were incorporated to control emissions also work to improve immunity to these phenomena. If an immunity problem shows up in your shack many of the same techniques discussed above to reduce emissions can be used to improve the situation. It is interesting to note that the 3 V/m radiated immunity test does not adequately demonstrate immunity to the field from a nearby handheld transceiver. The field from a Yaesu FT-727R dual band handheld transceiver was measured in an EMC facility a number of years ago and the field at a distance of 10 meters was found to be 1 V/m on both the 2 meter and 70 cm bands. Using the 1/R relationship this would predict a field of 10 V/m at 1 meter. Clearly this is a higher level than computers are tested to, but not one that has been shown to cause problems.

One part of your computer may react to fields from your transmitters more than any other part. This is the amplified speakers that are commonly part of a desktop computer system. Again, common-mode chokes will usually come to the rescue. Install them on the speaker wires as close to the amplifier(s) as possible. You may need to install a choke at each end, or just one. Experiment to see what works. As previously discussed in this chapter, use toroids for HF problems. Proper ferrite selection can be essential for optimum results. Several turns through the ferrite or around the toroid may be needed. Typically, six to eight turns on a torroid is often enough for an HF problem.

POWER LINE CONDUCTED EMISSIONS LIMITS

RF energy is put on the power cord from a variety of sources within the computer, but the main source is the switching power supply. This is an RF voltage that is conducted

Table 1
CE Mark Test Critera

Test	*Level*
Electrostatic Discharge	4 kV contact discharge (conductive surfaces) 8 kV air discharge (insulated or painted surfaces)
Radio frequency electromagnetic field, Amplitude modulated	80 to 1000 MHz, 3 V/m 1 kHz tone, 80% modulation
Electric fast transients	500 V (telecommunications ports) 1 kV (AC input)
Surge	1 kV line to line 2 kV line to earth
Radio frequency continuous conducted	150 kHz to 80 MHz, 3 V 1 kHz tone, 80% modulation
Power frequency magnetic field	1 A/m, 50 Hz
Voltage dips and dropouts	>95% reduction, ½ cycle 30% reduction, 25 cycles >95% reduction, 250 cycles

directly through the power cord to your house wiring. Once it reaches your house wiring, it can create an RF field which is picked up by nearby radio antennas, or be conducted directly from the computer to the receiver via the wiring. A combination of conducted and radiated paths is also possible.

The international standard for RF emissions from "Information Technology Equipment" (ITE), a fancy term for computers and their peripheral devices, is CISPR Publication 22. Both CISPR 22 and the FCC Rules provide the same limits for power line conducted emissions from ITE (computers). These limits are shown in Figure 15.7. They cover the frequency range of 150 kHz to 30 MHz. Note that there are two limit lines for Class A and Class B. Emissions are measured to these limits with two different detectors in the EMI receiver, a quasi-peak and an average detector. The quasi-peak detector is a peak detector with a reduced rise and fall time designed to simulate the response of the human ear to impulsive emissions as the repetition rate of the impulses increases. By 10 kHz or so, the quasi-peak and peak detectors read about the same. A single click isn't very annoying; your ear filters it out fairly well. However, as the repetition rate of those clicks increases the effect is far more annoying. This detector simulates the interference potential of a signal to an AM broadcast receiver (or television picture) fairly well and has been in use since the 1930s. The average detector takes a time average of the readings. If a peak detector shows readings below the average limit (the lower of the two) no testing with the other detectors is needed as the quasi-peak and average detectors will never read higher than the peak detector.

Radiated Emissions limits

Both the FCC and CISPR 22 provide limits for radiated emissions from computers. These limits start at 30 MHz. The FCC limits, depending on the maximum clock rate used in the product, go as high as 40 GHz (40,000 MHz). The CISPR limits were amended in 2005 to go as high as 6 GHz, having previously stopped at 1 GHz. These two sets of limits vary a bit with respect to each other from 30 MHz to 1 GHz, one being tighter than the other and swapping this condition at various frequencies. Figure 15.8 shows the FCC and CISPR 22 radiated emissions limits from 30 MHz to 1000 MHz. The FCC Class B limits are specified at a distance between the measurement antenna and the equipment under test (EUT) of 3 meters. The FCC Class A limits and both the Class A and Class B limits in CISPR 22 are specified at a distance of 10 meters. For the purpose of this Figure the FCC Class B limits have been adjusted downwards by 10.46 dB (20 Log (1/R)) to show them at the 10 meter test distance used for the other three limits. Figure 15.9 shows the FCC and CISPR 22 radiated emissions limits from 1 GHz to 40 GHz. The FCC Class A limits are specified at a distance of 10 meters, while the other three limits are specified at a distance of 3 meters. For the purpose of this Figure, the FCC Class A limits have been adjusted upwards by 10.46 dB to show them at the 3 meter test distance used for the other three limits. Testing to these limits between 30 MHz and 1 GHz is done with a quasi-peak detector and from 1 GHz to 6 GHz (CISPR 22) or 1 GHz to 4 0 GHz (FCC) testing is done with an average detector.

TESTING

A computer (or its peripheral devices) is tested as part of a representative system. Peripheral devices such as a keyboard, mouse, monitor and other devices are connected and the system is exercised with software th[illegible] communicates with the devices. For con[illegible] ducted emissions testing of table top equip[illegible] ment (such as personal computers) the sys[illegible] tem is placed on a non-conducting table, 8[illegible] cm in height and arranged in a controlle[illegible] manner. This is done with a vertical con[illegible] ducting plane (e.g. wall of a shielded room placed 40 cm behind the Equipment Unde[illegible] Test (EUT). Power is supplied through [illegible] device called a Line Impedance Stabilization Network (LISN) in the US and an Artificial Mains Network (AMN) in most of the rest of the world. Other than the name, the device is the same. Figure 15.10 shows a schematic diagram of an LISN for two leads, both the hot and the neutral. The LISN provides some filtering of RF energy coming from the mains, and in conjunction with the input impedance of the EMI receiver or spectrum analyzer present a 50 ohm load to RF current flowing from the EUT. The voltage generated across this 50 ohm load (input impedance of the receiver) is what is measured. The unused output is terminated in a 50 ohm load as shown. As long as the measured emissions are not above the stated limit, the EUT is deemed to pass this test.

Testing of radiated emissions is a bit more complex and takes significantly more time. The non-conducting table holding the EUT is placed on a turntable, either at an Open Area Test Site (OATS) or in an RF semi-anechoic chamber. The OATS is the "gold standard" test site for these measurements, but typically suffers from interference from ambient signals in the frequency range of interest such as FM and television broadcast stations (the primary source of interference to the measurements) and other sources. An OATS is simply a large clear area, free of objects above the ground that would reflect RF energy. The OATS is equipped with a flat metallic ground

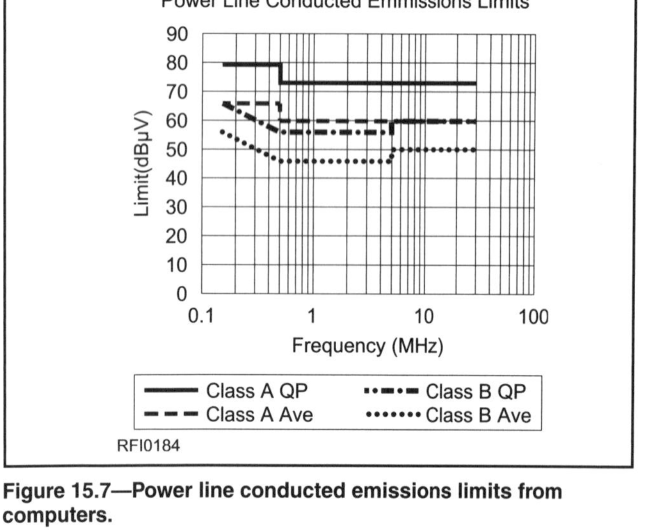

Figure 15.7—Power line conducted emissions limits from computers.

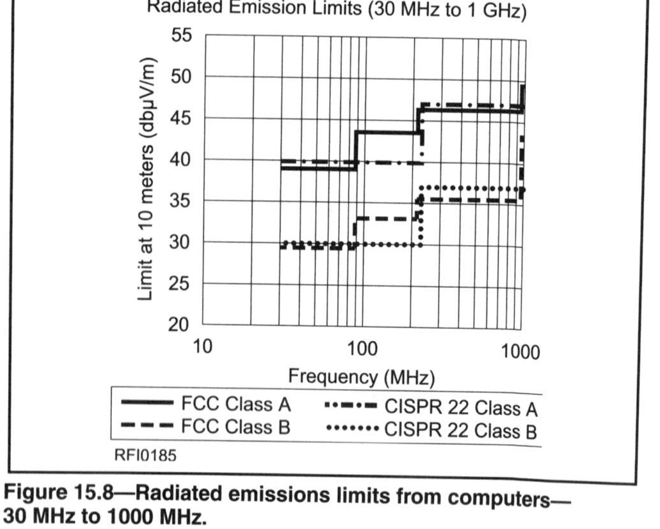

Figure 15.8—Radiated emissions limits from computers—30 MHz to 1000 MHz.

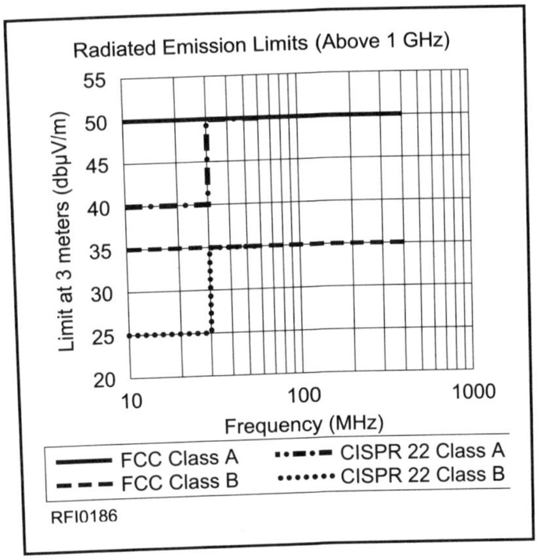

Figure 15.9—Radiated emissions from computers—above 1 GHz.

plane. Figure 15.11 is a photograph of a 30 meter OATS (30 meters from antenna to the center of the turntable) that was constructed in 1989. The turntable is located inside the RF transparent structure on the right of the picture. The test equipment is located in the building in the back of the picture. An RF semi-anechoic chamber solves the ambient signal problem by placing the measurement environment inside a Faraday cage[1], thus isolating it from the outside world. However, the walls and ceiling of a shielded room violate the requirement for an OATS that there be no reflecting objects above the level of the ground plane. The chamber gets around this problem by being built with RF absorbing material on the walls and ceiling, thus suppressing the reflections from these surfaces. A good semi-anechoic chamber is indistinguishable from an OATS in its performance against the qualification test for such facilities.

The EUT is rotated to find the azimuth which gives the highest reading at each frequency of interest. Both horizontal and vertical polarities of the receive antenna are investigated. The height of the antenna above the ground plane is varied from 1 meter to 4 meters above the ground to find the highest reading. This is done as there are two paths for RF energy to reach the antenna from the EUT. The first is the direct path, and the second is the reflected path via the ground plane. By varying the height of the antenna the relative lengths of these paths are varied and a point where the two waves add in phase can be found, giving a maximum reading. Height scan or turntable rotation may be done first at the discretion of the test laboratory. Many laboratories automate this process. At an OATS only the final measurements may be automated, but in a semi-anechoic chamber the entire process, including the initial scans to find signals, may be automated. This is a key advantage of RF semi-anechoic chambers. Another is that signals will not be missed by being masked by local broadcast radio transmitters.

Conformity Assessment Programs

After a laboratory has tested a computer and obtained data that shows it to be compliant with the applicable limits one step remains. Approvals from various nations' regulators must be obtained. These processes range from simply labeling the product and selling it without applying to the regulator to having to re-test the product in that country and following a formal certification process. A discussion of all of these processes is beyond the scope of this chapter, but we will discuss the two approval processes commonly used for computer products in the US.

Class A computing devices and some Class B devices (devices other that PCs or their peripherals) are subject to an approval process in the US called verification. This process requires that the manufacturer test the product in a laboratory that the manufacturer is convinced meets the requirements for the lab contained in the FCC Rules. Once the manufacturer has data that shows compliance with the limits, he labels the product, places certain required text in the user documentation and then sells the product. No prior notification to or approval from the FCC is required.

Class B computing devices and their peripherals are subject to an approval process called the Declaration of Conformity. Unlike verification, where any laboratory may be used, the DoC process requires that the laboratory be accredited by an approved accrediting body to the requirements contained in ISO/IEC 17025. This is a major undertaking for a laboratory, requiring the generation of large amounts of paperwork documenting the laboratory's processes. Formal assessments of the laboratory are performed by auditors from the accrediting body. In the US this body is either the American Association for Laboratory Accreditation (A2LA) or the National Institute of Standards and Technology National Voluntary Laboratory Accreditation Program (NIST NVLAP). Internal audits are also required. Once data showing compliance with the limits is obtained from an accredited laboratory, the manufacturer labels the product, places required text in the user documentation and sells the product. Like verification, no notification to or approval from the FCC is required. If your computer or peripheral was approved by the DoC pro-

[1]Faraday cage—a conductive enclosure surrounding the measurement area that isolates that area from the surroundings. RF energy inside the cage stays inside. RF energy outside the cage stays outside.

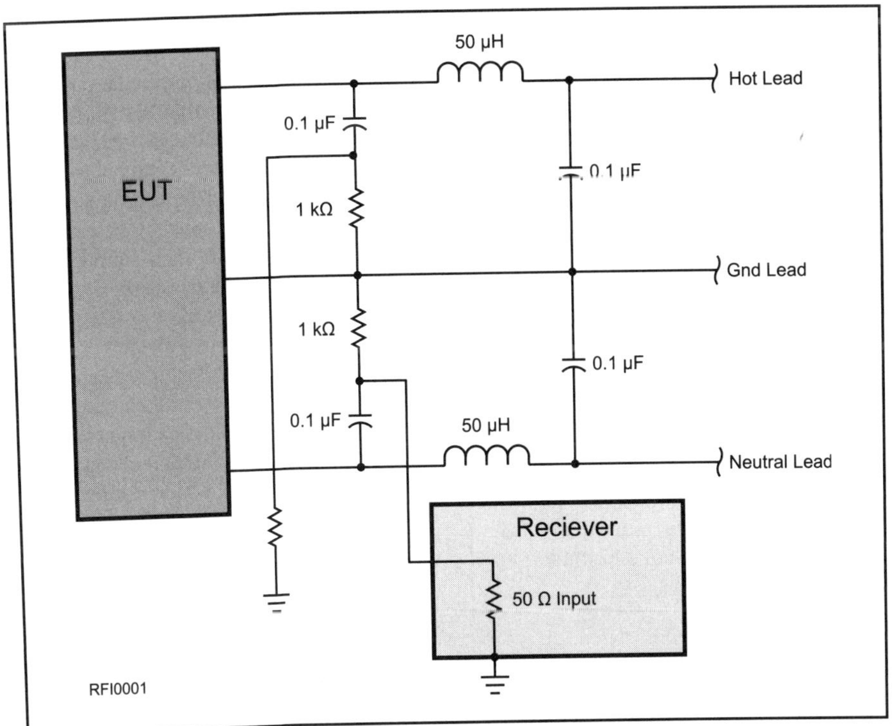

Figure 15.10—Schematic diagram of a LISN for 2 leads (hot and neutral).

Figure 15.11—A 30 meter Open Area Test Site (OATS).

§15.105—Information Required to be Included in User Documentation

(a) For a Class A digital device or peripheral, the instructions furnished the user shall include the following or similar statement, placed in a prominent location in the text of the manual:

Note: This equipment has been tested and found to comply with the limits for a Class A digital device, pursuant to part 15 of the FCC Rules. These limits are designed to provide reasonable protection against harmful interference when the equipment is operated in a commercial environment. This equipment generates, uses, and can radiate radio frequency energy and, if not installed and used in accordance with the instruction manual, may cause harmful interference to radio communications. Operation of this equipment in a residential area is likely to cause harmful interference in which case the user will be required to correct the interference at his own expense.

(b) For a Class B digital device or peripheral, the instructions furnished the user shall include the following or similar statement, placed in a prominent location in the text of the manual:

Note: This equipment has been tested and found to comply with the limits for a Class B digital device, pursuant to part 15 of the FCC Rules. These limits are designed to provide reasonable protection against harmful interference in a residential installation. This equipment generates, uses and can radiate radio frequency energy and, if not installed and used in accordance with the instructions, may cause harmful interference to radio communications. However, there is no guarantee that interference will not occur in a particular installation. If this equipment does cause harmful interference to radio or television reception, which can be determined by turning the equipment off and on, the user is encouraged to try to correct the interference by one or more of the following measures:

—Reorient or relocate the receiving antenna.

—Increase the separation between the equipment and receiver.

—Connect the equipment into an outlet on a circuit different from that to which the receiver is connected.

—Consult the dealer or an experienced radio/TV technician for help.

(c) The provisions of paragraphs (a) and (b) of this section do not apply to digital devices exempted from the technical standards under the provisions of §15.103.

(d) For systems incorporating several digital devices, the statement shown in paragraph (a) or (b) of this section needs to be contained only in the instruction manual for the main control unit.

(e) In cases where the manual is provided only in a form other than paper, such as on a computer disk or over the Internet, the information required by this section may be included in the manual in that alternative form, provided the user can reasonably be expected to have the capability to access information in that form.

cess, it will have a label similar to the one seen here.

FC Trade name
Model number

There is a special twist to the DoC process for personal computers in the US. If an integrator (e.g. your favorite local computer store) builds a PC from authorized parts (motherboard, power supply and peripheral devices—defined as internal components with connections for external cables) they simply have to maintain a record showing the authorization status of the parts, place a special label on the product, issue a DoC (usually included in the user documentation) and sell the product. No testing of the completed system is required under this process. The label on the PC would appear as:

FC Trade name
Model number
Assembled from tested components
Complete system not tested

The text called out for inclusion in the user manual is shown in the sidebar. In addition to this text the manufacturer must also provide a warning to the user that unauthorized (by the manufacturer) modifications to the product may void the user's authorization to operate the equipment.

As noted above, the limits that a computer must meet may not provide the level of protection needed when the computer is used in the shack.

SUMMARY

Interference in the ham shack from computers is much less of a problem today than it was when home computers first made their appearance. Due to the international standards and regulations dealing with emissions and immunity for these products they emit much less radio frequency energy than they did in the past and their immunity to various effects is demonstrated. However, in the case where you still have a problem, ferrite, copper tape and sand paper are still valuable tools.

EMC problems can be broken down into three basic parts—source, path and victim. Your computer may be the source, your radio the victim. You can fix the source, modify the path (increase separation) or fix the victim. However you attack the problem, have fun. It is another fascinating corner of this hobby we call Amateur Radio.

Chapter 16

Automobiles

From an RFI point of view, modern automobiles are a combination of mechanical and electronic technology. Most cars rely on computers to control many vehicle functions. In many cases, the only solution to automotive RFI problems is through the manufacturer.

By Terry Rybak, W8TR
PO Box 260, Pinckney, MI 48169-0260
and
Mark Steffka, WW8MS
1226 Hereford Ct, Canton, MI 48187-4676

About the Authors

Terry Rybak, W8TR, and Mark Steffka, WW8MS, wrote this chapter about automotive interference. Terry Rybak has been an Automotive EMC Engineer for 27 years. His previous experience was with the electrical utility industry. He is also a NARTE Certified EMC Engineer and was first licensed in1963.

Mark Steffka, WW8MS, has been in the automotive industry since 1991 and has been an Automotive EMC Engineer since 1997. His previous experience was with military and aerospace electronics. He is also an adjunct faculty member in the Electrical and Computer Engineering (ECE) department at the University of Michigan. He was first licensed in 1975.

BACKGROUND

This chapter describes the background and sources of noise, the diagnostic process that may be used to identify RFI problems and steps to eliminate the problems. This chapter is designed to assist in the diagnosis of interference problems on the vehicle, and to identify a "fix" to be made at the receiving end of the system (the mobile transceiver). Even if the "fixes" are not implemented, the diagnostic information should provide information to help your dealer or service shop locate a vehicle problem quickly and easily.

There are many urban legends about the ham that couldn't operate mobile because of the unsolvable RF noise emanating from his/her vehicle system—as well as an author of this chapter who had windshield wipers that turned themselves on during every transmission. While both of these situations make great stories, it turns out that these are very rare events. Some degree of EMC (Electromagnetic Compatibility) issues can occur in all vehicles, although severity varies with automobile make and model. This chapter will unravel some of the mystery about automotive EMC problems and attempt to answer hams' common questions: "Can I put a transceiver in my car?" and "What can I do if I encounter problems?"

In the years since this chapter was originally written, automotive systems have increased their functionality (and complexity).

The goal of this chapter is to relate some of the basics of RFI/EMC and automotive systems with respect to interactions with our amateur installations, as well as how vehicles have been designed and tested to ensure that properly installed amateur transmitters, of typical power levels and frequencies do not cause issues with vehicles.

Have you ever thought about the fact that today's vehicles represent three centuries of technology? It's true! They range from the internal combustion engine (19th century), to the addition of complex electrical systems and introduction of electronics (20th century), to today's (21st century) safety and comfort features (telematics, satellite radio, remote keyless entry, cruise control, antilock brakes, microprocessors, pulse width modulated motor speed controls, CDs, DVDs, "drive by wire"). Now we are witnessing the changing basics of the vehicle powertrain system itself with the introduction of today's gas/electric ("hybrid") vehicles and await tomorrow's fuel-cell powered vehicles. Your current vehicle may be only a few years old, but it has many features and electronics not found just a few model years before, and is very different from your friend's "well used" 1978 Chevrolet truck!

The installation of amateur equipment on a vehicle can result in very high RF fields around the vehicle, sometimes 10 to 100 V/m (Volts per meter), depending on the output power and/or antenna gain. Because of the wide range of bands available to amateurs, this can occur at many frequencies. This RF environment, caused by on-board amateur installations, can be nearly equal to some of the worst cases caused by off-board sources.

Automobiles—Then and Now

Automobiles of yesteryear used ignition systems identical in principle to modern gasoline powered vehicles. Ignition systems and other vital vehicle components such as the fuel pump, however, were all mechanical or electromechanical. This 1931 Ford Model A for example, uses mechanical switch contacts called "points" in the ignition system. The charging system for the 6 volt battery uses a generator with a commutator and brushes to provide direct current. It's also interesting to note that this vehicle did not have a fuel pump other than gravity.

Radio interference was typically of little or no concern in any of these vintage automobiles. As the ignition system photo shows, the ignition wires were simply bare copper straps from the distributor to the spark plugs. Today's modern microprocessor and electronic automobile systems would hardly have been recognizable in 1931!

The radio on the running board is a 1925 Thompson Neutrodyne with a Bristol loop antenna and speaker horn. Photos are courtesy of Marion Webster of *Marion's Attic*. Visit Marion's Web site at **members.aol.com/marionweb/**.

THE FIRST STEP: SHOPPING FOR A VEHICLE

Start with your dealer. The manufacturer of each vehicle is the expert on how that vehicle will perform. The dealer should have good communication with the manufacturer and should be able to answer your questions. Ask about service bulletins and installation guidelines. You can also ask your dealer about fleet models of their vehicles. Some manufacturers offer special modifications for vehicles intended for sale to police, taxicabs and other users who will be installing radios in their cars.

When shopping for a vehicle, it is useful to take along some portable (preferably battery operated) receivers or scanners and have a friend tune through your intended operating frequencies while you drive the vehicle. Tuning across various frequencies will help identify any "radiated" noise issues associated with that model vehicle. Since radiated noise can be more difficult to resolve than "conducted" noise, this process will give you an early indication of what to expect from that particular model. If you intend to make a permanent transceiver installation, give some consideration to how you will route the power and/or antenna cables. While looking for ways to run the wiring, keep in mind that some newer cars have the battery located in the trunk or under the rear seat, and that may make power wire routing easier.

Test the car before you buy it. A dealer expects you to take the car for a test ride; a cooperative dealer, who wants to make a sale, may let you test it for radio operation, too. Of course, the dealer can't permit you to do a full installation of your radio; the car would no longer be new when you were done. But you can do a fair amount of checking without digging too deeply into the car. You can check the vehicle for noise with a portable receiver. On VHF, your handheld transceiver will do the job nicely. On HF, you can usually locate a portable short-wave receiver, or operate your HF transceiver with a portable antenna and cigarette-lighter plug. With the engine running, tune across the bands of interest. You may hear some noise, or a few birdies, but if the birdies don't fall on your favorite net frequency, this is an encouraging sign.

To test the vehicle for susceptibility to your transmitted signal, you must transmit. To do this, you will need to bring a 12-volt capable radio, a separate battery[1], a power connector and a magnet-mount antenna (several *QST* advertisers sell mounts suitable for HF). Use the magnet-mount carefully; it is possible to scratch paint if any particles of dirt get on the bottom of the magnets. Use caution when drawing power through the cigarette lighter; they are not rated for the current some transmitters will draw. Transmit on each band you will use to see if the RF has any effect on the vehicle. This test is not an absolute guarantee because an installed transmitter connected to the vehicle's power source may act a bit differently, but if the car is basically immune to transmitter RF, any problem can usually be cured with a power-lead filter on the transmitter.

On both transmit and receive, you may want to experiment with the placement of the antenna. You may be able to find the magic spot that reveals the car is right for you.

Some dealers will be very cooperative (perhaps someone on their staff is a ham!); others will need a bit of persuasion and others may not permit these sorts of tests at all. Even that is important to know because if a dealer is not willing to work with you when you are standing in the parking lot with a pocketful of money, you have probably learned all you need to know about the level of support you might receive after the sale is complete.

[1]The recommended procedure is to connect directly to the battery terminals for a two-way radio power connection. This may not be possible for a test drive (How would you route the cables?); and the cigarette lighter/power port may not supply enough current.

Things to Know About Installing a Transceiver in Your Vehicle

While most amateurs are familiar with the process of installing a transceiver, there are some preferred practices that will help minimize potential problems. These include the support that may be expected from the automotive dealer, typical "best practices" installations, and considerations of special situations.

Manufacturer's Recommended Installation Requirements

The ultimate authority on whether a radio transmitter can be installed in cars is, naturally, the vehicle manufacturer. Surprisingly, vehicle manufacturers' policies about all types of aftermarket equipment, especially radio transceivers, vary widely. An article published in the September 1994 issue of *QST* detailed manufacturers' policies about installing transmitters in their vehicles. Some manufacturers indicated that they had published installation guidelines (information on how to obtain these is at the end of this chapter). Others, however, made it very clear that problems or damage caused by the installation of any aftermarket equipment would not be covered under warranty. One vehicle manufacturer even told us that the answers to the questions we asked were "proprietary"!

The first step is to ensure that your installation is in compliance with both the vehicle manufacturer and radio transceiver manufacturer's installation guidelines.

INTERFERENCE FROM AUTOMOBILE SYSTEMS TO AMATEUR RADIO

Vehicle system types

The application of many distributed electrical and electronic systems in vehicles can affect mobile amateur radio operators. Active electronic components have replaced many of the functions that were once controlled by mechanical linkages and systems. Not all noise is from the traditional sources such as ignition and motors. Some may be from various instruments and gauges, or from pulse-width modulated lighting systems or pulse-width modulated motor speed controllers on the vehicles. Most electronics modules today contain many active devices, including digital and analog components. These components and assemblies may emit RF noise, and they also need to be immune to external electromagnetic fields. Now add back to the mix the traditional noise sources, such as the numerous electric motors in today's vehicles, and the interference potential becomes much greater than even a few years ago.

There are many different types of systems in vehicles that may cause vehicle RFI issues. Some of the systems include:

- The high voltage system in the spark-ignited engine (traditional in the United States).
- The systems used in compression-ignited engines (commonly existing in diesel engines).
- Fuel injectors on both spark-ignited and combustion-ignited engines.
- Specialized systems in the newer hybrid vehicles that use both high power motors and internal combustion engines, as well as switching circuits for high voltage and high current.
- The microprocessor based engine control systems.
- Chassis control systems like HVAC, adaptive suspension, and anti-lock brakes.
- Driver information display systems (instrument panel).
- Adaptive cruise control.
- RADAR obstacle detection.
- Remote keyless entry.
- Key fob recognition systems.
- Pulse width modulated motor speed controls.

Noise Types

Most radio installations should result in no problems to either the vehicle systems or any issues with the transceiver. In those situations where issues do occur, the vast majority of issues are with interference to the receivers from the vehicle "on board" sources of energy that are creating emissions within the frequency band used by the receiver. In order to resolve these issues, it is useful to know about the nature of the noise types. The traditional way to describe the noise is to characterize it as one of two types:

- Broadband
- Narrowband

Broadband Noise

Broadband noise is defined as noise having a bandwidth much greater than the affected receiver's bandwidth and is reasonably uniform across a wide frequency range. For practical purposes, ignition and similar pulse-type noise, and brush-type motor noise, can be considered broadband. These systems are also referred to by their operation—many of them generate an electrical arc or a spark, and are called "arc and spark" devices. Not all broadband sources are "arc and spark", since today many devices and on-vehicle data communication signals may also consist of pulse-width modulated (PWM) signals. PWM signals exhibit a noise interference potential similar to broadband noise (see Figure 16.1).

Narrowband Noise

Narrowband noise is defined as noise having a bandwidth less than the affected receiver's bandwidth, shown in Figure 16.2. Narrowband noise consists of noise that is present on specific, discrete frequencies or groups of frequencies, with or without additional modulation. In other words, if you listened to it on a receiver, it would be tunable, just as if it were a desired signal. Microprocessor clock harmonics are an example of narrowband noise, and unlike today's PC clock speeds at or above the GHz range, automotive microprocessor clock speeds are still at approximately 50 MHz and below. These speeds are because of engineering issues involved in designing a computer that operates reliably over a wide temperature range and other harsh environmental factors. With these clock speeds, and from knowledge of clock speeds and harmonics, it's easy to see why there can be RFI across several amateur bands.

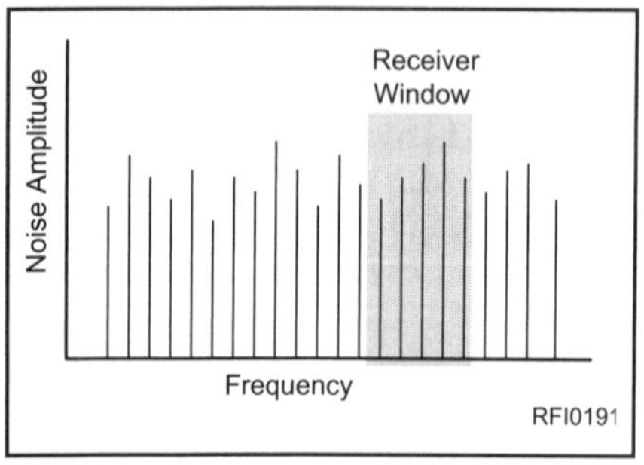

Figure 16.1—Broadband Noise

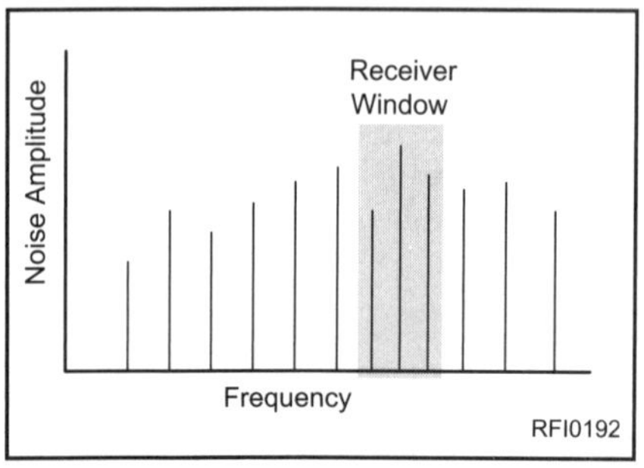

Figure 16.2—Narrowband Noise

Is it broadband or narrowband?

The first step in trying to identify the source of the noise is to determine its characteristics as described in the previous discussion. A straightforward and very effective way to do this is to use a battery operated portable broadcast receiver(s). As it turns out, an older "non-digital" receiver actually works best for this test for two reasons. First, since they have a continuous tuning capability, they will tune to frequencies between the allocated broadcast frequencies. Second, they do not have auto-mute or noise reduction circuits. If there is no noise heard in the following test, then it is almost a guarantee that any amateur transceiver will work very well in the vehicle.

1. Operate the vehicle systems and listen for noise on the radio audio. If RFI is detected by tuning across the band you can tell if it's "narrowband" or "broadband".

2. If it is *not* possible to "tune out" the noise no matter where the receiver dial is set, then it is probably broadband noise.

3. If only one or two stations are affected by the noise, then it is probably "narrowband" noise.

These steps can also be performed with handheld transceivers for any of the amateur bands, which will tell you specifically about the RFI that may be expected for actual installations.

EXAMPLES OF BROADBAND NOISE

Ignition Noise

A review of typical ignition systems and how they operate reveals how the various suppression schemes work. Figure 16.3 depicts systems present in most vehicles. The first

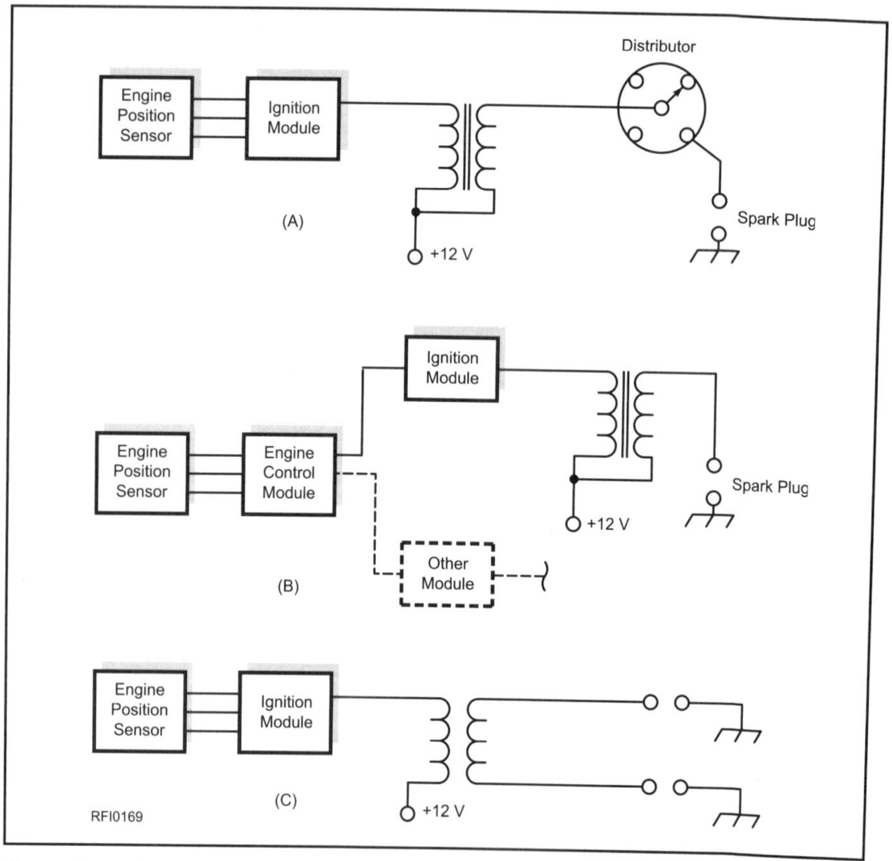

Figure 16.3—System diagrams representative of most vehicle ignition systems. At A, a distributor ignition system, which is common on older vehicles and some large-displacement contemporary engines. B shows an electronic-ignition system, found on most newer automobiles. C shows an electronic ignition system with two plugs per coil.

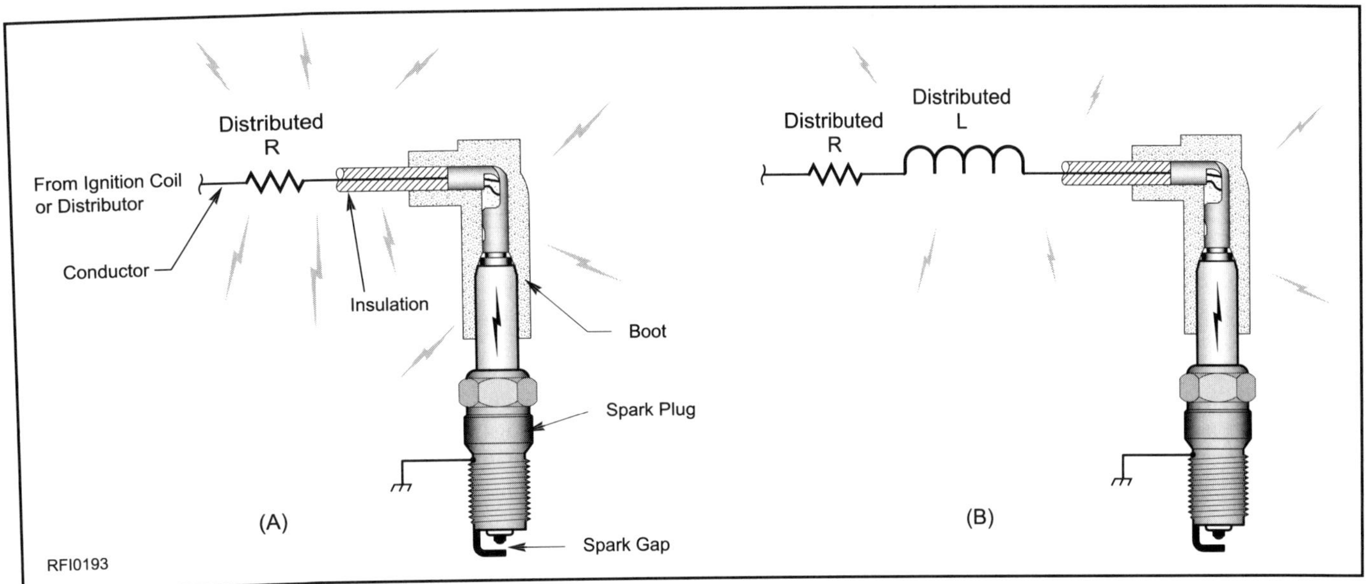

Figure 16.4—Ignition noise suppression methods.

system, Figure 16.3A, is used in older vehicles and a few new ones with large-displacement engines. Newer electronic ignition systems omit the distributor in favor of a coil per cylinder (Figure 16.3B) or coil per pair of cylinders (Figure 16.3C). During operation, a sensor in the distributor, or a crankshaft position sensor, sends a trigger signal to the engine computer and to an ignition module (essentially a high-power dc amplifier), which then drives a high-current pulse through the coil primary winding, T1. The secondary voltage can be 5 to 40 kV, depending on the engine operating conditions. The secondary current travels through the plug wire to the spark plug. For the configuration in Figure 16.3A, the secondary current also travels through the distributor rotor and its gap in the distributor cap before reaching the plug wire.

The fast-rise-time pulses of the coil current discharging across air gaps (distributor and spark plug) create broadband ignition noise. The theoretical models (zero rise time) of such pulses are called *impulse functions* in the time domain. When viewed in the frequency domain, the yield is a constant spectral energy level starting nearly at 0 Hz and theoretically extending up in frequency to infinity. In practice, real ignition pulses have a finite rise time, so the spectral-energy envelope decreases above some frequency.

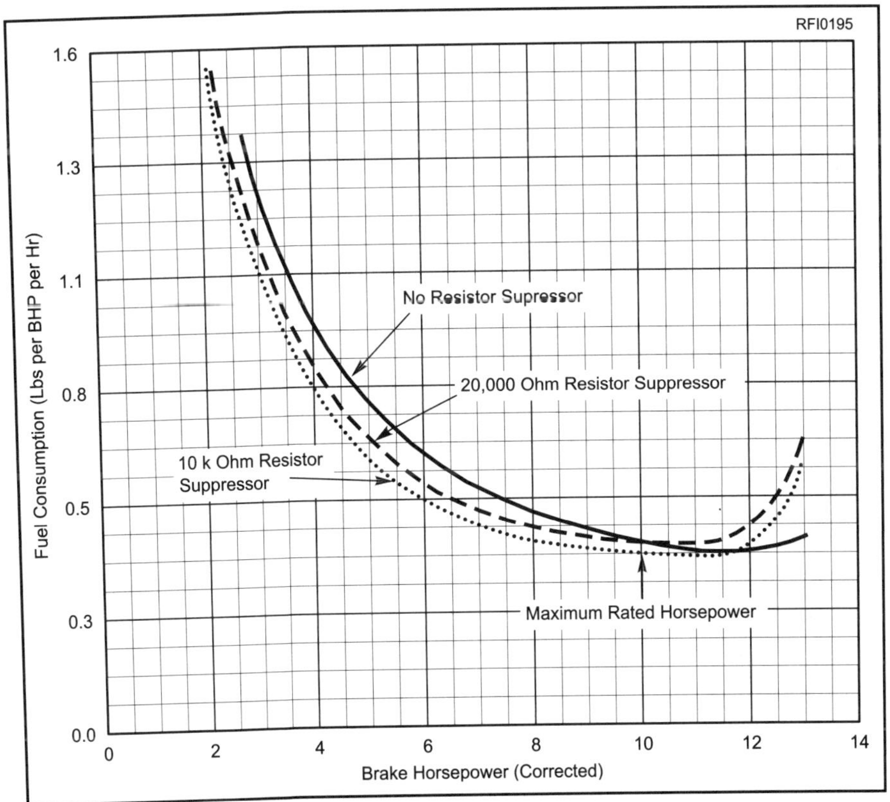

Figure 16.5—Fuel consumption versus brake horsepower for various ignition secondary impedances.

It turns out that both ignition spark generated noise and fuel injector noise manifest themselves as a regular, periodic "ticking" in the receiver audio output, which varies with engine RPM. If an oscilloscope were connected to the audio output, a series of distinct, separate pulses would appear. At higher speeds it sounds somewhat musical, like alternator whine, but with a harsher note. One distinguishing feature of ignition noise is that it increases in amplitude under acceleration. This results from the increase in the required firing voltage with higher cylinder pressure. Since ignition noise is usually radiated noise, it should disappear when the antenna element is disconnected from the antenna mount. The radiation may be from either the secondary parts of the system, or may couple from the secondary to the primary of the coil and be conducted for some distance along the primary wiring to the ignition system, and then radiated from the primary wiring.

Parameters in Secondary Ignition Systems

Radiated emissions occur typically above 30 MHz, and conducted emissions at lower frequencies because of the length of the path along the wire leads and cables con-

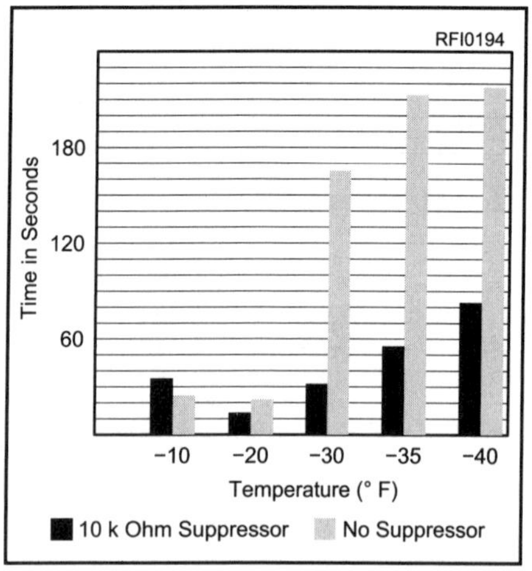

Figure 16.6—Time to start a cold engine versus temperature with and without a 10k Ω ignition secondary impedance.

nected to the various components. For the most part, because of the effect of the shielding that takes place by the body panels, adjacent vehicles or houses along the roadside are usually unaffected. On systems such as motorcycles that do not have an engine compartment to provide shielding, the spark plug may have a shield over it. The purpose of ground straps is to eliminate the potential across the gap. Otherwise, the gap may behave like a "slot" antenna. For years, there was also the common practice of adding ground straps from various vehicle body components to eliminate the voltage potential across the gap between the components. (This is not recommended, as it can defeat the rust prevention protection of body panels on today's vehicles by setting up a small electrical current flow.)

Specifics on the Source of RFI on Spark Ignited Engines

The source of the RFI in spark-ignited engines is the point of the breakdown of the mixture into combustion initiation. Ignition noise is caused by the noise present in the rise and fall times of the high-voltage ignition pulses. Most vehicles have low RFI emissions in the amateur bands, since there are industry standards that are used to control the levels of the emissions, primarily from 30 MHz to 1 GHz.

Two main methods are employed to suppress this noise—one involves adding an inductance, and the other involves adding a resistance, both in the secondary (high voltage) wiring. This is shown in Figure 16.4. It turns out that the addition of these elements does not have a measurable effect on the engine operation, because of the time constants involved in the combustion process. The way each method works is actually different. The resistance method suppresses RFI by dissipation of the energy that would have been radiated and/or conducted. Even though the amount of energy dissipated is small, it is still enough to cause interference to very sensitive amateur installations. The other method is via inductance, and even though the energy is not dissipated, the suppression takes place because the time constant resulting from the rise time of the secondary waveform is reduced. The inductor will store the energy for a short time and dissipate it into the ignition burn event, which is a low impedance path, reducing the RFI.

There have been a number of studies of the impact of both these suppression methods upon engine operation, as shown in Figures 16.5 and 16.6. As can be seen, there is no affect on fuel efficiency and, interestingly, engine starting may actually be enhanced at very low temperatures!

For traditional inductive discharge ignition systems, a value of about 5k Ω impedance (either real and/or reactive) in the spark plug provides effective suppression and, with this value, there is no detectable engine operation degradation. (Capacitor discharge systems, in comparison, are required to have very low impedance on the order of 10s of ohms in order to not affect the spark energy delivery amount, so they are not tolerant of series impedance). Most spark plug resistances are designed to operate with several kV across the plug gap, so a low voltage ohmmeter may not give proper resistance measurement results. Actually, the term "resistor wire" is somewhat misleading. High-voltage ignition wires usually contain both resistance and inductance. The resistance is usually built into suppressor spark plugs and wires, while there is some inductance with resistance in wires, rotors and connectors. The elements can be either distributed or lumped, depending on the brand, and each technique has its own merit. A side benefit of resistance in the spark plug is improved electrode wear.

Coil-on-Plug Ignition Noise

Potential fixes include:

1) Add a "bypass" capacitor in the harness near the coil assemblies (assuming there are two banks of 4 coils).

2) Try putting a "clamp on" ferrite on the harness that is attached to the coils to obtain suppression (sometimes 10 dB!). The key to success is to know where the noise "peaks" are and to select the correct ferrite for the frequency. Most ferrites are effective above 30 MHz (the FCC's Part 15 requirements for radiated emissions start at 30 MHz), so a specific type of ferrite may be required in the HF bands below 30 MHz.

The frequency of the noise may also sug-

Coil-on-Plug Ignition Noise

Many newer spark-ignition systems incorporate a "coil on plug" (COP) or "coil near plug" (CNP) approach. There are advantages to this from an engine operation standpoint, and this approach may actually reduce some of the traditional sources of ignition system RFI. This is because of the very short secondary wires that are employed (or perhaps there are no wires—the coil is directly attached to the spark plug). This reduces the likelihood of coupling from the secondary circuit to other wires or vehicle/engine conductive structures.

There will always be some amount of energy from the spark event that will be conducted along the lowest impedance path. It may mean that the energy that would have been in the secondary circuit will be coupled back on the primary wiring harness that is attached to the coils. This means that the problem may go from a radiated to a conducted phenomenon.

Two approaches that have worked are:

1) Placing a clamp-on ferrite on the 12 volt primary harness that is attached to the coils. Depending on the frequency of the noise and selection of the ferrite material, there can be significant improvement (sometimes 10 dB!). Key to this is knowing where the noise "peaks" and selecting the correct ferrite for the frequency. Most ferrites are effective above 30 MHz (the FCC lowest frequency for radiation emissions), and when people try to use those in the HF bands they conclude that it doesn't work. This may be only because of the incorrect ferrite material used.

Note: The frequency of the noise may also tend to suggest some of the suppression requirements. If the noise goes down to the AM band, for example, you may need to use more ferrites, or ferrites having a different mix.)

2) Add a "bypass" capacitor from the 12 volt coil primary wire to ground in the harness near the coil assemblies (there may be 2, 3, or 4 coils). This must be done carefully, however, because it could affect the functionality of the ignition system *and* perhaps most importantly, *may* void the vehicle warranty. This "bypass" capacitor performs the same function that bypass capacitors in any other application perform—separating the noise from the intended signal/power.

gest some of the suppression requirements. If the noise goes down to the AM band, for example, you may need to use more ferrites, or ferrites having a different core material. More on the use of ferrites can be found later in this chapter.

If ignition noise gets worse over time, replace worn ignition components; spark plugs, wires, distributor cap and rotor (if present).

Other Sources of Noise

Also watch for the "ignition noise" that is not caused by the ignition system. Many hams have been fooled into thinking that a noisy electric fuel pump, noisy pulsed fuel injectors, PWM (Pulse Width Modulated) light dimmers or motor speed controls, or some other source was an "ignition noise" problem. The ARRL Lab has received reports of "ignition noise" from diesel engines, which use no electrical ignition system at all.

In many cases, the problem extends beyond "ignition noise." Fuel injectors, also controlled by the engine controller, can mimic ignition noise. They produce a "pop-pop" sound at low speeds and change to a whine as speed increases. Today's alternators are often required to deliver higher current than their older counterparts. This also increases their noise-generating potential.

Identifying the noise source is crucial to fixing it.

Electric Motor Noise

The challenge in using electric motors on vehicles is that the motors can result in both radiated *and* conducted noise, depending on the frequency of the noise (see figures 16.7A and 16.7B).

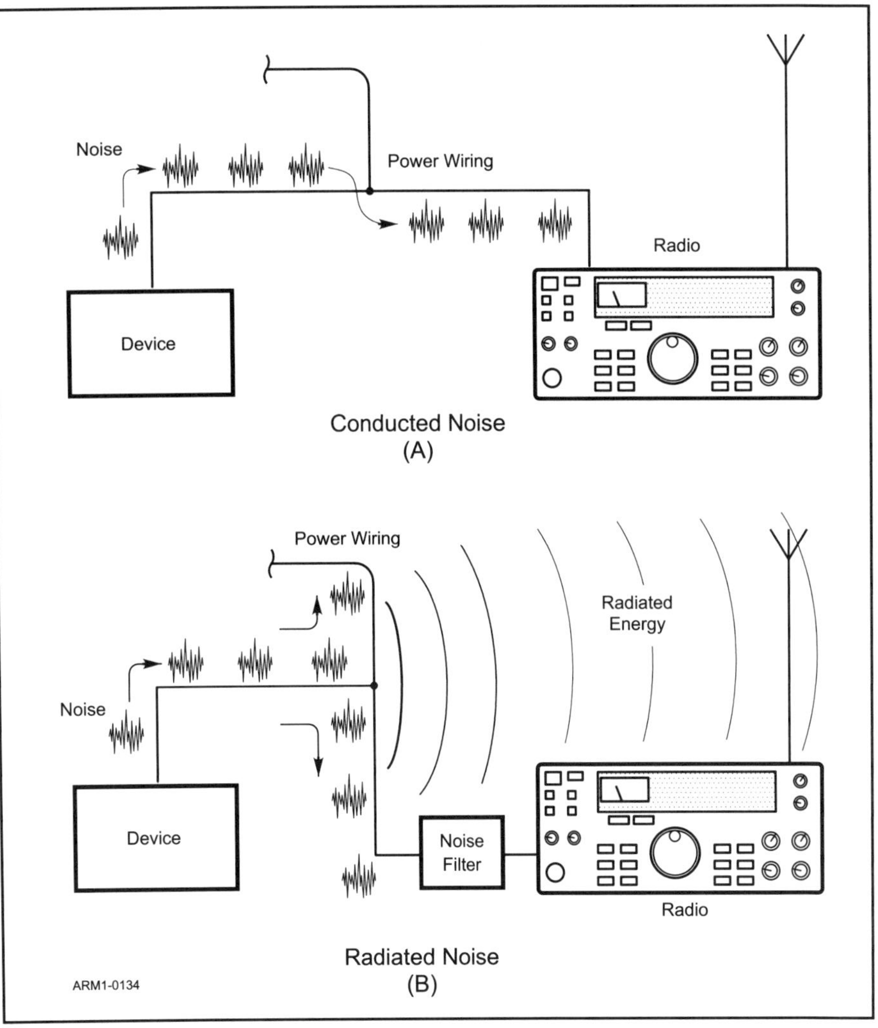

Figure 16.7—Noise from a device such as a motor can be conducted (at A) or radiated (at B) to a nearby radio. The noise filter shown at B is ineffective because noise energy has already been radiated by the quasi-antenna consisting of the long wire leads going to the filter.

EXAMPLES OF NARROWBAND NOISE

Narrowband noise is defined as noise having a bandwidth less than the affected receiver's bandwidth. For practical purposes, all noise other than ignition and similar pulse-type noise, and brush-type motor noise, can be considered narrowband. One characteristic of narrowband noise is that it usually contains discrete frequencies, with or without additional modulation. In other words, if you listened to it on a receiver, it would be tunable, just as if it were a real signal.

Harmonics from crystal oscillators exhibit good long-term stability even at frequencies as high as the two-meter band. Those from ceramic resonators may be less stable, especially with changes in ambient temperature. Stable oscillators may cause interference which always seems to be on your favorite

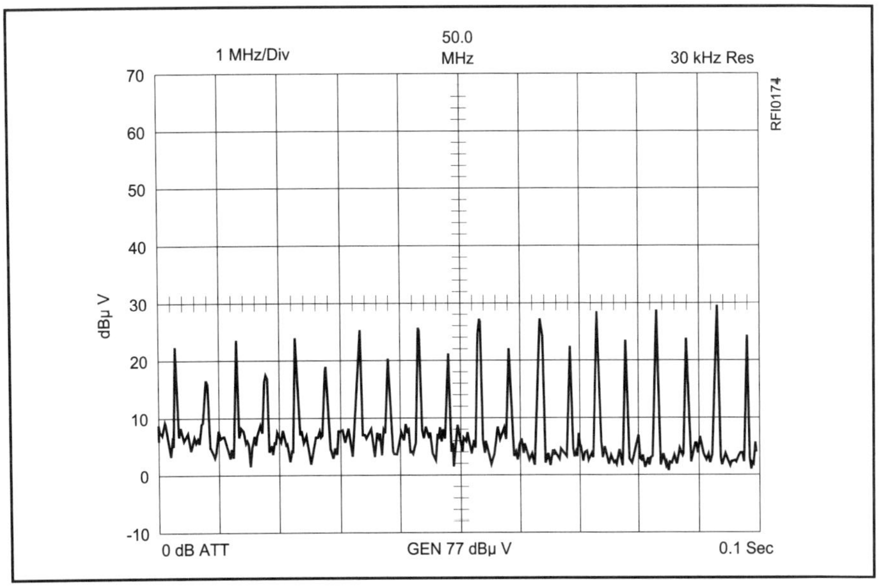

Figure 16.8—Spectral plot of a vehicle electronic module containing a microprocessor with a 1-MHz crystal oscillator.

repeater or simplex frequency, while less stable ones may cause interference only during a portion of the vehicle operation (such as warm-up). The passenger compartment and under-hood temperatures in a vehicle may change rapidly and may cause perplexing intermittent resonator-related EMI.

Microprocessors

A common source of narrowband noise is microprocessor based devices. These devices contain a master clock that generates square waves for timing. These square waves travel from the clock to the logic circuits. As these square waves switch from on-to-off and off-to-on, they draw a transient current from the power supply. Because microprocessors and other LSI chips contain many gates, the current pulses may reach large magnitudes and cause fluctuations in the Vcc power-supply bus. The fluctuations have a considerable number of HF components which may couple to other traces and modulate devices sharing the same power bus.

Clock signals cause the comb-like emissions spectrum characteristic of digital systems. A square-wave clock signal theoretically is composed of the fundamental and all odd harmonics. There may also be pulses below the fundamental if the oscillator signal is frequency divided inside the module to obtain a system clock. This is the case in Figure 16.8 where the crystal frequency is 1 MHz, and the separation between harmonics is 500 kHz. If you have the equipment to view the noise spectrum, such knowledge may help pinpoint the problem processor or module. If viewed on an oscilloscope connected to the receiver audio, the waveform may look like a sine wave, or, if modulated, like a complex wave (see Figure 16.9).

It is also possible to hear the various microprocessors. Vehicle manufacturers design their cars to industry standards and federal regulations. These standards, however, were not designed to offer absolute protection to sensitive amateur receivers. (Imagine the cost that would be involved to ensure that the vehicle emitted no noise on any frequency that might interfere with a QRP transmitter you are hearing from 10,000 miles away on HF!) It is almost certain that a sensitive amateur receiver will hear a number of microprocessor "birdies" on frequencies from HF up into the UHF range. Some of these can be rather strong, especially if your receiver antenna is located close to one of the microprocessors or its wiring. Unfortunately, it is possible that the umpteenth harmonic of one of the microprocessors will fall on your favorite local, 2-meter repeater.

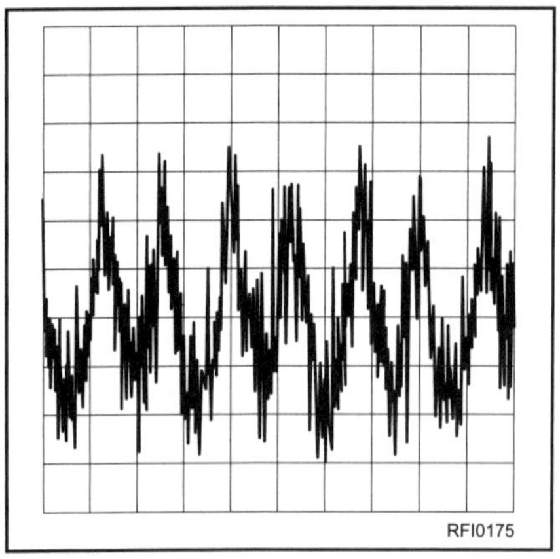

Figure 16.9—A time-domain plot of clock harmonics from the receiver audio output showing various modulation effects.

THE ROLE OF VEHICLE WIRING IN EMC

The module Vcc line is usually the most significant radiator, but other lines such as data, clock, and the I/O strobes may also be significant. The physical dimensions, layout, and filtering of the PC board traces play a large part in determining how much RF is generated. Energy may escape from the module directly by radiation, or by conduction onto the wiring harness.

Wiring harnesses may transport emissions from vehicle electronics. The term "common mode" refers to current that flows in phase (in the same direction) along all wires in a harness. "Differential-mode" currents flow 180 degrees out of phase (in opposite directions) in some wires of a harness. (See the EMC Fundamentals chapter for more information.)

Common-mode RF current is driven along the wiring by very small voltage differences between the module ground and the vehicle structure. The wiring may radiate energy, just like an antenna, and may couple to radio power supply wiring causing conducted noise, or may couple to the antenna causing radiated noise.

Besides being potential sources of radiated noise received by an on-vehicle antenna, all modules are powered from the vehicle power supply and may be a source of conducted noise to a receiver operating on the vehicle's 12 Volt supply.

As a rule of thumb, a conductor longer than 0.1 λ makes a good antenna. Traces on larger boards, like in an instrument cluster, may reach 60 cm (about 25 inches) in length. By the rule of thumb, the cluster could radiate noise above 50 MHz. This size board is a worst-case situation from a direct-radiation standpoint. Most modules are much smaller, with correspondingly shorter conductors (5 to 15 cm [2 to 6 inches]) and direct emissions are of most concern at VHF and UHF. If this is true, why do small modules cause EMI at frequencies well below those that could be radiated by their PC boards? The answer is *coupling*.

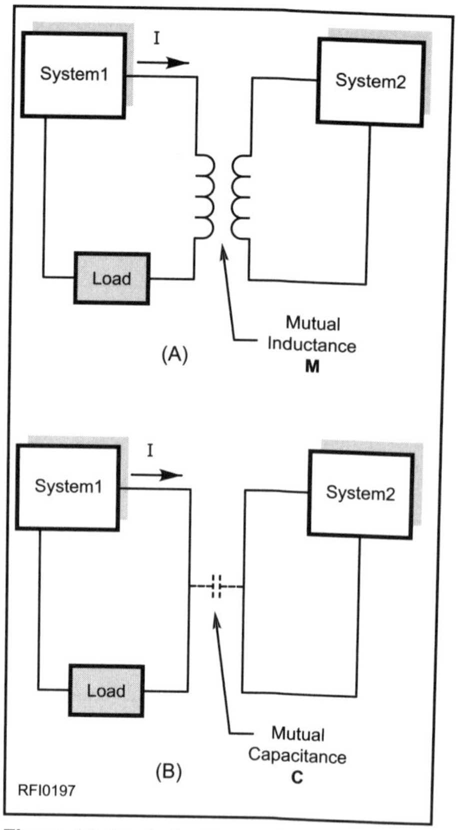

Figure 16.10—Inductive and capacitive coupling

To understand the coupling mechanism, visualize a noisy module as one or more noise-voltage sources that are magnetically and electrically coupled to the wiring harness via the PCB traces (see Figure 16.10). The coefficient of magnetic coupling, M, depends on the area of the loops formed by the two paths. The capacitive coupling depends on the length and separation of the traces. By minimizing the coupling parameters, the noise coupled out of a module and then radiated by the wiring harness may be reduced.

RESOLVING NOISE PROBLEMS IN GENERAL

Work with the Dealer

The dealer should be the first point of contact because the dealer should have access to information and factory help that may solve your problem. Much of this chapter is about how to work within that process, and how to hopefully make the process work if it breaks down. The manufacturer may have already found a fix for your problem (this applies to all problems, not just ignition noise) and may be able to save your mechanic a lot of time (saving you money in the process). If the process works properly, the dealer/customer-service network can be helpful.

Old time fixes are not always good today. Fewer specific vehicle fixes are mentioned in this chapter than in previous years. Simple fixes employed in earlier vehicles may produce unintended consequences with today's vehicles.

In amateur publications, many old-time RFI fixes involved bonding various body parts together, with the intent of improving the grounding. With today's vehicles, adding ground straps between exhaust systems and chassis, or between body panels, may defeat the electrical isolation designed in for corrosion control. Bypassing "noisy" wires could add enough capacitance to affect a data communication line. (Modules required for the vehicle to operate will no longer be able to communicate with one another.) Shielding the ignition system to reduce interference was once popular. With today's higher-energy ignition systems, shielding may greatly reduce spark energy without the benefit of lower noise. Bypassing motor power leads to ground to reduce motor noise has worked well with brush-type dc motors. However, if the motor speed is controlled by a PWM (Pulse-Width Modulated) signal, bypassing could detrimentally increase device dissipation. Fixes that solve problems at a specific frequency of concern to a vehicle owner (for example, adding or removing electronic module grounds or adding bypassing or series inductance at electronic module connectors) may simply move problems to other frequencies.

If you perform your own RFI work, in or out of warranty, you assume the same risks as you do when you perform any other type of automotive repair. Most state laws, and common sense, say that those who work on cars should be qualified to do so. In most cases, this means that work should be done either by a licensed dealer or automotive repair facility.

Future sections of this chapter deal with a general, educational overview of the types of EMC techniques that are designed into today's vehicles and an overview of the types of fixes that may be found in manufacturer's Technical Service Bulletins (TSBs) and used by their repair staff.

General Troubleshooting Techniques

An important aspect is the "source-path-victim" model. The path from the source to the receiver may be via radiation or conduction (as also discussed in detail in previous chapters). If the path is radiation, the field strength received is reduced as a function of the distance from the source to the receiver (in the "far field" it is a linear inverse relationship). If the source is in the "near field", as on most on-board amateur installations, then the relationship is non-linear. In addition, many automotive RFI issues are caused in the "near field reactive" region in which the fields can be very complex. The relationship between the near and far field is shown in Figure 16.11.

The best part of all this is that, with a general-coverage receiver or a spectrum analyzer, a fuse puller and a shop manual, the vehicle component needing attention may be identified using a few basic techniques. The only equipment needed could be as simple as:

- A mobile rig, scanner or handheld transceiver, or
- Any other receiver with good stability and an accurate readout, and
- An oscilloscope for viewing interference waveforms

Broadband Noise

If you suspect electric motor noise is the cause of the problem, one of the easiest steps is to obtain a portable AM receiver to check for this condition. Switch on the receiver, and then activate the electric motors one at a time. When a noisy motor is switched on, the background noise increases. It may be necessary to rotate the radio, since many radios use a highly directional ferrite rod antenna.

To check whether fuel pumps, cooling fans, and other vehicle-controlled motors are the source of noise, pull the appropriate fuse and see whether the noise disappears.

A note concerning fuel pumps: virtually every vehicle made since 1982 has an electric fuel pump, fed by long wires. It may be located inside the fuel tank. Don't overlook this motor as a source of interference just because it may not be visible. Electric fuel-pump noise often exhibits a characteristic time pattern. When the vehicle ignition switch is first turned on, without engaging the starter, the fuel pump will run for a few seconds, and then shut off when the fuel system is pressurized. The noise will follow this pattern, being present for a few seconds, and then stopping

Narrowband Noise

Once the noise source is tuned in and has stabilized, find the vehicle fuse panels and pull one fuse at a time until the noise disappears. If more than one module is fed by one fuse, locate each module and unplug it separately. Some modules may have a "keep-alive" memory that is not disabled by pulling the fuse. These modules may need to be unplugged to determine whether they are the noise source. Consult the shop manual for fuse location, module location, and any information concerning special procedures for disconnecting power.

A listening test may verify alternator noise, but if an oscilloscope is available, monitor the power line feeding the affected radio. Alternator whine appears as full-wave rectified ac, coupled with commutating pulses, superimposed on the dc level (see Figure 16.12).

Alternators rely on the low impedance of the battery for filtering. Check the wiring from the alternator output to the battery for corroded contacts and loose connectors when alternator noise is a problem.

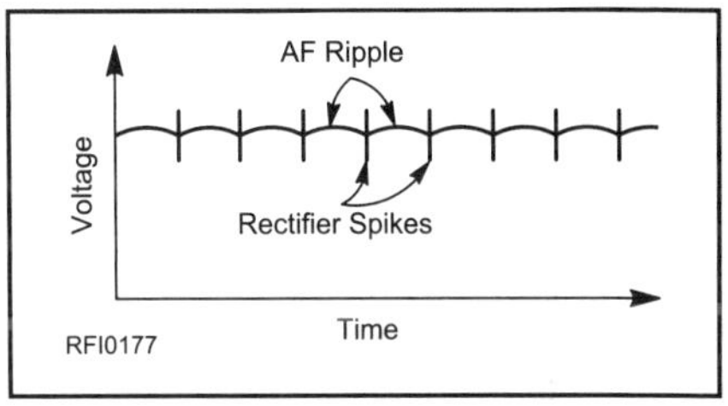

Figure 16.12—A time-domain plot of alternator whine contains full-wave rectified ac, along with rectified pulses.

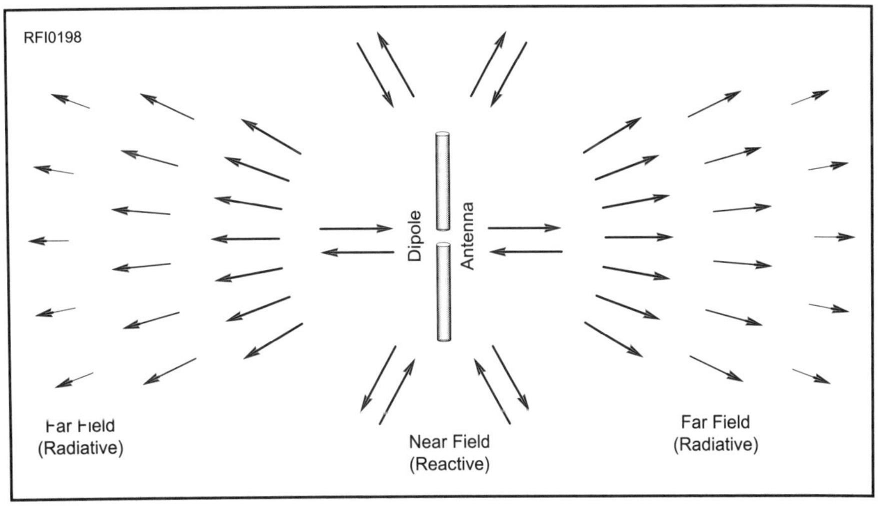

Figure 16.11—The flow of RF energy around a short dipole antenna.

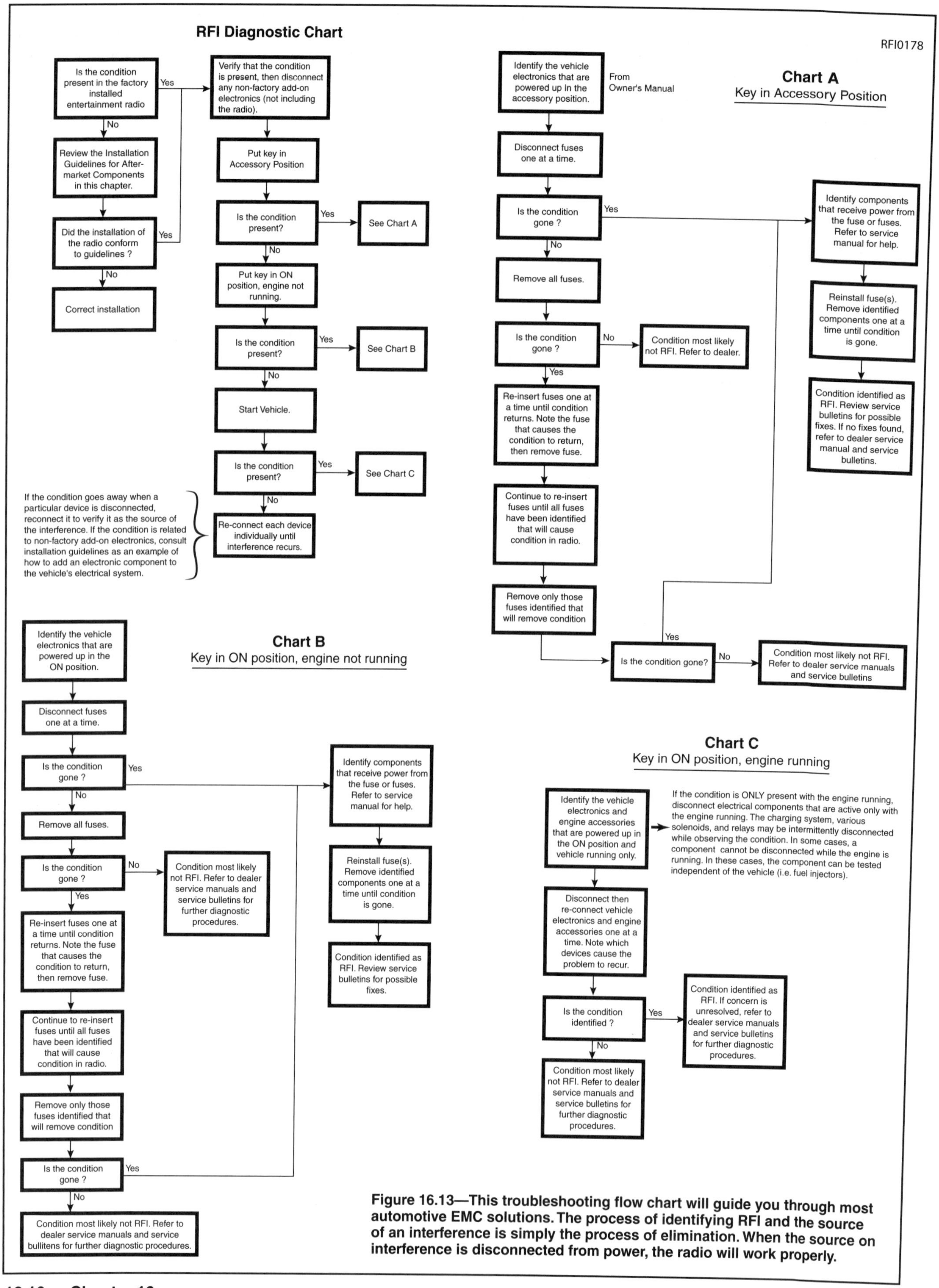

Figure 16.13—This troubleshooting flow chart will guide you through most automotive EMC solutions. The process of identifying RFI and the source of an interference is simply the process of elimination. When the source on interference is disconnected from power, the radio will work properly.

Receivers may allow conducted harness noise to enter the RF, IF, or audio sections (usually through the power leads), and interfere with desired signals. To ascertain this, check whether the interference is still present with the receiver powered from a battery or power supply instead of from the vehicle. If the interference is no longer present with the receiver operating from a battery or external supply, the interference is conducted via the radio power lead. Power line filters installed at the radio may resolve this problem.

The RFI diagnostic chart shown in Figure 16.13 should help guide you through this process.

Resolving Conducted Interference

To reduce common-mode current, impedance can be inserted in series with the wiring. The simplest and most widespread technique is the use of ferrite in the form of common-mode chokes. Ferrite works well in this application because its impedance is composed of resistive and reactive components that combine to yield significant impedance over a wide bandwidth. The result is broadband suppression.

Many companies make split ferrite cylinders to enable installation over entire wire bundles without cutting any conductors. Any number of cylinders may be added for increased attenuation. Bundles may also be wound around large toroids for the same effect. Figure 16.14 shows several ferrite suppressors and their application to wiring harnesses. Different ferrite mixes yield a wide range of permeabilities and cover frequencies in the HF and VHF spectrum. The choice of material depends on the range of frequencies that the choke must suppress. Curves showing effective permeability versus frequency are used to select the appropriate ferrite material. (See the EMC Fundamentals chapter.)

Ferrite chokes, however, do have their limitations. On HF, a single ferrite "bead" usually doesn't have enough inductance or resistance to be a good common-mode choke. Even more important are mechanical considerations. A motor vehicle is subject to a *lot* of vibration. If a choke is installed on a wire, this vibration may cause the choke to flex the wire, which may ultimately fail. It is critical that any additional shielding and/or chokes placed on wiring have been installed by qualified personnel who have considered these factors. These must be properly secured, and sometimes cable extenders are required to implement this fix.

Filtering Using Ferrites

Ferrite toroids have a nomenclature that can help with selecting a suitable core. Let's take the example of FT-240-73:

FT—"Ferrite Toroid". (Beads start with an "FB".)

240 is the outside diameter of the toroid in inches. The OD in this case is 2.4 inches.

73 is the ferrite material mix. A good mix for general purpose HF is type 73 material. Not all ferrite suppliers carry this mix. Use type 75 material for AM broadcast, 80 and/or 160-meter signals.

Select type-43 ferrite material for VHF signals, or HF and VHF signals.

Be sure to select toroids that will be large enough for your cables and connectors. You will need an ID that is large enough for several turns of cable—*plus the connector*. An alternative is to use a "split" toroid which has two halves that are held together by a clamp. You may want to favor 240 and 140 size toroids for most applications, which are the largest that are typically available. Smaller toroids can be used in tight spaces or in an "aesthetically sensitive" area as long as the wires and connectors are small enough.

To make a common mode choke, wrap 10 to 15 turns of the power cord or cable around the ferrite core. The common-mode chokes should be installed right at the source or (as near to as practical to) the device.

The ARRL RFI Web site (**www.arrl.org/tis/info/rfigen.html**) contains a list of EMI/RFI materials suppliers for ferrite chokes. You can also refer to the advertisements in *QST*—there are a few advertisers offering ferrite materials and chokes.

Filters for DC Motors

If the motor is a conventional brush- or commutator-type dc motor, the following cures shown in Figure 16.15 are those generally used, although as always, the mechanic should consult with the vehicle manufacturer. To diagnose motor noise, obtain an AM or SSB receiver to check the frequency or band of interest. Switch on the receiver, and then activate the electric motors one at a time. When a noisy motor is switched on, the background noise increases.

The pulses of current drawn by a brush-commutator motor generate broadband RFI that is similar to ignition noise. However, the receiver audio sounds more like bacon frying rather than popping. With an oscilloscope connected across the receiver audio, the noise appears as a series of pulses with no space between the pulses. Such broadband noise generally has a more pronounced affect on AM receivers than on FM because

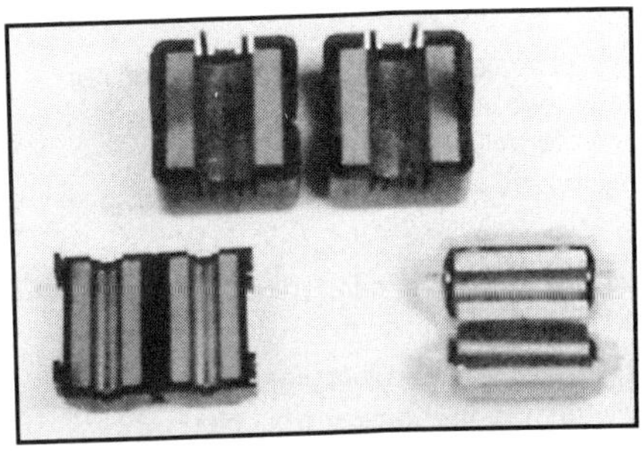

Figure 16.14—Examples of clamp-on ferrite cores for RFI suppression.

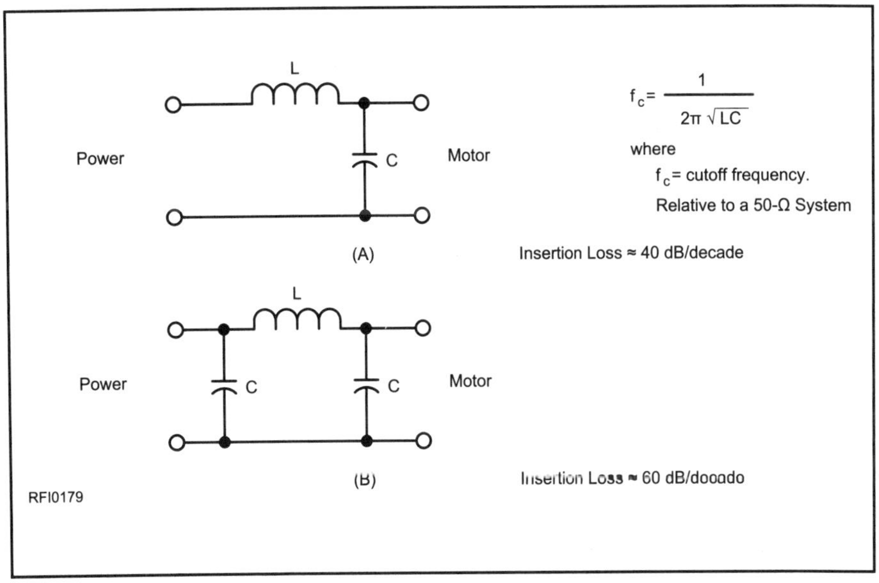

Figure 16.15—Filters for motors.

the noise consists of amplitude variations. Unfortunately, the pulses may affect FM receivers by increasing the "background noise level" and will reduce perceived receiver sensitivity because of the degraded signal-to-noise ratio.

Alternator and Generator Noise

As mentioned previously, brush-type motors employ sliding contacts which can generate noise. The resulting spark is primarily responsible for the "hash" noise associated with these devices. Hash noise appears as overlapping pulses on an oscilloscope connected to the receiver audio output. An alternator also has brushes, but they do not interrupt current. They ride on slip rings and supply a modest current, typically 4 A to the field winding. Hence, the hash noise produced by alternators is relatively minimal.

Generators use a relay regulator to control field current, and thus output voltage. The voltage regulator's continuous sparking creates broadband noise pulses that do not overlap in time. They are rarely found in modern automobiles.

Alternator or generator noise may be conducted through the vehicle wiring to the power input of mobile receivers and transmitters,

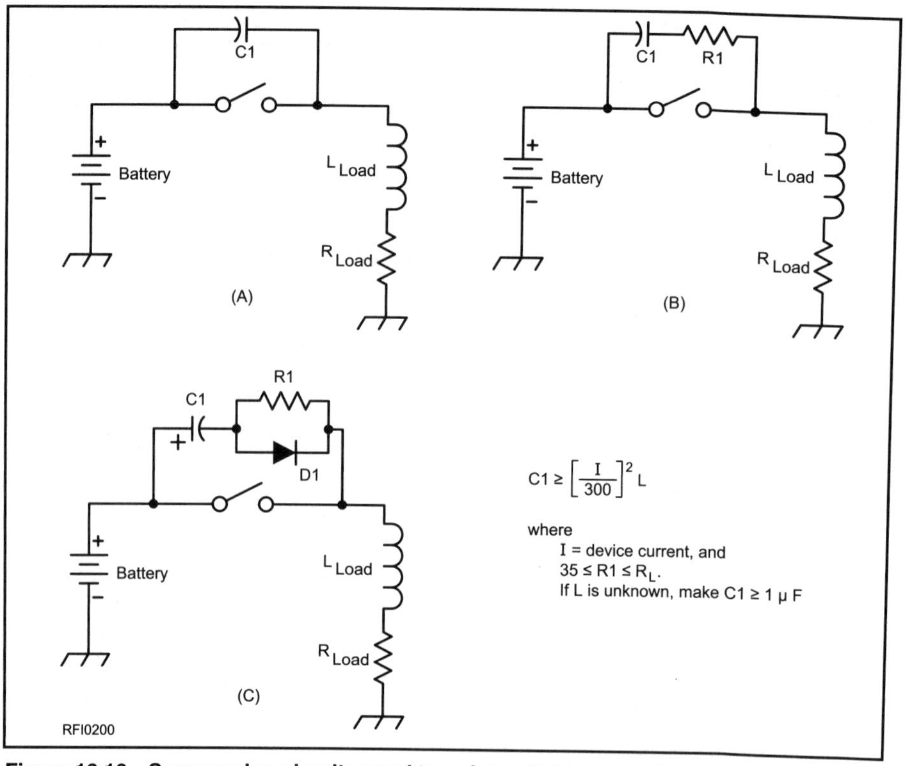

Figure 16.16—Suppression circuits used to quiet switching transients. A is the least effective and least costly. B is somewhat better. C is the most effective and costly. These schematics are informational only and do not necessarily apply to specific automotive designs and applications.

Table 1
General Wiring Practices

Recommendation	*Rationale*
Route wiring away from ignition system, spark plug wires and alternator wiring.	High energy noise may couple inductively or capacitively into wiring.
Don't bundle antenna, speaker or power wiring with vehicle wiring.	Noise may couple inductively or capacitively into wiring, even ground or shield wire.
Mount antenna away from vehicle wiring. Check antenna location with magnet-mount antenna before permanent installation.	Antenna is the most efficient interference receptor.
Run power and ground wires together as much as possible.	Minimizes magnetic coupling to other wiring.
Do not use vehicle frame for battery negative connection except as directed by the vehicle manufacturer.	Since receiver and antenna are both grounded separately, this avoids noise current flowing through the receiver case, and also avoids magnetic field coupling of interference.
Keep high-current leads as short as possible.	Reduces magnetic field coupling to other vehicle systems.
Do not run high current leads near cassette players or other devices using magnetic pickup coils. (CD players use a laser so may be less sensitive to current-carrying conductors).	Reduces magnetic field interference to tape heads.
Run wiring as close to metal vehicle body as possible.	Take advantage of vehicle metal body as a shield/filter. Wiring capacitance to ground, in combination with wiring inductance, will act as a low-pass filter.
Avoid making antenna coax splices in engine compartment.	Avoids noise pickup from splice, where coax shielding may be less effective.
Avoid connecting other accessory grounds to radio ground wire.	Eliminate noise current from other devices flowing on radio ground.
Fuse both positive and negative power leads on battery, unless vehicle is equipped with a ground block.	Prevents starter current from flowing through radio in the event battery-to-chassis connection becomes disconnected.
Secure wires physically to avoid movement.	Vibration may result in intermittent connections.

and may then be heard in the audio output.

Switches

Noisy switches may occur in vehicles. Unsuppressed switch contacts may generate fast, high-frequency voltage transients that travel through the vehicle wiring harness and radiate noise that sounds like "pops" in the receiver audio. These noise bursts may break receiver squelch and cause AGC-related dropouts. Problem switches may be easy to locate since many are readily accessible from the driver and passenger seats. In general, most hams don't bother curing these very transient pops—they may be an annoyance, but they don't usually disrupt communications. Figure 16.16 shows some of the design techniques used to suppress transients across switches. Exact component values are not shown because it is not intended that hams will apply these to the switches in their cars.

INTERFERENCE FROM AMATEUR RADIO TO AUTOMOBILE SYSTEMS

Just as vehicle electronic systems may interfere with mobile receivers, mobile transmitters may also affect the vehicle electronics if appropriate immunity measures are not in place. Electronic modules that radiate energy may also receive energy. This energy may come from on-board transmitters, nearby radio and TV stations, or any other device or event that generates an electric or magnetic field. Manufacturers use proven design techniques and run extensive tests to ensure RF immunity in their vehicles. Figure 16.17 shows an example of the EMC test facilities of different manufacturers.

Mobile Radio Installation Guidelines

The first step is to ensure that your installation is in compliance with both the vehicle manufacturer's and radio transceiver manufacturer's installation guidelines. Pay particular attention to the instructions about how to install the radio transceiver's power leads.

The installation guidelines of various manufacturers vary in their instructions about how to install radio transceiver's power leads. Most manufacturers recommend that the positive and negative leads from the radio be run directly to the battery. This is done to minimize the potential for the interaction between the radio's negative lead currents and vehicle electronics. If the manufacturer recommends that both wires be connected to the battery, they will also require that both wires be fused. This is necessary because, in the unlikely event that the connection between the battery and the engine block were to fail, excessive current could be drawn on the

(A)

(B)

(C)

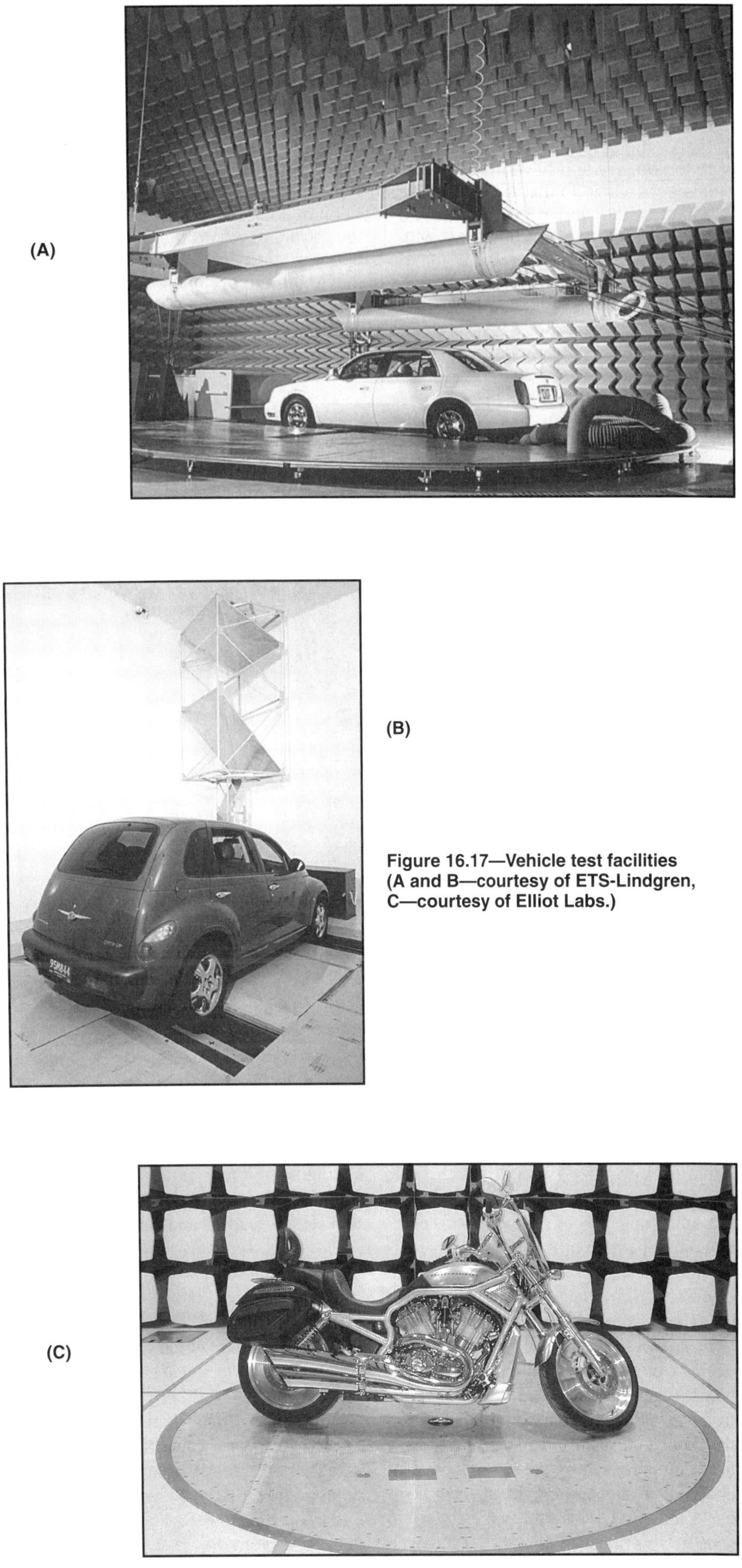

Figure 16.17—Vehicle test facilities (A and B—courtesy of ETS-Lindgren, C—courtesy of Elliot Labs.)

Use of Auxiliary Terminals for Radio Installation

Many mobile radio installation practices recommend directly connecting the power leads to the battery terminals. Over the last few years, some manufacturers have implemented an "auxiliary power terminal" or connection point on the battery positive (+) lead. These power terminals are typically located very close to the battery and have multiple connections to the vehicle electrical circuits. It is acceptable to utilize this connection instead of the previous recommendation of connecting at the battery. The intention of connecting to the battery is to eliminate the effect of vehicle wiring in supplying power to the transceivers. Since these auxiliary power terminals connect all the other vehicle systems to the battery, the wire from the power terminal to the battery has been sized to also minimize the effect of this short connection. In general, this approach will be included in subsequent revisions to the manufacturers' recommendations.

Connecting to this point does have other advantages in that they are normally much easier to access than the battery terminals and may eliminate the need for additional washers or hardware.

Figure A—Here's a good example of how to make your connection at the battery. Sandwich the ring terminals between two flat washers and use an additional nut to hold them in place. Coat the surface with plenty of conductive anti-seize compound.

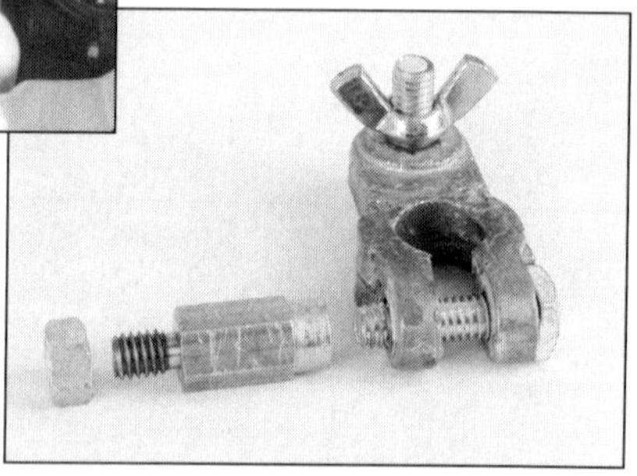

Figure B—These convenient and inexpensive terminal adapters were purchased at a local automotive supply retailer. They are ideal for connecting a mobile station directly to the automobile's battery terminals without compromising your battery's connections.

radio's negative lead when the vehicle starter is engaged. The disadvantage to this method is that, if the negative-lead fuse fails, the radio may continue to operate, using the shield of the radio's antenna coaxial cable as the negative return path. In most cases, this shield is not rated to carry the radio's full transmit dc current. Some vehicles provide a ground block near the battery for a negative cable to be connected. On these vehicles, run the negative power lead, unfused, to the ground block. When this technique is recommended by the manufacturer, the interaction between the ground currents and vehicle electronics has been evaluated by the manufacturer. In all cases, the most important rule to remember is this: If you want the manufacturer to support your installation, do it exactly the way the installation guidelines tell you to do it!

Following the practices outlined below will improve compatibility between on-vehicle transceivers and vehicle electronics:

1) The power leads should be twisted together (or run side by side) from the back of the rig all the way to the battery. Be sure to follow the manufacturer's instruction about whether to use the vehicle chassis as a power return. The power and antenna leads should be routed along the body structure, away from vehicle wiring harnesses and electronics. Any wires connected to the battery should be fused at the battery using fuses appropriate for the required current.

2) The coaxial feed line should have at least 95% braid coverage. The cable shield should be connected to every coaxial connector for the entire circumference (no "pigtails").

3) Antenna(s) should be mounted on a good ground plane, as far from the engine and the vehicle electronics as practical. Typical locations would be the rear deck lid or roof. Metal tape can be used to provide an antenna ground plane on non-metallic body panels.

4) Adjust the antenna for a low SWR.

5) Ensure that your installation is in compliance with both the vehicle manufacturer's and radio transceiver manufacturer's installation guidelines.

Refer to the wiring practices in Table 1 for a summary of these points.

Where to Ask for Help with Radio/Vehicle Interactions

RFI (Radio Frequency Interference) can occur in all vehicles, including motorcycles, boats, or anything with electronic control systems. Most vehicle manufacturers have EMC (Electromagnetic Compatibility) departments that resolve RFI problems in the design phases. EMC departments also verify the performance of their products and, in some cases, recommend specific RFI fixes and text content for their applicable factory service bulletins. The tests they use even include operation of on-board transmitters, purposefully using amateur frequencies and typical amateur power levels. Not surprisingly, a lot of EMC engineers are also hams.

Symptoms of susceptibility include display scrambling, erratic engine operation, false triggering of auxiliary systems and improper operation of entertainment radios. If you experience a susceptibility problem, first determine the interference source.

Problems associated with mobile transmitters come and go as the transmitter is keyed, but off-vehicle sources may switch on and off randomly. This makes them difficult to identify. Look for an EMI-trouble pattern with respect to location; there might be an antenna tower or a microwave dish nearby. If the problem is definitely a mobile transmitter, check through the steps in your manufacturer's installation guidelines to ensure that your installation follows "good engineering practice."

If a customer has any problem with a car, including EMC problems, they need to start with the dealer. The dealer may have some experience, and if not, should be able to locate the appropriate TSBs, (Technical Service Bulletins) using the same search methods that would be used to locate a fix for an ailing transmission. If the dealer has a problem they cannot resolve, the dealer may enlist the services of the engineers at the regional or zone offices. If the zone staff

Bonding Caveat

Many hams have traditionally bonded body panel and other parts together in an effort to reduce or eliminate automobile RFI. There is a significant caveat to this approach, however, especially when considering today's modern automobiles. It may compromise the manufacture's corrosion and rust proofing system, and ultimately lead to its failure.

The reason it may cause the failure is that the bonding can cause a galvanic reaction due to the introduction of a dissimilar metal. This creates galvanic corrosion, which starts at the connections point(s) and will expand over time. The worse the connection is, and the more the dissimilar metals are, the faster the corrosion will grow. Today's vehicles are comprised of metal with exacting characteristics and metallurgical properties to balance corrosion resistance, stamping and formability, strength, and weight. This is why there is a big problem in the auto repair business today when replacement body panels are purchased from off-shore sources whose primary concern is fit and form.

cannot help, the zone office has the factory contacts that will ultimately be in contact with the EMC labs, setting a process into motion to identify the problem and find a cure. The EMC lab will then disseminate that cure to the dealers. This process is designed to ensure that the customer has the best available local service and that the work of the engineers at the EMC lab will be made available worldwide to help others with the same problem. When this process works, it works well.

SUMMARY

Unfortunately, what most hams call "ignition noise" often originates from one or more of several sources in a modern automobile. There are many reasons for this. Modern vehicles must meet many regulatory objectives with respect to pollution control, fuel economy and vehicle safety. One approach taken by manufacturers to meet these guidelines has been to improve vehicle ignition systems. While increased spark plug voltage and decreased ignition pulse times can help meet these goals, both of those "improvements"—you guessed it—increase ignition noise. In many cases, the old coil and distributor system has been replaced by an ECM (Electronic Control Module) that sends signals to high-voltage modules at each plug.

In many cases, the problem extends beyond "ignition noise." Fuel injectors, also controlled by the ECM, can mimic ignition noise. They produce a "pop-pop" sound at low speeds and change to a whine as speed increases. Today's alternators are often required to deliver higher current than their older counterparts. This also increases their noise-generating potential.

Now add to the mix the traditional noise sources, such as the numerous electric motors in today's vehicles, and the interference potential becomes much greater than even a few years ago. It is hoped that the following frequently asked questions (FAQ's) would provide some assistance in understanding and identifying the source(s) of interference and possible solutions.

FAQ's about RFI and Mobile Radios:

Question: My vehicle is a few years old and there seems to be ignition noise across the HF band that is bad on some frequencies, yet okay on others. I verified that it is radiated noise by removing the antenna mast and the noise disappears. I've replaced the plugs, distributor cap and rotor. What could be happening? What should I try next?

Answer: It seems like there may be resonant conditions caused by the wiring. Eliminating these noise peaks may be as simple as re-routing wiring. However, it may only move the noise from one frequency to another, since if the energy is being emitted, it will find a way out at some frequency (ies)! Another suggestion is to replace the wires from the distributor to the plugs with ones specifically designed to reduce ignition noise.

Question: I once heard that I should "ground" the panels of a vehicle. What is that and what does it do?

Answer: "Grounding" panels is a technique of electrically connecting the panels together that can assist in the reduction of RFI due to "shorting" out the potential radiation. Sometimes this technique was also used on exhaust systems and worked to counteract the resonance of the exhaust pipes. As a general rule, this approach is not advised today as it may cause premature corrosion of those panels or exhaust components due to parasitic electrical current flow.

Question: I've got some RFI that I suspect is from the engine components. What's a good way to determine its source using commonly available items?

Answer: One method is to take a portable AM radio and tune it to the low end of the band. With the engine running, open the hood and slowly move the radio around the engine compartment to search for the location of maximum noise.

Question: I've got some RFI. How do I know if it's being conducted or radiated?

Answer: Figure out if the interference is coming through the antenna or the power feed wires. A simple test can tell you if it's radiated by the noise being eliminated when the antenna is disconnected. If it is not eliminated, it is being conducted by the power wiring. An in-line filter or perhaps a ferrite clamp on the wires can address this. Connecting directly to the battery with wires as short as possible is a good first step to ensure protection from conducted noise.

Question: I've heard that electrical fuel pumps can be sources of RFI. Is this true? How can this be fixed?

Answer: It may be true, since electrical fuel pumps are basically small electric motors connected to very long wires (antennas). The best solution for this is to ask your dealer for help to determine if there are any service bulletins specifically related to fuel pump RFI. If there are such bulletins, the

Hybrid Vehicles and RFI

Hybrid vehicles are becoming increasingly popular. Using both electric and gasoline power, they incorporate some fairly sophisticated electronics as well as high-power control circuitry and motors. Not surprisingly, development of hybrid vehicles could have an impact on mobile Amateur Radio installations. This is because of the extensive electronic systems required to operate a hybrid vehicle. In addition, high currents and high voltages may also be present. Switching of either can generate RFI. And since hybrid vehicles still utilize a gasoline engine, they still have all the traditional ignition and other systems that can cause problems in conventional vehicles. As the vehicle control system switches the vehicle between the gasoline engine and the electric motors, RFI can vary in intensity. RFI can also vary with acceleration and braking.

So how are mobile hams with hybrids doing? Based on a recent survey, it doesn't seem that they are any less compatible with VHF/UHF FM equipment than conventional vehicles. HF equipment, however, may present some additional challenges for the would-be mobile ham. Because of the small number of respondents and vehicles sampled in our survey, the jury is still out on this new technology.

Spark Event and Fuel Injector Noise— Identifying the Source is Crucial to Fixing It

Many times hams experience RFI that is related to engine RPM. Back in the "good ol' days" of carburetors, this noise would have been due to the ignition system. Now with almost every engine utilizing fuel injectors, noise that varies according to engine RPM may be either from the ignition system *or* fuel injection process.

Ignition noise manifests itself as a regular, periodic ticking sound that varies with engine RPM. At higher speeds, it sounds somewhat musical, like alternator whine, but has a harsher note. Fuel injection noise is similar and can easily be mistaken for ignition noise. They both vary with engine speed and can sound very similar to each other. There are some methods, however, to determine the source of the RFI.

The key to locating the source is to identify when the noise is most pronounced. One distinguishing feature of ignition noise is that it increases in amplitude under heavy acceleration. This results from the increase in the required firing voltage with higher cylinder pressure.

Ignition system noise is the result of the high voltage generated across the spark plug gap. As the demand upon the engine is increased (such as accelerating quickly from a low to a high speed), the RFI from the spark discharge will increase in magnitude. This results from an increase in firing voltage with higher cylinder pressure. Since more fuel is also being demanded, fuel injector pulses also get wider. The wider pulses have a higher duty cycle, which would mean less contribution to the RFI. At the other end of the engine operation, such as under light acceleration, the secondary voltage demand would be minimal and the injector duty cycle would also be minimal. This would result in the spark having less contribution to the RFI.

A good test can be conducted while driving on a flat road. Accelerate to approximately 50 MPH and provide just enough throttle in high gear to maintain a steady vehicle speed. There should be no road grade so as to provide a light engine loading condition. Apply a light throttle increase (to avoid the downshift that occurs when heavy throttle is applied, e.g., when passing a vehicle) and evaluate the noise during this condition. Repeat at three AM broadcast band frequencies at the low, middle and upper end of the band. Exact frequencies are not critical.

Ignition noise will typically increase in the AM broadcast band with an increase in engine load because of increased spark energy demand. Radiated spark ignition noise usually disappears when the receiver antenna is disconnected. If the noise still remains when the antenna is disconnected, then the high-current pulses in the ignition coil primary are generating noise that is conducted on the wiring harness to the receiver.

Another (and perhaps more direct) approach might be to make a "dipole" and "transmission line" out of a single pair of insulated wires that are several feet long. Make the "dipole" only a few inches long. Twist the remaining wires to form a "twisted pair" transmission line. Wrap the other end of the twisted pair around a transistor radio. (Alternately, use an appropriate length of coax with 1 inch loop and ferrite chokes at each end for the sniffer probe.)

By "sniffing" around with the "dipole" in the engine area, you may be able to determine the source of the noise. You can experiment with the length of the antenna and the number of turns around the radio.

If you determine the source is the fuel injection system, you may be able to use a series of clamp-on ferrite beads along the radiating conductors. (Cutting any wires is not recommended for obvious reasons.) If ferrites are used, be sure to secure them with strain relief as required.

Unfortunately, what most hams call "ignition noise" often originates from one or more of several sources in a modern automobile. If, however, you determine your problem to be ignition noise, one method is to visually inspect and verify that spark plugs have resistive elements in them. A resistive spark plug is typically 5k ohms (under operating conditions). Using an ohmmeter will give an indication if the plug is resistive or not; however, it may not be able to measure the actual value due to the low voltage used in most ohmmeters. Finally, be sure to check the ignition system in general. Some RFI investigators have induced ignition noise by partially removing a spark plug boot. An oscilloscope can now provide important clues as to the source using the "noise" wire as a reference.

dealer should be able to tell you what would be involved in fixing the problem.

Question: How can I tell the difference between injector and ignition noise? I think they both vary with engine speed.

Answer: There are some subtle differences. Ignition noise increases in amplitude under heavy acceleration. The difference, however, may not be discernable to an untrained ear. Use an AM radio to probe engine components and identify the offending item.

Question: I've done all you've suggested, I've found that it's radiated from the engine compartment and no matter what I do, the noise is still there. What else can I do?

Answer: A basic fact of all RFI problems is that if the receiver is moved away from the source, the problem will lessen. The bottom line—perhaps the cheap and easy fix—is to relocate the antenna from where it is installed now. Ask yourself if there is a way to move it farther away from the engine compartment. It should be clear by now that an antenna mounted on or near the rear of the vehicle picks up less engine noise than one mounted more towards the front of the vehicle.

ADDITIONAL RESOURCES

There are a number of Web pages that the League has created or identified that may help you.

"RFI—Automotive" **www.arrl.org/tis/info/rficar.html**—The automotive RFI "home page"—**www.arrl.org/tis/info/rficar.html**

Other pages that may help are:

Automotive Electric Motor and Fuel Pump Noise—**www.arrl.org/tis/info/fuel.html**—How to diagnose and cure noise from fuel pumps and other electric motors. This includes a reprint of the information in the Ford TSB on the subject.

Automotive Interference Problems: What the Manufacturers Say—**www.arrl.org/tis/info/carproblems.html**—This page gives the RFI policy statements of vehicle manufacturers.

Lab Notes—Mobile Installations and Electromagnetic Compatibility—**www.arrl.org/tis/info/pdf/39574.pdf**—Some general guidelines on various automotive RFI problems.

Off-site Web links:

NOISE and how to KILL it—**www.mindspring.com/~nx7u/mobile/noise.htm**—More information on automotive RFI and Ford fuel pump noise solutions.

Ford Explorer Radio Frequency Interference—**www.4x4central.com/tips.htm#rfi**—Still more information on Ford fuel pump noise solutions.

Radio Interference to/from two-way radio receivers.—**dodgeram.com/technical/tsb96/08_30_96.HTM**—Models: Dodge 1995-1997 BR Ram Truck.

Chapter 17

RFI From a Rules Point of View

Regulations help define acceptable conduct. The rules covering RFI help amateurs, their neighbors and equipment manufacturers understand and follow their responsibilities

By John Hennessee, N1KB (SK)
Dennis Bodson, W4PWF
ARRL Roanoke Division Vice Director
Paul C. Smith, KØPS
12328 Jasmine St, Brighton, CO 80601
Gary Hendrickson, W3DTN
1419 Larch Rd, Severn, MD 21144

ARRL Assistance

Few RFI problems involve only one person. In fact, most involve several people and, in some cases, whole armies of people! Your ARRL interference assistance team can include Technical Coordinators and Technical Specialists, Amateur Auxiliary members (Official Observers and Official Observer Coordinators), ARRL Volunteer Counsel and Volunteer Consulting Engineers, Local Interference Committees and RFI/TVI Committees as well as Section Managers and ARRL Headquarters staff. All of these volunteers work together to help *you* solve interference problems.

Members of the *Amateur Auxiliary to the FCC's Compliance and Information Bureau* are authorized to formally assist the FCC with data gathering for amateur-to-amateur interference problems. The Amateur Auxiliary is comprised of ARRL *Official Observers* who spend many hours of their time each month listening on the air for Part 97 rule violations. An ARRL *Official Observer Coordinator* oversees the Amateur Auxiliary in a particular section. The complaint must constitute a *specific* Part 97 rule violation to be actionable. Contact your ARRL *Section Manager* (see page 16 of any *QST*) for information about the Amateur Auxiliary and for the name of an OO or OOC in your section.

Some ARRL Affiliated Clubs sponsor *Local Interference Committees*. LICs deal only with amateur-to-amateur operational difficulties. LICs can often provide direction finding equipment and expertise to locate interference caused by amateurs. *ARRL HQ* can also refer you to the appropriate sources of assistance. Contact the ARRL by mail at 225 Main St, Newington, CT 06111; by telephone at 1-860-594-0200; by fax at 1-860-594-0259 or by e-mail at **reginfo@arrl.org**. The ARRL Web page can be found at **www.arrl.org**.

Don't contact the FCC first. Amateur-to-amateur interference problems are expected to be solved among the amateur community and through the Amateur Auxiliary. Sometimes the problem may involve a nonamateur or "bootlegger" who uses an amateur station illegally. The Amateur Auxiliary can help. This provision is meant to cover the *unauthorized* use of *amateur* frequencies by *amateurs*. See Chapter 1 for more information.

If you are faced with an interference problem and your neighbor is convinced that it's that "ham radio thing," help is available in evaluating the problem. It can be reported to your ARRL *Technical Coordinator* as suggested in the First Steps chapter. Technical assistance can also be sought from a club *RFI* or *TVI Committee.* In many cases, the technical cause of RFI is the device or system. Your ARRL Section Manager can refer you to your TC and to other local sources of assistance. The *ARRL Technical Information Service* can also refer you to appropriate sources of assistance. It can be contacted by mail at ARRL HQ (see contact information mentioned earlier) or by e-mail at **tis@arrl.org**. The ARRL TIS Web page at **www.arrl.org/tis/** contains a great deal of technical assistance.

Amateurs can sometimes receive interference from commercial radio services such as an improperly functioning pager or other transmitter which emits spurious emissions or other unwanted emissions. *It is the responsibility of the device operator to eliminate interference to other authorized radio services.* See the information on Parts 2 and 97 at the end of this chapter for a discussion of amateur responsibilities relating to non-amateur transmitter interference.

RFI FROM A RULES POINT OF VIEW

Amateurs tend to think of RFI as a technical problem, and it certainly is, but every interference problem also has social and regulatory components. In this chapter, we will look at how the rules affect amateurs. This discussion involves many sections of FCC regulations including Parts 0, 1, 2, 15, 18, 68, 76 and 97 of Title 47 of the *Code of Federal Regulations*. Sections of the Communications Act of 1934, as amended, also address interference. We will look at the regulatory responsibilities of amateurs, consumers, manufacturers and the FCC, as well as ways the ARRL and consumer groups can assist you in resolving an interference problem.

All of these regulations are tightly interwoven. How can we be expected to solve interference problems with all of this regulation? Generally, these regulations help both amateurs and consumers understand their responsibilities.

INTERFERENCE CLASSIFICATIONS

Although the focus of this book is primarily technical solutions to technical problems, some types of interference do not have technical solutions at all. Operational problems can cause significant difficulties for other amateurs and often require regulatory rather than technical solutions. RFI problems have many causes, but generally, they are caused by incompatibility between various electronic systems. RFI problems can be caused by inadequacies in the affected equipment, the proximity of the affected equipment to a

Power line Interference Problems—What Can I Do?

Power-line RFI problems can cause major headaches for amateurs. An entire chapter of this book is devoted to electrical and power-line interference problems. As with other Part 15 devices (a power line is an unintentional radiator) it is up to the system operator to resolve harmful interference problems to licensed radio services. If you are experiencing power-line noise, inform the appropriate individual at your power company of its obligation. Section 15.113 speaks of power-line carrier systems in the 9 to 490 kHz range, but this is of no significance to amateurs since there are no amateur bands in that segment. The provisions of Sections 15.5 and 15.13 apply. See the section in this chapter on Part 15.

Contact the power-line system operator through their usual RFI complaint process. If the interference hasn't been corrected after a reasonable time, write the General Manager in a registered, return receipt letter. Keep a copy of all correspondence. If you hear nothing after a reasonable time (usually 30 days), send a follow up letter to the General Manager of your power company with a copy sent to the President or Executive Vice President of your power company. If the situation is not corrected to your satisfaction after a reasonable time, send a registered letter to the Chairman or Commissioner of your state Public Utilities Commission (PUC). Your state PUC can usually be found in the "State Government" section of your telephone book. A list of PUC addresses can also be downloaded from **www.essential.org/**

strong RF signal or the radiation of unwanted signals by the transmitter that interfere with other radio services, such as spurious emissions.

Amateur-to-Amateur Interference

This topic may sound simple, but often it is not. Suppose, for example, that you're operating on your favorite 40-meter frequency, when you are suddenly plagued by severe splatter. Without the appropriate skills and equipment, it is not quite so simple to determine whether the source is an amateur or a nonamateur.

It is important for all amateurs to follow all FCC rules at all times. Specific technical standards can be found in Part 97 and they help amateurs avoid causing interference to other amateurs and nonamateurs. (See the Part 97 rules section later in this chapter.) The technical standards in the rules state that amateurs are not permitted to occupy more bandwidth than necessary for the information rate and type of emission; amateurs must use good engineering practice to reduce spurious emissions and amateurs must use the minimum power necessary to carry out the desired communications. The FCC may even impose "quiet hours" if the amateur station is found to be causing interference with receivers of good engineering design.

Nonamateur-to-Amateur Interference

Have you ever turned on your ham transceiver and encountered interference from an unknown source? Sometimes, amateurs receive interference caused by nonamateurs and these generally require technical *and* regulatory solutions as well as a good dose of common sense. Nonamateur sources could include other radio transmitters or non-transmitters, such as computers. In some cases of in-band interference, such as the 30-meter amateur band, Amateur Radio is *secondary* to the other radio services internationally. In that case, amateurs not only may need to accept the interference, but must be certain we do not cause interference to the primary users of the band. *The ARRL Operating Manual* gives details on how some ham bands are shared.

When faced with nonamateur RFI problems, it is important to point out to the owner of the interfering device what FCC rules say about causing interference to other services. It's no secret that as more poorly designed consumer devices enter the marketplace, the potential for RFI problems to (and from!) Amateur Radio equipment increases. These types of interference can involve many different government and non-government services. It isn't always easy to explain to the owners of the device or system that you are experiencing interference from *their* devices. It is important to conduct a technical investigation to determine whether the interference results from *noise*, *fundamental overload* or *unwanted emissions* such as *spurious emissions*. These terms are defined later in this chapter.

It is essential to have a good understanding of what FCC rules say about RF devices operating in more than a dozen authorized radio services. In most cases, RF devices aren't permitted to cause interference to other authorized radio services and if they do, the device operator is required to correct the interference.

The FCC also has rules that govern unlicensed emitters of RF energy. These can include computers, power supplies, motors and even some intentional emitters, such as wireless devices. Most unlicensed emitters are governed by Part 15 of the FCC's rules. Under Part 15, unlicensed devices are not in any sense a radio service. Unlicensed emitters are allowed to emit on licensed spectrum, subject to emissions limits and an unconditional requirement not to cause harmful interference.

Leaky Cable Television Systems—Help!

If you're operating on your favorite repeater on 145.25 MHz and a strong carrier shows up that you determine is emanating from a leaky cable television system, the cable company probably doesn't realize there is a problem. Part 76 regulations state, however, that cable companies can't exceed certain standards for leakage and if they do, it is their responsibility to correct the problem.

Inform the cable company of its obligations in resolving the interference. Even if the leak from the cable system is below the FCC leakage requirements, the interference could still be caused by a leak in the cable system. An interfering signal that's more than 40 dB below the video carrier level can still cause interference! If the leak is not severe, the cable company may tell you they don't have to fix it. The cable company must adhere to specific regulations concerning performance standards. See the discussion of Part 76 rules later in this chapter.

§76.605(a) states, among other things, that the carrier-to-noise ratio must be at least 43 dB and the carrier-to-coherent disturbances ratio must be at least 51 dB. Part 76, especially §§76.613 and 76.617 state clearly that interference from a cable system is the responsibility of the system operator. These regulations, and the ARRL Cable TV RFI web page, will help your cable operators to understand their responsibilities.

Remind the cable company of its obligation under the above provision in a certified, return-receipt letter. Keep copies of the correspondence and a record of all personal contacts. If you do not receive a satisfactory response within a reasonable amount of time (usually 30 days), send another certified letter to the General Manager, and send a copy to the National Cable Television Association (NCTA):

National Cable Television Assn
Director of Engineering
Science and Technology Dept
1724 Massachusetts Ave, NW
Washington, DC 20036
Tel: 202-775-3550
World Wide Web: **www.ncta.com/**

If your cable company needs some help understanding their responsibility, you or they should contact the ARRL Liaison to the NCTA. He is:

Robert V. C. Dickinson, W3HJ
Dovetail Surveys, Inc.
961 Marcon Blvd, Suite 450
Allentown, PA 18103
Tel: 610-264-0100
e-mail: rvcd@dovetailsci.com

Additional information on Cable leakage can be found on the ARRL Web at **www.arrl.org/tis/info/HTML/catvi/catvi-leakage.html**. If there is no solution after all recommended steps have been taken, contact the FCC in writing and explain the steps taken to solve the problem. Contact the FCC only as a last resort. Ask the FCC to assign a case number for later reference.—*John C. Hennessee, N1KB (SK)*

Part 15 goes into great detail about the interference potential of these unlicensed RF devices and the responsibility of the unlicensed device operator. Examples of Part 15 devices include touch lamps, computers, garage door openers and cordless microphones. They aren't *supposed* to cause interference to other services, but sometimes they *do*. In most cases, Part 15 device manufacturers must notify consumers of the interference potential. *It is important for amateurs to know that the manufacturer has told consumers about the interference potential.* This will be discussed in detail in the section of this chapter on Part 15 rules.

Amateurs are *not* protected from certain RF devices such as *Industrial, Scientific and Medical* devices which operate under Part 18. ISM devices can include kidney dialysis machines and devices used in heavy industries. This too will be discussed in detail later in this chapter.

Another common source of interference to amateurs is RFI from *cable television* and *power line* systems. See the accompanying text and sidebars.

Power line noise, another possible source of interference for amateurs, is regulated by Part 15 standards. Power line company systems must not cause interference to other licensed radio services, including the Amateur Radio Service. Of course, amateurs are cautioned against making any changes to cable or power-line wiring; that should only be done by an experienced employee of the cable or power-line company.

Amateur-to-Nonamateur Interference

This category of interference is what most hams mean when they say "RFI." Many amateurs are all too familiar with amateur-to-nonamateur interference; some have received the dreaded phone call or knock on the door from an irate neighbor advising them of an RFI problem. As with the other classifications of interference, there is a strong regulatory component. When faced with an RFI situation, the first reaction from the consumer is often "It's *your* fault!" The most helpful reply from an amateur is simply, "Not necessarily." As mentioned before, it is necessary to determine whether the interference is caused by electrical *noise, fundamental overload* or *unwanted emissions* such as *spurious emissions*. The other chapters in this book can be very helpful in evaluating such situations.

Amateur-to-Nonamateur Interference Inside the Ham Bands

Amateurs can potentially cause interference to users of other licensed radio services, and they can be across town or half a world away. Amateurs are prohibited from causing interference to users in other services (and vice versa) when the band is shared and the amateur allocation is the lower sharing status. This is discussed in great detail in Part 2 of the FCC rules. Services in a specific segment generally have *exclusive*, *primary* or *secondary* status. A station in a *secondary* service must not cause harmful interference to, and must accept interference from, stations in a *primary* service. Specific sharing

Causes of Interference

There are three types of interference:

Noise: This is interference caused by an electronic source such as a defective touch lamp or neon sign, computer systems, electric fence, power line noise and literally hundreds of other sources. Both amateurs and nonamateurs experience electrical noise. Noise caused by an electrical device can indicate an unsafe condition that needs to be corrected. Your neighbor may be surprised that he is the source of the interference, but pleased that you are his neighbor because you helped correct the problem for both of you!

Overload: *Fundamental overload* often results from the inability of a consumer device to reject strong and nearby signals. The device isn't *supposed* to receive your signal, but it does. *Amateurs are not legally responsible for interference to a consumer device resulting from your fundamental signal!* Consumer devices must be properly filtered and shielded in order to reject an amateur fundamental frequency where you are licensed to operate. Consumers must contact the device manufacturer for advice on shielding and filtering.

Unwanted emissions: Transmitters sometimes inadvertently transmit weak signals on frequencies for which the transmitter was not intended. These signals can have various regulatory classifications, but are most commonly referred to as *spurious emissions.* FCC regulations are quite clear about interference caused by spurious emissions: The operator of the transmitter must take whatever steps to reduce the "spurs" as required by FCC regulations. *Spurious emissions must not cause interference to other authorized radio services.* This point is echoed over and over in each of the FCC rule Parts mentioned in this chapter. If the interference is caused by spurious emissions from the transmitter, the transmitter operator *must* take whatever action is necessary to eliminate the interference. If the amateur is not causing interference to his or her own consumer devices, it's a pretty sure bet that he isn't transmitting spurious emissions.

RFI to Telephones and Audio Devices: Who is Responsible?

Interference to telephones and other audio devices is *not* the fault of the transmitter. A quick quote from the FCC's *Interference Handbook* should explain things clearly:

> Telephone interference generally happens because telephones are not designed to operate near radio transmitters and the telephone improperly functions as a radio receiver.

The FCC doesn't require that telephones include RFI protection and they don't offer legal protection to users of telephones that are susceptible to interference. One way of explaining this concept to consumers is to compare it to the difference between a *disease* and a *symptom* where the symptom is the interference and the disease is the susceptibility to interference. The disease has always been there, but has been lying dormant until a signal is present.

Cordless telephones are highly susceptible to interference; they are unlicensed low power transmitters (called Part 15 devices). Part 15 regulations require a label notifying consumers of the interference potential. Users are also required to discontinue operation if they cause interference to a licensed radio service, and they are required to eliminate any interference. Consumers must contact telephone manufacturers for technical assistance. If the manufacturer can't be contacted or isn't willing to help, contact the Consumer Electronics Association. Part 15 regulations will also be discussed later in the chapter.

arrangements are long and detailed. They can be found in §2.106. Complete details of the sharing status of all amateur bands can be found in *The ARRL Operating Manual*, available from ARRL HQ or Amateur Radio dealers.

Nonamateur-to-Nonamateur Interference

Nonamateurs sometimes blame hams for interference when the source actually originates from a nonamateur. After all, it's an RF jungle out there! Hams may be blamed for interference caused by electrical noise, another device inside a neighbor's own home, CB operators, defective consumer equipment or even another nearby amateur. Naturally, FCC regulations don't require you to be responsible for interference you are not causing. If this happens to you, as discussed in the First Steps chapter, your station log may help demonstrate to you and your neighbors that you are not the source of the interference.

Even though there is no obligation to do so, amateurs are encouraged to help solve the interference problem. Your own radio operation may also be affected by the interference. In addition, if you are able to help a neighbor track down a local interference problem, you may turn from the villain into a neighborhood hero. Your new reputation may prove especially helpful if you ever *are* involved in an RFI problem that involves your station.

Even the FCC doesn't have authority over certain types of interference that originate outside the US. One example: a strong foreign military radar signal that causes interference to the US Amateur Radio Service. The *ARRL Monitoring System* routinely reports instances of nonamateur intrusion on

WHAT IF . . .

What If . . . Your Landlord Threatens to Evict You Over RFI?

If a landlord threatens eviction over RFI, you need to obtain legal advice. Contact an ARRL Volunteer Counsel or other lawyer immediately. Lease agreements are private contractual agreements between the landowner and the person leasing the dwelling. Certain restrictions may prohibit antennas and RFI. Even if the agreement doesn't specifically state that antennas are prohibited, the landlord can do whatever he deems necessary to protect his property. PRB-1, the partial preemption of state and local regulations, offers no protection to amateurs who sign voluntary agreements.

What if . . . You are Faced with Restrictive RFI Covenants?

Amateurs are typically faced with two different types of covenant language in deed restrictions dealing with RFI incidents. The first is a provision that prohibits "radio or television transmitting facilities or equipment" within a dwelling or on the property. These are of questionable validity, since they have the effect of invalidating or restricting an amateur license at the station location. The second type is much more difficult to address. It would typically prohibit any "noxious or offensive activities" of a property owner. While this is a generalized provision, it has been applied in several instances to even public service activities such as amateur radio, if the activity results in RFI in consumer electronics of neighbors.

While the FCC indeed has exclusive jurisdiction over RFI, state courts have found that language in covenants, which are essentially private agreements between sellers and buyers of land, is to be enforced in any case. The courts are generally reluctant to interfere with private contractual agreements, and so they are usually upheld in case of challenges.

In situations where the covenant is deemed to "run with the land," covenant language need not even appear in the contract for the purchase of real property, if it appears in the subdivision plat and in the deed by which the property is conveyed to the buyer. This is the reason land records are checked to see if there are any type of restrictions from years past each time property is bought or sold. ARRL Volunteer Counsel members (hams who are also lawyers) can give you specific legal advice regarding covenants. To find the name and phone number of a VC, contact ARRL HQ or check the ARRL Web page at **www.arrl.org/field/regulations/local/vc.html**.

What If. . . Your Neighbor Threatens to Take You to Court Over RFI?

In rare instances, neighbors will threaten amateurs with court action over RFI matters. Even though the FCC has sole jurisdiction over interference, this can create complicated and often expensive legal problems as you defend yourself. If this happens to you, obtain the name of an ARRL Volunteer Counsel (VC) or other lawyer in your area. The initial consultation with an ARRL VC is always without charge.

What if . . . the Cops Show Up?

This may seem unlikely, but it could happen. After all, consumers don't usually like to hear that their device is at fault. They want the interference stopped and they want it stopped now! If local law enforcement officers are informed, they should know that only the FCC has jurisdiction over interference cases. This chapter clearly details the exclusive FCC jurisdiction. Owning a copy of *The RFI Book* can help you educate not only your neighbors, but also local law enforcement officials.

Even though you may not be at fault in an interference case, never argue with a law enforcement officer. Politely explain the FCC preemption. If an interference case becomes a dispute between two individuals, the interference problem may become secondary. ARRL Volunteer Counsel members can also help explain the legal basis for the interference preemption.

What If . . .

If you are involved in any of these "What If . . ." cases, it may be helpful to solicit the help of an attorney, even when the law is on the side of Amateur Radio. While the courts clearly do not have jurisdiction over RFI matters, preparing to appear before a judge is potentially expensive. A VC can often resolve the matter before the amateur is taken to court. Some neighbors call the police and try to have amateurs declared a "public nuisance" for causing interference. Only the FCC has jurisdiction over RFI cases!

For the most up-to-date RFI regulatory information you can check the ARRLWeb at **www.arrl.org/FandES/field/regulations/rfi-legal/index.html.**—*Chris Imlay, W3KD, ARRL General Counsel* amateur frequencies each month. This information is relayed to the FCC Notifications Branch in Washington, DC, for appropriate foreign governmental action.

As mentioned earlier, it isn't always possible to know whether over-the-air interference is being caused by an amateur or a nonamateur. If the interferer is found to be a nonamateur, the Amateur Auxiliary may continue to gather information on the matter since the Amateur Auxiliary has been expanded to cover these instances. This is sometimes the case when, for example, an unlicensed individual illegally uses the call of an amateur (a "bootlegger.") If the interference is caused by a *nonamateur* using a nonamateur device in the Amateur Radio Service, it is the responsibility of the Federal Communications Commission to solve these type problems. The FCC can be contacted by phone at 1-888-CALL FCC. The FCC Web page can be found at **www.fcc.gov**.

WHO'S RESPONSIBLE FOR RFI?

In one word—*everyone*. Everyone involved in an interference problem may have responsibilities and they must address those responsibilities fairly if a solution is to be found. Over and above the letter of the law, the FCC encourages an atmosphere of cooperation and trust when it comes to resolving RFI problems.

Responsibility may be shared between various people involved in the problem, but often to varying degrees. For example, if an electrical noise generator is the source of the interference, it is the responsibility of the device operator to rectify the problem. As another example, if an amateur transmitter is being operated in a completely legal manner using good engineering practice, but is still involved in an interference problem to a consumer device, the interference is probably caused by design deficiencies in the affected device, often fundamental overload.

Amateur Responsibilities

Let's cut to the chase—most hams would ask the question: "*When one of my neighbors has an interference problem, what do the rules say about whether it is my fault or not?*" This question is especially important if one of your neighbors raises it. The bottom line is simple: The amateur is responsible for the proper operation of his or her station, which means that all Part 97 rules must be followed at all times. If the amateur isn't experiencing a problem with his own consumer equipment, it's a pretty sure bet that his equipment isn't at fault.

As an amateur, you are responsible only for interference that results from FCC rules violations at your station. In the RFI world, this means that *if your station is transmitting signals outside the amateur band that cause interference to other radio services, it is your responsibility.* This is the *only* specific requirement under Part 97 rules.

The owner of a transmitter that emits spurious emissions is responsible for making sure that the transmitter meets all technical specifications of the service in which it operates. In the Amateur Radio Service, Subpart D of Part 97 gives technical standards for amateur transmissions.

Consumer Responsibilities

The *consumer* is responsible for cooperating with the amateur, the manufacturer and the FCC as a solution to the problem is sought. However, nonamateurs experiencing RFI often don't understand many of the technical concepts that are vital to resolving an RFI problem. When you first start talking about FCC rules with your neighbor, start by talking about what the rules require of you, as a radio operator. At this point, your neighbor will probably be listening, because you are talking about what you must do. You can then explain the other possible causes of interference, pointing out that the rules apply equally to everyone. They can *see* rules and you can gain middle ground in solving RFI problems.

In its *Interference Handbook*, the FCC is very clear in its advice to consumers faced with interference problems. This is very helpful in getting amateurs "off the hook" when the interference isn't caused by the amateur transceiver. Page 1 states:

> I. Check the Installation of Your Equipment
>
> Many interference problems are the direct result of poor equipment installation. Cost-cutting manufacturing techniques, such as insufficient shielding or inadequate filtering, may also cause your equipment to react to a nearby radio transmitter. This is not the fault of the transmitter and little can be done to the transmitter to correct the problem. If a correction cannot be made at the transmitter, actions must be taken to stop your equipment from reacting to the transmitter. These methods may be as easy as adjusting your equipment or replacing a broken wire. These and other simple corrections may be accomplished without the help of a service technician.

Never underestimate the importance of personal diplomacy when you're confronted with an RFI problem. The way you behave when your neighbor comes knocking sets the tone for everything that follows. No matter what you think of your neighbor, you have to remember that the best solutions are built on cooperation and trust. Knowing all the technical tricks in the book won't do you a bit of good if your neighbor won't even talk to you!

Manufacturer Responsibilities

The manufacturer of the consumer electronics device is responsible for meeting certain regulations and standards before the device can be marketed. Manufacturers are usually helpful with consumer RFI problems to their devices.

The manufacturers also shoulder some responsibility for RFI problems. Public Law 97-259, enacted in 1982, gave the FCC the authority to regulate the standards for the susceptibility of home entertainment equipment sold in the United States. The FCC, working with equipment manufacturers, decided to allow them to develop standards for RFI immunity and implement their own *voluntary* compliance programs. No system is perfect, especially a voluntary system, but the ARRL HQ Laboratory staff has noted that RFI complaints involving TVs, for example, seem to be decreasing.

The FCC has stated that many RFI problems aren't caused by a nearby transmitter, but by lack of RFI protection by the manufacturer. Most manufacturers respond appropriately if contacted about consumer RFI. The *Consumer Electronics Association* can help you find the right person to contact. Contact:

> Consumer Electronics Association (CEA)
> 2500 Wilson Blvd
> Arlington, VA 22205-3834
> Tel: 703-907-7626,
> **www.ce.org/**

Consumers should be encouraged to contact CEA if an interference problem with a piece of consumer electronic equipment occurs. They can determine who you should get in touch with for assistance. CEA prefers that you write rather than call. The details of a problem can often be communicated more clearly in written correspondence.

You may be surprised to know that the number of reported cases of interference to consumer electronic equipment in recent years has been very small. As a radio service, this is our fault! Amateurs are notorious for *not* reporting RFI problems to manufacturers. Contact the manufacturers! Working with manufacturers makes them aware of the need to continue to develop better shielding and filtering methods. It also demonstrates to your neighbor that the manufacturer should receive a little of his anger and frustration too.

The FCC and the ARRL are both active in an industry group (in cooperation with the American National Standards Institute, ANSI) that is studying RF immunity issues. This group, ANSI Accredited Standards Committee C63™, is working on voluntary industry standards for all types of EMC issues. One of its subcommittees, Subcommittee 5, Immunity, is chaired by Ed Hare, W1RFI, ARRL's Laboratory Manager. ARRL has long been an active and strong voice for Amateur Radio within industry committees, representing Amateur Radio's stake in a way that has secured ARRL staff leadership positions within these committees. (You can learn more about C63™ from its Web site at **www.c63.org**.)

FCC Responsibilities

The FCC will do what it can to assist consumers and amateurs, usually from an informational standpoint. The FCC can be contacted by phone at 1-888-CALL FCC, or by e-mail at **fccinfo@fcc.gov.** The role of the FCC has changed over the past decade from that of the sole problem solver for the masses experiencing RFI to that of a consultant and information source.

When faced with an amateur-to-nonamateur interference problem, encourage the complainant to obtain a copy of the FCC *Interference Handbook*. It will save you and your neighbor lots of time and undue frustration. A copy of the FCC's *Interference Handbook* can be found at the end of this book. The FCC *Interference Handbook* can be downloaded electronically via the FCC links at **www.arrl.org/tis/info/rfigen.html**.

FCC STRUCTURE AND REGULATORY PROCESS

To see how ham radio fits in to the rest of the radio world, let's consider the overall structure of the FCC and its rules. These are the rules that affect amateurs faced with RFI problems, either directly or indirectly.

The Amateur Radio Service is just one of many services administered by the FCC's Wireless Telecommunications Bureau. FCC regulation and treatment of RFI is discussed in greater detail later in this chapter.

The FCC does not function under any other Government department. It is an independent federal regulatory agency created by Congress and reports directly to Congress.

The FCC staff is organized on a functional basis. There are seven operating bureaus (Consumer & Governmental Affairs, Enforcement, International, Public Safety and Homeland Security, Wireline Competition, Wireless Telecommunications and the Media Bureau. There are ten staff offices: Communications Business Opportunities, Engineering and Technology, General Counsel, Strategic Planning and Policy Analysis, Workplace Diversity, Managing Director, Public Affairs, Legislative and Intergovernmental Affairs, Administrative Law Judges, and Inspector General. Its headquarters is in Washington, DC.

Amateur Radio Service policy and rules are the responsibility of the Wireless Telecommunications Bureau, which handles and implements amateur rule making proceedings. The FCC's Enforcement Bureau has the responsibility for monitoring and operations. In 1996, the FCC upgraded all of its monitoring stations around the country by installing new High Frequency Direction Finding (HFDF) equipment, which covers the frequency range of 100 kHz through 30 MHz. The 14 sites are now referred to as the National Automated Monitoring Network, are remotely controlled from a central office located in Columbia, Maryland.

Additionally, the FCC closed some of its 25 field offices, replacing them with two "resident agents" per office and consolidated its six Regional Offices into three (Chicago, Kansas City and San Francisco). A central, toll-free calling line provides an improved level of customer service. This telephone number, 1-888 CALL FCC (1-888-225-5322), should now be used for all contacts with the FCC.

FCC Preemption of RFI

Only the FCC has jurisdiction over interference. Municipal zoning authorities, including local law enforcement officials, do not have that authority.

As with restrictive antenna ordinances, the FCC, through a directive from Congress, has preempted any concurrent state or local regulation of RFI pursuant to the provisions of §302(a) of the Communications Act of 1934, as amended. The legal cite is: 47 USC §302(a) and it provides that the:

> Commission may, consistent with the public interest, convenience, and necessity, make reasonable regulations (1) governing the interference potential of devices which in their operation are capable of emitting radio frequency energy by radiation, conduction, or other means in sufficient degree to cause harmful interference to radio communications; and (2) establishing minimum performance standards for home electronic equipment and systems to reduce their susceptibility to interference from radio frequency energy. Such regulations shall be applicable to the manufacture, import, sale, offer for sale, or shipment of such devices and home electronic equipment and systems, and to the use of such devices.

The legislative history of §302(a) provides explicitly that the Commission has exclusive authority to regulate radio frequency interference (RFI). In its Conference Report No. 97-765, Congress declared:

> The Conference Substitute is further intended to clarify the reservation of exclusive jurisdiction to the Federal Communications Commission over matters involving RFI. Such matters shall not be regulated by local or state law, nor shall radio transmitting be subject to local or state regulation as part of any effort to resolve an RFI complaint.

The legal cite for this report is: *H.R. Report No. 765*, 97th Cong., 2d Sess. 33 (1982), reprinted at *1982 U.S. Code Cong. & Ad News* 2277.

State laws and local ordinances that require amateurs to cease operations or incur penalties as a consequence of radio interference thus have been entirely preempted by Congress. This was written by then-FCC General Counsel Robert L. Pettit in a letter dated February 14, 1990 to ARRL General Counsel Chris Imlay, W3KD. Amateurs experiencing difficulties convincing local zoning authorities and others of the FCC sole jurisdiction over RFI may obtain additional information from the ARRL Web Site at **www.arrl.org/FandES/field/regulations/rfi-legal/**.

FCC Preemption of Restrictive Antenna Ordinances

In many cases, neighbors who are concerned about amateur tower installations may focus on possible RFI problems as part of their concerns. Local zoning bodies sometimes try and overstep their legal bounds when it comes to zoning and this can cause headaches for amateurs. Local governments must reasonably accommodate amateur operations in zoning decisions as documented by the partial preemption called PRB-1. The legal cite for PRB-1 is: 101 FCC 2d 952 (1985). §97.15(e) provides that an amateur station antenna structure may be erected at heights and dimensions sufficient to accommodate effective amateur service communications. See the specific wording later in this chapter. Local authorities may adopt regulations pertaining to placement, screening or height of antennas, if such regulations are based on health, safety or aesthetic considerations and reasonably accommodate amateur communications.

Local governments may not, however,

base their regulation of amateur service antenna structures on the causation of interference to home electronic equipment—an area regulated exclusively by the Commission.

How does this affect RFI? Some local governments mistakenly believe that if an amateur antenna is lowered, the potential for interference decreases. The FCC has gone on record as stating that there is no reasonable connection between requiring an amateur to reduce the height of his tower and reducing the amount of interference to his neighbor's home electronic equipment. On the contrary, antenna height is inversely related to the strength, in the horizontal plane, of the radio signal that serves as a catalyst for interference in susceptible home electronic equipment. It is a matter of technical fact that the higher an amateur antenna, the less likely it is that RFI will appear in home electronic equipment. This statement was made in an October 25, 1994 letter from former FCC Private Radio Bureau Chief Ralph Haller.

For amateurs facing restrictions imposed by local ordinances, the ARRL Regulatory Branch offers a wide range of assistance. You will find current and useful information on PRB-1 online at: **www.arrl.org/FandES/field/regulations/PRB-1_Pkg/index.html**. Also note that besides Federal protection, 23 states to date have incorporated PRB-1 protections into state law. You can find a list of PRB-1 states and links to their specific statutes at: **www.arrl.org/FandES/field/regulations/statutes.html**. The ARRL also facilitates Volunteer Counsel (VC) and Volunteer Consulting Engineer (VCE) programs. These individuals have stepped up to assist amateurs in assessing their situations when being confronted by local regulations. A sortable list of VCs in your area can be found online at **www.arrl.org/FandES/field/regulations/local/vci.html**, while VCEs are listed at **www.arrl.org/FandES/field/regulations/local/vcei.html**.

Remember that PRB-1 is not a panacea to cure all antenna restrictions. It does not apply when you are facing covenants, conditions and restrictions (CC&Rs) placed on a property by deed or as part of a neighborhood home-owners association. PRB-1 also does not offer protection for those living in rental property—the property owner may always put restrictions or prohibitions on renters. But PRB-1 can be an effective tool when dealing with local ordinances.

FCC RULES, POLICY AND STANDARDS

As stated at the beginning of this chapter, the rules and regulations for telecommunications are very long and detailed.

The basis for all FCC rules and actions is the Communications Act of 1934, as amended. The FCC must follow a very specific set of directives from Congress when drafting rules. §302(a) is of particular interest to amateurs. It gives the FCC jurisdiction over interference and it allows the FCC authority to establish minimum performance standards for home entertainment equipment. This section states:

§302. [47 U.S.C. 302] DEVICES WHICH INTERFERE WITH RADIO RECEPTION.

(a) The Commission may, consistent with the public interest, convenience, and necessity, make reasonable regulations (1) governing the interference potential of devices which in their operation are capable of emitting radio frequency energy by radiation, conduction, or other means in sufficient degree to cause harmful interference to radio communications; and (2) establishing minimum performance standards for home electronic equipment and systems to reduce their susceptibility to interference from radio frequency energy. Such regulations shall be applicable to the manufacture, import, sale, offer for sale, or shipment of such devices and home electronic equipment and systems, and to the use of such devices.

§303(f) of the Communications Act of 1934, as amended, gives the FCC authority to prevent interference between stations:

§303. [47 U.S.C. 303] GENERAL POWERS OF COMMISSION.

Except as otherwise provided in this Act, the Commission from time to time, as public convenience, interest, or necessity requires shall—

(f) Make such regulations not inconsistent with law as it may deem necessary to prevent interference between stations and to carry out the provisions of this Act.

Another section of the Communications Act of 1934, as amended, prohibits willful and malicious interference. It states:

§333. [47 U.S.C. 333] WILLFUL OR MALICIOUS INTERFERENCE.

No person shall willfully or maliciously interfere with or cause interference to any radio communications of any station licensed or authorized by or under this Act or operated by the United States Government.

All telecommunications rules and regulations can be found in Title 47 of the *Code of Federal Regulations* (*CFR*). Each service has its own "part." As with most rule parts, other seemingly unrelated rules are, in fact, very important. The related parts include: Part 0 (FCC Organization), Part 1 (FCC Administrative Procedures), Part 2 (FCC Agreements, Procedures and Allocations), Part 15 (unlicensed RF devices), Part 18 (Industrial, Scientific, and Medical Equipment), Part 76 (Cable Television Service) and, of course, our own Part 97 (Amateur Service).

Part 2: FCC Agreements, Procedures and Allocations

Part 2 contains special requirements in international regulations, agreements, treaties and the table of frequency allocations. This part also contains requirements and procedures concerning the marketing, the equipment authorization and the importation of radio frequency devices capable of causing harmful interference. Parts 2 and 15 are closely related. Section 2.106 gives the specific and *very* detailed frequency allocations along with sharing arrangements. Other FCC rule parts refer to Part 2, which also

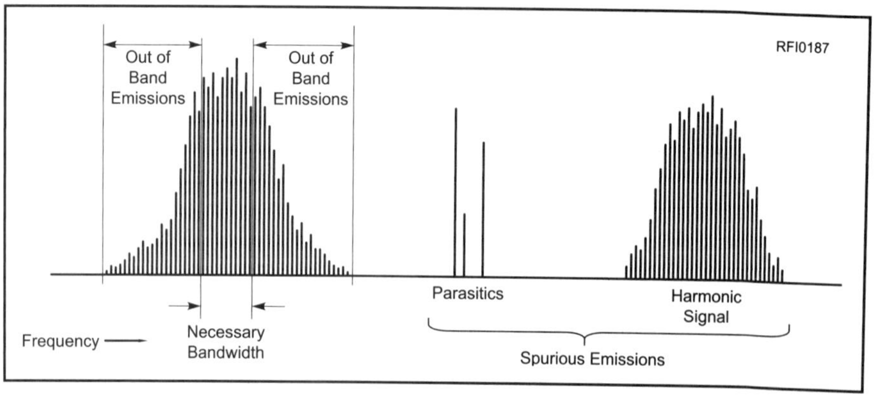

Figure 17.1—Some modulation sidebands are outside the necessary bandwidth. They are out-of-band emissions, but aren't considered spurious emissions. On the other hand, these sidebands must be reduced to the greatest extent practicable and must not interfere with other stations per §97.307(b). The harmonics and parasitics shown in this figure are spurious emissions and must be reduced to the levels specified in §97.307(c).

incorporates sections of the international Radio Regulations.

Part 2 is hundreds of pages long, but several important definitions are mentioned. Section 2.1 defines *harmful intereference* as:

> Interference which endangers the functioning of a radionavigation service or other safety services or seriously degrades, obstructs, or repeatedly interrupts a radiocommunications service operating in accordance with these [international] Radio Regulations.

The rules for nearly every licensed radio service prohibit harmful interference, and that includes the Amateur Radio Service.

Part 2 lists important definitions, some of which aren't mentioned in Part 97. It is important to have an understanding of these definitions as they relate to spurious emissions. As we mentioned before, if interference is caused by spurious emissions from an amateur station, it is the responsibility of the amateur station to reduce these emissions to those defined in Part 97. Amateurs need to be aware of the following Part 2 definitions:

Interference: The effect of unwanted energy due to one or a combination of emissions, radiations, or inductions upon reception in a radiocommunication system, manifested by any performance degradation, misinterpretation, or loss of information which could be extracted in the absence of such unwanted energy.

Necessary Bandwidth: For a given class of emission, the width of the frequency band which is just sufficient to ensure the transmission of information at the rate and with the quality required under specified conditions.

Occupied Bandwidth: The width of a frequency band such that, below the lower and above the upper frequency limits, the mean powers emitted are each equal to a specified percentage Beta/2 of the total mean power of a given emission.

Note: Unless otherwise specified by the CCIR for the appropriate class of emission, the value of Beta/2 should be taken as 0.5%.

Out-of-band Emission: Emission on a frequency or frequencies immediately outside the necessary bandwidth which results from the modulation process, but excluding spurious emissions.

Spurious Emission: Emission on a frequency or frequencies which are outside the necessary bandwidth and the level of which may be reduced without affecting the corresponding transmission of information. Spurious emissions include harmonic emissions, parasitic emissions, intermodulation products and frequency conversion products, but exclude out-of-band emissions.

Part 97 defines *bandwidth* in Section 97.3(a)(8) as: "The width of a frequency band outside of which the mean power of the transmitted signal is attenuated at least 26 dB below the mean power of the transmitted signal within the band." In engineering terms, this is usually referred to as the *occupied bandwidth*, a technical way of describing the amount of spectrum in which most of the transmitted signal is found. Bandwidth is the frequency range that a signal occupies around a center frequency. Bandwidth increases with the information rate.

Although it is not defined in Part 97, note that Part 2 also defines a term called *necessary bandwidth*. §2.202(b) states, "Necessary bandwidth. For a given class of emission, the minimum value of the occupied bandwidth sufficient to ensure the transmission of information at the rate and with the quality required for the system employed, under specified conditions." This essentially defines the minimum bandwidth necessary for a particular transmission—if any of this part of the signal were to be removed, the performance of the system would suffer. If part of the signal that extends past this bandwidth were removed, it would not have a detrimental affect on the performance of the system.

The *necessary bandwidth* is that which is required to transmit a certain emission, but *occupied bandwidth* is the *total* bandwidth including the upper and lower emissions. An *out of band emission* is *outside* the necessary bandwidth, but *within* the occupied bandwidth. Such emissions must be suppressed 26 dB down from the main signal. If they are not, they are the responsibility of the transmitter owner to correct.

Although most hams would interpret the term *out-of-band emissions* as being signals outside of the ham bands, this term actually has a different regulatory meaning. An out-of-band emission is an unwanted emission that results from the modulation process but is outside the necessary bandwidth.

For a better understanding of these definitions as they relate to spurious emissions, see Figure 17.1. Admittedly, this can get a bit confusing. A brief summary is that all signals outside the necessary bandwidth must be attenuated enough so they don't cause harmful interference.

We will discuss what the specific Part 97 rules say about the technical aspects of amateur operation later in this chapter.

Another important concept is that licensed stations operating in a primary service are offered protection from stations operating in a secondary service. A station in a secondary service must not cause harmful interference to, and must accept interference from, stations in a primary service. The amateur 160 and 30 meter bands and all amateur bands on and above 219 MHz are shared with other services. This section is very long and detailed. Complete details of the sharing status of all amateur bands can be found in *The ARRL Operating Manual*, available from ARRL HQ or Amateur Radio dealers

Part 97: Amateur Radio Regulations

Every ham should have a current copy of Part 97 on hand. Part 97 and other rule parts are updated frequently. Copies of any Title 47 part (or any other *CFR* title) can be ordered from the government or downloaded from the World Wide Web. See the sidebar titled "How to Obtain Copies of FCC Regulations."

Part 97 covers RFI thoroughly. To summarize the Part 97 regulations concerning RFI: (1) do not occupy more bandwidth then necessary for the information rate and type of emission; (2) use good engineering practice to reduce spurious emissions; (3) use the minimum power necessary to carry out the desired communications; and (4) the FCC may impose restricted hours of operation if the amateur station is found to be causing interference with receivers of good engineering design.

How to Obtain Copies of FCC Regulations

FCC Part 97 rules are available on the ARRL Web at **www.arrl.org/FandES/field/regulations/news/part97**.

If you need information on non-amateur rules, call the Government Printing Office Order Desk between 8 AM and 4 PM Eastern time, Monday through Friday, at 1-202-512-1800 or fax 1-202-512-2250. The Government Printing Office (GPO) operates 24 US Government Bookstores throughout the country. Each bookstore carries a selection of at least 1500 of the most popular Federal government publications, subscriptions and electronic products.

If you have access to the World Wide Web, the following page allows you to download needed parts of any section of the *Code of Federal Regulations* free of charge. See **www.gpoaccess.gov/cfr/index.html**.

Although §97.15 is not directly related to RFI, antenna location and height can contribute to RFI problems. §97.15 is of particular importance because it codifies PRB-1, the Federal preemption of overly restrictive state and local ordinances. Obviously, the higher the antenna, the less chance of causing RFI. §97.15 states:

> (b) Except as otherwise provided herein, a station antenna structure may be erected at heights and dimensions sufficient to accommodate amateur service communications. [State and local regulation of a station antenna structure must not preclude amateur service communications. Rather, it must reasonably accommodate such communications and must constitute the minimum practicable regulation to accomplish the state or local authority's legitimate purpose. [See PRB-1, 101 FCC 2d 952 (1985) for details.]

§97.101(d) states that no amateur operator shall willfully or maliciously interfere with or cause interference to any radio communication or signal. This point is echoed throughout the rules. Part 97 also speaks about the operational aspects and the specific responsibility of amateurs causing interference. §§97.201(c) and 97.205(c) state that when a repeater and/or auxiliary station causes harmful interference to another auxiliary station, the licensees are equally and fully responsible for resolving the interference unless one station's operation is recommended by a frequency coordinator and the other station's is not. In that case, the licensee of the non-coordinated auxiliary station has primary responsibility to resolve the interference. §97.205(g) goes on to say that the control operator of a repeater that inadvertently retransmits communications that violate the rules in this part is not accountable for the violative communications.

§97.121 offers some of the most direct wording the FCC has regarding RFI and Amateur Radio operation. It details restrictions the FCC may pose on the operation of an amateur station causing RFI while the problem is being resolved. The FCC can assign you "quiet hours" until an RFI investigation is completed. It is in the amateur's best interest to cooperate fully with the FCC. An amateur must be sure that transmissions are not causing interference to his or her own home electronic devices. It states:

§97.121 Restricted operation.

> (a) If the operation of an amateur station causes general interference to the reception of transmissions from stations operating in the domestic broadcast service when receivers of good engineering design, including adequate selectivity characteristics, are used to receive such transmissions, and this fact is made known to the amateur station licensee, the amateur station shall not be operated during the hours from 8 PM to 10:30 PM, local time, and on Sunday for the additional period from 10:30 AM until 1 PM, local time, upon the frequency or frequencies used when the interference is created.
>
> (b) In general, such steps as may be necessary to minimize interference to stations operating in other services may be required after investigation by the FCC.

The FCC's technical standards for amateurs aren't as rigid as for other services. Other specific technical standards apply directly to RFI. They include §97.307 Emission standards, and §97.313 Transmitter power standards.

§97.307 addresses the emission standards an amateur station must meet. Paragraphs (a) and (b) state that a signal can't occupy more bandwidth than necessary for the information rate and type of emission. Amateurs are prohibited from transmitting spurious emissions that are not suppressed according to the levels in the rules. It also states that splatter and keyclick interference isn't permitted. Paragraphs (c), (d) and (e) relate directly to RFI, specifying the action required to keep from interfering with other services, and the allowed levels of spurious emissions from amateur equipment. Always use good engineering practice to reduce spurious emissions (and to ensure that you comply with Part 97). The applicable Part 97 regulations relating to RFI are:

§97.307 Emission standards.

> (a) No amateur station transmission shall occupy more bandwidth than necessary for the information rate and emission type being transmitted, in accordance with good amateur practice.
>
> (b) Emissions resulting from modulation must be confined to the band or segment available to the control operator. Emissions outside the necessary bandwidth must not cause splatter or keyclick interference to operations on adjacent frequencies.
>
> (c) All spurious emissions from a station transmitter must be reduced to the greatest extent practicable. If any spurious emission, including chassis or power-line radiation, causes harmful interference to the reception of another radio station, the licensee of the interfering amateur station is required to take steps to eliminate the interference, in accordance with good engineering practice
>
> (d) For transmitters installed after January 1, 2003, the mean power of any spurious emission from a station transmitter or external RF power amplifier transmitting on a frequency below 30 MHz must be at least 43 dB below the mean power of the fundamental emission. For transmitters installed on or before January 1, 2003, the mean power of any spurious emission from a station transmitter or external RF power amplifier transmitting on a frequency below 30 MHz must not exceed 50 mW and must be at least 40 dB below the mean power of the fundamental emission. For a transmitter of mean power less than 5 W installed on or before January 1, 2003, the attenuation must be at least 30 dB. A transmitter built before April 15, 1977, or first marketed before January 1, 1978, is exempt from this requirement.
>
> (e) The mean power of any spurious emission from a station transmitter or external RF power amplifier transmitting on a frequency between 30-225 MHz must be at least 60 dB below the mean power of the fundamental. For a transmitter having a mean power of 25 W or less, the mean power of any spurious emission supplied to the antenna transmission line must not exceed 25 μW and must be at least 40 dB below the mean power of the fundamental emission, but need not be reduced below the power of 10 μW. A transmitter built before April 15, 1977, or first marketed before January 1, 1978, is exempt from this requirement.

Table 17.1
RFI Standards for Data Transmissions

Frequency (MHz)	*Baud Rate Limit*	*Frequency Shift Limit*
1.8-28.0	300	1 kHz
28.0-28.3	1200	1 kHz
50.1-51.0	1200	1 kHz
144.1-148	19600	20 kHz
222-225	56000	100 kHz
420-450	56000	100 kHz
902 and above	No limit	No limit

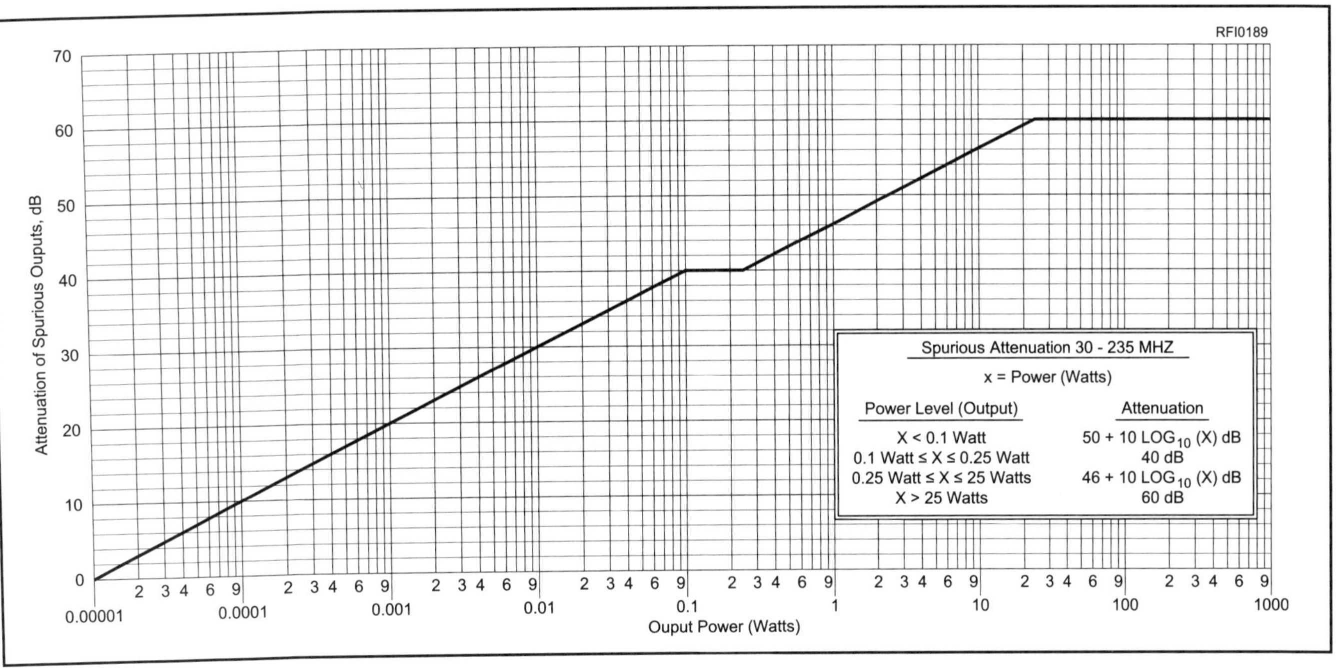

Figure 17.2—Required attenuation of spurious outputs in between 30 and 235 MHz.

The FCC is very specific about how far amateur spurs must be suppressed to avoid causing interference [97.307(d)]:

If your transmitter or RF power amplifier was installed after January 1, 2003 and transmits on frequencies below 30 MHz, the mean power of any spurious emissions must be at least 43 dB below the mean power of the emission.

If your transmitter or RF power amplifier was built after April 14, 1977, or first marketed after December 31, 1977, and transmits on frequencies below 30 MHz, the mean power of any spurious emissions must:

- never be more than 50 mW;
- be at least 30 dB below the mean power of the fundamental emission, if the mean power output is less than 5 W; and
- be at least 40 dB below the mean power of the fundamental emission, if the mean power output is 5 W or more. The requirement that no spurious emission exceed 50 milliwatts means that above 500 watts, the suppression must be greater than 40 dB. At 1500 watts, the suppression must be 44.77 dB.

The following requirements apply between 30 and 225 MHz [97.307(e)]:

- Transmitters with 25 W or less mean output power: spurs must be at least 40 dB below the mean power of the fundamental emission and never greater than 25 μW, but need not be reduced further than 10 μW. This means that the spurs from a 25 W transmitter must be at least 60 dB down to meet the 25 μW restriction.
- Transmitters with more than 25 W mean output power: spurious emissions must be at least 60 dB below the mean power of fundamental emission. The situation for transmitters operating between 30 and 225 MHz is more complex. The combination of the requirement that spurious emissions be less than 25 μW and the stipulation that they don't need to be reduced below 10 μW makes the requirements vary significantly with power level, ranging from 0 dB suppression required for a transmitter whose power is 10 μW to 60 dB of suppression required for power levels above 25 watts. The requirements for transmitters operating between 30 and 225 MHz are shown graphically in Figure 17.2. There are no absolute limits for transmitters operating above 225 MHz, although the general requirements for good engineering practice would still apply.

§97.307(f)(2) states that "No non-phone emission shall exceed the bandwidth of a communications quality phone emission of the same modulation type. The total bandwidth of an independent sideband emission (having B as the first symbol), or a multiplexed image and phone emission, shall not exceed that of a communications quality A3E emission." §97.307(f) details the specific standards for RTTY and data transmissions. See the summary in Table 17.1.

§97.313 is concerned with transmitter power standards and limits. Again, this section does not address RFI issues specifically. As with antenna location and height, however, excessive transmitter output power can aggravate RFI problems. Therefore, one of the cardinal rules of good amateur practice, "use the minimum power necessary to carry out the desired communications," directly applies.

Specific certification standards for amateur external RF power amplifiers can be found in §§97.315 and 97.317.

Part 15: Unlicensed RF Devices

Part 15 of Title 47 of the FCC Rules is very important to amateurs because it regulates low power, unlicensed devices that could cause interference to the Amateur Radio Service and vice versa. Part 15 covers an assortment of electronic equipment that generate RF energy whether it's *intentional*, *unintentional* or *incidental*. Amateurs will need to consider Part 15 as it relates to digital devices, computers, low-powered, unlicensed transmitters, electrical devices and any other "generic" device that might generate RF in the normal course of its operation.

Let's first define the most important terms:

An *intentional radiator* is a device that intentionally generates and emits radio frequency energy by radiation or induction. This term generally means "radio transmitter." Examples are cordless telephones, baby monitors and garage-door openers.

An *unintentional radiator* is a device that intentionally generates radio frequency energy for use within the device, or that sends radio frequency signals by conduction to associated equipment via connecting wiring, but which is not intended to emit RF energy by radiation or induction. Examples include computer systems and superheterodyne receivers.

An *incidental radiator* is a device that generates radio frequency energy during the course of its operation although the device is not intentionally designed to generate or emit radio frequency energy. Examples of incidental radiators are dc motors, power-line systems and mechanical light switches.

A *Class A digital device* is a digital device that is marketed for use solely in a commercial, industrial or business environment.

A *Class B digital device* is a digital device that is marketed for use in a residential environment. Examples of such devices include, but are not limited to, personal computers, calculators and similar electronic devices that are marketed for use by the general public. Class B equipment, intended for use in a residential environment where the likelihood of RFI is greater, must meet much stricter RF emission limits than the Class A devices.

The Scope of Part 15

Part 15 actually covers a *lot* of territory. Part 15 sets out the regulations under which an intentional, unintentional or incidental radiator may be operated without an individual license. It also contains the technical specifications for various types of devices. These technical specifications include absolute maximum radiated and conducted limits, in addition to the requirements stipulating that no harmful interference may result from the operation of a Part 15 device. In addition, the rules contain administrative requirements and other conditions relating to the marketing of Part 15 devices.

Because of space limitations, only the most applicable sections are included in this chapter. The sections of Part 15 that are most applicable to amateurs include: §§15.5(a), (b) and (c) Conditions of Operation, 15.13 Incidental Radiators, 15.17 Susceptibility to Interference and sections from Subpart B-Unintentional Radiators (15.101 Equipment authorization of unintentional radiators, 15.105 Information to the user, 15.107 Conduction limits, and 15.109 Radiated emission limits). The pertinent information from these sections follows.

The requirements for these unlicensed RF emitters are complex. To complicate the picture even more, not all unlicensed devices operate under Part 15; some operate under other FCC rule parts. Some of these devices must have a label stating that the device meets Part 15 specifications. The authorization procedures are outlined in Part 2.

In many cases, Part 15 devices use frequencies allocated to other radio services, including the Amateur Radio Service, on a non-interference basis. For example, some cordless telephones operate in the 902-928 MHz band, with the requirement that they not cause harmful interference to licensed users of that spectrum. Some frequency segments, including several in the amateur bands, have been approved for higher power Part 15 devices. In addition, some devices that do not specifically use any frequency, but still may radiate RF energy, are also covered in Part 15.

To help emphasize the unlicensed status of all devices operated under Part 15, the rules stipulate that the devices must not cause harmful interference to other radio services and must accept any interference caused by the legal operation of radio services. Amateurs need to know what the manufacturer has told the consumer and what the consumer is *supposed* to know about the interference potential. *Amateurs can often direct consumers to the owner's manual of the affected device for information on the potential for RFI and for its elimination.* These rules explain to the consumer whose responsibility it is to resolve the interference.

Summary of Equipment Authorization Procedures and General Information

No discussion of RFI would be complete without a discussion of FCC equipment authorization procedures, which are based on Parts 2, 15, 18, 68 and 76 of the FCC rules. Manufacturers can't simply market an RF device without the appropriate FCC approval; standards must be met to ensure that the device does not cause undue interference or constitute a hazard to users. These standards apply to a wide range of intentional, unintentional and incidental emitters of RF energy, ranging from unlicensed RF transmitters, such as baby monitors and low-power walkie-talkies to small digital devices to personal computers and peripherals.

Many people think that FCC regulations require these devices to be tested by the FCC, but in reality, few must actually undergo FCC testing. In most cases, the requirements are met by the manufacturer testing the device and either keeping the test results on file or by sending them to the FCC, depending on the type of device involved.

The following summary of Part 2, Subpart J, provided by the FCC Equipment Authorization Branch, provides some general information concerning various FCC approval processes for RF devices:

Certification: Requires submittal of an application that includes a complete technical description of the product and a measurement report showing compliance with the FCC technical standards. Devices subject to certification include: low power transmitters such as cordless telephones, security alarm systems, scanning receivers, superregenerative receivers, Amateur Radio external HF amplifiers and amplifier kits, and TV interface devices such as VCRs.

Declaration of Conformity: A Declaration of Conformity is the usual approval procedure for personal computers and personal computer peripherals. This authorization, based on a declaration that equipment complies with FCC requirements, it applies only to Class B personal computers and peripherals. A DoC is an alternative to Certification since no application to FCC is required, but the applicant must have the device tested at an accredited laboratory.

Notification: Requires submittal of an abbreviated application for equipment authorization, which does not include a measurement report, to the FCC. However, a measurement report showing compliance of the product with the FCC technical standards must be retained by the applicant and must be submitted upon request by the Commission. Devices subject to notification include: point-to-point microwave transmitters; AM, FM and TV Broadcast transmitters; certain microwave auxiliary broadcast transmitters and other receivers (except as noted elsewhere).

Verification: Verification is a self-approval process where the applicant performs the necessary tests and verifies that they have been done on the device to be authorized and that the device is in compliance with the technical standards. Devices subject to verification include: business computer equipment (Class A); TV and FM receivers; and, non-consumer Industrial, Scientific and Medical Equipment. Verified equipment requires that a compliance label be affixed to the device as well as information included in the operating manual regarding the interference potential of the device. The wording for the compliance label and the information statement regarding interference problems is included in Part 15 of the FCC Rules. Verified devices must be uniquely identified with a brand name and/or model number that cannot be confused with other devices on the market. However, they may not be labeled with an FCC identifier or in a manner that could be confused with an FCC identifier.

Type Acceptance: Type acceptance is no longer used by the FCC. Type acceptance procedures have been combined into certification procedures. The certification and type acceptance procedures have always been very similar. This report application required a complete technical description of the product and a test showing compliance with technical requirements. Traditionally, type acceptance has been applied to radio transmitters that are used in authorized radio services such as commercial and private mobile radio services. Type acceptance required and certification requires the filing of an application form and a technical report. Amateur Radio transceivers never required type acceptance, but external HF power amplifiers and amplifier kits did, and they now require FCC certification.

Although the FCC does still do some certification work directly, most of the approvals given to the FCC are done by FCC-approved Technical Certification Bodies (TSBs).

Specific information on obtaining an equipment authorization can be obtained from the FCC Web site at: **www. fcc.gov/oet/info/filing/ead/** or from the FCC Application Processing Branch in Columbia, Maryland. Individuals with questions concerning equipment authorization procedures should be addressed to:

Federal Communications Commission
Application Processing Branch
7435 Oakland Mills Rd
Columbia, MD 21046
Tel: 301-362-3000
Facsimile: 301-362-3000
E-mail: **labinfo@fcc.gov**

Interested amateurs can view the FCC equipment authorization database at: **https://gullfoss2.fcc.gov/prod/oet/cf/eas/reports/GenericSearch.cfm**

General Part 15 Technical Requirements

The following regulations apply to *all* Part 15 devices:

§15.5 General conditions of operation.

(a) Persons operating intentional or unintentional radiators shall not be deemed to have any vested or recognizable right to continued use of any given frequency by virtue of prior registration or certification of equipment, or, for power line carrier systems, on the basis of prior notification of use pursuant to §90.63(g) of this chapter.

(b) Operation of an intentional, unintentional, or incidental radiator is subject to the conditions that no harmful interference is caused and that interference must be accepted that may be caused by the operation of an authorized radio station, by another intentional or unintentional radiator, by industrial, scientific and medical (ISM) equipment, or by an incidental radiator.

(c) The operator of a radio frequency device shall be required to cease operating the device upon notification by a Commission representative that the device is causing harmful interference. Operation shall not resume until the condition causing the harmful interference has been corrected.

(d) Intentional radiators that produce Class B emissions (damped wave) are prohibited.

Table 17.2
Unlicensed Part 15 Bands

Band	*Typical Use*
160 - 190 kHz	Hobbyist
510 - 1705 kHz	Hobbyist
1.705 - 10 MHz	General
13.553 - 13.567 MHz	General
26.96 - 27.28 MHz	Hobbyist, walkie-talkie, baby monitor, etc.
40.66 - 40.70 MHz	Perimeter protection systems, control systems
43.71 - 44.49 MHz	Cordless telephones
46.60 - 46.98 MHz	Cordless telephones
48.75 - 49.51 MHz	Cordless telephones
49.66 - 50.0 MHz	Cordless telephones, walkie-talkie, baby monitor, etc.
72.0 - 73.0 MHz	Auditory assistance devices
74.6 - 74.8 MHz	Auditory assistance devices
75.2 - 76.0 MHz	Auditory assistance devices
88.0 - 108.0 MHz	Hobbyist, miscellaneous
174.0 - 216.0 MHz	Biomedical telemetry on unused TV channels
890.0 - 940.0 MHz	Measurement systems, **Amateur**
902.0 - 928.0 MHz	Multiple Part 15 uses, **Amateur**
1.91 - 1.93 GHz	Personal Communications Service (PCS)
2.39 - 2.4 GHz	PCS, **Amateur**
2.435 - 2.465 GHz	PCS, **Amateur**
2.9 - 3.26 GHz	Automatic vehicle identification systems (AVIS)
3.267 - 3.332 GHz	AVIS, **Amateur**
3.339 - 3.3458 GHz	AVIS, **Amateur**
3.358 - 3.6 GHz	AVIS, **Amateur**
5.15 - 5.35 GHz	Unlicensed National Infrastructure Devices (U-NII)
5.725 - 5.825 GHz	U-NII, other unspecified uses
10.500 - 10.550 GHz	Multiple Part 15 uses, **Amateur**
24.075 - 24.175 GHz	Multiple Part 15 uses, **Amateur**
46.7 - 46.9 GHz	Vehicular collision avoidance/radar systems
76.0 - 77.0 GHz	Vehicular collision avoidance/radar systems, **Amateur**

One point that echoes throughout Part 15 is that these devices can't cause harmful interference to other services. This point is illustrated below:

§15.13—Incidental radiators—states that "Manufacturers of these devices shall employ good engineering practices to minimize the risk of harmful interference."

Certain Part 15 technical regulations apply to all such devices:

§15.15 General technical requirements.

(a) An intentional or unintentional radiator shall be constructed in accordance with good engineering design and manufacturing practice. Emanations from the device shall be suppressed as much as practicable, but in no case shall the emanations exceed the levels specified in these rules.

(b) An intentional or unintentional radiator must be constructed such that the adjustments of any control that is readily accessible by or intended to be accessible to the user will not cause operation of the device in violation of the regulations.

(c) Parties responsible for equipment compliance should note that the limits specified in this part will not prevent harmful interference under all circumstances. Since the operators of Part 15 devices are required to cease operation should harmful interference occur to authorized users of the radio frequency spectrum, the parties responsible for equipment compliance are encouraged to employ the minimum field strength necessary for communications, to provide greater attenuation of unwanted emissions than required by these regulations, and to advise the user as to how to resolve harmful interference problems (for example, see §15.105(b)).

§15.17 is especially noteworthy because it advises designers and manufacturers of electronic devices that use RF energy to be aware of the potential for RFI from outside sources, such as Amateur Radio! It also advises manufacturers to take appropriate measures to control the susceptibility of their equipment to RFI:

§15.17 Susceptibility to interference.

(a) Parties responsible for equipment compliance are advised to consider the proximity and the high power of non-Government licensed radio stations, such as broadcast, amateur, land mobile, and non-geostationary mobile satellite feeder link earth stations, and of U.S. Government radio stations, which could include high-powered radar systems, when choosing operating frequencies during the design of their equipment so as to reduce the susceptibility for receiving harmful interference. Information on non-Government use of the spectrum can be obtained by consulting the Table of Frequency Allocations in §2.106 of this chapter.

Part 15 Label Requirements

Most Part 15 devices require a label attesting to the potential for interference and to the responsibility of the device operator. The following sections from the rules provide examples of the labeling requirements:

§15.19 Labeling requirements.

(a) In addition to the requirements in Part 2 of this chapter, a device subject to certification, notification, or verification shall be labeled as follows:

* * *

(3) All other devices shall bear the following statement in a conspicuous location on the device:

This device complies with part 15 of the FCC Rules. Operation is subject to the following two conditions: (1) This device may not cause harmful interference, and (2) this device must accept any interference received, including interference that may cause undesired operation.

Information to the User

Part 15 regulations are specific as to what the manufacturer must tell the consumer:

§15.105—Information to the user.

(a) For a Class A digital device or peripheral, the instructions furnished the user shall include the following or similar statement, placed in a prominent location in the text of the manual:

Note: This equipment has been tested and found to comply with the limits for a Class A digital device, pursuant to Part 15 of the FCC Rules. These limits are designed to provide reasonable protection against harmful interference when the equipment is operated in a commercial environment. This equipment generates, uses, and can radiate radio frequency energy and, if not installed and used in accordance with the instruction manual, may cause harmful interference to radio communications. Operation of this equipment in a residential area is likely to cause harmful interference in which case the user will be required to correct the interference at his own expense.

(b) For a Class B digital device or peripheral, the instructions furnished the user shall include the following or similar statement, placed in a prominent location in the text of the manual:

Note: This equipment has been tested and found to comply with the limits for a Class B digital device, pursuant to Part 15 of the FCC Rules. These limits are designed to provide reasonable protection against harmful interference in a residential installation. This equipment generates, uses and can radiate radio frequency energy and, if not installed and used in accordance with the instructions, may cause harmful interference to radio communications. However, there is no guarantee that interference will not occur in a particular installation. If this equipment does cause harmful interference to radio or television reception, which can be determined by turning the equipment off and on, the user is encouraged to try to correct the interference by one or more of the following measures:

—Reorient or relocate the receiving antenna.

—Increase the separation between the equipment and receiver.

—Connect the equipment into an outlet on a circuit different from that to which the receiver is connected.

—Consult the dealer or an experienced radio/TV technician for help.

Absolute Limits

As examples of the actual levels of the absolute limits in Part 15, the tables from Part 15 are reproduced here. These tables are for example only because each contains numerous notes that grant exceptions to these limits for various devices or explain how they are to be applied to different types of unintentional emitters. The notes also permit compliance with certain other regulations or standards in lieu of Part 15. Consult the actual text of Part 15 for more information. In general, the limits in Sections 15.107 and 15.109 apply to unintentional radiators such as computer systems and digital devices.

§15.107 Conducted limits.

(a) Except for Class A digital devices, for equipment that is designed to be connected to the public utility (ac) power line, the radio frequency voltage that is conducted back onto the ac power line on any frequency or frequencies within the band 150 kHz to 30 MHz shall not exceed the limits in the following table, as measured using a 50 μH/50 ohms line impedance stabilization network (LISN). Compliance with this provision shall be based on the measurement of the radio frequency voltage between each power line and ground at the power terminals. The lower limit applies at the band edges

Frequency of Emission (MHz)	*Conducted limit dBμV*	
	Quasi-peak	*Average*
0.15-0.5	66 to 56*	56 to 46
0.5-5	56	46
5-30	60	50

* Decreases with the logarithm of the frequency

(b) For a Class A digital device that is designed to be connected to the public utility (ac) power line, the radio frequency voltage that is conducted back onto the ac power line on any frequency or frequencies within the band 150 kHz to 30 MHz shall not exceed the limits in the following table, as measured using a 50μH/50 ohms line impedance stabilization network (LISN). Compliance with this provision shall be based on the measurement of the radio frequency voltage between each power line and ground at the power terminals. The lower limit applies at the band edges.

Frequency of Emission (MHz)	*Conducted limit dBμV*	
	Quasi-peak	*Average*
0.15-0.5	79	66
0.5-30	73	60

* * *

§15.109 Radiated emission limits.

(a) Except for Class A digital devices, the field strength of radiated emissions from unintentional radiators at a distance of 3 meters shall not exceed the following values:

Frequency of emission (MHz)	*Fieldstrength (microvolts/meter)*
30-88	100
88-216	150
216-960	200
Above 960	500

(b) The field strength of radiated emissions from a Class A digital device, as determined at a distance of 10 meters, shall not exceed the following:

Frequency of emission (MHz)	*Field strength (microvolts/meter)*
30-88	90
88-216	150
216-960	210
Above 960	300

Intentional Radiators

Part 15, Subparts C and D address RF emission limits from different types of *intentional radiators* such as remotely controlled garage door opener controls, R/C toy cars, hand-held automobile door lock and burglar alarm controls, cordless telephones (including both base and handset), wireless baby monitors, toy "walkie-talkies," dog trainers, carrier-current and perimeter-protection systems. Amateurs may encounter such devices, either as the source or the victim of an RFI problem.

The Part 15 regulations on intentional radiators are complex, taking page after page to explain all the nuances. Interested readers should refer to the FCC rules for further information about this equipment and its associated emission limits, since the Part 15 radiation limits contain many exceptions.

Part 15 essentially permits very low power, unlicensed intentional radiators on nearly any frequency, including all amateur bands. The following table, from §15.205, shows the frequencies for which intentional radiators cannot be operated. There are a number of notes in the actual text of the regulations; refer to Part 15 for more information.

§15.205 Restricted Bands of Operation

(a) Except as shown in paragraph (d) of this section, only spurious emissions are permitted in any of the frequency bands listed below:

MHz	*MHz*	*MHz*	*GHz*
0.090-0.110	16.42-16.423	399.9-410	4.5-5.15[1]
0.495-0.505	16.69475-16.69525	608-614	5.35-5.46
2.1735-2.1905	16.80425-16.80475	960-1240	7.25-7.75
4.125-4.128	25.5-25.67	1300-1427	8.025-8.5
4.17725-4.17775	37.5-38.25	1435-1626.5	9.0-9.2
4.20725-4.20775	73-74.6	1645.5-1646.5	9.3-9.5
6.215-6.218	74.8-75.2	1660-1710	10.6-12.7
6.26775-6.26825	108-121.94	1718.8-1722.2	13.25-13.4
6.31175-6.31225	123-138	2200-2300	14.47-14.5
8.291-8.294	149.9-150.05	2310-2390	15.35-16.2
8.362-8.366	156.52475-156.52525	2483.5-2500	17.7-21.4
8.37625-8.38675	156.7-156.9	2655-2900	22.01-23.12
8.41425-8.41475	162.0125-167.17	3260-3267	23.6-24.0
12.29-12.293	167.72-173.2	3332-3339	31.2-31.8
12.51975-12.52025	240-285	3345.8-3358	36.43-36.5
12.57675-12.57725	322-335.4	3600-4400	[2]
13.36-13.41			

[1] Until February 1, 1999, this restricted band shall be 0.490-0.510 MHz.

[2] Above 38.6

* * *

On other frequencies, §15.209 states that the radiated emission limits of intentional radiators generally can't exceed the field strength levels specified in the following table:

Frequency (MHz)	*Field Strength (microvolts/meter)*	*Measurement Distance (meters)*
0.009-0.490	2400/F(kHz)	300
0.490-1.705	24000/F(kHz)	30
1.705-30.0	30	30
30-88	100	3
88-216	150	3
216-960	200	3
Above 960	500	3

In the amateur 2-meter band, for example, these regulations would permit an intentional radiator to run a transmitter power of about 4 nanowatts into a half-wave dipole, resulting in a field strength of 150 microvolts/meter at a point 3 meters away from the dipole center.

This "legal" signal would be about S8 to a typical amateur station on that frequency. On HF, a "legal" unlicensed signal would typically be greater than S9 to nearby amateur stations. The provision in the Part 15 rules that requires that an unlicensed device not cause harmful interference to a licensed radio user is an important part of the rules to help control interference.

In addition, Part 15 permits a number of "periodic" devices to transmit on all permitted frequencies at a much higher power. Periodic devices are characterized as various control and signaling devices with very limited transmission times.

There are a number of bands set aside for higher power intentional radiators. These bands are used for various unlicensed walkie-talkies, baby monitors, cordless telephones and the like. See Table 17.2.

Part 18: Industrial, Scientific and Medical Regulations

The FCC rules in Part 18 deal with *industrial, scientific and medical* equipment (ISM) that emits electromagnetic energy in the RF spectrum. As with the equipment covered by Part 15, the ISM equipment addressed here is also a potential source of RFI. The most applicable sections of Part 18 follow.

The basic FCC definitions of ISM equipment are given in five paragraphs of §18.107. Industrial, scientific and medical equipment (ISM devices) generate and use locally RF energy for industrial, scientific, medical, domestic or similar purposes, excluding applications in the field of telecommunication. Typical ISM applications are the production of physical, biological or chemical effects such as heating, ionization of gases, mechanical vibrations, hair removal and acceleration of charged particles. They also include industrial heating equipment, medical diathermy equipment, ultrasonic equipment, consumer ISM equipment (domestic microwave ovens, jewelry cleaners for home use and ultrasonic humidifiers).

§18.109 directly addresses the interference issues of ISM equipment. Basically ISM equipment should be of "good engineering design" with adequate filtering to provide adequate suppression outside of the ISM bands. §18.111 outlines the basic RFI operating conditions for users of ISM equipment and §18.115 outlines procedures for eliminating and investigating harmful interference from ISM equipment. This section states that "(a) The operator of ISM equipment that causes harmful interference to radio services shall promptly take appropriate measures to correct the problem."

However, that provision "shall not apply in the case of interference to an authorized radio station or a radiocommunication device operating *in* an ISM frequency band." And, it "shall not apply in the case of interference to a receiver arising from direct intermediate frequency pickup by the receiver of the fundamental frequency emissions of ISM equipment operating in an ISM frequency band and otherwise complying with the requirements of this part." These conditions are very similar to the RFI operating conditions for Amateur Radio Service. In other words, if the ISM equipment is operating properly in its assigned band, any RFI it is causing must be accepted or corrected by the entity that is receiving the interference, including amateurs.

§18.213 is important because it alerts consumers in the operations manual or packaging material to the interference potential and simple measures that can be taken to correct the problem. §§18.301-311 deal with the operating frequencies and field-strength limits of the ISM bands. These bands are distributed across the spectrum in small segments from 6.78 MHz to 245 GHz.

The following ISM bands overlap all or portions of certain amateur bands:

ISM frequency	*Tolerance*	*Amateur Band*
915 MHz	±13.0 MHz	902-928 MHz
2,450 MHz	±50.0 MHz	2300-2310 and 2390-2450 MHz
5,800 MHz	±75.0 MHz	5.650-5.925 GHz
24,125 MHz	±125.0 MHz	24.0-24.25 GHz
245.00 GHz	±1.0 GHz	241-250 GHz

The field-strength and conduction limits depend on the band of operation and the equipment used. Interested readers should refer to these sections of the FCC rules and regulations for a complete listing of the bands and limits.

Part 76: Cable Television Regulations

The operation of the Cable Television Service is covered by Part 76. Because cable systems can, and do, operate in the amateur bands, there exists a potential for interaction between these systems and other radio services, including Amateur Radio. If they are working properly, cable systems operate as closed systems; that is, their transmissions must stay *inside* the cable. If old cable begins to break down, is installed improperly or is misused by the consumer, RFI can rear its ugly head!

Although most of Part 76 is of interest only to cable operators, part of Subpart K (Technical Standards) is of great interest to amateurs. These standards contain a number of provisions designed to protect over-the-air services from harmful interference from signals that might leak out of the cable. There are two provisions of interest: Part 76 sets an *absolute* limit on the amount of leakage that can occur.

In addition, Part 76 stipulates that *if harmful interference results from any leak, irrespective of its level, the cable operator must take whatever steps are reasonably necessary to resolve the interference.* There is often some disagreement about what constitutes harmful interference. Clearly, if a cable leak prevents a local amateur from accessing a local repeater, it is harmful interference. Simply hearing a squelch break while listening to an unused channel is probably not.

Cable System Interference

§76.613 contains specific obligations of cable companies when interference is caused to licensed radio services:

§76.613 Interference from a cable television system.

(a) Harmful interference is any emission, radiation or induction which endangers the functioning of a radionavigation service or of other safety services or seriously degrades, obstructs or repeatedly interrupts a radiocommunication service operating in accordance with this chapter.

(b) The operator of a cable television system that causes harmful interference shall promptly take appropriate measures to eliminate the harmful interference.

(c) If harmful interference to radio communications involving the safety of life and protection of property cannot be promptly eliminated by the application of suitable techniques, operation of the offending cable television system or appropriate elements thereof shall immediately be suspended upon notification by the Engineer in Charge (EIC) of the Commission's local field office, and shall not be resumed until the interference has been eliminated to the satisfaction of the EIC. When authorized by the EIC, short test operations may be made during the period of suspended operation to check the efficacy of remedial measures.

(d) The cable television system operator may be required by the EIC to prepare and submit a report regarding the cause(s) of the interference, corrective measures planned or taken, and the efficacy of the remedial measures.

In §76.617, the FCC clearly states that interference *from* a cable system is the responsibility of the cable operator to resolve:

§76.617 Responsibility for Interference

Interference resulting from the use of cable system terminal equipment (including subscriber terminal, input selector switch and any other accessories) shall be the responsibility of the cable system terminal equipment operator in accordance with the provisions of Part 15 of this chapter: provided, however, that the operator of a cable system to which the cable system terminal equipment is connected shall be responsible for detecting and eliminating any signal leakage where that leakage would cause interference outside the subscriber's premises and/or would cause the cable system to exceed the Part 76 signal leakage requirements. In cases where excessive signal leakage occurs, the cable operator shall be required only to discontinue service to the subscriber until the problem is corrected.

Part 76 Technical Standards

§76.601—Performance tests—states that the operator of cable systems "shall be responsible for ensuring that each such system is designed, installed, and operated in a manner that fully complies with the provisions of this subpart." In addition, they must be prepared to show the FCC, on request, that the system does, in fact, comply with the rules. Cable operators are required to conduct complete performance tests of that system at least twice each calendar year.

§76.605—Technical standards—details specific performance standards of systems.

§76.605 covers the technical standards for the operation of a cable system such as video-signal level and signal-to-noise ratio. This section also specifies the maximum radiation allowed from a cable system. Regarding harmful interference from a cable system, §76.613 clearly places responsibility on the cable company (see the sidebar "Leaky Cable Television Systems—Help!" earlier in this chapter).

In essence, Amateur Radio operators who receive interference from local cable television systems have legal recourse to cause the cable company to "clean-up their act." Nonetheless, if their signals can exit the system, amateur signals can penetrate it and interfere. As with all RFI problems, "cooperation" is a key word when solving cable interference problems.

§76.605 Technical standards.

(a) As of December 30, 1992, unless otherwise noted, the following requirements apply to the performance of a cable television system as measured at any subscriber terminal with a matched impedance at the termination point or at the output of the modulating or processing equipment (generally the headend) of the cable television system or otherwise as noted. The requirements are applicable to each NTSC or similar video downstream cable television channel in the system:

(7) The ratio of RF visual signal level to system noise shall be as follows:

(iii) As of June 30, 1995, shall not be less than 43 decibels.

(8) The ratio of visual signal level to the RMS amplitude of any coherent disturbances such as intermodulation products, second and third order distortions or discrete-frequency interfering signals not operating on proper offset assignments shall be as follows:

(i) The ratio of visual signal level to coherent disturbances shall not be less than 51 decibels for noncoherent channel cable television systems, when measured with modulated carriers and time averaged; and

(ii) The ratio of visual signal level to coherent disturbances which are frequency-coincident with the visual carrier shall not be less than 47 decibels for coherent channel cable systems, when measured with modulated carriers and time averaged.

Specific radiation limits are:

Frequency	*Signal leakage limits (microvolt/meter)*	*Distance in meters (m)*
Less than and including 54 MHz, and over 216 MHz	15	30
Over 54 up to and including 216 MHz	20	3

Most cable operators understand their requirements with respect to signals leaking out of a cable system. In some serious cases, the FCC can (and does) shut down cable channels if the leakage is persistent. This is especially true for leakage in the aircraft bands. (Hams can use this to good advantage. If a leak is heard on the amateur bands, it will almost always be heard equally well on the aircraft bands. If the latter is reported to the cable company along with the report of interference in the ham bands, it will certainly get their attention!) Additional information on solving cable leakage problem can be obtained from the ARRL Web site at **www.arrl.org/tis/info/catvi/index/html**.

However, in addition to mandating absolute leakage limits, §76.605 also mandates a specific level of signal quality that must be delivered to subscribers. There are specific required levels of video carrier to "coherent disturbance" and video carrier to noise. If a cable operator states that they are not responsible to leakage *into* the cable system, point to these regulations. It has never been established whether an amateur signal leaking in is a "coherent disturbance" or "noise," but this provision in the regulations should cover it in either case.

RFI and International Law

The regulations of United States Amateur Radio Service, as they relate to RFI, apply as long as the operator is within the territorial confines of this country or in international waters. One who wishes to operate within the confines of another country's territory may do so only with a guest or reciprocal-operating permit from that country if required.

The primary driving force for international RFI regulation outside the United States is the European Union (EU). On January 1, 1995, the 15 member states of the European Union adopted the European Union Directives for Electromagnetic Compatibility. The directives require the "CE" (Conformite' Europe'enne) mark to be on a wide range of electronic products sold within the Union. The intent of the EU directives is to eliminate non-tariff barriers to trade and to ensure that all products flowing through the EU comply with general safety and environmental standards. EMC/RFI are part of these directives. The member states of the EU are under legal obligation to implement the Directives.

The EU's Electromagnetic Compatibility (EMC) Directive defines two principal objectives that all electrical and electronic products must satisfy before they can be marketed: (1) the product should not be a source of interference to others, and (2) it should be immune to interference.

To demonstrate compliance with the EMC Directive, a manufacturer can use a number of routes. First, it can show compliance with the relevant technical standards, which are agreed upon throughout Europe. They are known as European Norms (ENs). If no ENs apply, he can use any member state's national standards that have been deemed to meet the requirements of the Directive (by a committee known as the Article 5 Committee). Where technical standards do not exist (for example, the immunity of computing equipment), he must keep a technical file that records the EMC performance of the product. The file can be inspected by the regulatory body should the need arise. If the device is a radio transmitting apparatus, an EC type examination certificate may be issued by a competent body.

History

Throughout the '60s and '70s, the growing problem of RFI continued to threaten the well being of the Amateur Radio Service. Amateurs were faced with more than just radio and television interference. Cable TV systems, power lines and a host of new consumer devices posed equal or greater threats. The Citizens Band (CB) radio boom increased the probability that a home-electronic device would be located near a transmitter. The ARRL tried to explain to disgruntled consumers that typical RFI situations involving radio amateurs result from design deficiencies in the affected device. Consumers, however, found it difficult to believe that a "passive" device, such as a TV or stereo, could be the "source" of interference. After all, when the ham was not transmitting, there was no problem! Of course, the perception was that "the ham down the street" caused the RFI.

Could the RFI situation become even worse? Yes, it could and would. During the '70s, some local governments began adopting ordinances making it "illegal" to interfere with television or radio reception. These laws were based on a "causing a public nuisance" concept. No one wanted to hear the ham explain. The situation became intolerable.

The Coming of the Law

Amateur Radio was being blamed for the inability of electronic devices to reject unwanted radio signals, and a solution was needed. At the time, the FCC did not have the authority to set minimum rejection standards for consumer electronic devices, and ARRL leaders knew the situation would only get worse. Amateur Radio needed a law that would amend the Communications Act of 1934 giving the FCC exclusive jurisdiction over RFI matters. This law would preempt regulation by state or local governments. During the "old days," amateurs did not want RFI law, but it became evident in the '70s that a law was needed if the Amateur Radio Service was to survive.

The ARRL took a leading role by lobbying Congress for RFI legislation. Early attempts were unsuccessful; many bills were introduced, but they all died. It became clear that not everyone shared the ARRL's enthusiasm for giving the FCC authority to set RFI-immunity standards.

The electronics industry came out against standards because they might be forced to add a $5 filter to the cost of a product. The Electronic Industries Alliance (EIA) lobbied fiercely against the bills.

Public Law 97-259

Several major changes in the regulations were made on September 14, 1982, when President Reagan signed into law a landmark measure that provided RFI protection to the amateur service.

The changes affect Amateur Radio and RFI in the following areas:

•The FCC was given the authority to regulate susceptibility of electronic equipment to RFI. In essence, this is an attempt to stem the flow of electronic devices that cannot function normally in the presence of RF energy.

•The FCC was authorized to use the services of volunteers in monitoring for rules violations. When assisting the FCC, the volunteers can issue advisory notices to apparent violators, but volunteers are not authorized to take enforcement actions.

The legislative history of the bill gave the FCC jurisdiction over RF susceptibility of home entertainment equipment. It states, in part:

> This law clarifies the reservation of exclusive jurisdiction over RFI matters to the Federal Communications Commission. Such matters shall not be regulated by local or state law, nor shall radio transmitting apparatus be subject to local or state regulation as a part of any effort to resolve an RFI complaint. The FCC believes that radio operators should not be subject to fines, forfeitures or other liability imposed by any local or state authority as a result of interference appearing in home electronic equipment or systems. Rather, the Commission's intent is that regulation of RFI phenomena shall be imposed only by the Commission.

This all sounds good on paper, but in practice the results have been mixed. The FCC has opted for voluntary standards, rather than formal rules and regulations. The FCC and the ARRL are both active in an industry group (in cooperation with the American National Standards Institute, ANSI) that is studying "RF immunity issues." Starting in the second half of the 1990s, consumers started to see some built-in immunity in consumer products. Although not a total solution, the voluntary standard is certainly a step forward as far as the amateur service is concerned.

Current and Future Progress on the Legal Front

The number of sources of RFI have gradually increased over the years. More and more electronic and electrical devices have been invented and sold to consumers. The good news is that these devices have made our life more efficient; the bad news: the devices create unwanted interference and spurious signals.

Amateur Radio is affected in two ways: the need to provide consumer equipment that is resistant to interference, and the need to reduce or control spurious and primary radiation of transmitting stations. Recent changes to the FCC regulations (that is, Public Law 97-259) have given the FCC authority to regulate equipment RFI susceptibility. Public Law 97-259 also attempts to ease the load on the FCC by allowing volunteers to assist in monitoring for rules violations. The FCC must then respond if enforcement is necessary.

Amateur Radio is on the upswing. There will probably be an increase in number of radio transmitters on the air (and energy radiated). Increased activity brings the possibility for more RFI and violations. It also increases the need for monitoring and possibly modification of regulations.

The development of features that allow for automated radio operation (for example, automatic linking) may affect methods of use and expand the use of Amateur Radio. Will this development affect RFI? Let's hope the results will be positive: greater understanding and increased RFI immunity. The potential for radiation of spurious signals (RFI), however, will obviously increase.

Chapter 18

How to Form a Local RFI Committee

In most cases, RFI problems are solved locally. Local hams can effectively help each other through local RFI committees. Although the best way to organize an RFI Committee is determined by local needs, this chapter offers an overview of the basics.

By Ed Hare, W1RFI
ARRL Laboratory Supervisor

Every ham, especially the new ham, dreads the knock on the door or the phone call from a neighbor concerned about an RFI problem. Even with the resources of the *ARRL RFI Book,* the technical and personal aspects can be overwhelming. Fortunately, the traditions of hams helping hams applies to RFI, too, and many areas have local RFI Committees (RFICs), usually organized through a local club. In many committees, there are some real technical "heavy hitters" *and* people skilled in handling the "personal diplomacy" aspects of RFI.

Although some clubs and areas have a committee, others do not. The question is often asked, "A few members of our club have been having RFI problems. How can we form an RFI committee to help our members in an organized way?" This chapter addresses that very question.

WHAT IS A LOCAL RFI COMMITTEE?

There has been some confusion between the concept of a "RFI Committee" (RFIC) and a "Local Interference Committee" (LIC), although in some cases, there can be some overlap in function and membership. The LIC is generally organized to deal with local cases of amateur-to-amateur interference. When someone starts jamming the local repeater, for example, the LIC will often track the offender down and try to resolve the problem by talking to the offender and explaining the reasons not to cause interference. An RFIC is generally formed to deal with two specific issues—interference to consumer equipment from the operation of amateur stations and interference to amateur stations from non-radio sources such as power lines or neighboring consumer equipment.

The keyword is "local." ARRL HQ does support the work of local RFICs, but these are not "formal" ARRL Field Organization positions, such as Technical Coordinators (TCs) or Affiliated Club Coordinators (ACCs). Even for a committee formed by an ARRL affiliated club, it is not necessary to be an ARRL member or to be appointed by the ARRL Section Manager. RFICs are organized and run locally. Most RFICs are organized through local clubs. This chapter will focus on clubs with the expectation that the methods described will also apply to non-club RFICs, too.

The makeup and duties of an RFIC can, and will, vary a lot, depending on local needs and resources. The RFIC of a large club of a few hundred members will generally be much different than the informal RFIC of a club consisting of a dozen members in a rural area. This chapter will describe typical ways to form and organize an RFIC in a larger club, with the expectation that the methods can be scaled down to fit smaller, less formal needs.

WHERE TO BEGIN

When the idea that a club needs an RFIC is brought up at a club meeting, the heads usually start nodding up and down, indicating general agreement. Most hams have had RFI problems and would certainly have appreciated some help in getting them resolved. However, it is sometimes hard to translate this general agreement into action (as most club members know from watching the results in other club areas). The first step, then, is to ask the club members how many would be interested in becoming active in a club RFIC.

Don't be surprised if no hands go up! Many hams fear RFI and feel that if you are not an EMC engineer, you won't be able to understand it or fix RFI problems. Although there *is* a lot to learn about RFI, or at least about all aspects of RFI, in most cases, the standard cures work. Once someone learns the basics, they can do a lot to identify RFI causes and suggest the correct cures. Nonetheless, the first step is going to be finding club members willing to do the job. If everyone in the club wants "George" to do it (probably because George never shows up at club meetings), then you can stop right here—your club will have to do without an RFIC.

Planning, Planning

If you are lucky and a few members of the club are willing to pitch in (where would clubs be without those who are willing to give so much of their time to their fellow amateurs?), their first task is to plan the organization and duties of the RFIC. In some clubs, for example, the RFIC may choose to limit its scope, perhaps dealing *only* wi consumer-equipment or power-line issue (This is fine—the RFIC will still be ve helpful in whatever they decide to do.) Th place to begin is to select either a committe chairman or a planning chairman—nearl every effort needs a leader. The chairma does *not* need to be the technical expert o the committee. In fact, in some cases, th best chairman is the one who is most skille at dealing with people, able to work with th rest of the committee, the involved hams an with the neighbors of hams who are often th hardest part of the problem to resolve. Th *most* important qualification of the chairma is that he or she is willing to do the job.

RFICs are comprised of volunteers. Volunteers have a funny habit; they tend to do those things they want to do and sometime get a bit testy if people start making demands of them. Keep this in mind as you set up the RFIC. First, assess the skills and interests o those who are willing to help. It is helpful to make a list of members' areas of expertise and their willingness to take on various tasks. This will help the committee to assign the right jobs to the right people.

The skills and interests may cover a wider range than you might think. When people first think about an RFIC, they think of a group of volunteers who visit hams and their neighbors and solve RFI problems in the field. While this certainly does happen, it doesn't work for all circumstances in all areas. In rural areas, for example, it is not often practical to get a large group together to drive many miles, several times, to resolve a problem. Even in more populated areas, you may find people willing to serve on the RFIC who will gladly talk to hams on the phone and offer their expertise, but do not have the

Putting a Committee Together

In general, the tasks that need to be formed to put an RFIC together are listed below. It is not necessary for every committee to follow every one of these steps, but they do represent the range of what might need to be done.

- Select an RFIC Chairman
- Agree on the scope of the committee
- Assess the skills and interests of the committee members
- Recruit new members
- Select volunteers to:
  - Document the committee scope and membership
  - Present information about the RFIC to the club
  - Present information about the RFIC to other clubs in the area
  - Act as liaison to the ARRL Field Organization (Section Manager, Technical Coordinator, Affiliated Club Coordinator, and so on)
  - Obtain and share ARRL information about RFI
  - Obtain the "RFIC Survival Kit" of filters, etc.
  - Function as point of first contact
- Personal Diplomacy Coordinator
- Technical leadership
- Reports and Documentation

time, or inclination, to do "house calls."

Assessing the willingness of the RFIC members to perform various tasks will often lead to a better definition of the scope of the committee. As the role of the committee is defined, it will ultimately be included in the written documentation of the committee organization, function and membership.

Tasks

Once the chairman is selected and the staff resources of the committee have been assessed, it is time to start organizing those who will get the committee off the ground. For larger, formally organized committees, there is a lot to do to get things rolling—more than will be necessary to keep it rolling once a bit of momentum is obtained. It may be possible to solicit the help of other club members who may not end up being active members of the RFIC (although once you have them hooked . . .). For example, it may be possible to find a club member willing to draft the document describing the scope and membership of the RFIC.

Skills and Interests

In most RFICs, what can be accomplished is determined as much by the skills and interests of the participants as it is by the needs of those the committee might help. Although the committee that has a really skilled EMC expert is truly blessed, a lot can be accomplished by less experienced members. There is plenty of non-technical work that needs to be done, too, so if that enthusiastic newcomer volunteers, snap that offer up fast! Most of the time, the "stock" cures for RFI fix the problem, so if someone has read through the chapters of this book, he or she is well qualified to serve as the technical wizard.

The Chairman

The person selected to chair the committee plays a special role. In most committees, the job of chairman involves leadership, not doing all of the work. The chairman will often serve as the first point of contact between the RFIC and club members or the general public. The chair will also usually run any formal meetings the RFIC may choose to have, keep track of the skills and availability of the RFIC members and be available to make decisions on behalf of the RFIC in those instances where it is not necessary to have a formal meeting and/or vote. It is important that the chairman have a broad-based general understanding of the issues surrounding RFI problems. Even more important, the chairman needs to be well organized and be able to maintain a cool head when approached by an irate neighbor of a ham.

The Peacekeeper

Most RFICs do have a technical wizard, but believe it or not, he or she may not be the most important member of the committee. Those who serve on the RFIC to help smooth out bad feelings between the ham and his or her neighbor may be more important; if everyone involved in the RFI problem is still fighting with each other, the services of the EMC experts might never be required. This role may be formal or informal, and often must be played by everyone on the committee, but as it becomes apparent that one or more individuals are successful in their negotiating with the neighbors of hams (or hams for that matter—some of them have contributed to those situations that have gotten out of hand), they will assume the role of peacekeeper.

Like the chairman, the peacekeeper must have a general understanding of RFI issues. He or she must be able to explain to a neighbor or ham the differences between interference that can be caused by improper operation of a transmitter (such as TVI to an antenna-connected TV) and that which is caused by fundamental overload, such as telephone interference.

Paperwork

It is important that the work of the committee be well documented. It doesn't happen very often, but the RFI case you are working on could end up at the FCC (or worse). The RFIC should document the entire chain of events relating to contact with the involved parties, the results of any tests or cures tried and any conclusions reached by the committee.

Public Relations

Related to all of this is the general topic of public relations. This topic covers a lot of area. Once the committee is up and running, someone needs to explain to the club what the committee is all about, what they can do for the club and what the RFIC expects and needs from the club members who will use its services. This will usually take about a half hour or so, plus time for questions—just about the length of time needed by the typical club speaker.

Some club RFICs are also willing to help out on a regional basis. In this case, it will be helpful to do some PR work with other clubs, or by posting information about the RFIC on local packet or land-line BBSs, and so on.

Last but not least, clubs are *always* looking for hot speakers, and RFI *is* a hot topic. If someone on the RFIC (preferably a member with a good understanding of the technical issues involved) is willing to put together a complete presentation on RFI, your club, or area clubs, will eat it up. This book provide *lots* of guidance on what to say. You may be able to convince your section's Technical Coordinator or Technical Specialists to do an RFI presentation for you.

Liaison

No organization works in a vacuum. Even if the RFIC is limited to serving one single club, it is important for hams in the area to know about it. Your RFIC may serve as an example to other clubs on forming their own, and you may be able to offer them some advice and help. You might also find that you can pick up a few new members from hams in your area who are willing to join your club to take advantage of the services provided by your RFIC.

Some of this will fall out of your RFIC "PR" activities, but there is another mechanism you can use—the ARRL and its Field Organization. You should appoint someone to serve as liaison to your ARRL Division Director and Vice Director and the ARRL Field Organization (if your club doesn't already have an ARRL liaison). Your ARRL Division Director, Vice Director, Section Manager, your section Technical Coordinator, any Technical Specialists that are near your club's "service area," your Affiliated Club Coordinator and perhaps others should be made aware of your RFIC and its activities so they can refer people to your club when questions about RFI come their way. Don't forget to liaison with ARRL HQ, too. If you run across a sticky RFI problem, you can contact the Technical Information Service in the ARRL Laboratory for some help.

Recruit New RFIC Members

With all of these non-technical jobs better defined, you may be able to recruit new members. In fact, one or more of these jobs may be just the ticket for someone who has been looking for the right role in your club. Don't be afraid to ask; in many clubs, most of the work is done by a handful of people, but you may be surprised at the number of others who are willing to help if they are given something *specific* to do.

Information

Every member of the RFIC should read a copy of the *ARRL RFI Book*. It may not be necessary for each to own his or her own copy, but if the club is going to publicize the RFIC,various members may occasionally be called upon to answer questions about RFI. For this reason, having a copy of the book handy would be helpful.

In addition to the book, ARRL HQ has some resources available, too. Our *RFI*

Consumer Pamphlet, written especially for the neighbors of hams, has been reprinted in the appendices. It is sometimes more effective to have a formally printed copy to hand to a neighbor. This pamphlet, by the way, is a joint effort by the ARRL and the Consumer Electronic Association. The ARRL solicited CEA's participation to help strengthen the credibility of the information in the pamphlet; it contains information that the electronic manufacturers and ham operators can agree on. The RFI package can be downloaded from the ARRL Web page (**www.arrl.org/tis/**).

Confidence

It is important that both the ham and his or her neighbor have confidence in the RFIC. After all, you may be working in their homes and with their equipment. If the RFIC is not able to work efficiently, and tries too many things that don't work, the neighbor may lose confidence and put a stop to the proceedings. This needs to be balanced, however, with *realistic* expectations. The RFIC is comprised of unpaid volunteers, and volunteerism does have its limits. All concerned have a right to expect a good faith effort, but there are no guarantees in the RFI business. The moral is: Don't make promises you can't keep. If you promise to cure the problem and it turns out to be direct pickup by an unshielded TV, you may be putting yourselves, and the ham you are helping, into a difficult situation.

RFI Survival Kit

To help the RFIC work efficiently, they should bring commonly needed items with them. The following items should be part of the RFI "Survival Kit:"

- Soldering iron and supplies
- Hand tools
- Dummy load
- HF low-pass filters (low and high power)
- VHF band-pass filters (2 meters, etc.) Ferrite toroid cores, assorted sizes (#75 and #43 material)
- Ferrite rods (assorted)
- "Split-bead" ferrites (assorted)
- AC-line filters
- High-pass filter
- Capacitors for bypassing
- Shielding spray
- Shielding material
- Copper or aluminum tape
- Audio (RCA) cables with ferrite attached
- Assorted lengths (non-resonant if possible) of telephone cable with modular plugs
- Cable ties
- Duct tape and electrical tape
- Clip leads
- 75-Ω cable TV terminators and barrel connectors
- Assorted known-good coax feed-line with male and female connectors (cable and amateur)
- Ground rod
- Ground wire
- VHF notch filter
- Telephone RFI filters (assorted)
- RFI-proof telephone
- Assorted wire
- RF "sniffer"
- A VOM (for checking integrity of connections)
- Hand-held scanner
- Known-good portable TV
- Handheld transceivers (for communications and signal reception)
- Field-strength meter
- AM broadcast receiver
- RDF equipment (optional)
- AC-convenience-outlet tester (looks like a three-prong plug; checks for polarity and proper ground)
- A list of consumer electronics manufacturer's addresses
- A copy of the FCC's interference handbook (to show the FCC's per-spective)
- Appropriate ARRL RFI information and RFI Pamphlet; see: **www.arrl.org/tis/info/rfigen.html**.
- *The ARRL RFI Book*

Having all of these items will enable the RFIC to diagnose and cure most RFI problems quickly.

Ready for Action

The RFIC is now ready to start tackling RFI problems. In most cases, they will hear first from the ham. The best way to handle things from there depends on many factors, including: the size of the RFIC, the skills of the individual members, and the previous relationship between the ham and the neighbor. The following example "case history" may provide some guidelines on how to proceed.

Case History

The Whoop and Holler Amateur Radio Club just formed an RFIC and they get their first call: Paul, W1EPG, has a neighbor with telephone and television interference. The RFIC Chairman, Mike, W1MG, is the RFIC contact person. He should start immediately by taking notes, asking Paul about the frequencies, power levels, and so on, that are causing the interference.

In this case, Paul knows that the problem is there when he operates 20 meter CW, but he rarely operates other bands, so he is not sure about the extent of the problem.

Paul and his neighbor have had a few discussions about the RFI, some of which have been a bit heated, so Mike, with Paul's concurrence, decides to ask Andy, W1FG, the RFIC "Personal Diplomacy" expert to give the neighbor a call. Andy does so, and he explains the complex issues of responsibility and cures to the neighbor. The neighbor doesn't really buy it 100%, but when Andy explains that Paul does want to see the problem get fixed and that the RFIC will be able to help, the neighbor is happy that forward progress will be made, so agrees to work together. (Andy is a diplomat extraordinaire!)

Paul, the RFIC, and neighbor arrange a convenient time for a test. Andy had suggested that the neighbor may want to involve a trusted friend, too, to watch the test from Paul's house. He selects Frank, another neighbor who has not had any RFI problems. Two members of the RFIC show up at Paul's house — Andy and Howard, W1HSR, a technical wizard of great renown. They give the neighbor a call on the phone, then walk next door to shake hands.

After everyone has a chance to meet and a few pleasantries are exchanged, Paul, Howard, and Frank go to Paul's house. They set up the station, pointing the beam right at the neighbor's house, and start testing. Paul operates band by band. They discover that there is RFI on the telephone from all HF bands. It is moderate on 20 meters, where Paul operates the most, but absolutely horrible on 40 meters. The television is connected to cable and seems to have RFI problems on 20 meters and up. When Paul transmits on 2 meters, the problem is occurring only on channel 18. A TV operating on rabbit ears gets interference on one channel when Paul transmits on 15 meters.

Paul and his neighbor agree to let the RFIC try a few stock cures. First, they put a low-pass filter on Paul's transmitter. This makes some difference on the antenna-connected TV, indicating that Paul *did* have some spurious emissions. (Under FCC rules, Paul must fix this, so Paul is going to be buying a low-pass filter.) They then try a high-pass filter on the antenna-connected TV. Bingo, the TVI to that TV is gone. One down, two to go.

The above cures made no difference to the cable-connected TV. The RFIC makes a common-mode choke from an FT-240-43 ferrite core and installs it on the cable, right where it connects to the set-top converter. Bingo, the TVI is gone.

Well, almost, anyway. When Paul transmits on 2 meters, the channel 18 TVI is still there. Channel 18 uses the 2 meter band, so this demonstrates that there is a leak in the cable system. The RFIC looks at some of the cable wiring and notes that the ground connection is quite corroded and that a number of the connections look suspect. They explain that the cable company should be able to diagnose the problem. The neighbor agrees to call the cable company; Paul agrees to be available for testing.

The telephone RFI problem is a bit more challenging. There are a number of telephones on the line, so the RFIC simplifies things by disconnecting all but one phone. A telephone-RFI filter cures it handily. Two other phones have no problem, *until* an answering machine is installed back on the line. Then, every phone in the house has RFI again. A telephone RFI filter on the answering machine line makes it quite a bit better on 20 meters, but on 10 meters, it is just as bad as ever. They then try a ferrite common-mode choke on the dc-power supply lead coming from the "wall cube" power supply. The problem is now gone. The last telephone also proves to be a bit troublesome, requiring both a line filter and a separate filter for the handset cord.

At this point, the RFIC feels it has helped all it can. Paul and his neighbor enthusiastically agree. They will call the cable company tomorrow and set up an appointment. The RFIC points out the Cable TVI chapter in this book; Paul will go to a nearby dealer and buy a copy tomorrow, so he can show the cable-TVI information to the cable company.

The RFIC asks Paul to keep them informed about the cable problem and offers that they will be willing to help out if Paul has any more RFI problems. Paul and his neighbor have both noted that the "stock" cures seemed to have worked and that if they have any more problems, they will "try the easy things first." In fact, Paul notes, one other neighbor has had problems, too, but Paul now feels he can give it a go, having watched the "pros" in action. The RFIC leaves, another one chalked up in the "win-win" column.

Over the next few days, the RFIC documents the entire sequence of events and Paul receives a copy of the report in the mail. He immediately makes a copy for his neighbor. They share it over a cup of coffee and wonder how they could have ever let things get out of hand.

Variations

This has been but an example. There are as many variations on this as there are RFI problems and individual people. In some cases, the RFIC may choose to do a lot more on the telephone, perhaps explaining things to Paul and his neighbor and leaving them to try the cures themselves.

Liability

No discussion would be complete without a mention of liability. In working with a neighbor, or the ham for that matter, there is a question of liability. In virtually no case should a ham or RFIC take apart consumer equipment. Hams should limit their involvement to helping a neighbor find solutions. Even trying external cures has its risks—what if the TV decides to die for other reasons the minute the filter is installed?

If the RFIC is being formed by an ARRL affiliated club already insured under the ARRL's club liability insurance, it would automatically be covered under that insurance plan. In general, if the activities of the RFIC are listed on the club's insurance application, the club and its members will be covered for claims alleging bodily injury and/or property damage subject to the policy terms and conditions. Specific questions should be directed to the ARRL insurance administrator, Marsh Affinity Group Services at 1-800-503-9230. For more information, see the ARRL Web site at **www.arrl.org/field/regulations/insurance/club_liability.html**.

In Conclusion

This section of the ARRL's *RFI Book* is intended to give general guidelines on how to set up an RFI Committee. There are a lot of good ideas out there from club RFI committees, and the ARRL is looking for any information clubs can provide to make this chapter a better tool. Contact the editor if you can help.

From here, it is up to you. These are guidelines, not hard and fast rules. You will set up an RFIC based on local needs and local talent. It is yet another example of hams helping hams—a proud tradition of the Amateur Radio Service.

Chapter 19

RFI and Unlicensed RF Noise Sources

Highlights, Review & Summary

We've already covered some of the more common sources and victims of RFI, as well as some basic troubleshooting techniques and cures. In this chapter, we will now attempt to review and summarize many of these concepts, and see how they might apply to a device not specifically covered elsewhere in this book. If the specific device associated with your problem is not covered in a previous chapter, this chapter should help!

By Mike Gruber, W1MG

INTRODUCTION

No single book about RFI could possibly cover every unlicensed device that can cause interference to radio services. Not only is every case different, but there are literally thousands of unlicensed devices with the potential to be at the heart of an RFI problem. In the United States, most unlicensed devices are regulated by Part 15 of the FCC's rules. Unlicensed devices include things like computers, switching mode power supplies and electric motors, to name just a few.

We've already covered some of the more common sources and victims of RFI, as well as some basic troubleshooting techniques and cures. In this chapter, we will now attempt to review and summarize many of these concepts, and see how they might apply to a device not specifically covered elsewhere in this book. If the specific device associated with your problem is not covered in a previous chapter, this chapter should help!

First, let's briefly review some of the Part 15 rules that pertain to RFI. While an understanding of these rules doesn't necessarily solve an RFI problem, they do provide important insight and background on interference both to and from a Part 15 device. More information about Part 15 is found in the RFI Regulations and Standards (Chapter 17), Power Lines and Electrical Devices (Chapter 11) and Computers (Chapter 15) chapters of this book.

FCC RULES AND PART 15 DEVICES

Part 15 of the FCC rules pertain to unlicensed devices. A Part 15 device can be almost anything not already covered in another Part of the FCC rules. It's also important to note that many Part 15 devices are not normally associated with electronics, RF or in some cases, even electricity. While televisions, radios, telephones and even computers obviously constitute a Part 15 device, the rules extend to *anything* that is capable of generating or responding to RF.

In an RFI case investigated by Mike Martin, K3RFI, for example, the source of a harmful interference complaint was traced to aluminum siding. A nearby transmission line induced a voltage in the siding, and arcing occurred as a result of a loose nail in the fascia board of the house. The aluminum siding then radiated the RFI. In this case, house siding acted as a Part 15 device! While this is obviously not a typical example of an RFI source, it does illustrate the scope of the Part 15 rules. It isn't surprising therefore that most RFI complaints reported to the ARRL involve Part 15 devices.

Part 15 describes three types of devices. The rules vary in each case. See Table 19.1 for a summary.

Let's now take a moment to review some of the more important highlights of the Part 15 rules—and perhaps debunk a few of the more persistent misunderstandings in the process:

- A Part 15 device can cause strong and harmful interference to a licensed radio service, such as Amateur Radio, and still meet FCC emissions limits. The device can be legally sold and purchased by consumers. The "legal limit" for a Part-15 device can result in S9+ noise levels nearby, depending on frequency, distance, etc. Therefore, *the fact that a particular device is causing harmful interference is not in itself evidence or proof of a rules violation with regard to emissions limits.*
- There are generally no FCC radiated emissions limits below 30 MHz.
- There are generally no FCC conducted emissions limits above 30 MHz.
- As Table 19.2 illustrates, all Part 15 devices are prohibited from causing harmful interference to a licensed radio service. *Most Part 15 interference cases are argued on this basis.*
- The FCC provides this specific definition of harmful interference:

  "Section 15.3 (m) Harmful interference. Any emission, radiation or induction that endangers the functioning of a radio navigation service or other safety services or seriously degrades, obstructs or repeatedly interrupts a radio communications service operating in accordance with this chapter." See sidebars "Where to Draw the Line" and "What is in the Public Interest" in the Electrical and Power-Line Interference chapter (Chapter 11) for more information on what might constitute harmful interference.
- There are no FCC rules or limits with regard to Part 15 device RFI immunity. Part 15 devices therefore receive no FCC protection from a legally licensed radio transmitter. There is a subtle caveat to this rule, however. Let's consider a case of TVI from an Amateur Radio transmitter as an example. While the TV itself may not be protected under the FCC rules—within its service area, the TV broadcast is protected from harmful interference caused by spurious emissions from other transmitters:
- If the interference is a result of a spurious emission on the television channel frequency, the fault is with the amateur transmitter. It is the responsibility of the amateur to correct the problem—*even if his or her transmitter otherwise meets FCC requirements for spectral purity.*
- If the interference is a result of fundamental overload, i.e., an improper and undesired response to the signal from the amateur transmitter on its intended frequency, *the fault is with the TV receiver.*

Hint: Knowing which device is at fault is often an important step toward knowing *where* to implement a cure. A low pass filter at the transmitter will not help in a TVI case that involves fundamental overload, for example. The cure in this case must be installed at the TV.

Conducted vs. Radiated Emissions

Notice the rules specify different absolute limits for two different types of emissions—conducted and radiated. Knowing the difference between conducted and radiated emissions is obviously important to understanding the rules, but can also be critical to developing and implementing an effective RFI cure. The general assumption of the rules is that below 30 MHz, devices are not good antennas, but the power lines they are connected to are, so conducted emissions limits are used to control the interference potential of those devices. Above 30 MHz, the power lines become more lossy and the devices become better radiators, so radiated emissions limits are used to control interference.

Let's now take a look at the difference:

- Radiated emissions are generated by and emanate directly from the Part 15 device. In this case, the Part 15 device acts as its own antenna. See Figure 19.1A

Table 19.1—Part 15 Device Types

	Intentional	*Unintentional*	*Incidental*
Definition	Intentionally generate RF energy and radiate it	Intentionally generate RF signals internatally, but do not intentionally radiate it	Generate RF energy only as an incidental part of its normal operation
Examples	Transmitters for garage door openers, cordless phones, baby monitors	Computers, superhetrodyne receivers, switching power supplies, TV sets	Power lines, arcing electric fence, arcing switch contacts

Table 19.2—Part 15 Interference Restrictions

Rules provision	*Intentional emitters*	*Unintentional emitters*	*Incidental emitters*
Absolute Conducted Emissions Limits	Apply below 30 MHz	Apply below 30 MHz	No specific limits, just a requirement to use good engineering practice
Absolute Radiated Emissions Limits	Apply on all frequencies, at various levels as defined in Part 15 and other rules	Apply above 30 MHz, except for carrier-current devices[1] that must for intentional emitters meet limits	No specific limits, just a requirement to use good engineering practice
Rules pertaining to harmful interference	Must not cause harmful interference to a licensed radio service	Must not cause harmful interference to a licensed radio service	Must not cause harmful interference to a licensed radio service

[1]Carrier-current devices intentionally conduct signals onto power lines for communications and other purposes.

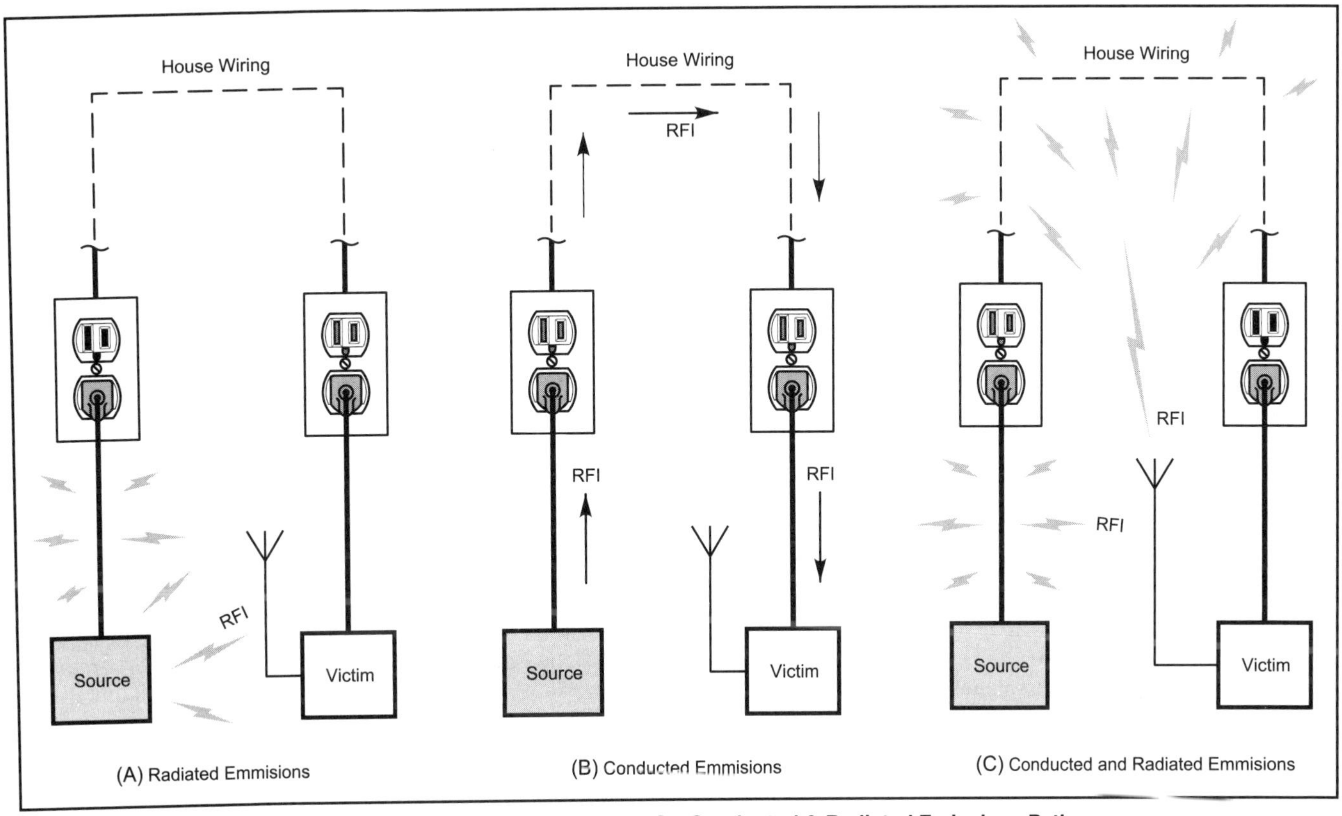

Figure 19.1 A—Radiated Emissions; B—Conducted Emissions; C—Conducted & Radiated Emissions Path

- Conducted emissions are generated by the Part 15 device but not radiated by it. In this case, the RF emissions are conducted to the power cable (or other wires connected to the Part 15 source device), which can then radiate the RF energy. The device itself does not serve as an antenna. The antenna is in fact the wires connected to the device, which in many cases can mean the power lines in a house. In some cases, the emissions can also be conducted (as opposed to being radiated) to its victim. See Figure 19.1B.

You may have been surprised to learn there are no radiated emissions limits below 30 MHz. There is a logical reason for this however. Consider the size of a typical HF antenna. Now compare its size to a typical Part 15 device. The Part 15 device is usually too small to be an efficient antenna at HF. Therefore, if there is interference at HF, the antenna is usually the wires or cable connected to it. RFI from a Part 15 device at HF is typically a conducted emissions problem.

At VHF and higher frequencies, the size of a typical Part 15 device becomes an appreciable part of a wavelength. The device itself can become an effective antenna—and the RF can therefore be radiated by the device itself. Power and other cables become less efficient transmission lines too, so conducted emissions are less of an issue at these frequencies. RFI from a Part 15 device at VHF and higher frequencies is typically a radiated emissions problem.

See the RFI Regulations and Standards chapter (Chapter 17) for both the conducted and radiated FCC emissions limits.

CARRIER-CURRENT DEVICES

Carrier current devices intentionally conduct signals onto power lines. They are still unintentional emitters for the most part, however. Examples of carrier-current devices are some of the "wireless telephone" jacks that use building electrical wiring to allow telephones to operate from rooms with no telephone wiring, using the electrical wiring in the building and broadband over power line (BPL) devices that use distribution and premise power wiring to deliver access to broadband services in homes and businesses.

In most cases, carrier-current devices would exceed the FCC's conducted emissions limits by tens of dB. To control interference from carrier-current devices, the FCC requires them to meet the general limits for intentional emitters. The RFI Regulations and Standards chapter (Chapter 17) contains more information about these limits.

PART 15 RULES—THE BOTTOM LINE

Part 15 of the FCC rules apply to unlicensed devices—which can be anything from house siding (as we saw in the case example earlier in this chapter) to an intentional transmitter such as a cordless telephone or garage door transmitter. Although reading and understanding Part 15 can appear rather formidable—especially at first glance—the rules pertaining to RFI can be roughly summarized as follows:

- A Part 15 device must not cause harmful interference to a licensed radio service, such as amateur radio. If it does, it is the operator of the device that is responsible for fixing it.
- A Part 15 device receives no protection from interference from a licensed radio service.

The Scope of the Part 15 Problem

There are literally thousands of Part 15 and potential Part 15 devices in a typical modern household. It's not surprising therefore that the vast majority of RFI cases reported to the ARRL involve Part 15 devices. As we've previously discussed, many of these are electronic devices and typically have semiconductor junctions in them. These devices can be both a source and a victim of RFI.

Not all Part 15 devices are electronic, however. Some are electrical but do not contain semiconductors or vacuum tubes. Typically, such electrical devices are more likely to be a source rather than a victim of RFI. The RF energy in this case is usually generated by an arc or spark. Typical examples include power lines, electric motors, and switch contacts. Some devices, such as the house siding discussed previously in this chapter, are not usually considered electrical, electronic or radio devices. It is important to understand, however, that *all* these devices are covered by Part 15!

Part 15 devices are ubiquitous, and locating an offending device might sometimes seem like trying to find a needle in a haystack. With a little patience, it is often possible to find the source of a problem in relatively short order. A little detective work is often required, and some cases require a little more perseverance than others. In either case, it is often easier to find an offending source than the first-time RFI investigator might think. A little knowledge of some procedures and techniques are required for efficient noise locating.

ON THE TRAIL

Whenever an unknown source of interference becomes an issue, and you suspect the problem could be your radio, remove the antenna connection as a preliminary check. Verify that the noise goes away, and therefore, is external to your radio. While this is not the usual case, it's a quick and easy test. Assuming the interference does go away, you are now ready to begin your hunt for a Part 15 device.

One of the first and most obvious clues is the sound and other characteristics of the RFI itself. Some RFI sounds are distinctive enough as to provide an important clue as to its likely source. The characteristic "snap—snap—snap" of an electric fence, for example, can be almost a smoking gun in that case. Frequencies at which the RFI appears can also be an important clue. Interference that occurs in discreet peaks and nulls, regularly spaced as you tune across the spectrum, is a characteristic of a switching mode power supply. Spacing between such noise peaks typically range from 20 to 120 kHz. The ARRL web site has a library of some common (and some not so common) RFI sounds. The URL is: **www.arrl.org/tis/info/HTML/rfi-noise/**.

Simply identifying an RFI source by sound can work in some cases. However, there are several significant caveats to using this approach. First, there are so many potential Part 15 RFI sources that it isn't possible to know them all. And, even if you could, many of them sound so similar that they are indistinguishable from each other. There are literally thousands of consumer electronic and electrical devices that could be located in

Part 15 Rules

The FCC's Part-15 rules contain two provisions to address interference. The first is an absolute maximum emissions limit that will not protect against all interference, but which is intended to minimize the likelihood of interference. Manufacturers of Part 15 equipment must meet these requirements. The second provision is that the equipment must be operated in a way that does not cause harmful interference. The operator of the Part-15 device is the one the rules make responsible for correcting any interference that does occur. If it does, the rules require that any steps necessary be taken, including shutting the device off.

Amateur radio has a similar requirement. Our HF transmitters, for example, are permitted to have as much as 75 milliwatts of harmonics. At that level, if the harmonic is on a TV channel, nearby TVs will experience interference. When that happens, there is no equipment violation, but there is still a requirement that the ham correct the interference.

Part 15 levels are pretty high. On VHF, that device is allowed to radiate a 150 μV/m field, 3 meters from the source. If you were to place a simple 2-meter ground plane in a 150 μV/m field, it would pick up a signal greater than S9 by the way most VHF-equipment S meters respond. That would be from a "legal" device, although if it is causing interference, the operator of the device would have to turn it off. Unfortunately, that is you in this case, and the Commission has never shown much sympathy for those who interfere with themselves.

Part 15 is not intended to prevent all interference, any more than our rules about harmonics are intended to prevent all interference from amateur radio. Both are intended to keep the interference potential local, so interference can be easily identified and case-by-case solutions applied only where necessary.

We will sometimes hear unlicensed emitters on our bands. This is not necessarily in and of itself a rules violation. After all, amateur radio is secondary to commercial operation on 30 meters and government operation on 70 cm. How would we feel if our primary sharing partners said that they could sometimes hear hams on their bands, so we had to get out? Part 15 is not really a sharing partner, but if the devices meet the Part-15 emissions limits, the operator of the device is violating the rules only if actual interference occurs. What constitutes interference can vary with circumstance. If a single "birdie" from an unlicensed device fell right on 144.2 MHz, for example, a 2-meter SSB/CW operator would experience interference, while if a weak birdie fell on the output of a local 2 meters repeater, the repeater could be much stronger in level, so in that case, there would be no interference.

More information on Part 15 rules is available at **www.arrl.org/tis/info/part15.html.**

a typical modern home. Simply attempting to know the source by its sound isn't practical. Perhaps even more importantly, however, even knowing what the source is or might be still may not solve the problem. You still would need to *locate* the source—and you don't need to know what the source might be in order to locate it. The easiest and best way to identify a Part 15 RFI source in many cases is to simply locate it!

Procedures for locating the offending device can roughly be broken down into two general categories:

- Sources inside of your house, and
- Sources outside of your house.

Locating an in-house source is so simple that it often makes sense to simply start an RFI investigation with this procedure. Professional RFI investigators, and experience at the ARRL RFI desk, indicate that most RFI sources are ultimately found right in the complainant's home! Starting in your own home makes sense for several reasons. See the "How To Locate Tips" sidebar in the Electrical and Power-Line Interference chapter (Chapter 11) for this step-by-step procedure.

Once you have verified the source is not inside your home, locating an outside source or residence obviously requires different techniques and procedures. Which one you use may depend on a number of factors, including available locating equipment, neighborhood layout, frequency and other interference characteristics, location of power lines, etc. In some cases, you may find you'll be using several techniques during your hunt.

Assuming an interference problem is *not* in your home, you must look outside. Power-line noise is the interference problem probably most frequently reported to the ARRL RFI desk. If you suspect the source of your interference is power-line noise, or not sure, see the "How To Resolve Power-Line Noise" section in Chapter 11 for information on how to locate the source of this type of noise. Although there really isn't a smoking gun, power-line noise tends to identify itself with several unique characteristics that are discussed in that chapter.

Remember, before taking your RFI hunt out of doors, you must verify to the best of your ability that your problem is not inside. Also, you can tell if it's likely to be power-line noise by examining its frequency range, relationship to changes in weather, the TV test, the "noise print" burst rate test, its general sound, and relationship to periods of human activity. All these things, which are discussed in Chapter 11, can be done before you open your door to leave your house. By knowing and performing these tests, you can better know if you are looking for something on a utility pole or your neighbor's house.

AN EASY FIRST STEP

Although power-line noise can propagate for several miles, as a general rule, interference from a Part 15 consumer device will not propagate nearly as far. In most cases, the source is located in a nearby residence. Mike Martin, K3RFI, a professional RFI investigator, reports most offending Part 15 consumer devices are located within 500 feet of the complainant and usually confined to the same secondary system. Due to the close proximity of most Part 15 sources, the number of possible residences is usually rather small, depending obviously on the type of neighborhood involved. See the "How to Locate the Residence" sidebar in the Electrical and Power-Line Interference chapter (Chapter 11) for an excellent procedure in this case. It's an easy first step—and often your last due to its effectiveness!

A NON-RDFing APPROACH

You don't need a beam antenna or other specialized equipment to expand your search beyond adjacent and nearby neighbors or secondary power distribution system. A battery operated portable radio capable of hearing the interference is all you need, along with some means of attenuating the signal as you get closer to the source. Walk or drive around in the general area of the interference to see if you can determine where it is the strongest. Follow power lines if, and as appropriate for the situation. Generally, you should head in the direction of increasing signal strength. When you come to a tee or intersection, try each direction one at a time to determine the correct direction.

If your radio does not have an RF gain control, an attenuator can be very helpful and possibly necessary when attempting to use this technique. Use minimum RF gain—or maximum attenuation—so that you just can barely hear the interference. You can also try removing or collapsing your receiver's antenna, depending on your particular radio's set-up. (A portable AM radio or your automobile's AM radio may be suitable, if you can hear the noise. An attenuator, however, may be difficult to implement.) Minimum RF reduces the effect of AGC action, thereby making it easier to determine where the interference is the strongest.

Note: In some cases, such as power-line noise, the noise can typically be heard at increasingly higher frequencies as you get closer to the source. The noise also tends to diminish in strength at higher frequencies. This phenomenon can often be used to your advantage whenever it occurs—especially if you do not otherwise have an RF gain control or attenuator. Since reduced noise levels can be obtained by listening at increasingly higher frequencies, you can tune upward in frequency to "attenuate" the noise to the desired level. As you get closer to the source, simply tune to progressively higher frequencies. Not only will this control the signal level, but shorter wavelengths can make it easier to pinpoint the source. Peaks and nulls often occur as a function of wavelength, and longer wavelengths have greater distance between peaks. A peak is, therefore, more likely to look like the source and fool the unwary RFI investigator. See Figure 11.11 and related text in the Electrical and Power-line Interference chapter (Chapter 11) for more information. Obviously, multiple

Body Fade RDFing

The simplest RDFing approach is the "body fade" technique—when you stand between the noise source and the handheld transceiver, the signal gets weaker. If the signal is too strong to use the S-meter, try removing the antenna. You may also be able to use a tiny antenna made out of a coaxial jumper cable cut in half. Fold back an inch of braid from the end of the cut cable. The body fade technique is most effective with single point sources.

If it appears you have more than one source or a distributed source, a much better technique is to use a directional antenna—and the more gain the better. One way of getting more gain is to move up in frequency—to where the noise is barely detectable. A 10 element 440 MHz beam has a lot more gain than a 4 element 146 MHz Yagi with the same size boom, even though the 2-meter Yagi is bigger and harder to handle. A 10 element 2-meter beam would be too big for most hams for use during neighborhood direction finding.

It helps to use a more effective detector—FM detectors are less effective than AM or SSB detectors. A stronger carrier causes more quieting with an FM detector—a stronger signal produces less noise! Many handheld transceivers feature an AM mode option, the best choice for pulse type noise, either by tuning below a certain frequency, typically around 137 MHz for the aircraft band, or by a menu selection. Their small size, tuning range and portability can make the ubiquitous HT a great RFI locating tool.—*Zack Lau, W1VT* sources can also be problematic. The signature method can help reduce confusion if multiple sources are involved. We'll discuss the signature method later in this chapter.

If possible, tune to the upper end of the band without a station. As a general rule, higher frequencies are better. VHF and UHF are best if you can hear the noise. (An air band radio is often ideal since it is both AM and VHF.) See if you can determine any noise hot spots as you continue your search. Again, adjust your attenuator or RF gain so that you can just barely hear the noise at any locations you determine are appropriate. You'll typically need to add more and more attenuation as you get closer to the source.

Once you know the approximate location of the noise, you may be able to further narrow it down. Go to a pole in the neighborhood and listen for the noise. Increase the attenuation or adjust the RF gain until you can just hear the noise. Move to the next pole and listen again. If you need to increase the attenuation to obtain the same level of noise, you are getting closer. If you need to decrease the attenuation, you are going away from the source.

This of course assumes all the poles are similar. It is important to note that a pole with a lot of "stuff" on it can radiate more noise than one with a minimal amount of hardware. Try to take measurements at or near identical types of poles. Be sure to use the minimum RF gain and maximum attenuation possible to just barely hear the noise. Once you think you have it narrowed down, continue on to a few more poles in each direction. The noise will always get stronger in the direction of the noise.

Finally, you need to narrow down the residence in which the noise is located. Look at the power lines to see which homes are connected to the loudest pole. Other nearby homes can also be suspect, depending on the particular power-line layout. Perform a "sniff test", described in the sidebar "How to "Locate the Residence" containing the RFI/TVI Source" in Chapter 11. Use power meters, lawn lights, etc. at private residences, if there are any accessible. Always use the same reference at each residence. Be sure to hold the radio the same distance away from the suspected source you are measuring for an accurate comparison.

While this approach may not always be the most efficient, especially in a complex environment or neighborhood, it will often work with a little patience. It is, therefore, usually worth a try before purchasing or building RDFing equipment.

RDFing THE SOURCE

If the source proves to be external to your home, RDFing techniques are typically the most efficient method to locate it—especially if it is located beyond your immediate area. While this approach may help save time during the noise search, it requires a directional antenna. A ferrite loop works at HF and is often found in many inexpensive portable AM radios. You can experiment a bit to learn the directional characteristics of a radio with known signal sources. A null often works best. A hand-held Yagi also works at VHF and UHF. As before, an attenuator is typically necessary. As with the non-RDFing approach, use as much attenuation as you can in order to get good readings. You'll typically need to add more and more attenuation as you get closer to the source.

See the Radio Direction Finding Chapter (Chapter 4), and Mike Martin's "Radio Direction Finding (RDF) Techniques" section in Chapter 11 for more information on RDFing.

THE SIGNATURE OR FINGERPRINT METHOD

Although most often associated with power-line noise investigations, the technique of fingerprint or signature matching can be a powerful tool when dealing with multiple RFI sources. In the case of power-line noise for example, multiple sources are often encountered during the investigation. To the ear, they may all sound the same or very similar. By looking at their waveforms however, it becomes possible to readily distinguish between them. The source or sources actually causing the RFI become apparent, and any sources not linked to the interference can be ignored.

This same technique may be useful in some non-power line related RFI investigations, although it is usually possible to simply turn off most consumer Part 15 devices to see if the interference goes away. If you think the signature or fingerprint method may be useful for your particular RFI investigation, see the Electrical and Power-Line Interference chapter (Chapter 11) for more information.

THE ANATOMY OF AN RFI PROBLEM

The ingredients of a typical RFI problem aren't much different than tuning into your favorite broadcast station. In order to hear a desired broadcast for example, there are fundamental requirements—a transmitter, a propagation path and a receiver. Although some terminology may be new—an RFI problem has these same basic requirements. The transmitter, path and receiver in an RFI case are typically referred to as the source, path and victim, respectively. Solving an RFI problem requires eliminating one or more of these three essential ingredients. See Figure 19.1.

In most RFI cases, the source and victim are obvious once known. The interference path, however, may not be so obvious, even when both source and victim have been located. An RFI path can be one or more of the following three possibilities:

– *Conducted*—The interference is conducted on wires or cables. A typical example is the power cord connected to an offending Part 15 device. This is one of the more likely paths by which HF RFI can propagate from or to a Part 15 device. For this reason, it is of particular interest when dealing with an HF RFI problem and most consumer household products. A filter or choke is installed where the cable or wires are connected to the Part 15 device.
– *Radiated*—In this case, the interference is a radio signal and propagates through space like any other radio wave. In a case where the interference is radiated directly by the Part 15 device itself, the problem is typically confined to VHF or higher frequencies. This is the point at which the size of the device becomes an appreciable portion of the wavelength.
– *Induced*—This interference path is less common than either the conducted or radiated path. An induced path typically requires proximity between the source and its victim. An induced path takes place as a result of magnetic field coupling. In some cases, the coupling can take place as a result of an electric field.

Propagation between a source and its victim is often a combination of radiated and conducted paths. In a typical example, the RF is conducted from a Part 15 source device via its power cord. The power cord then conducts the RF to the house wiring, and even the power distribution wires, where it can be radiated to its victim. See the EMC Fundamentals chapter (Chapter 2) for more information on RFI paths.

SOLUTIONS

Every RFI problem can be different. It may not be surprising, therefore, that effective solutions can vary too, depending on the source and specifics of a particular problem. In all cases, however, you should be looking at ways of eliminating one or more of the essential ingredients to an RFI problem. You must eliminate the source, break the path, or eliminate the victim. Some solutions can involve a combination of these three possibilities.

In some cases, a solution may be obvious. Consider the case of an arcing electric fence as an example. Not only is the arc generating unwanted RFI, but it provides an

Feedline Radiation from Balanced vs. Unbalanced Transmission Lines

Q. What is meant by the terms "balanced" and "unbalanced" when referring to transmission lines?

A. The physical differences between balanced and unbalanced feed lines are obvious. Balanced lines are parallel-type transmission lines, such as ladder line or twinlead. The two conductors that make up a balanced line run side-by-side, separated by an insulating material (plastic, air, whatever). Unbalanced lines, on the other hand, are coaxial-type feed lines. One of the conductors (the shield) completely surrounds the other (the center).

In an ideal world, both types of transmission lines would deliver RF power to the load (typically your antenna) without radiating any energy along the way. It is important to understand, however, that *both* types of transmission lines require a *balanced* condition in order to accomplish this feat. That is, the currents in each conductor must be equal in magnitude, but *opposite* in polarity.

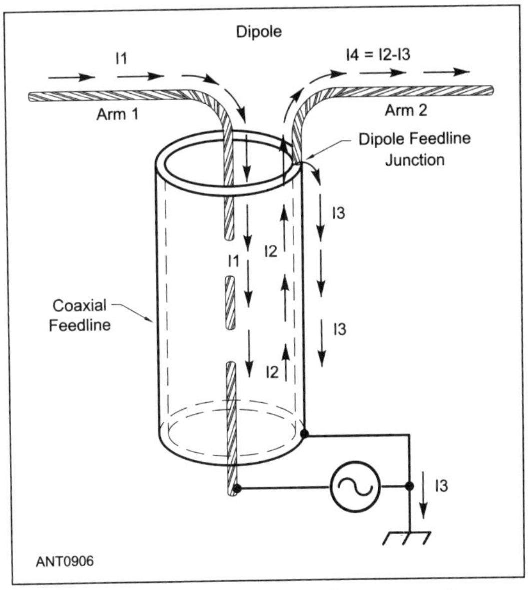

Figure A—Drawing showing various current paths at feed point of a balanced dipole fed with unbalanced coaxial cable. The diameter of the coax is exaggerated to show the currents clearly.

The classic definition of a balanced transmission line tells us that both conductors must be symmetrical (same length and separation distance) relative to a common reference point, usually ground. It's fairly easy to imagine the equal and opposite currents flowing through this type of feeder. When such a condition occurs, the fields generated by the currents cancel each other—hence, no radiation. An imbalance occurs when one of the conductors carries more current than the other. This additional "imbalance current" causes the feed line to radiate.

Things are a bit different when we consider a coaxial cable. Instead of its being a symmetrical line, one of its conductors (usually the shield), is grounded. In addition, the currents flowing in the coax are confined to the *outside* portion of the center conductor and the *inside* portion of the shield.

When a balanced load, such as a resonant dipole antenna, is connected to unbalanced coax, the outside of the shield can act as an electrical third conductor. This phantom third conductor can provide an alternate path for the imbalance current to flow. Whether the small amount of stray radiation that occurs is important or not is subject to debate. In fact, one of the purposes of a *balun* (a contraction of *balanced* to unbalanced) is to reduce or eliminate imbalance current flowing on the outside of the shield.

obvious and clear indication that the fence is broken. Furthermore, the function of the fence is compromised by this defect. The solution in this case is evident—repair the fence. Typically, this requires eliminating the gap across which the arc occurs. Repairing an arcing splice, for example, usually requires cleaning and tightening the connection. This simple (and typically zero-cost) repair not only eliminates the RFI problem, it simultaneously restores the functionality of the fence!

In some RFI cases, the solution may not be so obvious. In others it may not even be practical, at least to the average ham. Consider a case of direct radiation from the source device, especially one that is large and can't easily be moved. Let's now assume your lot is too small to move the receiver's antenna. Moving either the source or the receiver's antenna is not an option in this case. Shielding may offer some possibilities for an RFI cure, but may be difficult to implement in most cases. Other possibilities include internal circuit modification. Again, this may not be a practical option in most situations. Replacing the source device with one that doesn't create interference may work, but is likely to be expensive. Fortunately, this is not the typical case in most situations. If and when a difficult case does occur, hams are advised to start by contacting the manufacturer for help. (Fundamental overload to a Part 15 intentional receiver, such as a baby monitor, will also be discussed later in this chapter.)

Let's now look at a more typical Part 15 RFI example. Many cases involve RF that is conducted either into or from the device electronics by the wires and cables connected to it. As previously discussed, this is almost always true for HF, where the device itself is too small to be an efficient antenna. The solution in most of these cases is obvious – break the path at the point where the RF enters or leaves the device. Fortunately, the stock answer in this case will often fix the problem whether the Part 15 device is the source or the victim. Let's take a closer look.

A key ingredient to an effective RFI cure often involves understanding the difference between common mode vs. differential mode conduction. Although each mode requires a different cure, you can often make an educated guess as to which mode is most likely the problem. And, if you are not sure, you can always try each type of filter or combination of filters to see which works. Experimentation is often required. Let's now briefly review the difference between common vs. differential mode conduction.

- Differential mode currents have equal strength and opposite phase in some of the wires. There is no ground current. In a two wire transmission line, for example, the signal arrives on a pair of conductors with a 180° phase difference between the pair.
- Common Mode currents are equal in amplitude and phase in all of the wires. The return path for a common mode signal usually includes ground. The signal arrives on one or more of the conductors, in phase, and returns via some ground path, such as the power supply ground connection. In the case of a coaxial cable—the common mode currents flow on the *outside* of the cable shield. See the sidebar "Feedline Radiation from Balanced vs. Unbalanced Transmission Lines."

Applying the right filter or choke is essential for an effective cure. Some cases may require both a differential and common mode cure. Try the easy things first; but you may need to experiment a bit before you find a workable solution. One good rule of thumb is this: Parallel wires in close proximity to each other—relative to the wavelength of concern—tend to involve common mode signals. This is also true of coaxial cables. See the EMC Fundamentals chapter (Chapter 2) for more information on common mode vs. differential mode signals.

MEET THE CURE!

Filters and chokes are the number one weapon of choice for many Part 15 RFI problems. The same cure can work equally as well if the Part 15 device is the source or the victim. Both have been covered in other parts of this book, but let's review a few of the highlights.

– The *Common-Mode Choke*—Making a common-mode choke is usually pretty simple matter. Selecting the type of core and ferrite material can be crucial to an effective cure, however. Think toroids at HF. Wrap several turns of the cable or wire pairs around the toroid. The exact number isn't critical, but six to eight turns is a good place to start. Ten to fifteen turns is probably the practical upper limit in most cases. See sidebar for ferrite and size selection.

Although split ferrites, clamp-ons and ferrite beads are not as effective as toroids at HF, they can offer significant RFI suppression at VHF and higher frequencies. They also have the advantage of not requiring access to a cable end for installation. They also do not need to be sized for large connectors, just the cable diameter. Typically, the connector doesn't even need to be disconnected from the cable end. As with HF toroids, however, be sure to select an appropriate ferrite mix. Since these ferrite configurations are best suited for VHF and higher frequencies, type 43 mix is probably a good choice in most applications.

The common-mode chokes should be installed right at the affected or source device. The idea is to isolate the Part 15 device from its antenna, whether it is for receiving or transmitting the offending RFI signal. Do not assume that a ferrite common mode choke will not work unless you've tried a toroid core with a suggested ferrite mix. In some cases, additional chokes may be necessary.

Note: Some hams have reported success with some RFI problems by simply rerouting cables and/or wrapping a coil in the cable at the appropriate location. This can be worth a try, especially if you do not have any ferrite cores handy.

Optional: The ARRL "RFI Web Site" contains a list of EMI/RFI materials suppliers for ferrite chokes. You can also refer to the advertisements in *QST*—there are a few advertisers offering ferrite materials and chokes.

– *Differential Mode Filters*—RFI can enter and exit a Part 15 device via its ac power connection. The problem can be differential or common mode, so a differential mode filter may be in order. The so-called "brute-force" ac line filters are simple and easy to install—and commercially available. These filters can provide both common and differential mode suppression, depending on the filter. It is essential to be sure the filter is properly sized for the device current. If the device draws less than 300 watts (about 2.5 A), Radio Shack has a filter that is readily available and may help. See Radio Shack catalog #15-1111. Some of the filters sold by Industrial Communications Engineers, **www.arrl.org/cgi-bin/tisfind?patt=Industrial+Comm** should help if greater current capacity is required. Also, see the Short Takes review on page 48 of the March 2005 issue of *QST* for a review of an ICE filter.

See the EMC Fundamentals chapter (Chapter 2) for more information on common mode and differential mode filters. Common mode chokes are also covered in the Stereos and Other Audio Equipment (Chapter 10) and Automobiles (Chapter 16) chapters. See the Electrical and Power Line Interference chapter (Chapter 11) for information on brute force ac line filters.

THE FINAL APPROACH

Once you've found the source, you still will need to fix the problem. Depending on the source location, you may be dealing with neighbors. In some cases, especially if the situation has escalated, you may be dealing with companies, corporations, city or town officials, police or the FCC. The importance of diplomacy in these situations cannot be overstated. While it may seem that your people skills are taxed to their limit, it is important to handle these types of neighborhood situations in the most positive manner possible. For more information and suggestions on the people aspects of an RFI problem, see the First Steps chapter (Chapter 1). The ARRL web site also has a downloadable pamphlet that you can give your neighbor at **www.arrl.org/news/rfi/neighbors.html**.

Efficient and effective RFI resolution often requires a logical and methodical approach. Although every case can be different, there are some general guidelines that may help for most situations. From a technical standpoint, let's now take a closer look at a generic approach toward an RFI solution. Obviously, your mileage may vary and not all steps will apply in every situation:

1) Verify beyond any reasonable doubt that the suspect source is indeed the problem, whether it is an amateur transmitter or a Part 15 device. In most cases, the Part 15 device can simply be unplugged for this test.

2) Once you have located the RFI source, try to determine if the RFI is a result of a defect or malfunction unique to that particular source. If so, the solution may be obvious—repair or replace the broken device. As an example, consider an arcing splice on electric fence. Repairing a splice is not only simple and easy, but it's cost free and corrects deficiency in the operation of the fence. See the sidebar in the Electrical and Power-Line Interference chapter (Chapter 11) for more information on solving an electric fence problem.

Note: Just because the source device is generating interference doesn't necessarily mean it is broken or illegal under the FCC rules. As discussed earlier in this chapter, a Part 15 device can meet FCC absolute emissions limits and still generate strong interference to a nearby radio receiver.

3) If the RFI problem is *not* a result of a malfunction, try to determine if the cause of the problem is a result of either funda-

Ferrite Toroid Selection

Ferrite toroids have a nomenclature that can help with selecting a suitable core. Let's take the example of FT-240-73:

– FT—"Ferrite Toroid" (Note: Beads start with an "FB".)

– 240—This is the outside diameter of the toroid (donut shape) in inches. The OD in this case is 2.4 inches.

– 73—This is the ferrite material mix. A good mix for general purpose HF is type 73 material or its equivalent. Use 75 material for AM BC, 80 and/or 160-meter signals. Select –43 ferrite material for VHF signals.

Be sure to select toroids that will be large enough for your cables and connectors. You will need an ID that is large enough for several turns of cable—*plus* the connector. You may want to favor 240 and 140 size toroids for most applications, which are the largest that are typically available. Smaller toroids can be used in tight spaces or if in an "aesthetically sensitive" area as long as the wires and connectors are small enough.

mental overload or spurious emissions from the transmitter.

In the case of fundamental overload, the victim is incorrectly responding to the normal and desired output of the source. Fundamental overload is universally true for audio devices and non-receiver type devices. These are devices that should not respond to *any* radio signal. See the "Crystal Radio Effect" sidebar for information on audio devices.

In the case of a receiver device, fundamental overload typically occurs across a range of frequencies or spectrum, as opposed to a few frequencies or channels. In a case of TVI for example, fundamental overload is more likely to be a problem if all channels are affected by the interference, as opposed to just one or a few channels. *If fundamental overload is determined to be the cause of the problem, the cure will need to be applied to the victim.*

In the case of a spurious output from the source transmitter, the problem is obviously at the source, i.e., the transmitter itself. The victim device is not at fault. *If a spurious emission is determined to be the cause of the problem, the cure will need to be applied to the source transmitter.*

Spurious emissions are usually suspected when the interference occurs on one frequency or channel, or a limited number of frequencies or channels. In the case of TVI for example, if only one channel were being affected by interference from an amateur transmitter, a spurious emission would be the prime suspect. The stock cure in this case is a low pass filter installed at the transmitter. (Often, the frequency of a spurious emissions changes if the transmitter generating the emission changes frequency, although most transmitters, other than Amateur, that may generate spurious emissions operate on a fixed frequency.)

4) Tip: Try the easy things first—simple stock cures often work! Focus your initial attempts at a cure on the most likely cause of the problem—whether it is the source or the victim. Filters and chokes can prevent RFI from entering or leaving a Part 15 device. The same cures will work in both cases.

5) If the transmitter is the cause of the problem, try adding a low pass filter. Other possibilities include band-pass filters, but may not be as readily available. Statistically, RFI causing spurious emissions from an amateur transmitter is less typical than fundamental overload of an affected Part 15 device.

6) If the Part 15 device is the cause of the problem, try to isolate the path. A conducted emission path is typically the easiest to break:

- Simplify the problem! Remove as many devices, cables—and possibilities—as you can. See if you can isolate the RFI source to one or more cables connected to the Part 15 device. Filters and chokes only need to be installed where needed.
- Examine all cables and wires connected to the Part 15 device. Every one is suspect for conducted emissions—especially at HF. Try adding common mode chokes as appropriate. Proper selection and placement of the chokes can be important for a successful cure:
- Use toroids for HF.
- Use proper ferrite mix for frequency range of concern.
- A shotgun approach is okay, but not always necessary. If you fix the problem, you can try removing chokes, one at a time, to see if any are unnecessary. Keep only those required to fix problem.
- Several chokes may be required in some cases.
- Don't forget the power cord!
- See the EMC Fundamentals chapter (Chapter 2) for more information on common mode chokes, ferrites and brute force ac line filters. The Automobiles chapter (Chapter 16) also contains information on common-mode chokes.

7) If common-mode chokes don't cure the problem, try adding a brute-force ac line filter. These filters are best for a differential mode path via the power cord. In some cases, the problem may require both common mode chokes and a brute-force ac line filter. See the switching mode power supply sidebar on page 19.10.

8) If standard filters and chokes don't cure the problem, a radiated emissions path can be suspect—especially at VHF and higher frequencies. A direct radiated emissions path from or to a Part 15 device is usually more

The "Crystal Radio Effect"— Common Audio Interference Problem

A common interference problem involving audio devices is audio rectification. I sometimes refer to as "the crystal radio effect." The symptoms can often appear quite mysterious at first. The interference in this case can occur *even when the susceptible equipment is unplugged or turned off.* Just like the crystal radios you may remember from your youth, an external power source is not required to generate the audio, which in this case is interference, from an RF signal.

Not surprisingly, all the necessary components for a crystal radio to work must also be present for this phenomenon to occur. Specifically, these include an antenna, detector or crystal, and a transducer. While crystal radios typically used headphones, most of the interference cases you likely encounter will involve a speaker. Let's take a closer look.

Consider an audio amplifier connected to a speaker. In the case of many stereos for example, the speakers may be several feet away, making them an ideal "antenna" for strong signals from a nearby transmitter. (Note: As previously discussed in this chapter, the received signal in this example will typically be common mode.)

Now consider the output of the audio amplifier. It typically involves a semiconductor device of some sort, such as a transistor. Semiconductor junctions also make excellent detectors. In fact, many "modern" crystal radios often use a diode, such as a germanium 1N34A, for a detector. The RF is picked up by the speaker wire "antenna" and connected to the "detector" in the output of the audio amplifier. The detection process then produces an audio signal, which is conducted by the speaker wires to the speaker.

Whenever audio mysteriously emanates from an audio device, even when it's off or unplugged, the "crystal radio effect" is almost always the culprit. The output circuitry of the amplifier can act as diode, especially when the unit is turned off and the transistors have no forward bias. Fortunately the cure is usually pretty straightforward—simply install a common mode choke at the point the speaker wires are connected to the audio amplifier.

In cases where there is a second amplifier at the speaker itself, a second common mode choke may also be required. Most computer speakers, for example, have a power amplifier inside the speaker box. Always install the common mode chokes as close to the end of the speaker wires – and as close to the amplifier—as possible. You want to minimize the antenna length connected to the "detector" to the greatest extent possible. As with all common mode chokes, proper ferrite core and material selection can be crucial toward implementing a successful cure. See the Audio Rectification section in the "Stereos and Other Audio Equipment Chapter" for more information on this common problem.—*Mike Gruber, W1MG*

The Ubiquitous Switching Mode Power Supply

Switching mode power supplies are found everywhere these days. The reasons are simple. Not only can they be made small and lightweight, but inexpensively as well. Switching supplies also offer better efficiency over their linear counterparts, but that improved efficiency can come at a price – interference.

As its name implies, a switching mode power supply uses circuitry to switch – or chop – current. In order to achieve maximum power supply efficiency, the switching transition between "on" and "off" made by its circuitry must be made as fast as possible. Remember, a non-reactive circuit, power in watts is the product of the voltage times the current. In a perfect switch, as long as either the current through the switching device is zero, such as occurs during its "off" state, or the voltage across it is zero, such as occurs during its "on" state, no power is dissipated by the switching device.

In an ideal case, one in which no power was dissipated by the supply's circuitry; all input power to the supply is dissipated by its load. The supply's switching device would be 100% efficient. This imaginary supply would not require a heat sink either, since it would not generate any heat during its operation. Obviously, such an ideal power supply does not represent a practical real world device. Switching transitions for example are not instantaneous. It is during these transitions that switching mode power supplies are least efficient. Both current through and voltage across the switching device are simultaneously present, and power is therefore dissipated. One way to minimize loss, therefore, is to minimize the time during which transitions occur. Faster transitions, however, mean higher frequencies are generated in the process, and the resulting RF can be at the root of an interference problem.

Considering the economic and other advantages of a switching supply, it isn't surprising they continue finding their way into many consumer type products and gadgets. The proliferation of these products in typical households also tends to increase the probability that one of them will be an interference culprit. Many Part 15 interference complaints reported to the ARRL actually appear to be caused by a switching mode power supply in the culprit device.

RFI from a switching mode power supply is usually pretty easy to identify; the noise tends to repeat at regular and discreet intervals across large areas of spectrum. Typically, the peaks (and nulls) occur at intervals between 20 and 120 kHz, with the most common interval probably around 50 kHz +/- 10 kHz. The repetitive nature of the interference is often an important clue as to its source. Another typical characteristic of switching mode power supply interference is that the noise peaks (and nulls) tend to drift with respect to time. Most switching supplies' noise is also quite "buzzy," especially when listened to in the AM mode. See Figure 11.33.

Simply recognizing some of the characteristic symptoms of switching mode power supply interference won't necessarily provide much help with identifying the device causing the problem, or its location. Nearly every modern device in use today uses a switching power supply (often in the form of a "wall cube.") Follow the "how to locate" steps provided in the chapter and elsewhere in this book to locate an offending switcher. The cure in this case is also similar to other Part 15 devices – typically a common mode choke and/ or a brute force ac line filter. In the case of a wall cube, however, you may need to install an extension cord for the common mode choke. A choke may also be required on the output or dc cord.

Washing Machine RFI

The ARRL RFI Desk has received numerous reports of interference from washing machines that utilize pulsed dc motors. Stu Cohen, N1SC now tells us how he solved his Maytag Neptune washer RFI problems:

1) Add a common mode choke with a FT-240-73 or equivalent toroid core to the power cable where it enters the back of the machine. Wind approximately seven turns of the cable around the core. Keep the first and last turn on the core separated as much as possible. This is to minimize any capacitive coupling in the choke. (Note: The maximum number of turns possible is limited by the size of the plug. You may want to remove the molded plug, wind the toroid with as many non-overlapping turns as possible, and then re-attach a screw-on 3-wire plug.)

2) Add a second common mode choke about six inches up the cable.

3) Add a third and fourth common mode choke to the cable. This may require (as it did in my case) an extension cable that I made to accommodate the chokes. I used 12 gauge 3-wire cable and added the appropriate plug and socket to the extension. Be sure to observe proper polarity!

4) For additional suppression, plug the cable into a brute-force ac line filter, such as the ICE Model 475-3 ac Line filter that appears on the Short Takes column in the March 2005 issue of *QST*. Look for it on page 48. The filter wasn't necessary in my case; the four toroids were sufficient, but "your mileage may vary." (Note: If you use this filter, you may wish to experiment with the ground connection on the filter for best results.)

This cured my problem!—*Stu Cohen, N1SC*

difficult to break than a conducted one. Fortunately, this is less common than conducted emissions at HF. At best, a case of "direct radiation" can be a challenge. Suggestions in this case include:

- Consult manufacturer. Manufactures in some cases may have a standard cure for some RFI problems. In some cases, they may already have hardware and technical information to solve the problem.
- Solving a direct radiation problem to or from a typical Part 15 device often requires shielding or circuit modification. Such modifications may not always be practical.

9) Many Part 15 "intentional receivers," such as baby monitors, often have poor filtering and shielding. There simply may not be a practical cure in many cases. Nonetheless, Part 15 Rules still apply, and the device does not receive any protection from the FCC under its Rules. It may be easier to replace the Part 15 device in some cases. Some newer Part 15 receiving devices operate at much higher frequencies than older ones, for example.

FINAL COMMENTS

Most cases of RFI are curable. In this chapter, we've examined some common techniques used to suppress RFI. It is not only a summary of highlights contained in previous chapters, but serves as a guide toward applying some basic principles for devices not specifically covered elsewhere in this book. It provides some important key points and concepts, with references to other chapters for more information. These concepts are then applied in an example of a typical procedure that might be used to solve a generic RFI problem.

Good luck finding and fixing your RFI problems. I hope you have found this information useful.

Appendix A

Resources

This appendix contains address and contact information for government agencies and professional organizations. Also included are publishers and companies that provide materials and devices for controlling radio frequency interference.

AGENCIES, ORGANIZATIONS AND PUBLISHERS

American National Standards Institute (ANSI)
25 West 43rd St
(between 5th and 6th Avenues), 4 floor
New York, NY 10036
212-642.4900
Fax: 212-398.0023
Email: info@ansi.org
Web: www.web.ansi.org/

ARRL—The national association for Amateur Radio
225 Main St
Newington, CT 06111
860-594-0200
Fax: 860-594-0259
E-mail: rfi@arrl.org
Web: www.arrl.org/

Canadian Standards Institute (CSA)
178 Rexdale Blvd
Etobicoke (Toronto), ON M9W 1R3
Canada
416-747-4000
Fax: 416-747-4149
E-mail: sales@csa.ca
Web: www.csa.ca/

Consumer Electronics Association (CEA)
2500 Wilson Blvd
Arlington, VA 22205
703-907-7600
Fax: 703-907-7675
E-mail: cea@ce.org
Web: www.ce.org/

CQ Communications
25 Newbridge Rd
Hicksville, NY 11801
516-681-2922
Fax: 516-681-2926
E-mail: cqmagazine@aol.com
Web: www.cq-amateur-radio.com/

Electronic Industries Alliance (EIA)
2500 Wilson Blvd
Arlington, VA 22201
703-907-7500
E-mail: PublicAffairs@eia.org
Web: www.eia.org/

Federal Communications Commission (FCC)
1270 Fairfield Rd
Gettysburg, PA 17325
888-225-5322
717-338-2500
Fax: 717-338-2696
E-mail: callctr@nightwind.fcc.gov
Web: www.fcc.gov/

Federal Communications Commission Office of Engineering and Technology (OET)
445 12th St, SW
Washington, DC 20554
202-418-2470
E-mail: oetinfo@fcc.gov
Web: www.fcc.gov/oet/

Institute of Electrical and Electronics Engineers (IEEE) Headquarters
345 East 47th St
New York, NY 10017
212-705-7900
Fax: 212-705-4929
E-mail: member.services@ieee.org
Web: www.ieee.org/

IEEE Operations Center
445 Hoes Ln
PO Box 1331
Piscataway, NJ 08855
800-678-4333
732-981-0060
Fax: 732-981-9667
E-mail: customer.service@ieee.org
Web: www.ieee.org/

International Electrotechnical Commission (IEC)
3, rue de Varemb
PO Box 131
CH-1211 GENEVA 20
Switzerland
41 22 919 02 11
Fax: 41 22 919 03 00 (CCITT Type III)
E-mail: pubinfor@iec.ch
Web: www.iec.ch/

McGraw-Hill
1333 Burr Ridge Pkwy
Burr Ridge, IL 60521
630-789-4000
Web: www.mhhe.com/

National Association of Broadcasters (NAB)
1771 N St, NW
Washington, DC 20036
202-429-5300
Fax: 202-775-3520
Web: www.nab.org/

National Cable Television Association (NCTA)
25 Massachusetts Ave, NW
Suite 100
Washington, DC 20001-1431
202-222-2300
Web: www.ncta.com/

National Fire Protection Association (NFPA)
1 Batterymarch Park
Quincy, MA 02269
617-770-3000
Fax: 617-770-0700
E-mail: custserv@NFPA.org
Web: www.nfpa.org/
Publishers of the National Electrical Code (NEC).

Prentice-Hall
Division of Simon and Schuster
1 Lake St
Upper Saddle River, NJ 07458
201-236-7000
Fax: 201-236-7696
Web: www.prenhall.com/

Radio Society of Great Britain (RSGB)
Lambda House, Cranborne Rd
Potters Bar, Herts EN6 3JE
United Kingdom
44-1707-659015
Fax: 44-1707-645105
E-mail: RadCom@rsgb.org.uk
Web: www.rsgb.org/
The RSGB is the UK's internationally recognized national society for radio amateurs and the publishers of *RadCom* (*RC*) magazine.

Society of Automotive Engineers (SAE)
400 Commonwealth Dr
Warrendale, PA 15096
724-776-4841
Fax: 724/776-5760
E-mail: swiss@sae.org
Web: www.sae.org/

Society of Broadcast Engineers, Inc (SBE)
8445 Keystone Crossing, Suite 140
Indianapolis, IN 46240
317-253-1640
Fax: 317-253-0418
Web: www.sbe.org/

Telecommunications Industry Association (TIA)
2500 Wilson Blvd, Suite 300
Arlington, VA 22201
703-907-7700
Fax: 703-907-7727
E-mail: tia@tia.eia.org
Web: www.tiaonline.org/

Underwriters Laboratories (UL)
333 Pfingsten Rd
Northbrook, IL 60062
877-854-3577
Fax: 847-407-1395
E-mail: northbrook@ul.com
Web: www.ul.com/

US Department of the Interior
US Geological Survey (USGS)
509 National Center
Reston, VA 20192
888-275-8747
Fax: 703-648-4888
Web: www.usgs.gov/

Verband Deutscher Elektrotechniker (VDE)
VDE Testing and Certification Institute
Testing Laboratory for EMC Measurements
Merianstrasse 28
D-63069 Offenbach, Germany
49-69-8306-0
Fax: 49-69-8306-555
Web: www.vde.de/vde/html/e/home.htm

COMPANIES THAT SUPPLY MATERIALS AND EQUIPMENT FOR RFI SUPPRESSION

Allied Electronics
7410 Pebble Dr
Fort Worth, TX 76118
866-433-5722
(call for nearest distributor)
Web: www.alliedelect.com
Ferrite and ac power-line filters

Ameco Publishing Corp
253 Collins Avenue
Williston Park, NY 11596
516-248-7330
Fax: 516-248-7330
Web: www.amecocorp.com/
Low-pass filters for HF transceivers, high-pass filters for CATV and antenna connected televisions

Amidon, Inc
240 Briggs Ave
Costa Mesa, CA 92626
800-898-1883
714-850-4660
Fax: 714-850-1163
E-mail: sales@amidoncorp.com
Web: www.amidoncorp.com/
Ferrite cores, beads and rods. RFI and experimenters kits, containing a large assortment of ferrite beads and cores are also available

Barker and Williamson (B&W)
603 Cidco Rd
Cocoa, FL 32926
321-639-1510
Fax: 321-639-2545
Web: www.bwantennas.com/
Low-Pass filters for HF and 6-meter transceivers

Bencher, Inc
832 Anita St
Antioch, IL 60002
847-838-3195
(call for nearest distributor)
Fax: 847-838-3479
E-mail: bencher@bencher.com
Web: www.bencher.com/
Low-pass filters for HF transceivers

K-Y Filter Company
John K. Browne, KI6KY
3010 Grinnel Place
Davis, CA 95616
530-757-6873
Web: www.ky-filters.com
RFI filter construction, including RFI filters for modems and telephones

CWS ByteMark
1510 E Edinger Ave #B
Santa Ana, CA 92705
714-547-3276
Toll Free: 800 679-3184
Fax: 714-547-4433
e-mail:sales@cwsbytemark.com
Web: www.cwsbytemark.com/
Ferrite

Coilcraft
1102 Silver Lake Rd
Cary, IL 60013
Phone: 800-322-2645
847-639-6400
Toll Free: 800-322-2645
Fax: 847-639-1469
E-mail: info@coilcraft.com
Web: www.coilcraft.com/
Computer cable EMI filter modules, RJ11, RJ14, MOD4 handset and RJ45 modular jacks; EMI filters for 1, 2 and 4-line telephone systems

Communications and Energy Corp
7395 Taft Park Dr
East Syracuse, NY 13057
315-452-0709
Toll Free: 800-882-1587
Fax: 315-451-0732
E-mail: tech@cefilter.com
Web: www.cefilter.com/
CATV and satellite television interference filters

Corcom, Inc
844 E. Rockland Rd
Libertyville, Illinois 60048
847-680-7400
Toll Free: 800-522-6752
Fax: 847-680-8169
E-mail: info@cor.com
Web: www.cor.com/
AC power-line filters for installation inside equipment enclosures

Cornell-Dublier Electronics (CDE)
1605 E. Rodney French Blvd
New Bedford, MA 02744
508-996-8561
Fax: 508-996-3830
E-mail: lmiranda@cornell-dubilier.com
Web: www.cde.com
AC rated capacitors and ac power-line filters for installation inside equipment enclosures

DCI Digital Communications
20 South Plains Rd
Emerald Park, SK
Canada, S4L 1B7
800-563-5351
306-781-4451
Fax: 306-781-2008
E-mail: dci@dci.ca
Web: www.dci.ca/
Band-pass filters for VHF, 220 MHz, UHF and UHF ATV. Other frequencies are available by special order

Digi-Key Corp
701 Brooks Ave S
PO Box 677
Thief River Falls, MN 56701
800-344-4539
Fax: 218-681-3380
E-mail: webmaster@digikey.com
Web: www.digikey.com/
AC rated capacitors and ferrite

Down East Microwave
954 Rt 519
Frenchtown, NJ 08825
908-996-3584
Fax: 908-996-3702
Web: www.downeastmicrowave.com/
A wide variety of bandpass, low-pass and notch filters for HF, VHF, UHF and microwave applications

Electronic Rainbow
2238 West US 36
Danville, IN 46122
317-745-5602
Fax: 317-745-5604
Email: info@rainbowkits.com
Web: www.rainbowkits.com/
"Searcher" direction finder kit. (SDF-1)

Electronic Specialists, Inc
75 Middlesex Ave.
171 S Main St
Natick, MA 01760
800-225-4876
508-655-1532
Fax: 508-653 -0268
E-mail: esp@elec-spec.com
Web: www.elect-spec.com/index.htm
AC line filters

Elna Ferrite Laboratories, Inc
234 Tinker St
PO Box 395
Woodstock, NY 12498
800-553-2870
845-679-2497
Fax: 845-679-7010
E-mail: ferrite@mhv.net
Web: www.elna-ferrite.com/
Ferrite

EMR Corp
220-2 N 19th Ave
Phoenix, AZ 85027
800-796-2875
Fax: 623-584-9499
E-mail: info@emrcorp.com
Intermod Central Devices

Fair-Rite Products Corp
PO Box J, 1 Commercial Row
Wallkill, NY 12589
888 FAIRRITE (324-7748) or
845-895-2055
Fax: 888-FERRITE (337-7483) or
845-895-2629
E-mail: ferrites@fair-rite.com
Web: www.fair-rite.com
Ferrite

GC/Thorsen
1801 Morgan St
Rockford, IL 61105-1209
800-435-2931 (call for nearest distributor)
Fax: 815-968-9731
Conductive paint for RF shielding applications

Hamtronics, Inc
65-Q Moul Rd
Hilton, NY 14468-9535
585-392-9430
Fax: 585-392-9420
E-mail: sales@hamtronics.com
Web: www.hamtronics.com/
Helical resonator filters for VHF and UHF

Industrial Communications Engineers (ICE)
Array Solutions
350 Gloria Rd
Sunnyvale, TX 75182
USA
Tel: 972-203-2008
Fax: 972-203-8811
Email: wx0b@arraysolutions.com
www.iceradioproducts.com
www.arraysolutions.com
A wide variety of high-pass, band-pass, and low-pass filters for HF, VHF and UHF transceivers. BCB filters, CATV and antenna connected TV and VCR filters, telephone and stereo interference filters, dc line filters for mobile equipment and brute force ac line filters. Information bulletins on various RFI related topics are also available

Laird Technologies
Instrument Specialties Co
PO Box A
Delaware Water Gap, PA 18327-0136
570-424-8510 east
714-579-7100 west
800-843-4556
Web: www.instr.com/
Finger stock material and conductive tape for shielding

JPS Communications, Inc
5800 Departure Dr
Raleigh. NC 27616
800-533-3819
919-790-1011
Fax: 919-790-1456
E-mail: jps@jps.com
Web: www.jps.com/
Antenna noise canceller

K-Com
PO Box 82
Randolph OH 44265
330-325-2110
877-242-4540
Fax: 330-325-2525
E-mail: info@k-comfilters.com
Web: www.k-comfilters.com/
Telephone interference filters for 1 and 2-line phone systems and telephone handset cords

MFJ Enterprises
PO Box 494
Mississippi State, MS 39762
800-647-1800
662-323-0549
Fax: 662-323-6551
E-mail: mfj@mfjenterprises.com
Web: www.mfjenterprises.com/
Antenna noise canceling devices, low-pass filters for HF transceivers, high-pass filters for CATV and antenna-connected television interference and snap-together ferrite toroids.

Mouser Electronics
1000 N Main St
Mansfield, TX 76063
800-346-6873
Fax: 817-483-0931
E-mail: tech@mouser.com
Web: www.mouser.com/
AC rated capacitors and ferrite.

Newark Electronics
4801 N Ravenswood Ave
Chicago, IL 60640-4496
800-463-9275 (call for nearest distributor)
773-784-5100
Web: www.newark.com/
Computer cable and ac power connectors with built-in RFI suppression.
Ferrite

Ocean State Electronics
PO Box 1458
6 Industrial Dr
Westerly, RI 02891
401-596-3080
E-mail: ose@oselectronics.com
Web: www.oselectronics.com
Fax: 401-596-3590
Ferrite

Palomar Engineers
PO Box 462222
Escondido, CA 92046
760-747-3343
Fax: 760-747-3346
E-mail: Palomar@compuserve.com
Web: www.palomar-engineers.com
A wide variety of toroid cores and beads. RFI and experimenters kits, containing an assortment of ferrite beads and cores are also available. RFI tip sheets.

PAR Electronics
PO Box 645
Glenville, NC 28736
Tel: 828-743-1338
Fax: 828-743-1219
E-mail: www.parelectronics.com
www.rf-filters.com/
Notch filters for use on VHF and UHF transceivers. Typically used for notching out 152-153 MHz pager interference. Also available for other frequencies between 50 MHz and 1 GHz by special order

MtronPTI
100 Douglas Avenue
PO Box 640
Yankton, SD 57078 USA
605-665-9321
800-762-8800
Fax: 605-665-1709
Email: SalesYKT@mtronpti.com
Web: www.mtronpti.com/
Crystal band-pass filters

Radar Engineers
9535 NE Colfax St
Portland, OR 97220
503-256-3417
Fax: 503-256-1981
E-mail: radareng@teleport.com
Web: www.radarengineers.com/rfitvi.htm
"HotStick Line Sniffer" ultrasonic/VHF power-line interference location device

RadioShack
Riverfront Campus World Headquarters
300 RadioShack Circle
Fort Worth, TX 76102-1964
817-415-3700
(automated menu selection)
817-415-3011 (corporate switchboard)
Customer Relations
800-THE-SHACK (1-800-843-7422)
(or contact your local franchise)
Fax: 817-415-2303
E-mail: customer.relations@RadioShack.com
www.radioshack.com/
Split ferrite material, high-pass filters for CATV and antenna connected televisions and VCRs, ac line brute force filters and telephone interference filters. All telephones sold through Radio Shack are subjected to RFI testing

RF Parts Co
435 South Pacific St
San Marcos, CA 92069
800-737-2787 (orders only)
760-744-0700
Fax: 760-744-1943
E-mail: rfp@rfparts.com
Web: www.rfparts.com/
Low-pass filters for HF transceivers, band-pass filters by DCI, and ferrite

Richardson Electronics, Ltd
40W267 Keslinger Road
PO Box 393
LaFox, IL 60147-0393
630-208-2200
800-348-5580
Email: info@rell.com
Web: www.rell.com/
Finger stock

ROHN Industries
6718 W. Plank Road
Peoria, IL 61604
309-633-5607
Email: craig.ahlstrom@radiancorp.com
Web: www.radiancorp.com/ROHNNET/rohnnet2003/html/start.html
Communications towers

Schaffner EMC, Inc
52 Mayfield Ave
Edison, NJ 08837
732-225-9533
800-367-5566
Fax: 732-225-4789
Web: www.schaffner.com/
AC power-line filters for installation inside equipment enclosures

Sparrevohn Engineering
6911 E. 11th St.
Long Beach, CA 90815
Tel: 562-799-1577
Email: fred@sparrevohn.com
Web: www.sparrevohn.com/
Telephone interference filters for 1 and 2-line telephone systems

TTI Industries
1385 Akron St
Copiague, NY 11726
516-789-5020
Fax: 516-789-5063
E-mail: info@tti-industries.com
Residential telephone service entry panels with optional RFI filtering

Transtronics, Inc
3209 W 9th St
Lawrence, KS 66049
785-841-3089
Fax: 785-841-0434
E-mail: home@xtronics.com
Web: www.xtronics.com/
Ultrasonic translator hobbyist kit

Vectronics
300 Industrial Park Road
Starkville, MS 39759
800-363-2922
Fax: 662-323-6551
Email: mfjcustserv@vectronics.com
Web: www.vectronics.com/
Low-pass filters for HF transceivers, high-pass filters for CATV and antenna connected televisions, band-pass filters for VHF base, mobile and handheld applications

Vishay Intertechnology, Inc
One Greenwich Place
Shelton, CT 06484
402-563-6866
Fax: 1-402-563-6296
E-mail: business-americas@vishay.com
Web: www.vishay.com/
AC power-line filters for installation inside equipment enclosures

3M Electrical Specialties Division
6801 River Place Blvd
Austin, TX 78726-9000
512-984-1800
800-233-3636
Fax: 512-984-2010
Web: www.mmm.com/
Conductive tape for shielding applications

WEB RESOURCES

These Web sites contain contact information for a wide variety of consumer electronics manufacturers.

Circuit City Stores, Inc
www.circuitcity.com/help/mfrlist1.html

Consumer Electronics Association (CEA)
www.ce.org/

Electronics Manufacturers on the Net
www.electronics-oems.com/companies.html

Federal Communications Commission Compliance and Information Bureau Interference Handbook
www.arrl.org/fcc/tvibook.html

The ARRL maintains a Technical Information Service Web page. TIS provides a number of information packages and bibliographies to provide assistance for technical problems, including RFI, and information on other topics of interest to radio amateurs. Many of these packages are available at no charge from ARRLWeb, www.arrl.org/tis/

Appendix B

FCC Telephone Interference Bulletin CIB-10 August 1995

FEDERAL COMMUNICATIONS COMMISSION

COMPLIANCE & INFORMATION BUREAU

Telephone Interference
Bulletin CIB-10 August 1995

WHAT TO DO IF YOU HEAR RADIO COMMUNICATIONS ON YOUR TELEPHONE

Interference occurs when your telephone instrument fails to "block out" a nearby radio communication. Potential interference problems begin when the telephone is built at the factory. All telephones contain electronic components that are sensitive to radio. If the manufacturer does not build in interference protection, these components may react to nearby radio communications. Telephones with more features contain more electronic components and need greater interference protection. If you own an unprotected telephone, as the radio environment around you changes, you may sometimes hear unwanted radio communications. Presently, only a few telephones sold in the United States have built-in interference protection. Thus, hearing radio through your telephone is a sign that your phone lacks adequate interference protection. This is a technical problem, not a law enforcement problem. It is not a sign that the radio communication is not authorized, or that the radio transmitter is illegal.

Unprotected Telephone

Because interference problems begin at the factory, you should send your complaint to the manufacturer who built your telephone. A sample complaint letter is provided at the end of this bulletin.

You can also stop interference by using a specially designed "radio-proof" telephone, available by mail order. A recent FCC study found that these telephones, which have built-in interference protection, are a very effective remedy. A list of Radio-Proof telephones is provided at the end of this bulletin.

Radio–Proof Telephone

Interference problems in telephones can sometimes be stopped or greatly reduced with a radio filter. Install this filter at the back of the telephone, on the line cord, and/or at the telephone wall jack. Radio filters are available at local phone product stores and by mail order. (See attached list, Radio Interference Filters.) A list of Radio Interference filters is provided at the end of this bulletin.

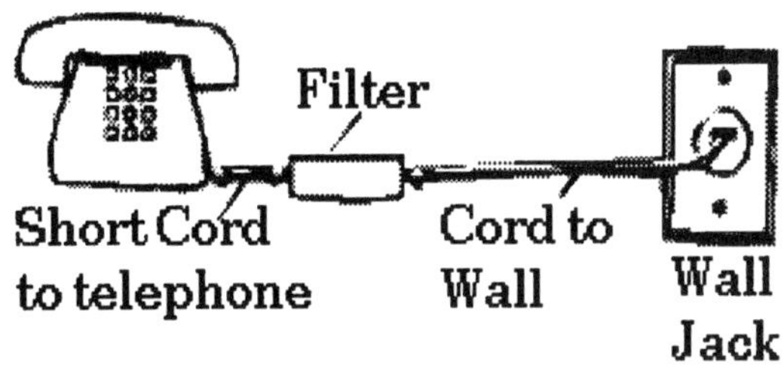

Next, it's important to follow through and contact the manufacturer. Telephone manufacturers need to know if consumers are unhappy about a product's failure to block out radio communications. Also, the manufacturer knows the design of the telephone and may recommend remedies for that particular phone.

To file a complaint, write a letter to the manufacturer, using the sample letter at the end of this document. To help the manufacturer select the right remedy, be sure to provide all the information in the sample, including the type of radio communication that the telephone equipment is receiving. You can identify the type of radio communication by listening to it. There are three common types:

(1) AM/FM broadcast radio stations—Music or continuous talk distinguishes this type of radio communication. The station identifies itself by its call letters at or near the top of each hour.

(2) Citizen's Band (CB) radio operators—These radio operators use nicknames or "handles" to identify themselves on the radio. Usually, the CB operator's voice is clearly heard. You may also hear sound effects or other noises.

(3) Amateur ("ham") radio operators—Amateur radio operators are licensed by the FCC. They use call letters to identify their communications. The amateur's voice can be heard but may be garbled or distorted.

Cordless telephones are low-power radio transmitters/receivers. They are highly sensitive to electrical noise, radio interference, and the communications of other nearby cordless phones. Contact the manufacturer for help in stopping interference to your cordless telephone.

Final note: Current FCC regulations do not address how well a telephone blocks out radio communications. At present, FCC service consists of the self-help information contained in this bulletin. A partial list of radio-proof telephones and radio filters is also attached.

The FCC strongly encourages manufacturers to include interference protection in their telephones as a benefit to consumers. The telephone manufacturing industry has begun to develop voluntary standards for interference protection. The FCC will continue regular meetings with manufacturers and will closely track the effectiveness of their voluntary efforts.

If you are not satisfied with the manufacturer's response, contact the Electronic Industries Association, 2500 Wilson Blvd., Arlington, Virginia 22201, phone: (703) 907-7500.

Dear Manufacturer: I am writing to register a complaint about telephone equipment I purchased (manufactured by your company). Unfortunately your product is receiving a nearby radio communication, making it difficult for me to complete phone calls. Please contact me within 30 days to discuss what steps your company will take to make my telephone work properly. Thank you for your help. I look forward to your prompt reply.

SOURCES OF RADIO-PROOF TELEPHONES AND RADIO FILTERS FOR TELEPHONES

The lists below show companies that sell radio-proof telephones and radio interference filters. If you would like to try a radio-proof telephone or radio interference filter, make sure that you can return it for a refund, and keep the purchase receipt.

THE FCC DOES NOT ENDORSE OR RECOMMEND THE USE OF ANY PARTICULAR GOODS OR SERVICES LISTED BELOW. SUCH GOODS OR SERVICES ARE LISTED FOR INFORMATION ONLY AND HAVE BEEN FURNISHED BY THE ORGANIZATIONS. FOR FURTHER INFORMATION, CONTACT THE INDIVIDUAL ORGANIZATIONS.

• RADIO-PROOF TELEPHONES:

TCE LABORATORIES, INC.
2365 Waterfront Park Dr.
Canyon Lake, TX 78133
(830) 899-4575
Note: Desk and wall models available. Will do custom orders for multiple-line phones, speaker phones, answering machines, etc. Advertises 30-day money-back guarantee.

PRO DISTRIBUTORS
2811 74th Street, Suite B
Lubbock, TX 79423
(800) 658-2027
Note: Desk and wall models available. Advertises 30-day money-back guarantee.

• RADIO INTERFERENCE FILTERS:

AT&T
(800) 222-3111
Notes: Also available at AT&T and GTE Phone Center stores.

COILCRAFT
1102 Silver Lake Road
Cary, IL 60013
(800) 322-2645
Note: Filters for computers and printers also available.

ENGINEERING CONSULTING
583 Candlewood Street
Brea, CA 92621
(714) 671-2009
Note: Also available filters for 2-line telephones.

INDUSTRIAL COMMUNICATIONS ENGINEERS (ICE)
350 Gloria Road
Sunnyvale, TX 75182
(972) 203-8811
Note: Also available hard-wired filters for wall-mount telephones.

K-COM
PO Box 82
Randolph, OH 44265
(216) 325-2110
Note: Also available filters for 2-line telephones.

KEPTEL, INC.
56 Park Road
Tinton Falls, NJ 07724
(908) 389-8800

KILO-TEC
PO Box 10
Oak View, CA 93022
(805) 646-9645

RICHARD MEASURES
6455 LaCumbre Road
Somis, CA 93066
(805) 386-3734

OPTO-TECH INDUSTRIES
PO Box 13330
Fort Pierce, FL 34979
(800) 334- 6786, or (407) 468- 6032

RADIOSHACK (ARCHER)
Available at nearest RadioShack store.
Catalog #273-104

SNC MANUFACTURING
101 W. Waukau Avenue
Oshkosh, WI 54901-7299
(800) 558-3325, or (414) 231-7370

SOUTHWESTERN BELL FREEDOM PHONE ACCESSORIES
7486 Shadeland Station Way
Indianapolis, IN 46256
(800) 255-8480, or (317) 841-8642

SPARREVOHN ENGINEERING
143 Nieto Avenue, Suite #1
Long Beach, CA 90803
(562) 799-1577
Note: Also available filters for 2-line telephones.

TCE LABORATORIES
2365 Waterfront park Drive
Canyon Lake, TX 78133
(830) 899-4575
Note: Also available filters for 2-line telephones.

[Editor's Note: Some of the information on this page is obsolete. It has been left unedited as part of FCC Bulletin CIB-10, August 1995. Please refer to Appendix A for current listings of Resources.]

• SAMPLE COMPLAINT LETTER:

Name
and address of
telephone manufacturer

Dear Manufacturer:

I am writing to register a complaint about telephone equipment I purchased (manufactured by your company). Unfortunately, your product is receiving a nearby radio communication, making it difficult for me to complete telephone calls. Please contact me within 30 days to discuss what steps your company will take to make my telephone work properly.

Thank you for your help.

I look forward to your prompt reply.

Name:__________________________ Telephone: ____

Address: ____

City/State/Zip Code: ____

Type of telephone equipment: ____

Model Number: ____

Description of Interference (AM/FM, CB, Amateur, etc):

____

____

Appendix C

Interference

To Home Electronic Entertainment Equipment

Handbook

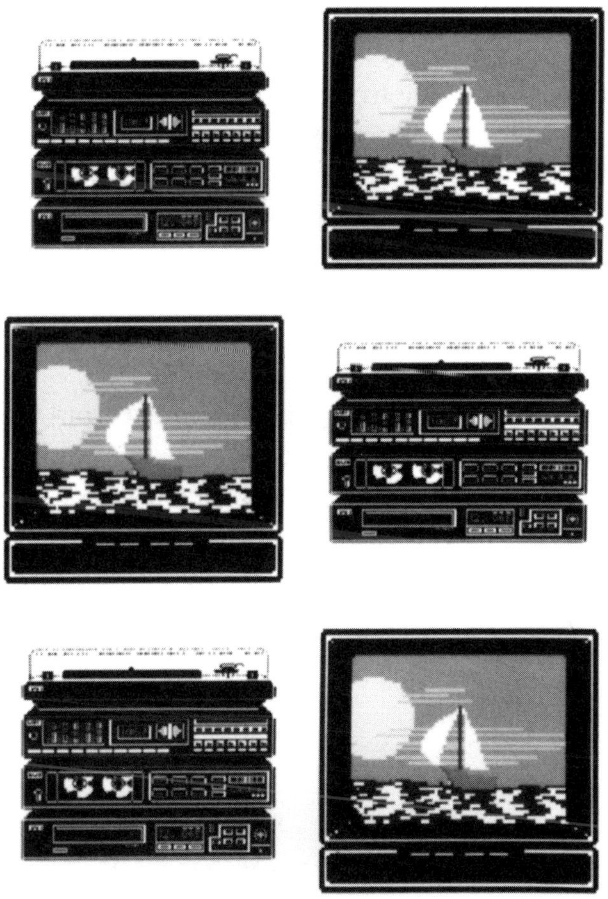

CIB
Compliance & Information Bureau

Bulletin CIB-2

May 1995

INTRODUCTION

Interference is any unwanted signal which precludes reception of the best possible signal from the source that you want to receive. Interference may prevent reception altogether, may cause only a temporary loss of the desired signal, or may affect the quality of the sound or picture produced by your equipment.

Interference to home electronic equipment is a frustrating problem; but, fortunately, there are several ways to deal with it. This handbook provides a step-by-step process for eliminating the interference.

If your problem is not eliminated by following the steps in this handbook, you should follow the instructions in the owner's manual of your equipment for contacting the manufacturer. We have provided a list of manufacturers. If the manufacturer of your equipment is not listed, look at the beginning of the list for additional help.

I. Check the Installation of Your Equipment

Many interference problems are the direct result of poor equipment installation. Cost-cutting manufacturing techniques, such as insufficient shielding or inadequate filtering, may also cause your equipment to react to a nearby radio transmitter. This is not the fault of the transmitter and little can be done to the transmitter to correct the problem. If a correction cannot be made at the transmitter, actions must be taken to stop your equipment from reacting to the transmitter. These methods may be as easy as adjusting your equipment or replacing a broken wire. These and other simple corrections may be accomplished without the help of a service technician.

A. Simplify the problem

Begin by disconnecting all equipment from the piece of equipment you are trying to fix. For example, if you are working with a television set, disconnect your VCR, stereo speaker wires and video game. Then, reconnect each of these additional devices individually to determine which device may be causing the interference to your television. Do the same thing for a telephone or stereo system. Disconnect all answering machines, telephones, CD players, facsimile machines, modems, etc. If the problem goes away when a device is not connected you have found the problem. It may be necessary to filter the device reacting to the transmitter.

For sale by the U.S. Government Printing Office
Superintendent of Documents, Mail Stop: SSOP, Washington, DC 20402-9328
ISBN 0 - 16 - 045542 - 1

B. Check your connections.

Make sure all cables are properly fastened and in good condition. Antenna wires, interconnecting cables and power cords often act as antennas and carry the interference into your system. All wires and cables should be as short as possible. If there are any loose connections or broken and damaged cables have them replaced or repaired. If you are using cable television services and have an in-house wire repair contract, contact your cable company for repair or replacement of the wires.

You should also test all splitters used in your system, if there are any. (A splitter is a device that provides a signal to more than one location.) To test the splitter, bypass it by hooking the antenna/cable connector directly to one TV. If the signal quality is improved or the interference goes away, the splitter is defective and should be replaced.

C. Check your amplifier.

Temporarily disconnect the amplifier and bypass it. By doing this, you allow the signal from the antenna to go directly to the TV or radio receiver. If the interference disappears, then the amplifier is causing the problem.

If your TV or radio receiver is connected to a master antenna television system (MATV), commonly used in large apartment complexes, you should contact the building management for assistance.

D. Check your antenna system.

Even though your antenna does not have moving parts, the wire and the antenna physically deteriorate due to the effects of time and harsh weather. Replace damaged or broken antennas. If the antenna is badly corroded clean or replace it. Check the incoming wire from your antenna for physical damage. If you are using twin-lead cable, replace it with coaxial cable. If you are currently using an inside antenna, try to replace it with an outside antenna to improve the signal.

Twin lead is usually a flat wire that connects the antenna to a receiving device. It is not shielded and the entire length of wire acts as an antenna. This may cause some receiving problems such as ghosting. Coaxial cable consists of two concentric conductors (an inner conductor and an outer braided sleeve) and is used in cable tv installations to prevent unwanted leakage from the cable system as well as entry of unwanted signals into the cable system.

If after following the steps described above your system continues to react to the interference, you should continue reading this booklet.

II. Identifying Other Sources

A. Simplify the problem

As a general rule, the more complex a system is, the more difficult it is to isolate a problem. Always start with the simplest system possible; one telephone, one television receiver, or just the stereo receiver.

For example, if your television is reacting to a nearby radio transmitter, remove all accessories, such as video games, VCRs, stereo system connections, booster amplifiers and even cable boxes. If you are connected to a cable system, connect the cable input directly to your TV set.

If the reaction stops when you disconnect any device, you have found the problem. You must now make a decision. You can attempt remedial action, contact the manufacturer for assistance, or replace the device with one that does not react to the nearby transmitter. If you choose remedial action continue with this section.

B. Collect Information about the Interference

Some basic information about the interference will help a lot in identifying its source.

1. When do you get the interference?

Keep track of the time of day you usually receive the interference. Do you get interference only at dinnertime? Does the interference occur day after day at the same time? Does the interference occur at all times or is it unpredictable?

If your equipment is reacting to the transmissions of a nearby radio operator, you will have the interference only when the radio operator is talking. The pattern will be much like that of a normal conversation except that you will hear only one half of the conversation. Usually the interference will occur for brief periods during specific times of the day.

If the interference is on constantly, it is not caused by a nearby radio operator. You may have electrical, broadcast, or paging interference.

2. What does the interference sound like?

Listen carefully to the interference. Read this section and see which part best describes the interference you are experiencing.

Do you hear music and voices from a broadcast station in the background? If so, try to identify which station you hear.

Stereo and other newer televisions may have separate audio channel capabilities. A foreign language that synchronizes with the video, a description of what is happening on the screen, or even continuous traffic reports may indicate that your television is switched to the separate audio channel. Refer to your operator's manual for instructions on how to change the audio channel.

Do you hear radio operator voices? Are the voices garbled? If the interference is intermittent and you hear clear or garbled voices, you are probably picking up the transmissions of a nearby CB or Amateur radio operator. If so, you will probably be able to see an antenna mounted on their house or car.

Do you hear static, or a constant buzzing sound like bacon frying? You may be picking up interference from an electrical device in either your house or your neighbor's house. You may also be picking up interference from electrical power lines. If the buzzing noise only occurs for short periods of time, you may be receiving interference from a household appliance, such as a hair dryer, vacuum cleaner, or electric drill.

Do you hear voices and tones? If so, you may be picking up a nearby paging system. Paging systems usually operate 24 hours a day, so your equipment may react to paging interference at any time.

3. What does the interference look like?

a. Electrical Interference

Electrical interference many times will be seen as two or three horizontal lines on the television screen and may be accompanied by a loud buzzing or sizzling sound through the TV speakers or stereo system. Many times the lines move upwards on the television screen and may be on for hours at a time or for a few seconds at a time. In severe cases, the entire screen may be covered with rolling horizontal lines.

A simple way to discover if the source of interference is in your home, is by going to the main fuse or circuit breaker box in your home. Using your TV set or a portable AM radio, determine whether the interference is active. Electrical interference will sound about the same on an AM radio as it does on your TV so you can use a portable AM radio as a detector. Assuming the interference is occurring, you should follow these steps to identify the circuit in your house that has the device causing the interference. Be very careful to avoid contact within anything in the box except the fuses or circuit breakers. Remove one fuse at a time, or shut off one circuit breaker at a time. If the interference stays on, turn the circuit breaker back on and try the next. If you are using your television to know that the interference is active, when you turn off the power to the circuit that supplies power to your TV to test that circuit, plug the TV into another circuit.

If the interference stops when a fuse is removed or a circuit breaker is turned off, go to the area that receives the electricity supplied by the disconnected circuit. Turn the power back on and wait until the interference is present. Next unplug each device on the circuit one at a time. If the interference stops after you unplug a device, you have found the culprit. The device causing the interference must be repaired or replaced. Remember that the device might be hidden. For example, you may have a bad amplifier in your attic, or a defective doorbell transformer that is connected directly to the power circuit.

An alternative method to locating electrical interference is to tune to a quiet frequency at the lower end of the AM dial. If you hear static or a buzzing sound, check to see if it corresponds with the interference to your TV or telephone. If it does, use the portable radio as a detection device to locate the source of the interference.

The noise will be loudest in the room where the interference is originating. Unplug each electrical device in the room one by one until the interference stops.

If you cannot locate the interference source in your own house, check with your neighbors to see if they also receive interference. The house that has the worst interference will most likely be the source of the interference. If your neighbor has strong interference, you may wish to try to track it down with a portable AM radio or run the circuit breaker test described above.

If you determine that the interference is not caused by any device in your home or that of your neighbors, contact the customer service department of your local power company.

b. Interference to Television (TVI)

The following pages illustrate what many common types of interference look like on a TV set. Find the one that best matches your interference.

Use this picture for comparison with the other pictures in this section.

Picture 1: Normal TV Reception

Picture 2: Poor TV Signal

TV stations are intended to serve viewers only within a limited distance of their transmitters. You can improve picture quality by raising your antenna or using a more directional antenna. Check your antenna line and connections. Try reorienting your antenna for better reception. If necessary, consider installing an outdoor antenna or a booster amplifier. Changing from flat twin lead to round, coaxial cable may also help.

Picture 3: Ghosting

Double images of a TV program, or "ghosting", is a common problem with off-air TV reception in urban areas. Ghosting may be caused by the TV signal being reflected off of a tall building or mountains. Ghosting may also indicate problems with the TV antenna or lead in wire. Reorient your antenna, or install a shielded (coaxial) lead in wire. You may need to install a directional, outdoor antenna.

Images from two different programs may appear on your TV screen when your set simultaneously receives two TV signals. Co-channel interference looks much like ghosting, except that the two images are different, as though one picture has been placed on top of the other. If the problem is caused by atmospheric conditions, it is usually temporary. Installing a directional, outdoor antenna, or relocating your indoor antenna may improve reception. You may also have similar interference if you are picking up signals leaking from a cable TV system. If you believe that you are picking up cable TV stations but you are not connected to the cable system, contact the cable TV company.

Picture 4: Co-channel Interference

Picture 5: Ham/CB Transmitter Interference

This picture may appear on your TV screen when your set is reacting to signals from a CB, amateur, police, or other two-way radio transmitter. The pattern will appear only when the operator transmits. The "lines" in the interference pattern may be wider, or may seem to "roll" up through the TV picture. If your TV is reacting to CB or amateur radio transmissions, you will often hear the operator's voice, although it may be garbled. If you are very close to the transmitting antenna, the TV screen may "black out" when the operator transmits. It is also very common to pick up the CB or amateur operator's voice on the telephone or on your radio or stereo system. For information about possible remedies see Section III.

Picture 6: Electrical Interference—very noisy (hair dryer)

This picture may appear on your TV screen when your set is reacting to an electrical device operating in or near your home. Home appliances and electrical equipment, such as hair dryers, electric razors and electric drills may cause temporary problems. You may choose to live with this type of interference as it is often expensive to correct. You may also experience intermittent problems from other home appliances, such as refrigerators and air-conditioners. For example, you may notice interference on your TV when your air conditioner cycles on or off. If the interference is on continuously, it may be caused by power line equipment. For information about possible remedies, see Section III.

Doorbell transformers are a common source of interference to TV reception. This interference looks like snow or bars on the TV screen. It looks very much like electrical interference, except it does not cover the whole screen. It appears as bars of electrical interference. A frying or sizzling sound comes and goes at about seven second intervals (the intervals may vary). The interference may also be received hundreds of feet from the source. If you suspect that you are experiencing interference from your own or your neighbor's doorbell transformer, and are unable to locate the transformer, you may wish to contact a service technician.

Picture 7: Electrical Interference—low noise

This pattern may appear on your TV screen if you are picking up signals from an FM broadcast transmitter. For information about possible remedies, see Section III.

Picture 9: Computer Interference

This type of picture may appear on your TV screen if you operate a computer in close proximity to the TV antenna. The interference may look like electrical interference or a series of diagonal dashed white lines among other patterns. Computer interference will usually occur only when the computer is very close to the TV antenna. You may try to reorient your antenna or computer cables, or move the computer or TV set to another location. For information about possible remedies, see Section III.

Low power radio devices, such as garage door openers also can cause interference. This pattern may appear on your screen, if you are picking up signals from a garage door opener or radio frequency doorbell.

Picture 10: Garage Door Opener

Booster amplifiers are often used to help receive weak or distant TV signals. Booster amplifiers may be installed at the TV set, at the TV antenna, or even in the attic. Although booster amplifiers do increase the TV station signal strength, they may also cause interference to your TV or even your neighbor's TV. A variety of patterns may appear on your TV screen because of booster amplifier interference. Sometimes a wavy pattern may appear, or the screen may black out for a short time. If your antenna system uses a booster amplifier, you should disconnect it and turn it off. Next connect your antenna directly to the TV. If the interference disappears, have your booster amplifier repaired or replaced. If you still have interference after disconnecting your booster amplifier, you may be receiving interference from a neighbor's amplifier. If several of your neighbors have interference, the one with the most severe interference is probably the one with the defective amplifier.

Amplifiers may also generate interference when used near strong signal sources, such as TV and radio broadcast stations, paging transmitters, or two-way radio stations.

c. Summary of Possible Interference Sources

1) Broadcast
   - AM Radio Station
   - FM Radio Station
   - TV Station

2) Two-way Radio Transmitters
   - Citizens Band (CB)
   - Amateur (Ham)
   - Taxi
   - Police
   - Business
   - Airport/Aircraft

3) Paging Transmitters

4) Cable TV

5) Electrical Devices
   - Doorbell transformers
   - Toaster Ovens
   - Electric Blankets
   - Ultrasonic pest controls (bug zappers)
   - Fans

- Refrigerators
- Heating pads
- Light dimmers
- Touch controlled lamps
- Fluorescent lights
- Aquarium or waterbed heaters
- Furnace controls
- Computers and video games
- Neon signs
- Power company electrical equipment
- Alarm systems
- Electric fences
- Loose fuses
- Sewing machines
- Hair dryers
- Electric toys
- Calculators
- Cash registers
- Lightning arresters
- Electric drills, saws, grinders, and other power tools
- Air conditioners
- TV/radio booster amplifiers
- TV sets
- Automobile ignition noise
- Sun lamps
- Smoke detectors

III. Remedies

WARNING — TO AVOID ELECTRICAL SHOCK, ELECTRIC OR ELECTRONIC EQUIPMENT SHOULD BE MODIFIED ONLY BY QUALIFIED SERVICE TECHNICIANS

Before you attempt any of the following solutions, you should see if moving or reorienting the affected device eliminates the interference. This often will prove to be a simple, but effective solution. For example, you may notice that your living room TV has perfect reception, while the bedroom TV has horrible interference. The farther away the affected device is from the interference source, the less severe the reaction will be.

If you believe that you are picking up interference through connecting cables or the antenna lead, you may wish to wrap several turns of the cable through a snap-together ferrite core.

It is always best if the affected device is modified in your home while it is reacting to the interference. This will enable the service technician to determine where the interfering signal is entering the equipment.

A. OFF-AIR or CABLE TV RECEPTION PROBLEMS

If you have reception problems such as a weak TV signal, ghosting, or co-channel interference, see Section II.B.3.b of this booklet.

B. TWO-WAY RADIO INTERFERENCE

The steps listed below may help you to eliminate TV interference that you experience from CB, amateur or other two-way radio stations. High-pass filters, common-mode chokes (choke filters), snap-together ferrite cores and ac-line filters are available from local electronics and department stores. A list of sources of filters is also provided at the end of this book. Remember, if your TV set or VCR has insufficient filtering or shielding, you may not be able to correct the interference yourself. You will have to obtain help from the manufacturer.

1. CB or Amateur Interference: Check to see if the TV volume control affects the interference level. If it does, install a "high-pass filter" at the antenna leads. This is a filter which will not allow signals in the low part of the radio spectrum to pass through it. The name "high pass filter" is all you need to know to purchase one. If the TV volume control has no effect on the interference level, go to step 4. If you still receive the interference after installing a high-pass filter, go to step 3.

2. Two-way FM Interference: If you are receiving two-way FM transmissions, such as police or taxi, you may need to install a "notch" or "band-reject" filter at the TV antenna leads. The filter must be designed to reject the specific transmissions that you are receiving. The electronics dealer where you purchase the filter should be able to assist you with your selection. If you still receive the interference after installing a notch filter, continue to the next step.

3. Install a common-mode filter (see list) and/or common-mode choke at the TV input. If you still receive the interference, go to the next step.

4. Disconnect the antenna lead from the TV set. If you still have the interference with the antenna disconnected, install an ac-line filter at the electrical outlet your TV is plugged into.

Try wrapping three or four turns of the TV set's power cord through a ferrite core. Do this as close to the TV set as possible. You may also install a ferrite core on the antenna coaxial cable where it enters the TV set. You local electronics store will know what a ferrite core is and will help you select one.

C. FM BROADCAST INTERFERENCE

There are three common interference problems that involve FM broadcast stations. The first problem may arise when you are receiving a distant TV signal and a new FM station begins operating in your area. The FM signal may overpower the weak TV signal. This problem often shows up as interference to TV channel 6 only. FM stations may also cause interference to radio receivers in your home.

The second problem is called "blanketing." It occurs when electrical or electronic devices are very close to the transmitter and where the FM signal is very strong. Operators of new FM stations are required to respond to all reasonable interference complaints in the immediate vicinity of the station. If the station is more than a year old or you live outside of the protected area, installation of an FM broadcast band rejection filter at your TV and/or stereo receiver antenna input leads, as well as use of a highly directional antenna may reduce your problem. You may also wish to reposition the location of your equipment to attempt to minimize or eliminate the problem(s).

Third, problems may occur when you are using an amplifier. Amplifiers are devices used to increase signals from distant stations and frequently react to strong nearby signals. If you suspect this is the case, you should install an FM band rejection filter or a tunable FM rejection trap in the antenna line between your antenna and booster. Some booster amplifiers have built in filters you simply switch on or off. Consult the instruction manual of your product. In extreme cases it may be necessary to install a second filter. Repair or replace your booster amplifier if it is defective.

D. PAGING SYSTEM INTERFERENCE

You may hear tones or voice transmissions from a one-way paging system coming through your TV set. The FCC requires paging system operators to assist in resolving interference caused by their transmissions only while the paging system is in the developmental licensing process. This is usually only within the first year of operation. Further, the FCC requires paging companies to resolve all interference complaints they may cause to television channels 4 and 5. If you know who the paging operator is, you may contact them directly. A band-reject, or "notch" filter designed to reject the paging frequency may be installed at the TV set input. If you need further assistance, you can contact your local FCC office listed in this handbook.

E. INTERFERENCE TO VIDEO CASSETTE RECORDERS (VCRs):

Your VCR is really a television receiver without a screen. If your VCR is receiving interference, replacing connecting wires with shielded wire (coaxial cable) or replacing push-on connectors with screw-on connectors may solve the problem. The solutions for interference from two-way radio, broadcasting, and paging, described for television interference above, also apply to VCR interference. If these do not work, contact the VCR manufacturer for alternative solutions.

F. COMPUTER INTERFERENCE:

Computer interference will normally disrupt only your TV reception. Check all computer connecting cables. You should especially be on the lookout for loose or unterminated cables. Greater distances between the TV and computer may solve the problem.

Occasionally, your computer may cause interference to your neighbor's TV reception. If this happens, you are required to discontinue using your computer system until you can eliminate the interference problem. Consult your owner's manual, the computer dealer, or the manufacturer for suggestions on how to resolve the interference.

G. TELEPHONE EQUIPMENT INTERFERENCE:

Telephone interference generally happens because telephones are not designed to operate near radio transmitters and the telephone improperly operates as a radio receiver.

You may contact the nearest FCC Office and request FOB Bulletin No. 10, "Telephone Interference". You may also try the following:

1. Contact the telephone company if you are using a leased phone. The telephone company may have responsibility for correcting interference to their leased phones.

2. Disconnect all of your telephones and accessories such as answering machines and take them to one telephone jack. Connect each instrument, one at a time, and listen for the interference. If you hear the interference through only one telephone, the interference is being generated in that unit.

3. Install a filter on the telephone line cord at the end nearest the telephone and/or at the telephone handset cord.

Filters are very selective. (See Section IV.) They must be designed for the type of interference your are experiencing or they will not work. For example, if your phone is reacting to an Amateur or a CB radio transmitter, install a filter designed for that purpose. FM broadcast interference requires a filter designed to reject FM broadcast stations. AM broadcast interference requires a filter designed to reject AM broadcast stations, etc.

4. Filter the incoming telephone line with ferrite beads and snap-together ferrite cores. You may need to experiment to find the best style of bead or core and the best location on the cord.

5. If you cannot eliminate the interference using the above techniques, you should consider purchasing a interference free telephone which has been specifically designed to be immune to interference.

Cordless telephones use radio frequencies and have no FCC protection from interference. If you are receiving nearby transmissions on a cordless phone your only recourse is to contact the manufacturer for assistance. The remedies above will not be of any use.

H. ELECTRICAL INTERFERENCE

If you have determined that the electrical interference is coming from within your home or one of your neighbor's homes, you should disconnect the defective equipment and replace it or have it repaired.

Devices such as electric razors, hair dryers, electric drills and saws can also cause temporary interference problems. You may choose to tolerate this type of interference since it is temporary and often expensive to eliminate. You may wish to contact the manufacturer for assistance.

If you determine that the interference is not caused by any device in your home, or a neighbor's home served by the same transformer, contact the service department of your local power company. Most power companies will investigate the problem and take steps to correct it.

I. INTERFERENCE TO OTHER EQUIPMENT

Stereos, electronic organs, and intercom devices, among others, can react to nearby radio transmitters. When this happens, the device improperly functions as a radio receiver. You should first determine what type of interference you are receiving. See Section II of this bulletin for assistance. You may try to relocate the device within your home. It is usually impractical or expensive to modify the affected device. Contact the manufacturer for assistance or consider changing to another brand or model.

IV. About filters:

Most filters are very selective. A filter designed to eliminate AM broadcast signals will have no effect upon an FM broadcast signal. Choose a filter designed for your needs. If you have doubts contact the manufacturers listed below to find a filter appropriate to your needs.

A. High pass filters

Andrew Corporation .... Several Models
1315 Industrial Park Dr
Smithfield NC 27577
(919) 934-9711

Cornel Dublier Electronics
1605 E. Rodney French Blvd
New Bedford, MA 02741
(508) 996-8561

Industrial Communications Engineers (ICE) .... Model 430 series high-pass
Array Solutions
350 Gloria Rd
Sunnyvale, TX 75182
(972) 203-2008

RadioShack .... Catalog: 15-579 for coaxial cable;
15-582 or 15-581 for twin lead systems

Ten-Tec, Inc .... Model 5060
1185 Dolly Parton Parkway
Sevierville, TN 37862
(865) 453-7172

Winegard..Model HP-2700
3000 Kirkwood Street
Burlington, IA 52601-1007
(800) 288-8094

B. Common mode filters

Industrial Communications Engineers (ICE)
Array Solutions
350 Gloria Rd
Sunnyvale, TX 75182
(972) 203-2008

Starplex Inc. ....Model Ghost Buster dual-ferrite Choke
1912 Falcon
Hanley Hills, MO 63133
(314) 423-5756

C. Notch filters (band reject filters)

Notch filters are very selective. To be effective notch filters must be tuned to the frequency of the offending transmitter. Therefore your service technician or the manufacturer should do the installation.

Communications and Energy Corporation ....channel reject traps
7395 Taft Park Dr
East Syracuse, NY 13057
(315) 452-0709 FAX (315) 451-0732

Microwave Filter Company
6743 Kinne St
E. Syracuse, NY 13057
(315) 438-4700
Fax: (315) 463-1467
Email: mfcsales@microwavefilter.com

Winegard..Model UT-2700 notched between 4700 and 800 MHz
3000 Kirkwood S
BF-1700 attenuate 110 -170 MHz
Burlington, IA 52601-1007

D. Ferrites and beads

Amidon
240 Briggs Ave
Costa Mesa, CA 92626
(800) 898-1883

Fair-Rite Products Corp .... Call to locate nearest distributor
PO Box J, 1 Commercial Row
Wallkill, NY 12589
888 FAIRRITE (324-7748) or (845) 895-2629
Web: http://www.fair-rite.com/

MFJ
PO Box 494
Mississippi State, MS 39762
(662) 323-0549

Ocean State Electronics
PO Box 1458
Westerly, RI 02991
(401) 596-3080

Palomar Engineers
PO Box 462222
Escondido, CA 92046
(760) 747-3343

RadioShack .... Catalog # 273-105 and 273-104

E. AC line filters

Industrial Communications Engineers (ICE) .... Model 472 isolator
Array Solutions .... Model 400 series
350 Gloria Rd
Sunnyvale, TX 75182
(972) 203-2008

RadioShack .... Catalog 15-1111 brute force AC line filter
.... 6-1395 surge-protection filter

F. Telephone filters

AT&T .... Model Z100B1
(800) 222-3111

Coilcraft
1102 Silver Lake Road
Cary, Il 60013
(800) 322-2645

Engineering Consulting .... Model TRIF
583 Candlewood St.
Brea, CA 92621
(714)671-2009

Industrial Communications Engineers (ICE)
Array Solutions
350 Gloria Rd
Sunnyvale, TX 75182
(972) 203-2008

K-Com .... Model RF-1 for modular phones;
PO Box 83.. RF-2 for wired phone connections
Randolph, OH 44265
(330) 325-2110
(877) 242-4540

Richard Measures
6455 LaCumbre Road
Somis, CA 93066
(805) 386-3734

Southwestern Bell Freedom Phone Accessories .... Model ZM04223
7486 Shadeland Station Way
Indianapolis, IN 46256
(317) 576-6847
(800) 366-0937

Sparrevohn Engineering .... Model F-1 Single line model
6911 E 11th St. .... F-2 two line model
Long Beach, CA 90815
(562) 799-1577

Telecom Industries.. Model 831-W1
1385 Akron St
Copiague, NY 11726
(631) 789-5020

V. MANUFACTURER LIST

Some manufacturer mailing addresses and contact points are listed below. It is suggested that you provide the model number of your device, the serial number, and any information concerning the specific problem.

For listings of consumer electronics manufacturers not provided in this book, you may want to contact the Electronics Industry Association/Consumer Electronics Group, 2500 Wilson Boulevard, Arlington, VA 22201 (703) 907-7600.

BANG AND OLUFSEN OF AMERICA
Customer Service Department, tel. 1 800 323-0378

CARVER CORPORATION, INC.
20121 48th Avenue, Lynnwood, WA 98046, tel. (206) 775-1202

CONCORD (A HARMAN INTERNATIONAL COMPANY)
80 Crossways Park West, Woodbury, NY 11797, attention: Customer Service, tel. (516) 496-3400

CONN KEYBOARDS, INC.
Contact your local Conn dealer or Organ Service Department, PO Box 352, French Lick, IN 47432, tel. (812) 836-9200

CONRAD-JOHNSON DESIGN, INC.
2733 Merrilee Drive, Fairfax, VA 22031, tel. (703) 698-8581

CROWN INTERNATIONAL
Technical Services Department, 1718 W. Mishawaka Rd, Elkhart, IN 46517, tel. (219) 294-5571

CURTIS MATHES
Contact your retail-dealer or write to Consumer Relations Division, 1220 Champion Circle, Suite 100, Carrolton, TX 75006.

DEFINITIVE TECHNOLOGY, INC.
Anne Conaway, Definitive Technology, Inc., tel. (410) 363-7148

DELCO ELECTRONICS (DIVISION OF GM CORPORATION) SEE GM CORP.
PO Box 9005, Kokomo, Indiana 46904-9005, Attention: Steve Butler
tel. (317) 451-3372

EMERSON RADIO CORPORATION
9 Entin Road, Parsippany, NJ 07054, Attention: Mr Jim Flatt, tel. (201) 428-2125.

EPI (A HARMAN INTERNATIONAL COMPANY)
80 Crossways Park West, Woodbury, NY 11797, Attention: Customer Service, tel. 1-800-343-9381.

EPICURE PRODUCTS, FORMERLY ELPA MARKETING INDUSTRIES, INC.
Mr. Jim Orman, National Service Manager, 25 Hale St, Newburyport, MA 01950, tel. 1 800 225-7932

FANNON COURIER CORPORATION
Mr. Mike Santana, Vice President of Engineering, 15200 San Fernando Mission Blvd, Mission Hill, CA 91345, tel. (213) 365-2531.

GENERAL ELECTRIC COMPANY
Consumer Relations Department, Mail Drop 43, Consumer Electronic Business Operations, Portsmouth, VA 23705, tel. (804) 483-5652.

GENERAL MOTORS CORPORATION
Seek assistance at your GM dealership or contact Steve Butler of Delco Electronics, Subsidiary of GM Hughes Electronics, PO Box 9005, Kokomo, IN 46904-9005, tel. (317) 451-3372.

H.H. SCOTT, INC.
9 Entin Road, Parsippany, NJ 07054. Attention: Mr. Jim Flatt, tel. (201) 428-2125

HARMAN/KARDON, INC.
80 Crossways Park West, Woodbury, NY 11797, Attention: Customer Service, tel. 1 800 422-8027 or 1 800 343-9381 (video division).

HITACHI HOME ELECTRONICS (AMERICA), INC.
Customer Service Relations, 3890 Steve Reynolds Blvd, Norcross, GA 30093, tel. (404) 279-5600, extension 22 or contact Mr. Chris Fabian, National Service Manager, ext. 145, 9am-5pm EST, M-F.

JBL CONSUMER PRODUCTS, INC.
80 Crossways Part West, Woodbury, NY 11797, Attention: Customer Service, tel. 1 800 336-4JBL (4525), or contact JBL Consumer Products, Inc., Mike Christian, tel. (516) 496-3400, Ext. 233.

LONESTAR TECHNOLOGIES, LTD.
920 South Oyster Bay Road, Hicksville, NY 11801, Attention: Arlene Vittore, tel. (516) 939-6116

MAGNAVOX (SEE PHILIPS CONSUMER ELECTRONICS COMPANY)

MARANTZ AMERICA, INC.
Marantz America Service Department, 440 Medinah Road, Roselle, IL 60172-2330, tel. (708) 307-3100, fax (708) 307-2687

MITSUBISHI ELECTRONICS AMERICA, INC.
Consumer Relations Department, tel. (714) 220-1464

NAIM AUDIO NORTH AMERICA, INC.
1748 North Sedgwick Street, Chicago, IL 60614, tel. (312) 944-0217, fax (312) 944-0255

NINTENDO OF AMERICA INC.
4820 150th Ave NE, Redmond, WA 98052, tel. 1 800 255-3700, hrs 4AM -Midnight PST, M-Sat, 6AM - 7PM, PST on Sundays. Contact is Dorothy Caravias tel. (206) 861-2796

PANASONIC (MATSUSHITA ELECTRIC CORPORATION OF AMERICA)
One Panasonic Way 2F-3, Secaucus, NJ 07094, Attn: Customer Satisfaction Center, tel. (201) 348-9090

PHILCO - (SEE PHILIPS CONSUMER ELECTRONICS COMPANY)

PHILIPS CONSUMER ELECTRONICS COMPANY/PHILIPS ELECTRONICS NORTH AMERICAN COMPANY
Dick Hovermale, Information and Consumers Affairs Center, 401 E. Old Andrew Johnson Highway, PO Box 555, Jefferson City, TN 37760, tel. (615) 475-8869 or (615) 475-0063, fax (615) 475-0203

QUASAR (MATSUSHITA ELECTRIC CORPORATION OF AMERICA)
50 Meadowland Parkway 2F-3, Secaucus, NJ 07094, Attention: Customer Satisfaction Center, tel. (201) 348-9090

ROTEL OF AMERICA
J. Laliberte, National Service Manager, PO Box 8, North Reading, MA 01864-0008, tel. (508) 664-3820

SONY ELECTRONICS INC.
Sony Customer Relations Center, One Sony Drive, Park Ridge, NJ 07656, tel. 1 800 282-2848 or Technical Services Department, tel. (201) 930-1000

THOMSON CONSUMER ELECTRONICS
Larry Elkin, Thomson Consumer Electronics, PO Box 6127, INH-315, Indianapolis, IN, tel. (317) 231-4151

TOSHIBA AMERICA CONSUMER PRODUCTS, INC.
Contact the nearest regional office listed below:

Mr. M. Pluchino, Eastern Regional Service Manager, 82 Totowa Road, Wayne, NJ 07470, tel. (201) 628-8000.

Mr. R. Fisher, Midwest Regional Service Manager, 1010 Johnson Dr, Buffalo Grove, IL 60089, tel. (708) 541-9400.

Mr. A. Marquez, Western Regional Service Manager, 5490A East Francis St, Ontario, CA 91761, tel. (909) 988-5303.

Service Division, 1420 Toshiba Drive, Lebanon, TN 37087, Facsimile: (615) 444-7520, tel. (615) 449-2360

VEHICLE SECURITY ELECTRONICS, INC.
Rudy Sanders, President, 21540-F Prairie Street, Chatsworth, CA 91311, tel. (818) 700-7900, fax (818) 700-7926

VIDEONICS, INC.
Videonics' Help Line, 1370 Dell Ave, Campbell, CA 95008, tel. (408) 370-9963; 8am - noon and 1pm - 5pm, Pacific Time

WADIA DIGITAL CORPORATION
Service Department, 624 Troy St., River Falls, WI 54022, tel. (715) 426-5900

YAMAHA ELECTRONICS CORPORATION, USA AND YAMAHA CORPORATION OF AMERICA
Mr. Bill Bower, Electronic Service Manager, Electronic Service Department, 6600 Orangethorpe Ave, Buena Park, CA 90620, tel. (714) 522-9100.

TELEPHONE MANUFACTURER CONTACTS

AT&T National Sales and Service Center
(800) 222-311

Cobra Comm
(800) 262 7222

Code-A-phone
(503) 655 8940

Fun Products
(415) 540-0334

General Electric
(800) 447-1700

ITT - Technical Assistance
(601) 287-5281

Lonestar
(516) 939-6116

Northern Telecom
(919) 992-5743

Pac Tel
(800) 426-2372

Panasonic Consumer Electronics
(800) 221-3438 ext. 6881

Phonemate
(800) 247-7889
(213) 320-9810

Plantronics
(800) 538 0748
(800) 662-3902 CA

Sanyo
(800) 524-0047 ext. 502 or 520

Soundesign Corporation
(201) 434-1050

Swisstel
(201) 470-5706

Toshiba
(714) 583-3700

TT system
(914) 968-2100

Uniden Corp.
(317) 842-2483

Appendix D

What to do if You Have an Electronic Interference Problem

This is a self-help guide for the consumer published jointly by the American Radio Relay League (ARRL), an organization representing Amateur Radio operators, and the Consumer Electronics Association (CEA).

Introduction

As our lives become filled with more technology, the likelihood of unwanted electronic interference increases. Every lamp dimmer, hair dryer, garage-door opener, radio transmitter, microprocessor-controlled appliance or remote-controlled new technical "toy" contributes to the electrical noise around us. Many of these devices also "listen" to that growing noise and may react unpredictably to their electronic neighbors.

Interference: What Is It?

Complex electronic circuitry is found in many devices used in the home. This creates a vast interference potential that didn't exist in earlier, simpler decades. Your own consumer electronics equipment can be a source of interference, or can be susceptible to interference from a nearby noise source. Interference can also result from the operation of nearby amateur, citizens band, police, broadcast or television transmitters.

The term "interference" should be defined without emotion. To some people, it implies action and intent. The statement, "You are interfering with my television" sounds like an outright accusation. It is better to define interference as *any unwanted interaction between electronic systems –period!* No fault. No blame. It's just a condition.

Personalities

You can't overestimate the importance of personal diplomacy when you're trying to solve a problem that involves two or more people! The way you react and behave when you first discuss the problem with other individuals, such as a neighbor, utility or cable company, or manufacturer, can set the tone for everything that follows. Everyone who is involved in an interference problem should remember that the best solutions are built on cooperation and trust. This is a view shared by electronic equipment manufacturers, the Federal Communications Commission (FCC) and the American Radio Relay League (ARRL).

Responsibilities

No amount of wishful thinking (or demands for the "other guy" to solve the problem) will result in a cure for interference. Each individual has a unique perspective on the situation —and a different degree of understanding of the technical and personal issues involved. On the other hand, each person may have certain responsibilities toward the other and should be prepared to address those responsibilities fairly.

Any individual who operates a radio transmitter, either commercial or private, is responsible for the proper operation of the radio station. All radio transmitters or sources are regulated by the FCC. The station should be properly designed and installed. It should have a good ground and use a low-pass filter, if needed. If consumer electronics equipment at the station is not suffering the effects of interference, you can be almost certain that the problem **does not** involve the radio station or its operation. However, if the interference is caused by a problem at the station, the operator must eliminate the problem there.

Manufacturers of consumer electronics equipment are competing in a difficult marketplace. To stay competitive, most of them place a high priority on service and customer satisfaction. For example, many manufacturers have service information that can be sent to a qualified service dealer. Most manufacturers are willing to assist you in resolving interference problems that involve their products. Over recent years, manufacturers have built up a good track record designing equipment that functions well in most electrically noisy environments.

The FCC will do what it can to help consumers and radio operators resolve their interference problems. They expect everyone involved to cooperate fully. Experience has taught them that solutions imposed from the outside are not usually the best solutions to local problems. Instead, they provide regulatory supervision of radio operators and manufacturers. To help consumers, basic information concerning interference solutions is [included in Appendix B and C of the ARRL RFI Book and is also available on the internet at www.arrl.org/tis/info/rfi-info-fcc.html.]

Finally, the consumer has responsibilities, too. You must cooperate with the manufacturer, the radio operator, and, if necessary, the FCC as they try to determine the cause of the problem. They need your help to find a solution.

What Causes Interference?

Interference occurs when undesired radio signals or electromagnetic "noise" sources are picked up by consumer electronics products –most often telephones, audio equipment, VCRs or TVs. It usually results in noise, unwanted voices or distorted TV pictures. In most cases, the source is nearby.

There are three common types of interference:

1) Noise: Interference can be caused by an electromagnetic noise source. Defective neon signs, bug zappers, thermostats, electrical appliances, switches or computer systems are just a few of the possible sources of this type of interference. Both you *and* your neighbors may be suffering from its effects. In some cases, the noise may be the result of a dangerous arc in electrical wiring or equipment and may provide warning of an unsafe condition that should be immediately located and corrected.

2) Overload: Even if a nearby radio signal is being transmitted on its assigned frequency, if it is strong your equipment may be unable to reject it. Your telephone, radio, stereo or TV should be able to separate the desired signal or sound from a large number of radio signals and electrical noises. This is shown in Figure 1. Consumer electronics equipment manufacturers have worked in cooperation with government regulators to set and meet voluntary standards of interference immunity. Modern equipment usually includes enough filtering and shielding to ensure proper performance under average conditions. Older equipment may not meet these standards, however, and even modern equipment can be affected if the interfering signal is particularly strong. In these cases, your equipment is working as designed, but it may need some additional filtering or shielding to function properly.

3) Spurious emissions: A nearby radio transmitter could be inadvertently transmitting weak signals on a frequency not assigned to that transmitter. These signals are called "spurious emissions." FCC regulations concerning spurious emissions are very clear. If interference is caused by spurious emissions, the operator of the transmitter must take whatever steps are necessary to reduce the spurious emissions as required by FCC regulations. Fortunately, modern transmitting equipment is manufactured to meet stringent regulations, and many radio operators are examined and licensed by the government. These federally licensed operators often have the technical skill to resolve interference problems that originate from their radio stations.

With all of these possibilities, it is difficult to **guess** which type of problem is causing your interference. Usually, only a technical investigation can pinpoint the cause and suggest a solution. This is where a spirit of cooperation and trust will pay off! If you believe your equipment is picking up signals from a nearby radio transmitter, the operator may be able to help you both find a solution to a mutual problem.

How to Find Help

Most consumers do not have the technical knowledge to resolve an interference problem. Even so, it's a comfort to know that help is available. Gather information about interference. The FCC and ARRL have self-help information packages and books. If the problem involves an electrical-power, telephone or cable-television system, contact the appropriate utility company. They usually have trained personnel who can help you and your neighbor pinpoint the cause of the problem.

Most consumer electronics manufacturers are willing to help you. Your owner's manual, or a label on your equipment, may give you information about interference immunity or tell you who to call about interference problems. If not, the Consumer Electronics Association will be able to give you the address of your equipment manufacturer's general customer service personnel. The manufacturers know their equipment better than anyone else and will usually be able to help you.

Operators of licensed amateur or commercial transmitters usually have some technical ability. These operators are the **nearest** source of help. Remember, the station operator may also be a neighbor! Use a polite approach to ensure that the relationship stays "neighborly." Licensed Amateur Radio operators have access to volunteers (Technical Coordinators and local interference committees) who are skilled at finding solutions for most interference problems.

Testing One, Two, Three . . .

If you think a neighbor's radio transmitter might be involved, you and your neighbor should arrange a test. It's important to determine whether the interference is (or is not) present when the radio station is "on the air." Your neighbor may want to ask another operator friend to participate in the test at your home. By the same token, you may want to invite a friend to attend the test at the radio operator's station. Having impartial witnesses will make you and your neighbor more comfortable with the outcome—whatever it may be. Be sure to choose your witness carefully. Select someone who is diplomatic and tactful.

The tests must be thorough. The transmitter operator must try all normally used frequencies, antenna directions and power levels. All results must be carefully written down. More than one set of tests may be needed. Once you and your neighbor have determined which frequencies and power levels cause the problem, you'll be one step closer to finding a solution.

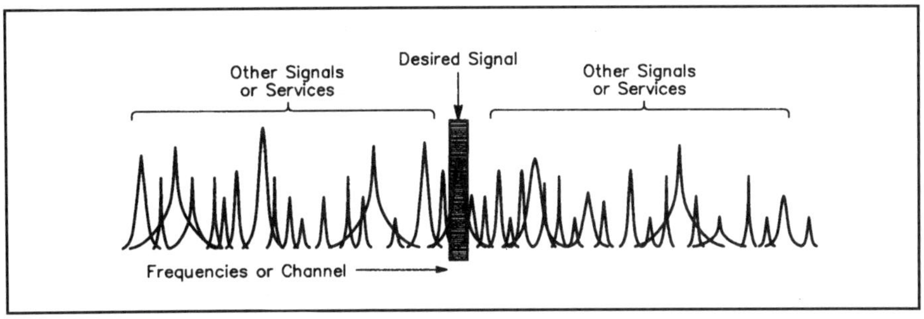

Figure 1—Every electronic appliance must select only the desired signal while rejecting all others.

Try the Easy Things First

Sometimes, the easiest solutions are the best. Many cases of interference can be resolved without the need for technical investigations or knowledge. As first steps, you might check your wiring for damage, for open outer wire shields, or for loose terminal connections. Try removing any added devices, such as video games, or even relocating the equipment or reorienting the device's antenna and power cord.

If you suspect that the problem is caused by electrical noise, check for overloaded circuits, frayed wires, loose sockets, etc. These types of problems should be fixed no matter what! Have your electrician shut off one breaker at a time, noting if this has any effect on the interference. If so, determine which devices are connected to that particular line, then remove the suspect devices one at a time. When the interference goes away, you've found the "culprit." Your electric utility company service department will offer assistance if the interference is coming from defective equipment on the power lines or distribution equipment.

Interference filters for your consumer electronics equipment can be purchased locally or by mail order. These filters usually eliminate unwanted interference if they are used properly on the equipment that is in need of additional filtering.

According to the FCC's Interference Handbook, telephones and other audio devices that pick up radio signals are improperly acting as radio receivers. The interference can usually be cured, but the necessary filtering must be applied to the affected device.

Several companies sell modular telephone interference filters that are very effective. Your telephone company service department also may be able to help.

A high-pass filter may reduce interference to an antenna-connected television or VCR. A common-mode filter should be tried first on TVs or VCRs connected to a cable system. An ac-line interference filter may help with electrical or radio interference. These items can be purchased locally or by mail order.

Some interference cures must be applied to the internal circuitry of the affected equipment. This should always be done by authorized service personnel.

The ARRL has an information package called "The RFI Package." They also sell a book, The *ARRL RFI Book*, that provides additional guidance and technical information. Although it was written for Amateur Radio operators, the book may be helpful to you, too. Contact the ARRL for information about their products and membership services.

Self-Help Cures

In some cases, when all else fails, you may need to resolve the problem yourself, or with the help of your electronic service person. It's impossible to use the remaining space to outline all of the possible cures for interference problems (the subject is quite complex). However, a few simple cures using commonly available parts can eliminate most problems. The self-help packages supplied by the ARRL and the FCC explain these cures in more detail.

For More Information...

The ARRL and the FCC have self-help packages available to help you resolve interference problems.

ARRL, Inc.
RFI Desk
225 Main Street
Newington, CT 06111
tel (860) 594-0214
Internet Web Site: **www.arrl.org**
E-mail: **rfi@arrl.org**

Federal Communications Commission
Consumer & Governmental Affairs Bureau
Consumer & Governmental Affairs Bureau
1270 Fairfield Rd
Gettysburg, PA 17326
tel (800) 225-5322 (CALL FCC)
Internet Web Site: **www.fcc.gov/cgb/**

Consumer Electronics Association
2500 Wilson Boulevard
Arlington, VA 22201-3834
tel (703) 907-7600
Internet Web Site: **www.ce.org**

For copies of U.S. Government publications, contact:

US Government Printing Office
710 North Capitol St, NW
Washington, DC 20401
tel (202) 512-0132
Internet Web Site: **www.access.gpo.gov**

SOURCES OF INTERFERENCE FILTER PRODUCTS:

High-pass filters:

Industrial Communication Engineers
RadioShack catalog #15-579, 15-577 (FM Trap)

Common-mode

Industrial Communication Engineers

Telephone-interference filters:

Industrial Communication Engineers
K-Com
RadioShack catalog #43-150

CONTACT:

Industrial Communication Engineers

350 Gloria Rd
Sunnyvale, TX 75182
tel (972) 203-2008

K-Com

PO Box 83
Randolph, OH 44265
tel (330) 325-2110

Interference Can Be Cured!

Remember, most cases of interference can be cured! It takes cooperation between the consumer, the manufacturer and the radio operator. With a little bit of work, you and your neighbor can both enjoy your favorite activities in peace.

Index

L

M

N

P

R

Notes

About the ARRL

The national association for Amateur Radio

The seed for Amateur Radio was planted in the 1890s, when Guglielmo Marconi began his experiments in wireless telegraphy. Soon he was joined by dozens, then hundreds, of others who were enthusiastic about sending and receiving messages through the air—some with a commercial interest, but others solely out of a love for this new communications medium. The United States government began licensing Amateur Radio operators in 1912.

By 1914, there were thousands of Amateur Radio operators—hams—in the United States. Hiram Percy Maxim, a leading Hartford, Connecticut inventor and industrialist, saw the need for an organization to band together this fledgling group of radio experimenters. In May 1914 he founded the American Radio Relay League (ARRL) to meet that need.

Today ARRL, with approximately 150,000 members, is the largest organization of radio amateurs in the United States. The ARRL is a not-for-profit organization that:

- promotes interest in Amateur Radio communications and experimentation
- represents US radio amateurs in legislative matters, and
- maintains fraternalism and a high standard of conduct among Amateur Radio operators.

At ARRL headquarters in the Hartford suburb of Newington, the staff helps serve the needs of members. ARRL is also International Secretariat for the International Amateur Radio Union, which is made up of similar societies in 150 countries around the world.

ARRL publishes the monthly journal *QST*, as well as newsletters and many publications covering all aspects of Amateur Radio. Its headquarters station, W1AW, transmits bulletins of interest to radio amateurs and Morse code practice sessions. The ARRL also coordinates an extensive field organization, which includes volunteers who provide technical information and other support services for radio amateurs as well as communications for public-service activities. In addition, ARRL represents US amateurs with the Federal Communications Commission and other government agencies in the US and abroad.

Membership in ARRL means much more than receiving *QST* each month. In addition to the services already described, ARRL offers membership services on a personal level, such as the ARRL Volunteer Examiner Coordinator Program and a QSL bureau.

Full ARRL membership (available only to licensed radio amateurs) gives you a voice in how the affairs of the organization are governed. ARRL policy is set by a Board of Directors (one from each of 15 Divisions). Each year, one-third of the ARRL Board of Directors stands for election by the full members they represent. The day-to-day operation of ARRL HQ is managed by a Chief Executive Officer.

No matter what aspect of Amateur Radio attracts you, ARRL membership is relevant and important. There would be no Amateur Radio as we know it today were it not for the ARRL. We would be happy to welcome you as a member! (An Amateur Radio license is not required for Associate Membership.) For more information about ARRL and answers to any questions you may have about Amateur Radio, write or call:

ARRL—The national association for Amateur Radio
225 Main Street
Newington CT 06111-1494
Voice: 860-594-0200
Fax: 860-594-0259
E-mail: **hq@arrl.org**
Internet: **www.arrl.org/**

Prospective new amateurs call (toll-free):
800-32-NEW HAM (800-326-3942)
You can also contact us via e-mail at **newham@arrl.org**
or check out *ARRLWeb* at **http://www.arrl.org/**

FEEDBACK

Please use this form to give us your comments on this book and what you'd like to see in future editions, or e-mail us at **pubsfdbk@arrl.org** (publications feedback). If you use e-mail, please include your name, call, e-mail address and the book title, edition and printing in the body of your message. Also indicate whether or not you are an ARRL member.

Where did you purchase this book?

☐ From ARRL directly ☐ From an ARRL dealer

Is there a dealer who carries ARRL publications within:

☐ 5 miles ☐ 15 miles ☐ 30 miles of your location? ☐ Not sure.

License class:

☐ Novice ☐ Technician ☐ Technician with code ☐ General ☐ Advanced ☐ Amateur Extra

---

Name ______________________ ARRL member? ☐ Yes ☐ No

Call Sign ______________________

Daytime Phone ( ) ______________________ Age ______________________

Address ______________________

City, State/Province, ZIP/Postal Code ______________________

If licensed, how long? ______________________ e-mail address: ______________________

Other hobbies ______________________

____

Occupation ______________________

For ARRL use only	RFI Book
Edition	3 4 5 6 7 8 9 10 11 12
Printing	1 2 3 4 5 6 7 8 9 10 11 12

From ______________________________

____

____

Please affix postage. Post Office will not deliver without postage.

EDITOR, RFI BOOK
ARRL—THE NATIONAL ASSOCIATION FOR AMATEUR RADIO
225 MAIN STREET
NEWINGTON CT 06111-1494

— — — — — — please fold and tape — — — — — —